Kurt Lang

Handbook of Signal Processing in Acoustics

Volume 1

Edited by

David Havelock

National Research Council, Canada

Sonoko Kuwano

Osaka University, Japan

Michael Vorländer

RWTH Aachen University, Germany

Handbook of Signal Processing in Acoustics

Volume 1

Editors
David Havelock
National Research Council
Institute for Microstructural
  Sciences
Acoustics and Signal
  Processing Group
1200 Montreal Road
Ottawa ON K1A 0R6
Canada

Sonoko Kuwano
Osaka University
Graduate School of Human
  Sciences
Department of Environmental
  Psychology
1-2 Yamadaok Suita
Osaka
Japan

Michael Vorländer
RWTH Aachen University
Institute of Technical Acoustics
Aachen
Germany

ISBN: 978-0-387-77698-9   e-ISBN: 978-0-387-30441-0

Library of Congress Control Number: 2008923573

© 2008 Springer Science+Business Media, LLC
All rights reserved. This work may not be translated or copied in whole or in part without the written permission of the publisher (Springer Science+Business Media, LLC, 233 Spring Street, New York, NY 10013, USA), except for brief excerpts in connection with reviews or scholarly analysis. Use in connection with any form of information storage and retrieval, electronic adaptation, computer software, or by similar or dissimilar methodology now known or hereafter developed is forbidden.
The use in this publication of trade names, trademarks, service marks, and similar terms, even if they are not identified as such, is not to be taken as an expression of opinion as to whether or not they are subject to proprietary rights.
Printed on acid-free paper

9 8 7 6 5 4 3 2 1

springer.com

Editor Biographies

David Havelock
National Research Council
Institute for Microstructural Sciences
Acoustics and Signal Processing Group
1200 Montreal Road
Ottawa ON K1A 0R6
Canada

Sonoko Kuwano
Osaka University
Graduate School of Human Sciences
Department of Environmental Psychology
1-2 Yamadaok Suita
Osaka
Japan

Michael Vorländer
RWTH Aachen University
Institute of Technical Acoustics
Aachen
Germany

Editorial Board

Last Name	Given Name(s)	Affiliation/Address
Beauchamp	James W.	University of Illinois Urbana-Champaign, School of Music, Dept. of ECE, Urbana, IL, USA
Christie	Douglas R.	Earth Physics, Research School of Earth Sciences, The Australian National University, Canberra, AUSTRALIA
Deffenbaugh	Max	ExxonMobil Research and Engineering Company, Annandale, NJ USA
Elliott	Stephen J.	Southampton University, Institute of Sound and Vibration Research, Southampton, ENGLAND
Fastl	Hugo	Technische Universitaet Muenchen AG Technische Akustik, MMK Muenchen, GERMANY
Gierlich	Hans Wilhelm	HEAD acoustics GmbH Telecom Division, Herzogenrath, GERMANY
Guyader	Jean-Louis	Labratory for Vibration Acoustics Villeurbanne, FRANCE
Jacobsen	Finn	Technical University of Denmark, Lyngby, DENMARK
Karjalainen	Matti	Helsinki University of Technology, FINLAND
Kollmeier	Birger	Universitat Oldenburg, Oldenburg, GERMANY
O'Shaughnessy	Douglas D.	INRS-EMT (Telecommunications), Montreal, QC, CANADA
Riquimaroux	Hiroshi	Doshisha University, Department of Knowledge Engineering & Computer Sciences, Sensory & Cognitive Neuroscience Research Laboratory, Kyotanabe, Kyoto, JAPAN
Sullivan	Edmund J.	EJS Consultants Portsmouth, RI, USA

Last Name	Given Name(s)	Affiliation/Address
Suzuki	**Hideo**	1-2-3-S2502, Utase, Mihama-ward, Chiba-city, JAPAN 261-0013
Taroudakis	**Michael**	University of Crete Department of Mathematics, and FORTH, Institute of Applied and Computational Mathematics, Heraklion, GREECE
Ueha	**Sadayuki**	Tokyo Institute of Technology, Director, Precision and Intelligence Lab, Yokohama, JAPAN
Verrillo	**Ronald T.**	(deceased) Institute for Sensory Research, Syracuse, NY, USA
Yamada	**Ichiro**	Aviation Environment Research Center, Airport Environment Improvement Foundation, Tokyo, JAPAN
Yamasaki	**Yoshio**	Waseda University, Graduate School of Global Information and Telecommunication Studies, Saitama, JAPAN
Tohyama	**Mikio**	Waseda University, JAPAN

Preface

Acoustics has a special relationship with signal processing. Many concepts in signal processing arise naturally from our general experience with sound and vibration and, more than in many other fields, acoustics is concerned with the acquisition, analysis, and synthesis of signals. Consequently, there is a rich resource of signal processing expertise within the acoustics community.

There are many excellent reference books devoted to signal processing but the objective of the *Handbook of Signal Processing in Acoustics* is to bring together the signal processing expertise specific to acoustics and to capture the interdisciplinary nature of signal processing within acoustics. It is also hoped that the handbook will promote networking and the interchange of ideas between technical areas in acoustics.

The handbook comprises 104 *Chapters* organized into 17 *Parts*. Each Part addresses a technical area of acoustics, reflecting the general demarcations of specialization within the acoustics community. An expert with broad knowledge of signal processing within their respective technical area was invited to act as a Section Leader for each Part of the handbook. These Section Leaders contributed substantially to the handbook project by helping to define the contents and scope of each chapter, finding an appropriate contributing expert author, and managing the review and revision of material. Collectively with the Editors, they form the Editorial Board for the handbook.

Planned sections on Architectural Acoustics, Nonlinear Acoustics, and Ultrasound are unfortunately omitted from the handbook; nevertheless, the handbook otherwise provides thorough coverage of the field of acoustics and we can hope that possible future editions might include these areas.

The handbook is written from the perspective of acoustics, by acousticians with signal processing expertise. Emphasis is placed in the description of acoustic problems and the signal processing related to their solutions. The reader is assumed to have basic knowledge of signal processing. Signal processing techniques are described but the reader is referred elsewhere for derivations and details.

The authors were not required to adhere to strict standards of style or notation, and were asked to prepare short, concise, self-sufficient chapters. This results in variations in style and notation throughout the handbook that reflects the diversity of perspectives within the acoustics community.

David Havelock

Acknowledgments

David Havelock gratefully acknowledges support from the Institute for Microstructural Sciences, of the National Research Council of Canada, for making time and resources available during the preparation of this handbook. He also thanks John C. Burgess (University of Hawaii, retired) for the encouragement to begin this project and expresses his gratitude to the co-Editors, Sonoko Kuwano and Michael Vorländer, who were a pleasure to work with throughout this project. The efforts of all of the Section Leaders are greatly appreciated and we thank each of them sincerely for their patience, perseverance, and faith that were required to see this project to completion. We thank each of the authors for their contributions.

Contents

Editor Biographies	v
Editorial Board	vii
Preface	ix
Acknowledgments	xi
Contributors	xvii

PART I   ACOUSTIC SIGNALS AND SYSTEMS 1

1	Signals and Systems	3
2	Acoustic Data Acquisition	17
3	Spectral Analysis and Correlation	33
4	The FFT and Tone Identification	53
5	Measuring Transfer-Functions and Impulse Responses	65
6	Digital Sequences	87
7	Filters	107
8	Adaptive Processing	125
9	Beamforming and Wavenumber Processing	131

PART II AUDITORY SYSTEM AND HEARING   145

10 Anatomy, Physiology and Function of the Auditory System   147

11 Physiological Measures of Auditory Function   159

12 Auditory Processing Models   175

13 Speech Intelligibility   197

14 Signal Processing in Hearing Aids   205

PART III PSYCHOACOUSTICS   213

15 Methods for Psychoacoustics in Relation to Long-Term Sounds   215

16 Masking and Critical Bands   229

17 Aspects of Modeling Pitch Perception   241

18 Calculation of Loudness for Normal and Hearing-Impaired Listeners   251

19 Psychoacoustical Roughness   263

PART IV MUSICAL ACOUSTICS   275

20 Automatic Music Transcription   277

21 Music Structure Analysis from Acoustic Signals   305

22 Computer Music Synthesis and Composition   333

23 Singing Voice Analysis, Synthesis, and Modeling   359

24 Instrument Modeling and Synthesis   375

25 Digital Waveguide Architectures for Virtual Musical Instruments   399

26 Modeling of Musical Instruments   419

PART V SPEECH   447

27 Display and Analysis of Speech   449

28 Estimation of Speech Intelligibility and Quality   483

29 Gaussian Models in Automatic Speech Recognition   521

30 Speech Synthesis   557

31 Speech Coders   587

PART VI AUDIO ENGINEERING   621

32 Transducer Models   623

33 Loudspeaker Design and Performance Evaluation   649

34 PA Systems for Indoor and Outdoor   669

35 Beamforming for Speech and Audio Signals   691

36 Digital Audio Recording Formats and Editing Principles   703

37 Audiovisual Interaction   731

38 Multichannel Sound Reproduction   747

39 Virtual Acoustics   761

40 Audio Restoration   773

41 Audio Effects Generation   785

42 Perceptually Based Audio Coding   797

PART VII TELECOMMUNICATIONS   819

43 Speech Communication and Telephone Networks   821

44 Methods of Determining the Communicational Quality of Speech Transmission Systems   831

45	Efficient Speech Coding and Transmission Over Noisy Channels	853
46	Echo Cancellation	883
47	Noise Reduction and Interference Cancellation	897
48	Terminals and Their Influence on Communication Quality	909
49	Networks and Their Influence on Communication Quality	915
50	Interaction of Terminals, Networks and Network Configurations	921
List of Important Abberviations		927

Contributors

Tomonari Akamatsu
National Research Institute of Fisheries Engineering, Fisheries Research Agency of Japan, Kamisu, Ibaraki, Japan

Benoit Alcoverro
CEA/DASE, BP 12, 91680 Bruyères Le Châtel, France

Lydia Ayers
Computer Science Department, Hong Kong University of Science and Technology, Clear Water Bay, Kowloon, Hong Kong, China, e-mail: layers@cs.ust.hk

Juha Reinhold Backman
Nokia, Espoo, Finland

Rolf Bader
Institute of Musicology, University of Hamburg, Hamburg, Germany

James W. Beauchamp
School and Music and Department of Electrical and Computer Engineering, University of Illinois at Urbana-Champaign, 2136 Music Building, 1114 West Nevada Street, Urbana, IL 61801, USA, e-mail: jwbeauch@uiuc.edu

Kim Benjamin
Naval Undersea Warfare Center, Newport, RI, USA

A.J. Berkhout
Delft University of Technology, Delft, The Netherlands

Jeff Bilmes
Department of Electrical Engineering, University of Washington, Seattle, WA, USA

Ole-Herman Bjor
Norsonic AS, Lierskogen, Norway, e-mail: ohbjor@norsonic.com

Stanley J. Bolanowski (deceased)
Institute for Sensory Research, Syracuse University, Syracuse, NY 13244, USA

Baris Bozkurt
Izmir Institute of Technology (IYTE), Izmir, Turkey

Nicolas Brachet
Provisional Technical Secretariat, CTBTO, Vienna International Centre, Vienna, Austria

Thomas Brand
University of Oldenburg, Oldenburg, Germany

David J. Brown
Geoscience Australia, Canberra, Australia, (Provisional Technical Secretariat, CTBTO, Vienna International Centre, Vienna, Austria), e-mail: David.Brown@CTBTO.org

John C. Burgess
Department of Mechanical Engineering, University of Hawaii, 2540 Dole Street, Honolulu, HI 968221, USA, e-mail: jcb@hawaii.edu

Paola Campus
Provisional Technical Secretariat, CTBTO, Vienna International Centre, Vienna, Austria

Yves Cansi
CEA/DASE/LDG, BP12, 91680 Bruyères-le-Châtel, France

Josef Chalupper
Siemens Audiological Engineering Group, Gebbertstrasse 125, Erlangen 91058, Germany, e-mail: Josef.Chalupper@siemens.com

N. Ross Chapman
University of Victoria, Victoria BC, Canada

Jakob Christensen-Dalsgaard
Institute of Biology, University of Southern Denmark, Campusvej 55, DK-5230 Odense M, Denmark, e-mail: jcd@biology.sdu.dk

Peter Daniel
Brüel & Kjær, GmbH, Bremen, Germany

Roger B. Dannenberg
Carnegie Mellon University, Pittsburgh, PA, USA

Torsten Dau
Technical University of Denmark, Lyngby, Denmark, e-mail: tda@elektro.dtu.dk

Hans-Elias de Bree
Microflown Technologies, The Netherlands, USA

Max Deffenbaugh
ExxonMobil Research and Engineering Company, Annandale, NJ, USA

Thierry Dutoit
Faculte Polytechnique de Mons, Mons, Belgium

Stephen J. Elliott
Institute of Sound and Vibration Research, Southampton, UK

Paulo A.A. Esquef
Nokia Institute of Technology, Rod. Torquato Tapajós, 7200, Tarumã 69048-660 Manaus-AM, Brazil, e-mail: paulo@esquef@indt.org.br

Richard R. Fay
Parmly Hearing Institute, Loyola University Chicago, 6525 N. Sheridan Rd., Chicago, IL 60626, USA, e-mail: rfay@luc.edu

Michael Fehler
Massachusetts Institute of Technology, Cambridge, USA

Sandy Fidell
Fidell Associates, Inc., Woodland Hills, CA, USA

Petr Firbas
Provisional Technical Secretariat, CTBTO, Vienna International Centre, Vienna, Austria (International Atomic Energy Agency, Vienna International Centre, Vienna, Austria)

Erling Frederiksen
Brüel & Kjær, Sound and Vibration Measurement A/S, Skodsborgvej 307, 2850 Nærum, Denmark

Milton A. Garcés
Infrasound Laboratory, University of Hawaii, Manoa, HI, USA, e-mail: milton@isla.hawaii.edu

H.W. Gierlich
HEAD acoustics GmbH, Herzogenrath, Germany

Norbert Goertz
Institute for Digital Communications, School of Engineering and Electronics, University of Edinburgh, King's Buildings, Mayfield Road, Edinburgh EH9 3JL, UK

Masataka Goto
National Institute of Advanced Industrial Science and Technology (AIST), Tokoyo, Japan

Philippe Gournay
Université de Sherbrooke, Sherbrooke, QC, Canada

Jørgen Hald
Brüel & Kjær, Sound & Vibration Measurements A/S, Nærum, Denmark

Joe Hammond
University of Southampton, Southampton, UK

Colin H. Hansen
School of Mechanical Engineering, University of Adelaide, Adelaide, SA 5005, Australia, e-mail: chanson@mecheng.adelaide.edu.au

Uwe Hansen
Department of Physics, Indiana State University, Terre Haute, IN, USA

Sabih I. Hayek
Department of Engineering Science and Mechanics, University Park, PA 16802, USA

Ulrich Heute
Institute for Circuit and System Theory, Faculty of Engineering, Christian-Albrecht, University, Kaiserstr. 2, D-24143 Kiel, Germany

Thomas Hoffmann
Provisional Technical Secretariat, CTBTO, Vienna International Centre, Vienna, Austria

Volker Hohmann
University of Oldenburg, Oldenburg, Germany

Masaaki Honda
Waseda University, Tokyo, Japan

Andrew B. Horner
Department of Computer Science, Hong Kong University of Science and Technology, Clear Water Bay, Kowloon, Hong Kong, China, e-mail: horner@cs.ust.hk

Adrianus J.M. Houtsma
Aircrew Protection Division, U.S. Army Aeromedical Research Laboratory, Fort Rucker, AL 36362-0577, USA, e-mail: adrian.houtsma@amedd.army.mil

Finn Jacobsen
Technical University of Denmark, Lyngby, Denmark, e-mail: fja@elektro.dtu.dk

Finn B. Jensen
NATO Undersea Research Centre, La Spezia, Italy, e-mail: jensen@nurc.nato.int

Walter Kellermann
Multimedia Communications and Signal Processing, University Erlanger-Nuremberg, Erlanger, Germany, e-mail: wk@lnt.de

Youngmoo E. Kim
Electrical & Computer Engineering, Drexel University, Philadelphia, PA, USA, e-mail: ykim@drexel.edu

Anssi Klapuri
Institute of Signal Processing, Tampere University of Technology, Korkeakoulunkatu 1, 33720 Tampere, Finland, e-mail: Anssi.Klapuri@tut.fi

Birger Kollmeier
University of Oldenburg, Oldenburg, Germany

Jan Felix Krebber
Institute of Communication Acoustics, Ruhr-University Bochum, Bochum, Germany, e-mail: jan.krebber@rub.de

Christine E. Krohn
ExxonMobil Upstream Research Company, Houston, TX, USA

Sonoko Kuwano
Osaka University, Osaka, Japan

Gerald C. Lauchle
State College, PA, USA

Walter Lauriks
Katholieke Universiteit Leuven, Heverlee, Belgium; Laboratorium voor Akoestick on Thermische Fysica, K.U. Leuven, Leuven, Belgium

Alexis Le Pichon
CEA/DASE/LDG, BP12, 91680 Bruyères-le-Châtel, France

Philippe Leclaire
Université de Bourgogne, Nevers, France; Laboratorium voor Akoestick on Thermische Fysica, K.U. Leuven, Leuven, Belgium

Roch Lefebvre
Université de Sherbrooke, Sherbrooke, QC, Canada

Cuiping Li
Department of Earth and Atmospheric Science, Purdue University, West Lafayette IN 47907, USA

Tapio Lokki
Department of Media Technology, Helsinki University of Technology, Helsinki, Finland, e-mail: Tapio.Lokki@tkk.fi

Aki Vihtori Mäkivirta
Genelec Oy, Iisalmi, Finland

Brian H. Maranda
Defence Research and Development Canada – Atlantic, Dartmouth, NS, Canada

James Mathews
PCB Piezoelectronics Inc., San Clemente, CA, USA

Manfred Mauermann
University of Oldenburg, Oldenburg, Germany

Walter Metzner
Department of Physiological Science, University of California, California, USA

Yasushi Miki
Computer Science, Faculty of Engineering, Takushoku University, Tokyo, Japan

Ben Milner
School of Computing Sciences, University of East Anglia, Norwich, Norfolk, UK

Riikka Möttönen
Department of Biomedical Engineering and Computational Science, Helsinki University of Technology, Helsinki, Finland

Swen Müller
National Institute of Metrology, Xerém, Brazil

Seiichiro Namba
Osaka University, Osaka, Japan

Ramesh Neelamani
ExxonMobil Upstream Research Company, Houston, TX, USA

David E. Norris
BBN Technologies, 1300 N. 17th St., Arlington, VA 22209, USA

Robert L. Nowack
Department of Earth and Atmospheric Science, Purdue University, West Lafayette, IN 47907, USA

Yasuhiro Oikawa
Waseda University, Tokyo, Japan

Kazuo Okanoya
Laboratory for Biolinguistics, Brain Science Institute, Riken, 2-1 Hirosawa, Saitama 351-0198, Japan, e-mail: okanoya@brain.riken.jp

John V. Olson
Geophysical Institute, University of Alaska, Fairbanks, USA

Thorkild Find Pedersen
Brüel & Kjær, Sound & Vibration Measurements A/S, Nærum, Denmark

Ville Pulkki
Department of Signal Processing and Acoustics, Helsinki University of Technology, Helsinki, Finland, e-mail: ville.pulkki@tkk.fi

Robert B. Randall
University of New South Wales, Sydney, Australia

Helmut Riedel
University of Oldenburg, Oldenburg, Germany

Francis Rumsey
Institute of Sound Recording, University of Surrey, Guildford, UK

Masahiko Sakai
Ono Sokki Co., Ltd., 1-16 1 Hakusan, Midori-ku, Yokohama, 226-8507 Japan, e-mail: msakai@onosokki.co.jp

Rebecca Saltzer
ExxonMobil Upstream Research Company, Houston, TX, USA

Mikko Sams
Department of Biomedical Engineering and Computational Science, Helsinki University of Technology, Helsinki, Finland

Lauri Savioja
Department of Media Technology, Helsinki University of Technology, Helsinki, Finland, e-mail: Lauri.Savioja@tkk.fi

Julius O. Smith
Center for Computer Research in Music and Acoustics (CCRMA), Stanford University, Stanford, CA 94305, USA, website: http://ccrma.stanford.edu/~jos/

Stefka Stefanova
Provisional Technical Secretariat, CTBTO, Vienna International Centre, Vienna, Austria

Edmund J. Sullivan
EJS Consultants, Portsmouth, RI, USA

David C. Swanson
The Applied Research Laboratory, The Pennsylvania State University, Philadelphia, PA, USA

Curt A.L. Szuberla
Geophysical Institute, University of Alaska, Fairbanks, USA

Hideki Tachibana
Institute of Industrial Science, University of Tokyo, Tokyo, Japan

Yasushi Takano
Department of Mechanical Engineering Research Laboratory, Hitachi, Ltd., Ibaraki, Japan

Ernst Terhardt
Technical University of Munich, Munich. Germany, e-mail: teirhardt@ei.tum.de

Christine Thomas
Department of Earth and Ocean Sciences, University of Liverpool, Liverpool, UK

Ippei Torigoe
Department of Mechanical Engineering and Materials, Science Faculty of Engineering Kumamoto University, Kumamoto, Japan

Stefan Uppenkamp
University of Oldenburg, Oldenburg, Germany

Ronald T. Verrillo (deceased)
Institute for Sensory Research, Syracuse University, Syracuse, NY 13244, USA

Tuomas Virtanen
Institute of Signal Processing, Tampere University of Technology, Korkeakoulunkatu 1, 33720 Tampere, Finland, e-mail: Tuomas.Virtanen@tut.fi

Stephen Voran
Institute for Telecommunication Sciences, Boulder, CO, USA

Erhard Werner
Tannenweg 16, D 29693 Hademstorf, Germany

Rodney W. Whitaker
Los Alamos National Laboratory, PO Box 1663, Los Alamos, NM 87544, USA

Paul White
University of Southampton, Southampton, UK

Ru-Shan Wu
University of California, Santa Cruz, CA, USA

Xianyn Wu
ExxonMobil Upstream Research Company, Houston, TX, USA

Ning Xiang
School of Architecture and Department of Electrical, Computer, and Systems Engineering, Rensselaer Polytechnic Institute, Troy, NY 12180, USA, e-mail: xiangn@rpi.edu

Ichiro Yamada
Aviation Environment Research Center, Airport Environment Improvement Foundation, Tokyo, Japan

Kohei Yamamoto
Kobayasi Institute of Physical Research, Tokyo, Japan

Yoshio Yamasaki
Waseda University, Tokyo, Japan

Nobutoshi Yoshida
Ono Sokki Co. Ltd., 1-16-1 Hakusan, Midori-ku, 226 8507 Yokohama, Japan, e-mail: yoshi@onosokki.co.jp

Udo Zölzer
Helmut Schmidt University, Hamburg, Germany

PART I
ACOUSTIC SIGNALS AND SYSTEMS

Finn Jacobsen

Technical University of Denmark, Lyngby, Denmark

1 Signals and Systems
Joe Hammond and Paul White 3
Temporal Signal Classification • System Definition and Classification

2 Acoustic Data Acquisition
David C. Swanson 17
Electronic Noise • Analog-to-Digital Conversion • Anti-aliasing Filters • Basic Digital Signal Calibration • Multichannel System Calibration • Data Validation and Diagnostic Techniques Summary

3 Spectral Analysis and Correlation
Robert B. Randall 33
Introduction • The Fourier Transform and Variants • Practical FFT Analysis • Spectral Analysis Using Filters • Correlation Functions • Time–Frequency Analysis

4 The FFT and Tone Identification
John C. Burgess 53
Introduction • The FFT (and DFT) • Leakage • Windows • Estimation of Amplitude, Frequency, and Phase • Applications

5 Measuring Transfer-Functions and Impulse Responses
Swen Müller 65
Introduction • Measurement Methods • Excitation Signals

6 Digital Sequences
Ning Xiang 87
Introduction • Golay Codes • Binary Maximum-Length Sequences • Fast MLS Transform (FMT) • Gold and Kasami Sequences • Legendre Sequences • Application Remarks

7 Filters
Ole-Herman Bjor 107
Introduction • Frequency Weightings • Octave- and Fractional-Octave-Bandpass Filters • Phase Linear Filters • In-Phase/Quadrature Filters • Perceptual Masking Filters

8 Adaptive Processing
Thorkild Find Pedersen 125
Introduction • The LMS Algorithm • Adaptive Applications

9 Beamforming and Wavenumber Processing
Jørgen Hald 131
Planar Nearfield Acoustical Holography • Beamforming

1
Signals and Systems

Joe Hammond and Paul White

University of Southampton, Southampton, UK

Signal processing is the science of applying transformations to measurements, to facilitate their use by an observer or a computer, and digital signal processing (DSP) is the enabling technology for applications across all disciplines and sectors.

The study of sound and vibration is highly dependent on the use of special purpose signal analysers or software packages. The accessibility and convenience of DSP analysis modules and procedures can sometimes create a deceptive air of simplicity in often complicated phenomena. It is important that practitioners, while availing themselves of the full range of DSP capabilities, should have a clear understanding of the fundamentals of the science of signal processing and so be fully aware of the assumptions, implications and limitations inherent in their analysis methods.

Signal processing and analysis involves the three phases of data acquisition, processing and interpretation (of the results of the processing) and, of course, all three are linked in any application. The last phase is naturally very much related to the subject under investigation, but the first two may be discussed independently of specific applications. A vast body of theory and methodology has been built up as a consequence of the problems raised by the need for data analysis; this is often referred to as "signal analysis" or "time-series analysis" depending on the context [1]. The choice of methodology is often reliant on some prior knowledge of the phenomenon being analysed. This usually relates to classifying the characteristics of the data and/or the way in which the data may be modelled owing to knowledge of (or assumptions about) the way in which the data may have been generated. This section considers signal and system characteristics that underpin signal processing.

In acoustics the signal in question is generally the output of a pressure transducer. Such a signal is a time history that depends on the spatial location of the transducer. We denote this by $p(t,\mathbf{r})$ where t denotes the time dependence and $\mathbf{r}$ is the vector representation for its spatial location (with Cartesian co-ordinates x, y, z). Multiple signals may be available from an array of sensors as required, e.g., in beamforming. For the present it is convenient to drop the spatial dependence and consider signals as evolving with time.

1 • Temporal Signal Classification

The physical phenomenon under investigation is often translated by a transducer into an electrical equivalent, and a single signal evolving in continuous time is denoted as $x(t)$. In many cases, data are *discrete* owing to some inherent or imposed sampling procedure. In this case the data might be characterised by a sequence of numbers. When derived from the continuous time process $x(t)$ we write $x(n\Delta)$ or x_n ($n = 0, 1, 2, \ldots$), where we have implied that the sampling interval Δ is constant (i.e. uniform sampling). The time histories that can occur are often very complex, and it is helpful to consider signals that exhibit particular characteristics. This allows us to relate appropriate analysis methods to those specific types of signal.

Figure 1 illustrates a broad categorisation of signal types.

A basic distinction is the designation of a signal as "random" or "deterministic" where by "random" we mean one that is not

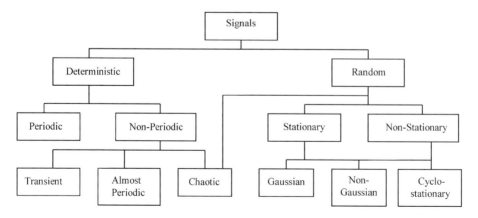

FIGURE 1 Classification of signals.

exactly predictable. Very often, processes are mixed and the demarcations shown in Figure 1 are not easily applied and consequently the analysis procedure to be used may not be apparent. We have included chaotic processes under both "deterministic" and "random" categories since such signals are generated from a deterministic non-linear phenomenon but nevertheless have an output with an unpredictable, random-like behaviour. The classification of data as being deterministic or random might be debatable in many cases and the choice must be made on the basis of the knowledge of the physical situation. Often signals may be modelled as being a mixture of both, e.g. a deterministic signal "embedded" in unwanted random disturbances (noise).

Many of the classes defined above, such as the stationary and periodic signals, are mathematical constructs to which no real-world signal can belong. However, these classes do provide one with a set of models which, in many cases, provide good approximations to measured processes and they suggest a suitable analysis framework.

A brief note for each of the categories is given below – continuous time is used throughout. An analysis method appropriate for each is also noted.

1.1 Periodic

A signal is periodic with period T_p if $x(t) = x(t + T_p)$.

Such a signal can be represented in the frequency domain as a Fourier series:

$$x(t) = a_0 + \sum_{k=1}^{\infty} a_k \cos(2\pi k f_0 t) + b_k \sin(2\pi k f_0 t) = \sum_{k=\infty}^{\infty} c_k e^{2\pi i k f_0 t}. \quad (1)$$

The feature of this is that the frequencies in representation are the fundamental, $f_0 = 1/T_p$, and multiples (harmonics) (plus a d.c. term), and the coefficients, a_k and b_k (or c_k), give the amplitude and phase of the components.

Figure 2a shows the time-series of a small section of voiced speech, the vowel /e/. The approximately periodic character of this signal is evident and suggests that it could be gainfully analysed by computing its Fourier series representation. This Fourier series is shown in Figure 2b, which has been computed based on the first period of the signal shown in Figure 2a. One should appreciate that the assumption of periodicity does not hold exactly for this example, as is commonly the case.

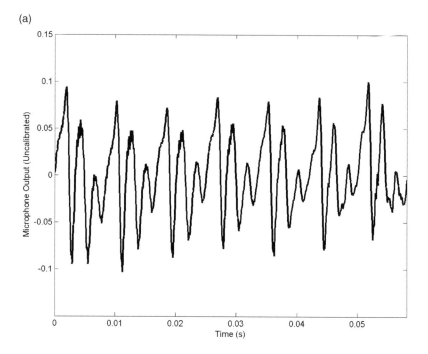
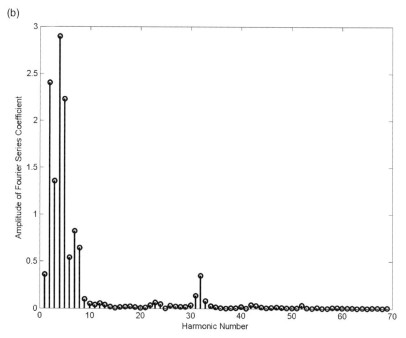

FIGURE 2 Example of voiced speech: (**a**) time-series, (**b**) Fourier coefficients.

1.2 Transient

A transient signal is one which is essentially localised, i.e. a signal that has a finite duration. Such a signal has a Fourier integral representation:

$$x(t) = \int_{-\infty}^{\infty} X(f) e^{2\pi i f t} df. \qquad (2)$$

This differs from the periodic case in that the frequency range becomes a continuum and amplitudes "in a band" are $X(f)df$ so $X(f)$ is now an amplitude density.

Figure 3a shows an example of the time-series of an unvoiced speech segment, in this case the sound /th/. The lack of a periodic structure in this signal is apparent. It can be considered as a transient signal and analysed by employing the Fourier transform defined in (2), which is shown in Figure 3b.

1.3 Almost Periodic

This could be regarded as a process where T_p in (1) varies with time. There is no natural generalisation of (1) for this.

An alternative way of constructing a simple almost periodic process is, e.g.,

$$x(t) = \sin(2\pi t) + \sin\left(2\pi\sqrt{2}t\right). \qquad (3)$$

This process whilst being the sum of two periodic signals is itself not periodic because the ratio of the two frequency components is not a rational number. In this case using the Fourier integral (2) is appropriate.

Figure 4a shows the time-series of a segment of a sustained piano note. These notes contain discrete frequencies that are approximately harmonically related. The inharmonicities present in piano string vibrations mean that such signals should be regarded as almost periodic and so the use of the Fourier transform (2) represents a suitable tool for analysis; the results of such an analysis are shown in Figure 4b.

1.4 Chaotic Process

The range of phenomena that lead to chaotic dynamics is wide, but a feature is that the describing equation is deceptively simple in its deterministic form, but generates intricate signals. An example is a second-order non-linear system driven by a sinusoidal excitation. The methodology of analysis uses topological feature analysis, by which one attempts to reveal the underlying simple generation mechanism.

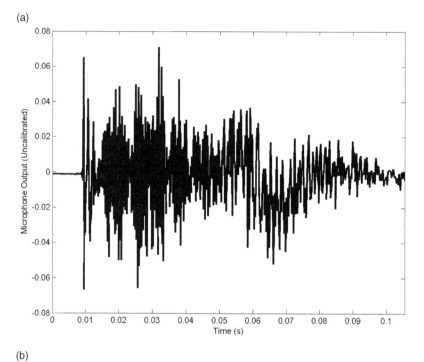
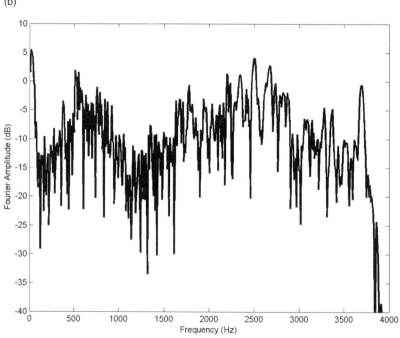

FIGURE 3 Example of unvoiced speech: (**a**) time-series, (**b**) magnitude of the Fourier transform.

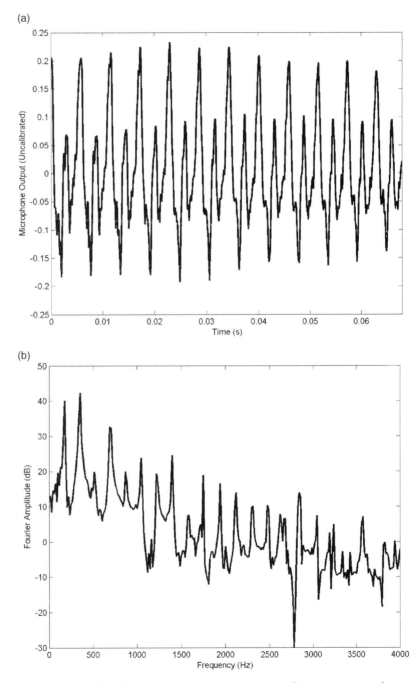

FIGURE 4 Example of a piano note: (**a**) time-series, (**b**) Fourier transform.

1.5 Stationary Random Processes

A stationary random process is one for which the statistical (probabilistic) structure is invariant under a shift in time. Strict stationarity requires that the joint probability density function

$$p_k(x_1, t_1; x_2, t_2; \ldots; x_k, t_k) = p_k(x_1, t_1 + T; x_2, t_2 + T; \ldots; x_k, t_k + T)$$

for all t_k. (4)

This very strong condition is often relaxed to its application to $k = 1$ and 2 only, i.e. stationary to second order. This is referred to as weak (or wide-sense) stationarity.

Stationarity can only be approximate, and judging when the assumption is valid relates to this specific application. The assumption of weak stationarity readily allows the study of the process in the frequency domain using the power (auto) spectral density and the cross-spectral density. In many applications signals are processed by invoking an assumption of stationarity over short timescales.

The autocorrelation function of a stationary signal $x(t)$ is the average of the product of $x(t)$ with a time-shifted version of itself, i.e. the average of $x(t)x(t+\tau)$, and is written as $R_{xx}(\tau)$. It is related to the power spectral density $S_{xx}(f)$ of $x(t)$ by

$$S_{xx}(f) = \int R_{xx}(\tau) e^{-2\pi i f \tau} d\tau. \qquad (5)$$

$S_{xx}(f)$ describes the decomposition of the power of the signal over frequency.

1.6 A Non-stationary Random Process

A non-stationary random signal is one for which p_k is not the same under a time shift.

It should be noted that the term "non-stationary" is now also used to denote any signal in which the "structure" changes as a function of time but which may not be immediately classified as random. An example is a deterministic chirp signal – for such signals the concept of a time–frequency analysis is often used.

For both classes of non-stationarity there are several methods that attempt to generalise the spectral representations [2]. Figure 5 shows an analysis of a section of the song of a nightingale (*Luscinia megarhynchos*).

1.7 Cyclo-stationary Processes

Cyclostationary processes are a class of non-stationary process for which there is an underlying periodic structure. The cyclostationary processes include periodic signals as a special case. An example of a more general cyclostationary process is a non-stationary process which is periodically correlated. A simple example of a cyclostationary process is modulated noise, e.g.

$$x(t) = \sin(2\pi f t) w(t), \qquad (6)$$

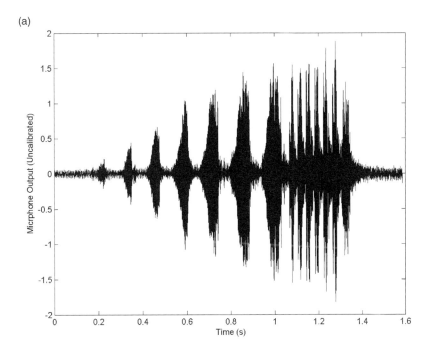

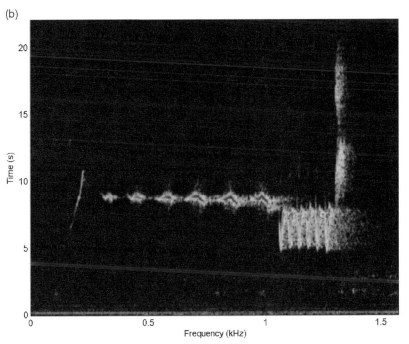

FIGURE 5 Example of a segment of the call of a nightingale: (**a**) time-series, (**b**) short-time Fourier transform.

where $w(t)$ is a broadband random process, with unit variance. The signal $x(t)$ has a structure (in this case a variance) which is periodic, whilst the signal itself is not periodic. Since it is the second-order statistics of the signal here which are the first to show periodicity the above signal is said to be second-order cyclostationary. Many acoustic signals associated with rotating systems, such as fans, contain components which are cyclostationary.

2 • System Definition and Classification

In signal processing the concept of a "system" is used alongside that of "signal". This may be because we wish to relate two (or more) signals that occur in the analysis of a system, or it is often convenient to regard a signal as having arisen as the output of a system (which may be "real" or simply a convenient model). This section describes some types of system.

2.1 Linear Systems

Figure 6 depicts a single input–single output system.

The link between input $x(t)$ and output $y(t)$ is linear if the concept of superposition applies – i.e. if $y_1(t)$ is the response to $x_1(t)$ and $y_2(t)$ is the response to $x_2(t)$ then the response to $ax_1(t) + bx_1(t)$ is $ay_1(t) + by_2(t)$.

Systems are often described by their response to a particular input – this provides both a characterisation of the system and the basis of how the response to more complex inputs may be built up. A fundamental description of a linear system is the response to a unit impulse (delta function). This is the impulse response function denoted $h(t, t_1)$ which is the response at time t due to the application of a unit impulse at time t_1.

Because the linearity of such a system is implied if an impulse of strength a is applied at t_1 and one of strength b is applied at t_2, the response to both is $ah(t, t_1) + bh(t, t_2)$.

The fact that the impulse response is a function of both t and t_1 indicates that the system, whilst being linear, may be *time variable*. The system is called *time invariant* if $h(t, t_1) = h(t - t_1)$, in which case what matters is the duration between t and t_1, not their explicit values.

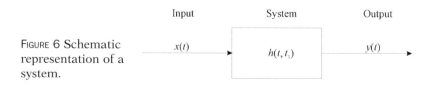

FIGURE 6 Schematic representation of a system.

A system is described as causal if $h(t,t_1) = 0$ for $t_1 > t$; in words the system only responds after the impact is applied. This clearly makes "obvious" sense for physical systems, but need not be so in mathematical models of data where the "past" and the "future" may be interpreted variously in, e.g., interpolation schemes for time histories or image processing.

A system is *stable* if $h(t,t_1)$ dies away "sufficiently quickly" as $t \to \infty$, i.e., the system "forgets" the impulse. There are several mathematical definitions of what constitutes a stable system. The most common definition is provided by the principle of bounded input bounded output (BIBO) stability, which states that a system is stable if for all bounded inputs, so $|x(t)| < M$ for all t, then one can find a value N such that $|y(t)| < N$. This is equivalent to requiring that

$$\int_{-\infty}^{\infty} |h(t,t_1)| \, dt < \infty \quad \forall t_1. \tag{7}$$

These concepts can be put together for a linear, time-invariant, causal, stable system to yield an expression relating the system's input, $x(t)$, to its output, $y(t)$, as follows:

$$y(t) = \int_{-\infty}^{t} h(t-\tau)x(\tau)\,d\tau = \int_{0}^{\infty} x(t-\tau)h(\tau)\,d\tau = h(t) * x(t), \tag{8}$$

where the operator $*$ is referred to as convolution which is defined via the above integral relationships. In the frequency domain the input/output relationship can be represented as

$$Y(f) = H(f)X(f), \tag{9}$$

where the function $H(f)$ is referred to as the transfer function and can be obtained by Fourier transforming the impulse response, $h(t)$.

These concepts extend simply to multi-input multi-output systems. If the system is excited by the P inputs $x_p(t)$ which generated the Q outputs $y_q(t)$ then the system can be characterised by the set of impulse responses (or transfer functions) that relates each of the individual outputs. Specifically

$$y_q(t) = \sum_{p=1}^{P} \int_0^{\infty} h_{p,q}(\tau) x_p(t-\tau)\,d\tau. \tag{10}$$

In the frequency domain this can be represented as a matrix product:

$$\mathbf{y}(t) = \mathbf{H}\mathbf{x}(t) \quad \mathbf{H} = \{H_{p,q}(f)\}_{p=1,\ldots,P, q=1,\ldots,Q}, \tag{11}$$

where $\mathbf{x}(t) = [x_1(t)\ldots x_P(t)]^t$ and similarly $\mathbf{y}(t) = [y_1(t)\ldots y_Q(t)]^t$.

2.2 Distributed Systems

In reality the concepts of single input–single output arise simply because we choose to look at signals aiming at single points in some field, e.g., $y(t)$ may be the output of a microphone measuring pressure at some specific spatial position:

$$y(t) = p(t; \mathbf{r}). \tag{12}$$

This means that the bigger picture relates some general excitation field x(t, r) (x ∈ X) to some response field y(t, r) (y ∈ Y) through a generalisation of a convolution, i.e.,

$$y(t, \mathbf{r}) = \iint h(t - t_1; \mathbf{r}, \mathbf{r}_1) x(t_1; \mathbf{r}_1) \, d\mathbf{r}_1 dt_1. \tag{13}$$

Every point of the input field X contributes to the response at the individual point $\mathbf{r}$ and this is contained in the spatial (three-dimensional) integral.

The impulse response function (often called the Green's function in this context) is $h(t, t_1; \mathbf{r}, \mathbf{r}_1)$, and if the system is time variant and spatially homogeneous it can be expressed as $h(t - t_1; \mathbf{r} - \mathbf{r}_1)$.

A classical example of the use of impulse responses of acoustic systems is found in the calculation of the reverberation time for an enclosure. The reverberation time is defined as the time taken for the reverberant energy to decay by a specified number of decibels, in this case 60 dB. There are a variety of methods that can be used to compute the reverberation time. For illustration the decay curve shown in Figure 7c is based on Schroeder integration [3,4], which is defined as a time-reversed integration of the instantaneous power of the signal. The Schroeder integral of an enclosure's impulse response, an example of which is shown in Figure 7a, can be used to compute the reverberation time. Note this is a very large room with an unusually long reverberation time.

2.3 Time-Varying Systems

For systems that are time-varying, and causal, the relationship between the input and output is given by a generalisation of the convolution integral, so that

$$y(t) = \int_{-\infty}^{t} h(t, t_1) x(t_1) \, dt_1. \tag{14}$$

2.4 Non-linear Systems

There is no simple relationship between the input and output of a general non-linear system. However, there are sub-classes of non-linear systems whose responses can be modelled through extensions of the linear convolution relationship. One of the most popular of which is the Volterra expansion:

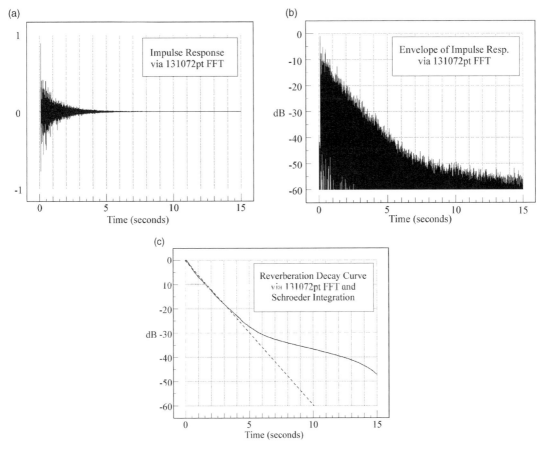

FIGURE 7 Computation of the reverberation time via the Schroeder integral (courtesy of K.R. Holland, ISVR, University of Southampton).

$$y(t) = \int_{-\infty}^{\infty} h_1(\tau) x(t-\tau) \, d\tau + \int_{-\infty}^{\infty} \int_{-\infty}^{\infty} h_2(\tau_1, \tau_2) x(t-\tau_1) x(t-\tau_2) \, d\tau_1 d\tau_2$$

$$+ \cdots + \int_{-\infty}^{\infty} \cdots \int_{-\infty}^{\infty} h_k(\tau_1, \ldots, \tau_k) x(t-\tau_1) \cdots x(t-\tau_k)$$

$$d\tau_1 \cdots d\tau_k + \cdots \tag{15}$$

This can be regarded as a generalised Taylor series expansion. To characterise the non-linear system one needs to define the set of kernel functions h_k. It is common practice to truncate this expansion at low order, e.g. the second or third terms. Such expansions have been used to describe the non-linear behaviour of loudspeakers [5].

References

1. F.J. Fahy and J.G. Walker (eds), *Advanced Applications in Acoustics in Noise and Vibration*, Spon Press, London (2004).
2. J.K. Hammond and P.R. White, "The Analysis of Non-stationary Signals Using Time Frequency Methods", *Journal of Sound and Vibration*, **190(3)** (1996), 419–447.
3. M.R. Schroeder, "New Method of Measuring Reverberation Time", *Journal of the Acoustical Society of America*, **37(6)** (1965), 1187–1188.
4. M.R. Schroeder "Integrated-Impulse Method Measuring Sound Decay Without Using Impulses", *Journal of the Acoustical Society of America*, **66(2)** (1979), 497–500.
5. A.J.M. Kaizer, "Modelling of the Nonlinear Response of an Electrodynamic Loudspeaker by a Volterra Series Expansion", *Journal of the Audio Engineering Society*, **35(6)** (1987), 421–433.

2
Acoustic Data Acquisition

David C. Swanson

The Applied Research Laboratory, The Pennsylvania State University, Philadelphia, PA, USA

By "acoustic" data we also mean vibrations in structures, internal stress waves, and structure–fluid interactions which involve acoustic radiation. There are a number of acoustic sensors that produce alternating voltage fluctuations that are then sampled at regular time intervals for analysis in the time, spatial, or frequency domains. The most essential part of data acquisition can be seen as objective signal calibration and characterization of signal-to-noise ratio (SNR). Calibration for multichannel systems involves not just signal amplitude, but relative phase or time delay between channels, as well as signal leakage, or cross-talk, between channels. Regardless of how expensive the data acquisition system is, these parameters of calibration, SNR, cross-talk, and relative phase are always an issue, albeit one that good engineering practice can manage.

1 • Electronic Noise

The electronics in the signal path from sensor to data acquisition all contain broadband noise in one form or another, and if the electrical impedance of the signal cables is high, noise interference from other electrical sources can be coupled into the sensor signals. Magnetic induction is very difficult to shield a cable from and can result in microamps of induced currents from nearby power lines. If the impedance of the signal path to ground in the sensor cable is high, say 50,000 ohms or greater, the signal current corresponding to 1 V RMS is also small (20 μA for 1 V over 50k ohms). For a high impedance signal path, it is very easy to have these weak induced currents appear as hundreds of millivolts along with the signal, especially if the cables are long. However, if one shunts the signal with say a 150 ohm resistor at the data acquisition end (assumed to have a high input impedance), and the sensor has a unity gain amplifier, the milliamps of sensor signal current driving the 150 ohm shunt resistor will be huge compared to the magnetically induced currents (still in the microamp range), and the noise interference disappears. Sometimes these magnetically coupled signals are called "ground loops", but this is actually a slightly different problem, although one also sees harmonics of the power line frequency (60 Hz in the USA, 50 Hz in Europe, etc.) which can be the same symptom.

Ground loops are caused from the sensor being locally grounded at a place some distance from the data acquisition system's ground. This is not only bad practice, but it can be a dangerous electric shock hazard if the sensor is located on equipment with high electric power such as motors, generators, electric heaters, or high-voltage systems. If a ground loop exists and an equipment electrical short occurs, the large short circuit current can make its way into the data acquisition system ground, electrifying anyone touching the chassis grounds as well as damaging the electronics. Removing all ground loops and using a single ground point not only keeps the sensor signals free of interference but also provides an added measure of safety.

If possible, one should listen to the electronic background noise, if not also examining it with an oscilloscope at every junction of the signal path from the sensor to the data acquisition system to identify and isolate grounding problems.

The following guidelines should be followed to avoid magnetic induction and ground loops in the sensor signal path to the data acquisition system:

1. Use a shielded twisted pair rather then a coaxial "BNC" cable, grounding the shield at the data acquisition system and floating the shield at the sensor.
2. Amplify the sensor signal as close to the sensor as possible. Place a resistor (150 to 2k ohm) across the twisted pair *at the data acquisition end* of a long cable run so that the current from the sensor amplifier drives a strong sensor signal current (well beyond 1 mA for a 1 V signal) in the cable relative to magnetically coupled interference, which is often in the microamp or

nano-amp range. This will greatly reduce power line and other electromagnetic interference. Most data acquisition systems have high input impedance, sometimes even in the giga-ohm range. One nano-amp of electromagnetic noise creates significant voltage interference relative to a typical sensor signal voltage when input into such a high impedance. By lowering the cable impedance and amplifying the sensor signal current, the sensor signal dominates the noise interference in the acquired data as desired.

3. Always amplify the sensor signal *before* filtering. Filters are designed for unity gain in the pass band and always have some added residual noise that should not be amplified. Many data acquisition systems have built-in signal amplifiers that may also be amplifying signal conditioner noise along with the sensor signal.
4. Calibration should be designed to be conveniently done in the field on the complete system, from sensor to digital data. This should be repeated before and after each data acquisition use to validate that nothing changed during the recording. The calibration signal should closely match the expected sensor signal in frequency and amplitude and should include the sensor. For example, devices called speakerphones or pistonphones are available for standard microphones for calibration purposes. Accelerometer calibrators are also available to provide a repeatable $1\,g$ RMS (one "g" is $9.8\,\text{m/s}^2$) acceleration at about $160\,\text{Hz}$. Calibration should be routine and records should be archived along with the sensor data.

2 • Analog-to-Digital Conversion

There are two main classes of analog-to-digital converters (ADC) called successive approximation (SA) and delta–sigma (DS) converters (also known as sigma–delta converters). Because the DS requires no external filters, it has become widespread particularly for audio-bandwidth applications. The type of built-in filtering in the DS converter can introduce signal delays as well as prevent DC coupling. However, the SA converter has an advantage of low latency in that the digital sample is available immediately and that you can select a specific external filter to suit the data acquisition requirements. This is very important for real-time control systems where acquired data must be processed to produce a control signal as rapidly as possible. Filters have a mathematical trade-off between transient response and roll-off in frequency response into the stop band. Sharper roll-off produces longer transient response "ringing". So for applications requiring real-time control and/or specific filter responses, the SA type of ADC is preferred, even though some DS converters have filters with very short delays. The DS converter typically internally samples the signal using only 1 bit at a very high rate with a modest prefilter, and then digitally filters and down-samples the data, increasing the sample word length during the processing. This processing requires processing a number of samples, thus causing a small latency which is not a problem if all one is doing in recording. Not all DS converters filter the same way, so it is very important to check the manufacturer's specifications in detail.

Digital signals are by definition band limited to a frequency bandwidth less then half the sampling frequency to insure all digital frequencies have corresponding analog frequencies. If a frequency outside the bandwidth defined by the sample rate and type is digitized, it will appear as an *aliased* frequency in the digital signal. The anti-aliasing filters used to limit the analog signal bandwidth prior to sampling generally should begin their roll-off at around one third of the sample rate. They should have a sharp roll-off of 48 dB/octave or greater so that any aliased frequencies are well attenuated in the digital signal. However, aliased frequency residuals can often be seen if there is low background noise and the digital signal has a bit depth (word length) sufficient to detect a wide dynamic range of signals. For 12-bit systems, 48 dB/octave filtering should be sufficient, while for 16- to 24-bit systems, a roll-off of 96 dB/octave is preferred. A good system test is to input a signal at the maximum undistorted amplitude and sweep the frequency from just below 1/2 the sample rate (called the Nyquist frequency) up to the sampling frequency slowly and use a Fourier transform to measure the aliasing filter leakage.

2.1 Successive Approximation

An SA converter uses a counter, a digital-to-analog converter, and a comparator to approximate a digital number output to the analog voltage of the input. A digital-to-analog converter is simply a bank of switches and a summing circuit to add the corresponding voltages. First, the analog voltage input is sampled and held on a capacitor during the digitization process. The voltage comparator signals the counter to count downward when the analog voltage is below the current digital approximation and to count upward when the analog voltage is above the digital approximation. The SA process converges very fast to approximate the sampled analog input voltage latched on the sample-and-hold capacitor.

For 16-bit samples, the number range is 0–65,535 for single-ended and −32,768 to +32,767 (90.3 dB dynamic range) for bi-polar input voltages. For 18-bit, it is $-131{,}072$ to $+131{,}071$ (102 dB), and for 24-bit, it is $-8{,}388{,}608$ to $+8{,}388{,}607$ (138 dB) because one binary number is reserved for zero. The maximum input voltage and the bit depth determine an effective "numeric gain" for the analog-to-digital converter of

$$G_{\text{ADC}} = \frac{2^{Q-1}}{V_{\max}}, \qquad (1)$$

where Q is the number of bits (for bi-polar voltages) and $V_{\max}$ is maximum input voltage (peak, not RMS). Equation (1) provides a very useful calibration metric for an analog-to-digital conversion system. Dividing the digital samples by this factor converts the integers to volts. Additional calibration factors for amplifier gains and transducer sensitivities can be applied to save the data in calibrated engineering units, which is always preferred unless only limited storage space is available.

2.2 Increasing Bit Depth Through Integration/Decimation

One can compensate for limited bit depth by sampling at a much higher rate and averaging successive groups of samples into fewer samples (decimation) where each sample has more bits. For example, summing pairs of 8-bit samples together gives 9-bit samples at half the sample rate. The summing operation is also an effective low-pass filter with a zero at the original Nyquist frequency. Suppose we wanted to measure vibrations at 1 kHz and below that had a dynamic range of over 90 dB, but only had a 12-bit SA converter that could sample at rates up to 128 kHz. If we sampled at 12-bits and 128 kHz, we could integrate and decimate to 18-bits at 2 kHz sample rate and capture signals with up to 102 dB dynamic range. One gains 6 dB of dynamic range (1-bit) with each halving of the decimated sample rate. This is a very simple and affordable solution that is often overlooked when specifying a data acquisition system.

2.3 Delta–Sigma Conversion

If we are not interested in DC signals (although some DS converters can handle them), we can do even better than the 6 dB per halving of the sample rate by differentiating the signal samples prior to summing and decimating. This is also called "noise shaping" because we are effectively filtering with a zero at 0 Hz and a +6 dB/octave roll-up across the signal bandwidth [1,2]. Summing the differences suppresses the low-frequency side by another 3 dB, thus giving a net dynamic range gain of 9 dB for each halving of the sample rate. The differencing is the "delta" part and the summing is the "sigma" part of the DS converter algorithm. Manufacturers' literature detail many variations and enhancements on this basic approach. The differences are seen mainly in the signal latency, phase responses near the Nyquist rate, and distortions that appear for low-level signals. For a DS converter to digitize an audio CD signal at 16-bits and 44.1 kHz, it begins with a 1-bit signal sampled at around 45 MHz and probably low-pass filtered at around 100 kHz and 12 dB/octave roll-off. About 10 "delta–sigma" operations would happen to enhance the bit depth to 16 bits at 44.1 kHz. Additional digital filtering may also be done to sharpen the roll-off near the Nyquist frequency, smooth the phase response to minimize audio distortion, and reduce ringing for transient signals. It is not uncommon for all this process to introduce a net latency of anywhere from 8 to 32 samples at the final decimated rate. This is only a problem for real-time control applications. Delta–sigma converters are quite good for digital recording of acoustic and vibration signals, although the DS sampling filter can cause problem if the analog signal is multiplexed. If a sample rate below around 8 kHz is desired, an external anti-aliasing filter is usually required. Again, since not all DS converters are alike, it is very important to check the specifications of the particular DS converter chip in use because the data acquisition system manufacturers often do not provide such detail.

3 • Anti-aliasing Filters

Figure 1 shows a typical collection of low-pass filter responses. While one might be tempted to always choose a filter type based on the sharpness of the roll-off, selection is even more critical considering stop-band leakage and phase response.

First, we briefly describe the filter types. The filters in Figure 1 are all pole-zero designs with a theoretically infinite impulse response [3]. In practice, filters with a sharper roll-off will tend to "ring" longer in response to a transient or impulsive signal. But as seen in Figure 1, the sharper roll-off comes at the expense of leakage in the stop band, which is unacceptable for anti-aliasing filters. The classic Butterworth filter frequency response is designed by placing the poles on a semicircle on the left-half complex s-plane ($s = \sigma + j\omega$), where the circle crosses the $j\omega$ axis at the positive and negative cut-off frequencies. The Chebyshev type I response places poles on a semiellipse in the left-hand s-plane. This gives a response with a slight ripple in the pass band and a faster roll-off in the stop band. The greater the eccentricity of the ellipse is, the larger the pass-band ripple and stop-band roll-off becomes. Generally, a ripple of about ± 0.5 dB or less is acceptable. The Chebyshev type II filter is derived from a type I by converting it from low pass to high pass and then inverting

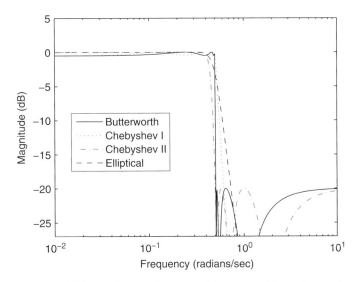

FIGURE 1 Several fifth order (36 dB/octave) low-pass filters (cutt-off at 0.5 rad/s) showing sharper roll-off for Chebyshev II and Elliptical, but also greater high frequency leakage. Butterworth or Chebyshev type I (0.5 dB ripple specified here) is most desirable as anti-aliasing filters.

the polynomial. The poles of the type I polynomial that caused pass band ripple are now zeros in the stop band for the type II filter. This leaves not only a flat response in the pass band, and a sharper roll-off, but also undesirable leakage bands in the stop band. The elliptical filter is similar to the Chebyshev type II except it is based on elliptical polynomials giving the sharpest roll-off, a flat pass band, but even greater leakage in the stop band. The sharp roll-off of the Chebyshev type II and elliptical filters also cause significant phase distortion near the cut-off frequency in the pass band. A fifth type of filter called a Bessel filter has a perfectly linear phase response (and the greatest group delay) but a very weak roll-off in the stop band [4]. While one should weigh all these factors in designing a data acquisition system, we recommend either Butterworth or Chebyshev type I anti-aliasing filters. Table 1 summarizes the filter characteristics as anti-aliasing filters showing why the Butterworth filter is desirable, especially if one samples over a wider bandwidth than needed.

Clearly, the octave from half the Nyquist frequency (one quarter of the sample rate) to the Nyquist frequency (half the sample rate) will have most of its response attenuated by the anti-aliasing filter roll-off. The sharper the roll-off, the more of this bandwidth can be used, which is why there is interest in sharp roll-off "brick

TABLE 1 Anti-aliasing filter characteristics

Filter type	Roll-off sharpness	Pass-band ripple	Stop-band leakage	Transient response
Butterworth	Weak	Excellent	Acceptable	Excellent
Chebyshev I	Acceptable	Poor	Good	Weak
Chebyshev II	Good	Weak	Poor	Acceptable
Elliptical	Excellent	Poor	Poor	Poor

wall" anti-aliasing filters. Using higher-order filters helps, but adds complexity, phase distortion, and time delay. So, in all cases, one has to deal with some level of leakage from the stop band into aliased frequencies in the pass band. It can be very difficult to recognize aliased interference, especially if it is broadband noise. Since it is always present to some degree, it is a good practice to measure the aliasing leakage by sweeping the frequency of a maximum amplitude sinusoid up past the Nyquist rate and measure the leakage directly using spectral analysis. Even delta–sigma converters with their sharp roll-off decimation filters can have leakage issues, especially if low sample rates are used.

4 • Basic Digital Signal Calibration

Calibration is an essential technique for validating the integrity of data. For example, in legal cases involving sound level measurements for noise mitigation, *one must calibrate both before and after a particular measurement* in order for the measurement to be admissible in the litigation. This is because there is no way to prove that the data acquisition system did not fail at some time during the measurement process if before and after calibrations are not available. One difficulty with electronics, battery, or connection failures when measuring noise is that the failure also produces noise. Even intermittent failures (say from a poor signal cable connection) can be identified using frequent calibration during a field test. However, to be practical, field calibration has to be fast and convenient. For microphones, products such as handheld speakerphone calibrators are an excellent choice for field work. They typically have selectable tones of 250 Hz, 1 kHz, and sound levels ranging from 94 dB (approximately 1 Pa RMS) to 124 dB. More expensive bench-type speakerphones cover many more frequencies and sound levels. The most conservative microphone calibrator is a mechanical pistonphone and barometer. The pistonphone has adapters to compensate for the air volume displaced by different-sized microphones and a barometer to compensate for the air density effect on the generated sound level in the chamber. For accelerometers, handheld calibrators typically produce 1g RMS (9.8 m/s²) at around 150 Hz as a reference and have a threaded hole for securely mounting the vibration sensor. Finally, a non-sensor calibration source is also quite convenient such as a 1 V RMS oscillator. These can be obtained with a convenient BNC-type connector for signal calibration of everything except the sensor.

4.1 Time Domain

Calibration of digital data is most often done using a simple "calibration constant" to convert the digital integers from the data acquisition system to floating point numbers in engineering units (Pascals, m/s², g's, etc.). One can estimate the "open-loop" calibration constant by collecting the advertised sensor voltage sensitivity (say S in mV/Pa for a microphone), the voltage gains of any amplifiers or attenuators in the signal path G_A, and the data acquisition "numerical gain" defined in Equation (1) as G_{ADC}. The calibration constant "C" is simply the inverse of the product of all the gains and the sensor sensitivity [5]:

$$C = \frac{1}{S \cdot G_A \cdot G_{ADC}}. \qquad (2)$$

Multiplying the integers from the data acquisition system by C in Equation (2) should convert the data into the appropriate engineering units. However, as a check, one could put a known calibration signal through the system and back out the "field" calibration factor to the "open-loop" calibration factor as a reference. A good way to do this in practice is to have handy a little notepad with the estimated "open-loop" factor, the actual measured "field" factor, and a date and time. If one uses a field calibrator such as a speakerphone, a simple list with calibration data, time, and level will suffice. These entries are usually RMS signal averages. If a battery is near failure, a sensor is damaged, or a connection is intermittent, the RMS levels will show more than a few percent change, giving you an important clue to check the system before proceeding. Pick a quiet place for calibration. One should not try to calibrate a microphone outdoors on a windy day, near loud noise sources, or using a speakerphone sound level significantly louder or quieter than the signals one expects to record. Likewise, one should not try to calibrate an accelerometer in a moving vehicle, a boat in high sea state, or on a factory floor with machinery running. Isolation of the calibration signal from background noise is essential.

4.2 Frequency Domain Calibration

Using fast Fourier transforms (FFT) and spectral analysis offers an even more precise calibration technique than the basic time-domain calibration factor. This is because spectral analysis can separate sinusoids from broadband background noise. The larger the size N of the FFT, the greater the signal-to-noise ratio (SNR) gain ($\sqrt{N/2}$) will be in the frequency domain. The estimated spectrum is often averaged over a period of time which averages variations in the inherent background noise, although this does not enhance the SNR. Finally, one often uses a data window, such as a Hanning (cosine tapered) window, on the data prior to an FFT to reduce spectral leakage. All of these factors must be taken into account when preparing a frequency domain calibration [6].

We begin with a known sinusoidal signal of amplitude A in appropriate engineering units. Calculation of an N-point FFT of the real-valued signal (we assume the imaginary part of the FFT input is zero) will yield an amplitude of $AN/2$ if the frequency of the sinusoid lies *exactly* on one of the FFT frequency bins, that is, some multiple of Fs/N Hz. If the sinusoid is in between adjacent FFT bins, the amplitude will be less than $AN/2$ by as much as 3 dB, depending on which data window is used. Clearly, spectral leakage must be managed for any meaningful frequency domain calibration. Unfortunately, this important detail has rarely been addressed in the literature. To produce a calibrated spectrum, we first choose the appropriate frequency domain units, either RMS units or spectral density (units-squared per Hz). RMS units are convenient when analyzing sinusoidal signals or acoustic responses with a few narrowband resonances. For broadband acoustic signals, the units for spectral density are preferred.

Table 2 summarizes frequency domain calibration and the effects of the data window on narrowband sinusoids and broadband noise. We assume an N-point FFT on a sinusoidal signal of amplitude A in Gaussian noise with variance σ^2 and a sample rate of Fs Hz. We use two-sided frequency domain calibration meaning that a real time-domain signal is assumed and the calibration is applied to the positive frequencies only. The "RMS"-calibrated FFT scales the raw FFT output so that a bin-aligned sinusoid with a rectangular window produces the bin amplitude in the correct RMS units. The "narrowband" correction factor is just a scale factor for the type of data window used. If the sinusoid frequency is not bin aligned, or if a non-rectangular data window is used, one must sum the squared RMS bin values over a frequency range capturing the spectral leakage and take the square root to produce the correct RMS level of the sinusoid. The power spectrum, often called the periodogram, produces the correct noise variance level in

TABLE 2 Frequency domain calibration techniques

Data type	Formula	Signal level	Noise level	Units
Time domain	$[y_1 y_2 \cdots y_N]$	$\frac{A}{\sqrt{2}}$	σ	RMS units
Basic FFT	$\eta^{-1} \sum_{n=1}^{N} W_n y_n e^{-j\omega n}$	$\frac{NA}{2}$	$\sigma\sqrt{N}$	units/bin
"RMS" FFT (two sided)	$\frac{\sqrt{2}}{N} \eta^{-1} \left\| \sum_{n=1}^{N} W_n y_n e^{-j\omega n} \right\|$	$\frac{A}{\sqrt{2}}$	$\frac{\sigma}{\sqrt{N/2}}$	RMS units/bin
"Power spectrum" (two-sided)	$\beta^{-1} \frac{2}{N} \left\| \sum_{n=1}^{N} W_n y_n e^{-j\omega n} \right\|^2$	$\frac{NA^2}{2}$	σ^2	Units^2/bin
"Spectral density" (two sided)	$\beta^{-1} \frac{2}{Fs} \left\| \sum_{n=1}^{N} W_n y_n e^{-j\omega n} \right\|^2$	$\frac{N^2 A^2}{2Fs}$	$\frac{\sigma^2}{Fs/N}$	Units^2/Hz
"Narrowband" scale factor η	$\eta = \frac{1}{N} \sum_{n=1}^{N} W_n$	Normalizes window amplitude to match rectangular window allowing narrowband signal calibration		
"Broadband" correction factor β	$\beta = \frac{1}{N} \sum_{n=1}^{N} |W_n|^2$	Normalizes window spectral leakage allowing broadband signal calibration		

every bin but only if the broadband correction factor is applied to account for spectral leakage. The spectral density, sometimes called the power spectral density (**PSD**), simply normalizes the power spectrum by the width of each bin in hertz. Many electronic component noise specifications are given as the square root of the PSD (i.e., units of $V/\sqrt{Hz}$). Calibration of broadband signals applies to the power spectrum and spectral density while narrowband signal calibration in the frequency domain should use the RMS FFT normalization and summing across the spectral leakage around the peak.

Averaging large numbers of non-overlapped spectral "snapshots" of sinusoids in Gaussian noise will yield a smoother looking spectrum but will not change the SNR. The noise statistics start out at the first average, with the mean-square and variance (over frequency bins) equal. After M spectral averages, the noise level mean stays the same while the standard deviation of the noise estimated over the frequency bins reduces by $\sqrt{M}$. Thus, the noise spectrum appears "smoother" over frequency, but its mean level is unchanged.

This implies that the relationship between the noise level and the noise standard deviation in the spectrum can tell you approximately how many averages are in the spectral estimate (if the underlying noise is a single Gaussian process). This is true only approximately but is a very useful relationship. It can be seen that each frequency bin of the averaged spectrum has Gaussian probability density with mean equal to the spectral noise mean and standard deviation equal to $1/\sqrt{M}$ of the noise mean. For a bin with a sinusoid plus the noise, the noise mean is now shifted up by the RMS sinusoid amplitude, but with the same standard deviations as the surrounding FFT bins. This is true for M greater than around 10 or for very high SNR. For low SNR or few averages, the noise-only bins have chi-square distribution of order $2M$, while the signal plus noise bin has a Rician distribution, which are more skewed toward zero [7].

Nonetheless, the simplified model of a shifted Gaussian for each spectral bin distribution is quite reasonable and implies that one can precisely specify a confidence interval about the averaged spectral level in each bin based on the estimated mean and standard deviation of the noise-only bins around the sinusoid bin. For example, if the RMS spectrum peak is 1 V (RMS) and the standard deviation of the surrounding noise bins is 100 mV (RMS), then the 68 % confidence interval for the spectral peak is from 1.1 to 0.9 V (RMS). For precision calibration, statistical metrics such as these are important to report explicitly along with the estimated calibration data. For high SNR and reasonable numbers of averages, the estimated spectrum is quite accurate. However, this conversely implies a minimum SNR for which the calibrations meet or exceed the defined confidence interval.

5 • Multichannel System Calibration

Many data acquisition systems for acoustic and vibrations measurements use an array of sensors at known locations to spatially process the waveforms for propagation direction, mode measurement or isolation, intensity flux, or wavenumber interference measurements for holographic field reconstructions. These systems require very precise spatial phase calibration and are intolerant of uncorrected time delays between channels. However, most data acquisition systems share the actual analog-to-digital converter device between channels using a multiplexing scheme. Signals can leak or "cross-talk" between channels and there can be a common noise signal across all the channels. Therefore, it is very important to characterize these effects through direct measurement, remove any biases such as multiplex time delays, and provide estimates of the single and cross-channel SNR and coherence.

5.1 Simultaneous Sample and Hold

This is a feature of many data acquisition systems designed for multichannel use. The sample-and-hold circuit is like a snapshot frame of a movie. The moving signal voltage is captured by quickly charging a capacitor (or a more sophisticated sample-and-hold circuit) to meet the signal voltage and then isolating the capacitor to hold the voltage until the analog-to-digital conversion is completed. Simultaneous sample and hold implies an array of these switched capacitors to take a snapshot of a group of channels where the analog-to-digital converter converts each voltage to a number very quickly so that all channel conversions in the group are completed prior to the next sample cycle. There is no time delay for each channel simultaneously sampled in the group, which is obviously what is preferred. However, for the capacitors to charge quickly and the captured voltages to have minimal drift, the electrical impedance is quite high, making these circuits susceptible to electronic noise interference and cross-talk, sometimes caused by a weak design element in the analog converter switching channels. A good way to measure cross-talk and noise is to connect a circuit short to every other channel, with maximal level different frequency sinusoids in the non-shorted channels. In a good data acquisition system, the cross-talk and noise in the shorted channels is quite small, but is still measurable using large FFTs and spectral averaging.

The specifications for sophisticated data acquisition systems can be misleading. For example, a PCI-bus data acquisition board may have eight simultaneously sampled channels each with a multiplexer for up to eight channels and advertise up to 64 data channels with simultaneous sample and hold. However, acquiring all 64 channels, one only has groups of eight that are simultaneously sampled where each group is delayed by 1/8 the sample time interval. This practical limitation introduces a time delay bias that can easily be removed in the frequency domain and verified using broadband frequency domain calibration.

5.2 Multiplexed Channel Phase Correction

For systems where a fixed time delay exists between channels due to a multiplexed converter, one can connect a broadband signal to the two channels simultaneously and measure the transfer function and coherence between the two channels. The broadband input can be either a natural Gaussian noise from an analog electronic circuit, a pseudo-random maximal length noise sequence, or a frequency chirp. Chirp signals should cycle synchronized with the FFT data blocks so that the chirp in each block is identical. The signal need not be spectrally white, but it should cover the spectrum of interest with a reasonable SNR. A reasonable number of spectra are averaged to maintain a small confidence interval for each spectral bin. With the same signal patched into both channels, an ideal intra-channel transfer function is unity amplitude with zero phase and unity coherence. The coherence $\gamma_{xy}^2(f)$ is a very useful quantity to measure because it provides a measure of the SNR and the transfer function errors [8]. The coherence is the magnitude-squared of the cross-spectrum average divided by the product of the input and output autospectra. One channel can be considered as the "input", while the other channel is considered the "output".

The SNR is calculated from coherence by

$$\text{SNR}(f) = \sqrt{\frac{\gamma_{xy}^2(f)}{1 - \gamma_{xy}^2(f)}}; \qquad (3)$$

so, as the coherence approaches unity, the SNR is approaching infinity. The transfer function amplitude error variance is approximated using

$$\sigma_{xy}^2(f) = \frac{1 - \gamma_{xy}^2(f)}{2M\gamma_{xy}^2(f)} |H(f)|^2, \qquad (4)$$

where $|H(f)|^2$ is the magnitude-squared of the transfer function and M is the number of averages used. The phase error standard deviation is estimated using

$$\sigma_{\theta_{xy}}(f) = \tan^{-1}\left[\sqrt{\frac{1 - \gamma_{xy}^2(f)}{2M\gamma_{xy}^2(f)}}\right]. \qquad (5)$$

The statistical metrics for M spectral averages of the transfer function in Equations (4) and (5) provide a means to define a confidence interval around the measured cross-channel amplitude or phase bias.

This cross-channel bias is corrected by simple complex spectral normalization. If the measured transfer function H is based on channel 2 as output (y) and 1 as input (x), channel 2 can be made electrically identical to channel 1 by dividing it (divide the amplitude and subtract the phase) by this transfer function. As such, every channel in the array can be normalized to channel 1 in this manner, thus removing any amplitude or phase variations across the array in the frequency domain. If time-domain transfer function corrected signals are needed, one would invert the transfer functions and inverse Fourier transform to create a non-causal FIR filter for each channel for postprocessing, which would be very computationally intensive. Non-causality is needed because some channels may require a time advance relative to the reference channel.

6 • Data Validation and Diagnostic Techniques Summary

This section has briefly outlined the most important considerations for data acquisition and calibration and how to validate digital data. Below is a summary checklist outline.

1. Signal integrity

    (a) Keep line impedances low. Amplify at the sensor and drive a strong current across a low impedance resistor at the data acquisition end.
    (b) Use twisted shielded pair cables with the shield floated at the sensor end.

2. Closed-loop calibration

    (a) Adopt or develop a convenient field calibration technique using a NIST-traceable source and the sensor in the loop.
    (b) Repeat calibration recordings frequently and always before and after critical recordings.
    (c) Keep a calibration log with date, time, and calibration factors or levels (for a specific calibration source).

3. SNR assessment

    (a) Measure the system SNR using a maximal amplitude sinusoid where the background noise is estimated away from obvious spectral leakage.
    (b) Record the environmental background noise in separate file as needed to describe the background in the absence of any signal.
    (c) Short the data acquisition input and measure the residual electronic noise spectrum using a large number (40–100) of spectral averages.

4. Multichannel issues

    (a) Short every other input channel and input channel specific frequencies at maximal amplitude to measure cross-talk. Repeat with the sinusoids and shorted channels switched.
    (b) Input a common broadband signal into all channels simultaneously and record the transfer functions and coherence from all channels (as outputs) relative to a reference channel input (say channel 1).
    (c) Repeat 4b with all channels shorted to measure the common noise levels and coherence.

References

1. P.M. Aziz et al., "An Overview of Sigma-Delta Converters", *IEEE Signal Processing Magazine* (Jan. 1996), 6183.
2. M.W. Hauser, "Principles of Oversampling A/D Conversion", *Journal of the Audio Engineering Society* **39(1/2)** (Jan. Feb. 1991).
3. A.V. Oppenheim, A.S. Willsky and Ian T. Young, *Signals and Systems*, Prentice – Hall International, London (1983).
4. T.W. Parks and C.S. Burrus, *Digital Filter Design*, Wiley, New York (1987).
5. D.C. Swanson, *Signal Processing for Intelligent Sensor Systems*, Chapter 1, Marcel Dekker, New York (2000).
6. D.C. Swanson, *Signal Processing for Intelligent Sensor Systems*, Chapter 6, Section 1, Marcel Dekker, New York (2000).
7. W.S. Burdic, *Underwater Acoustic System Analysis*, 2nd ed., Chapters 9 and 13, Prentice-Hall, Englewood Cliffs (1984).
8. J.S. Bendat and A.G. Piersol, *Engineering Applications of Correlation and Spectral Analysis*, Wiley, New York (1993).

3
Spectral Analysis and Correlation

Robert B. Randall

University of New South Wales, Sydney, Australia

1 • Introduction

Spectral analysis is the process of breaking down a signal into its components at various frequencies, and in the context of acoustics there are two very different ways of doing this, depending on whether the result is desired on a linear frequency scale with constant resolution (in Hz) or on a logarithmic frequency scale with constant percentage resolution. The fundamental connection between the time domain and the frequency domain, the Fourier transform, is most easily interpreted in terms of linear time and frequency scales, at least in the practical version now used to calculate it, the FFT (fast Fourier transform). However, expressing a spectrum on a linear scale automatically restricts its frequency range, since the upper frequency decade occupies 90% of the scale and the upper two decades 99% of the scale. The human ear has a frequency range of three decades (20 Hz–20 kHz) which can only be expressed on a logarithmic frequency scale with constant percentage bandwidth, i.e., where the bandwidth of a filter at any frequency is a fixed percentage of its center frequency, as explained in Chapter 7. Moreover, the human ear tends to interpret equal intervals on a logarithmic frequency scale as equal steps (e.g., octaves with a frequency ratio of 2 between the highest and lowest frequencies covered by the band, or 1/3-octaves with a ratio of $2^{1/3}$). For estimating spectra in terms of $1/n$-octaves, it is most efficient to pass the signal through a series of filters with constant percentage bandwidth and measure the power transmitted by each filter. The filters can be either analogue or digital, but there is an increasing tendency to use the latter, since, as explained below, by efficient "timesharing" it is possible to use a very limited number of actual filter units to filter in parallel over any number of octaves, simply by processing data streams with different sampling frequencies varying in octave (2:1) steps.

The "power" of a signal is directly related to the square of its instantaneous value, but is normally interpreted as its "mean square" value, obtained by averaging the instantaneous value over a defined averaging time (or over an ensemble of realizations for a random process). For continuous deterministic signals made up of sums of sinusoids, the mean square value of each sinusoid is half the square of its amplitude or peak value, and the total power or mean square value is simply the sum of the powers of the components. This is one version of Parseval's theorem. For stationary random signals, the power is continuously distributed with frequency (rather than being concentrated at a single frequency as for a sinusoid), and spectra are expressed in terms of "power spectral density" or PSD, which has to be integrated over a finite frequency band to give a finite power. The version of Parseval's theorem applicable to them is that the total mean square value as calculated in the time domain is equal to the integral of the PSD over all frequency. Power is meaningful for stationary or slowly varying signals, but for transient events, such as vibratory shocks or acoustic pulses, the fundamental quantity is their "energy" obtained by integrating their instantaneous "power" over their total length. The square of the magnitude of their Fourier transform gives the distribution of their energy with frequency, the so-called energy spectral density or ESD, and the version of Parseval's theorem applicable to them is that the integral of ESD over all frequency equals the total energy obtained by integrating the instantaneous power over all time. Attention is paid to these differences in dimensions and units in the following sections where the practical spectral analysis of different types of signals is discussed in detail.

As will be shown later, correlation functions are the equivalent of spectral functions in the time domain, being related to them by a Fourier transform. They express the relationship between signals (possibly the same signal) with various amounts of delay, so that for example periodic functions continue to correlate perfectly every period, whereas white noise does not correlate with itself for even the smallest delay, while narrow band noise does correlate over a number of periods of the basic oscillation, but the correlation deteriorates with increasing delay. The cross-spectrum (Fourier transform of the cross-correlation) between two signals and their two autospectra can be used to calculate their coherence, a measure of the degree of linear relationship between the two signals.

Theoretical frequency spectra represent averages over all time, but the ear is able to discern the way in which frequency patterns change with time, and so various time/frequency analyses have been developed to represent such changes in frequency patterns with time. A basic limitation here is the "Heisenberg uncertainty principle" which states that the product of the resolution in the time and frequency domains is a constant so that improvement in one comes at the cost of deterioration in the other. The various forms of time/frequency analyses discussed below represent different attempts to overcome this problem, for example to better represent what the human ear can achieve.

2 • The Fourier Transform and Variants

The basic connection between the time domain and the frequency domain is the Fourier transform, with the following equation [1]

$$G(f) = \int_{-\infty}^{\infty} g(t)e^{-j2\pi ft} dt, \qquad (1)$$

which transforms the time signal $g(t)$ into the frequency spectrum $G(f)$. For causal time signals (e.g., the impulse response function of all physical systems), the Fourier transform is equal to the Laplace transform evaluated along the imaginary axis of the Laplace plane (where the Laplace variable $s = j\omega = j2\pi f$) [2].

The time signal can be retrieved from its Fourier transform by the inverse transform

$$g(t) = \int_{-\infty}^{\infty} G(f)e^{j2\pi ft} df, \qquad (2)$$

which can be seen to be almost identical to (1), so that many properties of the forward transform apply equally to the inverse transform, for example the convolution theorem [2]. The latter states that a convolution in one domain corresponds to a multiplication in the other domain. For a convolution in the time domain represented by

$$g(t) = f(t) * h(t) = \int_{-\infty}^{\infty} f(\tau) h(t-\tau) d\tau \qquad (3)$$

where the asterisk is a shorthand way of representing the convolution operation, this corresponds to the following formula:

$$G(f) = \Im\{f(t) * h(t)\} = F(f) \cdot H(f), \qquad (4)$$

where $\Im\{.\}$ represents the Fourier transform of the argument, and the uppercase frequency variables represent the Fourier transforms of the corresponding lowercase time variables. Since the convolution operation of (3) represents the output $g(t)$ of any physical system with impulse response function $h(t)$ subject to the forcing function $f(t)$, it is evident that the corresponding multiplicative relationship of (4) is much simpler.

2.1 The Fourier Series

Strictly speaking, the Fourier transform of (1) only applies to transient functions (whose integral over all time is finite) [1], but using the convolution theorem it can be extended to the case of periodic functions as illustrated in Figure 1 [2]. Any periodic function can be generated by convolving one period (of length T) with an infinite train of unit delta functions with spacing T. Since the Fourier transform (actually Fourier series) of such a train of delta functions is another train of delta functions with scaling $1/T$ and spacing $1/T$, the spectrum of the periodically repeated transients is the product of the spectrum of the single transient with the scaled train of delta functions, thus sampling it at intervals of $1/T$, and multiplying these values by $1/T$. Thus, the Fourier series spectrum for a transient $g(t)$ with Fourier transform $G(f)$, repeated with a period of T, can be calculated as

$$G(f_k) = \frac{1}{T} G\left(\frac{k}{T}\right) = \frac{1}{T} \int_{-T/2}^{T/2} g(t) e^{-j2\pi kt/T} dt. \qquad (5)$$

This is the complex version of the better known expression where a periodic signal $g(t)$ is expressed in terms of sine and cosine functions:

$$g(t) = \frac{a_0}{2} + \sum_{k=1}^{\infty} a_k \cos(k\omega_0 t) + \sum_{k=1}^{\infty} b_k \sin(k\omega_0 t), \qquad (6)$$

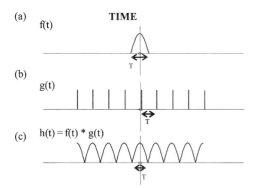
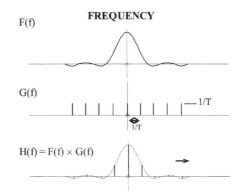

FIGURE 1 Illustration of how the Fourier series of a periodically repeated transient (a) can be obtained by sampling the Fourier transform of the transient.

where

$$a_k = \frac{2}{T} \int_{-T/2}^{T/2} g(t) \cos(k\omega_0 t) dt \qquad (7)$$

and

$$b_k = \frac{2}{T} \int_{-T/2}^{T/2} g(t) \sin(k\omega_0 t) dt, \qquad (8)$$

meaning that

$$G(f_k) = \frac{a_k}{2} - j\frac{b_k}{2} \qquad (9)$$

Note that (6) is for zero and positive frequencies only, whereas the frequency components of (5) extend from minus to plus infinity. Each sinusoidal component is made up of a sum of a positive and a negative frequency component, each with half the amplitude of the corresponding sinusoid (as indicated by (9)) since when they align, the total amplitude is double that of the individual components. This fact is important in scaling frequency spectra.

2.2 The Discrete Fourier Transform (DFT)

Equations (1) and (2) can be replaced by the following versions, when the signals are discretely sampled (in both time and frequency domains) and of finite length (implicitly one period of a periodic signal) [3,4]:

$$G(k) = \left(\frac{1}{N}\right) \sum_{n=0}^{N-1} g(n) e^{-j2\pi kn/N}, \qquad (10)$$

$$g(n) = \sum_{k=0}^{N-1} G(k)e^{j2\pi kn/N}, \qquad (11)$$

where (10) is the forward transform and (11) is the inverse transform. This version gives correctly scaled Fourier series spectra, and the scaling must be modified if other types of signals are processed, as described below in Section 3, "Practical FFT Analysis". The widely used signal processing package Matlab® makes the division by N in the inverse transform, giving a version more closely related to the Fourier transform pair of (1) and (2). However, the sum will then only represent the integral (over a finite length time signal) when the forward transform is multiplied by Δt, the time interval between samples.

Figure 2 shows the relationship between the Fourier transforms of (1) and (2) and the DFT of (10) and (11). It will be seen that the discrete sampling in each domain implies periodicity in the other domain so that the second half of the spectrum (from $f_s/2$ to f_s, where f_s is the sampling frequency) actually represents the negative frequency components from $-f_s/2$ to zero. This highlights the fact that a time signal sampled at f_s must not contain frequencies higher than half the sampling frequency before digitization, as discussed below.

2.3 The Fast Fourier Transform (FFT)

The FFT is simply a very efficient way of calculating the DFT using a number of operations of the order of $N \log_2 N$ instead of the N^2 required for a direct calculation [3–5]. For a typical transform size of 1,024 (2^{10}), this is more than 100 times faster. However, the FFT has all the properties of the DFT, and in particular the three "pitfalls" as described below in Section 3, "Practical FFT Analysis".

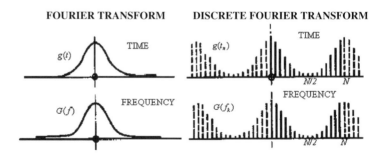

FIGURE 2 Relationship between the Fourier transform and the DFT.

2.4 Zoom FFT Analysis

As can be seen in Figure 2, the DFT always covers a frequency span from zero to half the sampling frequency. Sometimes it is desired to analyze a smaller range of frequency with high resolution, and this can be achieved by "zoom FFT". It is of course possible to achieve a high resolution by simply increasing transform size N, but a large transform size can be avoided if the signal can be preprocessed in real time by a zoom processor [4]. As illustrated in Figure 3, the real-time signal is first displaced in frequency by multiplying it by the (complex) unit vector $\exp(-j2\pi f_c t)$, which subtracts f_c from every frequency in the signal and shifts the frequency origin to f_c. The resulting signal is complex, but can be lowpass filtered by a digital filter to remove all frequency components outside the desired zoom band. The lowpass filtered signal can be resampled at a correspondingly lower sample rate without aliasing, meaning that a smaller number of samples N will correspond to a given record length T in seconds. Independent of the frequency shift, the frequency resolution is always given by $1/T$ so that a finer resolution is achieved in the zoom band. The FFT obtains the Fourier spectrum of the complex frequency-shifted time signal, which can simply be displayed as representing the original zoom band, rather than plus/minus half the new sampling frequency. The lowpass filtering and resampling can conveniently be done in octave steps, as explained below for digital filters, and so zoom is most commonly available in factors corresponding to a power of 2.

It is perhaps worth mentioning that the complex output of a zoom processor can also be used for demodulation, since its amplitude (modulus) represents the amplitude modulation of the original signal, and its phase the phase modulation signal,

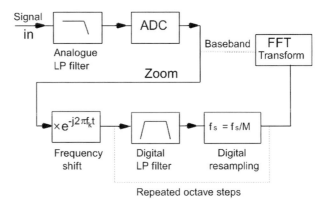

FIGURE 3 Schematic diagram of FFT zoom process.

for demodulation at the zoom center frequency [2,4]. The phase will have to be "unwrapped" to a continuous function of time (for example, using the Matlab® function UNWRAP) and can be differentiated to give the frequency modulation signal.

3 • Practical FFT Analysis

3.1 Aliasing, Windows, Leakage, and Picket Fence Effect

Aliasing, leakage, and picket fence effect are the three pitfalls of the FFT referred to above [4]. Figure 2 shows that before analogue to digital conversion (ADC) the signal must be lowpass filtered to exclude frequencies above half the sampling frequency because the periodic nature of the spectrum will otherwise cause "aliasing". To be able to use 80% of the calculated spectrum, the antialiasing filter must be very steep (120 dB/octave).

Because the signal record transformed by the FFT is effectively joined into a loop and repeated periodically, for general signals this can give a discontinuity at the join which is not part of the actual signal, and which modifies the results. The effects can be reduced by multiplying the signal record by a "data window", since otherwise the signal is effectively multiplied by a rectangular window of length equal to the record length. By the convolution theorem, this corresponds in the frequency domain to a convolution with the Fourier transform of the window function (for a rectangular window, a sin x/x function). The Fourier transform of the window function thus acts as a filter characteristic, with which each frequency component is replaced. This phenomenon is known as "leakage", since power originally at one frequency is leaked into adjacent frequencies in proportion to the square of this filter characteristic. The actual leakage depends not only on this filter characteristic but also on how it is sampled in the frequency domain. The sampling in the frequency domain results in another window property known as the "picket fence effect". If a sinusoidal frequency corresponds to an integer number of periods along the record length, it will coincide with a frequency line, and the sin x/x window characteristic will be centered on this line, and thus sampled at the zero crossings for all other frequency lines. In this special case, neither the leakage nor the picket fence effects will be visible, but in practice this can rarely be achieved for more than one periodicity at a time, and the frequency characteristic will be sampled at other than the zero crossings. The various window functions are usually compared for the worst case situation where the frequency is halfway between two lines (and, for example, the sidelobes of the sin x/x function sampled at their peaks).

Leakage is primarily indicated by three parameters, the highest sidelobe (−13.4 dB for the rectangular window), the rate of rolloff of the remaining sidelobes (−20 dB/decade for rectangular), and the noise bandwidth compared with the line spacing (1.0 for rectangular). The picket fence effect gives the maximum error of observed peak values compared with the actual peak (−3.9 dB for rectangular). The noise bandwidth is the width of ideal filter (i.e., rectangular in the frequency domain) that would transmit the same amount of power from a white noise source, i.e., with the same integral under the squared amplitude characteristic. Table 1 gives the properties of a number of window functions that are applied in practice to stationary signals.

The Hanning window is found in most FFT analyzers and is a useful general-purpose window with steep filter characteristic, small noise bandwidth, and reasonable picket fence effect. The flat top window is designed to eliminate the picket fence effect and is most useful when the spectrum is dominated by discrete frequency components, since their RMS values can be read directly without correction. This is particularly valuable when using a calibration signal to calibrate a frequency spectrum. However, it has a very large noise bandwidth, and so the discrete frequencies do not protrude so much from background noise in the spectrum (4 dB less than with Hanning). The Kaiser-Bessel window has the steepest filter characteristic in the sense of giving the best separation of closely spaced components of very different level, but such separation can also be achieved by using zoom.

Other windows are used for analyzing transient signals. For example, a decaying exponential window is used to force lightly damped responses to near zero at the end of the data record to avoid the discontinuity that would otherwise result from joining the signal into a loop. This corresponds to an added extra damping that is known exactly and can thus be subtracted from measured results.

TABLE 1 Properties of various windows

Window	Noise bandwidth	Highest sidelobe (dB)	Sidelobe rolloff (dB/decade)	Picket fence effect (dB)
Rectangular	1.0	−13	20	3.9
Hanning	1.5	−33	60	1.4
Kaiser-Bessel	1.8	−60	20	0.8
Flat top	3.8	−70	20	<0.1

3.2 Spectrum Averaging

After correction for picket fence effect, each estimate of a discrete frequency component will be the same, so little is achieved by averaging, except to confirm that a particular frequency component is deterministic.

For random signals, each FFT transform gives an estimate of the power or autospectrum, but a number of them must be averaged to obtain a valid estimate. The general formula for the standard deviation of the error in the RMS amplitude of an autospectrum estimate is [4,6]

$$\varepsilon = \frac{1}{2\sqrt{BT}}, \qquad (12)$$

where B is the bandwidth of each narrow band filter and T is the effective averaging time. In FFT analysis, the BT product given by each spectrum is 1, and thus for averaged FFT spectra, (12) can be used with BT replaced by n, the number of independent spectra averaged. For rectangular weighting, "independent" means from non-overlapping time records, but for the Hanning window most commonly used with random signals, up to 50% overlap of successive records can be used without significant loss of statistical reliability. Two-thirds and more overlap gives a temporally uniform time weighting, but n should still be estimated as for 50% overlap.

3.3 Spectrum Scaling

When the FFT according to (10) is used, and correction is made for picket fence effects, discrete frequency components will be scaled as half the amplitude of the equivalent sinusoidal components, as indicated in (9). Thus, the value must be multiplied by $\sqrt{2}$ to obtain the equivalent RMS value or by 2 to obtain the sinusoidal amplitude [2,4].

For random signals, the required averaging does not change the scaling, and thus the measured power obtained by squaring the value in each line must be divided by the equivalent filter bandwidth to give power spectral density (PSD). To include the power of the negative frequency components, the value obtained by squaring each positive frequency component should be multiplied by 2. The filter bandwidth by which this is to be divided is the line spacing $1/T$ multiplied by the relative noise bandwidth of the filter (as given in Table 1 for the most common windows) [2,4].

For transient signals which fit into the record length, no window need be applied, since there will be no discontinuity when the ends are joined, and so the noise bandwidth is 1. To convert to energy spectral density (ESD), the measured power in each line should be multiplied by T^2 [2,4].

4 • Spectral Analysis Using Filters

This will only be considered for stationary signals, since even though it is possible to analyze transients using filters it is quite complicated and beyond the scope of this section. An exception is the analysis of acoustic transients as expressed in terms of SEL (sound exposure level), since the result is still expressed in terms of power, as though the transient had constant power over 1 s. Thus the measurement is done in terms of power (e.g., L_{eq}) over whatever (linear) averaging time is necessary to capture the whole event, and then the L_{eq} can be converted to SEL by adding $10 \log_{10}(T_A)$ where T_A is the averaging time in seconds used for the L_{eq}. Note that the length of the event is not determined by the original signal alone. It may be dominated by the length of filter response (at low frequencies) which can be taken as $3/B$, where B is the filter bandwidth in hertz.

With $1/n$th-octave (constant percentage bandwidth) filters, the spectrum is usually expressed as power or RMS value in each band, though often expressed in dB re the appropriate reference level. The power is simply measured as the mean square value of the signal transmitted by each filter over the specified averaging time, or for stationary or slowly varying signals the equivalent obtained by exponential averaging with the corresponding time constant [4]. For exponential averaging, the equivalent linear averaging time T_A is given by $2RC$, where RC is the time constant of the electronic smoothing circuit used for the averaging, or the equivalent time constant for digital exponential averaging [4]. Thus for sound level meters set to "slow" (time constant 1 s), the equivalent averaging time is 2 s. For stationary random signals, the averaging time must be chosen in accordance with (12), keeping in mind that it will always be the lowest frequency filter that has the minimum bandwidth B, and thus the lowest BT product for a given averaging time [4].

$1/n$th-octave filters are efficiently generated digitally, since the characteristics of a digital filter are directly proportional to the sampling frequency, and so for 1/3-octave analysis, for example, only three sets of filter coefficients are required for an octave. If the sampling frequency is halved (by discarding every second sample), the same coefficients generate the 1/3-octave filters one octave below, and so on. Before reducing the sampling rate, the signals must be lowpass filtered to remove the information in the highest octave, and thus avoid aliasing, but this can also be done with just one set of filter coefficients. If M is the number of samples per second in the highest octave, because the sample rate is halved for each lower octave, the total number of samples to be processed per second for any number of octaves equals

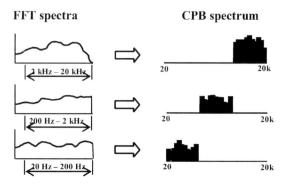

FIGURE 4 Conversion from FFT spectra to a 1/nth-octave spectrum.

$M(1+1/2+1/4+1/8+\cdots)=2M$. Thus if the processor is able to calculate twice as fast as the input sample rate, any number of octaves can be calculated in real time. The same procedure can be used for exponential averaging, as this is nothing other than a lowpass filtration by a first-order filter [4].

1/nth-octave spectra are sometimes calculated from FFT spectra by the process illustrated in Figure 4, where the upper decade of three separate FFT spectra is converted into constant percentage bandwidth and assembled into a single spectrum over three decades. The original FFT spectra are assumed to have constant PSD in each line, so the spectrum can be considered as a bar graph. The lower and upper cutoff frequencies of each 1/nth-octave filter are calculated, and the PSD in the corresponding lines (and parts of lines) of the FFT spectrum is integrated to give the total power in the 1/nth-octave filter. The problem with this approach is that the filter characteristic of the synthesized filter changes abruptly at the junctions between decades, since the slope changes by a factor of 10. The resulting filter characteristics may not meet the requirements of international and national standards [2].

To reduce this problem, some analyzers do the conversions over frequency bands of one octave rather than a decade.

5 • Correlation Functions

Correlation functions give a measure of how well one signal correlates with another as a function of the time displacement between them. The cross-correlation function between a signal $x(t)$ and a signal $y(t)$ is given in general by the expression [2]

$$R_{xy}(t,\tau) = E\left[x(t-\tau/2)y(t+\tau/2)\right], \qquad (13)$$

which gives the value centered on time t for a total displacement τ. The symbol $E[.]$ represents the expected value or statistical average over an ensemble of different realizations of the statistical process that produced the signals x and y. For stationary processes, there is no statistical variation with time, and the average can be taken over time, as represented by

$$R_{xy}(\tau) = \lim_{T \to \infty} \frac{1}{T} \int_{-T/2}^{T/2} x(t - \tau/2) y(t + \tau/2) \, dt. \qquad (14)$$

which is independent of time t, and a function of displacement τ only. Likewise because of the independence with respect to time t, the displacement does not have to be symmetric, and so the cross-correlation function is often represented as [7]

$$R_{xy}(\tau) = \lim_{T \to \infty} \frac{1}{T} \int_{-T/2}^{T/2} x(t) y(t + \tau) \, dt. \qquad (15)$$

When $x(t) = y(t)$, the so-called autocorrelation function results, giving the equivalent of (13), for non-stationary functions, as

$$R_{xx}(t, \tau) = E[x(t - \tau/2) x(t + \tau/2)] \qquad (16)$$

and of (15), for stationary functions, as

$$R_{xx}(\tau) = \lim_{T \to \infty} \frac{1}{T} \int_{-T/2}^{T/2} x(t) x(t + \tau) \, dt. \qquad (17)$$

The autocorrelation function gives a measure of how well a signal correlates with itself as a function of displacement. Any sinusoid correlates perfectly at zero displacement and for displacements corresponding to any exact number of periods. Thus, the autocorrelation function for any sinusoid (independent of initial phase) is a cosine function. Similarly, any periodic function has a periodic autocorrelation function consisting entirely of cosines so that it tends to be more impulsive than the original function (all harmonics are aligned at time zero).

It can be shown that the Fourier transform of the autocorrelation function is equal to the power spectrum, and before the advent of the FFT algorithm this was the normal means of calculating the power spectrum [6]. This gives a means of determining the autocorrelation function of noise signals as illustrated in Figure 5.

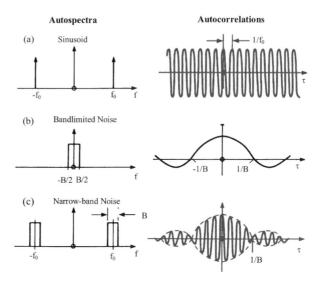

FIGURE 5 Autocorrelation vs autospectrum for three signals. Note that the spectrum of (**c**) is the convolution of (**a**) and (**b**).

Because the narrow band spectrum of Figure 5c is the convolution of a and b, the autocorrelation function of c is the product of the other two, and thus it can be seen that the "correlation length" of noise of bandwidth B is of the order of $1/B$ regardless of whether the noise is bandlimited or bandpass. This gives a limitation in the use of the autocorrelation function to detect echoes, since even if the echo is a perfect scaled copy of the original signal, there will only be a well-defined peak at the echo delay time if the bandwidth of the original signal is fairly large. The same applies when using the cross-correlation function to determine whether one signal is simply a scaled delayed version of another.

5.1 Cross-Spectrum

In the same way that the Fourier transform of the autocorrelation function is the autospectrum, or power spectrum, the Fourier transform of the cross-correlation function is the "cross-spectrum", which can also be calculated directly as [2,7]:

$$G_{xy}(f) = E\left[G_x^*(f) \cdot G_y(f)\right]. \quad (18)$$

This has an amplitude given by the product of the amplitudes of the two spectra (at each frequency) and a phase given by the difference of the two phases (i.e., the phase change from x to y). The effect of additive random noise tends to zero as a result of the averaging operation over a number of realizations.

5.2 Coherence

The coherence between two signals $x(t)$ and $x(t)$ is given by the formula

$$\gamma^2(f) = \frac{|G_{xy}(f)|^2}{G_{xx}G_{yy}} \quad (19)$$

which is effectively the (square of the) correlation coefficient between the frequency components as a function of frequency [4]. It has a value between zero and one and gives a measure of the degree of linear relationship between the two signals as a function of frequency. It is reduced by the addition of noise in either of the two signals, or if the relationship is non-linear. In practical terms, it will also be reduced by any delay between the equivalent parts of the two signals, if this is of the same order as the lengths of record used to generate the frequency spectra of the realizations. Note that the first estimate of coherence (for one realization only) will always be one, and it is only after averaging a number of realizations that the vector sum of the complex individual estimates of the cross-spectrum will be less than if they were aligned (as is the case when the relationship is fully linear).

As an example of the application of the coherence, the product of it with one of the autospectra used to calculate it gives the "coherent power", the amount of that autospectrum resulting from the linear part of the relationship between $x(t)$ and $y(t)$. This can be useful for identifying the sources of a measured spectrum [8].

6 • Time–Frequency Analysis

In principle, frequency analysis represents an average over all time, but it can be advantageous to investigate how the frequency content of a signal varies over time, as the human ear is certainly able to discern this.

6.1 Uniform Frequency Resolution

The most direct way of producing a time–frequency diagram is via the short time Fourier transform (**STFT**), which is an array of spectra obtained by moving a window (such as Hanning) along the time record and generating a spectrum for each position. This is illustrated in Figure 6, for a series of transients representing impulse responses. For some applications the individual spectra can be complex, but when they are autospectra, and particularly for sound signals, e.g., speech, the diagram is sometimes known as a sonogram.

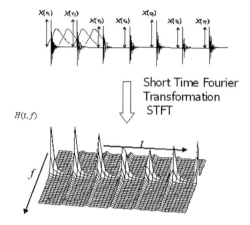

FIGURE 6 Generation of STFT.

The limitation with the STFT is that the product $\Delta t \times \Delta f =$ constant, where Δt is the resolution in time and Δf is the resolution in frequency (by the Heisenberg uncertainty principle). Thus, better resolution in one can only be obtained at the expense of poorer resolution in the other.

The Wigner–Ville distribution (WVD) was proposed as a means of obtaining better simultaneous resolution. It is defined as [9]

$$\mathrm{WVD}(t,f) = \int_{-\infty}^{\infty} x\left(t-\frac{\tau}{2}\right) x^*\left(t+\frac{\tau}{2}\right) e^{-j2\pi f \tau} d\tau. \tag{20}$$

The WVD appears to give better resolution but suffers from interference (cross) components that are difficult to distinguish from the real physical components in the signal. To overcome this problem, various smoothed versions have been proposed [3] that largely eliminate the interference at the expense of somewhat reduced resolution (but still better than the STFT).

Figure 7 gives an example comparing the STFT with the WVD, and a smoothed version, the so-called smoothed pseudo-Wigner–Ville distribution. It represents the time–frequency variations within a firing cycle of the vibration signal measured on the head of a diesel engine.

A distribution similar to the WVD is obtained by performing a Fourier transform with respect to time lag τ of the autocorrelation of (16). It is known as the Wigner–Ville spectrum (WVS). The difference is the averaging given by taking the expected value, instead of performing the Fourier transform on a single realization. Antoni [10] has shown that for second-order cyclostationary signals, the interference components average to zero,

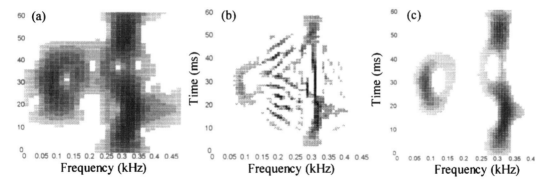

FIGURE 7 Comparison of time–frequency distributions for a diesel engine vibration signal: (**a**) STFT, (**b**) Wigner–Ville distribution, (**c**) smoothed pseudo Wigner–Ville distribution.

while the resolution remains unchanged. Figure 8 compares the WVS with the WVD for the impulse response of a non-linear single degree-of-freedom (SDOF) system. In the WVS, the decreasing natural frequency with amplitude is shown very clearly (with the same resolution as the WVD), whereas the interference terms in the latter cloud the issue. The system was excited by a burst random signal which is typically second-order cyclostationary because the mean value (first-order statistic) is zero, while the variance (second-order statistic) is a square wave (periodic).

6.2 Logarithmic Frequency Resolution

The STFT and other so-called Cohen class distributions have uniform resolution in both time and frequency, but another time–frequency representation with inherent logarithmic frequency resolution is given by wavelet analysis [11]. This is far too big a topic to be treated in detail here, but wavelet analysis can be said to be the result of filtering a signal with a number of matched filters obtained from a basic "mother wavelet" by dilation in geometric steps (thus giving a logarithmic frequency resolution), each dilation corresponding to a particular "scale", and displaced in time by amounts proportional to the wavelet length at each scale. Wavelets can be either continuous or discrete and either orthogonal or non-orthogonal.

Discrete orthogonal wavelets give the most efficient representation of a signal, with the possibility of analysis/synthesis and possible filtering in the wavelet domain. If the wavelets are chosen to be similar to features in the original signal which it is desired to enhance, then considerable denoising can be achieved by eliminating components below a defined value.

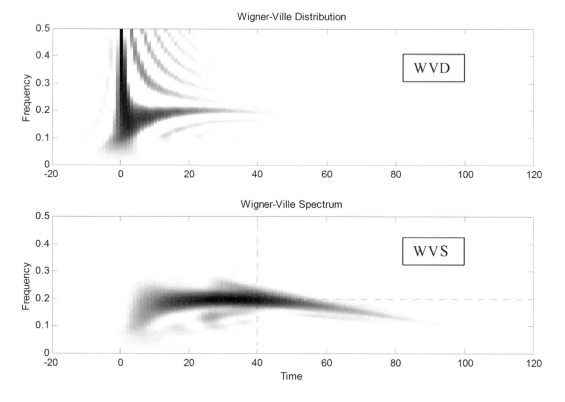

FIGURE 8 Impulse response of a non-linear SDOF system expressed as a Wigner–Ville spectrum (excitation by a burst random signal) and compared with the Wigner–Ville distribution with interference components [10].

The formula for performing a wavelet analysis is

$$W(a;b) = \frac{1}{\sqrt{a}} \int_{-\infty}^{\infty} x(t)\psi^*\left(\frac{(t-b)}{a}\right) dt, \qquad (21)$$

where $\psi(t)$ is the mother wavelet, translated by b and dilated by factor a, where a is the scale.

Continuous non-orthogonal wavelets, such as Morlet wavelets, which are Gaussian-windowed sinusoids, have properties very similar to $1/n$th-octave filters, except for being symmetrical and non-causal and thus with zero phase shift. The same can be achieved with digital $1/n$th-octave filtering. They have the property, which is advantageous in some cases, that even though they are subject to the uncertainty principle, the time resolution is better at high frequencies where the bandwidth is greater, and thus impulsive events can be well localized in time. Figure 9 (from [12]) illustrates this for a gear vibration signal where a local fault is better resolved at higher frequencies.

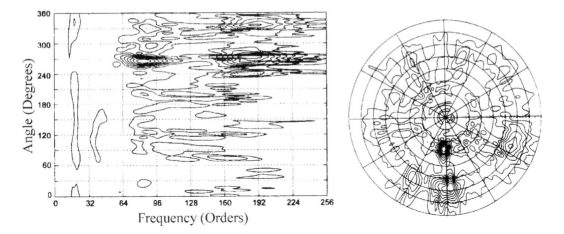

FIGURE 9 Use of wavelet transform to detect a local gear fault. (**a**) rectangular scales; (**b**) polar plot (frequency radial). From [6].

Lin and McFadden [12] proposed a radial display for this type of application (where the time axis can be related to rotation angle of a shaft) since it automatically gives better time resolution at higher frequencies when the frequency scale is radial.

References

1. Papoulis, P. (1962) *The Fourier Integral and Its Applications*. McGraw-Hill, New York.
2. Randall, R.B. (2001) "Signal Processing in Vibration Analysis". In *Structural Dynamics @ 2000: Current Status and Future Directions* (Eds. D.J. Ewins and D.J. Inman). Research Studies Press, Baldock, Hertfordshire, England.
3. Brigham, E.O. (1974) *The Fast Fourier Transform*. Prentice-Hall, New Jersey.
4. Randall, R.B. (revised ed. 1987) *Frequency Analysis*. Bruel & Kjaer, Copenhagen.
5. Cooley, J.W. and Tukey, J.W. (1965) "An Algorithm for the Machine Calculation of Complex Fourier Series". *Math. Comp.*, Vol. 19 No. 90, 297–301.
6. Bendat, J.S. and Piersol, A.G. (1971) *Random Data: Analysis and Measurement Procedures*. Wiley–Interscience, New York.
7. Bendat, J.S. and Piersol, A.G. (1980) *Engineering Applications of Correlation and Spectral Analysis*. Wiley, New York.
8. Halvorsen, W.G. and Bendat, J.S. (1975) "Noise Source Identification Using Coherent Output Power Spectra". *Sound Vibration*, Vol. 9, 15.
9. Cohen, L. (1995) *Time–Frequency Analysis*. Prentice-Hall, New Jersey.

10. Antoni, J. (2003) "On the Benefits of the Wigner–Ville Spectrum for Analysing Certain Types of Vibration Signals". Wespac8 Conference, Melbourne.
11. Newland, D.E. (1993) *An Introduction to Random Vibrations, Spectral and Wavelet Analysis*, 3rd ed. Pearson Education (formerly Longmans). Reprinted Dover, 2005.
12. Lin, S.T. and McFadden, P.D. (1995) "Vibration Analysis of Gearboxes by the Linear Wavelet Transform". *Second International Conference on Gearbox Noise, Vibration and Diagnostics*, IMechE, London.

4
The FFT and Tone Identification

John C. Burgess[1]

Department of Mechanical Engineering, University of Hawaii, 2540 Dole Street, Honolulu, HI 96822, USA, e-mail: jcb@hawaii.edu

1 • Introduction

Practical signal processing is based on the fast Fourier transform (FFT), an efficient algorithm for computing a discrete Fourier transform (DFT). While the FFT has a speed advantage, it is limited to data record lengths that are powers of 2. The slower DFT can operate on data records of any length. Except for the restriction on an FFT, *FFT* and *DFT* are interchangeable.

An FFT provides a spectral representation of a data record, a finite amount of sampled data taken from a stream of data. Because the data record has a finite length, the spectrum provided by an FFT is almost always corrupted by leakage. Data windows can reduce leakage.

When used with optimum data windows, an FFT can provide accurate estimates of sine wave parameters (tones). It can provide good estimates in the presence of moderate noise and moderate non-stationarity. This has practical application with

---

Copyright © 2004–2006 John C. Burgess. All rights reserved. No part of this manuscript may be published in any form without the prior written permission of the copyright owner.

[1]Emeritus. Home address: 1269 Mokulua Dr., Kailua, HI 96734, USA

quasi-stationary sounds, such as music, speech, bio-acoustics, and acoustic probes transmitted through the air, ocean, and biological specimens.

2 • The FFT (and DFT)

There are two distinct sets of equations. The set used here,

$$x(n) = \frac{1}{N}\sum_{k=0}^{N-1}\mathbf{X}(k)e^{j2\pi kn/N}, \quad 0 \leq n \leq N-1, \tag{1}$$

$$\mathbf{X}(k) = \sum_{n=0}^{N-1}x(n)e^{-j2\pi kn/N}, \quad 0 \leq k \leq N-1, \tag{2}$$

is based on a *z-transform* derivation (called "the discrete Fourier transform" by Oppenheim and Schafer [1]), where n is an integer time index, k an integer frequency index, $x(n)$ a discrete data record, $\mathbf{X}(k)$ a (complex) discrete spectrum, and N the length of a data record as well as the FFT (block) size. FFT designates the direct transform, Eq. (2), and IFFT designates the inverse transform, Eq. (1).

Another set of equations sometimes encountered is based on a *Fourier series* derivation [2] (Randall, Chapter 3, Eqs. (10), (11)), in which the scaling factor $1/N$ is applied to the FFT instead of the IFFT. *This slight difference can create difficulty in rationalizing published results.*

A Fourier series is defined as a one-sided, frequency-discrete function [2]. This deceptively implies that the discrete values are all that one has, and leads to terms such as the "picket-fence" effect.

2.1 An FFT Spectrum Is Not an Analog Spectrum

While an FFT spectrum has a superficial resemblance to an analog spectrum, the differences require careful interpretation. Leakage, data record, window, Nyquist frequency, and aliasing, important with digital spectra, have no meaning with analog spectra. Filter passband characteristics, important with analog spectra, have only limited meaning with digital spectra. *Reading a digital spectrum as if it were an analog spectrum can lead to significant interpretation error.*

2.2 The z-Transform and the FFT

The z-transform of a time-discrete data record is frequency-continuous, but periodic with period f_s, the sampling rate (or sampling frequency) [1]. This changes the signal analyzed into a new, periodic sequence. The FFT is defined as a sampled

z-transform. The signal "seen" by an FFT is a repeated data record, with period $T_r = N/f_s$, where T_r is the time length of a data record.

Both the z-transform and the FFT provide two-sided spectra, a feature essential to understanding leakage and its control. The limits 0 to $N-1$ on Eq. (2) are equivalent to $-N/2+1$ to $N/2$, meaning that half of the FFT represents positive frequencies and the other half negative frequencies.

These characteristics have practical effects:

- An FFT can be interpolated to whatever degree of spectral detail is desired (of the data record; seldom the true signal: Section 2.7 in [3]).
- If a data record includes an entire pulse (e.g., a finite impulse response), there is no leakage, and an FFT can provide a true spectrum of the pulse.

2.3 The Nyquist Frequency and Aliasing

The Nyquist frequency is defined as half the sampling rate, $f_{Ny} = f_s/2$. Data from an analog signal at frequencies higher than f_{Ny} are mirrored down in frequency and corrupt data at lower frequencies. For example, if the sampling rate is 10 kHz, the spectrum of a tone at 7 kHz in an analog signal will appear in the FFT spectrum centered at 3 kHz. *Once an analog signal is sampled, there is no way to separate aliased data from true data.*

There are only two ways to prevent aliasing: set the sampling rate high enough, $f_s > 2f_{max}$, or use low-pass (anti-aliasing) filters. Anti-aliasing filters have a disadvantage: the FFT spectrum will be corrupted by the filter's amplitude and phase distortion.

2.4 Frequency Relationships

Spectral plots can be presented as functions of frequency f, frequency index k, or normalized frequency, f_{norm}. Frequency and frequency index are related by $f(k) = kf_s/N$. Normalized frequency is defined by $f_{norm} = f(k)/f_s = k/N$. Corresponding frequency ranges are $-f_{Ny} \leq f \leq f_{Ny}$ for the z-transform, $-N/2 \leq k \leq N/2$ for an FFT, and $-0.5 \leq f_{norm} \leq 0.5$ for normalized frequency.

3 • Leakage

Leakage is a significant problem. It does not depend on whether the true signal is periodic or random. It can easily lead to misinterpretation.

For example, when a signal generator creates a pure tone, humans "hear" the tone and are trained to "see" a line spectrum, which has a single value at the tone frequency. But the z-transform of a data record shows tone energy spread over the entire spectrum. This is leakage. If a data record of a pure tone is not truncated exactly at an integer multiple of the tone period, an FFT will have N non-zero spectral values, none of which accurately represent tone amplitude, frequency, and phase (Section 5). Leakage can corrupt band-pass filtering (e.g., octave-band analysis) of signals containing tones.

3.1 What Causes Leakage?

An unmodified data record is "seen" through an open window (often called a rectangular window). But what about the unknown data outside the window? The misleading concept of a rectangular window implies that data outside the window are zero. This is usually incorrect in fact. It is also usually incorrect in a different way when an FFT is used. An FFT treats a data record as if it were periodically extended, not zero beyond its "ends" [3]. Almost always, this results in sharp discontinuities at the "ends" of the periodically extended data record. This results in leakage.

3.2 Can Leakage Be "Cured"?

Leakage cannot be "cured" with data records taken from extended signals. But it can be controlled by using a data window that depresses spectral window side lobes.

An attractive idea to "cure" leakage is to "taper the window gradually to zero at its ends". But this is not a necessary and sufficient requirement. While the Hann window (often called Hanning) does taper to zero at its ends, the Hamming and other optimum windows do not (Table 1).

TABLE 1 One-sided Fourier Series Coefficients (FSC) that define several data windows [1,12]

Name	$S(k)$	SAR (dB)	B_0	B_1	B_2	B_3
open	1.625	−13.3	1.0			
Hann(ing)	3.743	−31.5	0.5	−0.5		
Hamming	3.82	−42.7	0.54	−0.46		
opt43	3.946	−43.2	0.53836	−0.46164	0	
opt50	4.277	−50.0	0.50577	−0.47539	0.01565	−0.00319
Blackman	5.65	−58.1	0.42	−0.05	0.08	
opt59	5.075	−59.7	0.461445	−0.491728	0.046827	0
opt71	5.907	−71.5	0.424379	−0.497340	0.078281	0
opt80	6.644	−80.0	0.399770	−0.496413	0.101353	−0.002464
opt90	7.315	−90.0	0.379212	−0.493342	0.121205	−0.006241
opt98 [13]	7.929	−98.2	0.363579	−0.489179	0.136602	−0.010640

FSC are normalized to make data window center magnitudes 1.0. S is given in units of k as a real number. Values of B_i stated as 0 are zero in fact; those not stated are arbitrarily set to zero. The opt43 window is a more accurate Hamming window

4 • Windows

Practical windows are time-discrete functions of length N. Two kinds are used for spectrum analysis: data windows and lag windows. Data windows are used in the ordinary time domain, t, to modify a data record. Their two primary functions are to control leakage and provide insight into the parameters of tones (sinusoidal signal components). Data windows can also be used to design chirps for system identification [4] and for FIR filter design [5].

Lag windows [6] are used to modify correlation functions in the lag time domain, τ. Their primary function is to smooth fluctuations in the spectra of broadband signals. While data windows provide some smoothing, it is not significant.

The remainder of this section concentrates on the two primary functions of data windows.

4.1 Data Windows

The basic window is the open window, which is no window at all. It passes everything in a data record without change, altered only by leakage. All other windows, called data windows, modify an open window. There are a plethora of them [7].

Of all the possible applications for data windows, the two primary functions (Section 4) are the most useful. Attractive windows perform both functions. These windows are defined with one-sided Fourier series coefficients (FSC),

$$b(n) = \sum_{i=0}^{M-1} B_i \cos(2\pi i n/N), \ 0 \leq n \leq N-1, \quad (3)$$

where i is an integer, B_i the FSC, M their number, and N is even. Data windows are implemented by block multiplying a data record by the window,

$$\tilde{x}(n) = b(n)x(n), \quad (4)$$

where $\tilde{x}(n)$ represents a windowed data record.

4.2 Spectral Windows

The spectral result of Eq. (4) is mathematically equivalent to convolving the spectral window (spectrum of the data window) with the spectrum of the data record,

$$\tilde{\mathbf{X}}(k) = \mathbf{B}(k) * \mathbf{X}(k), \quad (5)$$

where $\tilde{\mathbf{X}}(k)$ represents the windowed spectrum, $\mathbf{B}(k)$ the spectral window, and $*$ represents convolution.

The practical meaning of Eq. (5) is this: the spectral features of a data window depend only on the FSC that define it, not on

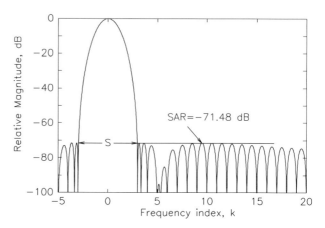

FIGURE 1 Spectrum ($20 \log_{10} |\mathbf{B}|$) of opt71 optimum data window (from [12]).

its use. The most important features of a spectral window are its selectivity, S (width of its main lobe), and its selectivity amplitude ratio, SAR (minimum depression of its side lobes) (Figure 1). Selecting a window is a trade-off between S and SAR, since it sets both.

There are two ways to define windows: with a few FSC or with equations. Table 1 lists S, SAR, and the FSC for several windows defined by a few FSC. Three windows defined by equations are the Dolph–Chebyshev [8–10], Kaiser [1,10,11], and Gaussian [10] windows.

4.3 Optimum Data Windows

All data window spectra are characterized by a main lobe and many side lobes (Figure 1). They are interdependent. Reducing side lobe amplitudes automatically increases main lobe width. An optimum data window must not only depress side lobes to control leakage, it must minimize main lobe width. Among the plethora of window shapes, only a few are optimum.

The Dolph–Chebyshev window is the best. It has the narrowest possible main lobe, and depresses all side lobes to the same amplitude. It has a computational disadvantage: $M = N$ different FSC are required for each side lobe amplitude depression and for each block size N.

The Kaiser and Gaussian windows not only have the same computational disadvantage as the Dolph–Chebyshev window, their main lobe widths S, like the Hann(ing) and Blackman windows (Figure 2), are greater than those for corresponding Dolph–Chebyshev windows at the same SAR.

Optimum data windows [12] depress side lobe amplitudes to a specified maximum value, have a main lobe width, S, nearly as narrow as a Dolph–Chebyshev window with the same SAR,

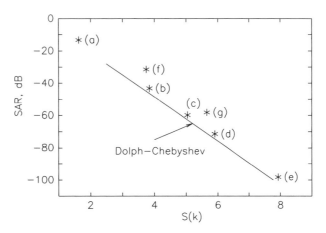

FIGURE 2 Selectivity S versus selectivity amplitude ratio SAR for various windows compared with Dolph–Chebyshev windows: (**a**) open; (**b**) Opt43 and Hamming; (**c**) opt59; (**d**) opt71; (**e**) opt98; (**f**) Hann(ing); (**g**) Blackman [12, Figure 2].

and the FSC are the same for any block size. Only four FSC are needed to depress side lobe amplitudes by 98 dB. While the FSC are intended for large N, they can provide useful results for $N \geq 16$. Figure 2 shows the relationship of various windows to Dolph–Chebyshev windows.

5 • Estimation of Amplitude, Frequency, and Phase

The amplitude, frequency, and phase angle of a pure tone are associated with the peak value of the main lobe in its spectrum (Figure 3). While discrete FFT values $X[k]$ (Fourier coefficients, often misleadingly called "bins") almost never provide this information, optimum data windows make it possible to extract accurate estimates.

5.1 Data Record

The data record for the mth single tone of frequency f_m can be expressed as

$$x_m(n) = A_m \cos(2\pi m n/N - \alpha_m), \quad 0 \leq n \leq N-1, \qquad (6)$$

where $m/N = f_m/f_s$, A_m is the signal amplitude, and α_m is the signal phase angle. The variable m is the (real) number of tone periods in the data record. It can be expressed by

$$m = \frac{T_r}{T_m} = p + q, \qquad (7)$$

where (Figure 3) p is the integer part of m, q is the fractional part of m, $T_r = N\Delta T$ is the data record period ($\Delta T = 1/f_s$ is the

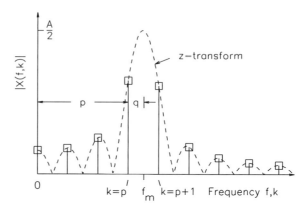

FIGURE 3 Pure tone spectrum for $m = 3.486$, showing p, q, and the principal values at $k = p$ and $k = p+1$. The *solid vertical lines* are FFT values. These and the z-transform illustrate leakage [3, Figure 12].

sampling interval), and $T_m = 1/f_m$ is the tone period. *Estimating q is the key to the procedure* (Section 5.3) [14].

5.2 Spectrum

The two-sided spectrum of a pure tone (z-transform or FFT) is mathematically the complex vector sum of two contributions, a direct contribution with its main lobe centered on the tone frequency f_m in the positive range and an image contribution centered on $-f_m$ in the negative range [14]. There appears to be no known way to separate the two after computation of an FFT.

The side lobes of each contribution corrupt the main lobes of the other. Optimum data windows provide a way around this effect by depressing the side lobes of one contribution within the main lobe of the other. When all side lobes are sufficiently depressed, the relatively simple direct contribution can provide a good approximation to the FFT [14, Eq. (3)]

$$\tilde{X}(k) \cong \tilde{X}(k)_{\text{direct}}. \tag{8}$$

Only two complex values of $\tilde{X}(k)$, called principal values, are required to obtain accurate estimates for tone amplitude, frequency, and phase angle. The principal values are the two values of $\tilde{X}(k)$ on either side of a tone frequency, at $k = p$ and $k = p+1$ (Figure 3),

$$\tilde{\mathbf{X}}(p) = \tilde{X}(p) e^{j\tilde{\phi}(p)}, \tag{9}$$

$$\tilde{\mathbf{X}}(p+1) = \tilde{X}(p+1) e^{j\tilde{\phi}(p+1)}. \tag{10}$$

A single tone has two main lobes, one in the direct contribution, one in the image contribution. When the tone frequency is near either 0 or f_{Ny}, the two main lobes may overlap. The following procedure applies as long as the main lobes are sufficiently separated (Figure 10 in [14]). The minimum required separation of the main lobe centers is

$$\Delta k_{\min} \geq \frac{S}{2} + 1 + \Delta k_{\text{shift}}, \qquad (11)$$

where S is the value of window *selectivity* (Table 1), and Δk_{shift} allows for quasi-stationarity and noise [14]. While $\Delta k_{\min}$ controls the minimum value of N required to properly analyze a single tone, values of N used in practice normally are much larger.

5.3 Procedure

The measured ratio of the two principal values is defined as

$$R = \left|\tilde{X}_p\right| \Big/ \left|\tilde{X}_{p+1}\right|. \qquad (12)$$

While small values of N can be used, increasing N decreases error by increasing the separation of direct and image main lobes. Error for a single tone signal is a minimum when the frequency is $f_m = f_s/4$, which positions it midway between zero and Nyquist frequencies. Error can be decreased significantly by using optimum windows (Table IV in [14]).

For a multi-coefficient window, R can also be specified as a function of q and the window coefficients B_i [14],

$$R = \left| \frac{q}{1-q} \sum_{i=0}^{M-1} \frac{B_i}{q^2 - i^2} \Big/ \sum_{i=0}^{M-1} \frac{B_i}{(q-1)^2 - i^2} \right|, \qquad (13)$$

which can be written as a polynomial in q,

$$\sum_{j=0}^{J-1} \sum_{i=0}^{M-1} a_{ji} B_i q^j = 0. \qquad (14)$$

Only one root of q will be in the range $0.0 \leq q < 1.0$. Optimum window coefficients B_i can be found in Table 1, and coefficients a_{ji} can be found in [14].

Accurate approximations for tone amplitude, phase angle, and frequency are given by

$$A \simeq \frac{2\pi \tilde{X}_p}{Nq \sin \pi q} \frac{1}{\sum_{i=0}^{M-1} B_i \frac{1}{(q^2 - i^2)}}, \qquad (15)$$

$$\alpha \simeq -\tilde{\phi}_p + \pi q \frac{N-1}{N} \simeq -\tilde{\phi}_p + \pi q, \qquad (16)$$

$$f \simeq \frac{p+q}{N} f_s. \qquad (17)$$

5.4 Multiple Tones

Multiple tones create multiple main lobes in the direct and image contributions. The single-tone procedure applies to each tone provided that all direct and image contributions meet the minimum separation requirement, $\Delta k_{\min}$, Eq. (11). Error may increase as a result of multiple side lobe interactions with principal values.

For each tone, determine $|m| = p$ by inspection of the spectrum, get the principal values, calculate the ratio R_m, determine q, and then the tone parameters.

5.5 Quasi-stationary Signals and Noise

When used with non-stationary signals or signals corrupted by noise, an FFT can provide only approximate values for the spectrum of the true, underlying signal. With quasi-stationary signals, there will be time intervals when the errors are small enough that results are useful. A reasonable estimate for Δk_{shift} (Eq. (11)) is necessary. Reference [14] includes discussion of expected errors resulting from moderate non-stationarity and noise.

6 • Applications

Two applications will illustrate how optimum windows can be used to provide useful data.

6.1 Probing Physical Systems

While one or more pure tones are useful as probes through physical media, the signal measured will almost always be quasi-stationary and corrupted by noise. An example concerning sound transmission through the atmosphere is given in [14].

Probe signals composed of one or more pure tones must have a minimum data record size N, a minimum sampling rate, $(f_s)_{\min}$, and are most efficiently analyzed with optimum data windows. Data record size N can be restricted by experimental factors, such as an upper limit on signal time length (e.g., the presence of echos), whether the signal is stationary or non-stationary, and the presence of unwanted experimental artifacts in the measured signal (e.g., power glitches [14]).

When the probe signal consists of a single tone, the minimum data record size is $N_{\min} = 2k_{\min}$. $N_{\min}$ varies from 4 for the open window to 10 for the opt98 window (N will normally be greater).

As an example, consider a single tone probe signal with the conditions $f_m = 1$ kHz, and $T_r \leq 0.1$ s. To minimize the effects of image side lobes, f_m should be placed at $f_{\text{Ny}}/2 = f_s/4$. This sets $f_s = 4$ kHz as the minimum sampling rate. The sampling interval $\Delta T = 1/f_s = 0.25$ ms. The condition $T_r \leq 0.1$ s sets $N_{\max} = 0.1/\Delta T = 400$. Thus $N = 256$ is the largest FFT block size meeting the test conditions. The requirement for Δk_{min} is more than met for any optimum window. A general method for designing multi-tone probe signals, with examples, is given in [14].

6.2 System Identification

Efficient waveforms for system identification must have a low peak factor and a broad spectrum. Maximum length sequences (MLSs) are well known. Linear chirps provide another. A linear chirp has the advantage over an MLS sequence that its power can be concentrated in a specified frequency range. But, when used with an FFT, it suffers from leakage. By using optimum windows, leakage in a repeated chirp, called a *choip*, can be minimized, and a choip becomes a useful tool for system identification. Details are given in [4] [and Chapter 4 by Ning Xiang].

References

1. A. Oppenheim and R. Schafer, *Discrete-Time Signal Processing*. Prentice-Hall, Englewood Cliffs, NJ, 1989.
2. R. Bracewell, *The Fast Fourier Transform and Its Applications*, 2nd Ed., Rev. McGraw-Hill Book Company, New York, NY, 1986.
3. J. C. Burgess, "Practical considerations in signal processing", in *Encyclopedia of Acoustics* (M. Crocker, ed.), vol. 3, ch. 101, pp. 1261–1279. Wiley, New York, 1997.
4. J. C. Burgess, "Chirp design for acoustical system identification", *J. Acoust. Soc. Am.*, vol. 91, pp. 1525–1530, 1992.
5. J. Proakis and D. Manolakis, *Introduction to Digital Signal Processing*. Macmillan, New York, 1988.
6. G. Jenkins and D. Watts, *Spectral Analysis and Its Applications*. Holden-Day, San Francisco, 1969.
7. F. Harris, "On the use of windows for harmonic analysis with the discrete Fourier transform", *Proc. IEEE*, vol. 66, pp. 51–83, 1978.
8. C. Dolph, "A current distribution for broadside arrays which optimizes the relationship between beam width and side-lobe level", *Proc. I.R.E*, vol. 34, pp. 335–348, June 1946.
9. H. Helms, "Nonrecursive digital filters: design methods for achieving specifications on frequency response", *IEEE Trans. Audio Electroacoust.*, vol. AU-16, pp. 336–342, 1968.
10. R. A. Roberts and C. T. Mullis, *Digital Signal Processing*. Addison-Wesley, Reading, MA, 1987.
11. L. Rabiner, B. Gold, and C. McGonegal, "An approach to the approximation problem for nonrecursive digital filters", *IEEE Trans. Audio Electroacoust.*, vol. AU-18, pp. 83–106, 1970.
12. J. C. Burgess, "Optimum approximations to Dolph–Chebyshev data windows", *IEEE Trans. Signal Process*, vol. 40, no. 10, pp. 2592–2594, 1992.

13. A. Nuttall, "Some windows with very good sidelobe behavior", *IEEE Trans. Acoust., Speech Signal Process*, vol. ASSP-29, pp. 84–91, 1981.
14. J. C. Burgess, "Accurate analysis of multitone signals using a DFT", *J. Acoust. Soc. Am.*, vol. 116, pp. 369–395, 2004.

5

Measuring Transfer-Functions and Impulse Responses

Swen Müller

National Institute of Metrology, Xerém, Brazil

1 • Introduction

The impulse response (IR) and its associated Fourier transform, the complex transfer function (TF), describe the linear transmission properties of any system able to transport or transform energy in a certain frequency range. As the name suggests, the IR is the response in time at the output of a system under test when an infinitely narrow impulse is fed into its input (Figure 1). Any other signal present at the input will be *convolved* with the system's IR $h(t)$:

$$o(t) = \int_{-\infty}^{\infty} h(\tau) \cdot i(t-\tau) \, d\tau = h(t) * i(t), \qquad (1)$$

with $i(t)$ = input signal, $h(t)$ = system IR, $o(t)$ = output signal, * = convolution symbol.

This corresponds to *multiplying* the signal's spectrum with the system's complex *transfer function* $H(f)$:

$$O(f) = H(f) \cdot I(f). \qquad (2)$$

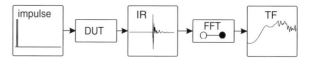

FIGURE 1 Basic setup to measure the IR of a device under test and derive its TF.

As the IR can be transformed into the TF via Fourier transform and back again into the IR via the inverse Fourier transform, both are equivalent and carry the same information, which can be extracted and visualized in different ways.

In the frequency domain, the best known is the *magnitude response*, which is the modulus of the complex TF and is most commonly plotted on a logarithmic (dB) *y*-scale. In acoustics, the frequencies on the *x*-axis are also normally plotted in a logarithmic fashion, which gives the well-known double-logarithmic display. The phase, which is the arc tangent of imaginary part divided by the real part of the complex TF, gives additional valuable information, for example to find the exact resonance frequency of a transducer. Sometimes even more important is the *group delay*, which is proportional to the negative derivative of the phase vs. frequency:

$$\tau_G = -\frac{d\varphi}{d\omega} = -\frac{d\varphi}{df \cdot 2\pi}. \quad (3)$$

The group delay is an estimate of the delay which each frequency suffers when traveling through the investigated system.

The transfer function often involves transformation from one physical entity to another, as is the case for transducers (voltage or current to sound pressure and vice versa) or in acoustics for solid materials (force to acceleration or velocity).

In room acoustics, the *room impulse response* between a sender (loudspeaker or impulsive noise source) and a receiver (microphone) plays a crucial role. It is used to evaluate a comprehensive set of parameters (most notably the frequency-dependent reverberation time) to characterize the acoustic properties of a room.

2 • Measurement Methods

Generally spoken, every kind of excitation signal can be used to determine the IR and the respective TF of a linear and time-invariant system, provided that it contains enough energy at every frequency of interest to overcome the noise floor. As can easily be seen from equation (2), the TF can be obtained by *dividing* the output spectrum of the system under test by the input spectrum:

$$H(f) = \frac{O(f)}{I(f)}. \qquad (4)$$

This usually implies Fourier-transforming input and output signals in order to perform the complex division in the spectral domain.

Alternatively to the processing in the spectral domain to obtain the TF, the IR can be retrieved directly in the time domain by convolving the DUT's response with the *inverse filter*. In case of a white excitation signal, this inverse filter simply is the *time-reversed* excitation signal (sometimes called "*matched filter*"; Figure 2). Convolution with a time-reversed signal is equal to *cross-correlation*. This plays an important role when using maximum-length sequences (MLS) as excitation signal, a special class of binary sequences with white spectrum presented in Chap. 6, because a very efficient cross-correlation algorithm operating in the time domain is available for them.

When using a non-white excitation signal, the general form to construct the inverse filter (sometimes misguidingly called "*mismatched filter*") is to Fourier-transform it, invert its spectrum (put the modulus upside down and negate the phases) and back transform it to the time domain (Figure 3). The convolution itself is also most efficiently performed by Fourier-transforming the DUT's response to the spectral domain where the convolution is replaced by the complex multiplication with the inverse filter spectrum and then performing an IFFT, which yields the IR.

All efficient modern methods for obtaining IR and TFs rely on the use of the fast Fourier transform (FFT) to switch between the

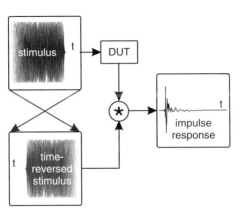

FIGURE 2 In the case of white excitation signals (here, a linear sweep has been used as example), the DUT's impulse response can be calculated by performing a cross-correlation, which is the convolution (denoted by the * *symbol*) with the time-reversed excitation signal.

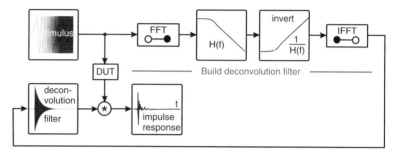

FIGURE 3 For any non-white excitation signal, the evaluation of the impulse response in the time domain can only be done by convolution with the inverse filter, which itself should be constructed in the frequency domain to yield exact results.

digitized time signal and its spectrum in the frequency domain. This has a few implications:

1. The analyzed signal must of course be band-limited to $f_S/2$ (f_S = sampling frequency) before digitalization. Care must be taken near the Nyquist frequency $f_S/2$ because the typical halfband digital filters used in modern oversampling AD converters only exhibit a fair attenuation (typically 6 dB) at $f_S/2$. Thus, a narrow frequency strip around $f_S/2$ usually contains invalid aliasing products.
2. Only signal parts with a length of 2^N samples (N = integer number) can be processed (at least for the commonly used decimation-in-2 algorithms).
3. The FFT only yields *exact* results if the signal part analyzed repeats periodically. This automatically means that the signal's spectrum becomes discrete and indeed only contains the frequencies given by $n \cdot f_S/2^N$, with N = FFT order and $n = 0, 1, 2, 3 \ldots, 2^{N/2}$.

As a consequence, the white noise constructed in the frequency domain shown later in Figure 10 will only exhibit the desired flat horizontal line in the magnitude spectrum if repeated periodically. It then consists of discrete frequencies with indeed all having exactly the same amplitude.

Conditions 2 and 3 frustrate the direct application of an FFT to maximum-length sequences (focused in Chap. 6), as these have a period length of $2^N - 1$. There is thus one value missing to complete the FFT block length and setting it to whatever value will lead to gross aberrations from the flat magnitude spectrum of a periodically repeated MLS.

While the periodic repetition of a deterministic noise signal is crucial to establish the desired spectral distribution, single pulses or sweeps do not rely on periodic repetition to establish a smooth spectrum. Thus, the latter two have to be ejected only once in a TF measurement, while deterministic periodic noise should be switched on at least one period before starting the actual measurement in order to establish a steady-state condition.

2.1 Dual Channel FFT Analysis

Instead of working with deterministic excitation signals, it is also possible to use an asynchronous signal source. As its output is not predictable, it is necessary to simultaneously capture the input and output signals of the DUT in each measurement cycle and process them in the frequency domain after FFT to obtain the TF. This leads to the classical two-channel FFT analyzer arrangement (see [1] for a comprehensive description) depicted in Figure 4.

A classical dual channel analyzer captures both the input (channel A in Figure 4) and the output signals of the DUT (channel B in Figure 4). The signal is cut in contiguous or overlapping segments which are windowed and then transformed to the spectral domain via FFT. The two resulting spectra are involved in three averaging processes: two for the autospectra

$$G_{AA}(f) = \frac{1}{n}\sum |A(f)|^2 \quad \text{and} \quad G_{BB}(f) = \frac{1}{n}\sum |B(f)|^2$$
$$(n - \text{number of averages}), \tag{5}$$

which are the squared modulus of the spectra for channels A and B, and one for the cross-spectrum

$$G_{AB}(f) = \frac{1}{n}\sum A^*(f) \cdot B(f), \tag{6}$$

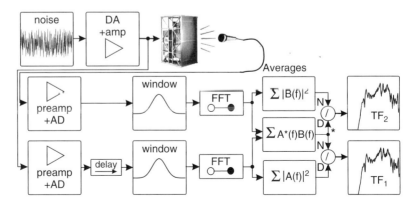

FIGURE 4 Simplified block diagram of classical two-channel FFT analysis.

which is the complex conjugate of the spectrum for channel A (spectrum A with its phases negated) multiplied with the spectrum for channel B.

While the two autospectra are real-valued and add up both signal and noise power in every measurement run, the cross-spectrum is complex, and uncorrelated noise with its random phase spectrum tends to average out with increasing number of averages.

The complex transfer function defined by

$$H(f) = \frac{B(f)}{A(f)} \tag{7}$$

can now be computed in two ways. Multiplying both the numerator and the denominator of the right side of the above equation with $A^*(f)$ yields

$$H_1(f) = \frac{A^*(f)B(f)}{A^*(f)A(f)} = \frac{G_{AB}(f)}{G_{AA}(f)}, \tag{8}$$

so $H(f)$ can be estimated by the division of the cross-spectrum with the autospectrum of the DUT's input. On the other hand, multiplying both the numerator and the denominator of the right side with $B^*(f)$ yields

$$H_2(f) = \frac{B^*(f)B(f)}{B^*(f)A(f)} = \frac{G_{BB}(f)}{G_{AB}^*(f)}, \tag{9}$$

so H(f) can also be computed by the division of the autospectrum of the DUT's output with the complex conjugated cross-spectrum. When noise is present neither in channel A nor in channel B, both methods obviously lead to identical results. With noise, however, the two results differ. $H_1(f)$ is a better estimate of the true TF when output noise prevails (the more typical case), while $H_2(f)$ comes closer to reality when input noise dominates.

To determine the amount of uncorrelated noise in the measurement, the two estimates of the transfer function can be divided:

$$\nu^2(f) = \frac{H_1(f)}{H_2(f)} = \frac{|G_{AB}|^2}{G_{AA} \cdot G_{BB}}. \tag{10}$$

This *coherence function* assumes values between 1 and 0. When no uncorrelated noise is present, the coherence function will become 1, as $H_1(f)$ and $H_2(f)$ are identical. When there is only uncorrelated noise without signal, the coherence function drops to near

0 because the noise tends to average out in the cross-spectrum, while it is being added up energetically in the autospectra.

With the help of the coherence function, the power of the autospectrum G_{BB} (derived from the DUT's output) can be split into the coherent power $\nu^2(f) \cdot G_{BB}(f)$, which originates from the excitation signal passed through the DUT, and the non-coherent power $(1 - \nu^2(f)) \cdot G_{BB}(f)$ originating from uncorrelated noise.

The signal-to-noise ratio becomes thus

$$\text{SNR} = \frac{\nu^2(f)}{1 - \nu^2(f)} \tag{11}$$

and it is an interesting feature of the dual channel analysis technique that this value is available without turning off the excitation signal.

Traditionally, FFT analyzers are operated with asynchronous noise sources. However, the use of an asynchronous excitation signal has some implications:

As already mentioned, the FFT yields correct results only for signals repeated with a period equal to the FFT block length. For non-periodic signals, start point and end point of the analyzed signal section generally do not match. This discontinuity introduces a considerable error, the famous *leakage*, which has to be lowered by windowing the analyzed sections prior to the FFT. However, windowing introduces a DC bias error and has a smoothing effect which reduces the linear spectral resolution, causing the *relative* spectral resolution (in fractions of an octave) to diminish toward low frequencies.

In measurement setups involving an acoustic transmission path, the propagation delay between sending and receiving transducer has to be carefully compensated with a digital delay of the same duration inserted in the analysis path for the direct signal. This is to guarantee that the same signal sections are submitted to the FFTs in every measurement cycle. The window attenuates all components outside the center, which can lead to errors when the DUT features a strong frequency-dependent group delay, as is the case for reflections or at frequencies near resonances.

The asynchronous noise generator commonly used as signal source has a white (or pink or other specified) spectrum when averaged over a long time, but a single snapshot of the noise signal has a very irregular spectral structure suffering from deep magnitude dips. Thus, a classical dual channel analyzer has to average over many individual measurements before being able to present a reliable result.

There is an interesting application which exploits the fact that any arbitrary signal source can be used, provided that at least in the long term, it covers all frequencies of interest: the unobtrusive measurement of a sound system during a performance, using the program material itself as the excitation signal. However, speech or music with their erratic spectral distribution require even longer averaging periods to achieve a meaningful result compared to asynchronous noise.

When the unobtrusiveness is not paramount, it is advantageous to replace the asynchronous noise (or other) signal source with a synchronous predetermined excitation signal that features a stable smooth spectrum over the whole frequency range of interest for the given FFT length. This can be a periodic noise or a zero-padded aperiodic sine sweep. Using these designed signals allows simplifying and considerably speeding up the entire analysis. First of all, due to the periodicity of the periodic noise (or the 0 start/end points for the sweep), there is no leakage and thus no need to window the DUT's input and output signals prior to the FFT. Furthermore, the inconvenient tunable delay to adjust the arrival time can be omitted (provided that the FFT block size accommodates all delayed components and, in case of the sweep, a sufficiently long gap of silence is introduced at the end). Any delay will be represented by the start position of the IR obtained by IFFT. Finally, as the excitation signal is smooth and has no dips at any frequency, a single measurement is sufficient to obtain a reliable result. Thus, the averaging process is normally not needed, provided that the SNR is not too poor. Without averages, however, the coherence function becomes meaningless.

Furthermore, using a synchronous deterministic excitation signal means that distortion components will always appear at the same locations in the estimated IR and hence do not diminish with increasing number of averages (Figure 5). In contrast, asynchronous noise gives the best linear fit of the transfer function in the long run (e.g., when performing many averages). Noise averages out, and so do all distortion products.

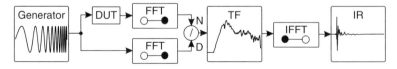

FIGURE 5 Simplified two-channel analysis to capture IR with deterministic excitation signal.

2.2 Single Channel FFT Analysis with Deterministic Excitation Signal

Employing a predetermined synchronous excitation signal that repeats exactly the same way in each measurement cycle means that the spectrum of the signal fed to the DUT will not change from measurement to measurement. In this case, it obviously has to be estimated only once at the beginning of a measurement session. The best way to do this *reference measurement* is to replace the DUT with a wire and pass the excitation signal over the same channel, as depicted in Figure 6. The captured signal is subjected to an FFT and the resulting spectrum inverted, yielding the deconvolution or *reference* spectrum. In the actual measurement of the DUT, this reference spectrum is applied as complex multiplication with the FFT of the DUT's output signal. The product will be the transfer function.

Using the same channel for both the reference and the actual measurement of the DUT has two advantages: It obviously eliminates the need for a second channel, and it leads to very high precision, as any gain error, any frequency response deviation and any delay of the components in the measurement chain (including amplifiers and converters, but not the microphone in the setup in Figure 6) are compensated. The resulting transfer function is actually a comparison to the frequency response of the flat wire used during the reference measurement.

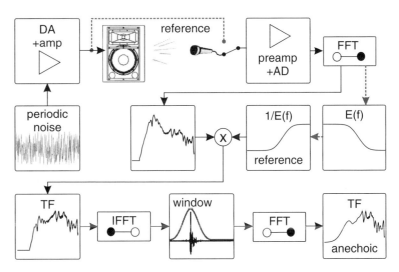

FIGURE 6 Single channel measurement, here with synchronous periodic excitation signal. The *dashed lines* depict the signal flow for the reference measurement. The processing also includes windowing in the time domain to reject unwanted reflection and noise, which allows quasi-anechoic measurements.

When working with a periodic excitation signal, the acquired impulse response will also be periodic. Care must be taken to guarantee that the chosen period is sufficiently long to accommodate all delayed components of the IR. In this case, one period of the periodic IR is virtually equal to the normal aperiodic IR. If the chosen period is too short, the IR's tail will smear into the beginning of the next period, an effect known as *time aliasing* [2].

2.3 Non-periodic Sweeps

Sweeps do not have to be periodically repeated, as their spectrum is similar for a single shot compared to periodical repetition. Thus, they can be used in a non-periodic fashion to perform TF measurements with the setup in Figure 6. In this case, a *linear deconvolution* is most appropriate to obtain the non-periodic IR. The linear deconvolution can be performed most efficiently by padding both the input signal and the output signal of the DUT with zeros in order to double the FFT block length to $2N$. Then both are Fourier-transformed and the output spectrum divided by the input spectrum. An IFFT will reveal the complete deconvolution result with positive and negative times. A noticeable amount of energy can show up in front of the rise of the actual impulse response. These are the distortion products, and it is a special feature of measurements with sweeps that they are all pulled to negative arrival times (relative to the direct sound), where they can be safely discarded, as no causal components of the IR can reside there (Figure 7). The IR itself stays totally free of harmonic distortion products, in contrast to a measurement with pseudo-random noise, where they are distributed as small IR repetitions over the whole period (*distortion peaks*).

For this reason, the dynamic range of an IR acquired with sweeps is virtually only limited by the background noise, whereas an IR obtained with a periodic pseudo-random noise stimulus

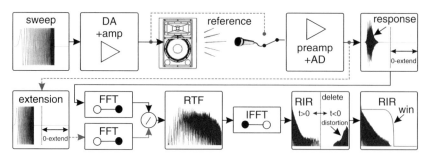

FIGURE 7 Linear deconvolution and separation of distortion products applied for room transfer function (RTF) and room impulse response (RIR) measurements with sweep.

(MLS including) is mostly restricted by the non-linearities of the transducers (most notably the sender). Reducing the level will reduce distortion, but when reducing it too much, background noise may dominate.

Performing synchronous averages helps to defeat the influence of uncorrelated background noise. Every doubling of the number of averages improves the SNR by 3 dB, provided that the undesired components are really of random nature. However, in an averaging process with synchronous periodic excitation, the distortion products are completely correlated with the excitation signal and thus averaging does not help to diminish their presence. This means that a compromise level (where distortion products and background noise have about equal contribution) has to been found by trial and error when working with synchronous noise excitation. In contrast, due to the ability to isolate the distortion products in the estimated IR, sweeps can be emitted with the maximum possible level, easily elevating the achievable SNR by 15–20 dB compared to a measurement with pseudo-random noise.

However, driving the transducer considerably into its non-linear range is only acceptable in a comparative measurement, in which the transducer itself is not the object of investigation. If the transducer's frequency response itself is to be estimated, operating in its non-linear range will of course lead to an apparent loss of sensitivity.

When a logarithmic (pink coloration) sweep is used to measure a transducer in anechoic conditions, a train of discrete spikes antecedes the actual IR. These are "harmonic impulse responses" (HIRs) in rising order from right to left. Due to the constant relative frequency increase (constant fraction of an octave per time unit) of the logarithmic sweep, the deconvolution process focuses the energy of each individual harmonic to a fixed arrival time over the whole frequency span, the same way as it happens for the fundamental (Figure 8). This makes it possible to determine the transfer function and the frequency-dependent level of all harmonics with one single log sweep and a bit of post-processing [3,4]. To do so, the fundamental and each individual HIR are isolated with appropriate windows and separately Fourier-transformed. The position of the HIRs can be found by

$$t_{\text{HIR}(k)} = t_{\text{FUND}} - \frac{\log_2(k)}{\text{Sweep\_rate}}, \qquad (12)$$

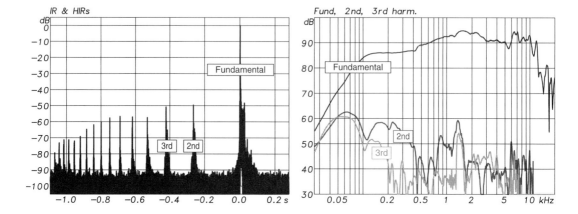

FIGURE 8 Extraction of just the second and third harmonics from a measurement with log sweep.

with $t_{\text{HIR}(k)}$ = arrival time of HIR of order k, t_{FUND} = arrival time of fundamental and

$$\text{Sweep\_rate} = \frac{\partial f}{f \cdot \partial t} \; (octaves/s).$$

The spectra of the HIRs are now compressed toward the left side of the frequency axis by a factor corresponding to their order. For the third-order harmonic, for example, the former 3 kHz component will be relocated to 1 kHz, as this is the frequency of the fundamental which caused the 3 kHz component of the third harmonic to appear. To calculate the frequency-dependent percentage of distortion for each harmonic, its relocated spectrum has only to be divided through the spectrum of the fundamental.

Loudspeakers, when pushed hard, are sometimes haunted by subharmonics, most notably half the fundamental. This component can also be detected by the method. Its HIR appears on the right side of the fundamental IR, at the same distance as the second harmonic on the left side. To isolate this component, good anechoic conditions and a slow sweep rate are vital in order to avoid smearing of reverberation into the subharmonic's IR.

Apart from the complete rejection of harmonic distortion, TF measurements with sweeps are also considerably less vulnerable to time variance compared to measurements with (pseudo-) random) noise. Time variance occurs when the sound medium is in movement (wind gusts, convection, water currents) or when the DUT itself is fairly time-variant (analogue recording gear). Breaking the exact periodicity of a periodic noise signal immediately disfigures its smooth spectrum, while a sweep suffering the same kind of time variance shows to be fairly immune.

2.4 Time Delay Spectrometry (TDS)

TDS is a vintage method to derive TF with the help of linear (white) sweeps (Figure 9). Devised by Richard Heyser specifically for measuring loudspeakers, it can be realized with purely analog means, which is the reason why it has been rather popular before the advent of powerful personal computers with FFT capabilities.

The mathematic framework of TDS has been extensively covered in [5]. To keep the description of the TDS principle and some practical aspects reasonably short, equations have been omitted here.

A TDS analyzer has a generator that produces both a swept sine and, simultaneously, a phase-locked swept cosine. The sine is fed to the DUT, and its captured response is multiplied separately by both the original sine (to get the transfer function's real part) and the 90° phase-shifted cosine (to get the imaginary part). The multiplier outputs are filtered by a low pass with fixed cutoff frequency. These multipliers work similar to the mixer circuits used in the intermediate frequency stages of radio frequency receivers based on the ubiquitous super-heterodyne principle, producing the sums and differences of the input frequencies. The sum terms of both multiplier outputs must be rejected by the low-pass filters, whereas the difference terms may pass up to a certain frequency. If both the generated and the captured frequencies are almost equal, the output difference frequency will be very low and thus will not be attenuated by the low-pass filters.

As the sound that propagates from the DUT arrives with a delay at the microphone, its instantaneous frequency will be lower than the one coming directly from the generator. This causes a higher mixer output difference frequency that, depending on the cutoff frequency, will be attenuated by the low-pass filters. For this reason, the generated signal must be *time-delayed* by the propagation time of the sound between loudspeaker and microphone before being multiplied with the DUT's response. In this way, the difference frequency will be near DC. In contrast to the direct sound, reflections will always take a longer way and thus arrive

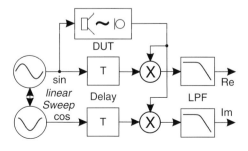

FIGURE 9 Basic TDS signal processing.

with a lower instantaneous frequency, causing higher frequency components in the multiplier outputs, which will be more attenuated by the low-pass filters, the longer their detour is. This is equivalent to windowing the respective IR. To obtain the latter, an IFFT of the complex TF has to be performed.

So, by proper selection of the sweep rate and the low-pass filter cutoff frequency, simulated quasi-free-field measurements are possible with TDS. In addition to the attenuation of unwanted reflections, distortion products, which arrive with a higher instantaneous frequency and thus cause high mixer output frequencies, are also suppressed very well. Extraneous noise in the wide band above the filter cutoff frequency will also be rejected.

The controlled suppression of reflections is the reason why TDS analyzers utilize a linear sweep ($df/dt = $ constant) as the excitation signal. The frequency difference between incoming direct sound and reflection stays constant over the whole sweep range, keeping the attenuation of each reflection frequency-independent.

For certain purposes, however, it could be interesting to use logarithmic sweeps with TDS – at least over some parts of the frequency span. This would lead to an effective IR window width that is inversely proportional to the frequency. This behavior corresponds to constant *relative*-bandwidth smoothing, a technique widely used for processing and displaying loudspeaker frequency responses. In contrast, standard TDS employing a linear sweep has a constant *absolute* bandwidth smoothing effect, just as windowing a broadband IR. Another important benefit of using a (semi-) log sweep would be to enhance the SNR at low frequencies where the linear (white) sweep is often not capable of injecting enough energy to sufficiently overcome background noise.

A common problem with TDS is ripple which occurs at low frequencies. As mentioned previously, the multipliers produce sum and difference terms of the "time-delayed" excitation signal and the incoming response. At higher instantaneous frequencies, the sum is sufficiently high to be attenuated by the output low-pass filter. But at the low end of the sweep range, when the sum is close to or lower than the low-pass cutoff frequency, "beating" will appear in the recovered magnitude response. To remedy this, the sweep can be made very long and the low-pass cutoff frequency reduced by the same factor.

Even then, some spectral ripple often appears at the beginning and at the end of the sweep frequency range because of the sudden onset and termination of the linear sweep. The abrupt switching corresponds to *multiplying* a continuous time signal with a rectangular window, which corresponds to *convoluting* the spectrum of the sweep by the rectangular window's spectrum (that is, the *sin(x)/x* function). A common way to circumvent this problem is to let the excitation sweep start well below the lowest frequency of interest. Synthesizing the excitation sweep in the spectral domain is a better solution, as will be shown in Sect. 3.2.

2.5 Noise Rejection

The frequency-dependent SNR of a non-windowed IR measurement depends on the spectral distribution and energy of both the background noise and the excitation signal during the data acquisition period, but not on the type (impulse, noise, sweep or other) of the excitation signal. However, differences exist in how the different types of interfering noise are distributed over the entire period of the estimated IR.

Monofrequent noise (such as hum) and uncorrelated broadband noise (for example, from air conditioning) will still appear as such in the IR, independently of the measurement method. This is due to the fact that their general character is not altered by manipulating the phase spectrum in the deconvolution process.

Short, impulsive noise events, such as clicks and pops, will be transformed into noise when using any noise stimulus. In contrast, they will be transformed into small time-inverted sweeps in a sweep measurement. Depending on the strength and time of occurrence, these time-reversed repetitions of the sweep can be masked by the main peak of the IR or become audible in its reverberant tail, giving it a bizarre melodic touch.

3 • Excitation Signals

Achieving a good signal-to-noise ratio (SNR) in impulse responses can be quite a challenge. In purely electrical environments, the measurement of the IR can be accomplished in direct form by injecting a pulse as shown in Figure 1. If the noise floor of the equipment under test is low, the SNR of the direct impulse response can be satisfactory. The situation is often different for acoustic ambiences, where the prevailing background noise, transducer non-linearities (loudspeaker distortion) and occasionally time variance (air movement) make the task difficult.

To improve the signal to background noise ratio, other excitation signals than the simple impulse are commonly used. The basic idea is to spread out the energy contained in a single impulse over a broader time interval.

Two fundamental types of excitation signals can be used: independent asynchronous signals (coming from analogue noise generators, for example) and custom-tailored deterministic signals which usually have the same length as the analysis time interval and are synchronously clocked with the acquisition of the DUT's output signal.

Synchronous deterministic signals can be conveniently created by defining their properties in the spectral domain and then getting the desired signal by IFFT. This is shown in Figure 10, where three basic types of excitation signals are created from exactly the same linear (white) magnitude spectrum. The difference of the three types lies in their phase spectrum.

For the impulse, all phases have to be set to 0°, which corresponds to an equal arrival time for all frequencies. For a pseudo-random noise signal, the phases must assume random values. More sophisticated algorithms exist to keep the crest factor of the noise signal low [6]. To create a white sweep (sometimes called *time stretched pulse*, TSP, see [7,8]), the group delay (which is proportional to the negative derivative of the phase) has to increase proportionally with the frequency.

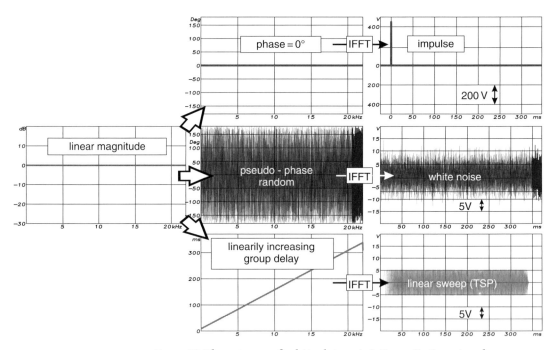

FIGURE 10 Three types of white deterministic excitation signals constructed in the frequency domain: impulse, pseudo-random noise and sweep.

In this example, the three excitation signals obtained after IFFT have identical energy. The sweep has the lowest peak value of all, about 6 dB lower than white noise. The impulse needs an amplitude of several hundred volts to concentrate the same energy as the sweep and noise signal, which shows that its use is unpractical.

3.1 Coloration of the Excitation Signal

In airborne and many other acoustical measurements, it is rarely useful to work with white excitation signals. The ambient noise floor almost always tends to increase considerably toward low frequencies. A white excitation signal often leads to insufficient SNR at the lower frequency bands [9,10]. In room acoustics, this may thwart the evaluation of parameters such as the reverberation time. A minimum SNR of at least 30 dB is desirable to obtain this value. For this reason, it is advisable to employ an excitation signal which strongly emphasizes the lower frequencies, for example by using pink (−3 dB/oct) or red (−6 dB/oct) coloration or a low-shelf filter, operating like the "bass" control in sound systems.

When a multi-way loudspeaker is used as sound source, the total energy fed to it can be increased considerably when working with pre-emphasis. The woofer often bears 10 times or more the power the tweeter can handle.

In many cases, a pink coloration is a reasonable choice to achieve sufficient SNR over the whole frequency range of interest. A low-shelf filter (see Figure 11) can be a better option in some cases. Above the upper corner frequency, its response gradually stops to decline and becomes linear. This can help overcome air absorption in long-distance open air measurements while maintaining substantial boost at low frequencies. This type of equalization can also be adequate if the background noise in a room is mainly caused by outside sources (such as traffic noise). In this case, the increased efficiency of the acoustical insulation of the walls can cause the acoustical noise level to drop below the microphone's or preamplifier's electrical noise level.

Additionally to the compensation of the background noise, the applied pre-emphasis can also include the loudspeaker's inverted frequency response. This is particularly useful when obtaining room IRs for auralization purposes. In this case, it is not acceptable that the loudspeaker's own response colors the results.

3.2 Generalized Sweep Synthesis

Sweeps used as excitation signal have a number of advantages presented earlier in this chapter. They can be created either directly in the time domain or indirectly in the frequency domain. In the latter case, their magnitude and group delay are synthesized and the sweep is obtained via IFFT of this artificial spectrum. This procedure avoids spectral ripple at the start and

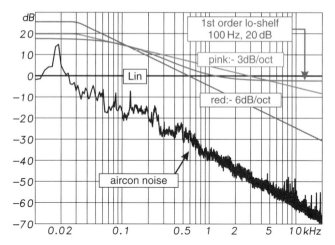

FIGURE 11 Typical background noise from an air condition unit (curve obtained from 30 energetic averages) and various pre-emphasis curves to overcome lacking SNR at low frequencies, compared to a white excitation signal of same total energy. To limit subsonic energy, the pink and red pre-emphasis curves have been chosen to not rise further below 30 Hz.

end frequencies which is mostly seen for sweeps built in the time domain. However, it can slightly worsen the crest factor, meaning that some cusps of the waveform can overshoot. Nevertheless, the generation in the frequency domain offers the crucial freedom to endow the sweep with a freely defined spectral distribution while almost perfectly maintaining its constant envelope (Figure 12).

This can be done by controlling the course of the group delay, which tells us the time at which each frequency is swept through. A steeply increasing group delay means that the corresponding frequency region is stretched out substantially in time, with the instantaneous frequency rising only slowly. This way, much energy is packed into this frequency region. On the other hand, an almost horizontal group delay course means that the corresponding frequency range is swept through in very short time, leaving little energy for it.

Thus, to construct a constant-envelope sweep with arbitrary spectral distribution, the group delay's inclination, or in other words, its derivative $\partial \tau_G(f)/\partial f$, has to be made proportional to the energy $|H(f)|^2$ at each frequency. For a discrete FFT spectrum, this means that the group delay τ_G has to increase according to

$$\tau_G(f) = \tau_G(f - \mathrm{d}f) + C \cdot |H(f)|^2, \qquad (13)$$

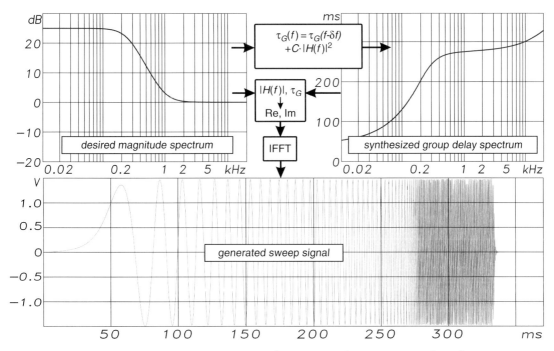

FIGURE 12 Generation of a sweep with constant envelope from an arbitrary magnitude spectrum.

with $\tau_G(f - df)$ being the predecessor (bin) of group delay $\tau_G(f)$ in the FFT spectrum, $|H(f)|^2$ the energy of the bin at frequency f, and C a normalization constant composed of the sweep's length divided by its total energy:

$$C = \frac{\tau_G(f_{\text{END}}) - \tau_G(f_{\text{START}})}{\sum_{f=0}^{f_S/2} |H(f)|^2}. \tag{14}$$

The sweep start time $\tau_G(f_{\text{START}})$ should be slightly above 0 s to allow the first half wave to evolve, the end time $\tau_G(f_{\text{END}})$ should be considerably smaller than the time interval $2^N/f_S$ (N = FFT order) to avoid smearing of the high-frequency tail into the sweep start and vice versa.

The phase is now easily calculated by integrating the group delay:

$$\varphi(f) = \varphi(f - df) - 2\pi \cdot df \cdot \tau_G(f) \quad \text{with} \quad df = f_S/2^N. \tag{15}$$

As a small detail, it is important that the phase resulting from the integration reaches exactly 0° or 180° at $f_S/2$. This condition generally must be fulfilled for spectra of real-time signals. It can be achieved easily by subtracting values from the phase spectrum that decreases linearly with frequency until exactly offsetting the former phase at $f_S/2$:

$$\varphi_{\text{NEW}}(f) = \varphi_{\text{OLD}}(f) - \frac{f}{f_S/2} \cdot \varphi_{\text{END\_OLD}}. \qquad (16)$$

This is equivalent to adding a minor constant group delay in the range of ± 0.5 samples over the whole frequency range.

An IFFT now reveals the synthesized sweep. Head and tail usually stretch out a little beyond the defined start and end time. Half-windows at the very beginning and end help to confine the sweep while unavoidably introducing a little deviation from the desired magnitude spectrum, although normally much smaller than the ripple of sweeps constructed by direct synthesis in the time domain.

The method can be easily extended to control the envelope in a frequency-dependent fashion instead of maintaining it constant. To do so, the desired magnitude spectrum is first *divided* by the *desired envelope spectrum* (which stipulates the frequency-dependent amplitude of the time signal). After synthesizing the group delay as previously, the desired magnitude spectrum is now *multiplied* with the desired envelope spectrum to re-establish the original spectral distribution. The final IFFT reveals a sweep which features both the desired spectral contents and the frequency-dependent amplitude. This generalized sweep generation method offers two crucial degrees of freedom: Any desired spectral distribution along with any freely definable frequency-dependent envelope will be transformed into a swept sine wave suitably warped in amplitude and time.

As a further sophistication, the sweep can also be split into various bands stored in separate channels to include the crossover functionality and equalization of a multi-way loudspeaker, this way replacing its passive or active crossover circuit. More details can be found in [4].

References

1. H. Herlufsen, "Dual Channel FFT Analysis (Part I, II)", Brüel & Kjær Technical Review No. 1, 1984, www.bksv.com/pdf/Bv0013.pdf
2. D. D. Rife, J. Vanderkooy, "Transfer Function Measurement with Maximum-Length Sequences", *J. Audio Eng. Soc.*, vol. 37, June 1989, pp. 419–444
3. A. Farina, "Simultaneous Measurement of Impulse Response and Distortion with a Swept-Sine Technique", *J. Audio Eng. Soc.*, vol. 48, 2000, p. 350, 108th AES Convention, Paris, Preprint 5093

4. S. Müller, P. Massarani, "Transfer-Function Measurements with Sweeps", *J. Audio Eng. Soc.*, vol. 49, June 2001, pp. 443–471
5. AES, R. C. Heyser, *Time Delay Spectrometry – An Anthology of the Works of Richard C. Heyser*, AES, New York, 1988
6. M. R. Schroeder, "Synthesis of Low-Peak-Factor Signals and Binary Sequences with Low Autocorrelation", *IEEE Trans. Inform. Theory*, 1970, pp. 85–89
7. N. Aoshima, "Computer-Generated Pulse Signal Applied for Sound Measurement", *J. Acoust. Soc. Am.*, vol. 69, May 1981, pp. 1484–1488
8. Y. Suzuki, F. Asano, H.-Y. Kim, T. Sone, "An Optimum Computer-Generated Pulse Signal Suitable for the Measurement of Very Long Impulse Responses", *J. Acoust. Soc. Am.*, vol. 97, February 1995, pp. 1119–1123
9. D. Griesinger, "Beyond MLS – Occupied Hall Measurement with FFT Techniques", *J. Audio Eng. Soc.*, vol. 44, p. 1174 (abstract), 101st AES Convention, Preprint 4403
10. E. Mommertz, S. Müller, "Measuring Impulse Responses with Preemphasized Pseudo Random Noise derived from Maximum Length Sequences", *Appl. Acoust.*, vol. 44, 1995, pp. 195–214

Further Reading

11. G.-B. Stan, J. J. Embrechts, D. Archambeau, "Comparison of Different Impulse Response Measurement Techniques", *J. Audio Eng. Soc.*, April 2002, pp. 249–262
12. J. Schoukens, R. Pintelon, E. van der Ouderaa, J. Renneboog, "Survey of Excitation Signals for FFT Based Signal Analyzers", *IEEE Trans. Instrum. Meas.*, vol. 37, September 1988, pp. 342–352
13. J. Shoukens, R. Pintelon, "Measurement of Frequency Response Functions in Noise Environments", *IEEE Trans. Instrum. Meas.*, vol. 39, December 1990
14. A. J. Berkhout, D. de Vries, M. M. Boone, "A New Method to Acquire Impulse Responses in Concert Halls", *J. Acoust. Soc. Am.*, vol. 68, 1980, pp. 179–183
15. J. C. Burgess, "Chirp Design for Acoustical System Identification", *J. Acoust. Soc. Am.*, vol. 91, 1992, pp. 1525–1530
16. M. Poletti, "Linearly Swept Frequency Measurements, Time-Delay Spectrometry, and the Wigner Distribution", *J. Audio Eng. Soc.*, vol. 36, June 1988, pp. 457–468
17. A. Lundeby, T. E. Vigran, H. Bietz, M. Vorländer, "Uncertainties of Measurements in Room Acoustics", *Acustica*, vol. 81, 1995, pp. 344–353

6

Digital Sequences

Ning Xiang

School of Architecture and Department of Electrical, Computer, and Systems Engineering, Rensselaer Polytechnic Institute, Troy, NY 12180, USA, e-mail: xiangn@rpi.edu

1 • Introduction

This section discusses the applications of digital sequences in acoustical system identification and characterization and describes Golay codes and binary maximum-length sequences (MLSs) in some detail. Legendre sequences and other coded signals are briefly described. Golay codes and MLS have been used for acoustic applications for years. Applications of Legendre sequences have also been reported. Digital sequences of other classes such as, e.g., binary Gold sequences and Kasami sequences have only recently found applications in acoustical system identification and characterization.

2 • Golay Codes

Golay codes [1] have been used as acoustic excitation signals for determination of head-related transfer functions (HRTFs) of human subjects [2–4]. Golay codes $\{a_n, b_n\}$, a pair of complementary binary sequences of length $L = 2^n$, with integer $n > 1$, can be recursively constructed from a pair of complementary codes

$$\begin{cases} a_1 = \{+1+1\} \\ b_1 = \{+1-1\} \end{cases} \quad (1)$$

using the following recursion:

$$\text{Append } b_{n-1} \text{ to } a_{n-1} \text{ to obtain } a_n$$
$$\text{Append } -b_{n-1} \text{ to } a_{n-1} \text{ to obtain } b_n.$$

For example, appending b_1 to a_1 and appending $-b_1$ to a_1 yield $a_2 = \{+1+1+1-1\}$ and $b_2 = \{+1+1-1+1\}$. The most important property of this pair of codes is that the sum of the periodic autocorrelations of both codes is a two-valued function [3]:

$$\phi(k) = \sum_j \{a(j)a(k+j) + b(j)b(k+j)\} = \begin{cases} 2L \text{ for } k=0 \\ 0 \text{ for } k \neq 0 \end{cases}, \quad (2)$$

where $a(i)$ or $b(i)$ is the ith element of a_n or b_n, respectively. For periodic (circular) correlation all the indices are calculated modulo L. Figure 1 illustrates normalized periodic autocorrelation functions (PACF) of a pair of Golay codes of length 16. In this example, all functions in Figure 1 have to be normalized by 2×16 which leads to a peak value 0.5 of individual PACFs in Figure 1a and b, while the sum of them provides a perfect PACF of $\delta(i)$ as shown in Figure 1c. Since the complementary pair of Golay codes are of radix two in length, the fast Fourier transform (FFT) operations can be directly applied for cross-correlations in the frequency domain. So the sum of the PACFs of a pair of Golay codes in the frequency domain can be expressed as [3]

$$\Phi(f) = \text{FFT}(a_n)\text{FFT}^*(a_n) + \text{FFT}(b_n)\text{FFT}^*(b_n) = 2L, \quad (3)$$

where FFT() and FFT*() stand for the fast Fourier Transform and its complex conjugates, respectively. In the context of acoustic transfer function measurements of a linear time-invariant (LTI) system under test as shown in Figure 2, $H(f)$ can be determined by exploiting the autocorrelation property expressed in Eq. (3) [3]:

$$H(f) = \frac{1}{2L}\{[\text{FFT}(a_n)H(f)]\text{FFT}^*(a_n) + [\text{FFT}(b_n)H(f)]\text{FFT}^*(b_n)\}. \quad (4)$$

A measurement procedure for a system under test (SUT) can be implemented by generating a pair of Golay codes of sufficient length, longer than the expected system impulse response, to avoid time aliasing. The measurement procedure illustrated in Figure 2 is summarized as follows [4]:

- Excitation of the SUT periodically using a_n code of $k+1$ periods. After the SUT reaches its steady state, record the system response to a_n, conceivably with synchronous averaging over k periods.

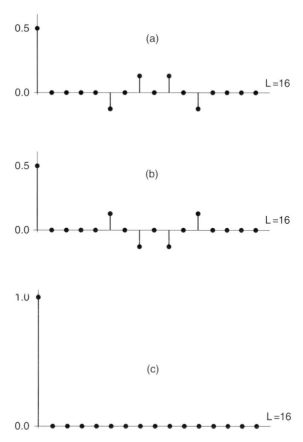

FIGURE 1 Normalized periodic autocorrelation functions of Golay codes of length $L = 16$ for the individual complementary a_4 and b_4 codes, shown in (**a**) and (**b**), respectively, along with the sum of both autocorrelation functions, shown in (**c**). The normalization factor of $2L$ leads to a peak value 0.5 of individual autocorrelation functions in (**a**) and (**b**), while the sum of them provides a perfect PACF of $\delta(i)$ as shown in (**c**) (from [4]).

- Wait until the system settles down, normally after one more period.
- Excitation of the SUT periodically using b_n code with $k+1$ periods. After the SUT reaches its steady state, record the system response to b_n, conceivably with synchronous averaging over k periods.
- Cross-correlation of one resulting period of response to a_n with the original code a_n in the frequency domain. In practice, its complex conjugate FFT$^*(a_n)$ is prepared in memory in advance for a direct multiplication as shown in Figure 2.

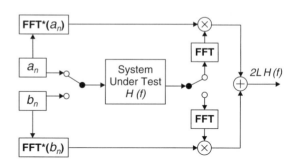

FIGURE 2 Flow diagram for determining the transfer function of a linear time-invariant system using a pair of Golay codes. FFT() and FFT*() stand for the fast Fourier Transform and its complex conjugates, respectively.

- Cross-correlation of one resulting period of response to b_n with the original code b_n in a similar fashion.
- Sum of the two correlation results and normalize it by $2L$, the transfer function of the SUT is determined.

Ease of generation using the recursive algorithm and the fact that FFT can be directly used for determining the transfer function of a SUT have been considered positive in the works by Zahorik [4]. Its shortcoming is that two sequences have to be used sequentially for acoustic excitation. The steady state of a LTI system has to be reached twice that makes any unavoidable time-variance effect more troublesome than those signals such as chirps [5], sweeps (see Chapter 5 by S. Müller), MLSs and other pseudorandom noise signals. The system characterization tasks, such as experimental determination of human HRTFs, are actually critical examples for the mentioned time variances [4].

3 • Binary Maximum-Length Sequences

Since the 1970s, binary maximum-length sequences (MLSs) have been introduced in acoustic applications [6] in shaping wall surfaces for diffuse sound reflections and in acoustic system identification, such as room impulse response measurements. This section briefs the acoustic system identifications using binary MLSs.

A n-stage linear feedback shift-register device (see Figure 3) can generate a binary periodic sequence $\{a_i\}$ with $a_i \in \{0, 1\}$, or equivalently $\{a_i\}$ can be derived by the linear recurrence [7]

$$a_i = \sum_{k=1}^{n} c_k a_{i-k}, \quad 0 \leq i < L, \tag{5}$$

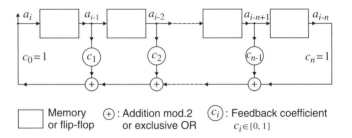

FIGURE 3 An n-stage linear shift register for generating binary periodic sequences.

with $c_k \in \{0, 1\}$ but $c_0 = c_n = 1$. The summation and indices are calculated modulo 2. Exclusive one all-zero state, there are $2^n - 1$ possible states for the shift register. When its feedback taps are appropriately connected, or equivalently an appropriate set of c_k is chosen, the periodic sequences arrive at their maximum period length of $L = 2^n - 1$. In this case, the sequences are referred to as maximum-length sequences (MLSs). A characteristic polynomial $f(x) = \sum_{k=0}^{n} c_k x^k$ expressing its feedback connection is then referred to as primitive. The positive integer n is said to be the degree of the MLS as well as its primitive polynomial (PP). A table of PPs for $n \leq 40$ can be found in [8]. In mathematical treatment MLSs are convenient in their binary form while a bipolar form $\{m_i\}$ is often used in practice to generate waveforms with $m_i = 1 - 2a_i \in \{-1, +1\}$.

MLSs enjoy a number of attractive properties that make them widely useful in broad scientific and engineering fields. They possess properties similar to those of random noise, but they are periodically deterministic and have a strict time structure within their periods. They are also termed pseudorandom sequence/noise. This section briefly reviews some basic properties pertaining to acoustic applications. Detailed description and definitions can be found in Refs. [6–11].

3.1 Autocorrelation Function

The most important property of binary MLS is that the normalized PACF of a bipolar MLS within one period is a two-valued function [12] with

$$\phi(i) = \frac{L+1}{L}\delta(i) - \frac{1}{L}. \qquad (6)$$

Figure 4 illustrates a normalized PACF of MLS of $L = 15$. When the period length $L = 2^n - 1$ is large enough the PACF approximates the unit sample sequence

$$\phi(i) \approx \delta(i). \qquad (7)$$

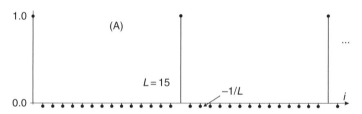

FIGURE 4 Normalized periodic autocorrelation function of an MLS of degree 4 ($L = 15$).

From its PACF, it is then readily deduced that the power spectrum of an MLS is completely flat, expect for a "dip" at the DC. One can also replace all -1s of a bipolar MLS $\{m_i\}$ by a value $-q$ with

$$q = \frac{1}{1 + 2/\sqrt{L+1}}, \qquad (8)$$

so that the slightly modified MLS possesses a perfect PACF of $\delta(i)$ [11].

3.2 Shift-and-Add

A (modulo 2) sum of an MLS $\{a_i\}$ with its phase-shifted version $\{a_{i+\tau}\}$ results in the same MLS with another phase shift

$$\{a_i\} \oplus \{a_{i+\tau}\} = \{a_{i+g(\tau)}\}. \qquad (9)$$

$g(\tau)$ is termed offset function, being a function of τ which can be explicitly determined in terms of Zech's logarithm [6]. The addition of modulo 2 becomes multiplication for bipolar MLS. The shift-and-add property leads directly to the proof of the two-valued autocorrelation function of bipolar MLSs (see Section 3.1). Along with Zech's logarithm it is also of practical significance for nonlinear system identification [13].

Generally a nonlinear system can be described by Volterra series

$$y(n) = \sum_{i=0}^{\infty} h(i) m(n-i)$$
$$+ \sum_{i=0}^{\infty} \sum_{j=0}^{\infty} h(i,j) m(n-i) m(n-j)$$
$$+ \sum_{i=0}^{\infty} \sum_{j=0}^{\infty} \sum_{k=0}^{\infty} h(i,j,k) m(n-i) m(n-j) m(n-k) \qquad (10)$$
$$+ \cdots,$$

where $h(i)$, $h(i,j)$ and $h(i,j,k)$ are first, second and third orders of Volterra kernels, respectively. For a linear system, in particular, second and all higher orders of Volterra kernels become zero. If the input signal of the nonlinear system $m(n)$ is an MLS, the periodic cross-correlation between $y(n)$ and $m(n)$ becomes

$$\phi(\tau) = \sum_{i=0}^{L-1} h(i) \sum_{n=0}^{L-1} m(n-i)m(n+\tau)$$
$$+ \sum_{i=0}^{L-1}\sum_{j=0}^{L-1} h(i,j) \sum_{n=0}^{L-1} [m(n-i)m(n-j)]m(n+\tau)$$
$$+ \sum_{i=0}^{L-1}\sum_{j=0}^{L-1}\sum_{k=0}^{L-1} h(i,j,k) \sum_{n=0}^{L-1} [m(n-i)m(n-j)m(n-k)]m(n+\tau)$$
$$+ \cdots . \tag{11}$$

According to the shift-and-add property, items in [.] of second and third line of Eq. (10) can be considered as phase-shifted MLS, while quantities following $\sum_{n=0}^{L-1}$ in each line represent shifted autocorrelation of the MLS as elaborated in Section 3.1. Equation (10) well explains "erroneous" behaviors in resulting impulse responses when the SUT contains weak nonlinearities, as reported via experimental observations [14,15]. For purpose of a linear system identification in practice, distinct spikes distributed along the impulse response are strong evidence and indication that the MLS excitation has already stretched beyond the linear range of the SUT. An amplitude reduction of the MLS excitation will often remove the erroneous nonlinear behaviors. Equation (10) also indicates an approximation scheme of nonlinear system identification relying on the determination of the offset function given the MLS. With some constrains, the higher order of Volterra kernels can be identified and separated from the linear response $h(n)$ of the SUT [13].

With respect to nonlinear immunity, inverse repeat sequences by appending an amplitude-inverse MLS to the original MLS can be used for improving even-order nonlinearity immunity [14, 16]. An overall high nonlinear immunity can be obtained using logarithmic sweep signals (see Chapter 5).

3.3 Characteristic MLS

An MLS $\{b_i\}$ can always be derived from a given one $\{a_i\}$ in terms of a circular phase shift such that $b_i = a_{i+\tau}$ so that an invariant decimation $b_i = b_{2i}$ can be satisfied. The MLS $\{b_i\}$ is designated as characteristic MLS and self-similar MLS [17], also termed idempotent MLS [6]. This invariant decimation holds only for characteristic MLSs with decimation factors of 2^k with k being a positive integer [7]. There exists a unique characteristic MLS for a given PP and a unique initial state of the shift register to generate it. An algorithm of deriving the required initial states given the PP has recently been documented in acoustics journals [17]. This special form of MLS is of practical significance in the interleaved sampling technique for ultrasonic applications where current analog-to-digital converters undesirably limit the upper limit of the sampling frequency in specific applications [17,18]. Ultrasonic non-destructive testing and evaluation generally work in frequency range of tens or hundreds of MHz. Figure 5 illustrates the interleaved sampling scheme. A self-similar MLS of 2^k periods or more, generated at an appropriate clock rate, is applied to an ultrasonic SUT given a LTI system. After the system arrives at its steady state a clearly lower sampling rate derived from the master clock can drive the digitizer to receive one period of the system response to the MLS. In this way a high-resolution, low-cost device can conveniently be employed [18]. Since at least 2^k periods have to be used to derive one period of the system response, a highly stationary (time-invariant) measurement environment is crucial to the success of this application.

3.4 Preferred MLS Pairs

With some proper decimation factors d, decimation of $\{a_i\}$ by d may yield distinctly different MLSs $\{b_i\} = \{a_{di}\}$ associated with a distinctly different PP of the same degree. When the decimation factor is properly chosen, the decimation will lead to a pair of MLSs, whose periodic cross-correlation function (PCCF) is of clearly lower value than the peak of the PACF. This cross-correlation property is well-known in the spread spectrum communication systems, in particular, in CDMA technology [19]. Recent research in outdoor sound propagation for acoustic atmospherical tomography [20] calls for a critical measurement

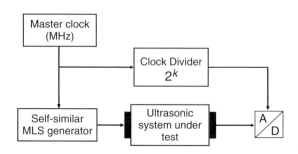

FIGURE 5 Characteristic MLS for ultrasonic applications. Interleaved sampling can take advantage of low sampling rate, high-resolution analog-to-digital converters [18].

technique for simultaneous sound speed determination using multiple sound sources through the atmosphere over the ground layer. The proper decimation of MLSs can make a significant contribution to the acoustic atmospherical tomography [21] (see Section 5).

If the degree of an MLS is a multiple of 4, then a decimation factor $d(n) = 2^{(n+2)/2} - 1$ will lead to an MLS pair whose PCCF takes on only four values. For such a decimation, the upper bound on the four-valued PCCF is

$$l_4(n) = \frac{2^{(n+2)/2} - 1}{2^n - 1}. \quad (12)$$

If the degree of an MLS is not a multiple of 4, then an appropriate decimation of that MLS will yield pairs of MLSs having relatively small three-valued PCCF [22]. For MLSs whose degrees are not a multiple of 4, some of decimation factors are of the form $d = 2^k + 1$ or $d = 2^{2k} - 2^k + 1$, where k is chosen such that $n/\gcd(n,k)$ is odd [22], with $\gcd(,)$ being the greatest common divider, the cross-correlation bounds take on preferred small values expressed by

$$l_3 = \frac{2^{\lfloor (n+2)/2 \rfloor} + 1}{2^n - 1}, \quad (13)$$

where $\lfloor x \rfloor$ denotes the integer part of the real number x. This section will refer to MLS pairs with both three-valued and four-valued PCCFs as preferred pairs. Figure 6 illustrates the PCCF of a preferred pair of degree 13 in comparison to the PACF of one of the MLSs. As one can see from the figure, the three-valued PCCF is bounded at a small value about 36 dB lower than the peak value of the PACF.

In addition to the preferred MLS pairs, the time-reversed sequence of every MLS and the MLS itself form a reciprocal pair. The reciprocal pair of every binary MLS of degree n also possesses small-valued PCCF [23] bounded by Eq. (12), yet not limited to three or four values.

Table 1 lists all bound values of preferred and reciprocal MLS pairs between degree 8 and 24 for an easy comparison. A careful comparison between columns in Table 1 reveals that the small-valued PCCF of the preferred MLS pairs are close to that of reciprocal ones for even-numbered degrees while for the odd-numbered degrees the preferred MLSs are approximately 1.4 times (3 dB) smaller than that of the reciprocal MLSs. Some acoustics applications require the bound values as small as possible, such as the acoustic atmospherical tomography [20]. When applying the cross-correlation technique in

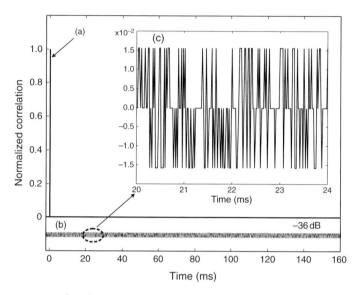

FIGURE 6 Normalized periodic correlation functions of a preferred MLS pair of degree 13 generated at sampling frequency of 50 kHz. (**a**) Autocorrelation function of an individual MLS. (**b**) Cross-correlation function between a preferred three-valued MLS pair (shifted downward beneath the autocorrelation function for a convenient comparison while keeping the same amplitude scale). The bound value of the cross-correlation amounts to 1.57×10^{-2}, 36 dB lower than the peak value of the autocorrelation. (**c**) Zoomed presentation of a segment from (**b**).

acoustic measurements using simultaneous sources driven by signals with above-mentioned correlation properties, individual channel impulse responses associated with each of individual sources can be straightforwardly resolved [21]. The smaller the bound values of PCCFs the less interference from other unwanted source channels.

While a pair of MLSs selected from two different preferred pairs of the same degree may not possess the preferred small-valued PCCFs as listed in Table 1, some pairs selected from different pairs do. A set of the MLSs for which each pair of the set has these preferred small PCCF values is referred to as connect set. The largest possible connected set is termed a maximum connect set (MCS) and the size of such a MCS is denoted by M_n. One can find the values of M_n for $n \leq 16$ with $M_n \leq 4$ [22]. The length of MLSs of these degrees and their reasonably low PCCF-bound values are of practical interest in a wide variety of acoustics measurements, such as in acoustic channel estimation using multiple simultaneous sources [21].

TABLE 1 Cross-correlation bound values of preferred and reciprocal MLS pairs, of Gold and Kasami sequences

Degree	Period length	Reciprocal MLS pairs	Preferred MLS pairs and Gold sequences	Kasami sequences
8	255	0.1216	0.1216	0.067
9	511	0.0881	0.0646	
10	1,023	0.0616	0.0635	0.0322
11	2,047	0.0437	0.0317	
12	4,095	0.0310	0.0310	0.0159
13	8,191	2.20E-2	1.57E-2	
14	16,383	1.56E-2	1.56E-2	7.87E-3
15	32,767	1.10E-2	7.84E-3	
16	65,535	7.82E-3	7.82E-3	3.92E-3
17	131,071	5.52E-3	3.91E-3	
18	262,143	3.90E-3	3.90E-3	1.96E-3
19	524,287	2.76E-3	1.95E-3	
20	1,048,575	1.95E-3	1.95E-3	9.77E-4
21	2,097,151	1.38E-3	9.77E-4	
22	4,194,303	9.76E-4	9.76E-4	4.89E-4
23	8,388,607	6.90E-4	4.88E-4	
24	16,777,215	4.88E-4	4.88E-4	2.44E-4

However, the number of MLSs in the MCS is strongly limited. One needs to use properly combined MLSs to form Gold or Kasami sequences (see Section 5) if specific applications require a large number of simultaneous sources.

4 • Fast MLS Transform (FMT)

For the purpose of experimental determination of impulse response of a linear time-invariant (LTI) system, we apply a periodic bipolar MLS $\{m_i\}$ to the LTI system under test. At the output of the system we receive one period or several periods of the system response to the MLS after the system arrives at its steady state. The receipt of several periods is for a synchronous averaging in order to further improve the signal-to-noise ratio of the measurement. Owing to the PACF of MLS in Eq. (7) the impulse response $h(i)$ can be determined in terms of cross-correlation between the excitation MLS and the system response to the MLS **Y** by

$$\mathbf{h} = \frac{1}{L+1}\mathbf{MY}, \qquad (14)$$

where **h**, **Y** are vectors of L elements. **M** represents the MLS matrix of dimension $L \times L$, its rows contain sequentially right-cyclically shifted MLS. Equation (14) is historically termed MLS transform [24].

4.1 Hadamard Transform-Based FMT

The MLS matrix $\mathbf{M}$ in Eq. (14) is permutationally similar to Hadamard matrix [25]

$$\mathbf{M} = \mathbf{P}_2 \mathbf{H}_N \mathbf{P}_1, \quad (15)$$

where $\mathbf{P}_1$, $\mathbf{P}_2$ denote the permutation and the repermutation matrix, respectively. $\mathbf{H}_N$ is the Sylvester-type Hadamard matrix, a square matrix of dimension N with $N = 2^n$, n being a positive integer, its row vectors contain mutually orthogonal elements being either -1 or $+1$. $\mathbf{H}_N$ can be recursively constructed by

$$\mathbf{H}_{2N} = \begin{pmatrix} +\mathbf{H}_N & +\mathbf{H}_N \\ +\mathbf{H}_N & -\mathbf{H}_N \end{pmatrix} \quad \text{with} \quad \mathbf{H}_1 = [1]. \quad (16)$$

For example,

$$\mathbf{H}_2 = \begin{pmatrix} +1 & +1 \\ +1 & -1 \end{pmatrix}, \quad \mathbf{H}_4 = \begin{pmatrix} +1 & +1 & +1 & +1 \\ +1 & -1 & +1 & -1 \\ +1 & +1 & -1 & -1 \\ +1 & -1 & -1 & +1 \end{pmatrix}. \quad (17)$$

This example indicates that the Hadamard transform (HT) consists of recursive butterfly structures, similar to the FFT, but requiring only addition operations. $\mathbf{H}_2$ makes its basic butterfly structure directly comprehensible. Figure 7 illustrates the fast Hadamard transform (FHT) of degree 3. The computation efficiency is $O(N \log_2 N)$ of add/subtraction.

Equation (15) implies a fast algorithm referred to as fast MLS transform (FMT) [24], since the FHT for Hadamard matrix $\mathbf{H}_N$ is adopted in the calculation. Figure 7 illustrates a flow diagram for the HT-based FMT of degree 3. It consists of three major steps:

- Permutation of the SUT response to the MLS being used as excitation ($\mathbf{P}_1\mathbf{Y}$).
- Fast Hadamard transform of the permuted vector ($\mathbf{H}_N(\mathbf{P}_1\mathbf{Y})$).
- Re-permutation of the transformed vector $\mathbf{P}_2[\mathbf{H}_N(\mathbf{P}_1\mathbf{Y})]$.

As illustrated in Figure 7, an impulse response in the time domain is directly obtained as a result right after the re-permutation except for a scale factor.

The Hadamard transform-based fast MLS transform generally requires two permutation matrices as indicated in Eq. (15). They contain only one non-zero element of value "1" in each row or column of the matrices. Instead of the matrix, the indices

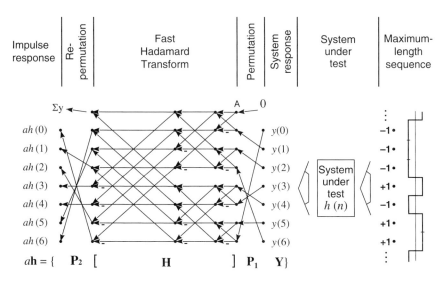

FIGURE 7 Flow diagram of the Hadamard transform based fast MLS transform with $n = 3$, $L = 2^3 - 1$ and a primitive polynomial $f(x) = 1 + x + x^3$. Apart from a scale factor $a = (L+1)$, the impulse response $h(i)$ of LTI system can be efficiently derived in time domain from the system response $y(i)$.

of non-zero elements from each row or column are usually stored [23]. Since the fast MLS transform performs inherently the cross-correlation between the MLS itself and the system response to the MLS, the two permutation matrices together can be considered as the original binary MLS in index form. There exist several different determination approaches to permutation matrices, some of them provide two required permutation matrices [12,24,26]. One of the algorithms derived from the characteristic MLS proved to require only single permutation matrix. Recently this algorithm has been extended for the single permutation matrix to work at the same time on a pair of reciprocal MLSs for the simultaneous dual source measurements in acoustic applications [23].

4.2 FFT-Based FMT

Due to odd length $L = 2^n - 1$ of MLS, the periodic cross-correlation expressed in Eq. (14) cannot be carried out simply using the fast Fourier transform (FFT) of 2^n points, since a one-point zero padding will dramatically destroy the nature of MLS (see also Chapter 5 by Müller). The augmented MLS by one zero or any value will totally destroy the PACF as shown in Figure 4, therefore totally destroy the flatness of the power spectra [27].

Recently, a dedicated algorithm has been reported for the FFT to work on the cross-correlation of coded signals not having the required length of 2^n [28]. Using this method the FMT can be performed economically with a computational efficiency of $O[2(n+1)2^{(n+1)}]$ with the following three major steps:

- Augmentation of the MLS $\{m_i\}$ of degree n to $\{\hat{m}_i\}$ and the response to the MLS $\{y_i\}$ to $\{\hat{y}_i\}$ by $2^n + 1$ zeros, respectively.
- Cross-correlation between two augmented sequences of length $M = 2^{(n+1)}$ using the FFT_M (M explicitly indicates $2^{(n+1)}$ points)

$$\hat{h}_i = \text{IFFT}_M\{\text{FFT}_M(\hat{m}_i)\text{FFT}_M^*(\hat{y}_i)\}. \tag{18}$$

- Addition of the first $L = 2^n - 1$ and the last L points in $\hat{h}_i$ to form the resulting h_i of L points

$$h_i = \hat{h}_i + \hat{h}_{i+M-L}, \quad 0 \leq i < L. \tag{19}$$

The algorithm summarized above is straightforwardly generalized to any periodic-coded signals whose period length is not a required length of p^n for FFT [28] with p being a prime, such as inverse repeat sequences, Legendre sequences as described in Section 6. The FFT-based FMT algorithm is computationally more expensive than the HT-based FMT, $O[2(n+1)2^{(n+1)}]$ multiplications and additions vs. $O[n2^n]$ additions [28]. However, the additional computation load will be less and less noticeable with the modern, fast developing computer technology. On the other hand, the FFT-based FMT is much easier to be implemented than the HT-based FMT. The FFT-based FMT is of practical significance when digital sequences other than MLSs have to be applied in acoustical experiments as the next section elaborates.

5 • Gold and Kasami Sequences

A binary Gold sequence $\{G_\tau(i)\}$ can be generated by combining a preferred or a reciprocal binary MLS pair $[\{a_i\}, \{b_i\}]$ as $G_\tau(i) = a_i \oplus b_{i+\tau}$, with $\oplus$ denoting addition modulo 2 [22]. In stepping τ point-by-point, a large number of Gold sequences result including the preferred MLS pair to yield a finite set $\{\{a\}, \{b\}, \{G_0\}, \{G_1\}, \ldots\}$ containing exactly $L+2$ sequences. Sequences in this finite set possess preferred PCCF properties. PACFs of bipolar Gold sequences associated with $\{G_\tau\}$ are peaked at zero lag, but they also possess small-valued side lobes. The PCCF of Gold sequences possesses the same bound value as that in the side lobes of the corresponding PACF. In fact, the bound values are the same as those of PCCF of preferred and reciprocal MLS pairs from which the Gold sequences are derived.

For even-numbered degree n, a decimation from an MLS $\{a_i\}$ with factor $\lambda(n) = 2^{n/2} + 1$ can lead to a pair $\{a_i\}$ and $\{e_i\}$ with $e_i = a_{\lambda i}$. In this case, $\{e_i\}$ is not an MLS of degree n. A binary Kasami sequence $\{K_\tau(i)\}$ can be generated by combining $\{a_i\}$ and $\{e_i\}$ as $K_\tau(i) = a_i \oplus e_{i+\tau}$. In similar fashion, a set of Kasami sequences $\{\{a\}, \{K_0\}, \{K_1\}, \ldots\}$ can be constructed [22] containing $2^{n/2}$ sequences. The side lobes of PACFs and the amplitude of PCCFs among these sequences present even lower bounds, approximately half of those of preferred pairs and Gold sequences:

$$l_k(n) = \frac{\lambda(n)}{2^n - 1}, \quad (20)$$

where n must be a positive even number. Table 1 also lists the bound values of Kasami sequences along with those of preferred and reciprocal MLS pairs for comparison. The power spectral density functions of Gold and Kasami sequences are the same as that of MLSs, being of broadband nature and covering the entire frequency range. In order to demonstrate the excellent correlation properties of Gold and Kasami sequences, Figure 8 shows

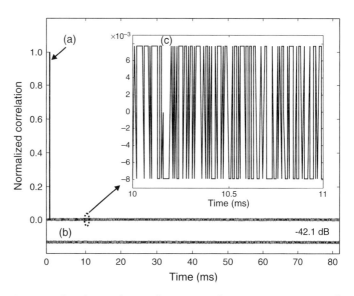

FIGURE 8 Normalized correlation functions of Kasami sequences of degree 14 generated at the sampling frequency of 200 kHz. (**a**) Periodic autocorrelation function (PACF) of one sequence. (**b**) Periodic cross-correlation function (PCCF) between two Kasami sequences (shifted downward beneath the autocorrelation function for a convenient comparison). The bound value of the PCCF is 42.1 dB lower than that of PACF. (**c**) Zoomed presentation of a segment from (**a**) showing three-valued side lobes of the PACF. Their peak values are the same as those of the PCCF in (**b**).

the PACF and PCCF of a pair of Kasami sequences, derived from an MLS of degree 14 decimated by a factor 129. In the example, the PCCF values and the side lobes of the PACF are bounded by 0.00787 or 42.1 dB below the peak value of the individual PACF. Unlike MLSs, the PACFs of Gold and Kasami sequences are pulse-like functions with small-valued side lobes. They are not MLSs any more by definition, but equivalent to the MLSs drawn from MCSs with regard to their usefulness in the simultaneous multiple acoustic source measurement technique [21], particularly when a large number of simultaneous acoustic sources have to be used. Kasami sequences have even lower bound values than that of Gold sequences, however, Kasami sequences are only available for even-numbered degrees. Moreover, the correlation measurement technique using these combined MLS sequences could not process the cross-correlation efficiently; only very recently, an efficient deconvolution algorithm [28] as described in Section 4.2 has become available.

6 • Legendre Sequences

Legendre sequences have been introduced as acoustic excitation signals for measurements of room impulse responses [9] since the late 1970s. Other applications have also been reported, such as measurements of brain-stem auditory-evoked responses (BAERs) [10].

Legendre sequences (LGSs) are periodic pseudorandom signals of period length L, where L is a prime of the form $4m-1$, with m being a positive integer. LGSs $\{a_i\}$ are defined as follows: $a_i = 1$, if i is a quadratic residue modulo L; $a_i = -1$, if i is a quadratic non-residue modulo L [9]. a_0 can be chosen arbitrarily as 0, 1 or -1. LGSs $\{a_i\}$ can also be generated by

$$a_i = i^{(L-1)/2}, \quad 0 < i < L, \tag{21}$$

where the result is taken as the least absolute remainder modulo L. LGSs are bipolar (binary) sequences for $|a_0| = 1$, while they are actually ternary sequences for $a_0 = 0$.

The PACF of LGSs normalized by its period length L is also a two-valued function

$$\phi(i) = \frac{L+|a_0|}{L}\delta(i) - \frac{1}{L}, \tag{22}$$

like that of MLSs as shown in Figure 4. The power spectrum of LGSs is therefore flat. It is this autocorrelation property that makes LGSs also useful for the correlation measurement technique. There exists a fast prime-length Fourier transform [29] so that the PCCFs can be efficiently processed. The algorithm described in Section 4.2 is also suitable for Legendre sequences for an efficient correlation processing.

Every LGS remains decimation invariant by a quadratic residue. Taking an example of $L = 7$, the LGS

$$a_i = 0, 1, 1, -1, 1, -1, -1; 0, \ldots . \tag{23}$$

A decimation by 2, being a quadratic residue given $L = 7$, results in,

$$a_{2i} = 0, 1, 1, -1, 1, -1, -1; 0, \ldots, \tag{24}$$

which exactly equals a_i. The invariant decimation can be exploited for interleaved sampling (in Section 3.3), a technique being applied in ultrasonic measurement techniques [18].

7 • Application Remarks

LGSs possess a number of intriguing properties, such as two-valued PACF, flat power spectra, pseudorandomness and invariant decimation, similar to MLSs. In addition, there are more period lengths of LGSs in the form of prime $4k - 1$ to choose from in practice than those of MLS-related sequences or Golay codes.

For LTI system identifications, both LGSs and MLS-related signals have a signal-to-noise ratio improvement proportional to their period length, since they possess a low peak factor, or high crest factor when compared to periodic impulsive excitations. In practice, these pseudorandom signals have to pass low/bandpass anti-aliasing filters, resulting in a crest factor of 10–11 dB [30,31] or lower depending on the filter characteristics. They can be used in noisy environments or at low excitation level when, e.g., the dynamic range of the SUT is severely limited or there are practical reasons for wanting a low excitation level [9], but with sufficient averaging the results can still be of high quality.

Due to excellent number-theory properties of MLSs, the MLS measurement technology has received wide acceptance in acoustic applications for over two decades. Relatively high noise immunity and an efficient correlation algorithm have significantly improved room impulse response measurements [9], acoustic impedance-tube measurements [32] and measurements of absorption coefficients in reverberation chamber even with a rotating diffuser [33]. In situ measurements of sound insulation of noise barriers [34], in situ measurements of acoustic surface reflection functions [35,36] both indoors and outdoors, measurements of head-related transfer functions [37] and hearing aids [31], applications in underwater acoustics [27] and acoustic technology of landmine detection [38] have been documented. Commercially available measurement systems based on MLS have already been serving audio engineers, acoustics researchers and consultants for quite some years. They are finding even more applications in architectural acoustics, audio engineering, physical acoustics, underwater acoustics and outdoor sound propagation.

The interest in MLS measurement technology has also stimulated investigations on its advantages, such as high-transient noise immunity [39]. However, pseudorandom signals in general, and MLSs in particular, are sensitive to time-variance components in the SUT [15,40] leading to low time-variance immunity. While some other methods with associated signals have been emerging for acoustical system identifications and characterizations with certain competing features (see Chapter 5 by Müller), such as even higher time-variance and nonlinear immunity, MLS-related signals and other digital sequences with their distinct advantages will still retain their practical significance in a wide variety of applications, particularly in nonlinearity detection/identification, and in challenging tasks using simultaneous multiple sources.

References

1. Golay, M. J. E.: "Complementary series", *IRE Trans. Inform. Theory* **7** (1961), pp. 82–87.
2. Foster, S.: "Impulse response measurement using Golay codes", *IEEE 1986, Conference on Acoustics, Speech, and Signal Processing* **2** (IEEE New York, 1986), pp. 929–932.
3. Zhou, B., Green D. M. and Middlebrooks J. C.: "Characterization of external ear impulse responses using Golay codes", *J. Acoust. Soc. Am.* **92** (1992), pp. 1169–1171.
4. Zahorik, P.: "Limitations in using Golay codes for head-related transfer function measurement", *J. Acoust. Soc. Am.* **107** (2000), pp. 1793–1796.
5. Burgess, J.: "Chirp design for acoustical system identification", *J. Acoust. Soc. Am.* **91** (1992), pp. 1525–1530.
6. Schroeder, M. R.: *Number Theory in Science and Communication*, 2nd enl. ed. Springer-Verlag, Berlin (1991).
7. Colomb, S. W.: *Shift Register Sequences*, Aegean Park Press, Laguna Hills, CA (1982).
8. MacWilliams, F. J. and Sloane, N. J.: "Pseudo-random Sequences and Arrays", *Proc. IEEE*, **64** (1976), pp. 1715–1729.

9. Schroeder, M. R.: "Integrated-impulse method measuring sound decay without using impulses", *J. Acoust. Soc. Am.* **66** (1979), pp. 497–500.
10. Burkard, R., Shi, Y. and Hecox K. E.: "A comparison of maximum length and Legendre sequences for the derivation of brain-stem auditory-evoked responses at rapid rates of stimulation", *J. Acoust. Soc. Am.* **87** (1990), pp. 1656–1664.
11. Lüke, H. D.: *Korrelationssignale*, Springer-Verlag, Berlin, New York (1992).
12. Borish, J. and Angell, J. B.: "An efficient algorithm for measuring the impulse response using pseudorandom noise", *J. Audio Eng. Soc.* **31** (1983), pp. 478–488.
13. Shi, Y. and Hecox, K. E.: "Nonlinear system identification by m-pulse sequences: application to brainstem auditory evoked responses", *IEEE Trans. Biomed Eng.*, **38** (1988) pp. 834–845.
14. Dunn, Ch. and Hawksford, M. O.: "Distortion immunity of MLS-derived impulse response measurements", *J. Audio Eng. Soc.* **41** (1993), pp. 314–335.
15. Vanderkooy, J.: "Aspects of MLS measuring systems", *J. Audio Eng. Soc.* **42** (1994), pp. 219–231.
16. Ream, N.: "Nonlinear identification using inverse repeat sequences", *J. Acoust. Soc. Am.* **76** (1984), pp. 475–478.
17. Xiang, N. and Genuit, K.: "Characteristic maximum-length sequences for the interleaved sampling method", *ACUSTICA* **82** (1996), pp. 905–907.
18. Mommertz, E. and Bayer, G.: "PC-based high frequency range M-sequence measurements using an interleaved sampling method", *ACUSTICA*, **81** (1995), pp. 80–83.
19. Simon, M. K., Omura, J. K., Scholtz, R. A. and Levitt, B. K.: *Spread Spectrum Communications Handbook*, McGraw-Hill (1994).
20. Wilson, D. K., Ziemann, A., Ostashev, V. E. and Voronovich, A. G.: "An overview of acoustic travel-time tomography in the atmosphere and its potential applications", *ACUSTICA* **87** (2001), pp. 721–730.
21. Xiang, N., Daigle, J. N. and Kleiner, M.: "Simultaneous acoustic channel measurement via maximal-length-related sequences", *J. Acoust. Soc. Am.* **117** (2005), pp. 1889–1894.
22. Sarwate, D. V. and Pursley, M. B.: "Cross-correlation properties of pseudorandom and related sequences", *Proc. IEEE* **68** (1980), pp. 593–619.
23. Xiang, N. and Schroeder, M. R.: "Reciprocal maximum-length sequence pairs for acoustical dual source measurements", *J. Acoust. Soc. Am.*, **113** (2003), pp. 2754–2761.
24. Cohn, M. and Lempel, A.: "On fast M-sequences transforms", *IEEE Trans. Inform. Theory* **23** (1977), pp. 135–137.
25. Lempel, A.: "Hadamard and M-sequences transforms are permutationally similar", *Appl. Optics*, **19** (1979), pp. 4064–4065.
26. Sutter, E. E.: "The fast m-transform: a fast computation of cross-correlations with binary m-sequences", *SIAM J. Comput.* **20** (1991), pp. 686–694.
27. Birdsall, T. G. and Metzger Jr., K.: "Factor inverse matched filtering", *J. Acoust. Soc. Am.* **79** (1986), pp. 91–99.

28. Daigle, J. N. and Xiang, N.: "A specialized fast cross-correlation for acoustical measurements using coded sequences", *J. Acoust. Soc. Am.* **119** (2006), pp. 330–335.
29. Rader, C. M.: "Discrete Fourier transforms when the number of data samples is prime", *Proc. IEEE* **56** (1999), pp. 1107–1108.
30. Rife, D. D. and Vanderkooy, J.: "Transfer-function measurement with maximum-length sequences", *J. Audio Eng. Soc.* **37** (1989), pp. 419–444.
31. Schneider, T. and Jamieson D. G.: "A dual-channel MLS-based test system for hearing-aid characterization", *J. Audio Eng. Soc.* **41** (1993), pp. 583–594.
32. Chu, W. T.: "Architectural acoustic measurements using periodic pseudo-random sequences and FFT", *J. Acoust. Soc. Am.* **76** (1984), pp. 475–478.
33. Chu, W. T.: "Room response measurements in a reverberation chamber containing a rotating diffuser", *J. Acoust. Soc. Am.* **77** (1985), pp. 1252–1256.
34. Garai, M. and Guidorzi P.: "European methodology for testing the airborne sound insulation characteristics of noise barriers in situ: experimental verification and comparison with laboratory data", *J. Acoust. Soc. Am.* **108** (2000), pp. 1054–1067.
35. Mommertz, E.: "Angle-dependent in-situ measurements of reflection coefficients using a subtraction technique", *Appl. Acoust.* **46** (1995), pp. 251–263.
36. Li, J. F. and Hodgson M.: "Use of pseudo-random sequences and a single microphone to measure surface impedance at oblique incidence", *J. Acoust. Soc. Am.*, **102** (1997), pp. 2200–2210.
37. Schmitz, A. and Vorländer, M.: "Messung von Aussenohrstossantworten mit Maximalfolgen-Hadamard-Transformation und deren Anwendung bei Inversionversuchen", *ACUSTICA*, **71** (1990), pp. 257–278.
38. Xiang, N. and Sabatier J. M.: "Laser-Doppler vibrometer-based acoustic landmine detection using the fast M-sequence transform", *IEEE Trans. Geosci. Remote Sens.* **1** (2004), pp. 292–294.
39. Ciric, D. G. and Milosevic M. A.: Transient noise influence in MLS measurement of room impulse response, *ACUSTICA*, **91** (2005), pp. 110–120.
40. Svensson, U. P. and Nielsen, J. L.: "Errors in MLS measurements caused by time variance in acoustic systems", *J. Audio Eng. Soc.* **47** (1999), pp. 907–927.

7
Filters

Ole-Herman Bjor

Norsonic AS, Lierskogen, Norway, e-mail: ohbjor@norsonic.com

1 • Introduction

Filters and frequency weightings are used to give emphasis or de-emphasis to parts of a signal based on its frequency content. The frequency range for filters is often separated in one or more passbands and one and more stopbands. The stopband consists of the band of those frequencies where we want the filter to remove the signal energy (high attenuation); the passband consists of the band of those frequencies where we want a low attenuation. An example is an octave-band filter where the passband is one octave wide. The filter will have low attenuation at frequencies within the octave band and high attenuation for signals outside the band.

The term "frequency weighting" is used when we want a more gradual emphasis/de-emphasis. An example is the A-weighting used in sound level meters introduced to mimic the variation in the sensitivity for the human auditory organ to sound with different frequencies.

The concept of frequency is strongly related to sinusoidal signal. Signals with a sinusoidal shape have a very special position in signal analysis. The reason is that this signal is the only signal that always retains its shape through a linear system. If a sinusoidal signal is used as an input to a linear system, the output will always be a signal with the same shape after the transient part due to the onset of the signal has decayed. The amplitude and phase may change, but not the sinusoidal shape and the frequency. This makes it easy to calculate the final response when a sinusoidal signal passes a sequence of linear elements. The link between time domain and frequency domain, as given by the Fourier transformation, gives one of more good reasons why we want to describe the signal by its frequency content.

Parseval's theorem states that [1]

$$\int_{-\infty}^{\infty} x^2(t)\,dt = \frac{1}{2\pi} \int_{-\infty}^{\infty} |X(\omega)|^2 \,d\omega, \quad (1)$$

where $x(t)$ is a signal as a function of time (assumed time limited) and $X(\omega)$ its Fourier transform, and $\omega = 2\pi f$ is the angular frequency. The theorem tells that the time integral of the squared signal may be obtained by integrating the squared magnitude of its Fourier-transformed function along the frequency axis. The integral on the right-hand side may be written as the sum of contiguous frequency bands:

$$\int_{-\infty}^{\infty} x^2(t)\,dt = \frac{1}{2\pi} \sum_n \left\{ \int_{-\omega_{n+1}}^{-\omega_n} |X(\omega)|^2 \,d\omega + \int_{\omega_n}^{\omega_{n+1}} |X(\omega)|^2 \,d\omega \right\}$$

$$= \sum_n \int_{-\infty}^{\infty} x^2_{w_n}(t)\,dt, \quad (2)$$

where the last equation is done by inverse Fourier transformation, and

$$x^2_{w_n}(t)$$

is the signal $x(t)$ filtered, so that only the frequencies between ω_n and ω_{n+1} are present. The equation tells that the total energy can be found as the sum of the energy in the different bands. It also tells that the phase response of the weighting is of no concern when we want to calculate the signal strength (energy) since the equation contains only the magnitude of the Fourier transformation.

2 • Frequency Weightings

2.1 Weightings for Sound Measurements

The human response to a continuous sound depends on the frequency. Traditionally, sound in the frequency range 20 Hz–20 kHz is regarded as audible. However, in reality there is a gradual change between audible and non-audible sound. A tone with a high or a low frequency outside this range may be audible if it has a sufficient strength, and a tone with a frequency inside this range may be inaudible if the strength is too low [2,3].

A network with a frequency-dependent gain may be used to mimic the frequency response of the human auditory organ. The most common network is the A-weighting. Historically, there were three common weightings for sound level measurements: A, B and C, to mimic the auditory response at low, medium and high sound levels. Today, we have better methods for describing the auditory response, and only A- and C-weightings are in general use. The A-weighting is mainly applied for general sound level measurement and to access the risk for hearing impairment due to loud noise. C-weighting is mainly used to access the risk of hearing impairment due to short time/high amplitude or peak values in the sound. Since the C-weighting is relatively flat within the normal audible frequency range, it is often used for indicating the non-weighted response. Recently, the Z-network has been introduced for this purpose.

The International standard for sound level meters IEC 61672-1 defines three frequency weightings for the measurement of sound level: A, C and Z [4]. The standard specifies the frequency response of the weightings as well as the applicable tolerances for two accuracy classes: Class 1 and Class 2. The design goal is similar for both classes, but Class 1 has the lowest tolerances.

A sound level meter consists in general of a sensor for the sound (microphone), a network for frequency weighting and a detector with an indicating device for measuring the strength of the sound. The detector shall indicate the square root of the time-averaged squared value of the frequency-weighted signal, abbreviated root mean square or rms value. As can be seen from the Parseval's theorem, the phase response of the network will not influence the measured rms value. However, other indicators of the signal strength like the peak value will be dependent on the phase.

The A-weighting is given by the following magnitude function:

$$w_A(f) = K_A \frac{f/f_1^2}{1+(f/f_1)^2} \frac{f/f_2}{\sqrt{1+(f/f_2)^2}} \frac{f/f_3}{\sqrt{1+(f/f_3)^2}} \frac{1}{1+(f/f_4)^2}, \quad (3)$$

where K_A, f_1, f_2, f_3 and f_4 are given by the following approximate values:

$$K_A = 1.258905$$
$$f_1 = 20.60 \text{Hz}$$
$$f_2 = 107.7 \text{Hz}$$
$$f_3 = 737.9 \text{Hz}$$
$$f_4 = 12194 \text{Hz}$$

The C-weighting is defined by the equation

$$w_C(f) = K_C \frac{(f/f_1)^2}{1+(f/f_1)^2} \frac{1}{1+(f/f_4)^2}, \quad (4)$$

where f_1 and f_4 are as defined above and

$$K_C = 1.007152$$

Both A- and C-weightings are defined to have unity gain at 1 kHz.

The Z-weighting is a weighting with the same tolerances as A- and C-weighting, but with a desired magnitude equal to one:

$$w_Z(f) = 1. \quad (5)$$

The corresponding gain/attenuation in decibel may be obtained by

$$L_{\text{weight}}(f) = 20 \, \log(w(f)), \quad (6)$$

where $w(f)$ are any of the weightings above (Table 1).

The responses at different frequencies are shown in Figure 1, plotted as the gain in decibel versus frequency in logarithmic scale.

TABLE 1 A-, C- and Z-weighting

Nominal frequency (Hz)	Frequency weightings (L_{weight}) (dB)		
	A	C	Z
10	−70.4	−14.3	0.0
12.5	−63.4	−11.2	0.0
16	−56.7	−8.5	0.0
20	−50.6	−6.2	0.0
25	−44.7	−4.4	0.0
31.5	−39.4	−3.0	0.0
40	−34.6	−2.0	0.0
50	−30.2	−1.3	0.0
63	−26.2	−0.8	0.0
80	−22.5	−0.5	0.0
100	−19.1	−0.3	0.0
125	−16.1	−0.2	0.0
160	−13.4	−0.1	0.0
200	−10.9	0.0	0.0
250	−8.6	0.0	0.0
315	−6.6	0.0	0.0
400	−4.8	0.0	0.0
500	−3.2	0.0	0.0
630	−1.9	0.0	0.0

800	−0.8	0.0	0.0
1,000	0	0	0
1,250	+0.6	0.0	0.0
1,600	+1.0	−0.1	0.0
2,000	+1.2	−0.2	0.0
2,500	+1.3	−0.3	0.0
3,150	+1.2	−0.5	0.0
4,000	+1.0	−0.8	0.0
5,000	+0.5	−1.3	0.0
6,300	−0.1	−2.0	0.0
8,000	−1.1	−3.0	0.0
10,000	−2.5	−4.4	0.0
12,500	−4.3	−6.2	0.0
16,000	−6.6	−8.5	0.0
20,000	−9.3	−11.2	0.0

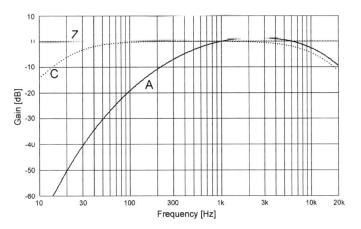

FIGURE 1 Frequency response of A-, C- and Z-weighting.

The weighting may be obtained by an analogue network or by digital signal processing. If a digital IIR network is used, some care is normally needed since the lowest characteristic frequency for the A-weighting is far below the typical sampling frequency [5]. Figure 2 shows a general second-order section for such applications with low sensitivity to truncation errors.

2.2 Networks for Vibration

A network with a gain proportional to $1/(2\pi f)$ may be used for transforming an acceleration signal to a signal proportional to the velocity. This is equivalent to an integration of the signal. This relationship can most easily be seen if an acceleration signal is assumed to be sinusoidal with amplitude A_0

$$a(t) = A_0 \sin(2\pi f t). \qquad (7)$$

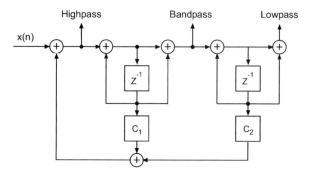

FIGURE 2 General second-order network with simultaneous outputs for lowpass, highpass and bandpass function. The network is well suited for filters with characteristic frequencies far below the sampling frequency (Courtesy: Norsonic AS).

The velocity may be obtained by time integration of the acceleration signal:

$$v(t) = \int_{-\infty}^{t} a(\tau)\mathrm{d}\tau = -\frac{A_0}{2\pi f}\cos(2\pi ft) = V_0\cos(2\pi ft). \qquad (8)$$

The change from sine to cosines indicates a 90° phase shift, which we have omitted in the network description since it will not influence the magnitude of the signal. The network may further be modified to take care of a requested frequency weighting like the weighting for whole body or hand-arm vibration. In a similar way, the displacement may be obtained from a velocity or an acceleration signal. Transformation in the opposite way is also possible by a suitable weighting.

3 • Octave- and Fractional-Octave-Bandpass Filters

For a long time, octave- and fractional-octave-band filters have been used in acoustics to measure the distribution of energy at different frequencies. The filters may be characterized as a constant relative bandwidth filter. This means that the bandwidth is a constant fraction of the midband or centre frequency of the filter. For octave filters, the upper bandedge frequency is an octave above the lower bandedge frequency.[1] Performance requirements for octave- and fractional-octave-band filters for use in acoustic analysis are specified in the International standard IEC61260 [6].

---

[1] See below for a definition of an octave in this context.

Octave- and fractional-octave-band filters are normally used in a set to cover a certain frequency range. This means that the bandedge frequency of neighbouring filters coincides so the total frequency range is covered by a number of adjacent filter bands. The levels in each band may then be used to describe the frequency distribution of the signal. Modern analyzers apply a set of parallel filters. Alternatively, one filter may sequentially be tuned to measure the level for each band. Figure 3 shows a typical set of octave-band filters with midband frequencies from 31.5 Hz to 16 kHz.

The most common bandwidths of these filters are octave and one-third octave, but 1/12 and 1/24 octave filters are also in common use.

The relation between the midband frequency and the bandwidth is chosen to make it possible to combine results obtained by a narrow filter set to results corresponding to a broader bandwidth. Three values based on 1/3-octave analysis may be converted to a 1/1-octave value. The accuracy is normally high for broadband signals, but for pure tone signals, some deviation may be observed in the transition range between the bands of the more narrow filters. Figure 4 shows the response of a typical octave-band filter and a filter synthesized by the summation of three bands with 1/3-octave width.

Although the name is octave- or fractional-octave-band filters, the characteristic frequencies for the filter are adjusted to fit the decade scale. If f_m is the midband frequency of a filter with a fractional-octave-bandwidth designator, b (b is whole number

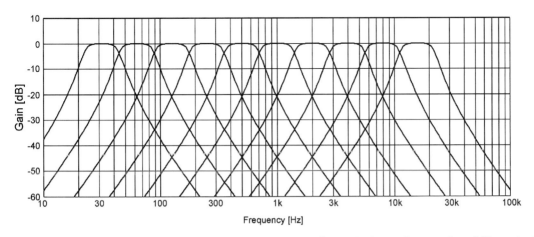

FIGURE 3 Frequency response of a typical set of octave-band filters in the range 31.5 Hz to 16 kHz. Filters may have steeper skirts in the transition between passband and stopband.

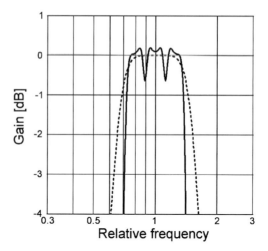

FIGURE 4 Comparison of a one-octave-band filter (- - - -) and a filter obtained as the sum of three one-third octave filters (———). Note that the sum is the sum of the squared signals (power) and thus the response is determined from the magnitude alone and is independent of the phase response of the filters. The diagram shows a typical response and may be different for other filter designs.

1, 2 or 3, etc.), the lower bandedge frequency f_L and the upper bandedge frequency f_H is given by

$$f_L = f_m \cdot 10^{3/20 \cdot b}, \tag{9a}$$

$$f_H = f_m \cdot 10^{-3/20 \cdot b}. \tag{9b}$$

For octave filters ($b = 1$), we have

$$\frac{f_H}{f_L} = 10^{3/10} = 1.995\ldots \approx 2, \tag{10}$$

which explains the designation for this filter since "octave" is used by musicians to describe tones for which the ratio between the pitches is two.

The midband frequency f_m is determined by the equation:

$$f_m = f_R \cdot 10^{3x/10 \cdot b} \quad \text{if } b \text{ is odd,} \tag{11a}$$

$$f_m = f_R \cdot 10^{3(2x+1)/20 \cdot b} \quad \text{if } b \text{ is even,} \tag{11b}$$

where f_R is the reference frequency 1,000 Hz, and x is any positive or negative whole number including zero. Rounded values for the

TABLE 2 Nominal and exact (rounded) frequencies in the range 1 kHz to 10 kHz for one-third octave-band filters

Designation (kHz)	x	Exact (kHz)
1	0	1.000000
1.25	1	1.258925
1.6	2	1.584893
2	3	1.995262
2.5	4	2.511886
3.15	5	3.162278
4	6	3.981072
5	7	5.011872
6.3	8	6.309573
8	9	7.943282
10	10	10.000000

frequency are in common use for designating the different filters for octave- and one-third octave-band filters. Table 2 shows the exact midband frequencies rounded to six digits together with the nominal values used for designation. The table applies to one-third octave-band filters in the frequency range from 1 kHz to 10 kHz. Similar figures, where the decimal point is shifted as appropriate, are used for any decade. A selection of every third number, starting at 1 kHz, is used for octave-band filters.

The relative bandwidth $(f_H - f_L)/f_m$ will be 70.5% for an octave filter, 23.1% for one-third octave and 2.88% for a 1/24th octave.

3.1 Power Spectral Density

A pure tone and a noise signal behave differently when they are analyzed with a filter with a certain bandwidth. The level of the pure tone will be almost independent of the bandwidth as long as the frequency is within the passband, whereas the level of a noise signal will be proportional to the square root of the bandwidth. Most commercial octave- and fractional-octave analyzers have a noise bandwidth close to the nominal value. The power spectral density (PSD) of the signal may be obtained by dividing the signal strength with the square root of the bandwidth [7]. Thus, if the fractional-octave value of the sound pressure is p, the mean PSD for the band will be

$$\text{PSD} = \frac{p}{\sqrt{f_m \cdot (10^{3/20 \cdot b} - 10^{-3/20 \cdot b})}}, \tag{12}$$

where f_m and b are as defined above.

3.2 Temporal Distribution

General signal theory states that the product of frequency resolution Δf and time resolution Δt, when describing the signal strength, cannot be selected below a certain limit [8]:

$$\Delta f \cdot \Delta t \geq 2\pi. \tag{13}$$

The implication of this is that the frequency resolution should not be too high (narrow bandwidth) if we want to have a high temporal resolution.

An application in acoustics where we often are restricted by this phenomenon is the measurement of short reverberation times. The reverberation time of a room, T_{60}, is the time, in seconds, that would be required for the sound pressure level to decrease by 60 dB after the sound source has stopped. Any operation to restrict the frequency content will create a virtual reverberation time. If this virtual reverberation time is not considerably less than the reverberation time for the room, the virtual reverberation will destroy the measurement accuracy.

Analogue and digital IIR fractional-octave-band filters with a bandwidth, B, in hertz, can normally be used for measurement of reverberation time without significant error if $T_{60} \geq 16/B$ [9]. However, if the impulse response of the analyzing filter is time reversed – as may be done by storing the signal – this limit can be reduced to one-fourth [10]. See Figure 5. The values are typical for filters found in commercial sound levels meters but will vary with the filter design. Filters with high selectivity may have even longer decay time. Digital FIR filters may have a limit close to the limit set by the time-reversed technique. In these cases, the lower limit is strongly dependent on which part of the decay is used for the extraction of the reverberation time. In general, the first part of the decay – the initial reverberation time or the early decay – is more influenced by the filter than the later part of the decay.

3.3 Weighted Response from Bandpass Filtered Values

It is often required to calculate weighted levels like the A-weighted level from the result of a fractional-octave-band analysis. This is done by obtaining the weighted sum of the power in each band. If the sound level in the fractional-octave-band number n with midband frequency $f_m(n)$ is L_n the weighted level is given by

$$L_w = 10 \log \left\{ \sum_n w^2[f_m(n)] \cdot 10^{L_n/10} \right\} = 10 \log \left\{ \sum_n 10^{(L_n + L_{\text{weight}}(f_m(n)))/10} \right\}, \tag{14}$$

where $w[f_m(n)]$ may be any of the previously defined weightings A, C or Z.

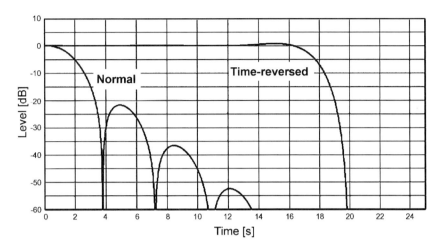

FIGURE 5 The envelope of the output from a 6-pole, Butterworth bandpass filter when a sinusoidal input signal with a frequency equal to the midband frequency of the filter is suddenly switched off is shown. The decay marked "Normal" is from an ordinary analogue filter (IIR). The decay marked "Time reversed" is when the impulse response for the "Normal" filter is time reversed. Note the delay needed to make the filter realizable. The difference in the decay rate is barely different in the first part of the decay (e.g. early decay time), but the rate is significantly higher for a larger dynamic range for the time-reversed filter. The "Normal" and the "Time-reversed" filter will have equal frequency response but a different phase response.

The accuracy of the calculation will in general be better the higher the resolution in the frequency analysis. One-third octave or more narrow analysis will normally be adequate for calculating the A- and C-weighted levels corresponding to the requirements for a Class 1 sound level meter. For sinusoidal signals, the largest deviations will typically be close to the transition frequencies between adjacent bands. The use of octave analysis for such calculation should not be recommended for other purposes than indication of weighted levels. Figure 6 shows the effective frequency response function obtained as the sum of typical octave-band analysis. Different filter shapes may make the deviation between the weighted sum and the ideal A-weighted response even larger.

3.4 Fractional-Octave-Band Analyzers

Fractional-octave-band analyzers contain a set of fractional-octave-band filters with a detector for each filter. The most common are octave-band and one-third octave-band analyzers. The signal is normally analyzed in real time, and such analyzers are therefore often called "real-time analyzers". Figure 7 shows

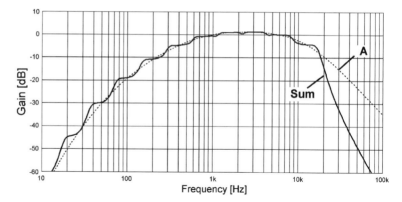

FIGURE 6 Comparison of the ideal A-weighting and the effective weighting obtained by summation of the response from a typical 1/1-octave-band analysis. For most filter designs, the result will not satisfy the requirement in the standard for sound level meters IEC 61672-1.

a typical display from a digital hand-held sound level meter with integrated octave band analyzer. The indicated level may be the instantaneous value like the *F*- or *S*-time weighted level or the mean level (equivalent value) for a specified measurement period.

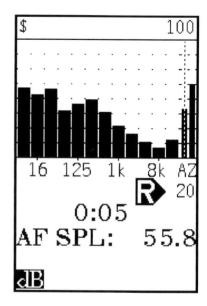

FIGURE 7 Display from a sound level meter with 1/1-octave-band analysis (Courtesy: Norsonic AS).

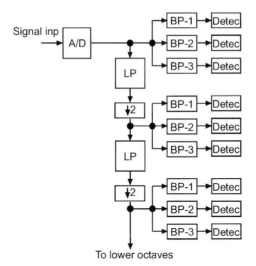

FIGURE 8 Analyzer for one-third octave. The analogue-to-digital converter converts the analogue input to a digital signal. Sampling frequency is 64 kHz. Each octave is analyzed by three 1/3-octave-band filters. The signal to the next lower octave is lowpass filtered, and the sampling is decimated to use every second sample. The block diagram corresponds to the filter implementation in the analyzer Nor830 (Courtesy: Norsonic AS).

Some digital analyzers reduce the sampling frequency for the lower octaves. According to the sampling theorem, each lower octave can use the half of the sampling frequency compared to the octave above. If a digital processor has adequate capacity for analyzing two octaves with the highest sampling frequency, it will have sufficient capacity to analyze all lower octaves if the sampling frequency is lowered for each octave. This can be shown from the sum property of geometric series (full capacity, half capacity, etc.):

$$C\left(1 + \frac{1}{2} + \frac{1}{4} + \frac{1}{8} + \cdots\right) = 2C. \quad (15)$$

The block diagram in Figure 8 shows a third-octave analyzer based on this method.

4 • Phase Linear Filters

Filters discriminate signals due to their frequency contents. A certain length of the signal is needed to determine the frequency, and the length needs to be longer for better frequency resolution.

All filters operating in real time therefore delay the signal. Most filters will have a delay that varies with the frequency. Filters having a frequency-independent delay are called linear phase filters. If the phase shift (in radians) through a filter is $\phi(f)$, the group delay is given by the equation [1]

$$\tau_{\text{group}} = -\frac{1}{2\pi}\frac{\partial \varphi(f)}{\partial f}. \qquad (16)$$

The group delay specifies the delay at frequencies around f.

Only finite impulse response filters (FIR) can have a true linear phase. The most common type is a digital filter with a structure as shown in Figure 9. Symmetry in the coefficients, a_n, around, a_0, will guarantee a symmetrical impulse response around the delay N.

Analogue, continuous time filters and digital IIR filters cannot have a true linear phase, but the response can approximate a linear phase response to any degree of accuracy.

For some application, it will be important that more filters have the same delay. A typical example is filters used for dividing the signal in different sub-band for later reconstruction. Applications may be transmission of the sub-bands in different transmission channels or in perceptual coding of sound.

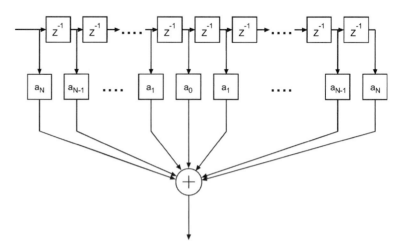

FIGURE 9 Finite impulse response (FIR) digital filter. Due to the symmetry in the filter coefficients, the filter will have a linear phase and a delay of N samples.

5 • In-Phase/Quadrature Filters

In-phase/quadrature filters are a special case of a polyphase filter. One application is to build narrow band filters like filters used in lock-in amplifiers. This is illustrated with an example in Figure 10. The signal to be filtered is modulated with two sinusoidal signals with angular frequency, ω_0, where the phase is in quadrature like a sine and a cosine signal. The spectrum of the signal will be frequency translated with the modulation frequency ω_0. The spectrum centred at ω_0 will, after modulation, be centred at the frequencies 0 (zero) and $2\omega_0$. In a similar way the spectrum centred at $-\omega_0$ will, after modulation, be centred at the frequencies 0 (zero) and $-2\omega_0$. A lowpass filter will remove the signal energy for those frequencies that are not close to zero. After a new modulation, the spectrum is again transferred to frequencies around ω_0. This will happen only for those frequency components that were not removed by the lowpass filtering. Thus the lowpass filters are performing an operation like a narrow bandpass filter.

Since the modulated signal around zero frequency is the result from the sum of two complex signals – positive and negative frequencies – at least two channels (paths) are needed to separate the two spectral components after the second modulation. If only one channel is used, the result will depend on the phase difference between the input signal and the modulation.

In a digital realization, the lowpass operation can often be performed at a reduced sampling frequency and thus reduce the needed capacity for processing power. A further decrease

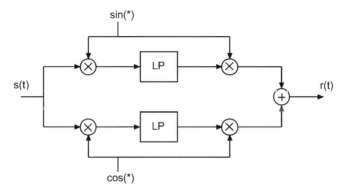

FIGURE 10 In-phase/quadrature filter where two lowpass filters are used to perform a bandpass operation.

in processing power can be gained in situation where the modulation can be substituted with two square waves in quadrature. In a digital design, this means that every second sample is routed to each of the channels.

In cases where only the filtered signal power shall be measured, the second modulation in each channel is not needed. The power can be obtained as the sum of the signal power at the output from each lowpass filter.

6 • Perceptual Masking Filters

Most of the sounds we observe have an envelope that changes slowly compared to the change in the sound pressure due to the frequency of the sound. If we sample such a signal, the information in the signal will then be less than if each of the samples were independent of the adjacent sample. This allows in most cases the signal to be coded in a way that reduces the needed capacity for transmission without loss of information (loss-less coding). A further reduction in the needed transmission capacity can be obtained if the perceptual limitation of our auditory system is utilized (perceptual coding). Perceptual coding is lossy; this means the original signal cannot be completely regenerated.

Sound with a level below a certain frequency-dependent limit cannot be heard. Furthermore, in the human auditory system, a strong frequency component will mask a neighbouring frequency component with a lower level. In particular, lower frequencies tend to mask higher frequencies. Spectral components that cannot be heard can be removed from the transmission without deteriorating the perceived sound quality. Use of this phenomenon is called perceptual coding. Masking can occur in the time domain (temporal masking) or in the frequency domain (frequency masking). A filter set is a major part of a perceptual coding algorithm [11]. The filter set may have a constant bandwidth or a bandwidth corresponding to the critical bandwidth of the auditory system. The critical bandwidth is assumed to be around 100 Hz for frequencies up to 500 Hz and more like a constant percentage bandwidth for higher frequencies (bandwidth about 15% of frequency). Examples of common perceptual coding formats are MP3, WMA and ATRAC, but more than 100 different formats are currently in use. The capacity for transmission or storing is typically reduced to 5–10% of the capacity needed for storing or transmitting the un-coded signal.

References

1. H.J. Blinchikoff, A.I. Zverev: *Filtering in the Time and Frequency Domains*. Wiley, 1976, ISBN 0-471-98679-8.
2. L.E. Kinsler, A.R. Frey: *Fundamental of Acoustics*. 2 edition. Wiley, 1962.
3. ISO 226 (2003): *Acoustics – Normal Equal-Loudness-Level Contours*.
4. IEC 61672-1 (2002–05): *Electroacoustics – Sound level meters – Part 1: Specifications*.
5. S.K. Mitra, J.F. Kaiser: *Handbook for Digital Signal Processing*. Wiley, 1993, ISBN 0-471-61995-7.
6. IEC 61260 (1995–08): *Electroacoustics – Octave-Band and Fractional-Octave-Band filters*.
7. J.S. Bendat, A.G. Piersol: *Engineering Applications of Correlation and Spectral Analysis*. Wiley, 1980, ISBN 0-471-05887-4.
8. J.B. Thomas: *An Introduction to Statistical Communication Theory*. Wiley, 1969.
9. F. Jacobsen: A note on acoustic decay measurements. *Journal of Sound and Vibration* (1987) **115**(1), 163–170.
10. F. Jacobsen, J.H. Rindel: Time reversed decay measurements. *Journal of Sound and Vibration* (1987) **117**(1), 187–190.
11. K. Brandenburg: *MP3 and AAC explained*. AES 17th International Conference on High Quality Audio Coding.

8

Adaptive Processing

Thorkild Find Pedersen

Brüel & Kjær, Sound & Vibration Measurements A/S, Nærum, Denmark

1 • Introduction

In acoustic signal processing, adaptive methods have found practical use in different applications, e.g. echo and noise cancelation, signal enhancement, system identification, adaptive equalization and speech coding.

Figure 1 shows the structure of an adaptive transversal filter. It is one of the simplest adaptive filter structures [1,2]. The filter consists of a Finite Impulse Response (FIR) filter of the order N with the adjustable coefficients $\mathbf{w} = [w_0 w_1 \cdots w_N]^T$.

The objective of the adaptive algorithm is to adjust the filter coefficients such that some predefined cost function is minimized. Typically, the cost function is the mean square error between the desired output $d(n)$ and $y(n)$ being the filtered reference signal $u(n)$. Using matrix notation, $y(n)$ is computed as the dot product of $\mathbf{w}$ and the reference vector $\mathbf{u}(n) = [u(n)u(n-1) \cdots u(n-N)]^T$.

In the ideal case, the filter output $y(n)$ will be the part of the reference signal that correlates with the desired output $d(n)$. The adaptive filter will thus attempt to cancel out the interference of the reference signal from the desired output resulting in the error signal $e(n)$. The adaptive filter is therefore suitable for interference canceling when a reference to the disturbing noise in the desired output signal is obtainable.

Adapting the filter coefficients is an iterative process where the coefficient $\mathbf{w}$ converges to an optimal value as the number of available samples increases. The notation $\mathbf{w}(n)$ is used to indicate

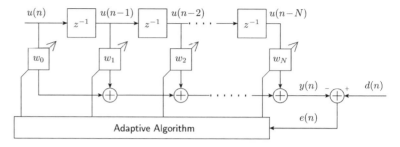

FIGURE 1 Adaptive transversal filter.

the estimates of the coefficients after the nth sample. There are numerous algorithms for adjusting the filter coefficients of which the Least Mean Square (LMS) algorithm described below is the most famous. The LMS algorithm uses the method of steepest descent to optimize the filter weights and converges relatively slowly. Faster convergence rates are obtainable with other algorithms, e.g. the Recursive Least Squares (RLS) algorithm but at the expense of higher computational complexity.

When comparing adaptive algorithms the important factors to consider are *rate of convergence*, the number of iterations it takes for the algorithm to converge to the optimum solution for a stationary input; *misadjustment*, the bias and variance of the error signal, i.e. the accuracy of filter; *tracking*, the ability of the filter to track nonstationary signals; and *robustness*, the sensitivity of the algorithm to internal or external disturbances and ability to avoid local optimum solutions.

2 • The LMS Algorithm

A good first choice algorithm for investigating adaptive processing is the LMS or the Normalized LMS (NLMS) algorithm. It is among the simplest adaptive algorithms and easy to implement. The general equations for the adaptive transversal filter is shown in (1) where the first step is to compute the difference of the current desired output $d(n)$ and the filtered output $y(n)$ based on the last estimate of the filter coefficients $\mathbf{w}(n-1)$. Depending on the algorithm type, an update step $\mathbf{\Delta}(n)$ is computed for updating the filter coefficients.

$$\begin{aligned} e(n) &= d(n) - \mathbf{w}^T(n-1)\mathbf{u}(n) \\ \mathbf{w}(n) &= \mathbf{w}(n-1) + \mathbf{\Delta}(n)e(n) \end{aligned} \quad (1)$$

For the LMS algorithm, the update step $\mathbf{\Delta}(n)$ is given by

$$\mathbf{\Delta}(n) = \mu \mathbf{u}(n) \quad (2)$$

where μ is a step-size parameter used to tune the adaption rate of the filter. The LMS algorithm converges for wide sense stationary signals when μ satisfies ($E\{\cdot\}$ is the expectation operator)

$$0 < \mu < \frac{2}{\sum_{k=0}^{N} E\{|u(n-k)|^2\}} \quad (3)$$

In the standard LMS algorithm, the update of the filter coefficients, $\mathbf{w}(n)$, is proportional to the input vector $\mathbf{u}(n) = [u(n)\ u(n-1)\cdots u(n-N)]^T$.

A problem with the LMS algorithm is *gradient noise amplification* caused by the fact that noise in the reference signal propagates directly into the filter coefficients. This is avoided in the NLMS where the update step is made time dependent by normalizing it with the reference vector, i.e.

$$\mathbf{\Delta}(n) = \frac{\mu \mathbf{u}(n)}{\delta + \|\mathbf{u}(n)\|^2} \quad (4)$$

where δ is a small positive constant to avoid numerical problems for small input signals. Generally, the NLMS algorithm converges faster than the LMS algorithm.

For large numbers of filter coefficients, it is beneficial to use the *Fast LMS* also called *Frequency Domain LMS* algorithm. The Fast LMS algorithm performs the filtering in the frequency domain using the fast Fourier transformation. The input data is therefore processed in blocks equivalent to the required FFT length, and the filter coefficients are only updated after each FFT block.

Table 1 summarizes the LMS, NLMS and RLS algorithms with respect to computing the update step in (1).

TABLE 1 Adaption algorithms for computing the update step in (1). For LMS and NLMS, the adaption rate is controlled by the step-size parameter μ. δ is a small positive constant. For RLS, it is controlled by the forgetting factor λ, $(0 \ll \lambda < 1)$. The closer λ is selected to 1, the slower the RLS algorithm adapts and also the more accurate it will be for stationary signals

Algorithm	Update step	Initialization
LMS	$\mathbf{\Delta}(n) = \mu \mathbf{u}(n)$	$\mathbf{w}(0) = \mathbf{0}$
NLMS	$\mathbf{\Delta}(n) = \frac{\mu \mathbf{u}(n)}{\delta + \|\mathbf{u}(n)\|^2}$	$\mathbf{w}(0) = \mathbf{0}$
RLS	$\mathbf{\Delta}(n) = \frac{\lambda^{-1} \mathbf{P}(n-1)\mathbf{u}(n)}{1 + \lambda^{-1}\mathbf{u}^T(n)\mathbf{P}(n-1)\mathbf{u}(n)}$	$\mathbf{w}(0) = \mathbf{0}$
	$\mathbf{P}(n) = \lambda^{-1}\mathbf{P}(n-1) - \lambda^{-1}\mathbf{\Delta}(n)\mathbf{u}^T(n)\mathbf{P}(n-1)$	$\mathbf{P}(0) = \delta \mathbf{I}$

3 • Adaptive Applications

Figure 2 shows four typical applications of the adaptive filter: (a) *system identification*, (b) *inverse modeling*, (c) *prediction* and (d) *interference canceling*.

When used for system identification, the filter is placed parallel with the unknown plant to be identified. The filters impulse response is then optimized to match that of the plant by using the driving signal of the plant as input and the plants output as the desired output.

For inverse modeling, the adaptive filter is placed in series with the output from the plant and uses the driving signal to the plant as the desired output. The adaptive filter will therefore attempt to cancel out the effect of the plant, hence the filters impulse response is optimized to match the inverse impulse response of the plant.

For prediction, the adaptive filter is optimized to predict future samples of the input signal. If the input signal is periodic with additive noise, then the periodic part can be separated from the noise if the delay M is larger than the correlation length of the noise. This use of the adaptive filter is also known as the Adaptive Line Enhancer (ALE) or Self-Adaptive Noise Cancelation (SANC). It is possible to cascade ALEs with different properties. In [3], Lee and White describe a two-stage ALE for enhancing impulsive noise from rotating and reciprocating machines. The first ALE

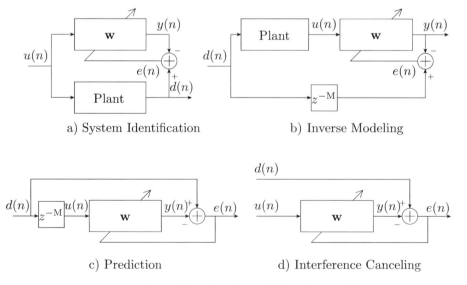

FIGURE 2 Applications of the adaptive filter.

serves to remove a strong periodic component from the signal using a relatively long delay and a transversal filter with many coefficients; the second ALE uses a shorter delay and less filter coefficients to enhance the transients in the error signal from the first ALE.

For interference canceling, the adaptive filter is used as mentioned previously, where the desired reference signal in some way is correlated with the additive noise in the desired output signal $d(n)$.

References

1. B. Widrow, S.D. Stearns: *Adaptive Signal Processing* (Prentice-Hall, Englewood Cliffs, NJ 1985).
2. S. Haykin: *Adaptive Filter Theory* 3rd edition (Prentice-Hall, Englewood Cliffs, NJ, 1996).
3. S.K. Lee, P.R. White: The enhancement of impulsive noise and vibration signals for fault detection in rotating and reciprocating machinery, *Journal of Sound and Vibration* (1998) **217**(3) 485–505.

9

Beamforming and Wavenumber Processing

Jørgen Hald

Brüel & Kjær, Sound & Vibration Measurement A/S, Nærum, Denmark

1 • Planar Nearfield Acoustical Holography

Planar nearfield acoustical holography (NAH) is a powerful experimental tool for studying and diagnosing noise radiation problems [1–11]. Based on measurements of typically sound pressure over a plane near the source, all sound field parameters can be calculated in any plane parallel with the measurement plane. In particular, the sound field can be mapped closer to the source than the measurement plane, which can provide very high spatial resolution of the source distribution. Since the calculation plane cannot in theory approach the source further than the nearest point on the source, planar NAH works best for but is not restricted to sources with a planar geometry. NAH formulations exist also for cylindrical, spherical and conformal geometries [3,5,6], but such formulations will not be treated here.

The geometry of the measurement problem to be considered is illustrated in Figure 1. All sound sources are assumed to be in the region $z < -d$, and the remaining half space $z \geq -d$ contains a homogeneous fluid with no viscosity. In that region,

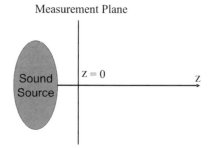

FIGURE 1 Measurement geometry.

the complex time-harmonic sound pressure $p(x,y,z)$ fulfils the Helmholtz equation [3],

$$\nabla^2 p + k^2 p = 0, \quad z \geq -d, \quad (1)$$

where $k = \omega/c = 2\pi/\lambda$ is the acoustic wavenumber, ω is the angular frequency, λ is the wavelength, c is the propagation speed of sound and ∇^2 is the Laplace operator. The implicit time factor is $e^{j\omega t}$.

For any given z-coordinate, $z \geq -d$, we now introduce the following Fourier transform pair of the sound pressure in the two dimensions (x, y):

$$P(k_x, k_y, z) = \int_{-\infty}^{\infty} \int_{-\infty}^{\infty} p(x, y, z) e^{j(k_x x + k_y y)} dx\, dy, \quad (2)$$

$$p(x, y, z) = \frac{1}{(2\pi)^2} \int_{-\infty}^{\infty} \int_{-\infty}^{\infty} P(k_x, k_y, z) e^{-j(k_x x + k_y y)} dk_x\, dk_y. \quad (3)$$

Here, the spatial angular frequencies k_x and k_y in the x- and y-directions, respectively, are also denoted as directional wavenumbers. Through insertion of Eq. (3) in Helmholtz equation (1) it can be verified that the following relation must hold:

$$P(k_x, k_y, z) = P(k_x, k_y, z_0)\, G(k_x, k_y, z - z_0) \quad z \geq -d, \quad z_0 \geq -d, \quad (4)$$

where

$$G(k_x, k_y, z) \equiv e^{-jk_z z} \quad (5)$$

with $k_x^2 + k_y^2 + k_z^2 = k^2$, and more precisely because all sources are in the region $z < -d$,

$$k_z \equiv \begin{cases} \sqrt{k^2 - (k_x^2 + k_y^2)} & \text{for } k_x^2 + k_y^2 \leq k^2 \\ -j\sqrt{(k_x^2 + k_y^2) - k^2} & \text{for } k_x^2 + k_y^2 > k^2 \end{cases}. \quad (6)$$

Thus, if the sound pressure p is known on the entire measurement plane $z = 0$, the wavenumber spectrum $P(k_x, k_y) \equiv P(k_x, k_y, 0)$ can be calculated using Eq. (2), and the pressure in any plane parallel with the measurement plane can then be obtained using a combination of Eqs. (3) and (4):

$$p(x, y, z) = \frac{1}{(2\pi)^2} \int_{-\infty}^{\infty} \int_{-\infty}^{\infty} P(k_x, k_y) e^{-j(k_x x + k_y y + k_z z)} dk_x \, dk_y \quad z \geq -d. \quad (7)$$

For $k_x^2 + k_y^2 \leq k^2$, i.e. inside the so-called radiation circle in the (k_x, k_y) wavenumber plane, the function $e^{-j(k_x x + k_y y + k_z z)}$ in Eq. (7) represents a plane wave propagating in the direction of the 3D wavenumber vector $\mathbf{k} \equiv (k_x, k_y, k_z)^T$, see Figure 2. For $k_x^2 + k_y^2 > k^2$, the same function represents a so-called evanescent wave that propagates along the xy-plane in the direction $(k_x, k_y, 0)^T$, but decays exponentially in the positive z-axis direction, i.e. away from the sources. Through the spatial Fourier transform in Eq. (2) we have therefore obtained a 3D representation of the sound pressure field in terms of plane propagating and plane evanescent waves. Based on the 3D pressure distribution in Eq. (7), Euler's

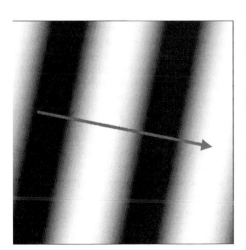
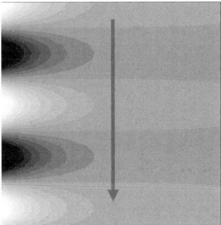

FIGURE 2 The two figures illustrate a plane propagating wave (*left*) and an evanescent wave with $k_x^2 + k_y^2 = (1.1k)^2$ (*right*). An arrow indicates the direction of wave propagation. The "source plane" $z = -d$ constitutes the leftmost limitation in both figures.

equation leads to the following expression for the particle velocity vector **u** [3]:

$$\mathbf{u}(x,y,z) = \frac{1}{(2\pi)^2} \int_{-\infty}^{\infty} \int_{-\infty}^{\infty} P(k_x, k_y) \frac{\mathbf{k}}{\rho_0 c k} e^{-j(k_x x + k_y y + k_z z)} dk_x \, dk_y, \quad (8)$$

where **k** is the wavenumber vector defined above and ρ_0 is the density of the homogeneous fluid. In particular, the normal component of the particle velocity, u_z, can be estimated on the source plane $z = -d$, and from the pressure and the particle velocity the sound intensity can be obtained.

1.1 Regularization

For $z \geq 0$ the sound field reconstruction based on Eqs. (7) and (8) is a forward simulation of wave propagation, which is numerically stable. But for $z < 0$ the reconstruction is an unstable inverse calculation, in which the evanescent wave components with $k_x^2 + k_y^2 > k^2$ must be exponentially amplified according to Eqs. (4)–(6). Measurement inaccuracies will therefore be amplified and totally dominate the reconstruction, unless some type of regularization is introduced.

The standard approach for regularizing the solution is to use a spatial low-pass filter in the form of a window function in the wavenumber plane (k_x, k_y). To prevent exponential amplification of error components, the window must cut off faster than the exponential amplification of the inverse calculation in Eq. (4). A widely used window function is the exponential window Π, defined as [3]

$$\Pi(k_x, k_y, k_c) \equiv \begin{cases} 1 - \frac{1}{2} e^{-(1-k_r/k_c)/\alpha} & k_r \equiv \sqrt{k_x^2 + k_y^2} \leq k_c \\ \frac{1}{2} e^{(1-k_r/k_c)/\alpha} & k_r \equiv \sqrt{k_x^2 + k_y^2} > k_c, \end{cases} \quad (9)$$

where k_r is the radial wavenumber, k_c is the radial cut-off wavenumber and α is a positive constant that determines how sharp the window cuts off. When α decreases towards zero, the window function Π approaches a rectangular window (Figure 3). Typically $\alpha = 0.1$ is used. A key problem is the identification of a cut-off wavenumber k_c that will prevent significant influence of measurement errors, but allow the largest possible part of the evanescent waves to be reconstructed. This can of course be done through a trial-and-error process, but typically it is done in the following way through specification of a signal-to-noise ratio SNR (dB): Considering reconstruction of the sound pressure in the source plane $z = -d$, the wavenumber spectrum of the pressure in that plane is assumed to be roughly constant, $|P(k_x, k_y, -d)| \approx P_s$.

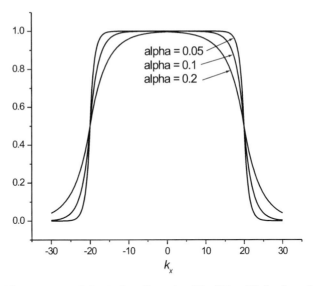

FIGURE 3 The exponential window function Π of Eq. (9) for $k_c = 20$, with α equal to 0.05, 0.1 and 0.2.

In the measurement plane, the propagating waves with $k_x^2 + k_y^2 \leq k^2$ will maintain the same amplitude P_S, while the evanescent waves will be attenuated by a factor $|G(k_x, k_y, d)|$. The second assumption is that the errors on the measured data will add a rather constant wavenumber spectrum, the level of which is approximately SNR (dB) lower than the level P_S of the plane wave spectrum. This assumption will hold if the measurement errors are spatially incoherent and incoherent with the sound sources. The radial cut-off wavenumber k_c is then chosen at the point where the level of the evanescent waves equals the level of the error spectrum:

$$e^{-\sqrt{k_c^2 - k^2}\,d} = 10^{-\text{SNR}/20} \Rightarrow$$

$$k_c = \sqrt{k^2 + \left(\frac{\text{SNR}}{20 d \log_{10}(e)}\right)^2} \simeq \sqrt{k^2 + \left(\frac{\text{SNR}}{8.7 d}\right)^2}. \quad (10)$$

Since data will be dominated by errors for radial wavenumbers larger than k_c, the filter can be used for any type of calculation.

The exponential window (also called the exponential filter) requires manual selection of the parameters α and k_c. A couple of other k-space filters have been described in the literature, for which there exist methods to determine the filter coefficients from a set of noisy measurement data: the standard Tikhonov filter and the high-pass Tikhonov filter [4,5].

1.2 Spatial Resolution

The spatial low-pass filter needed to regularize the reconstruction for $z < 0$ will of course introduce a smoothing and thereby limit the spatial resolution. The radial cut-off wavenumber k_c introduces a lower limiting wavelength of $\lambda_c = 2\pi/k_c$ on the reconstruction plane, implying that resolution cannot be better than approximately R_c given by [3]

$$R_c = \frac{\lambda_c}{2} = \frac{\pi}{k_c} \simeq \frac{1}{\sqrt{(2/\lambda)^2 + (\text{SNR}/27.3d)^2}}, \qquad (11)$$

where use has been made of Eq. (10). At high frequencies, where the wavelength λ is small, the resolution is approximately half a wavelength, but at low frequencies it approaches a constant equal to $27.3d/\text{SNR}$. So with 27.3 dB dynamic range the low-frequency resolution is approximately equal to the measurement distance.

1.3 Discrete Measurements

Until now we have assumed the pressure to be known at all points on the measurement plane $z = 0$. In practice measurements must be taken at discrete points over a finite area. Provided the pressure transducer is ideal, i.e. it measures the sound pressure at a point without disturbing the sound field, the discretization corresponds to a spatial sampling. In order to allow the spatial Fourier transforms to be evaluated using DFT algorithms, the measurement points must constitute a regular rectangular grid with constant spacing in the x- and y-directions. We will assume the same spacing a in both directions. As mentioned above, the propagation from the source plane $z = -d$ to the measurement plane $z = 0$ acts as a spatial low-pass filter, such that above the radial wavenumber k_c data will be dominated by errors. Since there is no reason to take into account the flat error spectrum in the spatial sampling, we just need at least two spatial samples per period up to the wavenumber k_c. This leads to the criterion

$$a \leq \lambda_c/2 = R_c, \qquad (12)$$

where R_c is the resolution defined in Eq. (11). In some practical cases the sample spacing (the microphone spacing) is given by a specific hardware configuration. Equations (11) and (12) can then be used to derive a lower limit on the measurement distance d. But notice that this minimum measurement distance and the maximum sample spacing in Eq. (12) are strict requirements based on the assumption that all wavenumber components have almost the same level in the source plane $z = -d$. If the pressure distribution on the source plane is "smooth", the requirements can be relaxed.

1.4 Finite Measurement Area

So far we have assumed the measurement grid to cover the entire measurement plane. The spatial sampling corresponds to multiplication with a 2D spatial comb function, which will introduce periodic replica in the wavenumber domain, but still the periodic wavenumber spectrum is in principle continuous. If we can assume the pressure samples to be almost zero outside a certain rectangular area, then we need only measure that area, and also the Fourier integral (2) needs only cover that area. If, however, the pressure is not zero outside the measurement window, then by restricting the area of the measurement and of the Fourier integration we implicitly assume the pressure to equal zero outside. The use of such a rectangular window function in the (x, y) plane corresponds to convolution with a sinc function in the wavenumber domain, which will cause components of the periodic repetitions to leak into the "baseband" and appear as aliasing. Another effect of the wavenumber leakage is that propagating waves will leak into the evanescent wave area and vice versa. The wavenumber leakage can be reduced through the use of a smooth window function instead of the rectangular window. To avoid loss of information over the central parts of the measurement area, a Tukey window is often used [3,4].

The fact that serious errors ("spatial window effects") occur if the measurement window does not extend into regions where the pressure has dropped off significantly is in many cases a serious limitation. A couple of the so-called *Patch holography* methods have been introduced to avoid this requirement. One method is to extrapolate outside the measurement area [6]. This method uses the measured data inside the measurement window and estimates the pressure in a region outside through an iterative process. Another method named SONAH (statistically optimal NAH) completely avoids the use of spatial Fourier transforms, and therefore the implicit assumption of zero pressure outside the measurement area is avoided [7,8]. In addition, since spatial DFT processing is not used, SONAH can operate with irregular measurement grids.

1.5 Spatial DFT Processing

With discrete measurement points and a finite measurement area it is attractive to perform the spatial Fourier transform with a DFT, which can be efficiently calculated using an FFT. The main implication on the result is associated with the fact that now data will be discrete in the wavenumber domain also, which is of course necessary for practical reasons: The discrete data set in the wavenumber domain represents a periodic pressure distribution in the measurement plane. When the discrete wavenumber data set is used to calculate for example the pressure in a parallel plane, then the periodic replica of the measurement area will contribute to the calculated pressure in principle over the entire calculation area. Errors introduced by these replicated measurement areas are called wrap-around errors [1]. The standard way to reduce wrap-around errors is to perform a zero-padding on the measured data before the spatial DFT. Each of the periodic repetitions in the represented pressure will then include both the measurement area and the zero-padding, which will push the first replica of the measurement area further away from the real measurement area.

Having described all the basic considerations and operations on the data, we can formulate mathematically the DFT-based holography calculation. The following formulation is very similar to the one given in [3]. The input data we will assume for the formulation is the measured pressure data after multiplication with a possible smooth window function and after zero-padding. We assume an even number N_x and N_y of data points in the x- and y-directions, respectively, located at positions $(m_x a, m_y a)$, where $-N_x/2 \leq m_x \leq N_x/2 - 1$ and $-N_y/2 \leq m_y \leq N_y/2 - 1$. The corresponding data set in the wavenumber domain has the same number of data points at the positions $(n_x \Delta k_x, n_y \Delta k_y)$, where $-N_x/2 \leq n_x \leq N_x/2 - 1$ and $-N_y/2 \leq n_y \leq N_y/2 - 1$ and where

$$\Delta k_x = \frac{2\pi}{N_x a}, \quad \Delta k_y = \frac{2\pi}{N_y a}. \tag{13}$$

Writing the discrete pressure data set in the measurement plane as $p_D(m_x, m_y)$ and the corresponding wavenumber data set as $P_D(n_x, n_y)$, the forward DFT transform corresponding to Eq. (2) becomes

$$P_D(n_x, n_y) = a^2 \sum_{m_x} \sum_{m_y} p_D(m_x, m_y) e^{2\pi j \left[\frac{n_x m_x}{N_x} + \frac{n_y m_y}{N_y}\right]}. \tag{14}$$

To calculate the pressure wavenumber data set in a parallel plane with $z \neq 0$, we must multiply with the transfer function G of Eq. (5), and to minimize the effects of measurement errors we multiply with the regularization window function Π of Eq. (9):

$$P_D(n_x, n_y, z) = P_D(n_x, n_y) \cdot G(n_x \Delta k_x, n_y \Delta k_y, z) \cdot \Pi(n_x \Delta k_x, n_y \Delta k_y, k_c), \tag{15}$$

where k_c is given in Eq. (10). The pressure data set in the calculation plane can now be obtained through inverse DFT:

$$p_D(m_x, m_y, z) = \frac{1}{N_x N_y a^2} \sum_{n_x} \sum_{n_y} P_D(n_x, n_y, z) e^{-2\pi j \left[\frac{n_x m_x}{N_x} + \frac{n_y m_y}{N_y}\right]} \tag{16}$$

The particle velocity in the same plane can be obtained by reusing Eqs. (15) and (16), just with a different transfer function G, which can be found from Eq. (8).

A lot of implementation details and considerations are given in [2,4].

1.6 Adaptations for Practical Application

So far we have considered only time-harmonic representation in the frequency domain. In practical measurement situations some effort is often needed to obtain such a representation of the sound field. As a first example, assume a rather large noise source radiating stationary, but only partially coherent noise. Partial coherence occurs because there are several independent (uncorrelated) excitation mechanisms inside the source. If the source is too large for a realistic microphone array to cover the necessary measurement area, then a smaller array must be scanned over that area. To obtain consistent phase information across the entire area, a reference signal is needed to provide a phase reference. But when averaging of typically cross-spectra is performed using the chosen reference signal, then only the part of the sound field coherent with that reference will be included in the averaged sound field representation. To obtain a complete representation we need a number of references at least equal to the number of independent source mechanisms, and from the measurement we then extract the same number of mutually incoherent time-harmonic sound fields that together represent the total sound field [9,10]. If the source is so small, however, that all positions can be measured simultaneously, then a time-harmonic representation can be obtained simply by doing a Fourier transform (using a full-record FFT) of the simultaneously recorded time data [11].

2 • Beamforming

Based on a planar measurement, NAH can reconstruct the near field of a sound source over a 3D region. By measuring a planar area that completely covers the sound source and by measuring with sufficient spatial sampling density (less than half a wavelength), all details of the sound field can basically be reconstructed, except for details that are lost in measurement errors. For large sources and/or at high frequencies, however, the required measurement area and sampling density make the method impractical. Here, beamforming is an attractive alternative based on a far-field array measurement: From typically a simultaneous recording with all array microphones the method determines, with a certain angular resolution, how much is incident from all directions. So where NAH provides a calibrated map of selected sound field parameters near the source, beamforming basically provides a directional map of contributions at the array position. The present section will give only a short introduction to the very basic delay-and-sum beamformer. A broad description of beamforming techniques is given in [12].

2.1 Introduction to Delay-and-Sum Beamforming

As illustrated in Figure 4, we consider a planar array of M microphones at locations $\mathbf{r}_m (m = 1, 2, \ldots, M)$ in the xy-plane of our coordinate system. When such an array is applied for delay-and-sum beamforming, the measured pressure time signals $\tilde{p}_m$ are individually weighted and delayed, and then all signals are averaged [12]:

$$\tilde{b}(\boldsymbol{\kappa}, t) = \frac{1}{M} \sum_{m=1}^{M} w_m \tilde{p}_m (t - \Delta_m(\boldsymbol{\kappa})). \qquad (17)$$

The individual time delays Δ_m are chosen with the aim of achieving selective directional sensitivity in a specific direction, characterized here by a unit vector $\boldsymbol{\kappa}$. This objective is met by adjusting the time delays in such a way that signals associated with a plane wave, incident from the direction $\boldsymbol{\kappa}$, will be aligned

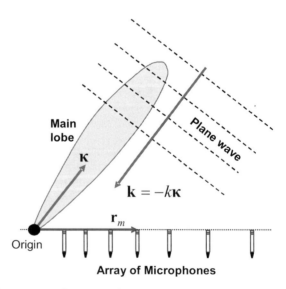

FIGURE 4 Illustration of a microphone array, a directional sensitivity represented by a mainlobe and a plane wave incident from the direction of the mainlobe.

in time before they are averaged. Geometrical considerations (see Figure 4) show that this can be obtained by choosing

$$\Delta_m = \frac{\kappa \cdot \mathbf{r}_m}{c}. \tag{18}$$

Signals arriving from other far-field directions will not be aligned before the averaging, and therefore they will not add up coherently. Thus, we have obtained a directional sensitivity as illustrated by the *mainlobe* in Figure 4.

The frequency domain version of expression (17) for the delay-and-sum beamformer output is

$$b(\kappa, \omega) = \frac{1}{M} \sum_{m=1}^{M} w_m \, p_m(\omega) e^{-j\omega \Delta_m(\kappa)} = \frac{1}{M} \sum_{m=1}^{M} w_m \, p_m(\omega) e^{j\mathbf{k} \cdot \mathbf{r}_m}, \tag{19}$$

where $\mathbf{k} \equiv -k\kappa$ is the wavenumber vector of a fictitious plane wave incident from the direction κ in which the array is focused, see Figure 4.

Through our choice of time delays $\Delta_m(\kappa)$, or equivalently of the "preferred" wavenumber vector $\mathbf{k} \equiv -k\kappa$, we have "tuned" the beamformer on the far-field direction κ. Ideally we would like to measure only signals arriving from that direction, in order to get a perfect localization of the sound source. To investigate how much "leakage" we will get from plane waves incident from other directions, we assume now a plane wave incident with a wavenumber vector $\mathbf{k}_0$ different from the preferred $\mathbf{k} \equiv -k\kappa$. The pressure measured by the microphones will then ideally be

$$p_m(\omega) = p_0 e^{-j\mathbf{k}_0 \cdot \mathbf{r}_m}, \tag{20}$$

which according to Eq. (19) will give the following output from the beamformer:

$$b(\kappa, \omega) = \frac{p_0}{M} \sum_{m=1}^{M} w_m e^{j(\mathbf{k}-\mathbf{k}_0) \cdot \mathbf{r}_m} \equiv p_0 W(\mathbf{k} - \mathbf{k}_0). \tag{21}$$

Here, the function W,

$$W(\mathbf{K}) \equiv \frac{1}{M} \sum_{m=1}^{M} w_m e^{j\mathbf{K} \cdot \mathbf{r}_m}, \tag{22}$$

is the so-called *array pattern*. It has the form of a 2D spatial Fourier transform of a weighting function w, which consists of delta functions at the microphone positions, refer Eq. (2). Because the microphone positions $\mathbf{r}_m$ have z-coordinate equal to zero, the array pattern is independent of K_z. We therefore consider the array pattern W only in the (K_x, K_y) plane, and when it is used as in Eq. (21) the 3D wavenumber vector is projected onto the (K_x, K_y) plane. In that plane, W has an area with high values around the origin with a peak at $(K_x, K_y) = (0, 0)$. According to Eq. (21), this peak represents the high sensitivity to plane waves coming from the direction κ, in which the array is focused. Figure 4 contains an illustration of that peak, which is called the *mainlobe*. Other directional peaks, which are called *sidelobes*, will cause waves from such directions to leak into the measurement of the mainlobe direction κ, creating the so-called *ghost sources* or *ghost images*.

From Eqs. (21) and (22) it is clear that the array pattern contains information about the sidelobe pattern for all frequencies. When the frequency increases, the wavenumber vectors **k** become longer, meaning that a larger area of the array pattern function W becomes "visible". For practical application of an array it is important to have low sidelobe levels and therefore good suppression of ghost sources over the frequency range of interest. This leads to the idea of optimizing an array design by adjusting the geometry and perhaps the weights w_m with the aim of minimizing the sidelobes of the array pattern over the area that is visible within a given frequency range [13]. In general, letting the weights go to zero towards the edges of the array can decrease the sidelobe level. This is, however, at the expense of a broadening of the mainlobe, which means poorer angular resolution [12]. The weights are therefore often kept equal to one, which is referred to as *uniform shading*. Irregular geometries are typically used, because they can provide fairly low sidelobe level over very broad frequency bands, avoiding the so-called *grating lobes* associated with strong spatial aliasing effects of regular geometries. Low sidelobe levels (at least 10 dB lower than the mainlobe) can then be maintained up to frequencies where the average element spacing in the array is several wavelengths, typically 3–4 wavelengths.

So far we have considered only resolution of incoming plane waves, corresponding to an infinite focus distance. To focus the array at a point (at a finite distance), the delays in Eq. (18) have to be changed in such a way that signals associated with a spherical wave radiated from the focus point are aligned in time before the averaging. But such a spherical wave will exhibit varying amplitudes across the array, which introduces the need to do amplitude compensation [12,13]. Focusing at a finite distance works well in general and does not significantly increase the sidelobe level if the measurement distance is not smaller than array diameter.

2.2 Resolution and Size of Mapping Area

The size and shape of the mainlobe of the array pattern is determined almost entirely by the size of the array [12]. The microphones just need to have a fairly uniform distribution across the array area, not for example being concentrated at the centre. Clearly, the mainlobe width will determine the angular resolution of the array: A broad mainlobe has the effect that a single incident plane wave will contribute to the beamformed map over a wide angular range. Resolution is here defined in connection with two incoherent monopoles of equal strength on a distant plane parallel with the array. It is defined as the smallest separation that allows them to be distinguished in a beamformed map. For an array with uniform shading that covers a circular area with diameter D, the resolution can be shown to be approximately

$$R(\theta) \simeq \frac{1.22}{\cos^3(\theta)} \frac{L}{D} \lambda, \qquad (23)$$

where L is the distance to the mapping plane and θ is the off-axis angle (from the z-axis) [13]. The resolution is seen to degrade very quickly beyond approximately 30° off-axis angle, so in practice mapping will be restricted to that angular range. Close to the axis, the best-case resolution is around one wavelength. This resolution is obtained when the array is used at the smallest recommended measurement distance, which is equal to the array diameter or slightly smaller. If this best possible resolution is required, then because of the 30° maximum off-axis angle the mapping area will be of approximately the same size as the array. As mentioned above, this area can be mapped with an array that has average microphone spacing up to 3–4 wavelengths. Compared to that, NAH requires less than half wavelength spacing. So to map a quadratic area of linear dimension one wavelength, NAH requires more than 4 measurement positions, whereas beamforming can do with down to 0.1 position.

The advantages of NAH are that scaled maps of any acoustic quantity are obtained, that the level of ghost sources can be kept considerably lower and that the resolution is half a wavelength or better, refer Eq. (11). If we look at low frequencies in particular, NAH can provide a resolution approximately equal to the typically very small measurement distance, whereas beamforming can still not do better than around one wavelength, which is very poor at low frequencies.

So at low frequencies, NAH is often required to get acceptable resolution, whereas at high frequencies beamforming is often required to map a given source area with an acceptable number of measurement positions. Reference [14] describes a solution where NAH and beamforming measurements can be performed with the same irregular array. The array is then optimized to provide low sidelobe level with beamforming, and at the same time it has a sufficiently uniform element density that it applies at low frequencies with the SONAH holography algorithm presented in [8]. Reference [14] also describes a method to scale the beamformed maps as sound intensity, which is a typical type of output from NAH.

References

1. J.D. Maynard, E.G. Williams and Y. Lee, "Nearfield acoustic holography: I. Theory of generalized holography and the development of NAH", *J. Acoust. Soc. Am.*, **78(4)**, 1395–1413 (1985).
2. W.A. Veronesi and J.D. Maynard, "Nearfield acoustic holography (NAH) II. Holographic reconstruction algorithms and computer implementation", *J. Acoust. Soc. Am.*, **81(5)**, 1307–1322 (1987).
3. E.G. Williams, *Fourier Acoustics – Sound Radiation and Nearfield Acoustical Holography*, Academic Press, (1999).
4. M.J. Crocker (editor), *Handbook of Noise and Vibration Control*, Chap. 48 by E.G. Wiliams, Wiley, New York (2005).
5. E.G. Williams, "Regularization methods for near-field acoustical holography", *J. Acoust. Soc. Am.*, **110(4)**, 1976–1988 (2001).
6. E.G. Williams, B.H. Houston and P.C. Herdic, "Fast Fourier transform and singular value decomposition formulations for patch nearfield acoustical holography", *J. Acoust. Soc. Am.*, **114(3)**, 1322–1333 (2003).
7. R. Steiner and J. Hald, "Near-field acoustical holography without the errors and limitations caused by the use of spatial DFT", *Proceedings of ICSV6* (1999).
8. J. Hald, "Patch near-field acoustical holography using a new statistically optimal method", *Proceedings of Inter-Noise* (2003).
9. J. Hald, "STSF – a unique technique for scan-based near-field acoustic holography without restrictions on coherence", *B&K Technical Review*, No. 1 (1989).
10. H.-S. Kwon, Y.-J. Kim and J.S. Bolton, "Compensation for source nonstationarity in multireference, scan-based nearfield acoustical holography", *J. Acoust. Soc. Am.*, **113(1)**, 360–368 (2003).
11. J. Hald, "Time domain acoustical holography and its applications", *Sound Vibr.*, 16–25 (February 2001).
12. D.H. Johnson and D.E. Dudgeon, *Array Signal Processing: Concepts and Techniques*, Prentice Hall, New Jersey (1993).
13. J.J. Christensen and J. Hald, "Beamforming", *B&K Technical Review*, No. 1 (2004).
14. J. Hald, "Combined NAH and Beamforming Using the Same Microphone Array", *Sound Vibr.*, 18–27 (December 2004).

PART II
AUDITORY SYSTEM AND HEARING

Birger Kollmeier

University of Oldenburg, Oldenburg, Germany

10 Anatomy, Physiology and Function of the Auditory System
 Birger Kollmeier 147
 External Ear • Middle Ear • Inner Ear • Auditory Pathway

11 Physiological Measures of Auditory Function
 *Birger Kollmeier, Helmut Riedel, Manfred Mauermann,
 and Stefan Uppenkamp* 159
 Head-Related Transfer Functions • Tympanometry • Acoustic Middle Ear Reflex •
 Otoacoustic Emissions • Evoked Potentials • Functional Imaging of the Auditory System

12 Auditory Processing Models
 Torsten Dau 175
 External and Middle Ear Transformation • Cochlear Transformation • Examples
 of Brainstem Processing Models • Models of Auditory Perception • Applications of
 Auditory Models

13 Speech Intelligibility
 Thomas Brand 197
 Speech Intelligibility • Measurement Methods • Factors Influencing Speech
 Intelligibility • Prediction Methods

14 Signal Processing in Hearing Aids
Volker Hohmann 205

Basic Approach to the Rehabilitation of Hearing Deficiencies • Factors of Hearing Impairment and Rehabilitation Strategies • Limitations of Current Rehabilitation Strategies • Conclusions

10

Anatomy, Physiology and Function of the Auditory System

Birger Kollmeier

University of Oldenburg, Oldenburg, Germany

The human ear consists of the outer ear (pinna or concha, outer ear canal, tympanic membrane), the middle ear (middle ear cavity with the three ossicles malleus, incus and stapes) and the inner ear (cochlea which is connected to the three semicircular canals by the vestibule, which provides the sense of balance). The cochlea is connected to the brain stem via the eighth brain nerve, i.e. the vestibular cochlear nerve or nervus statoacusticus. Subsequently, the acoustical information is processed by the brain at various levels of the auditory system. An overview about the anatomy of the auditory system is provided by Figure 1.

1 • External Ear

The pinna and the initial part of the external auditory canal are constructed from elastic cartilage and show a high interindividual variability. The more internal part of the external ear canal is surrounded by bone. In its rest position, the external ear canal is a bit winded so that the ear drum can only be inspected from outside if this winding is removed (e.g. by softly pulling the pinna to the back and upward). This is utilized when performing *otoscopy*: The permeability of the external auditory

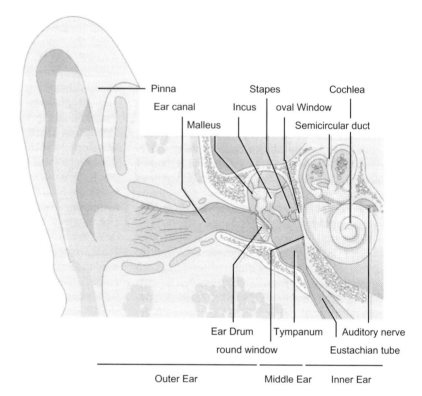

FIGURE 1 Overview of the anatomy of the outer ear, middle ear and inner ear (modified from Kollmeier, 1997).

canal, the shape and consistency of the ear drum, and any suspicious changes in the visible structures are considered by an inspection using a magnifying glass or a microscope. The function of the pinna is to sample the incident sound wave from a larger area into the smaller area of the ear canal (funnel principle, especially for high frequencies). Also, a spectral change (filter) is provided as a function of the incidence direction. The resulting change in timbre as a function of incidental direction can be utilized for localization of the sound source from where the sound is emitted. (Additionally, interaural differences in level and in arrival time occur that are utilized by the brain to perform a very exact localization in the horizontal plane by comparing the input to both ears. However, the interaural information has to be supplemented by the spectral changes provided by the pinna and other structures to avoid front/back confusions.) At low frequencies, the shape of the upper part of the body and the head is primarily responsible for

TABLE 1 Functional organization and possible pathologies of the auditory system as well as the range of audiological diagnostical methods to cover the function of the respective components

Part of the ear	Function	Dysfunction	Audiological diagnostics
Outer ear	Directional filtering	Malformation of pinna	Medical examination
Middle ear	Impedance transformation Air→Water	Conductive hearing loss (e.g., hydrotympanon, otitis media)	Otoscopy Impedance audiometry
Inner ear	Transformation water borne sound → neural excitation, frequency-place transformation	Cochlear hearing loss (e.g., noise-induced, age-related,….)	Tuning fork tests Tone audiogram Otoacoustic emissions
Auditory nerve/ brainstem	Coding of acoustical information, evaluation of interaural differences, modulation processing	Retrocochlear (neural) hearing loss, deterioration of localization	Suprathreshold tests Brain stem audiometry Speech audiometry
Cortex	Speech perception, complex perception	Central hearing disorders (e.g. aphasia)	Medical imaging Central speech tests

the exact direction-dependent filtering. The exact shape of the pinna is more relevant at high frequencies where the wavelength has the same order of magnitude as the dimensions of the pinna. These properties are utilized in the dummy-head technique which is termed "virtual acoustics" as soon as a computer gets involved. Basically, the recording can be made either with a dummy head or with a probe microphone in the individual's ear canal. If the sound is reproduced (for instance by equalized headphones) to the same subject, he or she has the impression to be acoustically located in the original recording site.

Most pathological changes of the outer ear and its function can be treated by otolaryngologists (i.e. physicians specialized in ear, nose and throat diseases). These changes can be observed either clinically (i.e. by the previous history and a physical examination) or by otoscopy (Table 1).

2 • Middle Ear

The middle ear is located in the air-filled tympanum which is connected via the Eustachian tube to the nose/throat cavity (Figure 1). This narrow tube can actively be opened by muscle tension when swallowing or yawning. The opening produces a pressure release on both sides of the tube which provides the same atmospheric pressure both from outside and inside the tympanic membrane. The maintenance of an equilibrium is important for the free movement of the tympanic membrane as well as the successive middle ear ossicles. A strong, static displacement of the tympanic membrane (e.g. by changes of the air pressure) can also be compensated by the joint between malleus head and incus that fixates itself into a new position (Hüttenbrink, 1988; Hudde and Engel, 1997).

The function of the middle ear is the impedance adjustment between the sound propagation in air (low impedance) and the quite high input impedance of the liquid-filled inner ear (high impedance). This is primarily achieved by the ratio between the large area of the tympanic membrane and the small area of the stapes footplate attached to the oval window. An additional contribution arises from the leverage of the long processus of the malleus connected to the ear drum and the shorter processus of the incus connected to the stapes. The total transformation effect produces an approximate increase of the force per unit of area by a factor of 50 at the stapes footplate. Even though the resulting impedance matching is not perfect (a factor of 3,000 would be required for a perfect impedance match between ear and water), it is sufficient to provide an optimum transmission efficiency with a minimum amount of biological effort (since larger membranes for a better impedance match would cause a higher susceptibility to possible damages).

At very high sound pressure levels, the stapedius muscle comes into action which is attached to the head of the stapes and produces a change of middle ear transmission efficiency that can be measured by acoustical impedance measurements (see below). The action of the second middle ear muscle, i.e. the tensor tympani cannot be recorded in audiological measurements.

A pathological change of the middle ear function results in a conductive hearing loss, i.e. an attenuation of an air-borne sound into the inner ear. This may be caused by insufficient ventilation through the Eustachian tube as well as fluid or infections in the middle ear cavity. Even though the type of the pathological change can be very different (i.e. otosclerosis with a fixation of the stapes footplate to the surrounding bone, disruption of the ossicular chain, defect in the tympanic membrane, etc.), the resulting dysfunction is very similar.

For a functional diagnostics of the middle ear, impedance audiometry is utilized. The acoustical reflection at the tympanic membrane is assessed with a microphone in the sealed ear canal in order to estimate the impedance and the sound transmission into the inner ear. Tympanometry tests the static air pressure delivered to the ear canal that yields the best transition (i.e. maximum acoustical compliance of the ear drum). If a middle ear pathology exists, the optimum static pressure value is displaced or no maximum value is found at all (Chapter 11). From the temporal change of the impedance after stimulating the ear with a high-intensity tone, the presence or the absence of the stapedius muscle reflex can be determined (reflex audiometry). Note that a conductive hearing loss is still connected to an intact bone conduction into the inner ear. This difference between air and bone conduction ("air–bone gap") can be assessed in clinical tuning fork tests (according to Weber and Rinne) and in audiometry with a bone conduction stimulator in comparison to thresholds obtained with an audiometric headphone.

3 • Inner Ear

The inner ear consists of a coiled shell-shaped tube which is embedded in the very hard temporal bone and consists of three compartments (Figure 2). The oval window connects to the scala vestibuli which is connected at the upper end of the cochlea, the so-called helicotrema, with the scala tympani. The intermediate compartment (scala media) is filled with endolymph and is separated from the scala tympani by the basilar membrane (Figure 2). The lateral width of the basilar membrane increases steadily from the oval window to the helicotrema and simultaneously decreases in its stiffness. If a sound excites the oval window, a pressure difference occurs orthogonal to the basilar membrane between the scala vestibuli and the scala tympani which produces a movement of the basilar membrane. Even though the water-borne sound travels with high velocity through all compartments of the cochlea, the location-dependent sound pressure difference across the basilar membrane leads to a peculiar wave pattern on the basilar membrane, the so-called travelling wave. It propagates with a comparatively slow velocity and a very high dispersion along the basilar membrane from stapes toward the helicotrema and displays maximum amplitude for high frequencies close to the stapes and for low frequencies close to the helicotrema (frequency-place transformation). Many theories have been developed in the past to explain this pattern, including a linear theory (de Boer, 1980) and three-dimensional finite element models (Böhnke and Arnold, 1999).

The displacement of the basilar membrane produces a shear movement in relation to the tectorial membrane which is detected by the inner and outer hair cells embedded on top of the basilar membrane (Figure 3). The stereocilia at the end of the hair cells

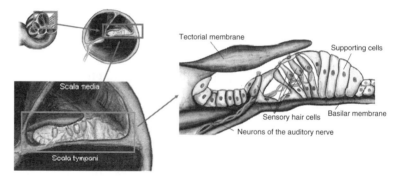

FIGURE 2 Schematical cross-section of the inner ear (modified after Silbernagl and Despopoulos, 2003).

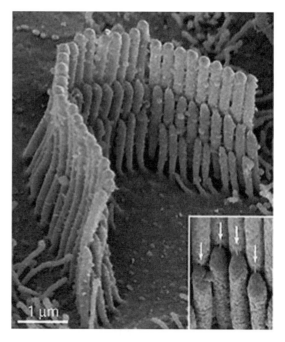

FIGURE 3 Scanning microscope photography of an intact hair cell bundle (stereocilia) located atop a hair cell. Tip links (*arrows*) connect shorter stereocilia to their taller neighbours. Reproduced with permission from Nature Reviews Genetics 5, 489–498, copyright 2004 Macmillan Magazines Ltd.

are connected with each other with tiny links ("tip links") that produce a change in ion permeability, if a mechanical stress is applied (Hudspeth et al., 1998, Dallos et al., 1991). Hence, the hair cells detect a shear movement by a change in their intracellular potential. While the inner hair cells are each connected to a large number of afferent nerve fibres (that transfer excitation from the receptors to the brain), the three rows of outer hair cells are primarily connected to efferent fibres (i.e. slowly changing adaptation information from the brain to the hair cells). The outer hair cells are capable of actively contracting themselves if an external voltage is applied. This is used to amplify vibrations at low acoustical input levels, so that an active feedback loop simultaneously increases the sensitivity and the frequency specificity of the basilar membrane response. Even though the exact mechanisms of these "active processes" in the inner ear is not yet fully understood (see Dallos et al., 1997; Gummer et al., 2002 for a review), they have a prominent importance for the normal auditory function especially when listening to low acoustical levels.

If the feedback from mechanical/acoustical energy via electrical voltage into the contraction of outer hair cells (which in turn is mechanical/acoustical energy) assumes a loop gain of greater than 1, the system becomes instable and begins to oscillate. These low-energetic oscillations are actually recordable with a sensitive microphone in the sealed ear canal when employing sophisticated noise reduction methods. They are called spontaneous *otoacoustic emissions* (SOAE). Such emissions can also be evoked by a short, transient stimulus (transient-evoked otoacoustic emissions, TEOAE) or with a continuous tone (simultaneous-evoked otoacoustic emissions). They can be applied to objectively test the auditory system (Table 1).

A damage to the inner ear causes a sensorineural hearing loss that may have various reasons (excessive noise exposure, age-related detriment of the hair cells, metabolic diseases, etc.). In most cases, the function of both the inner and the outer hair cells is decreased to a variable degree. The consequence is a reduced sensitivity of the ear for stimuli close to the respective frequency region. If primarily the inner hair cells are damaged, the active processes are still intact that still provide a kind of dynamic compression (i.e. amplification of low levels that saturates at high acoustic levels), while the overall sensitivity is reduced. Conversely, a partial or complete loss of outer hair cells produces both a loss in sensitivity and a loss in compression, i.e. no special adaptation occurs to amplify soft sounds. Consequently, a steep increase of perceived loudness can be observed when the stimulus level is increased (the so-called recruitment phenomenon, see Chapter 14 (Hohmann)).

The function of the inner ear can be tested with a tuning fork or with the audiogram using both bone-conduction transducers in comparison to measuring the sensitivity to air-borne sound with headphones. The presence of otoacoustical emissions is an indication of a (nearly) normal auditory system, while the threshold test and speech audiometry can be used to assess suprathreshold distortions of the auditory perception due to a hearing loss (i.e. recruitment phenomenon or decreased speech intelligibility in noise). As an "objective" auditory test which tests both the function of the inner ear and parts of the brain, the brain stem audiometry (brain stem-evoked response audiometry, BERA) is available for clinical diagnostics. With this method, the electrical voltage at the surface of the skull is recorded as response to an acoustical stimulus delivered to the patient's ear. To separate the stimulus-related reaction of the brain from the uncorrelated, random noise produced by the brain, an averaging method is used. If the recorded response for a given auditory stimulus complies with the normal value, one can objectively certify a normal functioning auditory system.

4 • Auditory Pathway

The auditory nerve originates in the middle of the cochlea and runs through the meatus acusticus internus (inner ear duct) to the brain stem where it reaches the cochlear nucleus. From this brain nucleus, several links exist to other nuclei in the brain stem (i.e. superior olive, nucleus accessories and to the nuclei of the lateral lemniscus). The auditory pathway goes up through the lateral lemniscus and the medial lemniscus into the inferior colliculus and the medial geniculate body until it finally reaches the primary auditory cortex which is located in area 41 of the temporal lobe in the cortex (Figure 4). Several linkages exist at each of these stages to the respective other side.

The function of the auditory nerve and the auditory pathway is the coding and processing of all acoustical information into neural excitation patterns in the appropriate structures that represent auditory information to the brain. The auditory nerve codes acoustical information by the synchronicity in the spiking pattern of different neural fibres as well as increasing the rate of the respective fibres with increasing stimulus level. In the brain stem, already complex auditory features are evaluated. For example, an interaural comparison to detect interaural arrival time and intensity differences is already performed in the superior olive complex as a basis for localization of sound sources (see above). In addition, a modulation frequency analysis is already performed in the inferior colliculus (Langner and Schreiner, 1988). The common modulation of the intensity in different frequency bands seems to be an important feature for

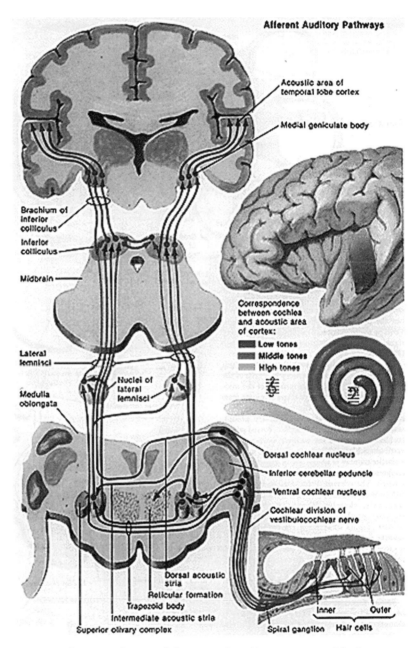

FIGURE 4 Schematical view of the central auditory system with the anatomical view (reprinted from the CIBA Collection of Medical Illustrations, Vol. 1).

an auditory object, so that a modulation analysis is assumed to take place already at a comparatively early processing stage of the auditory system (see Chapter 12 (Dau)). Furthermore, the whole auditory pathway is characterized by its tonotopic organization, i.e. neighbouring acoustical frequencies yield excitations in neighbouring structures within the whole auditory pathway. A similar spatial organization is supposed for the spatial organization of sound sources in space (spatiotopic organization) and a for a regular ordering of the different modulation frequencies in the brain (periodotopic mapping).

Pathological changes of the functions of the auditory nerves and the more peripheral parts of the central auditory system (e.g. acusticus neurinoma that mechanically damages the auditory nerve, or hearing deficits connected with other neurological symptoms as a consequence of a stroke or other neurological disease) are denoted as retrocochlear or neural hearing disorders. They can be detected with brain stem audiometry where a prolonged delay time is observed between the electrical waves originating from the cochlea (i.e. waves I and II) and the more centrally originating parts of the evoked response audiometry (ERA). Alternatively, imaging techniques like functional MR imaging or positron emission tomography (PET) can be used. More centrally located damages to the auditory system (for example due to blood circulation deficits or lesions) can result in various neurological symptoms, among which the aphasic syndromes are most prominent (i.e. the inability to speak correctly). Several central language tests have been developed and applied in order to improve the diagnostic properties here.

In a physical or communication theory sense, signal processing in the ear can be described by a succession of several signal processing stages that model the "effective" processing in the auditory system without taking the actual biological implementation into account. An important element of such a functional description is a filter bank as a first stage that distributes the sound according to its frequency into different frequency bands and hence mimics some aspects of the basilar membrane (Figure 5). It may be preceded by a broad band-pass filter that mimics the "effective" frequency shaping by the middle ear. In each frequency channel, the instantaneous energy is then obtained by extracting the temporal envelope which can be modelled by a half-wave rectifier with successive low-pass filtering. This process and the subsequent adaptation stage should simulate the function of the hair cell and the auditory nerve where the respective sensitivity is actually adapted to

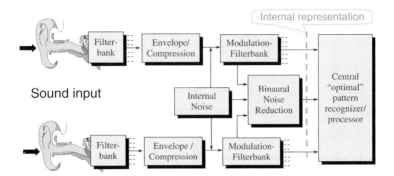

FIGURE 5 Schematical view of the "effective" signal processing in the auditory system.

the average of the respective input signal, and the exact fine structure of the input signal is lost for high-signal frequencies. The adaptation is followed by a binaural comparison as well as a separation into different modulation frequencies by a modulation filter bank. Hence, the output of these basic processing structures can be thought of as a two-dimensional pattern for each point in time that represents audio frequency on one axis and modulation frequency on the other axis. In addition, for each combination of modulation frequency and centre frequency, an interaural relation can be assumed to be present (Kollmeier and Koch, 1994).

The time course of this multidimensional pattern can be thought of as representing the "internal representation" of an acoustical stimulus presented to the auditory system. To account for processing inaccuracies of the auditory system, an additive processing noise ("internal noise") has to be added. Hence, most models of the "effective" signal processing in the auditory system assume that these most important processing steps of the auditory system can be represented with comparatively simple processing circuits that require very few parameters for their functioning. The actual detection of any signals in noise or of any changes in the acoustic input signal can be modeled then by an optimum detector or matched filter (Green and Swets, 1974; Dau et al., 1997). In other words, the current models assume that auditory detection and discrimination is not necessarily limited by the cognitive information processing and the previous knowledge about the stimulus and the training in the auditory task, but rather by any inaccuracies in representing the external acoustical stimulus in the internal representation. A more detailed review is provided in Chapter 12. Even though the current status of such models of the signal processing in the auditory system does not allow a very detailed description of the auditory system, it appears that a technical oriented way of interpreting the action of the auditory system seems to be very promising in order to "understand" the auditory system better and to derive applications from that (Kollmeier, 2003).

References

Böhnke F, Arnold W (1999) 3D-finite element model of the human cochlea including fluid-structure couplings. ORL 61, 305–310.

Dallos P, Evans BN, Hallworth R (1991) Nature of the motor element in electrokinetic shape changes of cochlear outer hair cells. Nature 350, 155–157.

Dallos P, He DZZ, Lin X, Sziklai I, Mehta S, Evans BN (1997) Acetylcholine, outer hair cell electromotility, and the cochlear amplifier. J Neurosci 17, 2212–2226.

Dau T, Kollmeier B, Kohlrausch A (1997) Modeling auditory processing of amplitude modulation: II. Spectral and temporal integration. J Acoust Soc Am 102, 2906–2919.

de Boer E (1980) Auditory physics. Physical principles in hearing theory I. Phys Rep 62, 87–174.

Green DM, Swets JA (1974) Signal detection theory and psychophysics, 2nd edition. Krieger, New York.

Gummer AW, Meyer J, Frank G, Scherer MP, Preyer S (2002) Mechanical transduction in outer hair cells. Audiol Neuro-Otol 7, 13–16.

Hudde H, Engel A (1997) Measuring and modeling basic properties of the human middle ear and ear canal. Part III: Eardrum impedances, transfer funtions and model calculations. Acustica/Acta Acustica 84, 1091–1108.

Hudspeth AJ, Choe Y, Mehta AD, Martin P (1998) Putting ion channels to work: Mechanoelectrical transduction, adaptation, and amplification by hair cells. Proc Nat Acad Sci USA 97, 11765–11772.

Hüttenbrink KB (1988) Die Mechanik der Gehörknöchelchen bei statischen Drucken, I. Normales Mittelohr Laryng Rhinol Otol 67, 45–52.

Kollmeier B (1997) Grundlagen. In: Kiessling J, Kollmeier B, Diller G, Versorgung und Rehabilitation mit Hörgeräten. Thieme, Stuttgart, 1–47.

Kollmeier B (2003) Auditory principles in speech processing: Do computers need silicon ears? In: Proceedings of the 8th European Conference on Speech Communication and Technology (Eurospeech), ICSA International Speech Communication Association, pp. 5–8.

Kollmeier B, Koch R (1994) Speech enhancement based on physiological and psychoacoustical models of modulation perception and binaural interaction. J Acoust Soc Am 95, 1593–1602.

Langner G, Schreiner CE (1988) Periodicity coding in the inferior colliculus of the cat. 1. Neuronal mechanisms. J Neurophysiol 60, 1799–1822.

Silbernagl S, Despopoulos A (2003) Taschenatlas der Physiologie, Thieme, Stuttgart.

11
Physiological Measures of Auditory Function

Birger Kollmeier, Helmut Riedel, Manfred Mauermann, and Stefan Uppenkamp

University of Oldenburg, Oldenburg, Germany

When acoustic signals enter the ears, they pass several processing stages of various complexities before they will be perceived. The auditory pathway can be separated into structures dealing with sound transmission in air (i.e. the outer ear, ear canal, and the vibration of tympanic membrane), structures dealing with the transformation of sound pressure waves into mechanical vibrations of the inner ear fluids (i.e. the tympanic membrane, ossicular chain, and the oval window), structures carrying mechanical vibrations in the fluid-filled inner ear (i.e. the cochlea with basilar membrane, tectorial membrane, and hair cells), structures that transform mechanical oscillations into a neural code, and finally several stages of neural processing in the brain along the pathway from the brainstem to the cortex. It is possible to investigate the function at nearly all of these stages by using methods tailored to exploit the particular physiological representation of interest. This chapter gives a brief overview of the main physiological measures that are currently in use to investigate the auditory system.

1 • Head-Related Transfer Functions

The main function of the outer ear and ear canal is to collect sound from the outside world and transmit it to the tympanic membrane. The shape of pinnae shows very large variability between subjects. An individual pinna can be characterized be its head-related transfer function (HRTF). This is generated from the Fourier transform of the impulse response from a sound played away from the listener and recorded at a point close to the tympanic membrane. The shape of the HRTF determines a "coloration" of the power spectrum of any perceived sound. It is mainly dependent on the angle between ear canal axis and the location of the external sound source. A certain amount of directional information is hence acoustically coded and can be centrally decoded at the output of the transduction mechanisms in the auditory system. A convolution of an acoustic signal with pre-recorded HRTFs can be used to simulate spatial listening conditions.

2 • Tympanometry

Functional diagnostics of the middle ear can be performed using a technique known as tympanometry that assesses the acoustic reflection at the tympanic membrane. The required apparatus to perform clinical tympanometry includes (1) a probe with a miniature loudspeaker (or tube connected to an insert earphone), (2) a precision measurement microphone, and (3) a connection to an external air pump. This probe is inserted in the ear canal using a sealing tip to allow pressurization of the ear canal. Using a probe tone typically of 220 Hz delivered to the loudspeaker, the sound pressure level measured with the microphone is kept constant as the static air pressure in the ear canal is varied. From the loudspeaker level that is needed to obtain the constant predefined level, the acoustical compliance of the middle ear can be obtained, i.e. the imaginary part of the inverse of the impedance. A review of models describing the middle ear input impedance, as a function of several structural parameters of the middle ear (including pathological cases), is provided by Hudde and Engel (1998). The compliance is expressed in millilitres of an equivalent air cushion. It usually exhibits a steep maximum if the static air pressure in the ear canal equals the static pressure in the tympanic cavity. This indicates the best sound transmission from the external ear to the ear drum and subsequent structures. If a middle ear pathology exists, the optimum static pressure is displaced or no maximum value is found at all. An overview of the effect of different middle ear pathologies on tympanometric results for clinical usage is given by Böhme and Welzl-Müller (1998) and Lehnhardt and Laszig (2000).

3 • Acoustic Middle Ear Reflex

The recording and interpretation of the acoustic middle ear reflex (AMER) is a powerful diagnostic technique in clinical audiology (overview in Møller, 2000). The contraction of the stapedius muscle in response to intense sounds decreases the sound transmission to the inner ear. The reflex is elicited in both ears, even with monaural stimulation, although it is slightly stronger in the ipsilateral ear where the threshold is also decreased. The absence of an acoustic reflex can be caused by conductive or sensorineural hearing loss, by disorders in the auditory nerve, or by disorders in the motor neurons in the facial nerve. The most important parameter recorded is the acoustic reflex threshold (ART). This is defined as the lowest stimulus intensity at which a contraction of the stapedius muscle can be detected (Clemis, 1984). The AMER is highly sensitive to middle ear pathology and can help to assess sensorineural hearing impairment. In most cases, however, a sensorineural hearing loss up to 60 dB does not alter the ART. The reflex threshold therefore appears to be more linked to the uncomfortable level (UCL) rather than absolute threshold.

In clinical usage, the acoustic reflex is most commonly detected as a change in the middle ear compliance by means of a low-frequency probe tone (Metz, 1951). A second short tone pulse presented at a high level elicits the reflex. The resulting decrease of acoustic compliance increases the level of the low probe tone, when measured with a microphone in the closed ear canal. When recorded in this way, the ART ranges between 80 and 100 dB hearing level for pure-tone stimulation in normal-hearing listeners. An alternative method to record the AMER was described by Neumann et al. (1996). They used a pair of successive short tone bursts rather than the combination of a single stimulus with the low-frequency probe tone. The method relies on the short latency, or time constant, of the acoustic reflex of the order of 25–100 ms. A change in the compliance of the middle ear results in a slight difference in the measured amplitudes for each of the two tone bursts. This can easily be detected by calculating the difference signal. The threshold criterion is based on the detection of phase coherence between successive measurements. With this method, the ART appears to be substantially lower than previously reported (Neumann et al., 1996). Acoustic reflex thresholds have also been linked to perceived loudness (Gorga et al., 1980; Kawase et al., 1998), which could substantially increase their clinical potential, e.g. aiding the fit of hearing aids based on physiological rather than psychophysical measures. This is currently still a controversial issue (Margolis and Popelka, 1975; Olsen and Rasmussen, 1999a, b).

4 • Otoacoustic Emissions

Otoacoustic emissions are acoustical signals produced in the cochlea, which can be recorded in the ear canal using a sensitive microphone and some noise cancellation techniques. They can be recorded from nearly all normal-hearing human listeners, and also from many different animal models, including nearly all vertebrates (Manley and Taschenberger, 1993). Otoacoustic emissions (OAE) are a consequence of the non-linear and active preprocessing of sound in the cochlea. People with a cochlear hearing loss in general show reduced or no emissions. This implies that OAE have the potential to act as an objective test of hearing function. The measurement of OAE is essentially the only non-invasive method to investigate physiological signals from the cochlea.

Different OAE types are classified according to the stimulus type and related details of the measurement paradigm. In this terminology, spontaneous OAE, i.e. narrowband signals emitted by about 40–60% of normal-hearing people at a very low level without any external stimulus, are separated from evoked OAE. These are responses from the cochlea to various types of acoustic stimulation.

In most cases, spontaneous OAE are very weak, with sound pressure levels of less than 5 dB SPL in an occluded ear canal. Usually they are recorded in the frequency domain by averaging several power spectra of the acoustic noise in the closed ear canal. So far, spontaneous OAE have little diagnostic value, as for the majority of people, neither the presence nor the absence of SOAE is linked to any audiological problem. Spontaneous OAE are simply a consequence of active processes in the inner ear. Many properties are in line with the idea of active, non-linear oscillators (e.g. van der Pol oscillator, van Dijk and Wit, 1990; Talmadge et al., 1991).

There are several types of evoked OAE. Transient-evoked emissions are weak sound emissions in response to a short, transient stimulus, like a click, a tone burst, or a chirp (see Figure 1a). In this case, stimulus and emission can be separated in time (Kemp, 1978). Click-evoked OAE are recorded with latencies between 3 and about 30 ms depending on the frequency

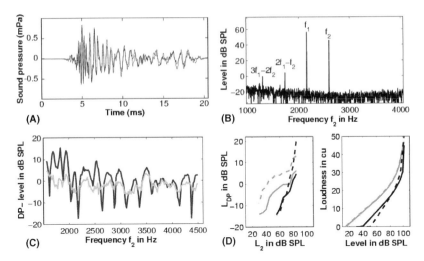

FIGURE 1 (a) Click-evoked otoacoustic emission for a normal-hearing listener. The two traces represent averaging in two separate buffers to allow for subsequent cross-correlation. (b) Spectrum of a single DPOAE measurement: Note the distortion products at $2f_1-f_2$ and $3f_1-2f_2$ in addition to the primary stimuli at f_1 and f_2. (c) High-resolution DP-Gram with and without additional suppressor tone near $2f_1-f_2$. Black line: DPOAE levels of the $2f_1-f_2$ distortion product for a fixed frequency ratio f_2/f_1 of 1.2; f_2 was varied in steps of 18 Hz. Grey line: DPOAE levels measured with the same paradigm, but with additional presentation of a suppressor tone 12 Hz below $2f_1-f_2$ at a level of 50 dB SPL. (d) Comparison of DPOAE I/O functions (*left panel*) and loudness growth (*right panel*) from categorical loudness scaling. The grey lines show the results from two normal-hearing subjects; the black lines in both panels show the respective results from two hearing-impaired subjects. Note the similarity of changes in both, indicating the loss of compression in hearing-impaired subjects.

of the transient stimulus. With broadband stimulation, nearly all normal-hearing listeners emit a broadband response. Click-evoked emissions are now widely used for screening of infant hearing, since they can sufficiently differentiate between populations of normal hearing and hearing-impaired subjects. However, the transition between normal hearing and hearing impairment is not very clear, with a large variability across subjects. A quantitative prediction of hearing thresholds for individual listeners based on click-evoked OAE is difficult.

Simultaneous-evoked OAE are emissions in response to stationary external stimulation, typically sinusoids (Kemp, 1979a; Wilson, 1980). One very important type is the distortion product

OAE (DPOAE), which allows a separation of stimulus and response in the frequency domain. DPOAE can be recorded as one or even several combination tones when stimulating with two sinusoids at different frequencies (Kemp, 1979b; Anderson and Kemp, 1979). The most prominent component is the distortion product at the frequency $2f_1 - f_2$, when the stimuli are sinusoids at f_1 and f_2 with an appropriate frequency ratio f_2/f_1 (see Figure 1b). A detailed review of the different types of OAE is given by Probst et al. (1991).

OAE classification related to the different stimulus and measurement paradigms should be clearly distinguished from a classification according to the underlying mechanisms generating the emissions in the inner ear (OAE source types). In general, two types of OAE sources are identified in the literature: (1) the so-called place-fixed or reflection emissions and (2) the wave-fixed or distortion emissions (see, e.g. Kemp, 2002; Shera and Guinan, 1999). Nearly "pure" reflection emissions can be recorded by the measurement of SFOAE or TEOAE at low levels. They are characterized by a strongly rotating phase of the emission across frequencies, i.e. a high group delay. These emissions in primates are assumed to be generated by a reflection, either at periodically distributed spatial irregularities across the cochlear partition (Strube, 1989) or as the coherent interference of wavelets scattered by randomly distributed perturbations located throughout the tall and broad peak of the travelling wave (Zweig and Shera, 1995). The latter is more consistent with physiological findings. Even spontaneous OAE in mammals can be explained as the result of multiple reflections between regions of coherent reflection and the oval window within the cochlea, resulting in amplitude-stabilized standing waves (Shera, 2003).

A non-linear transmission line model of the cochlea, allowing coherent reflection, can explain several properties seen in OAE, like the minimum spacing between SOAE, the fine structure in TEOAE and SFOAE, and also the fine structure observed in hearing thresholds and DPOAE (e.g. Talmadge et al., (1998)). Apical DPOAE ($2f_1 - f_2, 3f_1 - 2f_2, \ldots$, as can be seen in Figure 1b) measured in the sealed ear canal with a primary frequency ratio f_2/f_1 of about 1.2 are considered to be a mixture of two OAE sources with different underlying mechanisms. The first is a distortion source being in the region of maximal interaction of the two primaries close to the characteristic place of f_2. The second is a coherent reflection source at the characteristic place of the distortion product frequency itself at, e.g. $2f_1 - f_2$, excited by the travelling wave initiated by the distortion source. The distortion source can be extracted from DPOAE measurements either by using suppression and presenting a third tone close to the reflection site, typically $2f_1-f_2$ (e.g. Heitmann et al., 1998), or by an offline analysis called spectral smoothing (Kalluri and Shera, 2001) or latency windowing (Knight and Kemp, 2001; Mauermann and Kollmeier, 2004). Figure 1c shows one example of a DPOAE fine structure. This appears as a quasiperiodic variation of DPOAE level across frequencies of up to 15 dB, with a spacing between adjacent maxima of about 0.4 bark (see black line in the figure). The fine structure mirrors the interaction of the reflection source with its rotating phase and the distortion source, characterized by an almost constant phase across frequencies. The grey line in Figure 1c gives the almost separated contribution from the distortion source from the same subject. This was obtained by DPOAE measurements with an additional suppressor tone of 50 dB SPL at 15 Hz below the respective $2f_1-f_2$ frequencies. The pronounced spectral DPOAE fine structure related to the interaction of the two sources with their different phase characteristics is strongly reduced (compare black and grey lines in Figure 1c).

Input/output (I/O) functions for DPOAE, (i.e. the DP level as a function of the level of the primaries) appear to be a useful tool to assess hearing status (Figure 1d). The right panel of Figure 1d shows the results from categorical loudness scaling for two normal-hearing (grey lines) and two hearing-impaired listeners (black lines), indicating a clear recruitment in the hearing-impaired subjects. The psychoacoustically observed loss of compression ("recruitment") is clearly reflected by the I/O functions from DPOAE measurements. The data shown in Figure 1d were obtained by varying the primary level L_1 for a given primary level L_2 to get an optimized distortion product level. Furthermore, a suppressor tone was presented close to $2f_1-f_2$ to reduce the influence of the reflection source. The source separation appears to be essential for the usage of I/O functions from DPOAE measurements since DPOAE I/O functions are strongly influenced by the interaction of the two DPOAE sources (Mauermann and Kollmeier, 2004).

At present, most OAE recordings described in the literature are a mixture of the different sources, and clinically they mainly serve as an infant or neonatal screening tool. However, since the underlying OAE mechanisms are most probably differently affected by different cochlear pathologies, a clear separation of the contribution from different OAE sources and mechanisms may in future lead to a more detailed insight into cochlear status based on OAE measurements.

5 • Evoked Potentials

The neural activity along the auditory pathway and in the cortex can be observed by the recording of several types of far-field-evoked potentials. This is done using electrodes fixed to the head of the listeners, and utilizing various methods of noise cancellation during data acquisition, particularly averaging a large number of single recordings. When stimulated with a transient like a click, a characteristic pattern of several peaks and valleys is visible in the evoked potential, with latencies between 1 and more than 500 ms (Jewett and Williston, 1971). These potentials are classified according to their latency relative to stimulus onset, auditory brainstem responses (ABR) with latencies between 1 and 10 ms, middle latency auditory evoked potentials (MAEP) with latencies between 10 and approximately 50 ms, and finally late auditory evoked potentials (LAEP) with latencies of 100 ms and above. The most important type of far-field potential for clinical applications is the recording of ABR, usually summarized as brainstem evoked response audiometry (BERA).

The ABR is generated by the synchronized neural activity of many nerve fibres along the auditory pathway. The recorded electrical potential at the scalp represents the superposition of the activity of a very large number of electric dipole sources along all the stages of the pathway from the spiral ganglion up to the inferior colliculus (IC). Nevertheless, it is still possible to describe the peaks of the far-field potential by equivalent electric dipole sources, which can be linked to particular anatomical structures. For example, wave I of the ABR, at a latency of less than 1 ms, is closely related to the compound action potential which can also be recorded from the auditory nerve. Wave V of the ABR, with latencies varying between 5 and 8 ms as a function of stimulus level, is the most robust peak and can be recorded down to the absolute threshold of hearing. This part of the response appears to be generated by structures at the input to the IC (Scherg, 1991).

Electrophysiological responses to clicks are dominated by neural activity in the high-frequency channels. This is because high-frequency channels are synchronized, whereas the low-frequency channels are not, due to dispersion along the basilar membrane. One way to synchronize excitation in all frequency channels for ABR recording was described by Dau et al. (2000). They used optimized chirp stimuli with increasing instantaneous frequency to compensate for dispersion and demonstrated that these chirp signals enhance ABR wave V. This is probably due to increased synchronization of the low-frequency channels. An example of ABR recordings in response to a click and a synchronizing chirp stimulus is given in Figure 2. Two main effects can be observed. First, the amplitude of wave V is considerably increased

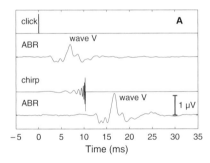 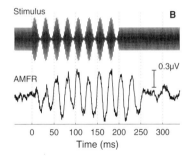

FIGURE 2 (**a**) Auditory brainstem responses to a click stimulus (*top*) and to a chirp stimulus (*bottom*) with increasing instantaneous frequency, designed to compensate for basilar membrane dispersion. The biggest peak, ABR wave V, has a latency of approximately 6 ms relative to stimulus offset and is significantly enhanced during chirp stimulation due to greater synchronization along the cochlear partition. Stimulus level: 40 dB nHL, diotic stimulation, recording bandwidth 100–1500 Hz, 10,000 averages. (**b**) Amplitude modulation following response. Stimulus: 60 dB SL, carrier frequency at 480 Hz, sinusoidally modulated at 40 Hz, monaural stimulation, recording bandwidth 15–1500 Hz. An evoked potential reflecting the amplitude modulation is clearly visible.

for the chirp in comparison to the click. Second, the position of wave V relative to the stimulus onset is delayed by an additional 10.5 ms for the chirp, while the delay relative to stimulus offset is almost identical to the click condition. This illustrates that the intended phase alignment between different BM channels is preserved at the level of IC. In a similar experiment with magnetoencephalography (MEG, i.e., the recording of the very weak magnetic field associated with brain activity), Rupp et al. (2002) have shown that this synchronization is even preserved in the middle latency response generated at the level of primary auditory cortex.

ABR responses and middle latency responses are dominated by cochlear filtering, compression, and temporal smoothing. Late auditory evoked potentials, however, are less stimulus specific. Also at the long latencies associated with cortical activity beyond primary auditory cortex, the response becomes more and more process specific. One famous example is the mismatch negativity response which arises from a perceptual contrast between an individual, deviant sound presented occasionally in a background series of standards that establish a context for the contrast (Näätänen et al., 1978). This response can be recorded reliably with both EEG and MEG.

Another important branch of evoked potentials are auditory steady-state responses (ASSR), such as frequency following responses (FFR), amplitude modulation following responses (AMFR) and sustained potentials or magnetic fields (Picton et al., 1978, 2003; Hall, 1979). All of these responses reflect one particular perceptual property of a stationary stimulus, like for example pitch, or modulation frequency. For ASSR, the stimulus is usually presented continuously to avoid onset effects. One example of an AMFR is given in the right panel of Figure 2. For a continuous 480 Hz tone, an amplitude modulation is switched on for eight full cycles. The response is dominated by a sinusoidal component at 40 Hz, corresponding to the modulation frequency of the stimulus. This 40 Hz response is assumed to be of cortical origin. It can be recorded at stimulus levels down to the subjective threshold of hearing for all audiometric carrier frequencies, which makes it potentially useful as an objective, frequency-specific audiometric tool.

Both EEG and MEG provide a temporal resolution of one millisecond or less and allow to study very fine details of temporal processing in the auditory pathway. However, the spatial resolution is inferior to the high resolution of modern neuroimaging methods like functional magnetic resonance imaging (see next section). In future, a combination of EEG and MEG with these methods therefore should substantially increase our knowledge of how the auditory brain is working.

6 • Functional Imaging of the Auditory System

The main neuroimaging methods to visualize the function of the brain are positron emission tomography (PET) and functional magnetic resonance imaging (fMRI). With PET, radioactive positron emitting tracer materials are injected into the subject, and the distribution of radiation from the brain tissue is observed, using a ring of gamma coincidence detectors. Tomographic 3D reconstruction of the distribution of activity within the brain is calculated by means of Fourier transformations (Cho et al., 1993), similar to computerized X-ray tomography. PET has been used for a big variety of auditory experiments, ranging from pitch perception to spoken language (overview in Johnsrude et al., 2002). PET has now been overtaken by functional MRI in terms of the number of studies performed, due to the wider distribution of MRI scanners and to the finer spatial and temporal resolution that can be achieved with modern MRI scanners. However, PET still has one crucial advantage, especially for auditory experiments, that is the absence of any loud distracting noise produced by the scanner.

Functional magnetic resonance imaging (fMRI) allows the exploration of neural activity in the brain at a spatial resolution of a few millimetres and a temporal resolution of several seconds. The combination of functional and anatomical images, acquired in the same listener, enables a direct relation of physiological processes to the underlying anatomical structures in the brain. In recent years, a great deal of progress has been made in utilizing this technique for auditory investigations.

The principle of MRI is based on radio frequency excitation and relaxation of nuclear magnetic moments that have been aligned with a strong external magnetic field. Functional MRI is usually based on the narrow magnetic resonance of proton spins. Spatial encoding is achieved using additional gradient coils that restrict RF excitation to predefined slices through the tissue. Phase and frequency encoding during RF excitation and measurement of the NMR signal are utilized for coding the other two spatial dimensions. The spatial resolution and signal-to-noise ratio are directly linked to the strength of the external magnetic field. For imaging human subjects, field strengths from 0.5 to 7 T are used.

There are several problems with fMRI for auditory experiments, for example, switching of the gradient coils in the strong external magnetic field results in very loud noises. This is due to mechanical vibrations caused by Lorentz forces acting on the coils. Precautions have to be introduced; the first one is to avoid damage to the subject's ears, and the second is to ensure that the recorded brain activity is caused by the actual experimental stimulus, and not by the scanner noise itself. There are now MR compatible, high-fidelity headphones available that have been fitted into conventional ear defenders. These give an attenuation of external noise between 15 and 40 dB. Additional damping of scanner noise can be achieved by sound absorbing material around the subject's head and also inside the walls of the scanner bore. In addition to these passive measures, active noise cancellation (ANC) has been utilized to compensate for the scanner noise (Goldmann et al., 1989; Chen et al., 1999; Chambers et al., 2001). By means of ANC, an additional reduction of the perceived scanner noise of 10–15 dB has currently been achieved.

With fMRI, a secondary, metabolic response is observed rather than the primary neural activation. The actual signal change of the measured NMR signal is based on the change of the oxygenation level of blood due to metabolic processes that are triggered by the neural activity. These changes result in a slight change of the relaxation times in activated regions of the brain and are the basis for an image contrast between different regions (overview in Buxton, 2002). As a consequence, the observed signal change has a comparatively long latency of several seconds relative to stimulus onset. This lack of immediate responses can be turned into an advantage for auditory fMRI by introduction of the paradigm of "sparse temporal sampling". It allows for a separation of the activation due to scanner noise from the one due to the actual stimulus of interest. Instead of continuous data acquisition during epochs of stimulus and epochs of rest, stimulation and data acquisition are separated in time, with periods of several seconds of stimulus presentation without any scanning (Edmister et al., 1999; Hall et al., 1999). A combination of sparse temporal sampling and cardiac gating, that is synchronization of the MR image acquisition with the cardiac cycle, has been shown to allow for imaging of not only the auditory cortex, but even the brainstem and thalamic structures with fMRI (Griffiths et al., 2001).

Since the observed NMR signal changes in activated regions of the brain are very small (in the order of 1 %), the analysis of fMRI data is heavily based on repeated measures and statistical models that test for significant differences between conditions (Worsley et al., 1992). A widespread approach is the method of statistical parametric mapping based on the theory of Gaussian random fields (Frackowiak et al., 2004; http://www.fil.ion.ucl.ac.uk/spm).

Figure 3 gives one example for the activation of auditory cortex in response to acoustic stimulation using sparse temporal sampling and statistical parametric mapping for data analysis.

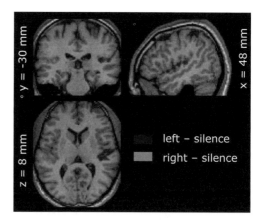

FIGURE 3 Activation of Heschl's gyrus for unilateral presentation of a broadband sound. The activation maps represent the T-statistics for significant differences between sound presentation and silence (i.e. no acoustic stimulus). Stimulation of the right ear largely activates the left hemisphere and vice versa. Activation maps are superimposed on an anatomical image of the same listener, in coronal (*top left*), saggital (*top right*), and axial view (*bottom left*). The coordinates are relative to the origin in commissura anterior.

The map shows significantly activated voxels in response to unilaterally presented broadband sound either to the left ear (in blue) or to the right ear (in red), when contrasted with the baseline condition (no sound presentation). The activation in these "sound–silence" contrasts is largely restricted to Heschl's gyrus in both hemispheres, and the crossing of the major projections on the way from the ears to cortex is clearly visible.

References

Anderson SD, Kemp DT (1979) The evoked cochlear mechanical response in laboratory primates: a preliminary report. Arch Otorhinolaryngol 224, 47–54.

Böhme G, Welzl-Müller K (1998) Audiometrie. Huber, Göttingen.

Buxton RB (2002) Introduction to Functional Magnetic Resonance Imaging. Cambridge University Press, Cambridge.

Chambers J, Akeroyd MA, Summerfield AQ, Palmer AR (2001) Active control of the volume acquisition noise in functional magnetic resonance imaging: method and psychoacoustical evaluation. J Acoust Soc Am 110, 3041–3054.

Chen CK, Chiueh TD, Chen JH (1999) Active cancellation system of acoustic noise in MR imaging. IEEE Transact Biomed Eng 46, 186–191.

Cho ZH, Jones JP, Singh M (1993) Foundations of Medical Imaging. Wiley, New York.

Clemis, JD (1984) Acoustic reflex testing in otoneurology. Otolaryngol Head Neck Surg 92, 141–144.

Dau T, Wegner O, Mellert V, Kollmeier B (2000) Auditory brainstem responses (ABR) with optimized chirp signals compensating basilar-membrane dispersion. J Acoust Soc Am 107, 1530–1540.

van Dijk P, Wit HP (1990) Amplitude and frequency fluctuation of spontaneous otoacoustic emissions. J Acoust Soc Am 88, 1779–1793.

Edmister WB, Talavage TM, Ledden PJ, Weisskoff RM (1999) Improved auditory cortex imaging using clustered volume acquisitions. Hum Brain Mapp 7, 89–97.

Frackowiak RSJ, Friston KJ, Frith CD, Dolan RJ, Price CJ, Zeki S, Ashburner J, Penny W (2004) Human brain function, 2nd edition. Elsevier Academic Press, London.

Goldmann AM, Gossmann WE, Friedlander PC (1989) Reduction of sound levels with antinoise in MR imaging. Radiology 173, 519–550.

Gorga MP, Lilly DJ, Lenth RV (1980) Effect of signal bandwidth upon threshold of the acoustic reflex and upon loudness. Audiology 19, 277–292.

Griffiths TD, Uppenkamp S, Johnsrude I, Josephs O, Patterson RD (2001) Encoding of the temporal regularity of sound in the human brainstem. Nat Neurosci 4, 633–637.

Hall JW (1979) Auditory brainstem frequency following responses to waveform envelope periodicity. Science 205, 1297–1299.

Hall DA, Haggard MP, Akeroyd MA, Palmer AR, Summerfield AQ, Elliott MR, Gurney EM, Bowtell RW (1999) "Sparse" temporal sampling in auditory fMRI. Hum Brain Mapp **7**, 213–223.

Heitmann J, Waldmann B, Schnitzler HU, Plinkert PK, Zenner HP (1998) Suppression of distortion product otoacoustic emissions (DPOAE) near $2f_1 - f_2$ removes DP-gram fine structure – evidence for a secondary generator. J Acoust Soc Am 103, 1527–1531.

Hudde H, Engel A (1998). Acoustomechanical human middle ear properties. Part III: eardrum impedances, transfer functions, and model calculations. ACUSTICA – Acta Acustica 84, 1091–1108.

Jewett DL, Williston JS (1971) Auditory-evoked far-fields averaged from the scalp of humans. Brain 94, 681–696.

Johnsrude IS, Giraud AL, Frackowiak RSJ (2002) Functional imaging of the auditory system: the use of positron emission tomography. Audiol Neurootol 7, 251–276.

Kalluri R, Shera CA (2001) Distortion-product source unmixing: a test of the two-mechanism model for DPOAE generation. J Acoust Soc Am 109, 622–637.

Kawase T, Hidaka H, Ikeda K, Hashimoto S, Takasaka T (1998) Acoustic reflex thresholds and loudness in patients with unilateral hearing losses. Eur Arch Otorhinolaryngol 255, 7–11.

Kemp DT (1978) Stimulated acoustic emissions from within the human auditory system. J Acoust Soc Am 64, 1386–1391.

Kemp DT (1979a) Evidence of mechanical nonlinearity and frequency selective wave amplification in the cochlea. Arch Otorhinolaryngol 224, 37–45.

Kemp DT (1979b) The evoked cochlear mechanical response and the auditory microstructure – evidence for a new element in cochlear mechanics. Scand Audiol Suppl 9, 35–47.

Kemp DT (2002) Exploring cochlear status with otoacoustic emissions. In: Otoacoustic Emissions – Clinical Applications, 2nd ed., edited by MS Robinette, TJ Glattke, Thieme, New York, pp. 1–47.

Knight RD, Kemp DT (2001) Wave and place fixed DPOAE maps of the human ear. J Acoust Soc Am 109, 1513–1525.

Lehnhardt E, Laszig R (2000) Praxis der Audiometrie. Thieme, Stuttgart.

Manley G, Taschenberger G (1993) Spontaneous otoacoustic emissions from a bird: a preliminary report. In: Biophysics of Hair Cell Sensory Systems, edited by H Duifhuis, JW Horst, P van Dijk, SM van Netten. World Scientific, Singapore, pp. 33–39.

Margolis RH, Popelka GR (1975) Loudness and the acoustic reflex. J Acoust Soc Am 58, 1330–1332.

Mauermann M, Kollmeier B (2004) Distortion product otoacoustic emission (DPOAE) input/output functions and the influence of the second DPOAE source. J Acoust Soc Am 116, 2199–2212.

Metz O (1951) Studies of the contraction of the tympanic muscles as indicated by changes in the impedance of the ear. Laryngoscope LXVIII(I), 48–62.

Møller AR (2000) Hearing – Its Physiology and Pathophysiology, Chapter 12. Academic Press, San Diego.

Näätänen R, Gaillard AWK, Mäntysalo S (1978) Early selective-attention effect on evoked potential reinterpreted. Acta Psychol 42, 313–329.

Neumann J, Uppenkamp S, Kollmeier B (1996) Detection of the acoustic reflex below 80 dB HL. Audiol Neurootol 1, 359–369.

Olsen SO, Rasmussen AN, Nielsen LH, Borgkvist BV (1999a) The acoustic reflex threshold: not predictive for loudness perception in normally-hearing listeners. Audiology 38, 303–307.

Olsen SO, Rasmussen AN, Nielsen LH, Borgkvist BV (1999b) The relationship between the acoustic reflex threshold and levels of loudness categories in hearing impaired listeners. Audiology 38, 308–311.

Picton TW, Woods DL, Proulx GB (1978) Human auditory sustained potentials. I. The nature of the response. Electroencephalogr Clin Neurophysiol 45, 186–197.

Picton TW, John MS, Dimitrijevic A, Purcell D (2003) Human auditory steady-state responses. Int J Audiol 42, 177–219.

Probst R, Lonsbury-Martin BL, Martin GK (1991) A review of otoacoustic emissions. J Acoust Soc Am 89, 2027–2067

Rupp A, Uppenkamp S, Gutschalk A, Beucker R, Patterson RD, Dau T, Scherg M (2002) The representation of peripheral neural activity in the middle-latency evoked field of primary auditory cortex in humans. Hear Res 174, 19–31.

Scherg M (1991) Akustisch evozierte Potentiale. Kohlhammer, Stuttgart.

Shera CA (2003) Mammalian spontaneous otoacoustic emissions are amplitude-stabilized cochlear standing waves. J Acoust Soc Am 114, 244–262.

Shera CA, Guinan Jr. JJ (1999) Evoked otoacoustic emissions arise by two fundamentally different mechanisms: a taxonomy for mammalian OAEs. J Acoust Soc Am 105, 782–798.

Strube HW (1989) Evoked otoacoustic emissions as cochlear Bragg reflections. Hear Res 38, 35–45.

Talmadge CL, Tubis A, Wit HP, Long GR (1991) Are spontaneous otoacoustic emissions generated by self-sustained cochlear oscillators? J Acoust Soc Am 89, 2391–2399.

Talmadge CL, Tubis A, Long GR, Piskorski P (1998) Modeling otoacoustic emission and hearing threshold fine structure. J Acoust Soc Am 104, 1517–1543.

Wilson JP (1980) Evidence for a cochlear origin for acoustic reemissions, threshold fine-structure and tonal tinnitus. Hear Res 2, 233–252.

Worsley KJ, Evans AC, Marrett S, Neelin P (1992) A three-dimensional statistical analysis for CBF activation studies in human brain. J Cereb Blood Flow Metab 12, 900–918.

Zweig G, Shera CA (1995) The origin of periodicity in the spectrum of evoked otoacoustic emissions. J Acoust Soc Am 98, 2018–2047.

12
Auditory Processing Models

Torsten Dau

*Technical University of Denmark, Lyngby, Denmark,
e-mail: tda@elektro.dtu.dk*

There are at least two main reasons why auditory processing models are constructed: to represent the results from a variety of experiments within one framework and to explain the functioning of the system. Specifically, processing models help generate hypotheses that can be explicitly stated and quantitatively tested for complex systems. The models can also help determine how a deficit in one or more components affects the overall operation of the system. The development of auditory models has been hampered by the complexity of the individual auditory processing stages and their interactions. This resulted in a multiplicity of auditory models described in the literature. The models differ in their degree of complexity and quantification. Models of auditory processing may be roughly classified into biophysical, physiological, mathematical (or statistical) and perceptual models depending on which aspects of processing are considered. Most of the models can be broadly referred to as functional models, that is, they simulate some experimentally observed input–output behavior of the auditory system without explicitly modeling the precise internal physical mechanisms involved. In the following, several models of auditory signal processing are summarized, including descriptions of the signal processing in the auditory periphery, examples of neural processing in the brainstem and models of auditory perception that often contain some decision stage after the assumed preprocessing. Some of the models can be useful for technical and clinical applications, such as improved man–machine communication by employing auditory-model-based processing techniques, or new processing strategies in digital hearing aids and cochlear implants.

1 • External and Middle Ear Transformation

The external ear is commonly subdivided into the external ear diffraction system and the pinna-ear canal resonator system. The outer ear enhances the sound pressure at the eardrum for frequencies in the range 1–9 kHz, with a maximum enhancement at 3 kHz of 10–15 dB. The transfer function of the external ear system is highly dependent on the direction of sound incidence (Shaw, 1980). Sound diffraction of the external ear system has been modeled as electrical network by Killion and Clemis (1981) and Giguère and Woodland (1994) using electroacoustic analogies. The sound propagation in the ear canal has been modeled as a transmission line by Gardner and Hawley (1972). The middle ear transfers airborne sound in the ear canal into fluid motion in the cochlea. The transmission in the middle ear is most efficient for midrange frequencies, with a maximum at 1 kHz, and drops off markedly for very low and very high frequencies. Zwislocki (1962) developed the first electroacoustic network model of the middle ear based on anatomical considerations and on input impedance data measured at the eardrum. The network was later adapted by Lutman and Martin (1979) to account for the action of the stapedius muscle. The latter model was later modified by Giguère and Woodland (1994) which allowed them to directly connect their cochlear network to the middle ear. An alternative approach has been provided by Moore et al. (1997) where the transfer of sound through the external and middle ear has been modeled effectively using fixed filters. In a recent implementation of the model (Glasberg and Moore, 2002), the combined effect of the outer and middle ear has been modeled by a single finite impulse response (FIR). The output was considered as representing the sound reaching the cochlea in a model of loudness of time-varying sounds.

2 • Cochlear Transformation

2.1 Basilar Membrane Processing

Stapes vibrations within the oval window set the cochlear fluid into motion. The response of the basilar membrane (BM) is a traveling wave propagating from the base to the apex of the cochlea. Models of the BM have been formulated in many ways. They range from mechanical structures, via electrical networks consisting of inductors, resistances, capacitors, diodes and amplifiers, to abstract structures that are put into a mathematical form. In biophysical models of BM processing, the network is typically realized as a transmission-line model, whereby the BM is spatially discretized into a series of segments of certain length. The transmission-line network elements are derived from electroacoustic analogies and from the assumption that the natural frequency of the resonant circuit in each segment is equal to the characteristic frequency of the BM at that place. These "classical" point impedance models of the cochlea are reviewed in, e.g., de Boer (1980, 1984, 1991), Giguère and Woodland (1994) and Hubbard and Mountain (1996). One of the characteristics of these (biophysical) models is that their impulse responses reflect not only the local mechanisms at the point of measurement but depend on the mechanics of the entire cochlea (e.g., Zweig, 1991; Neely, 1983; de Boer, 1995; Olson, 1999; Shera, 2001). This is fundamentally different from those models that treat the cochlea essentially as a bank of parallel bandpass filters. These "functional" models do not account for cochlear hydrodynamics; they simulate the response of the BM at a single site and the bandpass filters are instead suggested to originate via the differential build-up or decay of micromechanical resonances (e.g., Patterson et al., 1987; Irino and Patterson, 1997, 2000, 2001; Shekhter and Carney, 1997).

There has been evidence that the BM motion is highly nonlinear and is a major source of level compression (e.g., Rhode, 1971; Ruggero and Rich, 1991). Several functional modeling studies have explored combinations of linear filters and (memoryless) nonlinearities that provide a phenomenological representation of BM responses. Initial models of this type included the "bandpass nonlinearity" models of Pfeiffer (1970) and Duifhuis (1976). Goldstein (1990, 1995), Meddis et al. (2001), Lopez-Poveda and Meddis (2001) and Summer et al. (2003) extended their approach with a multiple bandpass nonlinear model, which included two interacting paths. One path in this dual resonance nonlinear (DRNL) filter configuration contains a lowpass filter followed by a memoryless nonlinearity and the other path contains a bandpass filter. The output of the system is the sum of the outputs of the linear and the nonlinear paths. The latest implementations of the model have been tested in large detail against a wide range of physical observations.

2.2 Transformation into Neural Activity

A strong correlation between the normal function of the outer hair cells and nonlinear BM responses has been demonstrated by Ruggero and Rich (1991). The outer hair cells have been shown to participate in the process of cochlear amplification as mechanical force generators acting on the cochlear partition (Kim, 1986). The inner hair cells (IHC) transduce the motion of the organ of Corti into firing activity in the auditory nerve (AN). Models of AN responses have been developed studying the representation of sounds at the first level of neural coding in the auditory system. Sachs and Abbas (1974) and Schoonhoven et al. (1997) investigated phenomenological models that explain rate-level functions for AN fibers in terms of the relation between AN threshold and BM compression. Also, detailed models of the IHC–AN synapse (e.g., Meddis, 1986, 1988; Geisler, 1990; Hewitt and Meddis, 1991) provide descriptions of several features of AN rate-level functions. Probably the most detailed descriptions of AN activity are given in the computational AN models of Heinz et al. (2001a,b) and Zhang et al. (2001) which are a modification of an earlier model by Carney (1993). The models describe many of the main response properties associated with the cochlear amplifier, including broadened tuning with increases in level, the associated compressive-magnitude and nonlinear phase responses, as well as two-tone suppression.

3 • Examples of Brainstem Processing Models

The activity of auditory brainstem and central neurons has not been studied in the same detail as the activity in the periphery. This is also reflected in a much smaller number of computational models of brainstem and central auditory processing. An exception might be the modeling of binaural processing and models of coding of amplitude modulation.

3.1 Neural Binaural Models

Several models of binaural interaction in brainstem neurons have been proposed. Most models are based on the coincidence counter hypothesis following an internal delay line as suggested by Jeffress (1948). The physiological basis for such coincidence counters are the excitation–excitation (EE) type cells (e.g., Goldberg and Brown, 1969; Joris and Yin, 1995) that are found in the superior olivary complex (SOC), representing the first neurons in the ascending auditory pathway exhibiting binaural interaction. Their discharge rate in response to binaural stimulation depends on the interaural time difference (ITD) and, at favorable ITDs, exceeds the sum of the responses for either ear alone (Goldberg and Brown, 1969). In models based on an array of EE-type cells with a range of "characteristic delays", the neural discharge rate resulting from the EE interaction is usually modeled as an interaural cross-correlation function (e.g., Jeffress, 1948; Shamma et al., 1989; Colburn et al., 1990). Within these models, an ITD results in a shift of the cross-correlation function along the interaural delay axis and hence leads to a predicted lateralization.

There has also been physiological evidence for excitation–inhibition type cells; corresponding EI-cells in the SOC and in the inferior colliculus (IC) are excited by the signals from one ear and inhibited by the signals from the other ear (e.g., Kuwada et al., 1984; Joris and Yin, 1995; Palmer et al., 1997; McAlpine et al., 1998). For neurons in the IC the exciting and inhibiting channels are typically reversed (IE-type cells). These observations have led to models consistent with the equalization cancellation (EC) theory by Durlach (1963, 1972). It has been shown in corresponding neural models that both ITD and IID sensitivities may be understood by considering the outcome of a subtractive mechanism for EI/IE-type neurons (e.g., Colburn and Moss, 1981; Zacksenhouse et al., 1992). A detailed review of the physiological binaural models can be found in Colburn (1996).

3.2 Neural Models of Amplitude-Modulation Processing

Another example has been the processing of amplitude modulations (AM). Envelope fluctuations are a common physical feature produced by many natural sound sources. While AM information is encoded in temporal activity patterns of neurons at the early stages of the auditory system, it is generally believed that this temporal periodicity code is transformed into a rate-based code at some higher stage of auditory processing. It has been suggested that arrays of neurons in the brainstem, each tuned to a different modulation rate, can be considered as a filter in the modulation domain, and the array of neurons is known collectively as a modulation filterbank. Neurons with appropriate properties have been found in the cochlear nucleus (Møller, 1976) and the inferior colliculus (Rees and Møller, 1983; Langner and Schreiner, 1988; Lorenzi et al., 1995). Detailed reviews can be found in Palmer (1995) and Joris et al. (2004).

The mechanisms by which the auditory system extracts, encodes and processes AM information are not yet fully understood. Several physiologically motivated models have been suggested, assuming different mechanisms of how temporal AM information can be transformed into a rate-based representation between CN and IC. The neural circuit suggested by Langner (1981) (see also Langner, 1997; Langer et al., 1997) consists of three functional units at the model CN stage, with each of these units being associated with a specific neural response type in the CN. At the model IC stage, a coincidence detector unit is located that represents the model output unit. In the model by Hewitt and Meddis (1994), the CN model stage employs populations of simulated chopper units that are morphologically associated with stellate cells in the ventral cochlear nucleus (VCN; Rhode et al., 1983; Smith and Rhode, 1989). The model by Hewitt and Meddis (1994) therefore transforms the bandpass-shaped temporal modulation transfer functions (tMTFs) observed in VCN chopper units into bandpass-shaped rate functions (rMTFs). The different best modulation frequencies (BMFs) of different IC coincidence detector units hereby result from different chopping frequencies among different populations of chopper units.

A third model describing the formation of bandpass-shaped rMTFs in IC neurons was suggested by Nelson and Carney (2004). The CN stage of their model employs a VCN bushy cell that receives undelayed excitatory AN input and delayed inhibitory AN input. The model bushy cell projects onto the model IC neuron in the same manner, i.e., it provides undelayed excitatory input and delayed inhibitory input. The interaction of excitatory and inhibitory synaptic inputs results in a bandpass-shaped rMTF at the model IC neuron if the excitatory synapses are assumed to be faster and weaker than the inhibitory synapses. Different BMFs within different IC model neurons are obtained by varying the time courses of the synaptic inputs.

4 • Models of Auditory Perception

Some perception models address the processing of the acoustical stimuli without much attention to underlying physiological processes (black box); others are inspired by neurophysiological findings. Typically, in the latter case, rather than trying to model many physiological details, the approach is to focus on the "effective" signal processing which uses as little physiological and physical parameters as necessary, but tries to predict as many perceptual data as possible. On the one hand, such a modeling strategy will never allow conclusions about the details of signal processing at a neuronal or single-unit level. On the other hand, if the effective model correctly describes the transformation of physical parameters in a large variety of experimental conditions, this strongly suggests certain general processing principles. These, in turn, may motivate the further study search for neural circuits in corresponding physiological studies.

4.1 Physiological Models for Basic Auditory Percepts

Explaining basic auditory perceptual phenomena in terms of physiological mechanisms has a long tradition. A nice review of this modeling approach is given in Delgutte (1996). There have been early systematic attempts at predicting psychophysical performance from the activity of AN fibers (e.g., Siebert, 1965, 1970). The same strategy has been used in a series of later modeling studies (e.g., Delgutte, 1990; Heinz et al., 2001a,b; Colburn et al., 2003) combining analytical and computational population models of the AN with statistical decision theory to evaluate performance limits in tasks like intensity discrimination and masking. A general result is that models incorporating a realistic cochlear frequency map, cochlear filters, suppression, and rate-level functions are reasonably successful to account for performance in basic detection, discrimination and masking tasks. One of the most recent implementations of this type of model (Heinz et al., 2001a,b, see Figure 1) quantifies the effects of the cochlear amplifier on temporal and average-rate based information given in the AN and evaluates the influences of compression and suppression on psychophysical measures of auditory frequency selectivity. In principle, the approach can also be used for more complex stimuli and tasks, although the increased importance of more central factors might limit the value of models based on AN activity. Figure 1 should be placed before Chapter 4.2 (It belongs to 4.1).

4.2 Binaural Perception Models

Most models of binaural perception are based on the overlapping concepts of interaural coincidence networks and interaural correlation functions applied to individual frequency bands. These models are all fundamentally similar, and the concept of an

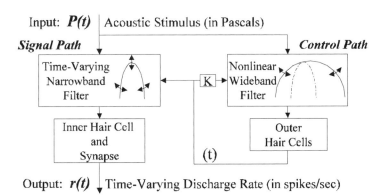

FIGURE 1 Schematical block diagram of the nonlinear auditory nerve (AN) model for a single characteristic frequency (CF) of Heinz et al. (2001a,b). The model uses a nonlinear control path to account for broadened tuning with increases in level, including the associated compressive-magnitude and nonlinear phase responses, suppression and the fast dynamics of these properties. The model was developed for use in modeling basic psychophysical experiments. The approach combines analytical and computational population models of the AN with statistical decision theory to evaluate performance limits in tasks like intensity discrimination and masking.

internal display of activity on a center frequency versus interaural time delay plane is almost universally used for insight about binaural phenomena (for a detailed review see Colburn, 1996). The models differ in their incorporation of interaural intensity differences (IIDs). Models based on interaural coincidence or correlation do not include the effects of IIDs such that additional assumptions and mechanisms need to be included to incorporate these effects. In some cases, the effects of IIDs are incorporated into a common mechanism (e.g., the EC model by Durlach, 1963, 1972; Lindemann, 1986; Gaik, 1993). In other cases (e.g., Jeffress, 1948; Colburn and Durlach, 1978; Colburn et al., 1990; Breebaart et al., 2001a,b,c), IID information is extracted separately and then combined with ITD information. The models of sound localization (e.g., Searle et al., 1976; Middlebrooks, 1992; Blauert, 1997) do not explicitly address mechanisms for the extraction of ITD or IID information; they are mainly concerned with the spectral variation in overall intensity and interaural differences that relate to source position. Concerning the ability of models to describe psychoacoustical data, most models have restricted their application only to a relatively small set of data. The binaural signal detection model by Breebaart et al. (2001a) probably represents the model that has been applied to the largest range of binaural detection conditions. Breebaart et al. (2001b,c) demonstrated that, for many experimental conditions, models based on the EC theory or the cross-correlation give similar results, but that in conditions where predictions differ, an EC-like mechanism may be favored over the cross-correlation.

4.3 Models of Loudness and Pitch

Relatively much is also known about the neural correlates of sensations like loudness and pitch in the early stages of processing. Accordingly, models of loudness and pitch have been commonly associated with mechanisms in these stages. For example, loudness models contain outer- and middle ear transformation, a calculation of the cochlear excitation pattern from the physical spectrum, a transformation of the excitation pattern to a specific loudness pattern and the calculation of the overall loudness. This type of model has been in most cases applied to steady sounds (e.g., Fletcher and Munson, 1937; Zwicker and Scharf, 1965; Zwicker and Fastl, 1999; Moore et al., 1997). Models of loudness of time-varying sounds, where the overall loudness is calculated from the short-term loudness using an averaging mechanism, were suggested by Zwicker (1965, 1977), Fastl (1993), Stone et al. (1997) and Glasberg and Moore (2002). The most recent model (Glasberg and Moore, 2002) predicts the loudness of brief sounds as a function of duration and the loudness of sounds that are amplitude modulated.

Pitch models have been somewhat more complex and none of them has been able to account for all of the experimental data. Detailed reviews about auditory representations of pitch can be found in Lyon and Shamma (1996) and de de Cheveigné (2005). For harmonic complex tones (such as voiced human speech), two main classes of models have been popular. Purely spectral (or spatial) models refer to algorithms that extract pitch from a representation of the spectral profile along the tonotopic axis. All these algorithms require that the input spectral profile be able to resolve (at least partially) several low-order harmonics which are the most important harmonics for pitch perception. There have been hypotheses to explain how pitch can be computed from resolved peaks in the spectral profile (e.g., Goldstein, 1973; Wightman, 1973; Terhardt, 1979; Cohen et al., 1995). A common principle in all these models is a pattern-matching stage which compares the resolved components spectrum to "internally" stored templates of various harmonic series in the central auditory system.

In the other class of models, the representation of pitch is derived exclusively from timing information. The pioneering study of Licklider (1951) about the duplex theory of pitch perception was the basis for later temporal models of pitch perception. For example, in the models of Meddis and Hewitt (1992) and Meddis and O'Mard (1997) the representation of pitch is derived exclusively from the phase-locked pattern of activity in each AN fiber (see Figure 2). It is assumed in the purely temporal models that the central auditory system makes no use of the spatial organization and tonotopic axis. The models "measure" the period or interval of the response of each AN channel and several ways have been suggested how this might be implemented in the central auditory system. Typically, the models form a running autocorrelation function (e.g., Brown and Puckette, 1989; Patterson et al., 1995). This makes it possible to estimate the perceived pitch from the peaks of the autocorrelation functions of the responses, as expressed for instance in a correlogram. An alternative formulation of a temporal mechanism as the basis for pitch has been suggested by de de Cheveigné (1998) where a cancellation is assumed instead of autocorrelation, thus assuming an inhibitory network instead of an excitatory one.

Some models of pitch reflect representations that utilize combinations of spectral and temporal cues. These spectro-temporal models assume, in different ways and implementations, that the tonotopic axis of the cochlea and the specific properties of the cochlear filters (like bandwidth and shape), as well as the information conveyed by phase-locking, are necessary for the representation of pitch (e.g., Shamma and Klein, 2000; Meddis, 1997;

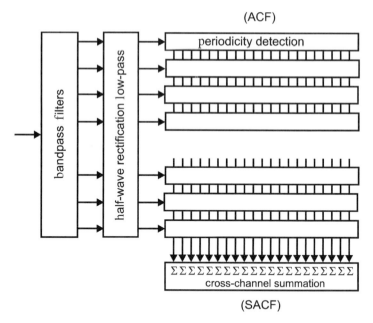

FIGURE 2 Computational model of pitch perception based on autocorrelation (Meddis, 1997). The model essentially consists of four main stages: (1) peripheral (mechanical) bandpass filtering, (2) half-wave rectification and lowpass filtering, (3) within-channel periodicity extraction via the autocorrelation function (ACF) and (4) across-channel aggregation of periodicity estimates via the summary autocorrelation function (SACF). The pitch decision algorithm is then applied to the result of the across-channel aggregation in stage 4.

Moore, 2003). Recently, experimental results by Oxenham et al. (2004) obtained with modulations imposed on high-frequency carrier tones strongly argued in favor of a fundamental role for tonotopic place in the computation of pitch and that, if temporal cues are involved, these must be intrinsically linked to the correct place (see also Shamma, 2004). These experiments provide constraints for future models of the representation of pitch in the auditory system.

4.4 Models of Temporal Resolution

Models of auditory temporal resolution typically have been described as consisting of an initial stage of bandpass filtering, reflecting a simplified action of BM filtering. Each filter is followed by a nonlinear device. In recent models, the nonlinear device typically includes two processes, half-wave rectification and a compressive nonlinearity, resembling the compressive input–output function on the BM (e.g., Ruggero and Rich, 1991; Oxenham and Moore, 1994, 1997; Plack and Oxenham, 1998).

The output is fed to a smoothing device implemented as a lowpass filter (Viemeister, 1979) or a sliding temporal integrator (e.g., Moore et al., 1988). This is followed by a decision device, typically modeled as the signal-to-noise ratio. In the model, the characteristics of the auditory filters are based on auditory filter shapes derived from masking experiments. The characteristics of the nonlinearity and the temporal integrator have been investigated extensively in conditions of nonsimultaneous masking. Forward and backward masking can be accounted for in terms of the build-up and decay processes at the output of the sliding temporal integrator. The same principle model structure has also been suggested to account for other phenomena associated with temporal resolution, such as gap detection and modulation detection (e.g., Viemeister, 1979).

An alternative way of describing forward masking is in terms of neural adaptation (e.g., Jesteadt et al., 1982; Nelson and Swain, 1996). Qualitatively, in an adaptation-based explanation, the portion of the signal furthest from the masker is most important in determining signal threshold, and that portions of the signal closer in time to the masker should contribute less to detection. This is because the response to portions closer in time to the masker will be more adapted, and hence less detectable, than later portions of the signal. Several processing models comprise adaptation and account for various aspects of forward masking (Dau et al., 1996a,b; Buchholz and Mourjoloulus, 2004a,b). For example, the model by Dau et al. (1996a) uses feedback loops to produce an effect similar to adaptation. The dynamic temporal properties of the feedback loops are combined with a logarithmic compression for stationary portions of the stimuli. This allows the model to also account for intensity discrimination data. The model assumes an optimal detector as the decision device, realized as a matched filter applied to the internal representations of the stimuli after the preprocessing stages. Both types of models, temporal integration and adaptation, can lead to similar results even though seemingly conceptually different (Oxenham, 2001).

4.5 Models of Spectro-temporal Processing and Modulation Perception

One of the models that include assumption about the processing in more central stages was suggested by Patterson et al. (1992, 1995). In their model, after the simulation BM motion and conversion into a neural activity pattern in the AN, a form of "strobed" temporal integration is applied to each channel of the neural activity pattern to stabilize any repeating patterns and convert it into a simulation of a stabilized "auditory image" of the sound. Sequences of auditory images were suggested to produce correlates of auditory perceptions that illustrate the dynamic response of the auditory system to everyday sounds.

Another model that includes explicit assumptions about the processing in central stages was suggested by Dau et al. (1997a,b) and in Jepsen et al. (2008). The key stage of these models is a modulation filterbank following the adaptation stage in each peripheral auditory filter (see Figure 3). The filterbank realizes a limited-resolution decomposition of the temporal modulations and was inspired by neurophysiological findings in the auditory brainstem (e.g., Langner and Schreiner, 1988; Palmer, 1995). The information at the output of the processing is combined by an optimal detector assuming independent observations across frequency. The parameters of the filterbank were obtained in simulations of perceptual modulation detection and masking data (e.g., Houtgast, 1989; Bacon and Grantham, 1989; Dau et al.

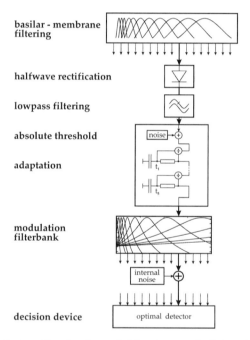

FIGURE 3 Modulation filterbank model by Dau et al. (1997a). The model consists of a simplified action of BM filtering and hair-cell transduction, an adaptation stage and a modulation filterbank. This processing transforms the signals into their internal representations. The signals are finally subjected to an optimal detector as the decision stage. The model describes modulation detection and masking data, spectro-temporal masking and forward masking. A modified version of it has also been used as (monaural) preprocessing for binaural detection (Breebaart et al., 2001a). The internal representation obtained with the model has been tested in several applications, such as in automatic speech recognition and objective speech quality assessment.

1997a,b; Ewert and Dau, 2000). The model has been tested in conditions of simultaneous and nonsimultaneous masking (Dau et al., 1996a,b; Derleth et al., 2001) and spectro-temporal masking (Derleth and Dau, 2000; Verhey, 2002; Verhey et al., 1999; Ewert et al., 2002). The original implementation of the model (assuming a lowpass filter instead of the modulation filterbank) was used as the preprocessing module in the binaural signal detection model by Breebaart et al. (2001a).

A conceptually similar model was presented by Shamma and colleagues (e.g., Chi et al., 1999; Elhilali et al., 2003). They suggested a spectro-temporal analysis of the envelope, motivated by neurophysiological findings in the auditory cortex (Calhoun and Schreiner, 1998; de Charms et al., 1998). In this model, a "spectral" modulation filterbank is integrated with the temporal modulation analysis resulting in two-dimensional spectro-temporal filters. Thus, in contrast to the current implementation of Dau et al.'s model, where spectral information is integrated subsequent to an independent temporal analysis, Shamma's model also contains joint (and inseparable) spectral–temporal modulations. In conditions where both temporal and spectral features of the input are manipulated, the two models respond differently. Shamma's model has been utilized for the assessment of speech intelligibility (Chi et al., 1999; Elhilali et al., 2003), the prediction of musical timbre (Ru and Shamma, 2003) and the perception of complex sounds (Carlyon and Shamma, 2003).

5 • Applications of Auditory Models

Auditory processing models have been utilized in several applications. For example, systems have been developed for digital recording and transmission of sound that can nearly give transparent high-fidelity reproduction quality but with a largely reduced bit rate (Brandenburg and Stoll, 1994; Gilchrist and Grewin, 1996). These perceptual coders (like, e.g., MP3) rely on the masking properties of the human auditory system, and thus are based on models of auditory signal processing. Listening tests with good coding systems have revealed that the great majority of human listeners cannot distinguish a perceptually coded signal from the original signal (Gilchrist and Grewin, 1996).

Loudness models have been used, for example, for fitting hearing aids. The models have been built into the hearing aids which is possible with modern digital hearing aids. One approach has been to attempt to restore loudness perception to "normal". Hereby, the loudness model helps to determine the frequency-dependent gain for a given sound such that the loudness perceived by the hearing-impaired listener is the same as would be perceived by a normal-hearing listener without amplification (e.g., Kollmeier and Hohmann, 1995; Launer and Moore, 2002).

In automatic speech recognition (ASR), front ends have been designed to transform the incoming speech signal into a representation which serves as input for later pattern recognition stages. Using the processing model of Dau et al. (1996a) as front end, Tchorz and Kollmeier (1999) and Kleinschmidt et al. (2001) showed that the robustness in noise is higher when compared to "classical" mel-scale cepstral feature extraction. It is mainly the temporal processing in the model that accounts for the robust behavior. The model reflects similar features as RASTA processing of speech (Hermansky and Morgan, 1994). Both techniques perform some kind of envelope bandpass filtering around 4 Hz and hence take roughly 200 ms of integration interval into account for feature calculation. The same preprocessing model has also been utilized for the objective prediction of the transmission quality of low-bit rate speech coding algorithms (Hansen and Kollmeier, 1999).

The computational AN model by Heinz et al. (2001a,b) has been utilized for the interpretation of auditory-evoked brainstem potentials (Dau, 2003). In particular, the model enables studying the effects of nonlinear cochlear processing on the evoked potential patterns generated in the brainstem. This is of interest for the clinical diagnostic of a (cochlear) hearing impairment, particularly in small children and newborns who cannot participate in perceptual listening tests.

References

Bacon SP, Grantham DW (1989) Modulation masking: effects of modulation frequency, depth, and phase. J Acoust Soc Am 85, 2575–2580.

Blauert J (1997) Spatial Hearing: The Psychophysics of Human Sound Localization (MIT, Cambridge, MA).

Brandenburg K, Stoll G (1994) ISO-MPEG-1 audio: a generic standard for coding of high-quality digital audio. J Audio Eng Soc 42, 780–792.

Breebaart J, van de Par S, Kohlrausch A (2001a) Binaural processing model based on contralateral inhibition. I. Model structure. J Acoust Soc Am 110, 1074–1088.

Breebaart J, van de Par S, Kohlrausch A (2001b) Binaural processing model based on contralateral inhibition. II. Dependence on spectral parameters. J Acoust Soc Am 110, 1089–1104.

Breebaart J, van de Par S, Kohlrausch A (2001c) Binaural processing model based on contralateral inhibition. III. Dependence on temporal parameters. J Acoust Soc Am 110, 1105–1117.

Brown JC, Puckette MS (1989) Calculation of a "narrowed" autocorrelation function. J Acoust Soc Am 85, 1595–1601.

Buchholz JM, Mourjoloulus J (2004a) A computational auditory masking model based on signal dependent compression. I. Model description and performance analysis. Acust Acta Acust 5, 873–886.

Buchholz JM, Mourjoloulus J (2004b) A computational auditory masking model based on signal dependent compression. II. Model simulations and analytical approximations. Acust Acta Acust 5, 887–900.

Calhoun BM, Schreiner CE (1998) Spectral envelope coding in cat primary auditory cortex: linear and non-linear effects of stimulus characteristics. Eur J Neurosci 10, 926–940.

Carlyon RP, Shamma S (2003) An account of monaural phase sensitivity. J Acoust Soc Am 114, 333–348.

Carney LH (1993) A model for the responses of low-frequency auditory-nerve fibers in cat. J Acoust Soc Am 93, 401–417.

Chi T, Gao Y, Guyton MC, Ru P, Shamma S (1999) Spectro-temporal modulation transfer functions and speech intelligibility. J Acoust Soc Am 106, 2719–2732.

Cohen MA, Grossberg S, Wysell LL (1995) A spectral network model of pitch perception. J Acoust Soc Am 98, 862–879.

Colburn HS (1996) Computational models of binaural processing. In Auditory Computation, edited by HL Hawkins, TA McMullen, AN Popper and RR Fay (Springer, New York), pp 332–400.

Colburn HS, Durlach NI (1978) Models of binaural interaction. In Handbook of Perception, edited by E Carterette and M Friedman (Academic, New York), Vol. IV, pp. 467–518.

Colburn HS, Moss PJ (1981) Binaural interaction models and mechanisms. In Neuronal Mechanisms of Hearing, edited by J Syka and L Aitkin (Plenum Press, New York), pp. 283–288.

Colburn HS, Han YA, Culotta CP (1990) Coincidence model of MSO responses. Hear Res 49, 335–346.

Colburn HS, Carney LH, Heinz MG (2003) Quantifying the information in auditory-nerve responses for level discrimination. J Assoc Res Otolaryngol 4, 294–311.

Dau T (2003) The importance of cochlear processing for the formation of auditory brainstem and frequency following responses. J Acoust Soc Am 113, 936–950.

Dau T, Püschel D, Kohlrausch A (1996a) A quantitative model of the "effective" signal processing in the auditory system. I. Model structure. J Acoust Soc Am 99, 3615–3622.

Dau T, Püschel D, Kohlrausch A (1996b) A quantitative model of the "effective" signal processing in the auditory system. II. Simulations and measurements. J Acoust Soc Am 99, 3623–3631.

Dau T, Kollmeier B, Kohlrausch A (1997a) Modeling auditory processing of amplitude modulation. I. Modulation detection and masking with narrowband carriers. J Acoust Soc Am 102, 2892–2905.

Dau T, Kollmeier B, Kohlrausch A (1997b) Modeling auditory processing of amplitude modulation. II. Spectral and temporal integration in modulation detection. J Acoust Soc Am 102, 2906–2919.

de Boer E (1980) Auditory physics. Physical principles in hearing theory. I. Phys Rep 62, 88–174.

de Boer E (1984) Auditory physics. Physical principles in hearing theory. II. Phys Rep 105, 142–226.

de Boer E (1991) Auditory physics. Physical principles in hearing theory. III. Phys Rep 203, 125–231.

de Boer E (1995) The inverse problem solved for a three-dimensional model of the cochlea. I. Analysis. J Acoust Soc Am 98, 896–903.

de Charms RC, Blake DT, Merzenich MM (1998) Optimizing sound features for cortical neurons. Science 280(5368), 1439–1443.

de Cheveigné A (1998) Cancellation model of pitch perception. J Acoust Soc Am 103, 1261–1271.

de Cheveigné A (2005) Pitch perception models. In Pitch: Neural Coding and Perception, edited by C Plack, AJ Oxenham, AN Popper and RR Fay (Springer, New York).

Delgutte B (1990) Physiological mechanisms of psychophysical masking: observations from auditory-nerve fibers. J Acoust Soc Am 87, 791–809.

Delgutte B (1996) Physiological models for basic auditory percepts. In Auditory Computation, edited by HL Hawkins, TA McMullen, AN Popper and RR Fay (Springer, New York), pp. 157–220.

Derleth RP, Dau T (2000) On the role of envelope fluctuation processing in spectral masking. J Acoust Soc Am 108, 285–296.

Derleth RP, Dau T, Kollmeier B (2001) Modeling temporal and compressive properties of the normal and impaired auditory system. Hear Res 159, 132–149.

Duifhuis H (1976) Cochlear nonlinearity and second filter: possible mechanism and implications. J Acoust Soc Am 59, 408–423.

Durlach NI (1963) Equalization and cancellation theory of binaural masking-level differences. J Acoust Soc Am 35, 1206–1218.

Durlach NI (1972) Binaural signal detection: equalization and cancellation theory. In Foundations of Modern Auditory Theory, edited by J Tobias (Academic, New York), pp. 369–462.

Elhilali M, Chi T, Shamma S (2003) A spectro-temporal modulation index (stmi) for assessment of speech intelligibility. Speech Commun 41, 331–348.

Ewert SD, Dau T (2000) Characterizing frequency selectivity for envelope fluctuations. J Acoust Soc Am 108, 1181–1196.

Ewert SD, Verhey JL, Dau T (2002) Spectro-temporal processing in the envelope-frequency domain. J Acoust Soc Am 112, 2921–2931.

Fastl H (1993) Loudness evaluation by subjects and by a loudness meter. In Sensory Research – Multimodal Perspectives, edited by RT Verrillo (Erlbaum, Hillsdale, NJ), pp. 199–210.

Fletcher H, Munson WA (1937) Relation between loudness and masking. J Acoust Soc Am 9, 82–108.

Gaik W (1993) Combined evaluation of interaural time and intensity differences: psychoacoustic results and computer modeling. J Acoust Soc Am 94, 98–110.

Gardner MB, Hawley MS (1972) Network representations of the external ear. J Acoust Soc Am 52, 1620–1628.

Geisler CD (1990) Evidence for expansive power functions in the generation of the discharges of 'low- and medium-spontaneous' auditory-nerve fibers. Hear Res 44, 1–12.

Giguère C, Woodland PC (1994) A computational model of the auditory periphery for speech and hearing research. I. Ascending path. J Acoust Soc Am 95, 331–342.

Gilchrist N, Grewin C (1996) Collected Papers on Digital Audio Bit Rate Reduction (Audio Engineering Society, New York).

Glasberg BR, Moore BCJ (2002) A model of loudness applicable to time-varying sounds. J Audio Eng Soc 50, 331–341.

Goldberg JM, Brown PB (1969) Response of binaural neurons of dog superior olivary complex to dichotic tonal stimuli: some physiological mechanisms of sound localization. J Neurophysiol 32, 613–636.

Goldstein JL (1973) An optimum processor theory for the central formation of the pitch of complex tones. J Acoust Soc Am 54, 1496–1516.

Goldstein JL (1990) Modeling rapid waveform compression on the basilar membrane as multiple-bandpass-nonlinearity filtering. Hear Res 49, 39–60.

Goldstein JL (1995) Relations among compression, suppression, and combination tones in mechanical responses of the basilar membrane: data and MBPNL model. Hear Res 89, 52–68.

Hansen M, Kollmeier B (1999) Continuous assessment of time-varying speech quality. J Acoust Soc Am 106, 2888–2899.

Heinz MG, Zhang X, Bruce IC, Carney LH (2001a) Auditory-nerve model for predicting performance limits of normal and impaired listeners. ARLO 5(3), 91–96.

Heinz MG, Colburn HS, Carney LH (2001b) Evaluating auditory performance limits: I. One-parameter discrimination using a computational model for the auditory nerve. Neural Comput 13, 2273–2316.

Hermansky H, Morgan N (1994) Rasta processing of speech. IEEE Trans Speech Audio Process 2, 578–589.

Hewitt MJ, Meddis R (1991) An evaluation of eight computer models of mammalian inner hair-cell function. J Acoust Soc Am 90, 904–917.

Hewitt MJ, Meddis R (1994) A computer model of amplitude-modulation sensitivity of single units in the inferior colliculus. J Acoust Soc Am 95, 2145–2159.

Houtgast T (1989) Frequency selectivity in amplitude-modulation detection. J Acoust Soc Am 85, 1676–1680.

Hubbard AE, Mountain DC (1996) Analysis and synthesis of cochlear mechanical function using models. In Auditory Computation, edited by HL Hawkins, TA McMullen, AN Popper and RR Fay (Springer, New York), pp. 62–120.

Irino T, Patterson RD (1997) A time-domain, level-dependent auditory filter: the gammachirp. J Acoust Soc Am 101, 412–419.

Irino T, Patterson RD (2000) A gammachirp perspective of cochlear mechanics that can also explain human auditory masking quantitatively. In Recent Developments in Auditory Mechanics, edited by H Wada, T Takasaka, K Ikeda, K Ohyama and T Koike (World Scientific, Singapore), pp. 230–236.

Irino T, Patterson RD (2001) A compressive gammachirp auditory filter for both physiological and psychophysical data. J Acoust Soc Am 109, 2008–2022.

Jeffress LA (1948) A place theory of sound localization. J Comp Physiol Psychol 41, 35–39.

Jepsen ML, Ewert SD, Dau T (2008) A computational model of human auditory signal processing and perception. J Acoust Soc AM in press, expected in issue 124(1) July 2008.

Jesteadt W, Bacon SP, Lehman JR (1982) Forward masking as a function of frequency, masker level, and signal delay. J Acoust Soc Am 71, 950–962.

Joris PX, Yin TCT (1995) Envelope coding in the lateral superior olive. I. Sensitivity to interaural time differences. J Neurophysiol 73, 1043–1062.

Joris PX, Schreiner CE, Rees A (2004) Neural processing of amplitude-modulated sounds. Physiol Rev 84, 541–577.

Killion MC, Clemis JD (1981) An engineering view of middle ear surgery. J Acoust Soc Am 69 (Suppl 1), S44.

Kim DO (1986) Active and nonlinear biomechanics and the role of the outer-hair-cell subsystem in the mammalian auditory system. Hear Res 22, 105–114.

Kleinschmidt M, Tchorz J, Kollmeier B (2001) Combining speech enhancement and auditory feature extraction for robust speech recognition. Speech Commun 32, 75–91.

Kollmeier B, Hohmann V (1995) Loudness estimation and compensation employing a categorical scale. In Advances in Hearing Research, edited by GA Manley, GM Klump, C Köppl, H Fastl and H Öckinghaus (World Scientific, Singapore).

Kuwada S, Yin TC, Syka J, Buunen TJ, Wickesberg RE (1984) Binaural interaction in low-frequency neurons in inferior colliculus of the cat. IV. Comparison of monaural and binaural response properties. J Neurophysiol 51, 1306–1325.

Langner G (1981) Neuronal mechanisms for pitch analysis in the time domain. Exp Brain Res 44, 450–454.

Langner G (1997) Neural processing and representation of periodicity pitch. Acta Otolaryngol Suppl 532, 68–76.

Langner G, Schreiner CE (1988) Periodicity coding in the inferior colliculus of the cat. I. Neuronal mechanisms. J Neurophysiol 60, 1799–1822.

Langner G, Sams M, Heil P, Schulze H (1997) Frequency and periodicity are represented in orthogonal maps in the human auditory cortex: evidence from magnetoencephalography. J Comp Physiol 181, 665–676.

Launer S, Moore BCJ (2002) Use of a loudness model for hearing aid fitting. V. On-line gain control in a digital hearing aid. Int J Audiol 42, 262–273.

Licklider JC (1951) A duplex theory of pitch perception. Experientia 7, 128–134.

Lindemann W (1986) Extension of a binaural cross-correlation model by contralateral inhibition. I. Simulation of lateralization for stationary signals. J Acoust Soc Am 80, 1608–1622.

Lopez-Poveda EA, Meddis R (2001) A human nonlinear cochlear filterbank. J Acoust Soc Am 110, 3107–3118.

Lorenzi C, Micheyl C, Berthommier F (1995) Neuronal correlates of perceptual amplitude-modulation detection. Hear Res 90, 219–227.

Lutman ME, Martin AM (1979) Development of an electroacoustic analogue model of the middle ear and acoustic reflex. J Sound Vibr 64, 133–157.

Lyon R, Shamma S (1996) Auditory representations of timbre and pitch. In Auditory Computation, edited by HL Hawkins, TA McMullen, AN Popper and RR Fay (Springer, New York), pp. 221–270.

McAlpine D, Jiang D, Shackleton TM, Palmer AR (1998) Convergent input from brainstem coincidence detectors onto delay-sensitive neurons in the inferior colliculus. J Neurosci 18, 6026–6039.

Meddis R (1986) Simulation of mechanical to neural transduction in the auditory receptor. J Acoust Soc Am 79, 702–711.

Meddis R (1988) Simulation of auditory-neural transduction: further studies. J Acoust Soc Am 83, 1056–1063.

Meddis R, Hewitt MJ (1992) Modeling the identification of concurrent vowels with different fundamental frequencies. J Acoust Soc Am 91, 233–245.

Meddis R, O'Mard L (1997) A unitary model of pitch perception. J Acoust Soc Am 102, 1811–1820.

Meddis R, O'Mard LP, Lopez-Poveda EA (2001) A computational algorithm for computing nonlinear auditory frequency selectivity. J Acoust Soc Am 109, 2852–2861.

Middlebrooks JC (1992) Narrow-band sound localization related to external ear acoustics. J Acoust Soc Am 92, 2607–2624.

Møller AR (1976) Dynamic properties of excitation and two-tone inhibition in the cochlear nucleus studied using amplitude-modulated tones. Exp Brain Res 25, 307–321.

Moore BCJ (2003) An Introduction to the Psychology of Hearing, 5th Ed. (Academic, Amsterdam).

Moore BCJ, Glasberg BR, Plack CJ, Biswas AK (1988) The shape of the ear's temporal window. J Acoust Soc Am 83, 1102–1116.

Moore BCJ, Glasberg BR, Baer T (1997) A model for the prediction of thresholds, loudness, and partial loudness. J Audio Eng Soc 45, 224–240.

Neely ST (1983) The cochlear amplifier. In Mechanics of Hearing, edited by E de Boer and MA Viergever (Martinus Nijhoff, The Hague), pp. 111–118.

Nelson PC, Carney LH (2004) A phenomenological model of peripheral and central neural responses to amplitude-modulated tones. J Acoust Soc Am 116, 2173–2186.

Nelson DA, Swain AC (1996) Temporal resolution within the upper accessory excitation of a masker. Acust Acta Acust 82, 328–334.

Olson ES (1999) Direct measurement of intracochlear pressure waves. Nature (London) 402, 526–529.

Oxenham AJ (2001) Forward masking: adaptation or integration? J Acoust Soc Am 109, 732–741.

Oxenham AJ, Moore BCJ (1994) Modeling the additivity of nonsimultaneous masking. Hear Res 80, 105–118.

Oxenham AJ, Moore BCJ (1997) Modeling the effects of peripheral nonlinearity in listeners with normal and impaired hearing. In Modeling Sensorineural Hearing Loss, edited by W Jesteadt (Erlbaum, Hillsdale, NJ).

Oxenham AJ, Bernstein JG, Penagos H (2004) Correct tonotopic representation is necessary for complex pitch perception. Proc Natl Acad Sci 101, 1421–1425.

Palmer AR (1995) Neural signal processing. In Hearing, edited by BCJ Moore (Academic, San Diego).

Palmer AR, McAlpine D, Jiang D (1997) Processing of interaural delay in the inferior colliculus. In Acoustical Signal Processing in the Central Auditory System, edited by J Syka (Plenum, New York), pp. 353–364.

Patterson RP, Nimmo-Smith I, Holdsworth J, Rice P (1987) An efficient auditory filterbank based on the gammatone function. In a paper presented at a meeting of the IOC Speech Group on Auditory Modeling at RSRE.

Patterson RD, Robinson K, Holdsworth J, McKeown D, Zhang C, Allerhand M (1992) Complex sounds and auditory images. In Auditory Physiology and Perception, Proceedings of the 9th International Symposium on Hearing, edited by Y Cazals, L Demany and K Horner (Pergamon, Oxford), pp. 429–446.

Patterson RD, Allerhand MH, Giguère C (1995) Time-domain modeling of peripheral auditory processing: a modular architecture and a software platform. J Acoust Soc Am 98, 1890–1894.

Pfeiffer RR (1970) A model for two-tone inhibition of single cochlear-nerve fibers. J Acoust Soc Am 48, 1373–1378.

Plack CJ, Oxenham AJ (1998) Basilar-membrane nonlinearity and the growth of forward masking. J Acoust Soc Am 103, 1598–1608.

Rees A, Møller AR (1983) Responses of neurons in the inferior colliculus of the rat to AM and FM tones. Hear Res 10, 301–330.

Rhode WS (1971) Observations of the vibration of the basilar membrane in squirrel monkeys using the Mössbauer technique. J Acoust Soc Am 49, 1218–1231.

Rhode WS, Oertel D, Smith PH (1983) Physiological response properties of cells labeled intracellularly with horseradish peroxidase in cat ventral cochlear nucleus. J Comp Neurol 213, 448–463.

Ru P, Shamma S (2003) Presentation of musical timbre in the auditory cortex. J New Music Res 26, 154–169.

Ruggero MA, Rich NC (1991) Furosemide alters organ of corti mechanics: evidence for feedback of outer hair cells upon the basilar membrane. J Neurosci 11, 1057–1067.

Sachs MB, Abbas PJ (1974) Rate versus level functions for auditory nerve fibers in cats: tone burst stimuli. J Acoust Soc Am 81, 680–691.

Schoonhoven R, Prijs VF, Frijns JH (1997) Transmitter release in inner hair cell synapses: a model analysis of spontaneous and driven rate properties of cochlear nerve fibres. Hear Res 113, 247–260.

Searle CL, Braida LD, Davis MF, Colburn HS (1976) Model for auditory localization. J Acoust Soc Am 60, 1164–1175.

Shamma S, Klein D (2000) The case of the missing pitch templates: how harmonic templates emerge in the early auditory system. J Acoust Soc Am 107, 2631–2644.

Shamma SA (2004) Topographic organization is essential for pitch perception. Proc Natl Acad Sci 101, 1114–1115.

Shamma SA, Shen NM, Gopalaswamy P (1989) Stereausis: binaural processing without neural delays. J Acoust Soc Am 86, 989–1006.

Shaw EAG (1980) The acoustics of the external ear. In Acoustical Factors Affecting Hearing Aid Performance, edited by GA Studebaker and IH Hochberg (University Park, Baltimore), pp. 109–125.

Shekhter I, Carney LH (1997) A nonlinear auditory nerve model for CF-dependent shift in tuning with sound level. Assoc Res Otolaryngol Abs 20, 670.

Shera CA (2001) Frequency glides in click responses of the basilar membrane and auditory nerve: their scaling behavior and origin in traveling-wave dispersion. J Acoust Soc Am 109, 2023–2034.

Siebert WM (1965) Some implications of the stochastic behavior of primary auditory neurons. Kybernetik 2, 206–215.

Siebert WM (1970) Frequency discrimination in the auditory system: place or periodicity mechanism. Proc IEEE 58, 723–730.

Smith PH, Rhode WS (1989) Structural and functional properties distinguish two types of multipolar cells in the ventral cochlear nucleus. J Comp Neurol 282, 595–616.

Stone MA, Moore BCJ, Glasberg BR (1997) A real-time DSP-based loudness meter. In Contributions to Psychological Acoustics, edited by A Schick and M Klatte (BIS Universität, Oldenburg), pp. 587–601.

Summer CJ, O'Mard LP, Lopez-Poveda EA, Meddis R (2003) A nonlinear filter-bank model of the guinea-pig cochlear nerve: Rate responses. J Acoust Soc Am 113, 3264–3274

Tchorz J, Kollmeier B (1999) A model of auditory perception as front end for automatic speech recognition. J Acoust Soc Am 106, 2040–2050.

Terhardt E (1979) Calculating virtual pitch. Hear Res 1, 155–182.

Verhey JL (2002) Modeling the influence of inherent envelope fluctuations in simultaneous masking experiments. J Acoust Soc Am 111, 1018–1025.

Verhey JL, Dau T, Kollmeier B (1999) Within-channel cues in comodulation masking release (CMR): experiments and model predictions using a modulation-filterbank model. J Acoust Soc Am 106, 2733–2745.

Viemeister NF (1979) Temporal modulation transfer functions based upon modulation thresholds. J Acoust Soc Am 66, 1364–1380.

Wightman FL (1973) The pattern-transformation model of pitch. J Acoust Soc Am 54, 407–416.

Zacksenhouse M, Johnson DH, Tsuchitani C (1992) Excitatory/inhibitory interaction in the LSO revealed by point process modeling. Hear Res 62, 105–123.

Zhang XD, Heinz MG, Bruce IC, Carney LH (2001) A phenomenological model for the responses of auditory-nerve fibers: I. Nonlinear tuning with compression and suppression. J Acoust Soc Am 109, 648–670.

Zweig G (1991) Finding the impedance of the organ of Corti. J Acoust Soc Am 89, 1229–1254.

Zwicker E (1965) Temporal effects in simultaneous masking and loudness. J Acoust Soc Am 38, 132–141.

Zwicker E (1977) Procedure for calculating loudness of temporally variable sounds. J Acoust Soc Am 62, 675–682.

Zwicker E, Fastl H (1999) Psychoacoustics – Facts and Models, 2nd Ed. (Springer, Berlin).

Zwicker E, Scharf B (1965) A model of loudness summation. Psychol Rev 72, 3–26.

Zwislocki JJ (1962) Analysis of the middle-ear function. Part I: input impedance. J Acoust Soc Am 34, 1514–1523.

13
Speech Intelligibility

Thomas Brand

University of Oldenburg, Oldenburg, Germany

1 • Speech Intelligibility

1.1 Definition

Speech intelligibility (SI) is important for different fields of research, engineering and diagnostics in order to quantify very different phenomena like the quality of recordings, communication and playback devices, the reverberation of auditoria, characteristics of hearing impairment, benefit using hearing aids or combinations of these things. A useful way to define SI so that it is directly and quantitatively measurable is: *"Speech intelligibility SI is the proportion of speech items (e.g., syllables, words or sentences) correctly repeated by (a) listener(s) for a given speech intelligibility test."*

1.2 Intelligibility Function and Speech Reception Threshold (SRT)

The Intelligibility function describes the listener's speech intelligibility SI as a function of speech level L (in dB), which may either refer to the sound pressure level of the speech signal or to the speech-to-noise ratio (SNR) if the test is performed with interfering noise (Figure 1).

In most cases it is possible to fit the function SI(L) to the empirical data using

$$\mathrm{SI}(L) = \frac{1}{A}\left(1 + \mathrm{SI}_{\max}\frac{A-1}{1+\exp\left(-\frac{L-L_{\mathrm{mid}}}{s}\right)}\right) \quad (1)$$

with

L_{mid}: speech level of the midpoint of the intelligibility function.
s: slope parameter. The slope at L_{mid} is given by $\frac{\mathrm{SI}_{\max}(A-1)}{4\,A\,s}$.

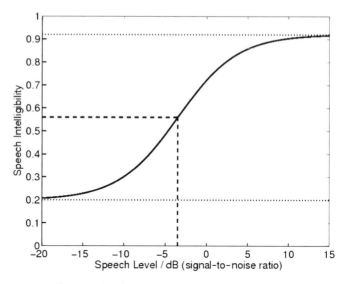

FIGURE 1 Typical example of SI function (*solid line*) for word intelligibility test (closed response format with five response alternatives). The *dashed line* denotes L_{mid}. The *dotted lines* denote the lower limit ($1/A$) and the upper limit (SI_{asymp}) of the SI function. Parameters: $L_{mid} = -3.5$ dB SNR, $SI_{max} = 0.9$ ($SI_{asymp} = 0.92$), $A = 5$, slope $= 0.05$/dB($s = 3.6$ dB).

SI_{max}: parameter for maximum intelligibility. This can be smaller than 1 in some cases (e.g., distorted speech signals or listeners with hearing impairment). The asymptotic maximum of SI is $SI_{max} + (1 - SI_{max})/A$.

A: Number of response alternatives. For example, $A = 10$ when the listener should respond in a closed response format using digits between "0" and "9". In SI tests with "open response format", like word tests without limiting the number of response alternatives, A is assumed to be infinite, which means

$$SI = SI_{max} \frac{1}{1 + \exp\left(-\frac{L-L_{mid}}{s}\right)} \quad \text{and slope} = \frac{SI_{max}}{4s}. \quad (2)$$

The primary interest of many applications is the speech reception threshold (SRT), which denotes the speech level (measured in dB) for a given intelligibility (e.g., $SI = 0.5$ or 0.7).

1.3 Accuracy of SI Measurements

The accuracy of SI measurements is given by the binomial distribution. Consequently, the standard error se(SI) of a SI estimate which is based on n items (e.g., words) is

$$\text{se}(\text{SI}) = \sqrt{\frac{\text{SI}(1-\text{SI})}{n}}. \tag{3}$$

A further increase of this standard error is caused by the fact that SI tests consist of several items (e.g., 50 words) which unavoidably differ in SI. Therefore, SI tests should be constructed in a way that the SI of all items is as homogeneous as possible.

To a first approximation, the standard error of the SRT is equal to se(SI_{SRT}) (the standard error of the SI estimate at the SRT) divided by the slope of the intelligibility function at the SRT:

$$\text{se}(\text{SRT}) = \frac{\text{se}(\text{SI}_{\text{SRT}})}{\text{slope}(\text{SRT})}. \tag{4}$$

2 • Measurement Methods

2.1 Speech Materials

A speech material (i.e., a set of speech items like words or sentences) is suitable for SI tests when different requirements are fulfilled: the different speech items have to be homogeneous in SI in order to yield high measurement accuracy and reproducibility in a limited measuring time. The distribution of phonemes should represent the respective language. Only speech materials which have been optimized properly by a large number of evaluation measurements can fulfill these requirements.

A large number of SI tests using different materials are available for different languages. An overview of American SI tests can be found in Penrod (1994). There are different formats, i.e., nonsense syllables, single words and sentences. Sentences represent a realistic communication situation best. Nonsense syllables and words allow for assessing confusion matrices and for analyzing transmission of information. Furthermore, the intelligibility functions of most sentence tests (Plomp and Mimpen, 1979; Hagerman, 1982; Nilsson et al., 1994; Kollmeier and Wesselkamp, 1997; Wagener et al., 1999, 2003) show slopes between 0.15 and 0.25 per dB which are considerably steeper than the values obtained with nonsense syllables or single-word tests.

Since the standard deviation of SRT estimates is inversely proportional to the slope of the intelligibility function, these sentence tests are better suited for efficient and reliable SRT measurements than single-word tests.

2.2 Presentation Modes

Signals can be presented via either loudspeakers (free field condition) or headphones. The free field condition is more natural but it has the drawbacks of requiring a larger experimental effort, being difficult to calibrate, and, especially for spatial speech/noise situations, small movements of the listeners head may influence the result of the SI measurement.

The advantages of presentation via headphones are: very good reproducibility for each individual listener; smaller experimental effort; and spatial speech/noise conditions can be easily realized using virtual acoustics. Drawbacks are: the individual calibrations are complicated because headphones produce different sound pressures in different ears; and measurements with hearing aids are not possible.

Adaptive procedures (Chapter 15) can be used to yield SRT estimates efficiently by concentrating the presentation levels near to the SRT. In sentence tests, each word can be scored independently which allows adaptive procedures to be designed which converge more efficiently than adaptive procedures usually used in psychoacoustics (Brand and Kollmeier, 2002).

3 • Factors Influencing Speech Intelligibility

3.1 Measuring Method

The different speech materials mentioned above generate different results. Therefore, only standardized speech materials or speech materials with well-known reference intelligibility functions should be used.

3.2 Noise and Room Acoustics

Noise and reverberation reduce SI. Therefore, if the intention is to measure SI in silence, environmental noise and reverberation have to be minimized using, for example, sound-insulated cabins and damping headphones. On the other hand, measurements can be used to specifically investigate the influence of different noises and room acoustics on SI, which is important for the so-called "cocktail party phenomenon".

3.3 Cocktail Party Phenomenon

The human auditory system has very impressing abilities in understanding a target talker even if maskers, i.e., competitive sound sources like different talkers, are present at the same time. This is the so-called "cocktail party phenomenon" and an interesting review of research on this topic can be found in Bronkhorst (2000). The SI in these multi-talker conditions is influenced by many masker properties like sound pressure level, frequency spectrum, amplitude modulations, spatial direction and number of maskers. The spatial configuration of target speaker and masker plays a very important role. Binaural hearing (hearing

3.4 Hearing Impairment

with both ears) enables a very effective release from masking (improvement of the SRT) of up to 12 dB compared to monaural hearing (hearing with one ear) (Bronkhorst, 2000).

An introduction to SI in clinical audiology can be found in Penrod (1994). Hearing impairment can lead to an increase of the SRT, a decrease of the maximum reachable intelligibility SI_{asymp} and a flatter slope of the intelligibility function. The most difficult situations for hearing impaired listeners are noisy environments with many interfering sound sources ("cocktail party situation"). Therefore, SI tests on noise are important diagnostic tools for assessing the daily life consequences of a hearing impairment and the benefit of a hearing aid. SI plays a very important role for the research on and the fitting of hearing aids (see Chapter 14).

4 • Prediction Methods

4.1 Articulation Index (AI), Speech Intelligibility Index (SII) and Speech Transmission Index (STI)

The most common methods for the prediction of speech intelligibility are the Articulation Index (AI) (French and Steinberg, 1947; Fletcher and Galt, 1950; ANSI, 1969), which was renamed Speech Intelligibility Index (SII) (ANSI, 1997), and the Speech Transmission Index (STI) (Houtgast and Steeneken, 1985; IEC, 1998) (Table 1). The strength of these models is the large amount of empirical knowledge they are based on. All these models assume that speech is coded by several frequency channels that carry independent information. This can be expressed by

$$AI = \sum_i AI_i, \qquad (5)$$

with AI denoting the cumulative Articulation Index of all channels and AI_i denoting the Articulation Index of a single channel.

AI and SII are derived from the speech signal by calculating the signal-to-noise ratio (SNR) in the different frequency channels:

$$AI = \sum_i \frac{W_i(SNR_i + 15)}{30}, \qquad (6)$$

with W_i denoting a frequency channel weighting factor and SNR_i denoting the signal-to-noise ratio in channel i. W_i depends on the speech material used and takes into account the fact that high frequencies are more important for the recognition of consonants than for the recognition of meaningful sentences. The main differences between the different versions of AI and SII are the way they include non-linearities like distortion, masking and broadening of frequency bands.

TABLE 1 Examples of methods for the prediction of speech intelligibility (SI)

Method (reference)	Signal parameters	Comments
Articulation Index, AI (French and Steinberg, 1947)	Levels and frequency spectra of speech and noise, kind of speech material	Macroscopic model that describes the influence of the frequency content of speech on intelligibility
Articulation Index, AI (Fletcher and Galt, 1950)	Levels and frequency spectra of speech and noise, kind of speech material	More complex than French and Steinberg version, describes more non-linear effects, seldom used
Articulation Index, AI (ANSI S3.5-1969)	Levels and frequency spectra of speech and noise	Simplified version based on French and Steinberg (1947), not in use anymore
Speech Intelligibility Index, SII (ANSI S3.5-1997)	Levels and frequency spectra of speech and noise, kind of speech material, hearing loss	Revision of ANSI S3.5-1969, includes spread of masking, standard speech spectra, relative importance of frequency bands
Speech Transmission Index, STI (IEC 60268-16)	Modulation transfer function	Predicts the change of intelligibility caused by a speech transmission system (e.g., an auditorium) based on the modulation transfer function of the system
Speech recognition sensitivity model, SRS (Müsch and Buus, 2001a,b)	Levels and frequency spectra of speech and noise, number of response alternatives	Alternative to SII, handles frequency band interactions and is better suited for unsteady frequency spectra
Holube and Kollmeier (1996)	Speech and noise signals, hearing loss	Microscopic modeling of signal processing of auditory system combined with simple automatic speech recognition

The Speech Transmission Index (STI) uses the modulation transfer function instead of the SNR and is especially successful for predicting SI in auditoria and rooms, because it explicitly takes into account the flattening of the information-carrying speech envelopes due to reverberation.

The transformation of AI, SII or STI, respectively, to speech intelligibility requires a non-linear transformation that has to be fitted to empirical data. The transformation depends on the kind

4.2 Statistical Methods

The assumption of independent information in different frequency channels does not hold in all situations because synergetic as well as redundant interactions between different channels occur. The speech recognition sensitivity model (Müsch and Buus, 2001a,b) takes these interactions into account using statistical decision theory in order to model the linguistic entropy of speech.

4.3 Functional Method

The methods mentioned above are based on relatively rough parameterizations of speech, such as long-term frequency spectrum or the modulation transfer function. The method proposed by Holube and Kollmeier (1996), however, is based on physiological and psychoacoustical data and is a combination of a functional model of the human auditory system [Chap. 12] and a simple automatic speech recognition system [Chapter 29]. A drawback of this approach is that there is still a large gap between recognition rates of humans and automatic speech recognition systems (for a review, see Lippmann, 1997).

of speech material used and is usually steeper for materials with context (e.g. sentences) than for materials with single words.

References

ANSI (1969) Methods for the Calculation of the Articulation Index, ANSI S3.5-1969, American National Standards Institute, New York.

ANSI (1997) Methods for Calculation of the Speech Intelligibility Index, ANSI S3.5-1997, American National Standards Institute, New York.

Brand T, Kollmeier B (2002) Efficient adaptive procedures for threshold and concurrent slope estimates for psychophysics and speech intelligibility tests, J Acoust Soc Am **111**(6), 2801–2810.

Bronkhorst A (2000) The cocktail party phenomenon: a review of research on speech intelligibility in multiple-talker conditions, Acustica **86**, 117–128.

Fletcher H, Galt RH (1950) The perception of speech and its relation to telephony, J Acoust Soc Am **22**, 89–151.

French NR, Steinberg JC (1947) Factors governing the intelligibility of speech sounds, J Acoust Soc Am **19**, 90–119.

Hagerman B (1982) Sentences for testing speech intelligibility in noise, Scand Audiol **11**, 79–87.

Holube I, Kollmeier B (1996) Speech intelligibility prediction in hearing-impaired listeners based on a psychoacoustically motivated perception model, J Acoust Soc Am **100**, 1703–1716.

Houtgast T, Steeneken HJM (1985) A review of the MTF concept in room acoustics and its use for estimating speech intelligibility in auditoria. J Acoust Soc Am **77**, 1069–1077.

IEC (1998) Sound System Equipment – Part 16: Objective Rating of Speech Intelligibility by Speech Transmission Index, INTERNATIONAL STANDARD 60268-16 Second edition 1998–03.

Kollmeier B, Wesselkamp M (1997) Development and evaluation of a German sentence test for objective and subjective speech intelligibility assessment, J Acoust Soc Am **102**, 2412–2421.

Lippmann R (1997) Speech recognition by machines and humans, Speech Commun **22**, 1–15.

Müsch H, Buus S (2001a) Using statistical decision theory to predict speech intelligibility. I. Model structure, J Acoust Soc Am **109**, 2896–2909.

Müsch H, Buus S (2001b) Using statistical decision theory to predict speech intelligibility. II. Measurement and prediction of consonant-discrimination performance, J Acoust Soc Am **109**, 2910–2920.

Nilsson M, Soli SD, Sullivan JA (1994) Development of the Hearing in Noise Test for the measurement of speech reception thresholds in quiet and in noise, J Acoust Soc Am **95**(2), 1085–1099.

Penrod JP (1994) Speech threshold and word recognition/discrimination testing. In Handbook of Clinical Audiology, edited by Katz J, Williams & Wilkins, Baltimore, MD, 4th ed., Chap. 10, 147–164.

Plomp R, Mimpen A (1979) Improving the reliability of testing the speech-reception threshold for sentences, Audiology **18**, 43–52.

Wagener K, Kühnel V, Kollmeier B (1999) Entwicklung und Evaluation eines Satztests für die deutsche Sprache I: Design des Oldenburger Satztests (Development and evaluation of a German sentence test I: Design of the Oldenburg sentence test), Z Audiol **38**, 4–15.

Wagener K, Josvassen JL, Ardenkjaer R (2003) Design, optimization and evaluation of a Danish sentence test in noise, Int J Audiol **42**(1), 10–17.

14

Signal Processing in Hearing Aids

Volker Hohmann

University of Oldenburg, Oldenburg, Germany

Since the introduction of the first commercial digital hearing aid in 1996, the possibilities of digital signal processing (DSP) have increasingly been exploited in hearing aids. DSP allows for the implementation of signal processing schemes (i.e., "algorithms") that have no counterpart in the analog domain and thus offers new, interesting perspectives for the rehabilitation of hearing impairment, only parts of which have been realized to date. This section reviews the processing schemes currently used in digital hearing aids and outlines the main lines of research toward improved processing schemes.

1 • Basic Approach to the Rehabilitation of Hearing Deficiencies

Processing schemes divide into two general categories. *Compensation schemes* compensate for altered aspects of auditory perception by preprocessing the acoustic signals in such a way that the combination of preprocessing and altered perception by the hearing impaired leads to an "aided" perception which is as close to the normal perception as possible. Amplification, dynamic range compression and several speech enhancement schemes fall into this category. A key problem is that the alteration of perception has to be described by nonlinear processes in many cases, which are not invertible in general: A combination of two nonlinear systems does not necessarily mimic the behavior of another (linear or nonlinear) system. Compensation schemes explicitly or implicitly consider a model of the altered perception (auditory processing models or perceptual models, see Chapter 12 and 15).

Substitution schemes, on the other hand, aim at substituting lost aspects of auditory perception by implementing these aspects fully by signal processing. Examples for substitution schemes are predominantly noise reduction schemes, which generally aim at improving the signal-to-noise ratio (SNR). They might be based fully or in parts on a simulation (model) of the principles of auditory perception, or might follow a purely technical-based approach.

All approaches have to obey the limits for the processing delay of the hearing aid between output and input signal (algorithmic filter delay plus hardware delay (e.g., for signal buffering)). Synchronization of audio information and visual information (lip reading) is maintained up to a delay of about 70 ms (see also Chapter 37). However, in case the hearing loss is small in specific frequency regions, both the direct sound and the processed signal are audible, and disturbing comb filter effects may occur. This effect is especially pronounced when listening to one's own voice. Based on the empirical results, it is generally agreed that the delay of a hearing aid should not exceed 10 ms to avoid all potentially detrimental effects [1]. This limit is reached in most FFT-based systems, whereas filterbank-based systems show lower delays between 6 and 2 ms, decreasing with frequency. An important consequence of the delay limit is that models of higher cognitive processes, e.g., processes involving short-term memory, cannot be used in the signal path in principle.

2 • Factors of Hearing Impairment and Rehabilitation Strategies

Table 1 indicates several factors of hearing impairment, which have been identified based on physiological and psychoacoustical evidence (Chapter 10 and [2–4]) and lists generic rehabilitation strategies applied to compensate for their perceptual consequences. For further reading, details on specific algorithms are given in [5,6] and in the references listed in Table 1.

The major components of current hearing aids are an amplification stage and a compressive automatic gain control (AGC) stage (or dynamic range compression stage) to compensate for the frequency-specific increase in hearing threshold level and the reduced dynamic range between threshold level and uncomfortable loudness level ("recruitment"). Several rules for

TABLE 1 Factors of hearing impairment and rehabilitation strategies

Factor	Perceptual consequence	Rehabilitation strategy	Refs.
Attenuation component	Loss of sensitivity; increased threshold level	Increase audibility by (a) frequency-specific amplification	[7,8]
		(b) frequency compression	[9,10]
Distortion component	(a) Loss of sensitivity; reduced dynamic range ("recruitment")	(a) Automatic gain control (AGC)	[11–13]
	(b) reduced frequency selectivity	(b) spectral enhancement	[14,15]
	(c) increased susceptibility to background noise	(c) noise reduction (see "neural component")	
Neural component	(a) increased susceptibility to background noise ("Cocktail-party effect")	(a) monaural noise reduction	[16,17]
		(b) directional microphones	[18]
	(b) impaired binaural capabilities	(c) beamformer	[19,20]
		(d) binaural noise reduction	[21,22]

individually fitting gain and compression parameters are used in practice [5]. Among the noise reduction schemes listed above, the first-order adaptive directional microphones [18] turned out to be the most successful strategies and are used in most current hearing aids for improved hearing in noise. Monaural noise reduction schemes [16] are also well established but are restricted to improving perceived noisiness of a noise condition, failing to improve speech intelligibility [23]. This limitation also holds for most published binaural noise reduction schemes [21].

In addition to the rehabilitation strategies, several technical limitations have to be addressed by DSP. In particular, feedback "howling" is a major problem because of the high acoustic gain and the close distance of microphone and speaker/receiver mounted in a single shell, giving rise to acoustic and mechanic feedback. DSP-based *feedback control* has been used to increase gain margins by about 10–15 dB in realistic conditions [24], but there is a demand for even more gain. As DSP increasingly offers specific processing schemes tailored to specific acoustic conditions (e.g., noise reduction for specific noise conditions like diffuse noise or stationary noise), an increasing demand for *automatic environment classification* [25] is observed.

Figure 1 shows a binaural hearing aid signal processing scheme, including the components listed above. Note that a fully binaural system which requires comparing signals from left and right ears is not commercially available to date because the

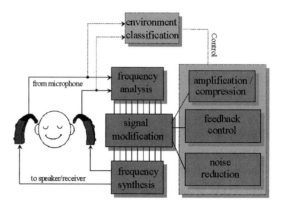

FIGURE 1 Block diagram of a generic binaural hearing aid signal processing scheme. Acoustic signals are picked up by microphones placed close to the ears, processed by a central processing device and played back at both ears. Processing includes frequency analysis and synthesis and several processing blocks (e.g., amplification and compression, feedback control, noise reduction schemes), which are controlled by automatic environment classification (e.g., activating noise reduction depending on estimated noise condition). Note that a fully binaural system as sketched here is currently not available in commercial hearing aids due to the lack of low-power wireless transmission.

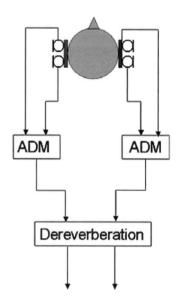

FIGURE 2 Generic binaural processing scheme combining two adaptive differential microphones (ADM), one at each ear, and a binaural dereverberation scheme, providing suppression of one directional noise source and of diffuse background noise and reverberation. The scheme provides a binaural output, i.e., the binaural cues are maintained and may be used by the residual binaural processing capabilities of the hearing impaired for localizing sound sources and exploiting binaural differences for noise reduction.

technology does not allow for signal transmission at the original data rate yet. Figure 2 sketches a promising combination of established processing schemes, i.e., the adaptive differential microphone [18] and a binaural dereverberation scheme [21], which possibly represents the current optimum that could be done in diffuse and reverberant noise conditions as soon as a binaural link is available.

3 • Limitations of Current Rehabilitation Strategies

Analyzing the performance of current hearing aids, further development should concentrate on two areas. First, AGC schemes use rather long adaptation time constants and a few frequency bands as compared to the normal hearing system, which provides a large number of overlapping bands and almost instantaneous amplitude compression [2]. Further research should be conducted to develop better simulations of the instantaneous spectro-temporal compressive processing implemented in the auditory system, which might be applied subsequently instead of the current AGC stages. Second, the benefit of noise reduction schemes, including published binaural schemes, is rather small and cannot compensate for degraded hearing in noise and reverberation associated with the most common types of hearing impairment. The normal hearing system, including the cognitive system, developed incredibly effective means of separating sound sources to ensure communication in difficult noise conditions. Some of the principles of the separation process have recently been experimentally revealed [26], but their simulation using computational auditory scene analysis (CASA) is still limited and not yet applicable to improve speech intelligibility in noise. Future research on noise reduction should therefore concentrate on exploiting CASA techniques. The properties of the hearing systems known so far suggest that the system exploits a priori knowledge on possible sound sources to identify sound objects and to extrapolate their temporal evolution in case part of the received acoustical information on these objects is ambiguous or even fully masked. Statistical filtering techniques (e.g., particle filters) incorporating dynamical models of possible sound sources [27] are therefore expected to play a major role in this line of research.

4 • Conclusions

Digital signal processing (DSP) is currently used to compensate for all aspects of hearing impairment and to overcome technical limitations of hearing aids. Increased user benefit of DSP has been shown in many studies. However, many of the digital processing schemes used in up-to-date hearing aids are not more than generalized versions of the analog schemes used before, and some hearing deficiencies are not satisfactorily rehabilitated by DSP yet, especially when hearing in noise and when reverberation is involved. Basic research needs to be pursued in order to reveal fundamentally new processing schemes for an improved rehabilitation of the aurally handicapped in difficult communication conditions, thereby fully exploiting the possibilities of DSP. The increasing knowledge on the cognitive processes involved in auditory processing is expected to play a major role in this line of research, requiring the development of algorithms which model the higher auditory functions and demanding much higher processing power of future speech processing devices.

Acknowledgments This preparation was supported by the Center of Excellence on Hearing Aid Technology (BMBF 01EZ0212). The author thanks B. Kollmeier and other members of the Medical Physics group for their helpful comments on earlier versions of this manuscript.

References

1. Agnew J and Thornton JM (2000). Just noticeable and objectionable group delays in digital hearing aids. J. Am. Acad. Audiol. 11, 330–336.
2. Moore BCJ (1995) Perceptual consequences of cochlear damage. Oxford University Press, Oxford.
3. Kollmeier B (1999) On the four factors involved in sensorineural hearing loss. In: Dau T, Hohmann V and Kollmeier B (Eds): Psychophysics, physiology and models of hearing. World Scientific, Singapore 211–218.
4. Kollmeier B (Ed) (1996) Psychoacoustics, speech and hearing aids. World Scientific, Singapore.
5. Dillon H (2001) Hearing aids. Boomerang Press, Turramurra, Australia.
6. Kates JM (1998) Signal processing for Hearing Aids. In: Kahrs M and Brandenburg K (Eds): Applications of digital signal processing to audio and acoustics. Kluwer, Dordrecht.
7. Plomp R (1994) Noise, amplification, and compression: Considerations of three main issues in hearing aid design. Ear Hear. 15, 2–12.
8. Turner C and Henry B (2002) Benefits of amplification for speech recognition in background noise. J. Acoust. Soc. Am. 112, 1675–1680.

9. Turner CW and Hurtig RR (1999) Proportional frequency compression of speech for listeners with sensorineural hearing loss. J. Acoust. Soc. Am. 106, 877–886.
10. Sakamoto S, Goto K, Tateno M and Kaga K (2000) Frequency compression hearing aid for severe-to-profound hearing impairments. Auris Nasus Larynx. 27(4), 327–334.
11. Stone MA, Moore BCJ, Alcantara JI and Glasberg BR (1999) Comparison of different forms of compression using wearable digital hearing aids. J. Acoust. Soc. Am. 106, 3603–3619.
12. Verschuure J, Maas AJJ, Stikvoort E, de Jong RM, Goedegebure A and Dreschler WA (1996) Compression and its effect on the speech signal. Ear Hear. 17, 162–175.
13. Jenstad LM, Seewald RC, Cornelisse LE and Shantz J. Comparison of linear gain and wide dynamic range compression hearing aid circuits: Aided speech perception measures. Ear Hear. 20(2), 117–126.
14. Baer T and Moore BCJ (1994) Spectral enhancement to compensate for reduced frequency selectivity. J. Acoust. Soc. Am. 95, 2992.
15. Yang J, Luo FL and Nehorai A (2003) Spectral contrast enhancement: Algorithms and comparisons. Speech Commun. 39(1–2), 33–46.
16. Ephraim Y and Malah D (1985) Speech enhancement using a minimum mean-square error log-spectral amplitude estimator. IEEE Trans. Acoust., Speech, Signal Process. ASSP 33(2), 443–445.
17. Tchorz J and Kollmeier B (2003) SNR estimation based on amplitude modulation analysis with applications to noise suppression. IEEE Trans. Speech Audio Process. 11(3), 184–192.
18. Elko GW and Pong A-TN (1995) A simple adaptive first-order differential microphone. IEEE ASSP Workshop on Applications of Signal Processing to Audio and Acoustics, New Paltz, NY, USA.
19. Campbell DR and Shields PW (2003) Speech enhancement using sub-band adaptive Griffiths – Jim signal processing. Speech Commun. 39(1–2), 97–110.
20. Widrow B and Luo FL (2003) Microphone arrays for hearing aids: An overview. Speech Commun. 39(1–2): 139–146.
21. Wittkop T and Hohmann V (2003) Strategy-selective noise reduction for binaural digital hearing aids. Speech Commun. 39, 111–138.
22. Bodden, M (1993) Modeling human sound-source localization and the cocktail-party-effect. Acta Acustica 1, 43–56.
23. Marzinzik M and Kollmeier B (2003) Predicting the subjective quality of noise reduction algorithms for hearing aids. Acta Acustica/Acustica, 89, 521–529.
24. Greenberg JE, Zurek PM and Brantley M (2000). Evaluation of feedback-reduction algorithms for hearing aids. J. Acoust. Soc. Am. 108, 2366–2376.
25. Nordqvist P and Leijon A (2004) An efficient robust sound classification algorithm for hearing aids. J. Acoust. Soc. Am.115(6), 3033–3041.

26. Bregman AS (1990). Auditory scene analysis: The perceptual organization of sound. MIT Press, Cambridge, MA.
27. Nix J, Hohmann V (2007) Combined estimation of spectral envelopes and sound source direction of concurrent voices by multidimensional statistical filtering. IEEE Trans. Audio Speech and Lang. Proc. 15, 995–1008.

PART III
PSYCHOACOUSTICS

Hugo Fastl

Department of Human Machine Communication, Munich University of Technology, Munich, Germany

15 Methods for Psychoacoustics in Relation to Long-Term Sounds
Seiichiro Namba and Sonoko Kuwano 215
Introduction • Methods for Psychoacoustics • Information Processing along the Temporal Stream – The Method of Continuous Judgment by Category • Information Processing in Hearing – Summary

16 Masking and Critical Bands
Bernhard U. Seeber 229
Introduction • Description of Masking Effects • Critical Bands in Psychoacoustical Experiments • Approximations of Auditory Filter Bandwidth • Limits of the Critical Band Concept • Modeling Auditory Masking • Conclusions

17 Aspects of Modeling Pitch Perception
Ernst Terhardt 241
Introduction • Basic Issues • Pitch of Tonal Stimuli • Pitch of Random Stimuli

18 Calculation of Loudness for Normal and Hearing-Impaired Listeners
Josef Chalupper 251
Introduction • Model • Conclusions

19 Psychoacoustical Roughness
Peter Daniel 263
Introduction • Roughness in Dependence of Signal Parameters • Roughness Calculation

15

Methods for Psychoacoustics in Relation to Long-Term Sounds

Seiichiro Namba and Sonoko Kuwano

Osaka University, Osaka, Japan

1 • Introduction

It is well known that auditory signals are effectively transferred to the inner ear as the acoustic energy through the outer ear canal, the ear drum, and the ossicles in the middle ear. It has been made clear using Moessbauer method or using laser radiation that sharp tuning of frequencies is made by active control system on the basilar membrane in the inner ear. Also each cochlear nerve fiber has its characteristic frequency to which it selectively responds. This makes it possible to transfer the selected clues of the frequency to the auditory cortex.

The information of the sound intensity is transferred in terms of the rate of generation of action potentials of auditory nerves. Superior olivary nucleus and inferior colliculus are considered to contribute to the sound localization by detecting the temporal difference in the information transmitted through both ears [1,2].

It is being made clear that the physical properties of sounds such as frequency and intensity and their difference between two ears are processed around peripheral areas in hearing system. However, the physiological mechanisms are not evident how the quantitative evaluation of loudness, the identification of pitch or tonality, or the recognition of special localization or movement of sound sources in our surroundings are achieved. Moreover, it is not explained yet how and where the complex information such as speech, music, and environmental noises is recognized in the auditory system.

Recent progress in brain science in visual perception shows how effectively the information of shapes, colors, and movements is processed and discriminated separately and individually in each specific area in the brain. However, it is still unclear how the information processed separately is integrated into a whole pattern in the central cortex [3].

Since even in visual perception, which is thought to have made a big progress in brain science, there is a long way to solve all the details of the mechanism of perception and cognition from physiological viewpoint, it will take much time to find how all the information in hearing is processed physiologically. At present when there is a black box between physiological findings and information processing of complex phenomena, the methods of psychoacoustics are useful tools to estimate the mechanism of information processing. Even if the brain science makes a drastic development in future, psychoacoustical approach is inevitable in order to find the laws in the subjective world in hearing.

2 • Methods for Psychoacoustics

2.1 Methods for Psychophysics

Psychophysical studies of pitch and loudness of steady-state sounds have a history for more than 100 years. Various thresholds such as absolute threshold of sounds and differential limens of frequency and intensity and point of subjective equality (PSE) of sounds of different frequencies have been measured using conventional psychophysical methods [4]. Psychophysical methods have also been used for the measurement of more complex phenomena than thresholds, such as signal detection under the presence of masking sounds. Information that seems to be processed in the peripheral auditory system can be measured using conventional psychophysical methods, and the relation is being made clear between physical values of sounds and the subjective responses to them. A brief outline of the conventional psychophysical methods is introduced taking an experiment to measure PSE of pitch or loudness as an example.

2.1.1 Method of Adjustment

This is the method where participants are required to control the frequency or the sound pressure level of the comparison stimulus so that both standard and comparison sounds are equally perceived. Sometimes the experimenter controls the stimulus according to the responses of participants. This method can easily be understood by participants, and usually it is fairly easy for participants to judge the stimuli with this method. On the other hand, it has a disadvantage that there is little reproducibility in this method.

2.1.2 Method of Limits

This is the method where the stimulus is presented by varying the stimulus parameter, such as frequency or sound pressure level, in a fixed step from low to high in ascending series or from high to low in descending series. Participants are required to compare the stimulus pair at each presentation and judge which sound is higher or louder in each pair. All the procedures are settled by the experimenter before starting the experiment and usually the experiment can be conducted in a short period. In experiments, participants can easily guess to which direction the stimulus is varied. This may possibly cause a bias in the judgment.

2.1.3 Method of Constant

This is the method where the stimulus pairs decided by a preliminary experiment are presented several times in random order and participants are required to judge each pair. This method has an advantage that it is free from various constant errors. At the same time, however, a disadvantage of this method is that it takes much time to obtain the data since many stimulus pairs must be judged.

2.1.4 Adaptive Procedure

In the above-mentioned psychophysical methods, the stimuli are presented in the same way as preliminarily decided. In the adaptive procedure, the stimuli to be presented are decided depending on the responses of the participants. The stimulus presentation is controlled by a computer. This procedure has a merit that useless trials can be skipped and it can shorten the time for experiments. On the other hand, experiments using adaptive procedures cannot be done without a computer for each participant and it is not easy to do experiments with a large group of participants together. There are several methods of adaptive procedure, such as "up and down method", "up and down transformed method", "parameter estimation by sequential testing", "maximum likelihood sequential procedure", and "hybrid adaptive procedure". There is a special issue of the journal *Perception & Psychophysics* on adaptive procedure [5].

2.1.5 Signal Detection Theory

There are sensory process and decision process when people respond to stimuli. These two processes are in the inner world of people and not visible from the outer world. It is considered that when a stimulus is given, sensory information is transferred to the decision process and a decision is made in reference to expectation, prediction, attitude, and knowledge of participants. In psychophysical methods, the decision process is controlled as much as possible in order to find the relation between stimuli and sensation. In the signal detection theory, it becomes possible to estimate the criterion of the participant's judgments through analysis of experimentally obtainable hit and false alarm rates [6].

2.2 Limitation of Psychophysical Methods

Psychophysical methods are a useful tool, but it has a disadvantage that in principle it is applied only to sounds of short duration. A pair of a standard stimulus and a comparison stimulus is prepared in order to measure threshold or PSE using psychophysical methods. A pair of the sounds is presented successively to participants, and after listening to the pair of the sounds, they are required to compare the sounds and judge whether they are subjectively equal or not. In this case, if the duration of the sounds is long, the impression of the standard sound may become vague after the presentation of the comparison sound and it may become difficult to compare the two sounds. The measurement of threshold is aimed to measure the sensitivity of auditory sensation. However, at the moment when the judgments are made after the presentation of the two sounds, the standard sound belongs to memory and the latter part of the comparison sound belongs to perception. This means that two phenomena that belong to different worlds are compared. This problem remains even if the orders of presentation of the standard and comparison stimuli are reversed. Therefore, generally, the duration of sounds used in psychophysical measurement is limited within about 1 s.

Such problems do not occur when psychophysical methods are applied to the measurement of visual perception. Both standard stimulus and comparison stimulus can be presented simultaneously in a visual field and compared. There is no effect of time order or the duration of stimuli. If necessary, successive comparison is also possible in visual perception taking the afterimage into consideration. It is necessary to control constant errors such as the effect of spatial position where the stimuli are presented, time order error, etc. If the experiment is conducted under the well-controlled condition, psychophysical methods can be applied to visual stimuli with complex patterns.

In hearing, the information is conveyed along the temporal stream. Since the information is expressed in the temporal variation, sounds have complex temporal patterns such as speech, music, and environmental noises and it takes time to be transmitted. The duration of an event sometimes becomes several 10 s as can be found in the case of a phrase of speech, a melody of music, or an occurrence of event such as a flyover of an aircraft.

There is a limitation in the application of conventional psychophysical methods to sounds of long duration since it is related to the interaction between perception and memory. Scaling methods such as rating method and magnitude estimation can be applied to sounds of fairly long duration.

2.3 Scaling Method

There are several scaling methods such as rating method, the method of paid comparison, the method of rank order, and magnitude estimation [7]. Among them, absolute judgments of a stimulus without comparing two stimuli can be made using rating method and magnitude estimation. This suggests that it is possible to measure the impression of fairly long-term sounds using these methods. The outlines of these scaling methods are introduced.

2.3.1 Rating Method

The judgment using rating method is made using category scales. The scales are indicated in various forms, e.g., numerical scales from 1 to 7 and graphical indication using the length of a line. Since the judgment using rating scale is easy, this method can be used for the measurement of wide aspects of our daily life experience; they are, for example, the judgments of aesthetic aspects in experimental aesthetics, the impression in daily life using questionnaires in social surveys and in clinical test methods as well as the impression of visual and auditory stimuli. It is also possible to obtain the judgment of long-term experience in questionnaires [8].

2.3.2 Magnitude Estimation

Magnitude estimation is a method that Stevens [9] used for verifying the power law between physical properties of stimuli and attributes of sensations. In the absolute magnitude estimation, participants are required to judge the impression quantitatively correlating their impression to randomly presented stimuli with numerals. This method can be applied not only to the measurement of sensations but to stimuli whose physical properties are not defined such as the preference of wrist watches [10].

It is true that scaling methods can be used for measuring the impression of long-term stimuli and experiences. However, in the case of long-term sounds with complex temporal variation, such as music, speech, and environmental noises, there is a problem that it is unclear whether the judgment is based on the overall impression of long-term sounds or whether some specific portions of the sounds have stronger effect on the overall judgment than the other portions.

In order to make clear the laws between factors involved in the stimuli and subjective responses to them, it is not easy to choose what kind of stimuli to be used: short-term sounds that include little information, but can be easily controlled, or long-term sounds that reflect our daily life experience and include much information, but may be difficult to obtain clear effect of each factor.

Scaling methods can also be applied to the judgments of comparison. The method of paired comparison and similarity judgment in multidimensional scaling can give us useful information. A slight difference among stimuli, especially, can be found and the sensory scale can be obtained by comparison. However, there is a limitation that they cannot be applied to stimuli of long duration.

3 • Information Processing along the Temporal Stream – The Method of Continuous Judgment by Category

As described above, sounds convey information along temporal stream. When the method of measurement of psychoacoustics is considered from the viewpoint of information processing, it is not desirable to divide the long-term event of sounds into short segments. This may destroy the frame of reference as a whole event. Conventional psychophysical methods are useful tools to find psychophysical laws between pitch and frequency or between loudness and intensity using short-term sounds. However, these laws show only a part of the information processing in peripheral regions of our auditory system.

It is possible to measure the impression of long-term temporally varying events such as a meaningful phrase of speech and music or a flyover of an airplane using conventional scaling methods. However, since components of these sounds are complex and varying with time, it is not easy to find a clear relationship between stimulus parameters and subjective responses.

The authors have developed a new method called "the method of continuous judgment by category" in order to obtain the instantaneous impression of temporally varying sounds [11–15]. This method enables us to measure the subjective responses along the temporal stream without destroying the natural steam and meaning of the sounds.

The idea of this method comes from piano performance. Piano performance itself is a kind of response by a pianist, and it is an expression of the musical image in the mind of the pianist by the motion of fingers. This suggests that it would be possible to measure the impression of temporally varying sounds continuously by pressing keys on a keyboard just as is done in the piano performance. It is not seldom that some passages of piano music are played in a speed faster than the critical identification rates [16], i.e., the shortest time where the temporal order of sounds can be identified. Therefore, after an exercise for the smooth motions of fingers such as "Hannon Album" in piano training, it would be possible to measure the impression of sounds continuously as quick as the critical identification rate using a computer keyboard as a response box.

As described before, when only the overall impression of long-term sound is measured using a conventional category rating scale, it is not clear which portions of the long-term sound contribute to the overall impression. In the method of continuous judgment by category, instantaneous impression is measured asking participants to respond continuously by pressing a key. In addition to the continuous judgment, participants are usually required to judge the overall impression of the sound in a questionnaire conducted after the continuous judgment. The relation between the instantaneous impression and the overall impression gives us a clue to find how the long-term sound is processed and the overall impression of the sound is formed from each portion.

Overall impression and instantaneous impression are not independent of each other. The overall impression may be determined by giving a kind of subjective weight to the impression of each portion of the whole stimulus. The impression of each portion is also affected by the frame of reference formed by the overall context and past experience [17].

In the design of conventional psychological experiments, stimuli are presented in random order in order to avoid constant errors. If we assume that a long-term sound, such as music, is divided into short segments and the segments are randomly arranged, it can be easily understood how strange it is to judge the randomly rearranged sound stream continuously. It is important to make clear in which condition the method of continuous judgment is effectively used. That is, this method is used for measuring the impression of the sounds that convey information along temporal steam in daily life such as speech, music, and environmental noises.

3.1 Introduction

When the instantaneous impression of loudness of a temporally varying sound is judged, for example, participants are required to judge the impression of the loudness at each moment using seven categories from very soft to very loud and press a corresponding key. They need not press the key if their impression does not change. The impression registered on the monitor will remain the same. When their impression of loudness changes, they have to press the appropriate key.

An example of the relation between instantaneous judgments and sound level in the case of traffic noises is shown in Figure 1 [18]. The instantaneous judgments were sampled every 100 ms and the judgments of eight participants were averaged taking the reaction time into account. High correlation can be seen between the instantaneous judgments and sound levels ($r = 0.969$). Another example of the analysis is shown in Figure 2. The instantaneous judgments for the sound level of every 1 dB step were averaged. The average of instantaneous judgments with the standard deviation is plotted against sound levels. Also high correlation can be seen ($r = 0.992$).

This method has many advantages as follows:

1. The judgment is easy for participants.
2. This method can be applied to a fairly long stimulus.
3. Since all the subjective responses are stored in a computer with the information of the timing they press the key, various analyses can be done after the experiment is over.

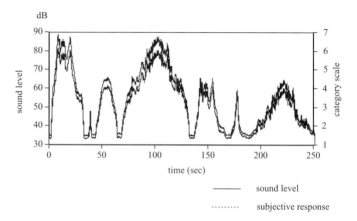

FIGURE 1 An example of the relation between sound level and instantaneous judgments sampled every 100 ms.

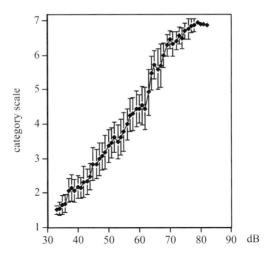

FIGURE 2 An example of the instantaneous judgments averaged for sound level of every 1 dB.

4. The relation between instantaneous judgments and physical values can give clues for understanding temporal characteristics of hearing.
5. The relation between overall impression and instantaneous impression can be compared and this makes it possible to investigate how the overall impression is determined by the impression of each portion.
6. In conventional psychophysical methods, participants have to pay attention to the stimulus to be judged. On the other hand, in experiments using this method, participants judge the impression along the temporal stream, i.e., in the same context, without paying attention to a specific portion of the long sound, and the impression of each portion of the sound can be obtained as in their daily life situations.
7. Multidimensional judgment is also possible using the method of continuous judgment by selected description [14,19].

3.2 Examples

3.2.1 Estimation of Psychological Present

In the continuous judgments, the judgment may possibly be affected by the preceding portions of the stimulus as well as the immediate portion. In order to find the duration of the preceding portions which may affect judgment at each moment, the sound energy was averaged for various durations and correlated with the instantaneous judgment [12]. From this it was found that the instantaneous judgment showed the highest correlation when the preceding sound energy was averaged for 2.5 s. This value may reflect the psychological present [20]. The physical present is a point, but the psychological present is not and may have a certain duration during which subjective impressions are integrated and perceived as a unit. If the coefficient of correlation between the integrated physical values and the instantaneous judgments is taken as an index of the psychological present, the integration time which shows the highest correlation with the instantaneous judgment may represent the duration of the psychological present.

3.2.2 Relation Between Overall Judgment and Instantaneous Judgment

For the evaluation of long-term temporally varying sounds, it would be helpful to measure instantaneous impression and overall impression as well. The time interval between the presentation of sounds and the overall judgment should be chosen carefully. As time passes, the impression of the prominent portion becomes stronger and that of the less prominent portion becomes weaker [12]. It is natural that loud sounds or loud portions easily become prominent, but it was found that even soft sounds such as bird twittering can be prominent and contribute to the overall impression [21]. It is interesting and important to find the factors which contribute to the overall impression.

3.2.3 Habituation

With the lapse of time, subjective impression of sounds may possibly change. Scharf [22] has indicated that with low-level sounds and high-frequency sounds loudness adaptation occurs, and the impression of loudness decreases when the sounds are presented for a longer time.

Habituation to sounds is the phenomenon where, when a certain stimulus is repeatedly presented to the auditory organ, the response to the stimulus gradually diminishes and finally no response will occur. Habituation is different from adaptation or fatigue in the point that it is easy to overcome habituation when attention is directed to the stimulus. Habituation has a positive meaning to help people cope with the environment. It is not easy to measure habituation by conventional psychological methods where participants have to pay attention in order to judge the stimulus. In experiments using the method of continuous judgment by category, participants are not necessarily forced to pay attention to the specific stimuli all the time. This makes it possible to approach a measurement of habituation.

In an experiment [13], participants were engaged on a mental task and they were highly motivated to do the task. While occupied with the mental task, they were asked to judge the noisiness of sounds continuously, using the method of continuous judgment by category. The sounds were of various kinds, such as road traffic noise, voices of street vendors, TV talk, and sounds from a washing machine and a vacuum cleaner. The duration of an experimental session was about one hour and a half. In this experiment when participants did not respond to the sounds for 30 s, they were given a signal on a monitor to respond. Participants did not always respond to the sounds in spite of the warning signal when they were devoted to the mental task. The length of time when they did not respond even if the warning signal was presented was used as an index of habituation. There was a high correlation between two trials in the habituation period for each participant. It was also found that the group of participants who showed long period of habituation said, in answer to a questionnaire, that they became habituated to noise fairly easily in daily life situations.

Another index of habituation is the number of responses made [23,24]. When participants pay attention to sounds, they respond to each change in their impression of sounds. On the other hand, when they are habituated to sounds, they are not conscious of the change and do not respond.

3.2.4 Trade-Off Effect in Subjective Responses among Stimulus Conditions

Taking the interior noise of a car, as an example, various sounds are audible in a car while driving, such as engine noise, air conditioner noise, winker signals, sounds from wipers, and road noise. The loudness of inside car noise while driving was judged using the method of continuous judgment by category [25]. The visual scenery while driving was also presented with the sound. This made the participants feel as if they were driving and could prevent participants from paying special attention to sounds. After the experiment, the averages of instantaneous judgments for each sound source were compared with background sounds. It was found that road noise when driving on a rough road is judged louder than the other portions in the same sound level conditions, which suggests that some countermeasures for noise due to road conditions are needed. It was also found in this experiment that air conditioner noise was judged louder than the other portions. This suggests that it is important to improve the air conditioner noise in order to make the cabin of a car comfortable in the case of the car used for the experiment.

3.3 Software

When the method of continuous judgment by category was developed in the 1970s, personal computers were not available. The responses of continuous judgments were recorded using a level recorder and the data at every 100 ms were measured manually. Then the data were punched on IBM cards, and using a large-size computer, the coefficient of correlation between physical parameters of stimuli and instantaneous responses was calculated. It was very laborious work.

Some years later, the experiments were conducted using a personal computer. The software was made by the authors with MS-DOS. This made it possible to analyze data from various viewpoints.

Now the new software for the method of continuous judgment by category for computers with OS of Windows is available [26]. The data can be analyzed very quickly, and the results can be obtained immediately after the experiment is over. The examples shown in Figures 1 and 2 were the results of the analysis using this software. It is also possible to analyze each portion of the sound separately. This software has Japanese and English versions and is commercially available.

4 • Information Processing in Hearing – Summary

In conventional psychophysical methods, usually a sound of short duration is used and absolute threshold or differential limen of frequency or intensity is measured. These measurements provide us valuable information about the basic characteristics of hearing. The signal processing in hearing in the peripheral area is becoming clear combining the results of psychophysical experiments and physiological findings.

However, temporal information processing is not considered in conventional psychophysical experiments, and important aspects of hearing to convey information along temporal stream are neglected. It is important to make clear how the information is processed in hearing along temporal stream in order to understand the mechanism of hearing to temporally varying sounds such as music, speech, and environmental sounds.

No one may deny the importance of ecological validity in hearing research. There is no one who underestimates the importance of precise control of stimulus conditions in experiments either. Though both aspects are important, it is difficult to fulfill both at the same time.

In this chapter, the method of continuous judgment by category is introduced that makes it possible to control physical parameters of sounds continuously and measure quantitatively the instantaneous impression of temporally varying sounds along the temporal stream. The main findings using this method are introduced in relation to information processing in hearing; they are the estimation of psychological present (i.e., psychological time window), the relation between overall impression and instantaneous impression, habituation to noise and trade-off effect among stimulus conditions. A newly developed software for the experiment using the method of continuous judgment by category is also introduced.

The information processing in hearing covers wide areas from short-term process to long-term experience by adjusting the frame of reference flexibly to the situation. It is expected that the information processing in hearing will be made clear in future investigations along the temporal stream.

References

1. K. Maki, K. Ito and M. Akagi, "Time–frequency response patterns of nerve firings in primary stage of auditory system", *The Journal of the Acoustical Society of Japan*, 59 (1), 52–58 (2003).
2. B. C. J. Moore, *Hearing*, 5th ed., Academic Press, 2003.
3. R. L. Gregory, *Eye and Brain: The Psychology of Seeing*, Oxford University Press, 1998.
4. G. A. Gescheider, *Psychophysics: Method, Theory, and Application*, 3rd ed., Lawrence Erlbaum Associates, Inc., 1997.
5. N. A. Macmillan (Ed.), "Psychometric functions and adaptive methods", special issue, *Perception & Psychophysics*, 63 (8), 1277–1455 (2001).
6. D. M. Green and J. A. Swets, *Signal Detection Theory and Psychophysics*, Wiley, 1966.
7. S. Namba and S. Kuwano, *Method of Psychological Measurement for Hearing Research*, Corona-sha, 1998.
8. S. Namba, J. Igarashi, S. Kuwano, K. Kuno, M. Sasaki, H. Tachibana, A. Tamura and Y. Mishina, "Report of the Committee of the Social Survey on Noise Problems", *The Journal of the Acoustical Society of Japan (E)*, 17, 109–113 (1996).
9. S. S. Stevens, *Psychophysics: Introduction to Its Perceptual, Neural and Social Prospects*, Wiley, 1975.
10. T. Indow, "An example of motivation research applied to product design", *Chosa to Gijutsu*, 102, 445–460 (1961).
11. S. Namba and S. Kuwano, "The relation between overall noisiness and instantaneous judgment of noise and the effect of background noise level on noisiness", *The Journal of the Acoustical Society of Japan (E)*, 1, 99–106 (1980).
12. S. Kuwano and S. Namba, "Continuous judgment of level-fluctuating sounds and the relationship between overall loudness and instantaneous loudness", *Psychological Research*, 47, 27–37 (1985).
13. S. Namba and S. Kuwano, "Measurement of habituation to noise using the method of continuous judgment by category", *Journal of Sound and Vibration*, 127, 507–511 (1988).

14. S. Namba, S. Kuwano and M. Koyasu, "The measurement of temporal stream of hearing by continuous judgments – in the case of the evaluation of helicopter noise", *The Journal of the Acoustical Society of Japan (E)*, 14, 341–352 (1993).
15. S. Kuwano, "Continuous judgment of temporally fluctuating sounds", in H. Fastl, S. Kuwano and A. Schick (Eds.), *Recent Trends in Hearing Research*, BIS, 1996, pp. 193–214.
16. R. Teranishi, "Critical rate for identification and information capacity in hearing system", *The Journal of the Acoustical Society of Japan*, 33, 136–143 (1977).
17. H. Helson, *Adaptation Level Theory*, Harper & Row, 1964.
18. S. Namba and S. Kuwano, "The continuous judgment of traffic noises from mixed sources" (unpublished data in 2003).
19. S. Namba, S. Kuwano, T. Hato and M. Kato, "Assessment of musical performance by using the method of continuous judgment by selected description", *Music Perception*, 8, 251–276 (1991).
20. P. Fraisse, *Psychologie du Temps*, Press University, 1957 (Japanese translation by Y. Hara, revised by K. Sato).
21. S. Kuwano, S. Namba, T. Kato and J. Hellbrueck, "Memory of the loudness of sounds in relation to overall impression", *Acoustical Science and Technology*, 24 (4), 194–196 (2003).
22. B. Scharf, "Loudness adaptation", in J. B. Tobias and E. D. Schubert (Eds.), *Hearing Research and Theory*, Vol. 2, Academic Press, 1983, pp. 1–57.
23. S. Namba, S. Kuwano and A. Kinoshita, "Measurement of habituation to noise", *Proceedings of the Autumn Meeting of the Acoustical Society of Japan*, 759–760 (1994).
24. N. Hato, S. Kuwano and S. Namba, "A measurement method of car interior noise habituation", *Transactions of Technical Committee on Noise, The Acoustical Society of Japan*, N95-48, 1–4 (1995).
25. S. Kuwano, S. Namba and Y. Hayakawa, "Comparison of the loudness of inside car noises from various sound sources in the same context", *The Journal of the Acoustical Society of Japan (E)*, 18, 189–193 (1997).
26. S. Namba, S. Kuwano, H. Fastl, T. Kato, J. Kaku and K. Nomachi, "Estimation of reaction time in continuous judgment", *Proceedings of the International Congress on Acoustics*, 1093–1096 (2004).

16

Masking and Critical Bands

Bernhard U. Seeber

Auditory Perception Lab, Department of Psychology, University of California at Berkeley, 3210 Tolman Hall #1650, Berkeley, CA 94720-1650, USA, e-mail: bernhard@bseeber.de

1 • Introduction

Auditory masking is often observed in everyday life: While walking on a street a passing truck can disrupt our conversation. Likewise, one would not want a cell phone to ring in quiet parts of a classical concert but it might not even be heard in a loud rock-concert. Components of one sound interact with components of another sound similar in frequency and time and render them inaudible. This is referred to as masking. Partial masking can also occur: Components are not inaudible, but their loudness is reduced (c.f. Chapter 18). This chapter first introduces the frequent case of masking by one tone on another for simultaneous and successive presentation.

Many spectral masking effects can be described by comparing signal power, integrated over a certain frequency region, with and without the probe present. Detection of changes in energy in the bandpass-filtered signal forms the basis of the critical band concept. The critical band (CB) is defined as the rectangular bandwidth of the filter. Section 3 presents a collection of experiments which reveal the influence of critical band filters.

It turns out that CBs observed in psychophysical masking experiments grow in the same way as the physiologically measured distance on the basilar membrane or the psychophysical scales of just-audible pitch steps or of ratio pitch. Scales of psychophysically derived CBs are presented in Section 4 whereas Section 5 discusses the limits of the CB-concept. The last section gives a brief introduction to models of masking.

2 • Description of Masking Effects

2.1 Simultaneous Masking

In a simultaneous masking experiment the level of a probe signal is varied in the presence of a masking sound to find the threshold at which the probe is just audible. Three variations for presenting the results are common: (1) *tuning curves* describe the level of a masker necessary to mask a probe of fixed frequency and low level, (2) *masking growth functions* show the dependence of probe level on masker level for fixed probe and masker frequencies, and (3) *masking patterns* depict the level of probes of different frequencies that are just audible in the presence of a masker of fixed level and frequency.

Figure 1 presents *masking patterns* of a tonal masker for simultaneously presented probe tones. It is immediately evident that the level of the test tone (L_T) has to be raised to assure audibility as its frequency (f_T) approaches that of the masker (f_M), i.e., a tonal masker produces more masking in its spectral vicinity. Masking even occurs for very low masker levels (L_M) but the effect is spread widely across frequencies for higher levels.

The slopes of the masking pattern depend heavily on masker level. To illustrate this non-linearity Figure 1 also shows masking patterns mirrored at the masker frequency. For tonal maskers of a level $L_M \approx 40$ dB SPL the pattern is almost symmetrical. If the masker level is lower than 40 dB SPL the pattern tilts so that the low-frequency slope is shallower than the high-frequency slope. For masker levels above 40 dB SPL the opposite occurs: The low-frequency slope becomes somewhat steeper whereas the slope on the high-frequency side becomes far more shallow. Masking spreads out to frequencies far above the masker frequency, which is called the *upward spread of masking*. The excess in masking between the mirrored high- and the low-frequency side is labeled "A" in Figure 1. Additionally, the frequency of maximum masking can shift upward from the masker frequency.

The way masking (L_T) grows with increasing masker level (L_M) can also be captured from Figure 1. At $f_T = f_M = 1,000$ Hz the different masking patterns are about 10 dB apart – equal to the increase in masker level. More precisely, masking increases from

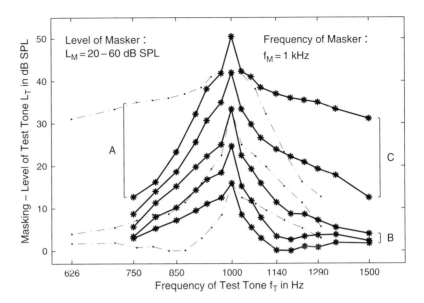

FIGURE 1 Tone-on-tone masking patterns for a simultaneous masker at $f_M = 1{,}000$ Hz with a level $L_M = 20\text{--}60$ dB SPL in 10 dB steps. The patterns for maskers at 20, 40 and 60 dB SPL are also plotted mirrored in frequency on the Bark scale (Section 4.1) to emphasize the non-linearity of the pattern (*dashed dotted line*). Data adopted from [28].

15.9 to 50.5 dB over a 40 dB change of L_M from 20 to 60 dB SPL. The small deviation from a linear increase is called the *"near miss" to Weber's law* and shows a ratio of 34.6 dB/40 dB ≈ 0.87.

On the high-frequency side masking shows a strong non-linear dependence on masker level. Unless the masker level exceeds a certain threshold no masking is present. This is labeled "B" in Figure 1 and evident for $f_T = 1{,}500$ Hz and $L_M \leq 40$ dB SPL. If the masker level exceeds this threshold masking occurs and it grows far stronger than linear with a ratio of 2–3 on average. In Figure 1, depicted with "C", masking increases from 12.5 to 31.1 dB SPL for a 10 dB change in masker level from 50 to 60 dB SPL. This equals a masking growth ratio of 1.9. Masking can even grow with a ratio of 6, i.e., for a 1 dB increase in masker level the probe must be raised by 6 dB.

The strong non-linear effect of the upward spread of masking can be exploited in audio compression systems. High compression ratios are possible for known playback levels as few strong components can produce masking over a wide frequency range. It should be mentioned that the measurement of tone-on-tone masking patterns is complicated by beats between the tones. Difference tones originating in the non-linearity of the basilar membrane can be masked with noise which was done to obtain the data in Figure 1 [13,24,27,28].

2.2 Simultaneous Masking by a Narrow Band of Noise

A narrow band of noise appears to produce stronger masking and more regular masking patterns than a single tone. Pronounced differences occur at the low-frequency slope of the masking pattern of the noise masker which appears less steep and less dependent on level with a constant slope of about 27 dB/Bark. This can be attributed to additional masking produced by difference tones in the auditory system. The tip region is broadened by the bandwidth of the noise and the influence of temporal beats is reduced. The high-frequency side shows similar upward spread of masking for noise and tonal maskers. Several tones spaced on and between the corner frequencies of the noise are sufficient to reproduce the masking patterns of the noise [5,13,27].

2.3 Temporal Masking Effects

Forward masking or *postmasking* is measured with a brief probe following the masker. It shows the decay of masking at the end of the masker. Forward masking is a non-linear effect depending on the masker frequency, its duration and level, the time between masker and probe and the probe frequency. Additional nearby components can reduce forward masking (suppression).

Tuning curves show the masker level necessary to mask a probe of fixed frequency and level. They are thought to resemble the shape of auditory filters as the listener is likely to always use the same filter to detect a faint probe of fixed frequency. Figure 2 compares simultaneous and forward masking tuning curves. At first sight forward masking tuning curves form a more regular pattern due to the absence of beats and difference tones. Except for the very tip region of the tuning curve the lower and upper slopes are steeper in forward masking. This indicates a higher frequency selectivity in forward than in simultaneous masking which is attributed mostly to suppression.

Forward masking will last only a few milliseconds for very short masker durations, but pronounced forward masking occurs if the masker duration is longer than 200 ms. Forward masking decays little within 5 ms after the end of the masker and ceases after about 100–200 ms relatively independent of masker level. This requires a faster decay of masking for higher masker levels which is reflected in a less than linear growth of masking with increasing masker level at the masker frequency ($f_M = f_T$). While masking increases about linearly with masker level in simultaneous masking, the growth ratio becomes smaller with

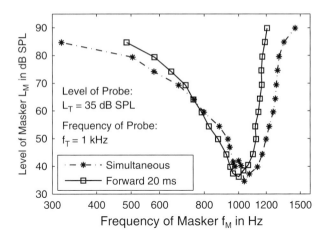

FIGURE 2 Psychophysical tuning curves for simultaneous and forward masking for a brief probe tone at $f_T = 1,000$ Hz with a level of $L_T = 35$ dB SPL. The tuning curves represent the level of a tonal masker required to keep the probe tone at the threshold of audibility. Data adopted from Figure 4 in [22].

increasing delay time in forward masking: 0.82, 0.5 and 0.29 for 0, 10, 30 ms delay. However, instead of showing the upward spread of masking for $f_T > f_M$, a linear growth of masking with increasing masker level is seen in forward masking [6,15].

In *premasking* or *backward masking* the probe precedes the masker. Although the masker appears later it can mask a preceding low-level probe within about 20 ms, but the effect is small and subject to high variance.

The tuning curves in Figure 2 show strong frequency selectivity. In the following section the idea of critical bands is developed based on the frequency selectivity seen in several psychoacoustical experiments.

3 • Critical Bands in Psychoacoustical Experiments

3.1 Band Widening

If the bandwidth of noise with constant spectral density is widened, the threshold of a tone masked by the noise will increase markedly less beyond a certain noise bandwidth, even though the overall level of the noise still increases. It seems that the noise power is spectrally integrated toward masking the tone up to this "critical" bandwidth and the energy outside this frequency band is disregarded. The band-widening experiment is the original measurement of the critical bandwidth [7]. Measurement difficulties arise from (1) the inherent modulation of the narrow-band noise masker which forms an additional detection cue besides energy, and (2) from the limited contribution of energy from the slopes of the non-rectangular auditory filters (e.g., [10,13,17]).

In a variation of the band-widening experiment a complex harmonic tone is detected as a function of it's number of harmonics. Since every doubling of the number of harmonics increases the level of the complex by 3 dB the detection threshold decreases by 3 dB if components fall within a CB, but it remains fairly constant if components are processed outside the CB. As CB-filters are not of rectangular shape it can be difficult to estimate the corner frequency. The experiment can be used to measure CBs at different levels by playing *uniform masking noise* as a background masker. Uniform masking noise is filtered to produce the same masked threshold at all frequencies [27].

The observation of critical bands likely stemmed from another effect: If the bandwidth of white noise is increased but the overall level is kept constant, the loudness will remain constant up to the critical bandwidth, but increase thereafter ([27], see Chapter 18).

3.2 Masking in Spectral Notches

If the ear separates signals into critical bands then the threshold of a tone should be less affected if flanking energy can be processed separately in adjacent CBs. In the *two-tone masking* experiment the spectral separation of two flanking tones is varied around a narrow band of noise [27]. The threshold for detecting the noise is low as long as the flanking tones are outside the CB and it rises when the tones enter the CB of the noise. Difference tones complicate the measurement.

One way to reduce the influence of difference tones is to use flanking bands of noise as a masker and a tone as the probe. The method produces slightly smaller CBs than the above methods. The *notched-noise method* reduces off-frequency listening since it forces the listener to base tone-detection on energy in the defined region of the spectral notch [17].

The influence of CBs can also be seen in the sensitivity to phase differences between tones which becomes apparent in AM and FM detection experiments. The isolated audibility of partials in harmonic complex tones can also be explained by the CB-concept.

4 • Approximations of Auditory Filter Bandwidth

4.1 Critical Bands

Zwicker averaged values of critical bandwidth obtained from several different methods and published them in tabulated form [25]. The data can be seen in Figure 3 along with their numerical approximation after Zwicker and Terhardt [29]:

$$\Delta f_G(f) \text{ [Hz]} = 25 + 75\left(1 + 1.4\left(f \text{ [kHz]}\right)^2\right)^{0.69}. \tag{1}$$

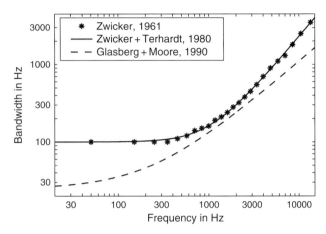

FIGURE 3 Bandwidth of psychophysically measured auditory filters: critical bandwidth according to the Bark-table published by Zwicker [25], "*", and according to the formula by Zwicker and Terhardt [29], "—", as well as the equivalent rectangular bandwidth (ERB) after a formula proposed by Glasberg and Moore [8], "- -".

The critical bands of Zwicker show a constant bandwidth of 100 Hz for $f < 500$ Hz and an approximately linear increase in bandwidth with $0.2f$ for $f > 500$ Hz. If 24 abutting CBs are lined up the *critical band rate scale* is formed which covers the audible frequency range. The unit is 1 Bark. An approximation of the critical band rate scale is given in [27,29]:

$$z(f) \text{ [Bark]} = 13 \arctan(0.76 f \text{ [kHz]}) + 3.5 \arctan\left(\frac{f \text{ [kHz]}}{7.5}\right)^2. \quad (2)$$

4.2 Equivalent Rectangular Bandwidth

Glasberg and Moore [8] suggested a formula for the auditory filter bandwidth based entirely on notched-noise masking data. This bandwidth definition is commonly referred to as the *equivalent rectangular bandwidth (ERB)*. The following formulas give the ERB-bandwidth $B_{ERB}(f)$ as well as the derived number ("rate") scale $ERB_N(f)$ [8,13]:

$$B_{ERB}(f) \text{ [Hz]} = 24.7(4.37f + 1), \quad (3)$$

$$ERB_N(f) \text{ [ERB]} = 21.4 \log_{10}(4.37f + 1). \quad (4)$$

Software that implements all formulas can be found in [20]. When comparing Zwicker's critical bands and ERBs in Figure 3 it becomes obvious that ERBs yield smaller values at all frequencies and that they decline for $f < 500\,\text{Hz}$. With respect to auditory models CBs tend to be somewhat large at low frequencies whereas ERBs appear to be slightly small, e.g., in comparison with the filter bandwidth necessary to model notched-noise masking data [18]. The differences stem largely and somewhat from the use of different methods and somewhat from measuring at different levels. As auditory filters widen with increasing level some methods show a level dependence of the measured critical bandwidth.

5 • Limits of the Critical Band Concept

5.1 Temporal Effects and Influence of Non-linearities

Since the critical band concept is based on energy detection in spectral channels it can be applied only if perception is not dominated by temporal effects. Specifically, the following temporal effects can influence detection of masked components: (1) Beats produced by the interaction of the probe tone with masker components can be detected as temporal fluctuations. (2) The phase of spectral components strongly influences the temporal waveform of the signal. The auditory system is able to listen into temporal gaps of signals with high peak factors (dip-listening). The long-term spectrum is not sufficient to predict masking of those signals and a short-term spectral analysis has to be used. (3) Masking shows an overshoot for probes presented at the onset of the masker, i.e., the probe is harder to detect if it is switched on in synchrony with the masker. The *overshoot effect* is maximal for wide-band maskers of medium levels. The additional masking can amount to 10–15 dB. This huge effect might be explained by a spectral sharpening of the active basilar membrane filtering process in combination with central effects [13,26].

The peripheral filtering process takes place on the basilar membrane and is clearly visible in the auditory nerve [19]. It forms the basis for CBs and a first stage of auditory models. To accommodate for many temporal masking effects the static energy detectors monitoring the output of CB-filters can be extended to follow dynamic changes in energy.

Although not a limitation of the CB-concept per se it should be noted that modeling peripheral frequency selection by linear filters becomes a simplification if the non-linearity of the basilar membrane response is involved. The non-linearity can introduce distortion products which may be heard before the probe tone itself becomes audible in masking experiments. A combination of peripheral and more central non-linear effects is likely responsible for non-linear addition of masking of two maskers in two-tone masking and suppression experiments. Finally, filter shape and bandwidth of peripheral filters are level dependent which somewhat influences effective critical bandwidth.

5.2 Comodulation Masking Release

If a masker is accompanied by flanking maskers in frequency bands away from the target band the detection threshold of a tone can decrease substantially if all maskers share the same low-frequency modulation. The effect of comodulation masking release (CMR) can be obtained over a wide frequency range and it can reach 1–6 dB for large and up to 16 dB for small spectral separations between the target and masker bands. The effect is maximal for low modulation frequencies and wide-band maskers [11]. CMR can be modeled either by comparing modulation across CBs or by assuming that the auditory system is able to listen into temporal gaps in the on-frequency band if the time instances for dip-listening are identified from flanking bands [2].

5.3 Profile Analysis

The detection of amplitude changes of a target component can improve in situations of stimulus uncertainty if the target is embedded in a background complex tone with multiple components [9,11]. It appears as if detection is facilitated by across-channel comparison of target level against background level. Profile analysis is linked to timbre perception because the effect degrades if the spectral extent of the background components is reduced or if the background deviates from a flat and dense spectral distribution.

5.4 Informational and Binaural Masking

Although this article is concerned with energetic masking it should be noted that several recent studies focused on the contribution of more central masking effects. Informational masking is commonly equated as masking that stems from attentional interference of sources or from stimulus uncertainty. It can be seen as masking that is not of peripheral, energetic origin [4]. Inter-subject variability is far higher in informational than energetic masking. Pronounced informational masking can be seen for speech sources (e.g., [1]).

When signals are played to both ears the masked threshold is dependent on the interaural phase difference between masker and probe, especially at low frequencies. If masker and probe are presented interaurally out of phase the threshold can be 15 dB lower than if they were presented in phase. The gain in detectability is called the *binaural masking level difference* [21].

6 • Modeling Auditory Masking

In most models of masking the detectability of signal components is derived from changes in the time- and frequency-dependent *excitation pattern*. Excitation patterns resemble a pattern close to masking patterns. They reflect the influence of the auditory periphery which analyzes the signal in frequency bands. Several functions for linear peripheral filters have been suggested. *Gammatone filters* consider the filter impulse response and are most widely used [16]. A simple, but powerful model to generate excitation patterns uses steep filters to compute the main excitation and applies slopes from normalized masking patterns to it. Frequency- and level-dependent non-linearities can be built into the lookup table [20,27].

Another approach builds on non-linear filterbanks which requires modeling the level-dependent compression and bandwidth in the filter center as well as the dependence of both tails on level (e.g., [12,18]). Temporal masking can be modeled by introducing compression and temporal integration (e.g., [3,14]). Inter-connectivity between channels for controlling channel-wise compression is necessary for simulating lateral suppression. Current modeling efforts aim to explain the non-linear behavior in simultaneous and forward masking. More general approaches will also model suppression effects, difference tones and the selection of spectral or temporal decision criteria.

7 • Conclusions

Since Helmholtz suggested in 1863 to view the cochlea as a series of resonators the filterbank approach to model spectral analysis became most widely accepted in auditory research [23]. The critical band concept adds the notion that masking can be predicted solely from energy comparisons at the output of these filters. The concept has proven successful in that numerous experimental results support this view. To predict results that cannot be explained directly by the CB concept, e.g., results on comodulation masking release or loudness, the critical band-based filterbank approach provides a starting point if it is followed by across-channel processes. Future models of masking have to be evaluated critically against the existing non-linear dynamic models to assess the contribution of other detection cues besides energy.

Acknowledgments I would like to thank Prof. H. Fastl, Dr. B. Edwards, and one anonymous reviewer for many helpful comments on previous versions of the manuscript.

References

1. D.S. Brungart, B.D. Simpson, C.J. Darwin, T.L. Arbogast, and G. Kidd. Across-ear interference from parametrically degraded synthetic speech signals in a dichotic cocktail-party listening task. *J. Acoust. Soc. Am.*, 118(5):292–304, 2005.
2. S. Buus. Release from masking caused by envelope fluctuations. *J. Acoust. Soc. Am.*, 78(6):1958–1965, 1985.
3. T. Dau, D. Püschel, and A. Kohlrausch. A quantitative model of the "effective" signal processing in the auditory system. I. Model structure. *J. Acoust. Soc. Am.*, 99(6):3615–3622, 1996.
4. N.I. Durlach, C.R. Mason, G. Kidd, T.L. Arbogast, H.S. Colburn, and B.G. Shinn-Cunningham. Note on informational masking. *J. Acoust. Soc. Am.*, 113(6):2984–2987, 2003.
5. J.P. Egan and H.W. Hake. On the masking patterns of a simple auditory stimulus. *J. Acoust. Soc. Am.*, 22(5):622–630, 1950.
6. H. Fastl. Temporal masking effects: III. Pure tone masker. *Acustica*, 43:282–294, 1979.
7. H. Fletcher. Auditory patterns. *Rev. Mod. Phys.*, 12:47–65, 1940.
8. B.R. Glasberg and B.C.J. Moore. Derivation of auditory filter shapes from notched-noise data. *Hear. Res.*, 47(5):103–138, 1990.
9. D.M. Green. *Profile Analysis*. Oxford University Press, Oxford, 1988.
10. D.D. Greenwood. Auditory masking and the critical band. *J. Acoust. Soc. Am.*, 33(4):484–502, 1961.
11. J.W. Hall III, J.H. Grose, and L. Mendoza. Across-channel processes in masking. In B.C.J Moore, editor, *Hearing*, Handbook of perception and cognition, pages 243–265. Academic Press, San Diego, 1995.
12. E.A. Lopez-Poveda and R. Meddis. A human nonlinear cochlear filterbank. *J. Acoust. Soc. Am.*, 110(6):3107–3118, 2001.
13. B.C.J. Moore. *An Introduction to the Psychology of Hearing*. Academic Press, San Diego, 5th edition, 2003.
14. A.J. Oxenham and B.C.J. Moore. Modeling the additivity of nonsimultaneous masking. *Hear. Res.*, 80:105–118, 1994.
15. A.J. Oxenham and C.J. Plack. Effects of masker frequency and duration in forward masking: further evidence for the influence of peripheral nonlinearity. *Hear. Res.*, 150:258–266, 2000.
16. R.D. Patterson, J. Holdsworth, I. Nimmo-Smith, and P. Rice. The auditory filterbank. SVOS final report, APU: 2341, 1988.
17. R.D. Patterson and B.C. Moore. Auditory filters and excitation patterns as representations of frequency resolution. In B.C. Moore, editor, *Frequency Selectivity in Hearing*, pages 123–177. Academic Press, London, 1986.
18. R.D. Patterson, M. Unoki, and T. Irino. Extending the domain of center frequencies for the compressive gammachirp auditory filter. *J. Acoust. Soc. Am.*, 114(3):1529–1542, 2003.
19. J.O. Pickles. The neurophysiological basis of frequency selectivity. In B.C. Moore, editor, *Frequency Selectivity in Hearing*, pages 51–121. Academic Press, London, 1986.

20. B. Seeber. Software on masking and critical bands. www.acoustics.bseeber.de.
21. S. van de Par and A. Kohlrausch. Dependence of binaural masking level differences on center frequency, masker bandwidth, and interaural parameters. *J. Acoust. Soc. Am.*, 106(4):1940–1947, 1999.
22. L.L.M. Vogten. Low-level pure-tone masking: a comparison of "tuning curves" obtained with simultaneous and forward masking. *J. Acoust. Soc. Am.*, 63(5):1520–1527, 1978.
23. H. von Helmholtz. *Die Lehre von der Tonempfindung als physiologische Grundlage für die Theorie der Musik*. Vieweg Verlag, Braunschweig, 1863.
24. R.L. Wegel and C.E. Lane. The auditory masking of one pure tone by another and its probable relation to the dynamics of the inner ear. *Phys. Rev.*, 23(2):43–56, 1924.
25. E. Zwicker. Subdivision of the audible frequency range into critical bands (Frequenzgruppen). *J. Acoust. Soc. Am.*, 33(3):248, 1961.
26. E. Zwicker. Temporal effects in simultaneous masking by white-noise bursts. *J. Acoust. Soc. Am.*, 37:653–663, 1965.
27. E. Zwicker and H. Fastl. *Psychoacoustics, Facts and Models*. Springer, Berlin Heidelberg New York, 2nd edition, 1999.
28. E. Zwicker and A. Jaroszewski. Inverse frequency dependence of simultaneous tone-on-tone masking patterns at low levels. *J. Acoust. Soc. Am.*, 71(6):1508–1512, 1982.
29. E. Zwicker and E. Terhardt. Analytical expressions for critical band rate and critical bandwidth as a function of frequency. *J. Acoust. Soc. Am.*, 68(6):1523–1525, 1980.

17
Aspects of Modeling Pitch Perception

Ernst Terhardt

Technical University of Munich, Munich, Germany,
e-mail: teirhardt@ei.tum.de

1 • Introduction

There is hardly any topic of psychoacoustics the approaches to which have been as diverse and controversial as the topic of modeling pitch perception. In the present contribution a number of basic aspects of this problem are discussed, and a framework for its solution is outlined.

2 • Basic Issues

Regarding psychophysical processes such as the perception of pitch, one can make a sensible distinction between a *theory* and a *model*. A theory is a formal description of the stimulus-to-sensation relationships that have become known from observations and experiments. The theory of pitch is – or should be – confined to an algorithmic description of the relationships that exist between the "observable variables" involved, i.e., the parameters of the stimulus, and the perceived pitches. Inclusion of hypothetical "mechanisms" should be avoided.

A *model* is supposed to simulate the perceptual process; it relates the stimulus itself, i.e., the sound wave, to the perceived pitches. If the theory exists, a model can be obtained by adding an algorithm that extracts from the sound wave the stimulus parameters which are defined and required by the theory. Any knowledge available about the perception of pitch – including the definition of the stimulus parameters – must be depicted by the theory, while extraction and measurement of the stimulus parameters as such is a purely physical problem whose solution does not require any knowledge about pitch perception.

An important practical consequence of this concept is that the theory – being self-contained – can be verified independently from the computationally expensive procedure of parameter extraction.

The theory of pitch perception can concisely be outlined by

$$P(t) = \Phi\{S_L(t), S_R(t), L\}, \tag{1}$$

where P denotes the set of perceived pitches; S_L, S_R the sets of stimulus parameters for the left and right ears; and L a set of listener-specific parameters. $\Phi\{..\}$ denotes an algorithm (in terms of computer language: a *procedure* or *function*) that includes specifications about which of the stimulus- and listener parameters are relevant and how they affect the set of pitches. For simplicity, in the following considerations the listener-parameter set L is mostly left out.

The pitch set P in general consists of several or many pitches, each of which in turn is a set of attributes of which the most important ones are h (height), σ (strength/pronouncedness), and λ (lateralization):

$$\begin{aligned} P(t) &= \langle p_1(t), p_2(t), \ldots \rangle \\ &= \langle \langle h_1(t), \sigma_1(t), \lambda_1(t) \rangle, \langle h_2(t), \sigma_2(t), \lambda_2(t) \rangle, \ldots \rangle. \end{aligned} \tag{2}$$

The two stimulus-parameter sets S_L, S_R may either include a complete description of the respective stimuli or may be confined to those parameters which are sufficient to determine the pitch set. In any case there is an inherent interdependency between the stimulus parameters and the algorithm Φ. The latter constitutes the core of the theory and implicitly defines which kind of stimulus parameters it operates on. The theory is complete only if it includes that definition. There is no apparent criterion according to which the stimulus parameters must be of either a "temporal" or "spectral", or any other kind. In general, the parameters must be presumed to be time variant.

For the evaluation of the implications of modeling pitch perception it is helpful to use a kind of standard description of the stimulus. A description which, on the one hand, enables specification of *any* type of stimulus and, on the other hand, is adapted to the fact that in pitch perception the sine wave plays a particular role, has the form

$$a(t) = \sum_{n=1}^{N} \hat{a}(t) \cos[2\pi f_n(t)t + \varphi_n]. \quad (3)$$

This is a *synthetic, parametric*, and universal description of the *audio signal a(t)*, i.e., the sound-pressure wave at each of the two ears.[1] As indicated, the sinusoidal components included in (3) are in general supposed to be time variant in frequency and amplitude. The time-variance of components ensures that *any* type of sound can be described in the form (3), i.e., with any number $N \geq 1$ of components.

By (3) the stimulus-parameter set S at each of the two ears is defined, i.e.,

$$S(t) = \langle\langle f_1(t), \hat{a}_1(t), \varphi_1 \rangle, \langle f_2(t), \hat{a}_2(t), \varphi_2 \rangle, \ldots, \langle f_N(t), \hat{a}_N(t), \varphi_N \rangle\rangle. \quad (4)$$

The actual number and values of the parameters may be different for the two ears, i.e., in the monaural and the dichotic conditions.

If only periodic sound waves would elicit pitch, and if there would occur only one pitch per periodic sound wave, specification of the stimulus-parameter set would be much simpler. For any singular periodic stimulus, S would not need to have the form (4) but would include only the type specification *periodic* and the length of the period or its reciprocal, i.e., oscillation frequency. The algorithm Φ would boil down to a mere formal frequency-to-pitch assignment. However, it is neither true that only periodic sound waves elicit pitch; nor is it true, for non-sinusoidal periodic sound waves, that there occurs only one pitch per periodic sound wave [25]. Therefore, any theory and/or model of pitch that rigorously depends on the stimulus' periodicity is inadequate.

The only type of periodic sound wave to which the above conditions apply – at least to a sufficient extent – is the singular sine wave. For singular sine waves the pitch theory is just as simple as was just hypothetically sketched for periodic signals in general: to a first approximation, there is a "one-to-one" relationship between a sine tone's oscillation frequency and pitch.

This provides for reliable and unambiguous measurement of the various pitches which in general are elicited by a complex stimulus: the test sound is presented to a listener in alternation

---

[1]This mathematical description must not be confused with a Fourier spectrum.

with a sine tone (the latter preferably presented diotically and with a definite SPL), and the sine tone's frequency is adjusted such that its pitch height matches the height of the test sound's particular pitch envisaged. The numerical value of any pitch can most conveniently be obtained by defining it to be identical to that of the match tone's frequency. Pitch height quantitatively expressed in this way is measured in *pitch units* (pu) [24–26]. The *strength* of any pitch can be determined by subjective assessment [9,29]. If required, also *lateralization* may be determined and included in the pitch set. In this way, the pitch set (2) of any complex sound can be obtained by fairly simple psychoacoustic experiments.

It has been proven helpful to make a distinction between *spectral pitch* and *virtual pitch* [22,25]. The former type of pitch is characterized by a "one-to-one" relationship to spectral singularities of any kind. Isolated sine tones are prominent examples of such singularity. However, spectral singularities may also occur in random and quasi-random types of stimuli. Any pitch which is not in a "one-to-one" relationship to such a singularity is termed virtual pitch. In general, virtual pitch is conceptualized as an entity that resides on the second or even a higher level of the perceptual hierarchy, while spectral pitch occurs on the first [22,25].

3 • Pitch of Tonal Stimuli

The tonal type of sound is characterized by the features that (a) the component frequencies included in (4) are to a large extent in a systematic relationship to one another (an objective criterion); and (b) a number of component frequencies elicit individual spectral pitches (an auditory criterion).

3.1 Spectral Pitch

Using (4), the formation of the two sets of spectral pitches $P_{s,L,R}$ elicited at the two ears can be outlined by

$$P_{s,L,R}(t) = \Phi_{s,L,R}\{S_{L,R}(t)\}, \qquad (5)$$

where $S_{L,R}$ corresponds to (4). Equation (5) actually depicts two independent equations, i.e., one for each ear. There is compelling evidence that spectral pitch is constituted by each ear alone [2, 25], though its height may be slightly affected by a contralateral stimulus [23].

Each of the two algorithms $\Phi_{s,L,R}$ must yield a set of spectral pitches each of which is characterized by a numerical value for its height, another for its strength, and an appropriate characterization of lateralization (which is left out of consideration here). To a first approximation, and by definition, the numerical value of pitch height can be set equal to that of the pertinent component's frequency, and the unit of height is pu [22,25]. As spectral-pitch height in general may also be affected by (a) SPL; (b) mutual interaction of adjacent components; (c) phase; and (d) interaural interaction, $\Phi_{s,L,R}$ should account for these effects. A preliminary algorithmic solution, which may serve as an example, was described in the virtual-pitch theory [24,26].

Pitch *strength* can be supposed to depend on (a) the pertinent component's amplitude (both absolute and relative to the threshold of hearing); (b) the degree to which the pertinent component can be assumed to stay unmasked by other adjacent components; (c) phase; (d) frequency-dependent prominence. (Components in the region around 700 Hz are most prominent, i.e., *dominant*, merely *because* they lie in that region [26].) As a result, each of the two spectral-pitch sets $P_{s,L,R}$ will include a number of pitches which is limited by the criterion that their strength must be $\sigma > 0$.

3.2 Virtual Pitch

One of the most striking features of virtual pitch is that it is not dependent on the presence of a sinusoidal component whose frequency corresponds to virtual-pitch height. However, the information about any virtual pitch surely is contained in the stimulus-parameter sets $S_{L,R}$.

Basically, several types of algorithm for the formation of virtual pitch are conceivable. In particular, it is tempting to suspect that virtual pitch may be obtained by an analysis of the temporal structure of the stimulus, i.e., essentially the structure created by those components which are not aurally "resolved" to such an extent that they can elicit individual spectral pitches [12,18]. However, already in the 1960s it was found that in fact virtual pitch essentially is dependent on components which evidently *are* resolved [22]. Moreover, it was found that the precise height of virtual pitches is in several ways correlated with the precise height of the spectral pitches which are elicited by the tonal stimulus [3,12,22,25,27]. These findings indicate that virtual pitch is not constituted by a mechanism which is independent of spectral pitch – irrespective of whether in the frequency- or time-domain – but is dependent on, and emerges from, the set of spectral pitches.

This conclusion is strongly supported by the observation that virtual pitch can even be elicited by a periodic stimulus whose set of harmonics is split, one subset being presented to the left ear, the other to the right [15]. While spectral pitch is monaural (Sect. 3.1), virtual pitch is binaural. Another compelling support comes from the observation that virtual pitches can be elicited by a harmonic complex tone whose harmonics are presented in a rapid temporal succession instead of simultaneously [11]. As this stimulus is a sequence of non-simultaneous sinewaves, those virtual pitches are not at all encoded in the time structure of the stimulus. As the auditory system evidently is equipped with a short-term memory for pitch, formation of virtual pitch from a rapid succession of spectral pitches is well conceivable.

Taking into account that any kind of perception is hierarchical, the above findings support the conclusion that virtual pitch is a perceptual entity that resides on a higher level of the hierarchy of pitches than spectral pitch and that virtual pitch is by an appropriate algorithm derived from the set of spectral pitches [22].

A number of authors have pointed out that a faint virtual pitch may be elicited by a monaural or diotic harmonic complex tone which is high-pass filtered such that it includes only quite high-order harmonics (e.g., above the 20th) [5,16,18]. Though it may appear questionable if such high harmonics can elicit individual "sine tone pitches", one cannot be sure that such a high-pass filtered complex tone does not elicit any spectral pitch at all. It is even likely that there is a spectral pitch corresponding to the cut-off frequency – and *one* spectral pitch can be sufficient to elicit faint subharmonic virtual pitches [14]. Moreover, examples such as *overtone singing*, and melodies played on the *Jew's harp*, demonstrate that both spectral pitch and virtual pitch can be elicited by very-high-order harmonics of a periodic stimulus with a very low oscillation frequency. Whatsoever, the very faintness and singularity of the virtual pitch of "unresolved harmonics" render its potential theoretical implications overruled by the wealth of compelling evidence just outlined.

Hence, an adequate outline of the formation of virtual pitch can be denoted by the formula

$$P_v(t) = \Phi_v\{P_{s,L}(t), P_{s,R}(t)\}, \tag{6}$$

where $P_{s,L}$ and $P_{s,R}$ are obtained from (5).

A guideline for the development of the algorithm Φ_v is provided by the eminent *robustness* of the pitch elicited by complex stimuli. In real life, the auditory stimulus routinely suffers drastic linear distortion, such that even for systematic and steady types of sound the amplitude- and phase-distribution of components at the ear is variable and in several respects unpredictable. It is the spectral frequencies – and, therefore, the spectral pitches – which provide reliable clues to the characteristics of the sounds emitted by the sound sources. The evident perceptual constancy of the pitch percept suggests that, on a second or higher level of the auditory hierarchy, a *completion mechanism* must be involved, and it makes sense to assume that virtual pitches are the result of such kind of pitch-pattern completion. The completion algorithm appears to operate on the presumption that in real life any set of spectral pitches emerges from the presence of one or several *harmonic* complex tones. In fact, harmonicity of components is a strong indicator for those components to be pertinent to one and the same periodically oscillating source. The principle of *subharmonic coincidence detection* applied in the virtual-pitch theory [24,26] may serve as an example for the way in which the above notions may be cast into the algorithm Φ_v.

3.3 The Tonal Percept

The ultimate pitch set P, which consists of both spectral and virtual pitches, can be assumed to emerge by evaluation and interpretation of the monaural pitch sets $P_{s,L}, P_{s,R}$, and the binaural set P_v, i.e., on a higher level of the perceptual hierarchy. For the description of the ultimate pitch set another algorithm Φ_p is required. This algorithm may also account for aspects such as formation of multiple sensory objects and contextual effects [12,18,25]. When listener-specific influences are also envisaged, one eventually obtains the formula

$$P(t) = \Phi_p\{P_{s,L}(t), P_{s,R}(t), P_v(t), L\}, \tag{7}$$

where $P_{s,L}$ and $P_{s,R}$ are obtained from (5), and P_v from (6).

3.4 Conclusion

The outline (7) of the pitch theory for tonal sounds is meant to account for those basic aspects and conditions which can be regarded as evident, while there remains plenty of space for specification of details and variants. Indeed, many details – in particular, where high-level pitch evaluation is concerned – remain to be specified. Several basic algorithms implied in (7) have been worked out and verified in the virtual-pitch theory [24–26]. Those algorithms may serve as examples.[2]

As the theory as such does not require processing of the audio signal, it can be tested and verified with moderate computational burden.

Extension of the theory into a *model* requires inclusion of an algorithm that extracts the component parameters (4) from the audio signal. A considerable number of such algorithms are available [4,10,13,17,21]. The model thus obtained provides a basis for applications such as pitch extraction from speech; automatic transcription of music; root-finding of musical chords.

---

[2]Software for testing the virtual-pitch theory may be downloaded from ftp://ftp.mmk.e-technik.tu-muenchen.de/pub/akom/ter/.

4 • Pitch of Random Stimuli

4.1 Pitch of Band-Limited Random Noise

When presented monaurally or diotically, narrow-band random noise typically elicits a pitch that essentially corresponds to its center frequency. When the bandwidth is increased and the spectral slopes are steep, the pitch splits into two pitches which approach the edges of the spectral contour [20,29]. It is conceivable that this behavior can be accounted for by a generalized form of the spectral-pitch algorithm Φ_s.

4.2 Pitch of Wideband Random Noise with Spectral Peaks

Equidistant peaks and gaps do occur in the power spectrum of wideband random noise when a truly random noise after a brief delay is superimposed to itself. For both monaural and diotic presentation, one or several pitches may be heard that are related to the reciprocal of the time delay [1,28,29]. It is conceivable that the phenomenon can be explained on the basis of a theory of the spectral pitch of narrow-band random noise (Section 4.1), and probably virtual pitches become involved in a way equivalent to that of tonal sounds (Section 3.2) [1].

4.3 Dichotic Pitch Phenomena

A number of dichotic pitch phenomena have been described which typically are elicited by dichotic presentation of two wideband random-noise stimuli, one of which was somehow manipulated [1,7,8]. Ordinarily, no pitch is heard when either of the stimuli is presented alone, i.e., monaurally. For example, the phase of the Fourier spectrum of a wideband random noise was within a narrow frequency interval continually shifted by 360°. When the noise thus manipulated is presented to one ear and the original noise to the other, a faint pitch can be heard that resembles that of a narrow-band noise situated in the frequency interval of phase transition [6,7]. As this type of pitch obviously is dependent on a kind of spectral singularity, the phenomenon demonstrates an exception from the rule that spectral pitch is basically monaural (cf. Section 3.1). To explain it, the algorithm for spectral-pitch formation needs to be generalized, and an account of binaural interaction must be included [1,7].

4.4 Pitch of Amplitude-Modulated Wideband Noise

When wideband random noise gets "pulsed", i.e., periodically switched off and on, a faint pitch corresponding to the modulation frequency may be heard for both monaural and diotic presentations [19,29]. The very faintness of the pitch excludes the possibility that periodicity of the temporal envelope as such can be the determining clue. The power spectrum of the modulated noise fails to reveal any clue. Hence, the faint pitch may be determined by the phase spectrum. As yet there does not appear to exist a theoretical description of that relationship [12].

4.5 Conclusion

For several of the pitches of random stimuli it is not readily apparent if and how they can be accounted for by a theory akin to the one for tonal sounds. Yet it is conceivable that by appropriate generalization of the algorithm for formation of spectral pitch a unified theory and model can be achieved.

References

1. F.A. Bilsen: J. Acoust. Soc. Am. **61**, 150 (1977)
2. G. van den Brink: Acustica **32**, 159 (1975)
3. G. van den Brink: Acustica **32**, 167 (1975)
4. J.C. Brown, H.S. Puckette: J. Acoust. Soc. Am. **94**, 662 (1993)
5. R.P. Carlyon, T.M. Shackleton: J. Acoust. Soc. Am. **95**, 3541 (1994)
6. E.M. Cramer, W.H. Huggins: J. Acoust. Soc. Am. **30**, 413 (1958)
7. J.F. Culling, A.Q. Summerfield, D.H. Marshall: J. Acoust. Soc. Am. **103**, 3509 (1998)
8. J.F. Culling, D.H. Marshall, A.Q. Summerfield: J. Acoust. Soc. Am. **103**, 3527 (1998)
9. H. Fastl, G. Stoll: Hear. Res. **1**, 293 (1979)
10. T. Funada: Signal Process. **13**, 15 (1987)
11. J.W. Hall, R.W. Peters: J. Acoust. Soc. Am. **69**, 509 (1981)
12. W.M. Hartmann: J. Acoust. Soc. Am. **100**, 3491 (1996)
13. W. Heinbach: Acustica **67**, 113 (1988)
14. T. Houtgast: J. Acoust. Soc. Am. **60**, 405 (1976)
15. A.J.M. Houtsma, J.L. Goldstein: J. Acoust. Soc. Am. **51**, 520 (1972)
16. A.J.M. Houtsma, J. Smurzynski: J. Acoust. Soc. Am. **87**, 304 (1990)
17. R.J. McAulay, T.F. Quatieri: IEEE Trans. ASSP **34**, 744 (1986)
18. B.C.J. Moore: *An Introduction to the Psychology of Hearing* (Academic, New York 1989)
19. I. Pollack: J. Acoust. Soc. Am. **45**, 237 (1969)
20. A. Rakowski: Pitch of filtered noise. In: *Proc. 6 Int. Congr. Acoust.* (Tokio 1968) pp. 105–108
21. S. Seneff: IEEE Trans. ASSP **26**, 358 (1978)
22. E. Terhardt: J. Acoust. Soc. Am. **55**, 1061 (1974)
23. E. Terhardt: Acustica **37**, 56 (1977)
24. E. Terhardt: Hear. Res. **1**, 155 (1979)
25. E. Terhardt: *Akustische Kommunikation* (Springer, Berlin Heidelberg New York 1998)
26. E. Terhardt, G. Stoll, M. Seewann: J. Acoust. Soc. Am. **71**, 679 (1982)
27. K. Walliser: Acustica **21**, 319 (1969)
28. W.A. Yost: J. Acoust. Soc. Am. **100**, 511 (1996)
29. E. Zwicker, H. Fastl: *Psychoacoustics*, 2nd edn (Springer, Berlin Heidelberg New York 1999)

18

Calculation of Loudness for Normal and Hearing-Impaired Listeners

Josef Chalupper

Siemens Audiological Engineering Group, Gebbertstrasse 125, Erlangen 91058, Germany, e-mail: Josef.Chalupper@siemens.com

1 • Introduction

Loudness is a hearing sensation which is almost ubiquitous in everyday life: traffic is *too loud*, announcements are *too soft*, children are *very loud* and a stereo can be adjusted to a *comfortable loudness*. Thus, it is not surprising that scientists started very early to carry out psychoacoustic experiments to investigate loudness. In 1933, Fletcher and Munson [1] measured "equal-loudness contours". These curves are obtained if a reference sound (e.g. a 1,000-Hz sinusoid) is fixed in level and the test sound is adjusted to give a loudness match. Some typical results are shown in Figure 1 [2]. Figure 1 shows equal-loudness contours for loudness levels from 3 to 100 phon.[1]

---

[1] "phon" is the unit of the loudness level and is defined as the level (dB SPL) of a 1-kHz sinusoid with the same loudness as the test sound.

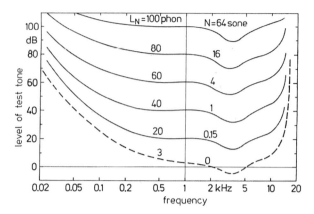

FIGURE 1 Equal-loudness contours for pure tones in a free sound field. Taken from [2].

Obviously, loudness of pure tones depends on frequency. Moreover, this dependency varies with level, as the dip between 2 and 5 kHz becomes deeper with increasing levels. The remarkable differences between the slopes of the equal-loudness contours below 1 kHz resulted in the definition of level-dependent weighting curves ("dB A", "dB C") which are often applied in technical acoustics [3]. In many other listening experiments (for comprehensive overviews see [2,4]), it was shown that loudness is affected by a large variety of signal parameters: frequency, bandwidth, duration, level and temporal structure. In addition, early experiments by Steinberg and Gardner [5] indicated that loudness perception also strongly depends on properties of the human listener. In particular, the rate of growth of loudness with increasing level is greater for listeners with cochlear hearing impairment than for normal listeners. This phenomenon is called "recruitment" (Figures 2 and 3).

In Figure 2, median loudness scaling data (circles) for 57 normal hearing subjects using the WHF method ("Würzburger Hörfeld", see [6]) are shown [7]. With this method, the loudness of 1/3-octave-band noises is measured by a combination of verbal categories and an underlying scale of "Categorical Units" (CU) ranging from 0 ("inaudible") to 50 ("too loud"). In general, the curves in Figure 2 are very similar across frequencies: starting with 0 CU for a sound level of 10 dB; values close to 50 CU are obtained for levels around 90 dB. Interestingly, the

---

"sone" is the unit of loudness. A 1-kHz sinusoid with a level of 40 dB SPL by definition elicits a loudness of 1 sone.

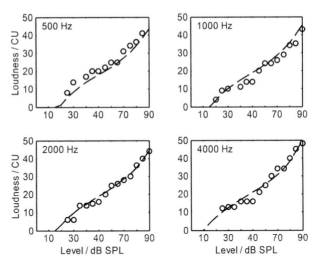

FIGURE 2 Loudness scaling by normal hearing subjects. Data measured by the WIII method (*circles*) [6] and predictions by the DLM (*broken curves*). From [7].

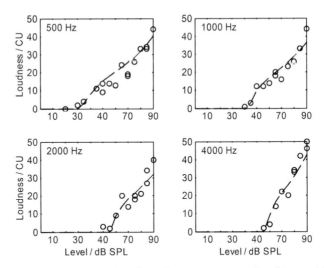

FIGURE 3 Same as Figure 2, but for a hearing impaired subject with presbyacusis. From [7].

curves of the hearing-impaired subject (Figure 3) have nearly the same loudness at high levels, whereas the starting point of the loudness functions is shifted to higher levels due to the individual hearing loss. Thus, the loudness functions at 2 and 4 kHz, where hearing loss is about 50 dB HL, are nearly twice as steep as the respective curves of the normal hearing listeners in Figure 2. If loudness functions are measured by assigning numbers to sounds according to their relative or absolute loudness ("magnitude estimation"), typically 10 dB increase in level is required to increase loudness by a factor of 2 [2].

As a consequence of this large number of complex dependencies of loudness on stimulus parameters, methodology and properties of the human hearing system, various computational models of loudness were developed to calculate the loudness of a given sound [1,8,9]. The loudness model of Zwicker [8,10] was originally designed for calculating the loudness of steady-state sounds for normal hearing listeners. In brief, first the excitation pattern is calculated from a spectral analysis based on properties of the human hearing system (as revealed mainly in masking), and then the excitation pattern is transformed into the "specific loudness pattern". The area under the specific loudness pattern indicates the overall loudness. Later on [11], this model was extended to account also for the loudness of time-varying sounds and was used as the basis for the calculation of other hearing sensations like roughness and sharpness [2]. In 1991, Zwicker's procedure for calculating of loudness became a normative standard [12]. New psychoacoustic data for loudness of nonstationary stimuli allowed for further refinement of the original dynamic model [2,13,14]. At present, it is widely used in many practical applications, in particular for sound quality engineering tasks [2]. In 2002, Zwicker's model was revised and extended to hearing-impaired listeners by Chalupper and Fastl [7]. The structure of this model (DLM – dynamic loudness model), specific aspects and some predictions for important test cases will be presented in more detail in the following section.

2 • Model

Figure 4 shows the basic structure of the dynamic loudness model.

The first stage is a filter, which takes into account the transmission through outer and middle ear. The frequency response of this filter depends on the acoustical conditions for the input signal. Currently, freefield conditions are assumed, but also diffuse field sounds can be used [2]. Moreover, if the loudness of in situ or KEMAR recordings shall be assessed, appropriate equalizing can be accomplished in this stage.

The next block is an aurally adequate spectro-temporal analysis consisting of a critical band filterbank. This filterbank was designed according to the Fourier-t transformation (FTT) [15].

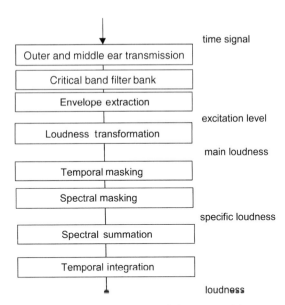

FIGURE 4 Dynamic loudness model (DLM) for normal hearing and hearing-impaired listeners.

Analysis frequencies are the centre frequencies of critical bands as described by Zwicker [16]. Essentially, the FTT analysis is identical to a Gammatone filterbank [17] with an equivalent rectangular bandwidth of 1 Bark [18].

The following block "envelope extraction" contains a window to extract the temporal envelopes of the outputs from the bandpass filters according to a proposal of Plack and Moore [19]. At the output of this block, the excitation level $L_E(z,t)$ is available, which is related to excitation E by

$$L_E = 10 \cdot \log\left(\frac{E}{E_0}\right) \text{dB}, \tag{1}$$

with E_0 the excitation corresponding to the reference value of sound intensity $I_0 = 10^{-12} \text{W}/\text{m}^2$.

In contrast to other loudness models for hearing-impaired listeners [20–22], it is not required to adapt the spectro-temporal analysis to the reduced spectral and temporal resolution typically seen in cochlear hearing impairment [23,24]. This is possible because in the DLM, the effects of spectro-temporal resolution are splitted in a linear and a nonlinear part. Whereas the linear part ("critical band filterbank" and "envelope extraction") takes into account the maximum possible resolution of the human hearing system, the nonlinear part ("temporal masking" and

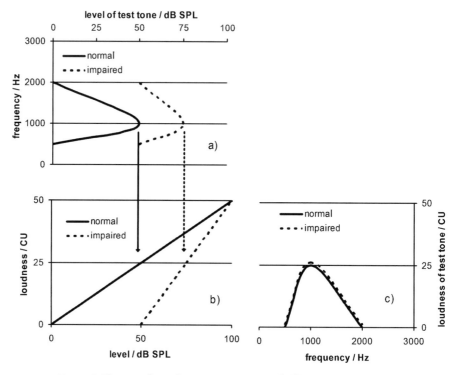

FIGURE 5 Slopes of masking patterns are shallower in hearing-impaired listeners (**a**). If the masking patterns are transformed to loudness by the individual loudness functions (**b**), the steepness of the slopes is the same for normal and hearing-impaired listeners (schematic sketch).

"spectral masking") is primarily designed to model dependencies on level and duration. Due to the position of these blocks after the loudness transformation, they also account for the effects of hearing loss on masking patterns, and thus on spectro-temporal resolution. This interesting behaviour can be explained best with the schematic sketch in Figure 5.

Figure 5a shows typical spectral masking patterns obtained with a narrow band noise. The levels of the masking noise have been adjusted to the same loudness for the normal and hearing-impaired listener. The slopes of the masking patterns of the normal listener are steeper by factor of 2 than the respective slopes of the hearing-impaired listener. If each data point of these masking patterns is transformed from level (in dB SPL) to loudness (in CU) by the respective loudness functions (Figure 5b), the slopes of the resulting patterns in the loudness domain have the same steepness for normal and hearing-impaired listeners. The same considerations also can be applied to temporal masking patterns by simply changing the labelling of the *x*-axes in Figure 5a and c from "frequency" to "time". This means that reduced spectral and temporal resolution of hearing-impaired listeners is not required to be adapted explicitly if masking effects are modelled in terms of loudness. Since masking is taken into account after the loudness transformation in the DLM, the only adjustment required to adopt the model to an individual hearing-impaired listener is to modify the steepness of the loudness transformation.

The loudness transformation as proposed by Zwicker [8] transforms excitation into specific loudness:

$$N'(z) = N_0 \left(\frac{E_{\text{THQN}}(z)}{s(z)E_0}\right)^\alpha \left[\left(1 - s(z) + s(z)\frac{E(z)}{E_{\text{THQN}}(z)}\right)^\alpha - 1\right]$$

$$\text{sone / Bark}, \quad (2)$$

where $E_{\text{THQN}}(z)$ denotes the excitation at threshold in quiet, in case of normal hearing, and N_0 is a constant, which is chosen to achieve a loudness of 1 sone for a 1 kHz sinusoid with a level of 40 dB. The threshold factor s is the ratio between the intensity of the just-audible test tone and the intensity of the internal noise appearing within the critical band at the test tone's frequency. The exponent $\alpha = 0.23$ indicates that specific loudness is related to the fourth root of intensity. For details, the reader is referred to [10].

In order to account for individual differences in the loudness transformation for people with hearing impairment, a so-called two component approach [21,25] is used. In essence, the total hearing loss HTL is divided into a component which accounts for the attenuation HTL_a (seen in elevated thresholds) and another component realizing an expansion HTL_e (seen in the recruitment phenomenon). The following equations apply:

$$\text{HTL} = \text{HTL}_a + \text{HTL}_e, \quad (3)$$

$$\text{HTL}_e = k \times \text{HTL} \quad (0 \leq k \leq 1), \quad (4)$$

$$\text{HTL}_a = (1-k)\text{HTL}. \quad (5)$$

In order to incorporate both components into the model, $E(z)$ in (2) is replaced by

$$E(z) = E_0 \cdot 10^{(L_E(z) - \text{HTL}_a/10)} \quad (6)$$

and $E_{\text{THQ}}(z)$ by

$$E_{\text{THQ}}(z) = E_{\text{THQN}} \cdot 10^{(\text{HTL}_e/10)}. \tag{7}$$

The factor k in (4) and (5) is determined by minimizing the difference between measured loudness scaling data and calculated loudness. This means that besides the hearing loss, loudness scaling data are required to adapt the loudness model to an individual hearing-impaired listener. If no loudness scaling data are available, these are either estimated from the uncomfortable level (UCL) or k is set to 0.8, which has been found empirically to be a good estimate [22]. In order to transform loudness N in sone into category units CU, the following formula was derived from empirical data [26]:

$$\text{CU} = 0.6 \, N/\text{sone} + 11.8 \log(16.5 \, N/\text{sone}). \tag{8}$$

Figure 6 shows the resulting specific loudness functions for $k = 0$ and $k = 1$ for a hearing loss of 50 dB in comparison to a normal hearing listener. If k is optimized individually, results of loudness scaling of narrow band noises are predicted with a high accuracy for both normal and hearing-impaired listeners (see Figure 2).

After the loudness transformation, the nonlinear aspects of temporal and spectral masking effects are modelled. At the output of these blocks, the specific loudness–time pattern is available with a resolution of 0.1 Bark. By integration across

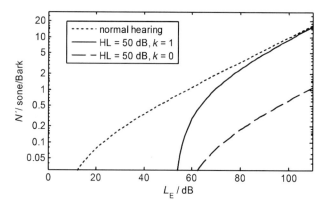

FIGURE 6 Specific loudness as a function of excitation level at 1 kHz. *Dotted*: normal hearing subject, *solid*: hearing-impaired subject with 50 dB hearing loss and $k = 1$. *Dashed*: hearing-impaired subject with 50 dB hearing loss and $k = 0$.

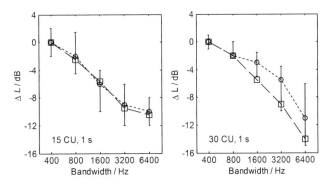

FIGURE 7 Level difference at same loudness between noises with various bandwidths and a narrow band noise of 400 Hz at 2 kHz. *Circles*: median and quartiles (*vertical lines*) of data from eight normal hearing subjects. *Squares*: prediction by the DLM. *Left panel*: loudness of the 400 Hz wide noise is 15 CU (45 dB). *Right panel*: 30 CU (70 dB). From [7].

frequency and time, the dependencies of loudness on bandwidth and duration are taken into account. If the level of narrow band noise is adjusted to match the loudness of a broad band noise, the resulting level is higher than the level of the broad band noise. This effect is called "loudness summation" and amounts to about 10–15 dB (depending on bandwidth and level) for normal hearing listeners (Figure 7). In subjects with sensorineural hearing loss, loudness summation is reduced (Figure 8). Reduced spectral resolution and recruitment in literature are often considered to be the reasons for reduced loudness summation [20,24]. As in Figures 7 and 8, the DLM is capable of predicting the amount of loudness summation both for normal and hearing-impaired listeners – without explicitly modifying spectral resolution (remember Figure 5).

In the last block of the DLM, the temporal integration of loudness is addressed. Experimental data have shown that loudness decreases for signal durations smaller than 100–200 ms. If the loudness of a 10 ms tone is matched to the loudness of a 100 ms tone, typically an increase in level of 10 dB is required for normal hearing listeners [2]. Similar to loudness summation, also the amount for temporal integration of loudness is reduced in hearing-impaired listeners [27]. Following the line of reasoning for reduced spectral resolution (Figure 5), in the DLM, the same low pass is sufficient to successfully model temporal integration for normal as well as hearing-impaired listeners [7].

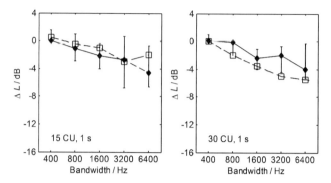

FIGURE 8 Same as Figure 7, but for subjects with cochlear impairment. *Filled symbols*: data from seven subjects with cochlear impairment (medians and quartiles). *Open squares*: predictions by the DLM. From [7].

3 • Conclusions

Loudness is a hearing sensation which depends on many properties of the assessed signal (stimulus) and the human hearing system (receiver). Since loudness has been of interest for science as well as industry (e.g. hearing aid manufacturers, sound quality labs), the most important effects are well understood today and computational models are available for calculating loudness of arbitrary sounds – not only for normal listeners but also for listeners with hearing impairment.

References

1. Fletcher, H., Munson, W.A., "Loudness, its definition, measurement and calculation", J. Acoust. Soc. Am., 5, 82–108, 1933.
2. Zwicker, E., Fastl, H., "Psychoacoustics: Facts and Models", 2nd ed., Berlin, Heidelberg, New York: Springer, 1999.
3. Beranek, L., "Acoustical measurements", J. Acoust. Soc. Am., 1988.
4. Moore, B.C.J., "An Introduction to the Psychology of Hearing", 4th ed., London: Academic Press, 1997.
5. Steinberg, J., Gardner, M., "Dependence of hearing impairment on sound intensity", J. Acoust. Soc. Am., 9, 11–23, 1937.
6. Heller, O., "Orientated category scaling of loudness and speech-audiometric validation", In: Contributions to Psychological Acoustics, Vol. 5, A. Schick, Ed., Bibliotheks- und Informationssystem der Univ. Oldenburg, Oldenburg, 135–159, 1991.
7. Chalupper, J., Fastl, H., "Dynamic loudness model (DLM) for normal and hearing-impaired listeners", Acustica – Acta Acustica, 88, 378–386, 2002.

8. Zwicker, E., "Über Psychologische und Methodische Grundlagen der Lautheit", Acustica, 8, 237–258, 1958.
9. Zwicker, E., Scharf, B., "A model of loudness summation", Psychol. Rev., 72, 3–26, 1965.
10. Zwicker, E., "Ein Verfahren zur Berechnung der Lautstärke", Acustica, 10, 304–308, 1960.
11. Zwicker, E., "Procedure for calculating loudness of temporally variable sounds", J. Acoust. Soc. Am., 62, 675–682, 1977.
12. DIN45631, "Berechnung des Lautstärkepegels und der Lautheit aus dem Geräuschspektrum, Verfahren nach E. Zwicker", DIN 45631, 1991.
13. Zwicker, E., "Dependence of post-masking on masker duration and its relation to temporal effects in loudness", J. Acoust. Soc. Am., 75, 219–223, 1984.
14. Widmann, U., Lippold, R., Fastl, H., "Ein Computerprogramm zur Simulation der Nachverdeckung für Anwendungen in Akustischen Meßsystemen", In: Fortschritte der Akustik, DAGA'98, Dt. Gesell. für Akustik e. V., Oldenburg, 96–97, 1998.
15. Terhardt, E., "Fourier transformation of time signals: conceptual revision", Acustica, 57, 242–256, 1985.
16. Zwicker, E., "Subdivision of the audible frequency range into critical bands (Frequenzgruppen)", J. Acoust. Soc. Am., 33, 248, 1961.
17. Patterson, R.D., Holdsworth, J., Nimmo-Smith, I., Rice, P., "SVOS final report: the auditory filterbank", MRC Applied Psychology Unit, Cambridge, APU report 2341, 1986.
18. Chalupper, J., Fastl, H., "Simulation of hearing impairment based on the Fourier Time Transformation", Proceedings ICASSP 2000, Istanbul, Turkey, 857–860, 2000.
19. Plack, C.J., Moore, B.C.J., "Temporal window shape as a function of frequency and level", J. Acoust. Soc. Am., 87, 2178–2187, 1990.
20. Florentine, M., Zwicker, E., "A model of loudness summation applied to noise-induced hearing loss", Hear. Res., 1, 121–132, 1979.
21. Moore, B.C.J., Glasberg, B.R., "A model of loudness perception applied to cochlear hearing loss." Aud. Neurosci., 3, 289–311, 1997.
22. Moore, B.C.J., Glasberg, B.R., Vickers, D.A., "Further evaluation of a model of loudness perception applied to cochlear hearing." J. Acoust. Soc. Am., 106, 898–907, 1997.
23. Tyler, R., "Frequency resolution in hearing-impaired listeners", In: Frequency Selectivity in Hearing, B.C.J. Moore, Ed., Academic Press, London, 309–371, 1986.
24. Moore, B.C.J., "Perceptual Consequences of Cochlear Damage", London: Oxford University Press, 1995.
25. Launer, S., Hohmann, V., Kollmeier, B., "Modeling loudness growth and loudness summation in hearing-impaired listeners" In: Modeling Sensorineural Hearing Loss, W. Jestaedt, Lawrence Erlbaum Mahwah, NJ: 175–185, 1997.

26. Baumann, U., Stemplinger, I., Arnold, B., Schorn, K., "Kategoriale Lautstärkeskalie-rung in der klinischen Anwendung", In: Fortschritte der Akustik, DAGA'96, Dt. Ge-sell. für Akustik e. V., Oldenburg, 128–129, 1996.
27. Buus, S., Florentine, M., Poulsen, T., "Temporal integration of loudness in listeners with hearing losses of primarily cochlear origin", J. Acoust. Soc. Am., 105, 3464–3480, 1999.

19

Psychoacoustical Roughness

Peter Daniel

Brüel & Kjær, GmbH, Bremen, Germany

1 • Introduction

Roughness is a fundamental hearing sensation for fast amplitude modulations. Fast means in this sense that modulation is not any longer perceived as loudness fluctuation but instead as a new quality of the sound: the sound becomes rough. The transition between the sensation of loudness fluctuation and roughness occurs at about 12 Hz for sinusoidal amplitude-modulated sounds.[1]

The sensation of roughness is similar to the sensation of dissonance caused by the beating of tones. It seems to be produced at early stages of the hearing process. If the two beating tones are presented separately to both ears, roughness disappears [9]. As basic listening sensation, roughness was extensively investigated by numerous researchers back from the nineteenth century. Detailed description can be found, e.g., already in Hermann von Helmholtz fundamental book *Die Lehre von den Tonempfindungen* dating from 1863 [15].

---

[1] The perception of loudness fluctuation – called fluctuation strength – is maximal at a modulation frequency of about 4 Hz, correlating well with the average rate of syllables in speech.

In general, roughness correlates highly with the perceived unpleasantness and annoyance of sounds [8,4,16]. Thus, it is an important parameter for the assessment of the perceived quality of sounds, mainly of technical sounds. Deviations in gears cause, e.g., increased friction with dominating rough components.

2 • Roughness in Dependence of Signal Parameters

The dependency of roughness on signal parameters is very complex and was thoroughly investigated in jury tests [2,6,7, 10–14]. A comprehensive survey and summary of these studies is given in [4,5,17].

2.1 Quantitative Definition of Psychoacoustical Roughness

In order to quantify roughness perception, *the roughness produced by a 1-kHz tone sinusoidal modulated with a modulation frequency of 70 Hz and a degree of modulation of 1 is defined to be one asper.*

By this definition pychoacoustical roughness is linked to the character of this reference sound. In the following the term roughness is used in this sense!

Its dependencies on the different signal parameters can be summarized as follows.

Threshold of roughness. The lower limit of roughness perception equals 0.07 asper for a 1-kHz tone modulated with 70 Hz.

Just noticeable difference. In order to perceive roughness differences, the degree of modulation of a sinusoidal amplitude-modulated sound has to be changed by about 10 % resulting in a just noticeable roughness difference of about 17 %.

Degree of modulation. Main factor influencing roughness perception is the degree of modulation m. For sinusoidal amplitude-modulated sounds perceived roughness R is on average proportional to $m^{1.6}$. (Figure 1)

Modulation frequency and carrier frequency. Roughness shows a bandpass characteristic in dependence on modulation frequency and carrier frequency. The bandpass ranges from about 12 to 300 Hz, with its maximum depending on the carrier frequency. It varies from a modulation frequency of about 30 Hz for a carrier frequency of 125 to 70 Hz for carrier frequencies of 1 kHz and higher. (Figure 2)

Sound pressure level. Sound pressure level has a minor but significant influence on roughness. Roughness increases by about 25 % when the SPL is increased by 10 dB. This implies on the

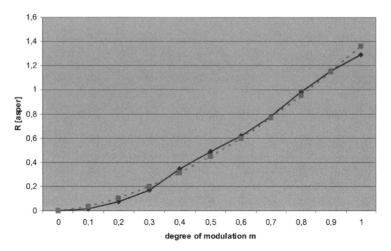

FIGURE 1 Roughness in dependence of the degree of modulation for a 1-kHz tone sinusoidal modulated at 70 Hz (SPL = 70 dB): jury test data (*dotted line*) from [17] approximated in very good agreement by the power function $m^{1.6}$, *solid line* model data.

other side that a reduction of the loudness of a rough component is often not solving sufficiently a sound quality problem.

Specific roughness. Investigations on roughness summation by Aures, Terhardt and Vogel [2,11,13] showed that total roughness sensation is composed of specific roughness, a concept which is similar to the composition of total loudness. The specific roughnesses of the critical bands simply add up to overall roughness if the contributing modulated parts of the sound do not interfere with the basilar membrane. (Figure 3)

Roughness of noise. Modulated broadband noise attains considerably higher roughness values than amplitude-modulated tones. The maximal roughness of amplitude-modulated broadband noise is about 4 asper. Even if noise is unmodulated, a considerable roughness sensation may occur for narrow band noises due to random envelope fluctuations (Figure 5). Unmodulated broadband noise however is not perceived as rough. As most environmental and technical sounds contain noise, it is very important to model the roughness of modulated and unmodulated noise correctly.

3 • Roughness Calculation

Two main approaches can be distinguished for the calculation of roughness: (1) calculating roughness out of an effective degree of modulation [2,5] and (2) calculating roughness out

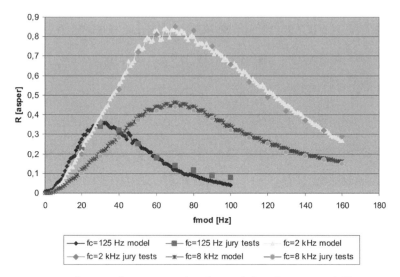

FIGURE 2 Roughness of 100% amplitude-modulated tones at different carrier frequencies in dependence of modulation frequency: model data (*solid lines*) and data from jury tests from [17] (bigger symbols, see legend).

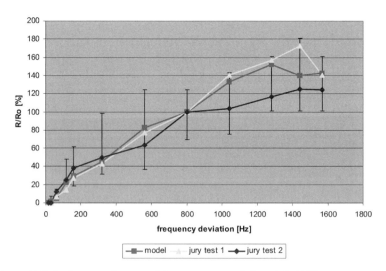

FIGURE 3 Relative roughness of frequency-modulated 1.6 kHz tones ($f_{\mathrm{mod}} = 70\,\mathrm{Hz}$, $L = 60\,\mathrm{dB}$) in dependence of frequency deviation. Two different standards were used in the jury tests from [7] (*black* and *white curve*). Quartiles are plotted for better visibility only for the jury test with weaker standard (an amplitude modulated 1.6 kHz tone). Data from model (*gray curve*) are within the jury test data.

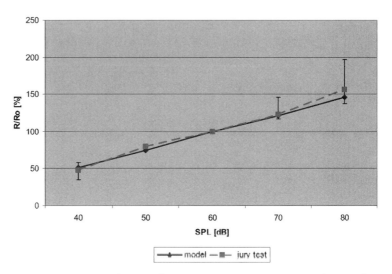

FIGURE 4 Relative roughness of a 1.6-kHz FM tone in dependence of level: *solid line* data from model; *dotted line* data from jury test of [7].

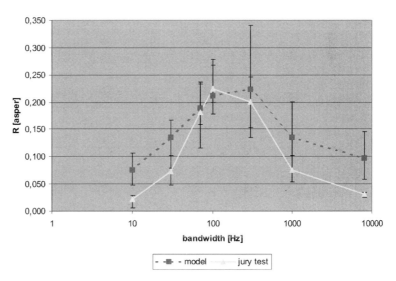

FIGURE 5 Roughness of unmodulated bandpass noise in dependence of bandwidth. Model data (*solid line* with quartiles of 21 calculated values) and jury test data from [1] (*dotted line* with quartiles).

of the product of modulation frequency and excitation level differences [6,16,17]. There are several different implementations available on the market including combinations of the two approaches. Their results differ often considerably for the same sound, and especially for low roughness values, the precision is often not sufficient to decide whether a sound contains rough components or not. Due to overestimation of specific roughness of uncorrelated noise events in the critical bands by various roughness models, calculated roughness of technical sounds often exceeds, e.g., the roughness of 0.1 asper by far, although they are not perceived as rough by experienced test persons. On a 5-point equally distant category scale ranging from not to very rough, category 3 (medium roughness) ranged, e.g., already at 0.1 asper, whereas a value of 0.05 asper near the threshold of roughness is ranked to be slightly rough [4].

3.1 Model Description

The goal of the described model is to provide calculated roughness values with a precision within the range of just noticeable difference of psychoacoustical roughness. As reference values for psychoacoustical roughness, the data of [17] are used. As a basis for this approach, the model of Aures [2] was selected in [4,5] due its flexibility. Several modifications of the model of Aures have been necessary in order to reduce deviations between the calculated roughness and data from jury tests below the just noticeable difference of 17%. The main changes are doubling of the analysis window from 100 ms in the model of Aures to 200 ms and a doubling of the analysis channels to model psychoacoustical roughness of amplitude-modulated tones in dependence of modulation and carrier frequency correctly. Although the correlation of the envelopes of neighboring critical bands is already taken into account in the model of Aures, a stronger reduction of roughness of uncorrelated narrow band noises had to be introduced in addition.

The structure of the model is shown in Figure 6 and it is summarized in the following (a more detailed description is given in [4,5]).

The roughness is calculated from time frames of 200 ms weighted by a Blackman window. After a discrete Fourier transformation, the spectrum is multiplied by the factor $a_0(f)$ (representing the transmission between free field and the peripheral hearing system) [17]. Then each frequency component is transformed into a triangular excitation pattern on the bark scale, modeling its excitation at the different places on the basilar membrane. The excitation level of the component is set to the level of the Fourier component at the place on the bark scale which corresponds to the frequency of the component.

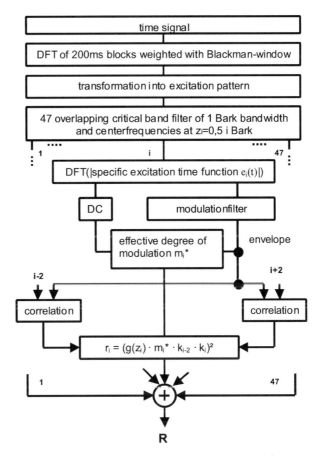

FIGURE 6 Structure of the roughness model.

The excitation decreases with a constant slope toward lower frequencies and with a level dependent slope toward higher frequencies. The slopes are chosen in accordance with the proposal of Terhardt [3]:

$S_1 = -27 \, \text{dB} \, /\text{Bark}$
for the lower slopes and
$S_2 = [-24 - 0.23 \text{kHz}/f + 0.2L/\text{dB}] \, \text{dB/Bark}$
for the level-dependent upper slopes, with f and L denoting frequency and level of the spectral component concerned.

The changes in levels and slopes between different frames result in a time-dependent filtering characteristic.

After the transformation into an excitation pattern, specific excitations are calculated within 47 overlapping 1-Bark-wide channels with equally spaced center frequencies z_i at

$$z_i = 0.5 \cdot i \text{ [Bark]}, \quad i = 1, 2, \ldots, 47.$$

A specific excitation spectrum in a channel i is obtained by linearly superimposing the excitation contributions in an interval $[z_i - 0.5, z_i + 0.5]$. The contribution of a spectral component is determined as follows: If the Bark value $z(f)$ of the frequency f of a component is higher than the upper limit of the interval $z(f) > z_i + 0.5$ Bark, the excitation of its lower slope S_1 at $z_i + 0.5$ Bark is taken as the contribution of the component to the specific excitation spectrum of the interval i (for the transformations from the frequency domain into the Bark domain $f \rightarrow z(f)$ the table in [17] is used with linear interpolations). If $z(f) < z_i - 0.5$ Bark, the contribution is the excitation of its upper slope S_2 at $z_i - 0.5$ Bark, and if the component f, i.e., $z(f)$ falls in the interval $z_i - 0.5$ Bark $\leq z(f) \leq z_i + 0.5$ Bark, its unchanged level L fully contributes to the specific excitation spectrum of the channel i. Excitation contributions which are less than the threshold in quiet L_{TQ} minus the logarithmic transmission factor a_0 in a given channel z_i are omitted (see also [17]). This way of calculating specific excitation spectra in each of the 47 bark channels allows the incorporation of the essential level-dependent nonlinear behavior of the excitation pattern into the model and is therefore superior to a constant linear filterbank as input stage. It is equivalent to the use of trapezoidal critical-band filters with level-dependent lower slopes.

The original phases of the frame spectrum are used in a subsequent inverse discrete Fourier transform of each of the 47 specific excitation spectra, yielding 200-ms-long specific excitation time functions $e_i(t)$.

The fluctuations of the envelope are contained in the low-frequency part of the spectrum $F(|e_i(t)|)$ of the absolute value of the specific excitations. This spectrum is appropriately weighted in order to model the bandpass characteristic of the roughness on modulation frequency. The weighting functions $H_i(f_{\text{mod}})$ have been optimized for that purpose.

The shapes of the weighting functions $H_i(f_{\text{mod}})$ of the channels i are described in detail as they have a decisive influence on the final relation between the total roughness R and the modulation frequency f_{mod} for AM tones. The spectral weighting functions $H_4(f_{\text{mod}})$ at $z_4 = 2$ Bark, $H_{16}(f_{\text{mod}})$ at $z_{16} = 8$ Bark and $H_{42}(f_{\text{mod}})$ at $z_{42} = 21$ Bark are drawn in Figure 7. They are basic in the sense that the weighting functions in the other channels can be derived from them in the following way:

for $i = 1, 2, \ldots, 4$ ($z_i = 0.5, 1, \ldots, 2$ Bark)
$H_1 = H_2 = H_3 = H_4$

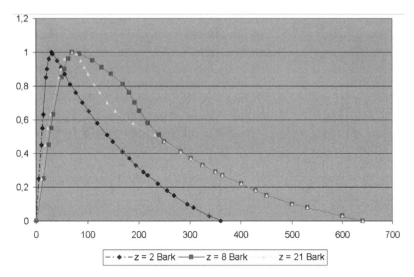

FIGURE 7 Weighting functions $H_i(f_{\mathrm{mod}})$ of the channels $i = 2, 16, 42$.

for $i = 6, 8, \ldots, 16$ ($z_i = 3, 4, \ldots, 8$ Bark)
$H_{i-1} = H_i$ and H_i is linearly interpolated between H_4 and H_{16}
for $i = 17, 18, 19, 20$ ($z_i = 7.5, 8, \ldots, 10$ Bark) $H_i = H_{16}$
for $i = 22, 24, \ldots, 42$ ($z_i = 11, 12, \ldots, 21$)
$H_{i-1} = H_i$ and H_i is linearly interpolated between H_{20} and H_{42}
for $i = 43, 44, \ldots, 47$ ($z_i = 21.5, 22, \ldots, 23.5$ Bark) $H_i = H_{42}$

The time functions of the bandpass-filtered envelopes $h_{\mathrm{BP},i}(t)$ are calculated via inverse Fourier transform.

A generalized degree of modulation m_i^* is calculated by

$$m_i^* = \frac{\tilde{h}_{\mathrm{BP}i}}{\mathrm{DC}_i} \quad \text{if} \quad \frac{\tilde{h}_{\mathrm{BP}i}}{\mathrm{DC}_i} \leq 1$$
$$\text{else} \quad m_i^* = 1.$$

with $\tilde{h}_{\mathrm{BP}i}$ equal to the rms value of the filtered envelope $h_{\mathrm{BP}}(t)$ and its DC value before the modulation filtering:

$$DC = \overline{|e_i(t)|}.$$

The roughness dependency on the frequency of the carrier is introduced into the model by multiplying the calculated m_i^* by a weighting function $g(z_i)$ with factors from 0.6 to 1.1 according to the dependence of roughness on the carrier frequency of amplitude-modulated tones.

A simple addition of the squares of $g(z_i) \cdot m_i^*$ would overestimate the total roughness if the modulations in the different channels are uncorrelated as in the case of noise events. Thus, the weighted generalized degree of modulation $g(z_i) \cdot m_i^*$ of each band is multiplied by the crosscorrelation coefficients k_{i-2} between the envelopes of the channels k_{i-2} and i and k_i calculated from channels i and $i+2$ with $\Delta z = 1$ Bark in both cases.[2] The squared product of the crosscorrelation factors instead of their average as used by Aures [1] has proved to be more effective in reducing roughness of unmodulated noise. The specific roughness is given by

$$r_i = (g(z_i) \cdot m_i^* \cdot k_{i-2} \cdot k_i)^2.$$

The use of the exponent 2 for defining the specific roughness does not necessarily mean that the overall relation between the degree of modulation and the entire calculated roughness will also be of power 2 as can be seen in Figure 1. The total roughness is given by

$$R = 0.25 \sum_{i=1}^{47} r_i \text{ [asper]}.$$

The calibration factor 0.25 is chosen such that the model produces a roughness of 1 asper for a 1-kHz tone at 60 dB SPL that is 100% amplitude modulated with a modulation frequency of 70 Hz.

3.2 Discussion

The model is fitted to the roughness of amplitude-modulated tones in dependence on modulation frequency and carrier frequency. Results for the degree of modulation and modulation frequency are shown for typical carrier frequencies in Figures 1–3 together with jury test data from [17]. A more detailed analysis can be found in [4,5]. The differences between results from jury tests and the model are in general within the just noticeable difference of 17%.

The roughness of unmodulated bandpass noise is also predicted fairly well by the model. Typical examples are shown in Figure 4 for bandpass noises centered at 4 kHz with varying bandwidth. Model data (solid line) with quartiles for 21 calculated values are plotted together with data from jury tests from [1]. Data for other center frequencies fit even better to the

---

[2] In the regions of $z = 0.5$, 1 Bark and $z = 23$, 23.5 Bark only the possible crosscorrelation factors k_i and k_{i-2}, respectively, are used. A compensation for the reduced correlation is given by the enlarged threshold of hearing in these regions.

data from jury tests [5]. For a bandwidth of 10 Hz the random fluctuation is about 6 Hz and thus calculated roughness near 0 asper. The jury test data are also near or below the threshold of roughness of 0.07 asper.

The model was verified by several tests, especially by comparing calculated roughness of frequency-modulated sounds with data from jury tests. Figure 7 shows the relative roughness of frequency-modulated 1.6 kHz tones ($f_{\mathrm{mod}} = 70$ Hz, $L = 60$ dB) in dependence of frequency deviation. Two different standards were used in the jury tests from [7] (black and white curves). Quartiles are plotted for better visibility only for the jury test with weaker standard (an amplitude-modulated 1.6 kHz tone). Data from model (gray curve) are within the jury test data. Figure 6 shows the dependence of roughness of frequency-modulated tones on sound pressure level. The data from jury tests from [7] are very well predicted by the model without being fitted to this parameter.

References

1. Aures, W.: Berechnungsverfahren für den Wohcklang Selicdiger Schallsignale, ein Beitrog zurigehörsezogenen Schall analyse. PhD thesis, Tu München (1984).
2. Aures, W.: Ein Berechnungsverfahren der Rauhigkeit. Acustica 59, 130–141 (1985).
3. Terhardt, E.: Calculating virtual pitch. Hearing Research 1, 155–182 (1979).
4. Daniel, P.: Berechnung und kategoriale Beurteilung der Rauhigkeit und Unangenehmheit von synthetischen und technischen Schallen. Fortschritt-Berichte VDI Reihe 10 Nr. 465. VDI Verlag, Düsseldorf (1997).
5. Daniel, P. and Weber, R.: Psychoacoustical roughness: implementation of an optimized model. Acustica – Acta Acustica 81, 1–12 (1995).
6. Fastl, H.: Roughness and temporal masking patterns of sinusoidal amplitude modulated broadband noise. Psychophysics and Physiology of hearing, Academic Press, London, 403–414 (1977).
7. Kemp, S.: Roughness of frequency-modulated tones. Acustica 50, 126–133 (1992).
8. Mellert, V. and Weber, R.: Gehörbezogene Verfahren zur Lärmsenrteilung. Walcher, K.P. and Schick, A., editors, Bedentungslehre des Schalls, Bern, Verlag Peter lang, 183 (1984).
9. Roederer, J.: Physikalische und psychoakustische Grundlagen der Musik. Springer, Berlin, Heidelberg (1977).
10. Terhardt, E.: Über akustische Rauhigkeit und Schwankungsstärke. Acustica 20, 215–224 (1968).
11. Terhardt, E.: On the perception of periodic sound fluctuation (roughness). Acustica 30, 201–213 (1974).

12. Über ein Äguivalenzgesetz für Intervalle akustischer Empfindungsgrößen. Kybernetik 5, 127–133 (1968).
13. Vogel, A.: Roughness and its relation to the time-pattern of psychoacoustical excitation, 241. Springer, Berlin, Heidelberg, New York (1974).
14. Vogel, A.: Über den Zusammenhang zwischen Rauhigkeit und Modulationsgrad. Acustica 32, 300–306 (1975).
15. von Helmholtz, H.: die Lehre von den Tonempfindungen. Wissenschaftliche Buchgesellschaft, Darmstadt, 7. unveränderte Auflage (1. Auflage Braunschweig 1863), (1968).
16. Widmann, U.: Ein Modell der Psychoakustischen Lästigkeit von Schallen und seine Anwendung in der Praxis der Lärmbeurteilung. PhD thesis, TU München (1992).
17. Zwicker, E. and Fastl, H.: Psychoacoustics. Springer, Berlin, Heidelberg (1990).

PART IV
MUSICAL ACOUSTICS

James W. Beauchamp

University of Illinois at Urbana-Champaign, Urbana, IL, USA

20 Automatic Music Transcription
Anssi Klapuri and Tuomas Virtanen 277

Introduction • Musical Sounds and F0 Estimation • Auditory Model-Based Multiple-F0 Analysis • Sound Separation Using Sinusoidal Modeling • Statistical Inference Within Parametric Signal Models • Unsupervised Learning Techniques • Summary and Conclusions

21 Music Structure Analysis from Acoustic Signals
Roger B. Dannenberg and Masataka Goto 305

Introduction • Features and Similarity Measures • Segmentation • The Similarity Matrix • Finding Repeating Sequences • Music Summary • Evaluation • Summary and Conclusions

22 Computer Music Synthesis and Composition
Lydia Ayers 333

Introduction • Music with Live Performers • Form, Structure and Time • Timbre and Texture • Tuning Systems • Musique Concrète and Collage • Environments • Conclusion

23 Singing Voice Analysis, Synthesis, and Modeling
Youngmoo E. Kim 359

Introduction • Principles of Singing Voice Modeling • Models of the Glottal Source • Models of the Vocal Tract • Synthesis of the Singing Voice • Conclusion

24 Instrument Modeling and Synthesis
Andrew B. Horner and James W. Beauchamp 375

Introduction • Spectral Analysis • Wavetable Synthesis • Wavetable Matching • Frequency Modulation (FM) Synthesis Models • Frequency Modulation Matching • Conclusions

25 Digital Waveguide Architectures for Virtual Musical Instruments
Julius O. Smith 399
Introduction • Vibrating Strings • Wind Instruments • Recommended Reading

26 Modeling of Musical Instruments
Rolf Bader and Uwe Hansen 419
Introduction • Finite-Element Formulation • Finite-Difference Formulation • Turbulent $k - \epsilon$ Model of Flute-Like Instruments • Multiphonics in Blown Instruments in Terms of a Split-Phase Pressure Description • Time-Dependent Model of a Classical Guitar Body • Guitar Back-Plate Timbre Brightening by Stress Bending • Church Bell Clapper Time-Dependent Striking Force

20
Automatic Music Transcription

Anssi Klapuri and Tuomas Virtanen

Institute of Signal Processing, Tampere University of Technology, Tampere, Finland

1 • Introduction

Written musical notation describes music in a symbolic form that is suitable for performing a piece using the available musical instruments. Traditionally, musical notation indicates the pitch, target instrument, timing, and duration of each sound to be played. The aim of music transcription either by humans or by a machine is to infer these musical parameters, given only the acoustic recording of a performance. In terms of data representations, this can be seen as transforming an audio signal into a MIDI[1] file. Signals of particular interest here are *polyphonic* music signals where several sounds are playing simultaneously.

Automatic recovery of the musical notation of an audio signal allows modifying, rearranging, and processing music at a high abstraction level and then resynthesizing it again. Structured audio coding is another important application: For example, a MIDI-like representation is extremely compact yet retains the characteristics of a piece of music to an important degree. Other uses of music transcription comprise information retrieval,

---

[1] Musical instrument digital interface (MIDI) is a standard for exchanging performance data and parameters between electronic musical devices.

musicological analysis of improvised and ethnic music, and interactive music systems which generate an accompaniment to the singing or playing of a soloist.

Attempts to automatically transcribe polyphonic music have been reported over a timescale of about 30 years, starting from the work of Moorer in 1970s [50].

State-of-the-art transcription systems are still clearly behind human musicians in accuracy and flexibility, but considerable progress has been made during the last 10 years, and it is the purpose of this chapter to survey selected literature representing these advances. The main emphasis will be on signal processing methods for resolving mixtures of pitched musical instrument sounds. Methods will be discussed both for estimating the fundamental frequencies (F0s) of concurrent sounds and for separating component sounds from a mixture signal. Other subtopics of music transcription, in particular beat tracking, percussion transcription, musical instrument classification, and music structure analysis, will not be discussed here, but an overview can be found in [37].

This chapter is organized as follows. Section 2 will introduce musical sounds and the basic principles of F0 estimation. Section 3 will discuss multiple-F0 estimators that follow principles of human hearing and pitch perception. Section 4 will describe sound separation methods that are based on the sinusoidal representation of music signals. Section 5 will discuss statistical inferences from parametric signal models. Section 6 describes unsupervised learning methods which make no assumptions about the nature of the sound sources and are thus suitable for percussion transcription, too. Concluding remarks are made in Section 7.

2 • Musical Sounds and F0 Estimation

In the majority of Western music, melody and harmony are communicated by *harmonic* sounds. These are sounds that are nearly periodic in the time domain and show a regular spacing between the significant spectral components in the frequency domain (Figure 1). An ideal harmonic sound consists of frequency components at integer multiples of its F0. In the case of plucked and struck string instruments, however, the partial frequencies are not in exact integral ratios but obey the formula $f_j = jF\sqrt{1+\beta(j^2-1)}$, where F is the fundamental frequency, j is the partial index, and β is an inharmonicity factor due to the stiffness of real strings [20]. Despite this imperfection, the general structure of the spectrum is similar to that in Figure 1.

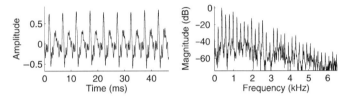

FIGURE 1 A harmonic sound in the time and frequency domains. The example represents a violin sound with fundamental frequency 196 Hz and fundamental period 5.1 ms.

Some of the methods discussed later in this chapter assume harmonic sounds as input. This limitation is not very severe in Western music where harmonic sounds are produced by most instrument families, namely by bowed and plucked string instruments, brass and reed instruments, flutes, pipe organs, and the human voice. Outside this category are mallet percussion instruments (e.g., vibraphone, xylophone, and marimba) whose waveforms are nearly periodic but whose spectra are definitely inharmonic.

A large number of different methods are available for monophonic F0 estimation. For speech signals, comparative evaluations of these can be found in [56,26,13]. Here, only the main ideas of different algorithms are introduced so as to provide a background for the analysis of polyphonic music signals described in detail in subsequent sections.

Algorithms that measure the periodicity of a time-domain signal have been among the most frequently used F0 estimators [64,14]. As noted in [13], quite accurate F0 estimation can be achieved simply by an appropriate normalization of the autocorrelation function (ACF) $r(\tau)$. Then, the F0 can be computed as the inverse of the lag τ that corresponds to the maximum of $r(\tau)$ within a predefined lag range. The ACF of a signal $x(n)$ within a K-length analysis frame is given by

$$r(\tau) = \frac{1}{K} \sum_{k=0}^{K-1-\tau} x(k)x(k+\tau). \qquad (1)$$

Figure 2 illustrates the ACF for the violin waveform shown in Figure 1.

An indirect way of measuring time-domain periodicity is to match a harmonic pattern to the signal in the frequency domain. According to the Fourier theorem, a periodic signal with period τ can be represented with a series of sinusoidal components at frequencies j/τ, where j is a positive integer.

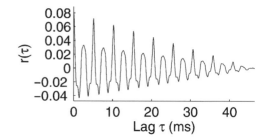

FIGURE 2 Autocorrelation function calculated within a 46 ms time frame for the violin waveform of Figure 1.

As an example, Doval and Rodet [16] performed maximum-likelihood spectral pattern matching to find the F0 which best explained the observed frequency partials. Other strategies for spectral pattern matching have been proposed by Brown [6] and Maher and Beauchamp [44]. The time-domain ACF calculation is closely related to these methods since, according to the Wiener–Khintchine theorem [25, p. 334], the ACF of a signal $x(n)$ is the inverse Fourier transform (IDFT) of its power spectrum $|X(k)|^2$. Writing out the IDFT we get

$$r(\tau) = \frac{1}{K} \sum_{k=0}^{K-1} \cos\left(\frac{2\pi\tau k}{K}\right) |X(k)|^2, \qquad (2)$$

where K is the size of the analysis frame. It is easy to see that Eq. (2) emphasizes frequency components at harmonic spectral locations (integer multiples of $k = K/\tau$) because $r(\tau)$ is maximized when the cosine maxima line up with the frequencies of the harmonics (Figure 4).

Another class of F0 estimators measures the periodicity of the Fourier spectrum of a sound [40]. These methods are based on the observation that harmonic sounds have a quasi-periodic magnitude spectrum, the period of which is the F0. In its simplest form, the autocorrelation function $\tilde{r}(m)$ of a K-length magnitude spectrum is calculated as

$$\tilde{r}(m) = \frac{2}{K} \sum_{k=0}^{K/2-m-1} |X(k)||X(k+m)|. \qquad (3)$$

The above formula bases F0 calculations on a fundamentally different type of information than the time-domain ACF: Here, any two frequency components with a certain spectral interval support the corresponding F0. An interesting difference between this method and the time-domain ACF is that measuring the periodicity of the time-domain signal is prone to F0 halving since the signal is periodic at twice the fundamental period too, whereas the methods that measure the periodicity of the magnitude spectrum are prone to F0 doubling since the spectrum is periodic at twice the F0 rate, too. There are ways to combine these complementary approaches to improve the overall result, as will be discussed in the next section.

3 • Auditory Model-Based Multiple-F0 Analysis

The human auditory system is very efficient at analyzing sound mixtures. It is therefore reasonable to learn from its function as much as possible, especially since the peripheral parts of hearing are relatively well known, and precise auditory models exist which are able to approximate the signal in the auditory nerve [49]. This enables the computation of a data representation similar to that used by the central auditory system.

3.1 Model of the Auditory Periphery

Computational models of the peripheral auditory system comprise two main parts which can be summarized as follows:

(1) An acoustic input signal is passed through a bank of bandpass filters (aka *channels*) which represents the frequency selectivity of the inner ear. Typically, about 100 filters are used with center frequencies f_c uniformly distributed on a critical-band scale,

$$f_c = 229(10^{\xi/21.4} - 1), \qquad (4)$$

where ξ is the critical band number. Usually *gammatone* filters are used, with the bandwidths b_c of the filters obeying $b_c = 0.108 f_c + 24.7$ Hz.

(2) The signal at each band is processed to simulate the transform characteristics of *hair cells* that produce neural impulses to the auditory nerve [27]. In signal processing terms, this involves compression, half-wave rectification, and lowpass filtering.

These seemingly very simple steps account for some important properties in pitch perception. In particular, the half-wave rectification operation in subbands allows a synthesis of the time-domain and frequency-domain periodicity analysis mechanisms discussed in Section 2. The half-wave rectification (HWR) is defined as

$$HWR(x) = \max(x, 0) = \frac{1}{2}(|x| + x). \qquad (5)$$

Figure 3 illustrates the HWR operation for a subband signal which consists of five overtones of a harmonic sound. What is

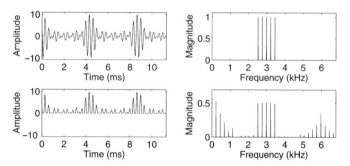

FIGURE 3 *Upper* panels show a signal consisting of the overtone partials 11–15 of a sound with F0 230 Hz in the time and the frequency domains. Lower panels illustrate the signal after half-wave rectification.

important is that the rectification generates spectral components that correspond to the frequency intervals between the input partials. The components below 1 kHz in the figure represent the amplitude envelope of the input signal. Any signal that consists of more than one frequency component exhibits periodic fluctuations, beating, in its time-domain amplitude envelope. That is, the partials alternatingly amplify and cancel out each other, depending on their phase. The rate of the beating caused by each pair of two frequency components depends on their frequency difference and, for a harmonic sound, the frequency interval corresponding to the F0 dominates.

After the rectification, periodicity analysis of the subband signals using, for instance, the ACF, would make use of both time and frequency domain periodicity by matching a harmonic pattern to both the input partials and the beating partials, which leads to more reliable F0 analysis [39]. Balance between the two types of information can be determined by applying a lowpass filter which partly suppresses the original passband at higher subbands but leaves the spectrum of the amplitude envelope intact.

3.2 Pitch Perception Models

The above-described two steps produce a simulation of the signal that travels in the different fibers ("channels") of the auditory nerve. Pitch perception models try to explain how this signal is further processed to derive a stationary percept of pitch. The processing mechanisms in the brain are not accurately known, but the prevailing view is that periodicity analysis of some form takes place for the signals within each auditory channel, and the results are then combined across channels [11]. Meddis and Hewitt [48] implemented these two steps as follows:

3) ACF estimates $r_c(\tau)$ are computed within channels.
4) The ACFs are summed across channels to obtain a summary autocorrelation function $s(\tau) = \sum_c r_c(\tau)$. The maximum of $s(\tau)$ within a predefined lag range is used to predict the perceived pitch.

Together with the above-mentioned peripheral processing stages, this became known as the "unitary model" of pitch perception, and the authors showed that it can reproduce many important phenomena in pitch perception, such as missing fundamental, repetition pitch, and pitch shift of equally spaced inharmonic components [48].

Some music transcription systems have applied an auditory model to compute an intermediate data representation that is then used by a higher level inference procedure. Martin [46] proposed a system for transcribing piano performances of four-voice Bach chorales. His system used the log–lag correlogram model of Ellis [17] (similar to the unitary model) as a front end to an inference architecture, where knowledge about sound production was integrated with rules governing tonal music. More recently, Marolt [45] used adaptive oscillators and neural networks to detect notes at the output of the peripheral hearing model described above.

3.3 Extensions of the Unitary Pitch Model

The pitch perception model outlined above is not sufficient as such for accurate multiple-F0 estimation. It suffers from certain shortcomings, and the following described methods can be seen as attempts to alleviate these problems. Most importantly, the model accounts only for a single simultaneous pitch. Several pitches in a mixture signal cannot be detected simply by picking several local maxima in the summary ACF. De Cheveigné and Kawahara [12] addressed this problem by proposing a system where pitch estimation was followed by the cancellation of the detected sound in order to reveal the other sounds in the mixture. The cancellation was performed either by subband selection or by performing within-band cancellation filtering, and the estimation step was then iteratively repeated for the residual signal.

Tolonen and Karjalainen [65] addressed the computational complexity of the unitary model by first prewhitening the input signal and then dividing it into only two subbands, below and above 1 kHz. A "generalized ACF" was then computed for the lower channel signal and for the amplitude envelope of the higher channel signal, and the two ACFs were summed. Despite the drastic reduction in computation compared to the original unitary model, many important characteristics of the model were preserved. Extension to multiple-F0 estimation was

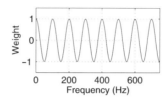 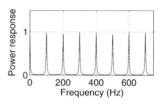

FIGURE 4 The *left* panel shows the weights $\cos(2\pi\tau k/K)$ of ACF calculation in (2), when the lag τ is 10 ms (corresponding to 100 Hz). The *right* panel shows the power response of a comb filter with a feedback lag of 10 ms.

achieved by cancelling subharmonics in the summary ACF, by clipping the summary ACF to positive values, timescaling it to twice its length, and by subtracting the result from the original clipped summary ACF. This cancellation operation was repeated for scaling factors up to about five. From the resulting enhanced summary autocorrelation function, all F0s were picked without iterative estimation and cancellation. The method is relatively easy to implement and produces good results when the component F0s are clearly below 1 kHz.

The generally weak robustness of the unitary model in polyphonic music was addressed by Klapuri [39]. He used a conventional peripheral hearing model (Section 3.1) but replaced the ACF by a periodicity analysis mechanism where a bank of comb filters was simulated in the frequency domain. Figure 4 illustrates the power response of a comb filter for a time lag candidate τ and, for comparison, the corresponding weights $\cos(2\pi\tau k/K)$ of the ACF calculation in (2). Giving less emphasis to the spectrum between the harmonic partials of an F0 candidate alleviates the interference of concurrent sounds. Estimation of multiple F0s was achieved by canceling each detected sound from the mixture and repeating the estimation for the residual, similarly to [12]. Quite accurate multiple-F0 estimation was achieved in the pitch range 60–2100 Hz.

4 • Sound Separation Using Sinusoidal Modeling

Sound separation refers to the task of estimating the signal produced by an individual sound source from a mixture. Multiple-F0 estimation and sound separation are closely related: An algorithm which achieves sound separation facilitates multiple-F0 estimation, and vice versa. Even when the F0s of musical tones have already been found, additional information obtained using sound separation is useful for classifying the tones to sources. For example, similar F0s and spectra in adjacent frames can be concatenated to form notes, and notes with different F0s can be grouped to obtain the whole passage of an instrument.

An efficient initial decomposition for the sounds produced by musical instruments is afforded by the sinusoids plus noise model, which represents the signal as a sum of deterministic and stochastic parts, or, as a sum of a set of sinusoids plus a noise residual [4,58]. Sinusoidal components represent the vibration modes of sound sources, and the residual contains the energy produced by the excitation mechanisms and other components which are not a result of these modes. The deterministic part of the model, which is called the sinusoidal model, has been used widely in audio signal processing, for example in speech coding by McAulay and Quatieri [47]. In musical signal processing, it became known by the work of Smith and Serra [61].

The usefulness of the sinusoidal model in sound source separation and automatic music transcription stems from its physical plausibility. Since all sinusoids correspond to the vibrating modes of the sources in a mixture, it is plausible to map the sinusoids into the individual sound sources and to estimate higher level information such as notes played by each source. Sound source separation algorithms which apply the sinusoidal model can be roughly divided into three categories: (1) methods which first estimate sinusoids and then group them into sound sources, (2) methods which first estimate F0s of the sources and then estimate sinusoids using partial frequencies predicted by the F0s, and (3) methods, which jointly estimate the number of sources, their F0s, and parameters of the sinusoids. The general signal model is discussed in Section 4.1, and the three separation approaches are discussed in Sections 4.2, 4.3, and 5, respectively.

4.1 Signal Model

The parameters of natural sound sources are usually slowly varying; therefore, the signal is analyzed in short segments or frames. In general, the parameters are assumed to be fixed during each frame, although continuous time-varying parameters are used in some systems. The sinusoidal model for one frame of a music signal can be written as

$$x(n) = \sum_{j=1}^{J} a_j \cos(2\pi f_j n + \theta_j) + e(n), \qquad (6)$$

where the J sinusoids represent the harmonic partials of all sources, n is the time index, a_j, f_j, and θ_j are the amplitude, frequency, and phase of the jth sinusoid, respectively, and $e(n)$ is the residual.

Basic algorithms for the estimation of sinusoids from music signals have been reviewed by Rodet [57] and by Serra [58], and a theorethical framework for the parameter estimation is discussed by Kay [35]. Usually the estimation is done in the frequency domain. The frequencies of sinusoids can be estimated by picking the most prominent peaks from the magnitude spectrum. Many useful practical details for estimating the peaks are given in [3]. Also, matching pursuit algorithms have been used, which use a dictionary of time-domain elements to decompose the signal. For harmonic sounds, the use of harmonic atoms [24] provides a decomposition which is a good basis for the analysis of music signals.

4.2 Grouping Sinusoids to Sources

Psychoacoustically motivated methods have been among the most widely used in sound source separation. The cognitive ability of humans to perceive and recognize individual sound sources in mixture signals is called *auditory scene analysis* [5]. Computational models of this function typically consist of two main stages where an incoming signal is first decomposed into its elementary time–frequency components and these are then organized to their respective sound sources.

Bregman pointed out a number of measurable acoustic "cues" which promote the grouping of time–frequency components to a common sound source as perceived by human listeners [5]. Among these are proximity in time–frequency, harmonic frequency relationships, synchronous changes in the frequency or amplitude of the components, and spatial proximity (i.e., the same direction of arrival). In the computational modeling of auditory scene analysis, the most widely used cues are the proximity in time and harmonic frequency relationships; usually the signals are analyzed in short frames, and they are assumed to be harmonic or close to harmonic.

Kashino and Tanaka [34] implemented a subset of Bregman's cues for the purpose of sound source separation in music signals. Using the sinusoidal model as a decomposition, the authors viewed the source separation task as two grouping problems where sinusoids were first clustered to sound events which were then grouped to particular sound sources (instruments). Harmonic mistuning and onset asynchrony were used as the cues to initialize new sound events, and sinusoids were then grouped to these. Grouping the sound events to their respective sources was achieved by using timbre models and an "old-plus-new" heuristic. The latter means that a complex sound is interpreted as a combination of "old sounds" as much as possible, and the remainder is perceived as a "new sound". Evaluation results were shown for polyphonies up to three simultaneous sounds and for several different instruments. In a subsequent paper, the authors implemented the system in the framework of a Bayesian probability network and integrated musical knowledge to it [33].

The two-way mismatch procedure proposed by Maher and Beauchamp implements grouping based on harmonic frequency relationships [44]. Their method can be used to find the most likely F0s of harmonic sounds within one frame, given the frequency and amplitude estimates of prominent sinusoids. The fundamental frequencies are chosen so that the mismatch error between the estimated sinusoids and the partials predicted from trial F0 values is minimized. The name two-way mismatch stems from a procedure in which each estimated sinusoid is matched to the closest predicted partial, and each predicted partial is matched to the closest estimated sinusoid, and the total mismatch is measured as the average discrepancy in all the matches. The exact procedure and mismatch function is explained in [44]. The method requires a search for possible F0s within a designated range, and the number of sources has to be set manually. The algorithm is relatively straightforward to implement and has been used for the separation of duet signals [43] and to analyze the melody line in a polyphonic recording [53].

Sterian [62] implemented perceptual grouping rules as a set of likelihood functions, each of which evaluated the likelihood of the observed sinusoids given a hypothesized grouping of the sinusoids to note candidates. Distinct likelihood functions were defined to take into account onset and offset timing, harmonicity, low partial support, partial gap, and partial density (see the reference for the definitions of the latter concepts). The product of all the likelihood functions was used as a criterion for optimal grouping.

Godsmark and Brown [21] applied an auditory model to estimate dominant time–frequency components at different bands. These were mapped to sound sources by performing grouping according to onset and offset synchrony, temporal and frequency proximity, harmonicity, and common frequency movement.

4.3 Sound Separation Given the Fundamental Frequencies

When the number of concurrent sounds is high, the robustness of the grouping methods presented in the previous section decreases mainly because overlapping partials (partials whose frequencies are very close or the same) are difficult to resolve at the grouping stage. Estimation of the partials can be done more robustly if some higher level information is available, such as the F0s of the sources, which can usually be estimated more accurately than individual sinusoids. This section deals with the estimation of sinusoids assuming that the F0s are estimated in advance using some other method.

Usually sinusoidal parameters are estimated by minimizing the reconstruction error between the sinusoidal model and the observed signal. Since the sinusoidal model is nonlinear with respect to frequencies, in general, a global solution for optimal parameters cannot be guaranteed. However, estimated F0s can be used to generate rough predictions of the partial frequencies, and iterative procedures which alternately estimate amplitudes and phases while frequencies are fixed and then update the frequencies, repeating until convergence is achieved, can be used. Most authors have used the energy of the residual to measure the reconstruction error, which leads to least-squares estimation. A general procedure, using a nonlinear least-squares method has been described in [63].

For fixed frequencies, amplitudes and phases can be estimated using the standard least-squares approach [36] which usually produces good results, even when the frequencies of the partials are close to each other. However, when the frequencies of two or more partials are equal, the exact amplitudes and phases of these partials cannot be resolved based on spectral peaks. This phenomenon is very common in musical signals, since F0s are often in harmonic relationships. However, an estimate of the amplitudes can be obtained by interpolating from adjacent frames or partials. The shape of a typical instrument spectrum is slowly varying with respect to time and frequency, so that, in general, interpolation of amplitudes produces tolerable results. The phases are perceptually less important, so they can be easily interpolated to produce smooth transitions between frames.

Quatieri and Danisewicz [55] used adjacent frame interpolation for separation of speech voices. The method is not as useful for musical signals since F0s are often in harmonic relationship for relatively long durations. Maher [43] suggested using interpolation from adjacent partials. However, a problem arises if several adjacent partials overlap. The method proposed by Virtanen and Klapuri [71] performs interpolation implicitly, since it forces smooth amplitude spectra of sources by means of linear overtone series models.

Using the estimated amplitudes and phases, frequencies can be updated, for example, by using an algorithm proposed by Depalle and Hélie [15]. Each partial's frequency can be separately estimated, or frequency ratios can be fixed, so that only the fundamental frequency is updated, as done by Quatieri and Danisewicz [55] and by Virtanen and Klapuri [70].

If an F0 estimate fails, the signal $x(n)$ will contain components for which no sinusoids have been assigned in the model given by Eq. (6). The additional components may affect the estimate, since both the target and the interfering sounds have a harmonic spectral structure. The effect of harmonic interference can be decreased, for example, by postprocessing the estimated parameters using a perceptually motivated spectral smoothing method proposed by Klapuri [38].

5 • Statistical Inference Within Parametric Signal Models

By making certain assumptions about the component sounds in a music signal, the multiple-F0 estimation problem can be viewed as estimating the parameters of a chosen signal model. In this section, statistical methods to this end are discussed.

5.1 Frequency-Domain Models

Let us first consider a model for the short-term power spectrum of a signal. Here it is useful to apply a logarithmic frequency scale and to measure frequency in cent units. One cent is 1/100 of a semitone in equal temperament and there are 1,200 cents in an octave. The relationship between cents and frequencies in Hertz is given by

$$z = 1200 \log_2(f_{Hz}/440) + 6900, \quad (7)$$

where the reference frequency 440 Hz is fixed to 6,900 cents.

To enable statistical modeling, the power spectrum $\Psi(z)$ of a signal is considered to be a distribution of small units of energy along the frequency axis. In terms of an underlying random variable ζ, a realization $\zeta = z$ is thought to assign a unit of spectral energy to cents-frequency z. Consider the following parametric model for the spectral energy distribution proposed by Goto [22]:

$$p(\zeta = z | \theta) = \sum_{i=1}^{I} \alpha_i \sum_{j=1}^{J_i} a_{i,j} G(z; \omega_i + 1200 \log_2(j), \sigma), \quad (8)$$

where $p(\zeta = z | \theta)$ represents the probability of observing energy at cents-frequency z given the model parameters θ. The function $G(z; z_0, \sigma)$ is a Gaussian distribution with mean z_0 and standard deviation σ. The above formula assumes a mixture of I harmonic sounds where ω_i is the fundamental frequency of sound i in cents,

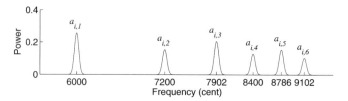

FIGURE 5 Illustration of the parametric model for spectral energy distribution in Eq. (8). Here ω_i is 6,000 cents (260 Hz).

α_i is the gain of sound i, and $a_{i,j}$ are the powers of the partials j of sound i. The shorthand notation θ represents all the parameters, $\theta = \{\omega_i, \alpha_i, a_{i,j}, I, J_i\}$.

Figure 5 illustrates the above model. Intuitively, each of the component sounds is modeled as a mixture of Gaussian distributions centered at integer multiples of F0. The variance around each partial models spreading of spectral energy due to time-domain windowing, F0 variations, and the inharmonicity phenomenon mentioned in Section 2. A nice feature of the model is that, due to the logarithmic frequency scale applied, the actual variance is larger for the higher order partials although σ is common for all.

Goto [23] extended the above model by introducing one additional dimension, so that multiple adaptive tone models were estimated for each F0. In practice, the weights became two-dimensional ($\alpha_{i,m}$) and the adaptive tone models three-dimensional ($a_{i,m,j}$), where m denotes the model index. This was done in order to allow various kinds of harmonic structures to appear at each F0 and to enable the use of several alternative tone model priors (spectral templates) side by side.

The problem to be solved, then, is to infer the model parameters θ given an observed short-time spectrum and possibly some prior information about the parameter distributions. Goto derived a computationally feasible expectation-maximization (EM) algorithm which iteratively updates the tone models and their weights, leading to maximum *a posteriori* parameter estimates. He avoided the explicit estimation of the F0 parameters in Eq. (8) by distributing a large number of F0s uniformly on the log-frequency scale and by estimating only the weights for each F0. The algorithm requires only a few iterations to converge and is suitable for real-time implementation. Goto used the method to detect the melody and the bass lines in real-world CD recordings. Temporal continuity of the estimates was considered by framewise tracking of the F0 weights within a multiple-agent architecture. Although the overall method in [23] is quite complex, the core EM algorithm is straightforward to implement and particularly suitable for detecting one (predominant) F0 in polyphonic signals.

Kameoka et al. [32] applied the signal model (8) and performed maximum-likelihood estimation of the F0s ω_i, weights α_i, and tone model shapes $a_{i,j}$ using another EM algorithm. In addition, they used an information theoretical criterion to estimate the number of concurrent sounds I. They proposed a two-stage estimation procedure where F0 values were first constrained within a one-octave range and equal amplitudes were used for all harmonics; then the true F0s were found among the harmonics and subharmonics of each candidate and the spectral shape of the sounds was estimated. Promising results were reported for polyphonic music transcription and for pitch estimation of two simultaneous speakers.

5.2. Time-Domain Models

In time-domain models, the main difference compared to frequency-domain models is that the phases of partials have to be taken into account. Walmsley [72] performed parameter estimation in a Bayesian framework using a reformulated sinusoidal model for the time-domain signal:

$$y(t) = \sum_{i=1}^{I} \gamma_i \sum_{j=1}^{J_i} \left[a_{i,j} \cos(j\omega_i t) + b_{i,j} \sin(j\omega_i t) \right] + e(t), \qquad (9)$$

where γ_i is a binary indicator variable switching the sound i on or off, ω_i is the F0 of sound i, and $a_{i,j}$, $b_{i,j}$ together encode the amplitude and phase of individual partials. The term $e(t)$ is a residual noise component. Similarly to Eq. (8), the above model assumes a mixture of harmonic signals, but a significant difference is that here the phases of the partials are also taken into account and estimated.

Another novelty in Walmsley's method was that the parameters were estimated jointly across a number of adjacent frames to increase robustness against transient events. A joint posterior distribution for all the parameters (given an observed signal segment) was defined which took into account the modeling error $e(t)$, prior distributions of the parameters, and the dependencies of the parameters on longer term hyperparameters that modeled frequency variation over time. Optimal parameter estimates were produced by locating regions of high probability in the joint posterior distribution. Whereas this method proved intractable for analytical optimization, Walmsley was able to use Markov chain Monte Carlo (MCMC) methods to generate samples from the posterior. A transition kernel was proposed which consisted of heuristic rules for the fast exploration of the parameter space. For example, sounds were switched on and off, F0 values were switched between their harmonics and subharmonics, and the residual was analyzed to detect additional notes.

Davy and Godsill [10] extended Walmsley's model to accommodate time-varying amplitudes, nonideal harmonicity, and nonwhite residual noise. They also improved the MCMC algorithm and reconsidered the prior structure. The resulting system was reported to work robustly for polyphonies up to three simultaneous sounds.

Another interesting time-domain model has been proposed by Cemgil et al. [9] who placed emphasis on explicitly modeling the sound generation process. They modeled musical sounds as a sum of harmonically related and damped "noisy" sinusoids drawn from a stochastic process. The damping factors were tied to a global factor which was assumed to be trained in advance. The sound generators were controlled by a collection of binary indicator variables $\gamma_{i,t}$, a "piano roll", which represented the activity of different notes i as a function of time t. A matrix $[\Gamma]_{i,t} = \gamma_{i,t}$ was used to denote the entire piano roll.

The transcription problem was then viewed as the task of finding a piano-roll Γ^* which maximizes the posterior probability $p(\Gamma|\mathbf{y})$ given a time-domain signal $\mathbf{y}$. Since this optimization task was analytically intractable, Cemgil et al. developed inference procedures for certain special cases. Chords were identified by a greedy algorithm which started from an initial configuration and iteratively added or removed single notes until the probability of the mixture converged. Piano-roll inference was achieved by analyzing the signal in short time-windows and by assuming that only one note may switch on or off within a window.

6 • Unsupervised Learning Techniques

Recently, information theoretical methods such as independent component analysis (ICA) have been successfully used to solve blind source separation problems in several application areas [30]. ICA is based on the assumption of statistical independence of sources, and in certain conditions, it enables the blind estimation of sources from mixed observations. The term blind means that there is no prior knowledge of the sources. Basic ICA cannot be directly used to separate one-channel time-domain signals, but this becomes possible with a suitable signal representation, as described in Section 6.1. In addition to ICA, other unsupervised learning algorithms such as nonnegative matrix factorization (NMF) [41] and sparse coding have been used in the analysis and separation of monaural (single-channel) music signals. Unlike ICA, NMF and sparse coding do not aim at statistical independence but just at representing the observed data efficiently. In the case of music signals, this usually results in the separation of sources, at least to some degree. The basic principles of ICA, NMF, and sparse coding are described in Sections 6.2, 6.3, and 6.4, respectively.

6.1 Signal Model for Monaural Signals

Basic ICA and many other blind source separation algorithms require that the number of sensors is greater than or equal to the number of sources. In multichannel sound separation, this means that there should be at least as many microphones as there are sources. However, automatic music transcription usually aims at finding notes in monaural or stereo signals, for which basic ICA methods may not be adequate. The most common approach to overcome this limitation is to represent the input signal in the time-frequency domain, for example by using the short-time Fourier transform (STFT). Basic generative models of ICA, NMF, and sparse coding are linear: For each analysis frame m, the short-time spectrum vector $\mathbf{x}_m$ can be expressed as a weighted sum of basis spectra $\mathbf{b}_j$. The model is not necessarily noise free. Thus, with a residual term $\mathbf{e}_m$, it can be written as

$$\mathbf{x}_m = \sum_{j=1}^{J} g_{j,m} \mathbf{b}_j + \mathbf{e}_m, \qquad (10)$$

where J is the number of basis spectra and $g_{j,m}$ is the gain of the jth basis spectrum for the mth frame. For M frames, the model can be written in a matrix form as

$$\mathbf{X} = \mathbf{BG} + \mathbf{E}, \qquad (11)$$

where $\mathbf{X} = [\mathbf{x}_1, \ldots \mathbf{x}_M]$ is the spectrogram matrix, $\mathbf{B} = [\mathbf{b}_1, \ldots, \mathbf{b}_J]$ is the basis matrix, $[\mathbf{G}]_{j,m}$ is the gain matrix, and $\mathbf{E} = [\mathbf{e}_1, \ldots, \mathbf{e}_M]$ is the residual matrix. Usually $\mathbf{X}$ is given as the input and $\mathbf{B}$, $\mathbf{G}$, and $\mathbf{E}$ are unknown. Each $\mathbf{b}_j$ corresponds to a short-time spectrum, which has a time-varying gain $g_{j,m}$. The term *component* is used to refer to one basis spectrum and its time-varying gain. Each sound source is usually modeled as a sum of one or more components. Multiple components per source are used since each pitch value of a harmonic source corresponds to a different spectrum. Also, even one note of an instrument may not have a stationary spectrum over time, so that multiple components may be needed. However, the model is flexible in the sense that it is suitable for representing both harmonic and percussive sounds. A simple

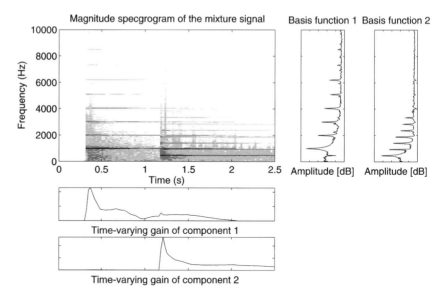

FIGURE 6 A mixture signal consisting of two piano notes (B5 and A#4, starting at times $t = 0.3$ and $t = 1.2$, respectively) and two separated components. Both components have a fixed spectrum with a time-varying gain.

two-note example is illustrated in Figure 6 where the components were estimated using the nonnegative matrix factorization algorithm described in [60] and [41].

The phase spectra of natural sound sources are very unpredictable. Therefore, the phases are often discarded and the estimation is done using the magnitude or power spectra. Linear superposition of time-domain signals does not imply linear superposition of the magnitude or power spectra. However, a linear superposition of incoherent time-domain signals can be approximated as a linear superposition of short-time power spectra.

In most systems, a discrete Fourier transform of fixed window size is used, but, in general, estimation algorithms allow the use of any time-frequency representation. For example, short-time signal processing has been used without an explicit frequency transform by Abdallah [1] and Jang and Lee [31]. It turns out that their systems learned basis functions from time-domain music and speech signals which were very similar to those used in wavelet and STFT analysis.

There are several alternative criteria for the estimation of the unknown parameters, the basic principles of which are discussed in the next three sections. In many cases, the number of basis functions, J, is unknown, and principal component analysis (PCA) can be used to estimate it. However, in the case of monaural sound source separation, the problem of estimating the number of sources J has not received much attention yet.

Once the short-time spectrum of the input signal is separated into components, the components are further analyzed to obtain musically important information. This includes, for example, the onset and offset times and F0s of each component. Ideally, a component is active when its gain $g_{j,m}$ is nonzero, but in practice, activity detection has to be made using a threshold which is larger than zero. There are some alternative possibilities for the estimation of the F0. For example, prominent peaks can be located from the spectrum and the two-way mismatch procedure [44] can be used, or the fundamental period can be estimated from an autocorrelation estimate. Sometimes a component may represent more than one pitch. This happens especially when notes occur consistently simultaneously, as in the case of chords.

6.2 Independent Subspace Analysis

In monaural sound source separation, the term independent subspace analysis (ISA) has been used to denote the ICA of the spectrogram of a time-domain signal [8]. In ISA, either the time-varying gains or the spectra of the components are assumed to be statistically independent. There are several criteria and algorithms for obtaining the statistical independency, but they all have a common mathematical background [42]. The criteria include, for example, nongaussianity and negentropy, which are usually measured using high-order cumulants such as kurtosis [30]. As a preprocessing for ICA, the observation matrix is usually whitened and the dimensions are reduced by PCA. The core of ICA is the estimation of an unmixing matrix. Independent components are obtained by multiplying the whitened observation matrix by the estimate of the unmixing matrix. There are several ICA algorithms, some of which are freely available, for example, FastICA [29] and JADE [7].

ISA has been used in several automatic music transcription and sound source separation systems, for example, by Casey and Westner in general audio source separation [8] and by FitzGerald in percussion transcription [18].

6.3 Nonnegative Matrix Factorization

If short-time power spectra are used as observations, each component has a fixed power spectrum with a time-varying gain. It is reasonable to restrict these to be entry-wise nonnegative, so that the components are purely additive and the spectra do not have negative values. A problem with ISA is that the standard ICA algorithms do not allow non-negativity restrictions, and in practice, the algorithms produce also negative values. This problem has been addressed, e.g., by Plumbley and Oja [54], whose nonnegative PCA algorithm solves the problem with some limitations.

Nonnegative matrix factorization (NMF) has been successfully used in unsupervised learning with the non-negativity restrictions [41]. When $\mathbf{X} \approx \mathbf{BG}$, the non-negativity of $\mathbf{B}$ and $\mathbf{G}$ alone seems to be a sufficient condition for the blind estimation of sources in many cases. In the analysis of music signals, NMF has been used, for example, by Smaragdis and Brown [60].

Lee and Seung proposed two cost functions and estimation algorithms for obtaining $\mathbf{X} \approx \mathbf{BG}$ [41]. The cost functions are the square of the Euclidean distance, or, the Frobenius norm of the error between the observation matrix $\mathbf{X}$ and the model $\mathbf{BG}$, given by

$$\|\mathbf{X} - \mathbf{BG}\|_F^2 = \sum_{k,m} \left([\mathbf{X}]_{k,m} - [\mathbf{BG}]_{k,m}\right)^2 \tag{12}$$

and divergence D, defined as

$$D(\mathbf{X}\|\mathbf{BG}) = \sum_{k,m} d\left([\mathbf{X}]_{k,m}, [\mathbf{BG}]_{k,m}\right), \tag{13}$$

where the function d is given by

$$d(p,q) = p \log\left(\frac{p}{q}\right) - p + q. \tag{14}$$

Both cost functions are lower bounded by zero, which is obtained only when $\mathbf{X} = \mathbf{BG}$. The estimation algorithms presented in [41] initialize $\mathbf{B}$ and $\mathbf{G}$ with random values, and then update them iteratively, so that the value of the cost funtion is nonincreasing at each update.

Since the NMF aims only at representing the observed spectrogram with nonnegative components, it does not guarantee the separation of sources. Especially when the sources are always present simultaneously, the algorithm tends to represent them with a single component. However, it has turned out that the factorization of a magnitude spectrogram by minimizing the divergence given by Eq. (13) produces better separation results than ISA in most cases [69].

6.4 Sparse Coding

A technique called sparse coding has been successfully used to model the early stages of vision [51]. The term sparse refers to a signal model in which the data are represented in terms of a small number of active elements chosen out of a larger set.

The sparseness restriction is usually placed for the gains $[\mathbf{G}]_{j,m}$ in Eq. (11). Sparseness of $\mathbf{G}$ means that the probability of an element of $\mathbf{G}$ being zero is high, so that only a few components are active at a time, and each component is active only in a small number of frames. For musical sources, this is usually a valid assumption, since each component typically corresponds to a single pitch value or to a percussive source.

A sparse representation is obtained by minimizing a cost function which is the sum of a reconstruction error term and a term which incurs a penalty on the nonzero elements of $\mathbf{G}$. An example of such a cost function c is given by

$$c(\mathbf{B},\mathbf{G}) = ||\mathbf{X}-\mathbf{B}\mathbf{G}||_F^2 + \lambda \sum_{j,m} f([\mathbf{G}]_{j,m}). \tag{15}$$

The function f is used to penalize nonzero entries of $\mathbf{G}$, and the scalar $\lambda > 0$ is used to balance the reconstruction error cost and the sparseness cost. For example, Olshausen and Field [51] used $f(x) = \log(1+x^2)$, and Hoyer [28] used $f(x) = |x|$. This approach requires that the scale of either $\mathbf{B}$ or $\mathbf{G}$ is fixed, for example to unity variance.

As in NMF, the parameters are usually solved using iterative algorithms. Hoyer [28] proposed a nonnegative sparse coding algorithm by combining NMF and sparse coding. For musical signal analysis, sparse coding has been used for example by Abdallah and Plumbley [1,2] to analyze pitched sounds and by Virtanen [67] to transcribe drums from synthesized MIDI signals.

It is not clear whether the explicit assumption of sparseness ($\lambda > 0$) really increases the quality of the separation. In [69] the sparseness criterion did not improve the average SNR of the separated sources, although it may be useful for musical transcription.

6.5 Discussion

Manual music transcription requires a lot of prior knowledge and training. It is not known whether automatic transcription of music is possible without prior information of the sources, for example the knowledge that they are harmonic. In some simple cases, it is possible to estimate the components without prior information, but this may not always be the case. Some attempts to utilize prior information have been made. Usually they are based on supervised learning since it is difficult to constrain harmonic basis functions in the model given by Eq. (10). Vincent and Rodet proposed a polyphonic transcription system based on ISA and hidden Markov models which were trained using monophonic material [66]. FitzGerald [18] used spectral templates of instruments as an initialization of his prior subspace analysis based drum transcription system. The time-varying gains obtained for the components were further processed using ICA. All these systems use a limited instrument set. In general, all the possible instruments cannot be trained in advance but some kind of model adaptation is needed.

The linear model (10) is not well suited for separating singing voice signals, since different phonemes have different spectra. Also, other instruments with a strongly varying spectral shape are problematic. Researchers have tried to overcome the limitations of the model by using a more complex model which includes a two-dimensional time–frequency basis function for each component instead of a static spectrum. Initial experiments with this kind of approach have been presented by Smaragdis [59] and Virtanen [68]. The model can also be used to represent time-varying fundamental frequencies [19].

The dynamic range of music signals is wide, and low-intensity observations may be perceptually important. The power spectrum domain is problematic in the sense that it causes separation algorithms to concentrate on high-intensity observations, thus failing to separate low-energy sources. This has been addressed by Vincent and Rodet [66], who used a specific algorithm in the log-power spectral domain, and by Virtanen [68], who used perceptually motivated weights to mimic human loudness perception.

Automatic music transcription and sound source separation using unsupervised learning techniques is currently an active research topic, and none of the proposed methods is clearly better than others. The existing algorithms are most successful in cases where instrument sets and polyphony are limited. For example, FitzGerald [18] reported good results for an algorithm extended from ISA, and Paulus and Virtanen [52] successfully used NMF for the transcription of drum patterns consisting of bass, snare, and hi-hat drums.

7 • Summary and Conclusions

As the preceding sections have shown, music transcription can be performed using very different kinds of methods and assumptions. The underlying assumptions made for unsupervised learning methods are completely different than those made for signal-model-based statistical inference methods, yet both approaches can yield meaningful results. At the present time, none of the described main approaches stands out as clearly the most promising. Instead, one of the decisive factors determining the popularity of different methods is their conceptual simplicity vs. performance. A problem with many transcription systems is that they are very complex entities. However, there are methods in all the main categories that are quite straightforward to implement and lead to good analysis results. Among these are, for example, the auditorily oriented method of Tolonen and Karjalainen [65], the sinusoidal-modeling method of Maher and Beauchamp [44], the expectation-maximization algorithm of Goto [23] for detecting the most predominant F0 in polyphonic signals, and the non-negative matrix factorization algorithm of Smaragdis and Brown [60].

Research is being carried out to combine the advantages of different approaches, and in many cases, this has produced the most successful results. Also, it should be remembered that this chapter has focused primarily on acoustic signal analysis methods without addressing the use of musicological information or the larger scale structure of music signals.

References

1. S. A. Abdallah. *Towards Music Perception by Redundancy Reduction and Unsupervised Learning in Probabilistic Models*. PhD thesis, Dept. Electronic Eng., King's College London, 2002.
2. S. A. Abdallah and M. D. Plumbley. An independent component analysis approach to automatic music transcription. *Audio Eng. Soc. 114th Convention*, Preprint No. 5754, Amsterdam, Netherlands, March 2003.
3. M. Abe and J. O. Smith. Design criteria for simple sinusoidal parameter estimation based on quadratic interpolation of FFT magnitude peaks. *Audio Eng. Soc. 117th Convention*, Preprint No. 6256, San Francisco, CA, 2004.
4. X. Amatriain, J. Bonada, A. Loscos, and X. Serra. Spectral processing. In U. Zölzer, editor, *DAFX – Digital Audio Effects*. Wiley New York, 2002.
5. A. S. Bregman. *Auditory Scene Analysis*. MIT Press, Cambridge, MA, 1990.
6. J. C. Brown. Musical fundamental frequency tracking using a pattern recognition method. *J. Acoust. Soc. Am.*, 92(3):1394–1402, 1992.
7. J.-F. Cardoso. High-order contrasts for independent component analysis. *Neural Computation*, 11(1), 157–192, 1999.
8. M. A. Casey and A. Westner. Separation of mixed audio sources by independent subspace analysis. *Proc. 2000 Int. Computer Music Conf.*, Berlin, Germany, 2000, pp. 154–161.
9. A. T. Cemgil, B. Kappen, and D. Barber. A generative model for music transcription. *IEEE Trans on Speech and Audio Processing*, 14(2):679–694, 2006.

10. M. Davy and S. Godsill. Bayesian harmonic models for musical signal analysis. *Proc. Seventh Valencia Int. Meeting on Bayesian Statistics 7*, pp. 105–124, Tenerife, Spain, June 2002.
11. A. de Cheveigné. Pitch perception models. In C. J. Plack, A. J. Oxenham, R. R. Fay, and A. N. Popper, editors, *Pitch*. Springer, New York, 2005, pp. 169–233.
12. A. de Cheveigné and H. Kawahara. Multiple period estimation and pitch perception model. *Speech Communication*, 27:175–185, 1999.
13. A. de Cheveigné and H. Kawahara. Comparative evaluation of F0 estimation algorithms. *Proc. 7th European Conf. Speech Communication and Technology*, pp. 2451–2454, Aalborg, Denmark, 2001.
14. A. de Cheveigné and H. Kawahara. YIN, a fundamental frequency estimator for speech and music. *J. Acoust Soc. of Am.*, 111(4): 1917–1930, 2002.
15. Ph. Depalle and T. Hélie. Extraction of spectral peak parameters using a short-time fourier transform modeling and no sidelobe windows. *Proc. 1997 IEEE Workshop on Applications of Signal Processing to Audio and Acoustics*, pp. 19–22, New Palz, NY, 1997.
16. B. Doval and X. Rodet. Estimation of fundamental frequency of musical sound signals. *Proc. 1991 IEEE International Conference on Acoustics, Speech, and Signal Processing*, pp. 3657–3660, Toronto, Canada, 1991.
17. D. P. W. Ellis. *Prediction-Driven Computational Auditory Scene Analysis*. PhD thesis, Massachusetts Institute of Technology, 1996.
18. D. FitzGerald. *Automatic Drum Transcription and Source Separation*. PhD thesis, Dublin Institute of Technology, 2004.
19. D. FitzGerald, M. Cranitch, and E. Coyle. Generalised prior subspace analysis for polyphonic pitch transcription. *Proc. Int. Conf. Digital Audio Effects*, pp. 77–81, Madrid, Spain, 2005.
20. N. H. Fletcher and T. D. Rossing. *The Physics of Musical Instruments*. Springer, Berlin, Germany, 2nd edition, 1998.
21. D. Godsmark and G. J. Brown. A blackboard architecture for computational auditory scene analysis. *Speech Communication*, 27(3): 351–366, 1999.
22. M. Goto. A robust predominant-F0 estimation method for real-time detection of melody and bass lines in cd recordings. *Proc. IEEE Int. Conf. Acoustics, Speech and Signal Processing*, pp. 757–760, Istanbul, Turkey, June 2000.
23. M. Goto. A real-time music scene description system: predominant-F0 estimation for detecting melody and bass lines in real-world audio signals. *Speech Communication*, 43(4):311–329, 2004.
24. R. Gribonval and E. Bacry. Harmonic decomposition of audio signals with matching pursuit. *IEEE Trans. Signal Processing*, 51(1):101–111, 2003.
25. W. M. Hartmann. *Signals, Sound, and Sensation*. Springer, New York, 1998.
26. W. J. Hess. Pitch and voicing determination. In S. Furui and M. M. Sondhi, editors, *Advances in Speech Signal Processing*, pp. 3–48. Marcel Dekker, New York, 1991.

27. M. J. Hewitt and R. Meddis. An evaluation of eight computer models of mammalian inner hair-cell function. *J. Acoust Soc. Am.*, 90(2):904–917, 1991.
28. P. Hoyer. Non-negative sparse coding. In *IEEE Workshop on Networks for Signal Processing XII*, Martigny, Switzerland, 2002.
29. A. Hyvärinen. Fast and robust fixed-point algorithms for independent component analysis. *IEEE Trans. Neural Networks*, 10(3):626–634, 1999.
30. A. Hyvärinen, J. Karhunen, and E. Oja. *Independent Component Analysis*. Wiley, New York, 2001.
31. G.-J. Jang and T.-W. Lee. A maximum likelihood approach to single channel source separation. *J. Machine Learning Research*, 23: 1365–1392, 2003.
32. H. Kameoka, T. Nishimoto, and S. Sagayama. Separation of harmonic structures based on tied gaussian mixture model and information criterion for concurrent sounds. In *IEEE Int. Conf. on Acoustics, Speech, and Signal Processing*, (ICASSP 2004), Montreal, 2004.
33. K. Kashino, K. Nakadai, T. Kinoshita, and H. Tanaka. Organisation of hierarchical perceptual sounds: music scene analysis with autonomous processing modules and a quantitative information integration mechanism. In *Int. Joint Conf. Artificial Intelligence*, pp. 158–164, Montreal, 1995.
34. K. Kashino and H. Tanaka. A sound source separation system with the ability of automatic tone modeling. *Proc. 1993 Int. Computer Music Conf.*, pp. 248–255, Hong Kong, China, 1993.
35. M. Kay. *Modern Spectral Estimation*. Prentice Hall, Englewood Cliffs, NJ, 1988.
36. S. M. Kay. *Fundamentals of Statistical Signal Processing: Estimation Theory*. Prentice Hall, Englewood Cliffs, NJ 1993.
37. A. Klapuri and M. Davy, editors. *Signal Processing Methods for Music Transcription*. Springer, New York, 2006.
38. A. P. Klapuri. Multiple fundamental frequency estimation based on harmonicity and spectral smoothness. *IEEE Trans. Speech and Audio Processing*, 11(6):804–815, 2003.
39. A. Klapuri. Multipitch analysis of polyphonic music and speech signals using an auditory model. *IEEE Trans. Audio, Speech and Language Processing*, 16(2):255–266, 2008.
40. M. Lahat, R. Niederjohn, and D. A. Krubsack. Spectral autocorrelation method for measurement of the fundamental frequency of noise-corrupted speech. *IEEE Trans. Acoustics, Speech, and Signal Processing*, 6:741–750, June 1987.
41. D. D. Lee and H. S. Seung. Algorithms for non-negative matrix factorization. In *Neural Information Processing Systems*, pp. 556–562, Denver, CO, 2001.
42. T.-W. Lee, M. Girolami, A. J. Bell, and T. J. Sejnowski. A unifying information-theoretic framework for independent component analysis. *Computers and Mathematics with Applications*, 31(11): 1–21, 2000.

43. R. C. Maher. Evaluation of a method for separating digitized duet signals. *J. Audio Eng. Soc.*, 38(12):956–979, 1990.
44. R. C. Maher and J. W. Beauchamp. Fundamental frequency estimation of musical signals using a two-way mismatch procedure. *J. Acoust. Soc. Am.*, 95(4):2254–2263, 1994.
45. M. Marolt. SONIC: transcription of polyphonic piano music with neural networks. *Proc. MOSART Workshop on Current Research Directions in Computer Music*, pp. 217–224, Barcelona, Spain, November 2001.
46. K. D. Martin. Automatic transcription of simple polyphonic music: robust front end processing. Technical Report 399, MIT Media Laboratory, Perceptual Computing Section, 1996.
47. R. J. McAulay and T. F. Quatieri. Speech analysis/synthesis based on a sinusoidal representation. *IEEE Trans. on Speech and Audio Processing*, 34(4):744–754, 1986.
48. R. Meddis and M. J. Hewitt. Virtual pitch and phase sensitivity of a computer model of the auditory periphery. I: pitch identification. *J. Acousti. Soc. Am.*, 89(6):2866–2882, 1991.
49. B. C. J. Moore, editor. *Hearing – Handbook of Perception and Cognition*. Academic Press, San Diego, CA, 2nd edition, 1995.
50. J. A. Moorer. *On the Segmentation and Analysis of Continuous Musical Sound by Digital Computer*. PhD thesis, Department of Music, Stanford University, 1975. Distributed as Department of Music report No. STAN-M-3.
51. B. A. Olshausen and D. F. Field. Sparse coding with an overcomplete basis set: a strategy employed by V1? *Vision Research*, 37:3311–3325, 1997.
52. J. Paulus and T. Virtanen. Drum transcription with non-negative spectrogram factorisation. *European Signal Processing Conf.*, Antalya, Turkey, 2005.
53. G. Peterschmitt, E. Gómez, and P. Herrera. Pitch-based solo location. *Proc. MOSART Workshop on Current Research Directions in Computer Music*, Barcelona, Spain, 2001.
54. M. D. Plumbley and E. Oja. A "non-negative PCA" algorithm for independent component analysis. *IEEE Trans. on Neural Networks*, 15(1):66–67, 2004.
55. T. F. Quatieri and R. G. Danisewicz. An approach to co-channel talker interference suppression using a sinusoidal model for speech. *IEEE Trans. Acoustics, Speech, and Signal Processing*, 38(1):56–69, 1990.
56. L. R. Rabiner, M. J. Cheng, A. E. Rosenberg, and C. A. McGonegal. A comparative performance study of several pitch detection algorithms. *IEEE Trans. Acoustics, Speech, and Signal Processing*, 24(5):399–418, 1976.
57. X. Rodet. Musical sound signal analysis/synthesis: Sinusoidal+ residual and elementary waveform models. *Proc. IEEE Time-Frequency and Time-Scale Workshop*, Coventry, UK, 1997.
58. X. Serra. Musical sound modeling with sinusoids plus noise. In C. Roads, S. Pope, A. Picialli, and G. De Poli, editors, *Musical Signal Processing*. pp. 91–122, Swets & Zeitlinger, 1997.

59. P. Smaragdis. Discovering auditory objects through non-negativity constraints. *Proc. ISCA Tutorial and Research Workshop on Statistical and Perceptual Audio Processing*, Jeju, Korea, 2004.
60. P. Smaragdis and J. C. Brown. Non-negative matrix factorization for polyphonic music transcription. *Proc. IEEE Workshop on Applications of Signal Processing to Audio and Acoustics*, pp. 177–180, New Palz, NY, 2003.
61. J. O. Smith and X. Serra. Parshl: an analysis/synthesis program for Proc. non-harmonic sounds based on a sinusoidal representation. *Int. Computer Music Conf.*, pp. 290–297, Urbana, IL, 1987.
62. A. D. Sterian. *Model-Based Segmentation of Time–Frequency Images for Musical transcription*. PhD thesis, University of Michigan, 1999.
63. P. Stoica and R. L. Moses. *Introduction to Spectral Analysis*. Prentice Hall, Englwood Cliffs, NJ, 1997.
64. D. Talkin. A robust algorithm for pitch tracking. In W. B. Kleijn and K. K. Paliwal, editors, *Speech Coding and Synthesis*, pp. 495–517. Elsevier, Amsterdam, 1995.
65. T. Tolonen and M. Karjalainen. A computationally efficient multi-pitch analysis model. *IEEE Trans. on Speech and Audio Processing*, 8(6):708–716, 2000.
66. E. Vincent and X. Rodet. Music transcription with ISA and HMM. *Proc. 5th Int. Symp. on Independent Component Analysis and Blind Signal Separation*, pp. 1119–1204, London, U.K., 2004.
67. T. Virtanen. Sound source separation using sparse coding with temporal continuity objective. *Proc. 2003 Int. Computer Music Conference*, pp. 231–234, Singapore, 2003.
68. T. Virtanen. Separation of sound sources by convolutive sparse coding. *Proc. ISCA Tutorial and Research Workshop on Statistical and Perceptual Audio Processing*, Jeju, Korea, 2004.
69. T. Virtanen, Monaural Sound Source Separation by Non-Negative Matrix Factorization with Temporal Continuity and Sparseness Criteria. *IEEE Trans. Audio, Speech, and Language Processing*, 15(3):1066–1074, 2007
70. T. Virtanen and A. Klapuri. Separation of harmonic sounds using multipitch analysis and iterative parameter estimation. *Proc. IEEE Workshop on Applications of Signal Processing to Audio and Acoustics*, pp. 83–86, New Paltz, NY, 2001.
71. T. Virtanen and A. Klapuri. Separation of harmonic sounds using linear models for the overtone series. *Proc. IEEE Int. Conf. Acoustics, Speech and Signal Processing*, pp. 1747–1760, Orlando, FL, 2002.
72. P. J. Walmsley. *Signal Separation of Musical Instruments. Simulation-Based Methods for Musical Signal Decomposition and Transcription*. PhD thesis, Department of Engineering, University of Cambridge, September 2000.

21
Music Structure Analysis from Acoustic Signals

Roger B. Dannenberg[1] and Masataka Goto[2]

[1] *Carhegie Mellon University, Pittsburgh, PA, USA*
[2] *National Institute of Advanced Industrial Science and Technology (AIST), Tsukuba, Japan*

1 • Introduction

Music is full of structure, including sections, sequences of distinct musical textures, and the repetition of phrases or entire sections. The analysis of music audio relies upon feature vectors that convey information about music texture or pitch content. Texture generally refers to the average spectral shape and statistical fluctuation, often reflecting the set of sounding instruments, e.g., strings, vocal, or drums. Pitch content reflects melody and harmony, which is often independent of texture. Structure is found in several ways. Segment boundaries can be detected by observing marked changes in locally averaged texture. Similar sections of music can be detected by clustering segments with similar average textures. The repetition of a sequence of music often marks a logical segment. Repeated phrases and hierarchical structures can be discovered by finding similar sequences of feature vectors within a piece of music. Structure analysis can be used to construct music summaries and to assist music browsing.

Probably everyone would agree that music has structure, but most of the interesting musical information that we perceive lies hidden below the complex surface of the audio signal. From this signal, human listeners perceive vocal and instrumental lines, orchestration, rhythm, harmony, bass lines, and other features. Unfortunately, music audio signals have resisted our attempts to extract this kind of information. Researchers are making progress, but so far, computers have not come near to human levels of performance in detecting notes, processing rhythms, or identifying instruments in a typical (polyphonic) music audio texture.

On a longer time scale, listeners can hear structure including the chorus and verse in songs, sections in other types of music, repetition, and other patterns. One might think that without the reliable detection and identification of short-term features such as notes and their sources, it would be impossible to deduce any information whatsoever about even higher levels of abstraction. Surprisingly, it *is* possible to automatically detect a great deal of information concerning music structure. For example, it is possible to label the structure of a song as AABA, meaning that opening material (the "A" part) is repeated once, then contrasting material (the "B" part) is played, and then the opening material is played again at the end. This structural description may be deduced from low-level audio signals. Consequently, a computer might locate the "chorus" of a song without having any representation of the melody or rhythm that characterizes the chorus.

Underlying almost all work in this area is the concept that structure is induced by the repetition of similar material. This is in contrast to, say, speech recognition, where there is a common understanding of words, their structure, and their meaning. A string of unique words can be understood using prior knowledge of the language. Music, however, has no language or dictionary (although there are certainly known forms and conventions). In general, structure can only arise in music through repetition or systematic transformations of some kind.

Repetition implies there is some notion of similarity. Similarity can exist between two points in time (or at least two very short time intervals), similarity can exist between two sequences over longer time intervals, and similarity can exist between the longer-term statistical behaviors of acoustical features. Different approaches to similarity will be described.

Similarity can be used to segment music: contiguous regions of similar music can be grouped together into segments. Segments can then be grouped into clusters. The segmentation of a musical work and the grouping of these segments into clusters is a form of analysis or "explanation" of the music.

2 • Features and Similarity Measures

A variety of approaches are used to measure similarity, but it should be clear that a direct comparison of the waveform data or individual samples will not be useful. Large differences in waveforms can be imperceptible, so we need to derive features of waveform data that are more perceptually meaningful and compare these features with an appropriate measure of similarity.

2.1 Feature Vectors for Spectrum, Texture, and Pitch

Different features emphasize different aspects of the music. For example, Mel-frequency cepstral coefficients (MFCCs) seem to work well when the general shape of the spectrum but not necessarily pitch information is important. MFCCs generally capture overall "texture" or timbral information (what instruments are playing in what general pitch range), but some pitch information is captured, and results depend upon the number of coefficients used as well as the underlying musical signal.

When pitch is important, e.g., when searching for similar harmonic sequences, the chromagram is effective. The chromagram is based on the idea that tones separated by octaves have the same perceived value of *chroma* (Shepard, 1964). Just as we can describe the *chroma* aspect of pitch, the short-term frequency spectrum can be restructured into the *chroma spectrum* by combining energy at different octaves into just one octave. The *chroma vector* is a discretized version of the *chroma spectrum* where energy is summed into 12 log-spaced divisions of the octave corresponding to pitch classes (C, C#, D, ...B). By analogy to the spectrogram, the *discrete chromagram* is a sequence of chroma vectors.

It should be noted that there are several variations of the chromagram. The computation typically begins with a short-term Fourier transform (STFT) which is used to compute the magnitude spectrum. There are different ways to "project" this onto the 12-element chroma vector. Each STFT bin can be mapped directly to the most appropriate chroma vector element (Bartsch and Wakefield, 2001), or the STFT bin data can be interpolated or windowed to divide the bin value among two neighboring vector elements (Goto, 2003a). Log magnitude values can be used to emphasize the presence of low-energy harmonics. Values can also be averaged, summed, or the vector can be computed to conserve the total energy. The chromagram can also be computed by using the wavelet transform.

Regardless of the exact details, the primary attraction of the chroma vector is that, by ignoring octaves, the vector is relatively insensitive to overall spectral energy distribution and thus to timbral variations. However, since fundamental frequencies and lower harmonics of tones feature prominently in the calculation of the chroma vector, it is quite sensitive to pitch class content, making it ideal for the detection of similar harmonic sequences in music.

While MFCCs and chroma vectors can be calculated from a single short-term Fourier transform, features can also be obtained from longer sequences of spectral frames. Tzanetakis and Cook (1999) use means and variances of a variety of features in a 1-s window. The features include the spectral centroid, spectral rolloff, spectral flux, and RMS energy.

Peeters et al. (2002) describe "dynamic" features, which model the variation of the short-term spectrum over windows of about 1 s. In this approach, the audio signal is passed through a bank of Mel filters. The time-varying magnitudes of these filter outputs are each analyzed by a short-term Fourier transform. The resulting set of features, the Fourier coefficients from each Mel filter output, is large, so a supervised learning scheme is used to find features that maximize the mutual information between feature values and hand-labeled music structures.

2.2 Measures of Similarity

Given a feature vector such as the MFCC or chroma vector, some measure of similarity is needed. One possibility is to compute the (dis)similarity using the Euclidean distance between feature vectors. Euclidean distance will be dependent upon feature magnitude, which is often a measure of the overall music signal energy. To avoid giving more weight to the louder moments of music, feature vectors can be normalized, for example, to a mean of zero and a standard deviation of one or to a maximum element of one.

Alternatively, similarity can be measured using the scalar (dot) product of the feature vectors. This measure will be larger when feature vectors have a similar direction. As with Euclidean distance, the scalar product will also vary as a function of the overall magnitude of the feature vectors. If the dot product is normalized by the feature vector magnitudes, the result is equal to the cosine of the angle between the vectors. If the feature vectors are first normalized to have a mean of zero, the cosine angle is equivalent to the correlation, another measure that has been used with success.

Lu et al. (2004) use a constant-Q transform (CQT), and found that CQT outperforms chroma and MFCC features using a cosine distance measure. They also introduce a "structure-based" distance measure that takes into account the harmonic structure of spectra to emphasize pitch similarity over timbral similarity, resulting in additional improvement in a music structure analysis task.

Similarity can be calculated between individual feature vectors, as suggested above, but can also be computed over a window of feature vectors. The measure suggested by Foote (1999) is vector correlation:

$$S_w(i,j) = \frac{1}{w} \sum_{k=0}^{w-1} (V_{i+k} \bullet V_{j+k}), \qquad (1)$$

where w is the window size. This measure is appropriate when feature vectors vary with time, forming significant temporal patterns. In some of the work that will be described below, the detection of temporal patterns is viewed as a processing step that takes place after the determination of similarity.

2.3 Evaluation of Features and Similarity Measures

Linear prediction coefficients (LPC) offer another low-dimensional approximation to spectral shape, and other encodings such as moments (centroid, standard deviation, skewness, etc.) are possible. Aucouturier and Sandler (2001) compare various approaches and representations. Their ultimate goal is to segment music according to texture, which they define as the combination of instruments that are playing together. This requires sensitivity to the general spectral shape, and insensitivity to the spectral details that vary according to pitch. They conclude that a vector of about 10 MFCCs is superior to LPC and discrete cepstrum coefficients (Galas and Rodet, 1990).

On the other hand, Hu et al. (2003) compare features for detecting similarity between acoustic and synthesized realizations of a single work of music. In this case, the goal is to ignore timbral differences between acoustic and synthetic instruments, but to achieve fine discrimination of pitches and harmonies. They conclude that the chroma vector is superior to pitch histograms and MFCCs.

3 • Segmentation

One approach to discovering structure in music is to locate segments of similar musical material and the boundaries between them. Segmentation does not rely on classification or the discovery of higher order structure in music. However, one can envision using segmentation as a starting point for a number of more complicated tasks, including music summarization, music analysis, music search, and genre classification. Segmentation can also assist in audio browsing, a task that can be enhanced through some sort of visual summary of music and audio segments.

3.1 Segmentation Using Texture Change

Tzanetakis and Cook (1999) perform segmentation as follows: Feature vectors V_i are computed as described above. A feature time differential, Δ_i, is defined as the Mahalanobis distance:

$$\Delta_i = (V_i - V_{i-1})^T \Sigma^{-1} (V_i - V_{i-1}), \tag{2}$$

where Σ is an estimate of the feature covariance matrix, calculated from the training data, and i is the frame number (time). This measure is related to the Euclidean distance but takes into account the variance and correlations among features. Next, the first order differences of the distance, $\Delta_i - \Delta_{i-1}$, are computed. A large difference indicates a sudden transition. Peaks are picked, beginning with the maximum. After a peak is selected, the peak and its neighborhood are zeroed to avoid picking another peak within the same neighborhood. Assuming the total number of segments is given a priori, the neighborhood is 20% of the average segment size. Additional peaks are selected and zeroed until the desired number of peaks (segment boundaries) has been obtained.

3.2 Segmentation by Clustering

Logan and Chu (2000) describe a clustering technique for discovering music structure. The goal is to label each frame of audio so that frames within similar sections of music will have the same labels. For example, all frames within all occurrences of the chorus should have the same label. This can be accomplished using bottom-up clustering to merge clusters that are similar. Initially, the feature vectors are divided into fixed-length contiguous segments and each segment receives a different label. The following clustering step is iterated:

Calculate the mean μ and covariance Σ of the feature vectors within each cluster. Compute a modified Kullback–Leibler (KL) distance between each pair of clusters, as described below. Find the pair of clusters with the minimum KL2 distance, and if this distance is below a threshold, combine the clusters. Repeat this step until no distance is below the threshold.

The KL2 distance between two Gaussian distributions A and B is given by

$$KL2(A,B) = KL(A;B) + KL(B;A) \tag{3}$$

$$= \frac{\Sigma_A}{\Sigma_B} + \frac{\Sigma_B}{\Sigma_A} + (\mu_A - \mu_B) \cdot \left(\frac{1}{\Sigma_A} + \frac{1}{\Sigma_B}\right). \tag{4}$$

3.3 Segmentation and Hidden Markov Models

Another approach to segmentation uses a hidden Markov model (HMM). In this approach, segments of music correspond to discrete states Q and segment transitions correspond to state changes. Time advances in discrete steps corresponding to feature vectors and transitions from one state to the next are modeled by a probability distribution that depends only on the current state. This forms a Markov model that generates a sequence of states. Note that states are "hidden" because only feature vectors are observable. Another probability distribution, $p(V_i|q_i)$, models the generation of feature vector V_i from state q_i. The left side of Figure 1 illustrates a four-state ergodic Markov model, where arrows represent state transition probabilities. The right side of the figure illustrates the observation generation process, where arrows denote conditional probabilities between variables.

The HMM has advantages for segmentation. In general, feature vectors do not indicate the current state (segment class) unambiguously, so when a single feature vector is observed, one cannot assume that it was generated by particular state. However, some features are more likely to occur in one state than another, so one can observe the *trend* of feature vectors, ignoring the unlikely outliers and guessing the state that is most consistent with the observations. If transitions are very unlikely, one may have to assume many outliers occur. On the other hand, if transitions are common and segments are short, one can change states rapidly to account for different feature vectors. The HMM formalism can determine the segmentation (the hidden state sequence) with the maximum likelihood given a set of transition probabilities and observations, thus the model can formalize the trade-offs between minimizing transitions and matching features to states. Furthermore, HMM transition probabilities can be estimated from unlabeled training data, eliminating the need to guess transition probabilities manually.

Aucouturier and Sandler (2001) model the observation probability distribution $P(V_i|q_j)$ as a mixture of Gaussian distributions over the feature space:

$$P(V_i|q_j) = \sum_{m=1}^{M} c_{j,m} \cdot \mathbf{N}(V_i, \mu_{j,m}, \Gamma_{j,m}), \qquad (5)$$

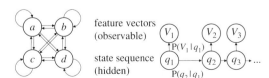

FIGURE 1 Hidden Markov model with four hidden states a, b, c, and d. As shown, feature vectors depend only upon the current state, which depends only upon the previous state.

where **N** is a Gaussian probability density function with mean $\mu_{i,m}$, covariance matrix $\Gamma_{j,m}$, and $c_{j,m}$ is a mixture coefficient. Here, i indexes time and j indexes state. They train the HMM using the Baum–Welch algorithm using the sequence vectors from the single song chosen for analysis. The Viterbi algorithm is then used to find the sequence of hidden states with the maximum likelihood, given the observed feature vectors.

One potential drawback of this approach is that the HMM will segment the signal according to fine-grain changes in spectral content rather than long-term elements of musical form. For example, in one of Aucoturier and Sandler's test cases (see Figure 2), the HMM segmentation appears to isolate individual words of a singer rather than divide the song according to verses and instrumental interludes (Aucouturier et al., 2005). In other words, the segments can be quite short when there are rapid changes in the music. Although this might be the desired result, it seems likely that one could detect longer-term, higher-level music structure by averaging features over a longer time span or applying further processing to the state sequence obtained from an HMM.

Peeters et al. (2002) approach the problem of clustering with a two-pass algorithm. Imagine a human listener hearing a piece of music for the first time. The range of variation of music features becomes apparent, and templates or classes of music are formed. In the second hearing, the structure of the music can be identified in terms of the previously identified templates.

An automated system is inspired by this two-pass model. In the first pass, texture change indicates segment boundaries, and "potential" states are formed from the mean values of feature vectors within segments. In the second pass, potential states that are highly similar are merged by using the K-means algorithm. The resulting K states are called the "middle" states. Because they represent clusters with no regard for temporal contiguity, a hidden Markov model initialized with these "middle" states is then used to inhibit rapid inappropriate state transitions by penalizing them. The Baum–Welch algorithm is used to train the model on the sequence of feature vectors from the song. Viterbi decoding is used to obtain a state sequence. Figure 3 shows the result of an analysis using this smoothing technique.

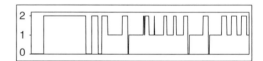

FIGURE 2 Segmentation of 20 s of a song. State 0 is silence, State 1 is voice, accordion, and accompaniment, and State 2 is accordion and accompaniment.

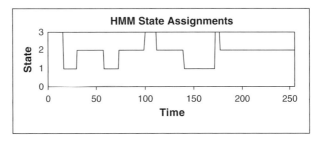

FIGURE 3 Classification of states in "Head Over Feet" from artist Alanis Morisette. (Adapted from Peeters et al. 2002).

4 • The Similarity Matrix

A concept used by many researchers is the similarity matrix. Given a sequence of feature vectors V_i and a measure of similarity $S(i,j)$, one can simply view $S(i,j)$ as a matrix. The matrix can be visualized using a grayscale image where black represents dissimilar vectors and white represents similar vectors. Shades of gray represent intermediate values. Since any vector is similar to itself, the diagonal of the similarity matrix will be white. Also, assuming the similarity measure is symmetric, the matrix will be symmetric about the diagonal. The interesting information in the matrix is in the patterns formed off the diagonal.

In very general terms, there are two interesting sorts of patterns that appear in the similarity matrix, depending on the nature of the features. The first of these appears when features correspond to relatively long-term textures. The second appears when features correspond to detailed short-term features such as pitch or harmony and where similar sequences of features can be observed. These two types of patterns are considered in the next two sections.

4.1 Texture Patterns

First, consider the case where features represent the general texture of the music, for example whether the music is primarily vocal, drum solo, or guitar solo. Figure 4 shows an idealized similarity matrix for this case. The white diagonal appears because feature vectors along the diagonal are identical. Notice that wherever there are similar textures, the matrix is lighter in color (more similar), so for example, all of the feature vectors for the vocals (V) are similar to one another, resulting in large light-colored square patterns both on and off the diagonal. Where two feature vectors correspond to different textures, for example drums and vocals, the matrix is dark.

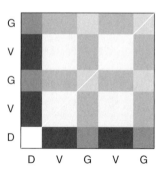

FIGURE 4 An idealized similarity matrix for segments of drum (D), vocal (V), and guitar (G) texture.

Notice that along the diagonal, a checkerboard pattern appears at segment boundaries, with darker regions to the upper left and lower right, and lighter regions to the lower left and upper right. Foote (2000) proposes the correlation of the similarity matrix S with a kernel based on this checkerboard pattern in order to detect segment boundaries. The general form of the kernel is

$$C = \begin{bmatrix} 1 & 1 & -1 & -1 \\ 1 & 1 & -1 & -1 \\ -1 & -1 & 1 & 1 \\ -1 & -1 & 1 & 1 \end{bmatrix}. \qquad (6)$$

(Note that in Eq. (6), row numbers increase in the downward direction whereas in the similarity matrix images, the row number increases in the upward direction. Therefore the diagonal in Eq. (6) runs from upper left to lower right.) The kernel image in Figure 5 represents a larger checkerboard pattern with radial smoothing. The correlation $N(i)$ of this kernel along the diagonal

Checkerboard Kernel

Checkerboard Kernel with Radial Smoothing

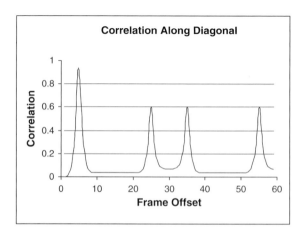

FIGURE 5 The correlation of the kernel shown at *lower left* with a similarity matrix.

of a similarity matrix S can be considered to be a measure of novelty (Foote, 2000):

$$N(i) = \sum_{m=-L/2}^{L/2} \sum_{n=-L/2}^{L/2} C(m,n) S(i+m, i+n). \qquad (7)$$

A graph of $N(i)$ for the similarity matrix in Figure 4 is shown in Figure 5. A peak occurs at each transition because transition boundaries have the highest correlation to the checkerboard pattern.

Cooper and Foote (2003) extend this technique for finding segment boundaries with a statistical method for clustering segments.

4.2 Repeating Sequence Patterns

While the texture patterns described above are most useful for detecting transitions between segments, the second kind of pattern can be used to discover repetition within a song. For these patterns to appear, it is important that features reflect short-term changes. Generally, features should vary significantly with changes in the pitch of a melody or with changes in harmony. If this condition is satisfied, then there will not be great similarity within a segment and there will not be a clear pattern of light-colored squares as seen in Figure 4. However, if a segment of music repeats with an offset of j, then $S(i,i)$ will equal $S(i, i+j)$, generating a diagonal line segment at an offset of j from the central diagonal. This is illustrated schematically in Figure 6, where it is assumed that the vocal sections (V) constitute three repetitions of very similar music, whereas the two guitar sections (G) are not so similar. Notice that each non-central diagonal line segment indicates the starting times and the duration of two similar sequences of features. Also, notice that since each pair of similar sequences is represented by two diagonal line segments, there are a total of six (6) off-central line segments in Figure 6.

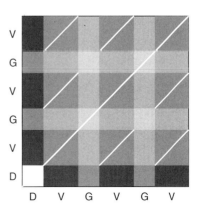

FIGURE 6 When sections of music are repeated, a pattern of diagonal line segments is generated.

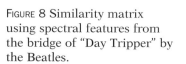

FIGURE 7 The vocal segments (V) in this similarity matrix contain a repetition, generating additional pattern that is characteristic of music structure.

Although not shown in Figure 6, the similarity matrix can also illustrate hierarchical relationships. For example, if each vocal section (V) consists of a phrase that is repeated, the similarity matrix would look like the one in Figure 7.

Figure 8 illustrates both texture patterns and repeated sequence patterns from the song "Day Tripper" by the Beatles. The bridge is displayed, starting with three repetitions of a two-measure guitar phrase in the first 11 s, followed by six measures of vocals. Notice how a checkerboard pattern appears due to the timbral self-similarity of the guitar section (0–11 s) and the vocal section (11–21 s). Finer structure is also visible. A repeated sequence pattern appears within the guitar section as parallel diagonal lines. This figure uses the power spectrum below 5.5 kHz as the feature vector, and uses the cosine of the angle between vectors as a measure of similarity.

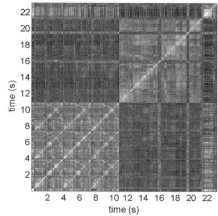

FIGURE 8 Similarity matrix using spectral features from the bridge of "Day Tripper" by the Beatles.

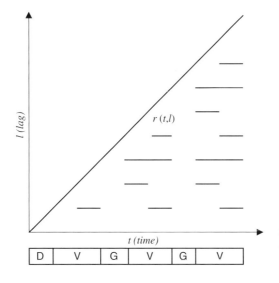

FIGURE 9 Time-lag matrix representation of the similarity matrix in Figure 7.

4.3 The Time-Lag Matrix

When the goal is to find repeating sequence patterns, it is sometimes simpler to change coordinate systems so that patterns appear as horizontal or vertical lines. The *time-lag* matrix r is defined by

$$r(t, l) = S(t, t - l), \text{ where } t - l \geq 0. \tag{8}$$

Thus, if there is repetition, there will be a sequence of similar frames with a constant lag. Since lag is represented by the vertical axis, a constant lag implies a horizontal line. The time-lag version of Figure 7 is shown in Figure 9. Only the lines representing similar sequences are shown, and the grayscale has been reversed, so that similarity is indicated by black lines.

5 • Finding Repeating Sequences

Of course, with audio data obtained from real audio recordings, the similarity or time-lag matrix will be full of noise and ambiguity arising from spurious similarity between different frames. Furthermore, repetitions in music are rarely exact; variation of melodic, harmonic, and rhythmic themes is an essential characteristic of music. In order to automate the discovery of musical structure, algorithms must be developed to identify the structure that lies within the similarity matrix.

5.1 Melodic Sequence Matching

One way to find repetition is to transcribe the melody and perform matching on the resulting symbolic transcription. While extracting the melody from a polyphonic recording (Goto, 2004) is very difficult in general, an approximate transcription from an instrumental recording or from a monophonic (melody only) recording is relatively easy. Dannenberg (2002) describes a simple transcription system based on the enhanced autocorrelation algorithm (Tolonen and Karjalainen, 2000) applied to a ballad recorded by John Coltrane. The transcription results in a quantized integer pitch value p_i and real inter-onset interval d_i for each note (inter-onset intervals are typically preferred over note duration in music processing). This sequence is processed as follows:

1. First, a similarity matrix is constructed where rows and columns correspond to notes. This differs from the similarity matrix described above where rows and columns correspond to feature vectors with a fixed duration.
2. Each cell of the similarity matrix $S(i,j)$ represents the duration of similar melodic sequences starting at notes i and j. A simple "greedy" algorithm is used to match these two sequences. If note i does not match note j, $S(i,j) = 0$.
3. Simplify the matrix by removing redundant entries. If a sequence beginning at i matches one at j, then there should be another match at $i+1$ and $j+1$. To simplify the matrix, find the submatrix $S(i:u, j:v)$ where the matching sequences at i and j end at u and v. Zero every entry in the submatrix except $S(i,j)$. Also, zero all entries for matching sequences of length 1.
4. Now, any non-zero entry in S represents a pair of matching sequences. By scanning across rows of S we can locate all similar sequences. Sequences are clustered: The first non-zero element in a row represents a cluster of two sequences. Any other non-zero entry in the row that roughly matches the durations of the clustered sequences is added to the cluster. After scanning the row, all pair-wise matches are zeroed so they will not be considered again.

The result of this step is a set of clusters of similar melodic segments. Because repetitions in the music are not exact, there can be considerable overlap between clusters. It is possible for a long segment to be repeated exactly at one offset, and for a portion of that same segment to be repeated several times at other offsets. It may be desirable to simplify the analysis by labeling each note with a particular cluster. This simplification is described in the next section.

5.2 Simplification or Music Explanation

The goal of the simplification step is to produce *one* possible set of labels for notes. Ideally, the labels should offer a simple "explanation" of the music that highlights repetition within the music. The AABA structure common in songs is a typical explanation. In general, longer sequences of notes are preferable because they explain more, but when sequences are too long, interesting substructure may be lost. For example, the structure AABAAABA could also be represented as AABA repeated, i.e., the structure could be labeled AA, but most theorists would consider this to be a poor explanation. Hierarchical explanations offer a solution, but there is no formal notion as yet of the optimal simplification or explanation.

Dannenberg uses a "greedy" algorithm to produce reasonable explanations from first note to last (Dannenberg and Hu, 2002). Notes are initially unlabeled. As each unlabeled note is encountered, search the clusters from the previous section to find one that includes the unlabeled note. If a cluster is found, allocate a new label, e.g., "A", and label every note included in the cluster accordingly. Continue labeling with the next unlabeled note until all notes are processed.

Figure 10 illustrates output from this process. Notice that the program discovered a substructure within what would normally be considered the "bridge" or the B part, but this substructure is "real" in the sense that one can see it and hear it. The gap in the middle of the piece is a piano solo where transcription failed. Notice that the program correctly determines that the saxophone enters on the bridge (the B part) after the piano solo. The program also identifies the repeated two-measure phrase at the end. It fails to notice the structure of ascending pitches at the very end because, while this is a clear musical gesture, it is not based on the repetition of a note sequence.

FIGURE 10 A computer analysis of "Naima" by John Coltrane. The automatic transcription appears as a *"piano roll"* at the top, the computer analysis appears as *shaded bars*, where *similar shading* indicates similar sequences, and conventional labels appear at the bottom.

5.3 Finding Similar Sequences in the Similarity Matrix

Typically, transcription of a music signal into a sequence of notes is not possible, so similar sequences must be detected as patterns in the similarity or time-lag matrix. For example, Bartsch and Wakefield (2001) filter along diagonals of a similarity matrix to detect similarity. This assumes nearly constant tempo, but that is a good assumption for the popular music used in their study. Their objective was not to identify the beginnings and endings of repeating sequences but to find the chorus of a popular song for use as an "audio thumbnail" or summary. The thumbnail is selected as the maximum element of the filtered similarity matrix, with the additional constraints that the lag is at least one-tenth of the length of the song and the thumbnail does not appear in the last quarter of the song.

Peeters and Rodet (2003) suggest using a 2D structuring filter on the lag matrix to detect similar sequences. Their filter counts the number of values in the neighborhood to the left and right of a point that are above a threshold. To allow for slight changes in tempo, which results in lines that are not perfectly horizontal, neighbor cells above and below are also considered. Lu et al. (2004) suggest erosion and dilation operations on the lag matrix to enhance and detect significant similar sequences.

Dannenberg and Hu (2003) use a discrete algorithm to find similar sequences which is based on the idea that a path from cell to cell through the similarity matrix specifies an alignment between two subsequences of the feature vectors. If the path goes through $S(i,j)$, then vector i is aligned with j. This suggests using a dynamic time warping (DTW) algorithm (Rabiner and Juang, 1993), and the actual algorithm is related to DTW.

The goal is to find alignment paths that maximize the average similarity of the aligned features. A partial or complete path P is defined as a set of pairs of locations and is rated by the average similarity along the path:

$$q(P) = \frac{1}{|P|} \sum_{(i,j) \in P} S(i,j), \qquad (9)$$

where $|P|$ is the path length using Euclidean distance. Paths are extended as long as the rating remains above a threshold. Paths are constrained to move up one cell, right one cell, or diagonally to the upper right as shown in Figure 11 (and adopting the orientation of the similarity matrix visualizations where time increases upward and to the right). Therefore, every point that is on a path can be reached from below, from the left, or from the lower left. Each cell (i,j) of an array is computed by looking at the cell below, left, and below left to find the (previously calculated) best path (highest $q(P)$) passing through those cells. Three new

FIGURE 11 Extending a path from $S(i,j)$.

ratings of r are computed by extending each of the three paths to include (i,j). The path with the highest rating is remembered as the one passing through (i,j).

Because cells depend on previously computed values to the lower left, cells are computed along diagonals of constant $i+j$, from lower left to upper right (increasing $i+j$). When no path has a rating above some fixed threshold, the path ends. A path may begin wherever $S(i,j)$ is above threshold and no previous paths exist to be extended.

5.4 Forming Clusters

After alignment paths are found, they are grouped into clusters. So if sequence A aligns to sequence B, and sequence A also aligns to sequence C, then A, B, and C should be grouped in a single cluster. Unfortunately, it is unlikely that the alignments of A to B and A to C use exactly the same frames. It is more likely that A aligns to B and A′ aligns to C, where A and A′ are mostly overlapping. This can be handled simply by considering A to equal A′ when they start and end within some fraction of their total length, for example within 10%. Once clusters are formed, further simplification and explanation steps can be performed as described above.

5.5 Isolating Line Segments from the Time-Lag Matrix

If nearly constant tempo can be assumed, the alignment path is highly constrained and the alignment path approach may not work well. Taking advantage of the fact that similar sequences are represented by horizontal lines in the time-lag matrix, Goto (2003a) describes an alternative approach to detecting music structure. In this work, the time-lag matrix is first normalized by subtracting a local mean value while emphasizing horizontal lines. In more detail, given a point $r(t,l)$ in the time-lag matrix, six-directional local mean values along the right, left, upper, lower, upper right, and lower left directions starting from $r(t,l)$ are calculated, and the maximum and minimum are obtained. If the local mean along the right or left direction takes the maximum, $r(t,l)$ is considered a part of a horizontal line and emphasized by subtracting the minimum from $r(t,l)$. Otherwise, $r(t,l)$ is considered a noise and suppressed by subtracting the maximum from $r(t,l)$; noises tend to appear as lines along the

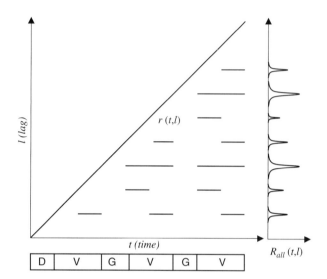

FIGURE 12 The summary $R_{\text{all}}(t,l)$ indicates the possibility that there are similar segments at a lag of l.

upper, lower, upper right, and lower left directions. Then, a summary is constructed by integrating over time:

$$R_{\text{all}}(t,l) = \int_{l}^{t} \frac{r(\tau,l)}{t-l} \, d\tau. \qquad (10)$$

R_{all} is then smoothed by a moving average filter along the lag. The result is sketched in Figure 12. R_{all} is used to decide which lag values should be considered when searching for line segments in the time-lag matrix. A thresholding scheme based on a discriminant criterion is used. The threshold is automatically set to maximize the following between-class variance of the two classes established by the threshold:

$$\sigma_B^2 = \omega_1 \omega_2 (\mu_1 - \mu_2)^2, \qquad (11)$$

where ω_1 and ω_2 are the probabilities of class occurrence (the fraction of peaks in each class), and μ_1 and μ_2 are the means of peak heights in each class.

Each peak above threshold determines a lag value, l_p. For each peak, the one-dimensional function $r(\tau, l_p)$ is searched over $l_p \leq \tau \leq t$. A smoothing operation is applied to this function and the discriminant criterion of Eq. (11) is again used to set the threshold. The result is the beginning and ending points of line segments that indicate repeated sections of music.

5.6 Modulation Detection

A common technique in pop music when repeating a chorus is to change the key, typically modulating upward by half-steps.

(Note that "modulation" in music is not related to amplitude modulation or frequency modulation in the signal processing sense.) Since modulation changes all the pitches, it is unlikely that a feature vector that is sensitive to pitch sequences could detect any similarity between musical passages in different keys. To a first approximation, a modulation in music corresponds to frequency scaling, as if changing the speed of a vinyl record turntable or changing the sample rate of a digital recording. On a logarithmic frequency scale, modulation is simply an offset, and when the scale is circular as with pitch classes and chroma, modulation is a rotation. To rotate a vector by ζ, the value of the ith feature is moved to become feature $(i+\zeta)$ mod 12. One would expect the chroma vectors for a modulated passage of music to be quite similar to a rotation of the chroma vectors of the unmodulated version.

Goto (2003a) exploits this property of the chroma vector by extending the time-lag matrix to incorporate chroma vector rotation by a transposition amount ζ. Denoting V_t^ζ as a transposed (rotated) version of a chroma vector V_t, $r_\zeta(t,l)$ is the similarity between V_t^ζ and the untransposed vector V_{t-l}^0. Since we cannot assume the number of semitones at the modulation in general, the line segment detection is performed on each of 12 versions of $r_\zeta(t,l)$ corresponding to the 12 possible transpositions (this usually does not increase harmful false matches). The segments from all 12 versions are combined to form the set of repeated sections of music, and the transposition information can be saved to form a more complete explanation of the music structure.

5.7 Chorus Selection After Grouping Line Segments

Since each line segment indicates just a pair of repeated contiguous segments, it is necessary to organize into a cluster the line segments that have mostly overlapping frames. When a segment is repeated n times ($n \geq 3$), the number of line segments to be grouped in a cluster should theoretically be $n(n-1)/2$ in the time-lag matrix. Aiming to exhaustively detect all the repeated segments (choruses) appearing in a song, Goto (2003a) describes an algorithm that redetects missing (hidden) line segments to be grouped by top-down processing using information on other detected line segments. The algorithm also appropriately adjusts the start and end times of line segments in each cluster because they are sometimes inconsistent in the bottom-up line segment detection. Lu et al. (2004) describe another approach to obtain the best overall combination of segment similarity and duration by adjusting segment boundaries.

A cluster corresponding to the chorus can be selected from those clusters. In general, a cluster that has many and long segments tends to be the chorus. In addition to this property, Goto (2003a) uses heuristic rules to select the chorus with a focus on popular music; for example, when a segment has half-length repeated sub-segments, it is likely to be the chorus. The *choruslikeness* (chorus possibility) of each cluster is computed by taking these rules into account, and the cluster that maximizes the choruslikeness is finally selected.

5.8 Texture Sequences

Detecting repeating patterns in the similarity matrix is equivalent to finding sequences of similar feature vectors. An alternative is to find sequences of similar texture classes. Aucouturier and Sandler (2002) perform a segmentation using hidden Markov models as described earlier. The result is a "texture score", a sequence of states, e.g., 11222112200, in which patterns can be discovered. They explore two methods for detecting diagonal lines in the similarity matrix. The first is kernel convolution, similar to the filter method of Bartsch and Wakefield (2001). The second uses the Hough transform (Leavers, 1992), a common technique for detecting lines in images. The Hough transform uses the familiar equation for a line: $y = mx + b$. A line passing through the point (x, y) must obey the equation $b = -mx + y$, which forms a line in the (m, b) space. A series of points along the line $y = m_0 x + b_0$ can be transformed to a series of lines in (m, b) space such that all intersect at (m_0, b_0). Thus, the problem becomes one of finding the intersection of lines. This can be accomplished, for example, by making a sampled two-dimensional image of the (m, b) space and searching for local maxima. It appears that the Hough transform could be used to find patterns in the similarity matrix as well as in the "texture score" representation.

One of the interesting features of the texture score representation is that it ignores pitch to a large extent. Thus, music segments that are similar in rhythm and instrumentation can be detected even if the pitches do not match. For example, "Happy Birthday" contains four phrases of six or seven notes. There are obvious parallels between these phrases, yet they contain four distinct pitch sequences. It seems likely that pitch sequences, texture sequences, rhythmic sequences, and other feature sequences can provide complementary views that will facilitate structure analysis in future systems.

6 • Music Summary

Browsing images or text is facilitated by the fact that people can shift their gaze from one place to another. The amount of material that is skipped can be controlled by the viewer, and in some cases, the viewer can make a quick scan to search for a particular image or to read headlines. Music, on the other hand, exists in time rather than space. Listeners cannot time-travel to scan a music performance, or experience time more quickly to search for musical "headlines". At best, one can skip songs or use fast-forward controls with recorded music, but even this is confusing and time-consuming.

One application of music structure analysis is to enable the construction of musical "summaries" that give a short overview of the main elements of a musical work. Summaries can help people search for a particular piece of music they know or locate unfamiliar music they might like to hear in full. By analogy to low-resolution versions of images often used to save space or bandwidth, summaries of music are sometimes called "music thumbnails".

Cooper and Foote describe a simple criterion for a music summary of length L: the summary should be maximally similar to the whole. In other words, a summary can be rated by summing the similarity between each feature vector in the summary with each feature vector in the complete work. The rating for the summary beginning at feature vector i is

$$Q_L(i) = \frac{1}{NL} \sum_{m=i}^{i+L} \sum_{n=1}^{N} S(m,n). \qquad (12)$$

The best summary is then the one starting at the value of i that maximizes $Q_L(i)$. The formula can be extended by weighting $S(m,n)$ to emphasize earlier or louder sections of the song.

Other approaches to summary construction are outlined by Peeters et al. (2002). Assume that music has been segmented using one of the techniques described above, resulting in three classes or labels A, B, and C. Some of the interesting approaches to musical summary are:

- Use the most common class, which in popular music is often the chorus. Some research specifically aims to determine the chorus as described earlier (Bartsch and Wakefield, 2001; Goto, 2003a).
- Use a sample of music from each class, i.e., A, B, C.
- Use examples of each class transition, i.e., A → B, B → A, A → C.

In all cases, audio segments are extracted from the original music recording. Unfortunately, artificially and automatically generated transitions can be jarring to listeners. Music structure analysis can help to pick logical points for transitions. In particular, a cut from one phrase of music to a repetition of that phrase can be inaudible. When a cut must be made to a very different texture, it is generally best to make the cut at an existing point of strong textural change. In most music, tempo and meter create a framework that is important for listening. Cuts that jump from the end of one measure to the beginning of another preserve the short-term metrical structure of the original music and help listeners grasp the harmonic and melodic structure more easily. Segments that last two, four, or eight measures (or some duration that relates to the music structure) are more likely to seem "logical" and less disruptive. Thus, music structure analysis is not only important to determine what sections of music to include in a summary, but also to organize those sections in a way that is "musical" and easy for the listener to comprehend.

An alternative to the construction of "music thumbnails" is to provide a "smart" interface that facilitates manual browsing of entire songs. The SmartMusicKIOSK music listening station (Goto, 2003b) displays a time line with the results of an automatic music structure analysis. In addition to the common stop, pause, play, rewind, and fast-forward controls, the SmartMusicKIOSK has controls labeled "next chorus", "next section", and "prev section" (see Figure 13). These content-based controls allow users to skim rapidly through music and give a graphical overview of the entire music structure, which can be understood without listening to the entire song.

FIGURE 13 The SmartMusicKIOSK user interface showing music structure and structure-related controls.

7 • Evaluation

Most research in this area has been exploratory, with no means to evaluate whether computer-generated structures and segments are "correct". In most cases, it is simply interesting to explore what types of structures can be uncovered and what methods can be used. Quantitative evaluations will become more important as problems are better understood and when competing methods need to be compared.

Tzanetakis and Cook (1999) conducted a pilot study to compare their automatic segmentation with human segmentation. They found most human subjects agreed on more than half of the segments, and their machine segmentation found more than half of the segments that humans agreed upon.

Bartsch and Wakefield (2001) hand-selected "true audio thumbnails" from 93 popular songs and measured "recall", the fraction of true frames labeled by their program as the chorus, and "precision," the fraction of labeled chorus frames that are true frames. With the chorus length set to around 20–25 s, the average recall and precision is about 70%, compared to about 30% for a chorus interval selected at random.

Goto (2003a) also used hand-labeled choruses in 100 popular songs from the RWC Music Database, a source that enables researchers to work with common test data (Goto et al., 2002). Goto judged the system output to be correct if the F-measure was more than 0.75. The F-measure is the harmonic mean of recall rate (R) and precision rate (P): F-measure $= 2RP/(R+P)$. The system dealt correctly with 80 of 100 songs.

Evaluating music structure descriptions is difficult. Structure exists at many levels and often exhibits hierarchy. The structure intended by the composer and perhaps determined by a music theorist may not correspond to the perception of the typical listener. Nevertheless, one can ask human subjects to identify pattern and structure in music, look for consistency between subjects, and then compare human descriptions to machine descriptions of music. One can also evaluate the impact of music structure detection upon tasks such a browsing, as in SmartMusicKIOSK (Goto, 2003b).

8 • Summary and Conclusions

Knowledge of musical structure can be used to construct music summaries, assist with music classification, provide high-level interfaces for music browsing, and offer high-level top-down guidance for further analysis. Automatic analysis of music structure is one source of music meta-data, which is important for digital music libraries.

High-level music structure is generally represented by partitioning the music into segments. Sometimes, segments are labeled to indicate similarity to other segments. There are two main principles used to detect high-level music structure. First, segment boundaries tend to occur when there is a substantial change in musical texture. In other words, this is where the music on either side of the boundary is self-similar, but the two regions differ from each other. Secondly, segments can be located by detecting patterns of repetition within a musical work.

It should be noted that the music signal, viewed as a time-domain waveform, is not directly useful for analysis because repetition in music is never exact enough to reproduce phase and amplitude relationships. Therefore, the signal is processed to obtain features that capture useful and more-or-less invariant properties. In the case of texture analysis, features should capture the overall spectral shape and be relatively insensitive to specific pitches. Low-order MFCCs are often used to measure texture similarity. To detect music repetition, features should capture changes in pitch and harmony, ignoring texture which may change from one repetition to the next. The chroma vector is often used in this case.

The similarity matrix results from a comparison of all feature vector pairs. The similarity matrix offers an interesting visualization of music, and it has inspired the application of various image-processing techniques to detect music structure. Computing the correlation with a "checkerboard" kernel is one method for detecting texture boundaries. Using filters to detect diagonal lines is one method for detecting repetition.

Detecting segment boundaries or music repetition generates individual segments or pairs of segments. Further processing can be used to merge segments into clusters. Hidden Markov models, where each hidden state corresponds to a distinct texture, have been applied to this problem. When music is analyzed using repetitions, the structure can be hierarchical, and the structure is often ambiguous. Standard clustering algorithms assume a set of distinct, fixed items, but with music analysis, the items to be clustered are possibly overlapping segments whose start and end times might be adjustable.

Music structure analysis is a rapidly evolving field of study. Future work will likely explore the integration of existing techniques, combining texture-based with repetition-based segmentation. More sophisticated features including music transcription will offer alternative representations for analysis. Finally, there is the possibility to detect richer structures, including hierarchical patterns of repetition, rhythmic motives, harmonic progressions and key changes, and melodic phrases related by transposition.

Acknowledgments The authors wish to thank Jonathan Foote for Figure 8. Jean-Julien Aucotourier, Mark Bartsch, Jonathan Foote, Geoffroy Peeters, Ning Hu, Xavier Rodet, Mark Sandler, George Tzanetakis, and Greg Wakefield have made contributions through their work, discussions, and correspondence.

References

Aucouturier, J.-J., F. Pachet, and M. Sandler. 2005. "'The Way It Sounds': Timbre Models for Structural Analysis and Retrieval of Music Signals". *IEEE Trans. on Multimedia* 7(6), 1028–1035.

Aucouturier, J.-J., and M. Sandler. 2001. "Segmentation of Musical Signals Using Hidden Markov Models". *110th Convention of the Audio Engineering Society*, Preprint No. 5379.

Aucouturier, J.-J., and M. Sandler. 2002. "Finding Repeating Patterns in Acoustic Musical Signals: Applications for Audio Thumbnailing". In *AES22 Int. Conf. on Virtual, Synthetic and Entertainment Audio*. Audio Engineering Society, pp. 412–421.

Bartsch, M., and G. H. Wakefield. 2001. "To Catch a Chorus: Using Chroma-Based Representations for Audio Thumbnailing". *Proc. the 2001 IEEE Workshop on Applications of Signal Processing to Audio and Acoustics (WASPAA 2001)*. New York: IEEE, pp. 15–18.

Cooper, M., and J. Foote. 2003. "Summarizing Popular Music via Structural Similarity Analysis". *Proc. 2003 IEEE Workshop on Applications of Signal Processing to Audio and Acoustics (WASPAA 2003)*. New York: IEEE, pp. 127–130.

Dannenberg, R. B. 2002. "Listening to 'Naima': An Automated Structural Analysis from Recorded Audio". *Proc. 2002 Int. Computer Music Conf. (ICMC 2002)*. San Francisco: International Computer Music Association, pp. 28–34.

Dannenberg, R. B., and N. Hu. 2002. "Discovering Musical Structure in Audio Recordings". In Anagnostopoulou, C., et al. eds. *Music and Artificial Intelligence, Second International Conference, (ICMAI 2002)*, Berlin: Springer-Verlag, pp. 43–57

Dannenberg, R. B., and N. Hu. 2003. "Pattern Discovery Techniques for Music Audio". *J. of New Music Research*, 32(2), 153–164.

Foote, J. 1999. "Visualizing Music and Audio Using Self-Similarity". *Proc. ACM Multimedia '99*. New York: Association for Computing Machinery, pp. 77–80.

Foote, J. 2000. "Automatic Audio Segmentation Using a Measure of Audio Novelty". *Proc. Int. Conf. on Multimedia and Expo (ICME 2000)*. New York: IEEE, pp. 452–455.

Galas, T., and X. Rodet. 1990. "An Improved Cepstral Method for Deconvolution of Source–Filter Systems with Discrete Speara: Application to Musical Sounds". *Proc. 1990 Int. Computer Music Conf. (ICMC 1990)*. San Francisco: International Computer Music Association, pp. 82–84.

Goto, M. 2003a. "A Chorus-Section Detecting Method for Musical Audio Signals". *Proc. 2003 IEEE Int. Conf. on Acoustics, Speech, and Signal Processing (ICASSP 2003)*. New York: IEEE, pp. V-437–V-440.

Goto, M. 2003b. "SmartMusicKIOSK: Music Listening Station with Chorus-Search Function". *Proc. 16th Annual ACM Symposium on User Interface Software and Technology (UIST 2003)*. New York: Association for Computing Machinery, pp. 31–40.

Goto, M. 2004. "A Real-Time Music Scene Description System: Predominant-F0 Estimation for Detecting Melody and Bass Lines in Real-World Audio Signals". *Speech Communication (ISCA Journal)*, 43(4), 311–329.

Goto, M., T. Nishimura, H. Hashiguchi, and R. Oka. 2002. "RWC Music Database: Popular, Classical, and Jazz Music Databases". *Proc. 3rd Int. conf. on Music Information Retrieval (ISMIR 2002)*. Paris: IRCAM, pp. 287–288.

Hu, N., R. B. Dannenberg, and G. Tzanetakis. 2003. "Polyphonic Audio Matching and Alignment for Music Retrieval". *Proc. 2003 IEEE Workshop on Applications of Signal Processing to Audio and Acoustics (WASPAA 2003)*. New York: IEEE, pp. 185–188.

Leavers, V. F. 1992. *Shape Detection in Computer Vision Using the Hough Transform*. Berlin: Springer-Verlag.

Logan, B., and S. Chu. 2000. "Music Summarization Using Key Phrases". *Proc. 2000 IEEE International Conference on Acoustics, Speech, and Signal Processing Proceedings (ICASSP 2000)*. New York: IEEE, pp. II-749–II-752.

Lu, L., M. Wang, and H.-J. Zhang. 2004. "Repeating Pattern Discovery and Structure Analysis from Acoustic Music Data". *Proc. 6th ACM SIGMM Int. Workshop on Multimedia Information Retrieval*. New York: Association for Computing Machinery, pp. 275–282.

Peeters, G., A. L. Burthe, and X. Rodet. 2002. "Toward Automatic Audio Summary Generation from Signal Analysis". *Proc. 3rd Int. conf. on Music Information Retrieval (ISMIR 2002)*. Paris: IRCAM, pp. 94–100.

Peeters, G., and X. Rodet. 2003. "Signal-Based Music Structure Discovery for Music Audio Summary Generation". *Proc. 2003 Int. Computer Music Conf. (ICMC 2003)*. San Francisco: Int. Computer Music Association, pp. 15–22.

Rabiner, L., and B.-H. Juang. 1993. *Fundamentals of Speech Recognition*. Englewood Cliffs, NJ: Prentice Hall.

Shepard, R. 1964. "Circularity in Judgements of Relative Pitch". *J. Acoust. Soc. Am.*, 36(12), 2346–2353.

Tolonen, T., and M. Karjalainen. 2000. "A Computationally Efficient Multi-pitch Analysis Model". *IEEE Trans. on Speech and Audio Processing, 8*(6), 708–716.

Tzanetakis, G., and P. Cook. 1999. "Multifeature Audio Segmentation for Browsing and Annotation". In *Proc. 1999 IEEE Workshop on Applications of Signal Processing to Audio and Acoustics (WASPAA 1999)*. New York: IEEE.

22

Computer Music Synthesis and Composition

Lydia Ayers

Clear Water Bay, Kowloon, Hong Kong, China,
e-mail: layers@cse.ust.hk

1 • Introduction

What is computer music composition? Composers are using the computer for everything from MIDI instruments communicating with computer sequencers, pitch trackers analyzing the sounds of acoustic instruments and converting them to pitch information, live performers with recorded music, performers with interactive computer programs, computer music produced by dancers using sensors, automatic music composition with the computer programs composing the music, composing with sounds or parts of sounds rather than notes, how to structure the use of time, composing with timbres, or the colors of sounds, and timbre morphing, such as a gong morphing to a voice, composing with textures and texture morphing, such as flutter-tonguing morphing to pitch, granular synthesis, trills and convolution. Computers are especially accurate in tuning, such as just intonation, Pythagorean tuning, Indonesian gamelan tuning, many kinds of equal temperaments and pitch bend, which can make any tuning even more expressive. The computer can control manipulation of samples in musique concrète and collage. Many composers use computer speech to tell stories, which can take place in fantasy environments with the movements of the sounds in the aural landscapes part of the composition.

What role does synthesis play in computer music composition? While it is true that composers used some of these techniques on acoustic instruments before they composed for computers, the computer has made it possible to use these techniques in new ways and with new sounds.

What makes a performance of computer music dramatic? One of the issues in computer music composition is whether the audience can watch someone perform the music. While we can watch the actions that produce the sounds on acoustical musical instruments in concerts, composers often synthesize computer music in studios using processes that don't occur in real time. Then they save the final mix of the music in a digital medium, such as a binary computer file, tape or CD recording. Some composers then play the recordings in concerts. But how can music with no human performer be interesting to "watch"? Some composers perform their compositions on mixing consoles, moving the sound around on "symphonies" of loudspeakers. Concerts may have such colorful names as the "Concerts of Nothing to See" in Montreal, the "Concerts for Closed Eyes" in Hong Kong or John Cousins' "Cinema for the Ears" at the Australasian Computer Music Conference in Wellington, New Zealand. Other composers team up with dancers or graphic artists to give the audience something to watch. This chapter gives brief descriptions of these aspects of computer music.

2 • Music with Live Performers

2.1 Compositional Interface

The compositional interface is the communication channel between the composer and the computer. The most suitable compositional interface to use for a particular composition depends, in part, on the type of performance. Writing computer code for synthesis languages such as Csound, or programming languages, such as C++, is suitable for recorded music. It is also possible to mix samples in sound editors. Music with live performers, however, requires a program that reacts to instructions quickly enough to perform music in "real time". The interface may include desktop objects in a program such as Max, which a computer performer can manipulate, and even improvise with, in real time. Other interfaces use hardware input devices, such as sensors or electronic musical instruments, to send signals to the computer which cause the computer to produce sounds. Although some useful commercial electronic musical instruments, such as the Yamaha wind controller, already exist, many composers are inventing their own hardware devices and programming the related software.

2.2 Notation Systems

When live performers are involved in computer music, they need to know what to do, and when to do it, just as performers on acoustic instruments do. Notation should handle performer parts, and it may be helpful if it can also handle reference parts describing the computer music. Notation can also be useful for pitch input for recorded compositions. Computer notation programs, such as Finale and Sibelius, handle notation more or less suitable for traditional western instruments, but their graphics are limited in describing computer sounds. Many computer notation programs, such as Finale and Cubase Score, contain sequencers which can play back the score using synthesized instruments. These programs can store the score information as MIDI files, which can be used in other computer music systems.

Chinese and Indonesian instruments normally use a cipher notation with numbers from 1 to 7. In Chinese music, pitch 1 is the tonic of a major key. In Indonesian music, pitch 1 is the first tone of a five-tone scale which is tuned to the taste of the owner of a gamelan. One computer input method uses the numbers for the pitches, and then transposes the scale up and down by octaves. The octave is set by multiplying the pitch by another power of 2 for each higher octave, so the original octave is 1, the next octave is 2 and the next octave is 4.

2.3 Pitch Input Methods

Composers use several methods to get their musical ideas into computers. Composers using the computer as a notation tool can use the MIDI sequencers in music notation programs, such as Finale, Sibelius and Cubase, to synthesize demos of pieces composed for acoustic instruments, but the quality of the synthesis is limited to the quality of the MIDI instruments available on the computer's sound card. The musical result tends to be more about the notes than the characteristics of the sound, and it doesn't exploit many of the characteristics of computer music. Other composers play MIDI instruments to communicate with computer sequencers, or use pitch trackers to analyze the sound of acoustic instruments and convert it to pitch information. However, the MIDI instruments do not have to be commercial instruments, such as wind controllers and keyboards. Many commercial instruments arose from hardware tinkering, and such tinkering is still going on. Using MIDI instruments to input pitch information can result in interactive music, where a computer program processes the input information in real time, possibly mixing it with sound synthesis and sending the result to the playback system.

Another computer performance method involves the composer connecting objects on the screen. Using such a system, a composer can create music in a computer environment similar to those on analog synthesizers, such as the Buchla. Max/MSP is one of the most well-known graphical programming environments for music, audio and multimedia. Max users create their own software using a visual toolkit of objects and connect them together with patch cords. The Max environment includes a user interface, MIDI control and timing objects. Hundreds of objects work on top of Max. Jitter is a set of real-time video and image processing and algorithmic image generation objects integrated with the musical capabilities of Max/MSP (Cycling 74 2004).

In Christopher Keyes' *Touched by Space & Time*, microphones capture the sounds of percussion instruments, including circular saw blades, and send them to a computer which alters and spatializes them over the multichannel sound system. Played like an "air marimba", the infrared batons trigger a note when they pass a certain point in the vertical plane, with pitch relative to the

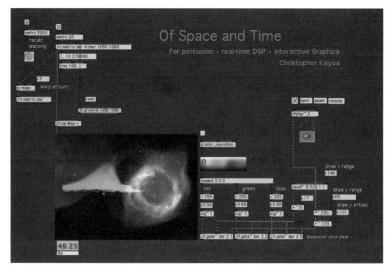

FIGURE 1 Screen shot of Jitter patch for Keyes' *Of Space and Time*.

horizontal plane and dynamics relative to the velocity at the point the baton passes the plane. Keyes wrote a Max/MSP/Jitter patch for the piece to process the incoming percussion audio signals from two microphones and output them over four channels, route MIDI messages to a sampler patch triggering notes in 24- and 36-tone equal temperament, play a four-channel prerecorded accompaniment and send the incoming audio signal to a Jitter patch which generates oscilloscope-like images (Figure 1).

2.4 Live Performers with Recorded Music

Recording a digital computer file and playing it back to accompany live performers is one of the most portable solutions to the problem of mixing performers with computer music. The live performers can play without requiring a computer operator to travel with them. However, the live performers must play in time with the accompanying recording because the recording cannot "follow" them.

2.5 Performers with Interactive Computer Programs

What if the computer could follow the performers? Max Matthews' Radio Baton (Figure 2) is a three-dimensional MIDI controller that tracks the X, Y and Z motions of the ends of two batons held in a performer's hands. At the end of each baton is a small radio-transmitting antenna that transmits a signal to five antennas arranged around the receiving surface. The batons can also send MIDI triggers by crossing invisible

FIGURE 2 Max Matthews and Radio Baton.
http://csounds.com/mathews

planes at various heights above the antenna board. Using a real-time Csound interface on the PC or the Max/MSP interface on the MacIntosh, the performer's movements can control the computer to follow other performers (Mathews). David Jaffe's piece, *Racing Against Time*, uses a similar method with drum sticks that can perform on percussion instruments while they transmit motion data (Jaffe 2001). Todd Winkler (2001) discusses many issues related to interactive music systems in *Composing Interactive Music: Techniques and Ideas Using Max*.

2.6 Computer Music Produced by Dancers Using Sensors

Systems ranging from simple on/off triggers placed on the floor to sophisticated systems incorporating ultrasound, radio and real-time visual analysis convert the movements of dancers into signals to control musical performances. Dancers "play" "active control" systems expressively like musical instruments. With the "passive" approach, the system responds to dancers' normal movements. The audience is usually unaware of how the movements trigger the computer – dancing to a recording can produce the same effect. However, performers like the sense of control that interactive devices provide. Russell Pinkston has used the University of Texas MIDI Dance Floor (Figure 3) in interactive dance compositions, including *Dervish Sketches*, *Memory of Absence* and *Song for the Living/Dance for the Dead*. The dance floor uses force-sensing resistors which transmit the position coordinates, velocity and pressure information as MIDI messages. These are attached to heavy-duty plastic sheeting

FIGURE 3 Dancers on MIDI Dance Floor.

with contact cement and covered with polyethylene foam. Used with an "intelligent" external MIDI processing system, it permits dancers to affect music and lighting with their movements and positions on its surface (Pinkston et al. 2001).

Chris van Raalte wears a BodySynth to control music, video and animation. Electrodes on different parts of his body control different synthetic and instrumental sounds. His muscle movements produce electrical energy, which a wireless transmitter sends to a computer which controls a MIDI synthesizer. In *Hondo Butoh*, after dancing covered by a large pink fabric that looks like a bedspread, van Raalte emerges looking geeky in glasses and a lab coat. A provocative "striptease" gradually reveals the BodySynth (Figure 4) (Van Raalte and Zane-Cheong 1995).

2.7 Automatic Music Composition

Automatic music composition, also called algorithmic composition or computer-assisted composition, uses computer programs to compose music, and in some cases such automatic music composition produces music for live performers without any computer synthesis at all. Using an Illiac computer at the University of Illinois, Lejaren Hiller and Leonard Isaacson created the first computer-composed composition, *The Illiac Suite for String Quartet*, in 1956 (Roads 1996). Iannis Xenakis described the use of stochastic procedures in *Formalized Music* (1963, 1992) and used such procedures in works ranging from *Eonta* (1964) to *Gendy 3* (1991) and *S.709* (1994).

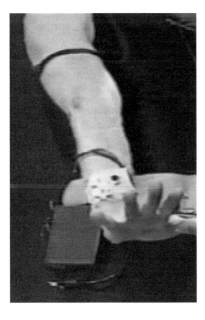

FIGURE 4 Chris van Raalte wearing the BodySynth.

Xenakis created the latter works using his GENDYN program which functions as a "black box" that can create an entire composition using a small amount of user input. Barry Truax developed a PODX (Poisson Distribution) system, which he used to compose with sounds and parts of sounds, rather than notes, in pieces such as *Inside* (1995) and *Beauty and the Beast* (1989). In my pieces, *Balinese* and *Pendopos* (Ayers 2006), I wrote C++ computer programs which composed Csound scores algorithmically, loosely within the Balinese gamelan style. Some elements of this style are the rhythmic structure and pentatonic scales using Balinese tunings (McPhee 1966). *Pendopos* is a walk up an imaginary mountain past open pavilions playing gamelan music.

3 • Form, Structure and Time

How to structure the use of time is an interesting question in computer music composition. While it is possible to create a symphony in sonata form, it seems more idiomatic to use experimental forms. In concert performances, the audience cannot tell when an invisible piece will begin, nor is it obvious whether a pause in the sound is the end of the piece or a break between movements. This characteristic of computer music has encouraged one-movement works, or works where sections flow together. Everything can seem quite static, or the movement can have contrasting sections.

Composers often play with the way we perceive time. Time can move disconnectedly as points of sound or smoothly as a slowly evolving drone. Pierre Boulez coined the term, "striated time", to describe music with meter, beats or at least events that can be counted, even if they are irregularly spaced. "Smooth time" occurs in compositions in which it is difficult to count time (Boulez 1963, 1971). Spectrally evolving drones may be completely smooth, with no events marking time. In smooth time, listeners may even lose their awareness of how much (or how little) time is passing. Computer music contains many shades between completely smooth time and completely striated time. Curtis Roads' *Field* (Roads 1981) contrasts smooth time with striated time in "a flowing musical process". The piece begins with long filtered noise sounds, which seem to be nearly a textured drone. The spectra evolve as the composition unfolds. He combined individual sound objects into compound sound mixtures, using computer synthesis techniques, such as frequency modulation, waveshaping, additive and subtractive synthesis and granular synthesis, or mixing them with digitized natural sounds. He processed the compound sounds and linked

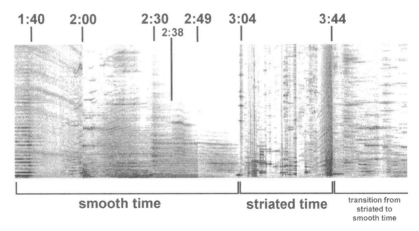

FIGURE 5 Smooth and striated time in Roads' *Field*.

them into layered melodies. Interest focuses on the qualities of the sounds as they slowly morph into other sounds. Figure 5 shows an episode of striated time interrupting the smooth time.

Striated time layers can move at different speeds. I used this technique in *Nostalgia Strata*, which remembers some familiar twentieth century melodies using some of the woodwind designs from our book, *Cooking with Csound* (Horner and Ayers 2002). Slices of the famous melodies form a "strata", which suggests both fugal layers and the name of a layered dish. Heavy-effects processing masks a lot of the melody, but the layers in this piece form a distorted fugue, with the "subject" reappearing at different pitch levels in simultaneously different tempi, stretching and contracting the stretto effect.

4 • Timbre and Texture

Timbre is the color of the sound, the quality that distinguishes between, say, a trumpet and a violin, and in computer music composition it ranges from pitched musical instrument tones to environmental sounds, such as wind, water and birds, to artificial sounds. Timbre may be independent of and more important than rhythm and pitch. Timbre morphing, such as a gong morphing to a voice, slows down spectral interpolation to the point that the focus is on the spectral change process rather than on the resulting sound goal.

Larger scale texture has been described as relationships between monophonic, homophonic, polyphonic and heterophonic melodies and harmonies. Another way to use texture is on the surface of a sound, as silk usually has a smoother texture than wool fabric. This texture may be smooth versus striated or continuous versus discontinuous.

4.1 Granular Synthesis

Granular synthesis uses many short overlapping grains of sound. Multiplying samples or waveforms by amplitude envelopes creates grains. Mara Helmuth's StochGran stochastic granular synthesis program generates complex sounds from tiny grains of recorded sounds, which maintain some relationship to the original source (Helmuth 2002). She used granular synthesis to produce clouds of sound in *Loonspace* and *Abandoned Lake in Maine*, and at the end of *Implements of Actuation* (co-composed with Allen Otte) (1998).

Barry Truax used granular synthesis in pieces such as *Inside* (Truax 1995) and *Beauty and the Beast* (Truax 1989).

4.2 Flutter Tonguing

Flutter tonguing is a grainy texture on wind instrument tones that is accomplished by rolling the "r" at the front of the tongue or gargling in the back of the throat. The movement of the performer's throat or tongue produces amplitude modulation near 20 Hz, the border frequency between clicks and tone. We modeled it using a function that represents the average amplitude envelope of the individual modulations (Ayers 2004). Changing the speed of amplitude modulation on a synthesized bassoon tone morphs the texture from flutter tonguing to pitch (Figure 6).

4.3 Trills

Trills are a special case of a slur and are similar to flutter tonguing as well. A repeating function controlling the frequency of a single tone produces smoother trilled notes than overlapping tones does, and we have used amplitude and frequency modulation functions to cross-fade the trills in wavetable instrument designs

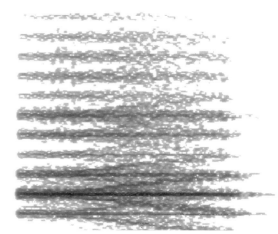

FIGURE 6 Synthesized bassoon tone with flutter tonguing morphing to pitch.

(Ayers 2003, Ayers 2004, Ayers and Horner 2004a and b). An additive synthesis design improves the transitions between the trilled notes. The additive synthesis instrument uses a single signal with a line segment on all the parameters for each of its harmonics. That is, each harmonic is one component signal for the spectrum, with a frequency and an amplitude line segment handling a varying number of notes. Then all the harmonics are added together. The repeating trill function models the frequency change for the trill. The function does not need to model pitch variation of the average tone, change of speed or jitter, so it can represent one average cycle of the trill, and adjusting the parameters randomly within their typical ranges can vary each cycle. The amplitude function modulates the difference between the harmonic amplitudes of the lower and higher notes to give each note in the trill the correct spectrum. The design controls an initial, middle and final tremolo rate and the time required to change from the first tremolo rate to the second. Figure 7 shows an example of speeding up the trill until it becomes frequency modulation. A separate score parameter controls the changing amplitude of the trill as it would for a single sustained tone (Ayers 2005, 2006, Ayers and Horner 2006, 2007).

I am using these designs in a number of compositions. I first used it for the Chinese *erhu* (Figure 8) in *Why Ask Questions*, which was composed as a background music for an installation art exhibition. At the time, the 1997 Handover was on the horizon, and many in Hong Kong were questioning what the future would bring. Others asked, "Why ask questions? Why not just accept fate?" I composed the original 30-min microtonal piece algorithmically (using the computer to make some of the compositional decisions). Then I added multi-effect processing to Chinese voices

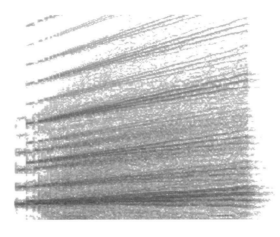

Figure 7 Changing speed from trill to frequency modulation.

FIGURE 8 *Erhu* slurs, pitch bend and trills in Ayers' *Why Ask Questions*.

and Chinese computer instrument designs for *dizi, xiao, erhu, pipa* and gong.

4.4 Convolution

Convolution, a process of multiplying all the samples in one signal by all the samples in another signal, is often used in cross-synthesis, filters and reverb (Boulanger 2000, Roads 1996). John Rimmer's *Ancestral Voices* (Rimmer 2000) explores sampled sounds, including the Maori trumpet-like, male *putorino*, and the flute-like, female *koauau*, the Australian didgeridoo, three Asian instruments and two European instruments. Rimmer colored the samples with convolution and contrasted the active gestures and granulations of the "male" sections with the "delicate, colored and highly resonant sonorities" of the "female" sections. That is, the female sections are in smooth time, and the male sections are more striated.

5 • Tuning Systems

Computers are especially accurate with tuning. Virtually any kind of tuning is possible, paired with any timbre, so it is possible to have real or imaginary computer musical instruments play in tunings that acoustic instruments can only wish for.

5.1 Just Intonation

The just intonation tuning system uses integer ratios to calculate frequencies at intervals from an originating frequency. Opinion differs about which ratios fall into the category of just intonation. Some people limit it to only the simplest integer ratios, such as those needed for the major and minor scales. However, the major scale can require two ratios for the major second in order to have consonant intervals in typical chords. The major second, 9/8, is typically used in the just major scale, but harmonic music may add 10/9 to form a consonant minor third (6/5) interval with the perfect fourth (4/3) and major sixth (5/3) in the *ii* chord

(10/9 × 6/5 = 4/3). *The Just Intionation Primer* (Doty 1993) gives an introduction to just intonation, and *Genesis of a Music* (Partch 1974) gives a very detailed discussion.

Other people prefer a wider view of just intonation, such as simple or small integer ratios. It is a matter of opinion how small "simple" integers should be. The most unlimited definition is any integer ratios, but some intervals formed that way are quite dissonant. Depending on the individual definition, just intonation may or may not include dissonant intervals. However it is defined, the ratios should contain integer fractions, so real numbers such as the 12th root of 2 are not included.

Bill Alves piece, *In–Yo* (Alves 1999), uses just intonation with a live shakuhachi player and computer accompaniment. *In* and *yo*, the Japanese equivalent to the Chinese *yin* and *yang*, are also the names of two Japanese scales. *In*, a scale of large and small intervals, is associated with meditation, darkness and melancholy. In contrast, *yo* has no very small intervals, and expresses the extroverted joy of folk music. Alves represents these perspectives with two systems of just intonation. The *in* section uses "utonal" intervals in which the pitches are related consonantly to each other, but not necessarily to a common fundamental. In the *yo* section, Alves used "otonal" intervals, where all the pitches are harmonics related to a single fundamental (Partch 1974). Alves used Csound to manipulate the breath sounds of a bamboo flute, the deep tolling of a Buddhist temple bell (in which the partials have been altered to match the tuning system) and the human voice.

5.2 Pythagorean Tuning

The "Pythagorean" tuning method was actually first discovered by the Chinese musician Ling Lun (Partch 1974) and was later named after Pythagoras in Greece. On acoustic instruments, the first interval is tuned as a perfect 5th, then another interval is tuned to a perfect 5th from the upper note and so on until all the notes are tuned. Transposing tones up or down by octaves organizes them into scales. The computer produces Pythagorean tuning by multiplying 3/2 by 3/2 and then repeating until it produces all the desired notes.

My piece, *Bioluminescense* (Ayers 2006), uses Indian tunings from ragas of the deep night, including *Paraj*, *Sohini*, *Lalita*, *Vibhasa*, *Durga*, *Hambir* and *Khammaja*. According to Daniélou (1980), some of these ragas, along with many others, use Pythagorean tuning. *Bioluminescense* also explores the boundary between tuning and timbre. Gamelan-like timbres result from a tuning using groups of 17-denominator ratios, generally with the higher frequencies dropping out before the lower frequencies

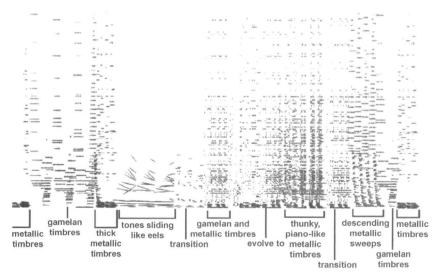

FIGURE 9 End of Ayers' *Bioluminescense*.

(Figure 9). Metallic timbres result from a relationship among high harmonics, the quality of which changes when it is detuned. In this piece, I also flirted with just intonation utonalities and combined aggregates of ratios which are not closely related to each other, such as a section combining ratios of 7, 8, 13, 19 and 23. Some of the thickest aggregates give metallic and strange, thunky timbres resulting from detuning the high harmonics.

5.3 Equal Temperaments

Computers can easily produce accurate 12-tone equal temperament, the division of the octave into 12 equal pieces, which is the hoped-for tuning in western music for about the last 150 years. And they can go beyond it into many other equal temperaments. Equal divisions of 19 and 31 are especially popular. Equal divisions of 13 sound especially strident. Bill Sethares has exploited a number of octave divisions in his *Xentonality* CD (Sethares 1997), which includes, among others, *Ten Fingers* in 10-tone equal temperament, *Seventeen Strings* in 17-tone equal temperament and *Unlucky Flutes* in 13-tone equal temperament. *Chaco Canyon Rock*, included on the CD in his book, *Tuning, Timbre, Spectrum, Scale* (Sethares 1998), is an example of Sethares' technique mapping the spectrum of a sound to the tuning of the scale he uses in the piece. He used a fast Fourier transform to analyze 12 samples of a circular rock from Chaco Canyon. Striking different positions on the rock produced different sounds, each with a unique spectrum. All of the sounds have a resonance near 4,070 Hz, and other resonances occur in

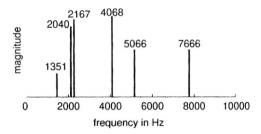

FIGURE 10 Sethares' combined line spectrum for *Chaco Canyon Rock*.

just one or two samples. Sethares combined the most prominent partials into a composite line spectrum that captures much of the acoustic behavior of the samples (Figure 10). Sethares then used a program to draw the dissonance curve for the tone. He derived his scale (Table 1) for the rock instrument from the most consonant points on the dissonance curve. The *Chaco Canyon Rock* piece uses sampled tones tuned to this scale and electronic rocks produced with additive synthesis of the inharmonic partials shown in Table 1.

5.4 Indonesian Tunings

While Bill Alves has composed music for Indonesian gamelan instruments, and Lou Harrison composed for justly tuned gamelan instruments that Bill Colvig built for him, these pieces depend on the physical instruments for which they are composed. Playing them on other instruments requires changing their tunings, which readily occurs in Indonesian gamelan music. Intonational differences, however, may not always please contemporary American composers. Virtual gamelan instruments, on the other hand, can stay in a single desired tuning or play simultaneously in multiple tunings. In *Merapi* (Ayers 2006), inspired by the name of Java's most active volcano, I used the first approach, a single hypothetical tuning derived from the tuning of the Berkeley Gamelan (Ayers 1996). The computer program randomly generated the piece within the style of Indonesian gamelan music. In addition to making Risset's bell design (Risset 1969, 1996) less inharmonic for the *bonang* (a group of tuned horizontal gongs) by tuning its partials to harmonics, I adapted Judy Klein's design for the sound of struck crystal glasses. "Waiting" opens the piece with a fairly tradi-

TABLE 1 Scale used by Sethores for the rock instrument

Ratio	1	1.063	1.17	1.24	1.25	1.37	1.385	1.507	1.612	1.69	1.77	1.88	2
Cents	0	105	272	372	386	545	564	710	827	908	988	1,093	1,200

tional rhythmic structure, "Simmering" is a slow section with strange chords, the "Eruption" is a meterless tape interlude in which the tuning gradually rises to symbolize the eruption of the volcano and the "Bubaran" is a short closing piece. I used a variety of Balinese tunings (McPhee 1966) in two other algorithmically composed pieces, *Balinese* and *Pendopos* (Ayers 2006), a walk up an imaginary mountain past open pavilions playing gamelan music. The computer program generated several short segments using Balinese tunings (McPhee 1966), loosely within the Balinese gamelan style. The synthesized gamelan reuses the *bonang* adaptation of Risset's bell design and the Woodstock gamelan design (Ayers and Horner 1999). I also synthesized my voice and a Balinese *suling* flute. Each section of *Pendopos* represents a pavilion. As the trail curves around, you hear the first two gamelan groups again, playing the same fragments transposed to different pitch centers with a changing background full of chattering birds, including a few ducks, that I recorded in Bali. Suddenly, passing the tree line, the gamelan layer changes from the heavier Risset gongs (Risset 1969, 1996) to the bright and sparkly Woodstock gamelan, which accompanies the voice. At the summit, the voice swirls around everywhere as the bird calls fade into the distance.

5.5 Overview of How to Tune a Computer

The following example shows one way to tune a computer using Csound. First, set a global tuning frequency. One number in the note statement can control all the instruments in the orchestra. For example, for C major, middle C, at 261.6 Hz, is a good note for a tuning frequency, and for A minor, A at 440 Hz is a good tuning frequency. To modulate to another key in the middle of a piece, simply set a new global tuning frequency. The Csound orchestra code for instrument 1 uses the global frequency and attenuation values from the score. As these values are initialized at the start time of the note, the duration is irrelevant as long as it is greater than 0, so I just use a short value such as 1.

```
;--------------------------------------------------------------------------------
instr 1
gifreq      =     p4      ; set global tuning frequency
giattn      =     p5      ; set global attenuation scaling
endin
;--------------------------------------------------------------------------------
```

Once the global frequency is set, a few simple lines of code in the orchestra set the frequency for the individual notes.

```
;----------------------------------------------------------------
instr 25
. . .
ipitch       =     p8
ioct         =     p9
ipitchtable  =     9                    ; table of indexes for
                                          table 10
iratiotable  =    10                    ; table of frequency ratios
iprange   table  ipitch*100, ipitchtable  ; pitch range of note
inum      table  (iprange*2), iratiotable    ; numerator of note
iden      table  (iprange*2)+1, iratiotable  ; denominator of note
ifreq     =      inum/iden * gifreq     ; ratio x global freq
ifreq     =      (ioct = 0 ? ifreq/2 : ifreq * ioct)
. . .
;----------------------------------------------------------------
```

Function tables 9 and 10 control the tuning of the pitches. Function table 10 sets just intonation in the following example. The input pitches for a major scale consist of the integers 1–7. The input method is more intuitive if the chromatic pitches occur between the integers. For example, the input for pitch 3b, the minor third, would be 2.5. As the indexes to table 10 must be integers, table 9 provides the conversion. Csound then reads the pair of numbers occurring at the indexed position and the following position in table 10 as the numerator and denominator of the frequency ratio for the tone.

```
;----------------------------------------------------------------
; function table for pitch to choose ratios
f9 0 16384 -17 0 0 105 1 205 2 255 3 305 4 405 5 505 6 605 7 705 8
; function table for just intonation ratios (commented out)
f10 0 64 -2 1 1 9 8 6 5 5 4 4 3 3 2 5 3 15 8
;----------------------------------------------------------------
```

5.6 Pitch Bend

Pitch bend often occurs in natural sounds, such as bird songs and cat meows. My piece, *Yazi's Dream*, celebrates the life of Yazi, my first cat. "Looking at Birds" represents the birds she watched and tried to talk to through the window. The synthesized trills on the *dizi*, a Chinese membrane flute which resembles a keyless bamboo piccolo, represent Yazi's bird fantasies. The excerpt shown in Figure 11 includes a *dizi* melody and trill, lowered cat sounds with natural pitch bend, stretched (slowed) and granulated lowered cat sounds, and normal and lowered crow sounds with natural pitch bend.

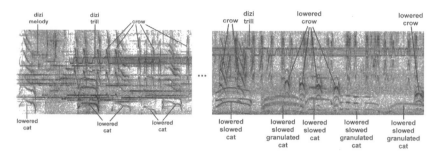

FIGURE 11 Pitch bend in Ayers' *Yazi's Dream*.

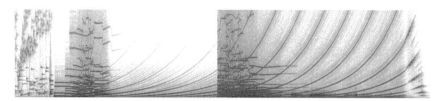

FIGURE 12 Endless ascending glissandi in Risset's *Mutations*.

The endless glissando, an illusion in which the pitch seems to keep going up (or down) but never gets anywhere (Dodge and Jerse 1984), is another use of pitch bend. French composer Jean-Claude Risset used the endless glissando in *Mutations*, synthesized using Max Matthews' Music V program at Bell Labs in 1969. The piece uses mostly additive synthesis and is the first work using FM, based on John Chowning's work. A melodic motive transforms into a chord, which stretches into a timbre. The density of the frequencies increases and gradually mutates to an endless ever-ascending glissando (Figure 12).

6 • Musique Concrète and Collage

Although most people barely notice the sounds surrounding them, some have found those sounds musical. For example, Benjamin Franklin assembled a glass harmonica from wine glasses, and, in the middle of the twentieth century, such composers as Pierre Schaeffer and Luciano Berio recorded collages of such concrete, or real, sounds as locomotives (Schaeffer 2000a,b) and voices (Berio 1999). Their compositions often took months of cutting and splicing taped sounds together. Schaeffer described these techniques extensively in the *Traité des objets musicaux* (Schaeffer 1966). Now, it is much easier to cut and paste recorded samples together to create music with computer "instruments".

6.1 Sampling

One evening after dinner, Paul Lansky's sons played everything in their kitchen which would make noise. In *Table's Clear*, Lansky makes syncopated gamelan music mixed with a jazz chorus or two from such sounds as pots and pans, glasses, tupperware drums, water drops and the boys talking, belching, burping and clapping. In the middle, the gamelan-like patterns twist as they modulate from one pitch center to another (Lansky 1992).

Joyce Tang derived the title *Cympergon* (1996) from the names of its cymbal, paper and gong sound sources. She then transformed these "sonic seeds" into usable computer "instruments" for sampling and sound-wave editing. The imagined motions of striking, scratching, tearing and rubbing these instruments are an important part of the piece's character. Combining the gong timbres with polymetric and polyrhythmic patterns gave her piece a gamelan-like flavor, but pitch bending and time stretching transform the gongs and other timbres into an imaginary landscape (Figure 13).

In *Kasumi*, Ian Whalley sampled seven New Zealand Mãori instruments, played by Richard Nunns, and Chikako Komaki reciting a Japanese poem. The samples of the Mãori wind instruments are melody fragments, rather than isolated tones, and Whalley repeats these fragments and eventually disintegrates them. The composition is constructed in levels like a folk tale. Figure 14 illustrates the beginning of the piece, which begins with a *tumutumu* (a pilot whale jawbone struck with wood), a *pahü pounamu* (small Greenstone gong) and two higher layers of melody fragments on the *kōauau kōiwi kuri* and *kōauau pongäihu*

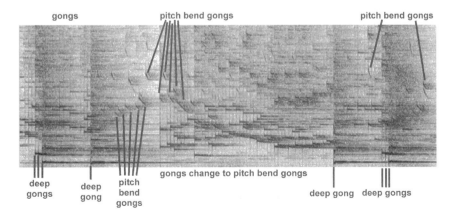

FIGURE 13 Gongs from Tang's *Cympergon*.

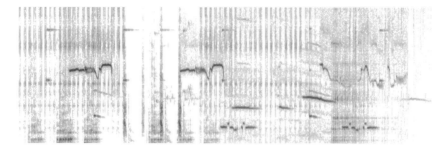

FIGURE 14 Beginning of Whalley's *Kasumi*.

(dog bone and nose flutes). Whalley manipulates the sample fragments later in the piece with a great deal of processing, and also uses many synthetic instruments (hybrids of electronics and real instruments) (Whalley 2002).

In *Implements of Actuation* (1998), Mara Helmuth and Allen Otte transformed the sounds of an electric mbira and a kid's rusty bicycle wheel into a delicate texture (Figure 15) using a Javanese scale and a computer. Implements applied to the mbira and bicycle wheel, such as a Chinese wok brush, red lentils, comb, credit card and baseball trading card, became subjects to be acted upon by the computer applications used both as a mixing tool for all of the collected fragments which became the tape part and also as an algorithmic generator of both the tape and the live performance score.

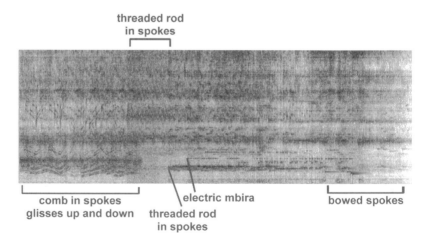

FIGURE 15 Sampling in Helmuth's *Implements of Actuation*.

6.2 Speech, the Voice and Program Music

Many composers have been inspired by the possibilities of manipulating voices on the computer. For example, in *Idle Chatter*, Paul Lansky transformed unintelligible chattering speech into rhythmic chanting (Lansky 1985). Charles Dodge was attracted to "making music out of the nature of speech itself", to "changing the timbre of the voice" in *He Destroyed Her Image*. Dodge tried to depict the reversal, from looking outside to being inwardly confused in a poem by Mark Strand, with the changes of tone quality in the voice, back and forth between a speechlike electronic phrase (you can understand the words) and a less speechlike electronic phrase (you can't understand the words). This happens even though the two have the same pitch patterns (Dodge 1972).

Program music paints a picture or tells a story. Anna Rubin's *Family Stories*, co-composed with Laurie Hollander, used Paul Koonce's phase vocoder instruments (Koonce http://silvertone.princeton.edu/winham/PPSK/koonce.html) in a text/sound piece using narrative, sampled ambient sounds and computer-generated music to tell the story of Rubin's mother, Sophie, the child of Russian Jewish immigrants in Atlanta. At seven, Sophie lost her mother and was left in the care of an African-American woman, Sally Johnson. Sophie's unbearable loss is framed by the atmosphere of racism and anti-Semitism in early twentieth-century Atlanta (Rubin and Hollander 2004).

7 • Environments

Soundscapes are landscapes "painted" with sounds. For example, while "on the ice" in Antartica, composer Chris Cree Brown (NZ) recorded the sounds of walking on snow, skua birds, polar wind, radio communications, Weddell seals, an Adelie penguin rookery and Antarctic white out. In his piece, *Under Erebus*, he sculpted his recordings into a soundscape reflecting his experiences of the Antartic environment (Brown 2000).

7.1 Spatialization and Composing an Imaginary Space

Jean-Claude Risset's *Invisible* is inspired by Calvino's book, *Le città invisibili*, in which Marco Polo describes to Kublai Khan cities from his dreams, desires, fantasies, fears and utopias. With digital synthesis and processing using MUSIC V, SYTER and Sound Mutations, Risset produced harmonically composed timbres, voice and wind hybrid sounds and illusory bells, gongs and voices that dialog with the *Invisible* real voice of soprano Irene Jarsky in imagined spaces similar to Calvino's invisible cities. Through spatial movement (Figure 16), Risset transforms the voice into the butterfly described in the texts of Taoist Chinese poet Zhuangzi (Risset 1996).

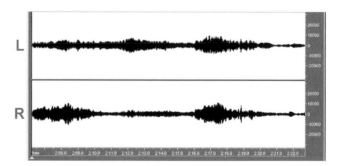

FIGURE 16 Stereo spatialization in Risset's *Invisible*.

Mark Applebaum composed *Pre-Composition* (2002) using only his voice and put each of the eight characterizations of himself in a different speaker. The result is a humorous piece about his internal dialog whenever he plans a piece of computer music.

Australian composer Jim Barbour recreated an outback waterhole in *Outback* (Barbour 2004). After converting the subwoofer into a regular channel, he used four horizontal loudspeakers and two overhead loudspeakers in the 5.1 surround sound format to create a three-dimensional sound environment. The sounds move over and around listeners with reverberation reflections from all over the vertical hemisphere. Some sources sound at ground level, some in trees while others are flying around and over a listener, as they would in their original environment. The wild budgerigar birds, for example, fly and sing in the trees only from the elevated loudspeakers (Barbour 2004, 2006).

7.2 Composing for a Space

An installation is a continuous sound design that fits a place rather than a composition for a set length of time. Installations include realistic or imaginary sound scenes all around the space. French composers Michel Redolfi, Luc Martinez and Michel Pascal composed pieces for an installation for the Nausicaa Exposition. They mixed sounds recorded in the Pacific Ocean and South China Sea into music in their studios. A central computer controls three-dimensional sound circulation around the public through all levels of the museum. The Chalutier room simulates the fury of a fish in the high sea in a sort of eight-channel acoustic "holograph". As people walk through the Remontée, their random movements interactively recompose the sound. Synthesized music and biological sounds evoke a scientific sea where 20,000 known sonorous species murmur the last siren songs of a strange opera (Redolfi 1990–91).

8 • Conclusion

This chapter has given brief examples of some of the major aspects of computer music composition. Computer music takes place, as all music does, during time and within the mind. I have described computer music which plays purely in the sonic dimension and works using additional "dimensions", such as spatialization and the visual dimension in multimedia works. Some composers perform their compositions on mixing consoles, moving the sound around on "symphonies" of loudspeakers. Other composers team up with performers with interactive computer programs, dancers using sensors or graphic artists to give the audience something to watch. I have described the use of computers in various types of tuning systems, automatic music composition with the computer programs composing the music, and how the computer has made it possible to use synthesis techniques in new ways and with new sounds. Composers are using the computer for composing with sounds or parts of sounds rather than notes, composing with timbres and timbre morphing, composing with textures and texture morphing, granular synthesis, trills and convolution. Many composers use computer speech to tell stories in fantasy environments with sounds dancing toward the horizon in the aural landscapes.

Acknowledgments Many thanks to the composers whose works are cited in this chapter. Details in many of the piece descriptions have been summarized from their program notes.

References
Publications

Ayers, L. 1996. "A Composition for Gamelan and Computer-Generated Tape: *Merapi*", *Leonardo Music Journal*. **6**: 7–14.

Ayers, L. 2003. "Synthesizing Trills for the Chinese Dizi", *Proceedings of the 2003 International Computer Music Conference*. Singapore, pp. 227–230.

Ayers, L. 2004. "Synthesizing Timbre Tremolos and Flutter Tonguing on Wind Instruments", *Proceedings of the 2004 International Computer Music Conference*. Miami, FL, pp. 390–393.

Ayers, L. 2005. "Synthesizing Chinese Flutes Using Csound", *Organised Sound*. **10**(1): 37–49.

Ayers, L. 2006. "Fantasy Birds in Yazi's Dream", *Proceedings of the 2006 International Computer Music Conference*, New Orleans, LA, p. 33.

Ayers, L. and Horner, A. 1999. "Modeling the Woodstock Gamelan for Synthesis", *Journal of the Audio Engineering Society*. **47**(10): 813–823.

Ayers, L. and Horner, A. 2004a. "Expressive Modeling of Chinese Folk Wind Instruments", *Journal of the Acoustical Society of America*. **116**(4): 2619.

Ayers, L. and Horner, A. 2004b. "Synthesis of Chinese *Dizi* Ornaments", *Ghost in the Machine: Performance Practice in Electronic Music, Australasian Computer Music Association Conference*. Wellington, New Zealand.

Ayers, L. and Horner, A. 2006. "Synthesizing the Chinese *Suona*", *Medi(t)ations, Australasian Computer Music Conference*, Adelaide, Australia, pp. 15–22.

Ayers, L. and Horner, A. 2007. "Synthesizing the Chinese *Suona*", *Asia Pacific Festival and Conference*. Wellington, New Zealand.

Barbour, J. 2004. "Exploration of the Height Dimension in Audio Reproduction", *Australasian Computer Music Conference*, Wellington, New Zealand.

Barbour, J. 2006. "Creating Acoustic Space", *Australasian Computer Music Conference*, Adelaide, Australia.

Boulanger, R., Ed. 2000. *The Csound Book: Perspectives in Software Synthesis, Sound Design, Signal Processing and Programming*. MIT Press, Cambridge, MA.

Boulez, P. 1963. *Penser la Musique Aujourd'hui*. Editions Gonthier, Geneve.

Boulez, P. 1971. *Boulez on Music Today*. Bradshaw, S. and Bennett, R.R., trans. Faber, London.

Cycling 74. 2004. http://www.cycling74.com/products/maxmsp.html.

Daniélou, A. 1980. *The Ragas of North Indian Music*. New Delhi: Munshiram Manoharlal.

Dodge, C. and Jerse, T. 1984. *Computer Music*. 1st ed., Schirmer, New York.

Dolson, M. 1986. "The Phase Vocoder: A Tutorial", *Computer Music Journal*. **10**(4): 14–27.

Doty, D. 1993. *The Just Intonation Primer: An Introduction to the Theory and Practice of Just Intonation*. The Just Intonation Network, San Francisco, CA.

Helmholtz, H. 1954. *On the Sensations of Tone* 2nd ed., Dover Publications, New York, trans. Alexander J. Ellis. p. 90.

Helmuth, M. 2002. "StochGran on OSX", *Proceedings of the 2002 International Computer Music Conference*, Gotheborg, pp. 77–78.

Helmuth, M. StochGran available at: http://meowing.ccm.uc.edu/softmus.htm.

Horner, A. and Ayers, L. 1998a. "Modeling Acoustic Wind Instruments with Contiguous Group Synthesis", *Journal of the Audio Engineering Society*. **46**(10): 868–879.

Horner, A. and Ayers, L. 1998b. "Modeling Chinese Musical Instruments", *Proceedings of the 135th Meeting of the Acoustical Society of America*. Seattle, WA, **4**: 2541–2542.

Horner, A. and Ayers, L. 2002. *Cooking with CSound, Part 1: Woodwind and Brass Recipes*. AR Editions.

Horner, A., Ayers, L. and Law, D. 1999. "Synthesis Modeling of the Chinese *Dizi*, *Bawu*, and *Sheng*". *Journal of the Audio Engineering Society*. **47**(12): 1076–1087.

Koonce, P. "PVC", http://silvertone.princeton.edu/winham/PPSK/koonce.html.

Mathews, M. *The Radio Baton*. http://csounds.com/mathews/

McPhee, C. 1966. *Music in Bali*. Yale University Press, New Haven.
Partch, H. 1974. *Genesis of a Music*, 2nd ed. Da Capo, New York.
Pinkston, R., Kerkhoff, J. and McQuilken, M. 2001. *The U. T. Touch-Sensitive Dance Floor and MIDI Controller*. http://www.utexas.edu/cofa/music/ems/research/midifloor/midifloor.html.
Risset, J-C. 1996. "Introductory Catalogue of Computer Synthesized Sounds" and "My 1969 Sound Catalogue: Looking back from 1992", *Historical CD of Digital Sound Synthesis (Computer Music Currents 13)*. Wergo, Germany.
Roads, C. 1996. *The Computer Music Tutorial*. MIT Press, Cambridge, MA.
Schaeffer, P. 1966. *Traité des Objets Musicaux*, Seuil, Paris, France.
Sethares, W. 1998. *Tuning, Timbre, Spectrum, Scale*. Springer, London.
Winkler, T. 2001. Composing Interactive Music. MIT Press, Cambridge, MA.
Xenakis, I. 1963. *Musiques Formelles: Nouveaux Principes Formels de Composition Musicale*. (original French edition), republished by Stock, Paris, 1981.
Xenakis, I. 1992. *Formalized Music* (updated from first English edition, 1971). Pendragon Press, Stuyvesant, NY.
Zakros InterArts, http://www.zakros.com/archives/Echoes.html.

Music References

Alves, B. 1999. "In-Yo" on *ICMC 99* conference disk, Tsinghua University, Beijing China.
Applebaum, M. 2002. "Precomposition" on *Music from SEAMUS*, Vol. 13, SEAMUS, Los Angeles, CA.
Ayers, L. 2006a. "Balinese" on *Virtual Gamelan*, Albany Records, Albany, NY, TROY874.
Ayers, L. 2006b. "Bioluminescense" on *Virtual Gamelan*, Albany Records, Albany, NY, TROY874.
Ayers, L. 2006c. "Merapi" on *Virtual Gamelan*, Albany Records, Albany, NY, TROY874.
Ayers, L. 2006d. "Pendopos" on *Virtual Gamelan*, Albany Records, Albany, NY, TROY874.
Barbour, J. 2004. "Outback", diffused at *Australasian Computer Music Conference*, Concert 7, Wellington, New Zealand.
Berio, L. 1999. "Visage" on *Passaggi/Visage*, Bmg Ricordi, B000005SSX.
Brown, C.C. 2000. "Under Erebus" on *New Zealand Sonic Art*, Vol. II (2001), The University of Waikato, Hamilton, New Zealand, UWMD1201.
Dodge, C. 1972. "He Destroyed Her Image" on *Any Resemblance Is Purely Coincidental*. New Albion, San Francisco, CA.
Helmuth, M. 1998. *Implements of Actuation*, Electronic Music Foundation, Albany, NY, EMF 023.
Hiller, L. and Isaacson, L. 1956. *The Illiac Suite for String Quartet*, excerpt in *The Voice of the IBM 7090 Computer*.

Jaffe, D. 2001. *Racing Against Time* performed at the 2004 International Computer Music Conference, Miami, Florida

Lansky, P. 1985. *Idle Chatter*. Wergo Compact Disc 2010–50, Bridge Records, 9050CD.

Lansky, P. 1992. "Table's Clear" on *Home Brew*, Bridge Records, BCD 9035.

Norris, M. 1998. "Aquarelle" on *New Zealand Sonic Art III* (2002), The University of Waikato, Hamilton, New Zealand.

Redolfi, M. 1990–91. *Nausicaa, Bande Originale de L'exposition*. Nausicaa, Boulogne-sur-Mer, France.

Rimmer, J. 2000. "Ancestral Voices" on *New Zealand Sonic Art*, Vol. II (2001), The University of Waikato, Hamilton, New Zealand, UWMD1201.

Risset, J-C. 1969. "Mutations" on *OHM: The Early Gurus of Electronic Music: 1948–1980*. Elipsis Arts, Roslyn, New York, CD3670.

Risset, J-C. 1996. *Invisible*. GMEM, Marseille, France.

Roads, C. 1981. "Field" on *Digital Rewind 25: Music for Instruments and Computer, Twenty-Fifth Anniversary of The MIT Experimental Music Studio, 1974–1999*. The Media Laboratory, Massachusettes Institute of Technology.

Rubin, A. and Hollander, L. 2004, "Family Stories: Sophie, Sally" on *Electric Music*, Capstone Records, Broolyn, NY, CPS-8743.

Schaeffer, P. 2000a. "Etude aux Chemins de Fer" on *Pierre Schaeffer: L'Oeuvre Musicale*, INA/EMF Media EM155–3.

Schaeffer, P. 2000b. "Etude aux Chemins de Fer" on *OHM: The Early Gurus of Electronic Music*, Ellipsis Arts, B00004T0FZ.

Sethares, W. 1997. *Xentonality*. Frog Peak Music. Lebanon, NH.

Tang, J.W.C. 1996. "Cympergon" on 1996 *International Computer Music Conference CD*, Hong Kong. October Music, Taiwan, 96081–10004.

Truax, B. 1989. "Beauty and the Beast" on *Inside*, Cambridge Street Records, Burnaby, BC, Canada.

Truax, B. 1995. "Inside" on *Inside*, Cambridge Street Records, Burnaby, BC, Canada.

Whalley, I. 2002. "Kasumi" on *New Zealand Sonic Art III* (2002), The University of Waikato, Hamilton, New Zealand.

Xenakis, I. 1964. "Eonta" on *Metastasis*, Chant du Monde, France.

Xenakis, I. 1991. "Gendy 3" on *Xenakis*. Harmonia Mundi, Los Angeles, CA; Neuma Records, Chelmsford, MA.

Xenakis, I. 1994. "S.709" on *Electronic Music Foundation*. EMF, Albany, NY.

Video References

Van Raalte, C. and Zane-Cheong, J. 1995. "The BodySynth" on *International Computer Music Association Video Research Reports*. V. II, The International Computer Music Association, San Francisco, California.

23
Singing Voice Analysis, Synthesis, and Modeling

Youngmoo E. Kim

Electrical & Computer Engineering, Drexel University, Philadelphia, PA, USA, e-mail: ykim@drexel.edu

1 • Introduction

The singing voice is the oldest musical instrument, but its versatility and emotional power are unmatched. Through the combination of music, lyrics, and expression, the voice is able to affect us in ways that no other instrument can. The fact that vocal music is prevalent in almost all cultures is indicative of its innate appeal to the human aesthetic. Singing also permeates most genres of music, attesting to the wide range of sounds the human voice is capable of producing. As listeners we are naturally drawn to the sound of the human voice, and, when present, it immediately becomes the focus of our attention.

To meet the demands of vocal communication, the vocal apparatus is extremely flexible, yet also remarkably self-consistent. An endless variety of sounds can be produced by a fairly simple physical system of vibration and resonance. Describing the distinctive character of a voice, however, is difficult without resorting to vague and subjective terms (e.g., "squeaky") that have no objective correlates. These qualities are believed to be a combination of innate physical characteristics and learned patterns of expression, but quantifying, extracting, and modeling these features have proven to be a difficult task. Likewise, the singing voice has also proven difficult to simulate convincingly because of its greater physical variation compared to other instruments. In order to pronounce different words, a singer must move his or her jaw, tongue, teeth, etc., changing the shape and thus the acoustic properties of the vocal mechanism. This range of acoustic variation is difficult to capture in a low-dimensional model, and analysis and synthesis techniques which have been successful for other musical instruments often do not apply well to speech or singing.

Numerous models for analysis and synthesis of singing have been designed for a variety of applications. Some applications, such as singing detection and singer identification, focus only on analysis, while others, such as singing voice coding, require both analysis and synthesis components. This chapter explores the prevalent models used in modern singing voice analysis and synthesis and highlights the features and targeted applications of each.

Section 2 discusses background principles of singing voice modeling, such as vocal physiology and the source–filter formulation. Sections 3 and 4 examine multiple models used for the glottal source and vocal tract filter, respectively. This is followed by a description of various approaches to overall singing synthesis in Section 5.

2 • Principles of Singing Voice Modeling

2.1 The Anatomy of the Singing Voice

The anatomy of the voice consists of three primary collections of organs: the *lungs*, the *larynx*, and the *oropharynx*. Exhalation from the lungs results in the air pressure changes necessary for singing. The larynx consists of a skeleton of cartilage enclosing and supporting two structures of muscle and ligaments covered by mucous membranes known as the *vocal folds*. The oropharynx is more commonly known as the *vocal tract*. The key characteristic of the vocal tract is its ability to assume a wide range of different shapes, which are easily altered by articulating the positions of the jaw, tongue, and lips.

The process of singing begins with breath pressure produced by the lungs. In the case of *voiced* sounds (vowels, e.g., [a], [i], and semivowels, e.g., [m], [r]), muscles initially adduct (close) the vocal folds, but the breath pressure forces them to open. The airflow through the opening is uneven, with the air adjacent to the folds traveling a greater distance than the unimpeded air flowing through the opening. The result is a pressure differential, which causes the vocal folds to be sucked back together by the Bernoulli force. Rapid repetition of this process is called *phonation*, and the frequency of this repetition correlates to our perception of the pitch. In voiced sounds, phonation results in a largely harmonic sound source. For *unvoiced* sounds (e.g., [f], [t]), the vocal folds remain open and the breath pressure results in free airflow through the larynx into the mouth, where it is impeded by a constriction (caused by the tongue, soft palate, teeth, or lips), resulting in a sound source caused by air turbulence. Some vocal sounds require both phonation and turbulence sound sources and are referred to as *mixed* sounds (e.g., [v], [z]). For a detailed description of voice anatomy and the process of singing, see [49].

In all three cases, the *source* (phonation, turbulence, or both) is modified by the shape of the vocal tract (throat, mouth, nose, tongue, teeth, and lips). Since the acoustic properties of an enclosed space follow directly from the shape of that space, the physical flexibility of the vocal tract lends itself to tremendous acoustic flexibility. From a signal processing standpoint, the vocal tract represents a time-varying acoustic *filter*, which colors an input signal. This description of the human voice is called the linear *source–filter* model (also known as the *excitation-resonance* model), which is the basis of the majority of voice models.

2.2 Speech Versus Singing

Historically, speech and singing research have been closely linked, but there are important differences between the two methods of vocal production. The vast majority of sounds generated during singing are voiced (approximately 90%), whereas speech contains a much larger percentage of unvoiced sounds (about 60% voiced and 40% unvoiced for English) [8]. In the most common classical singing technique, known as *bel canto*, singers are taught to sustain vowels as long as possible between other phonemes because they are the most efficient and audible sounds. Classical singers usually employ a technique in which they lower the larynx, creating an additional high-frequency resonance (around 2–4 kHz) not present in other types of vocal production. This resonance, known as the *singer's formant*, is especially important for being heard in the presence of other instruments, for example allowing an opera singer to be heard over an entire orchestra [49]. Because of these differences, speech models do not always translate directly into accurate models for singing.

2.3 The Signal View of the Singing Voice

The process of singing results in a time-varying acoustic signal that is periodic over short intervals. The short-term periodicity relates to the pitch, and the amplitude envelope roughly corresponds to the perceived dynamic level. The more slowly varying spectral features, such as the spectral envelope, contribute to the

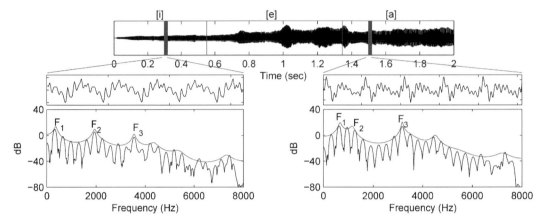

FIGURE 1 *Top*: time-domain signal of sung vowel sequence [i][e][a]. *Bottom*: magnified time-domain signals and short-time frequency analysis of phonemes [i] and [a]. The spectral envelope is depicted by the gray line, and the first three formants are labeled F_1, F_2, and F_3.

other components of singing (e.g., words, timbre, and musical expression). The relationship between spectral features and perceived expressive features is much less robust, but some general associations emerge. The spectral envelope contains several areas of increased energy (resonances) which may span across a range of harmonics. These resonances are called formants and have a great effect on vocal timbre [49]. The general position of the formants determines the phoneme of the singing. More subtle variations in formant frequency and bandwidth reflect the individual characteristics of the singer. An illustration of the singing signal in the time domain and frequency domain for two different phonemes is shown in Figure 1. Models of the singing voice have approached the problem from varying levels of abstraction: from physical models of vocal physiology to signal models of the acoustic waveform to symbolic or semantic models based on phonemes and musical notation. Several models are hybrids between different abstraction levels.

3 • Models of the Glottal Source

Because of the convenience of the linear source–filter decomposition, many computational representations of singing employ separate models for the source excitation signal. The vast majority of singing is voiced, and much of the work on excitation modeling has therefore focused on accurate modeling of the glottal flow. In voicing, the periodic but abrupt opening and closing of the vocal folds produces a glottal flow waveform rich in harmonics. Spectrally, the glottal flow tends to fall off smoothly as frequency increases, and this slope is known as *spectral tilt*.

For convenience in representation, models often instead target the derivative of the glottal flow, which has a generally flat spectrum. This accommodation is often made because some vocal tract models can easily incorporate the spectral tilt parameter into their representation. The abrupt opening and closing of the folds leads to an impulsive quality in the derivative glottal waveform, a characteristic that has been used to great advantage in early source models. The glottal flow and its derivative have been shown to have significant correlation to an individual's vocal quality, and glottal models have been used in systems for individual voice identification [35].

3.1 Signal Models

The earliest computational voice model, Homer Dudley's *Vocoder* (a contraction of voice coder) in 1936 [9], used excitations of a periodic impulse train to approximate the derivative glottal waveform and white noise to approximate the air turbulence of unvoiced sounds. These excitations are still used in some modern very low bitrate speech codecs because of the small number of parameters needed. Reconstructed speech and singing using the impulse train excitation, however, tend to sound overly buzzy, and more recent work has moved toward more complex models.

Modern signal models for the glottal flow include the KLGLOTT88 model [18] and the transformed Liljencrants–Fant (LF) model [12], each of which model a single period of the derivative glottal waveform. The LF model has gained wide acceptance because of its accuracy while still employing a reasonably small number of parameters.

3.2 Physical Models of the Vocal Folds

Researchers have also explored physical models of the vocal folds in order to accurately synthesize glottal flow. Initial models used a fixed number of lumped masses and springs to represent the mechanics of the vocal folds. The first models [15] used two masses and springs per fold, but were limited in their flexibility. More recently, the 16-mass model proposed by Titze [50] with extensions by Kob [19] has found wider acceptance. These models are able to simulate wide-range vocal-fold mechanics, and the computational requirements have become less of a restriction on modern computers. Recent work by Hunter et al. [14] pursues the even more complex approach of a finite element model of the vocal folds. This model no longer assumes lumped masses and springs but uses a large number (1,721) of very small elements to calculate the geometry and movement of the folds. Real-time calculation of this model, however, is still out of reach of today's computers.

These physical models can exhibit great authenticity when compared to the actual movement of the vocal folds and have been used to great effect in simulations, gaining traction in research on voice pathologies. Another benefit of these physical models is that they can be combined with physical models of the vocal tract in a nonlinear manner to model source–filter coupling and boundary effects. The greatest deficiency of physical models lies in the difficulty of parameterizing them especially from measured data. The mass values and spring constants in these models do not directly map to perceptually intuitive controls, though this is an area of continuing research [51].

4 • Models of the Vocal Tract

Dudley's Vocoder initially modeled the vocal tract response using a set of bandpass filters with amplitude controls to roughly approximate the spectral envelope of a voice signal. A modern extension of this principle is the *phase vocoder* (a short-time Fourier transform decomposition of the input signal), which has also found widespread use beyond voice coding in applications such as pitch- and time-scaling of audio [33]. Other modeling techniques have taken varied approaches to modeling the vocal tract, and some combine both source and filter elements to model the overall signal.

4.1 Formant Vocoder

A significant branch of vocoder research has focused on modeling the individual formants resulting from the shape of the vocal tract. The formant vocoder models each formant using a resonant filter (typically a second-order complex pole pair), resulting in a compact and efficient parameter set (see Figure 2b for an illustration). The formant filters are applied to an excitation signal to produce a resynthesized signal [39]. Difficulties are encountered, however, in the accurate estimation of the formant parameters (center frequency, amplitude, and bandwidth). In coding speech, the overall sound quality proved to be fairly low, and eventually the formant vocoder was eclipsed by other coding techniques that provided better sound quality at comparable bitrates.

The heightened role of formants in singing, however, has led to greater success for the formant vocoder in parametric singing synthesis. One of the earliest synthesizers designed specifically for singing voice is the Music and Singing Synthesis Equipment (MUSSE) developed at the Royal Institute of Technology (KTH) in Stockholm [21]. In this system, a simulated glottal pulse is sent

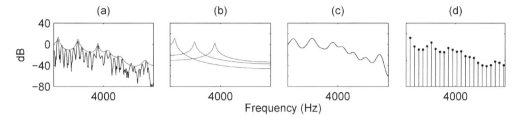

FIGURE 2 Graphical depiction in the frequency domain of several vocal tract models: (**a**) input signal, (**b**) formant vocoder, (**c**) cepstral model, and (**d**) sinusoidal model.

through a series of resonant filters to model the formants characteristic of a bass/baritone singer. Johan Sundberg and his group at KTH have determined general formant frequencies for all of the vowels and consonants. The most recent version of this synthesizer is MUSSE DIG, a digital implementation of the same structure that includes improvements to the glottal pulse model [4].

4.2 Homomorphic Vocoder

The homomorphic vocoder, based on an analysis by Alan Oppenheim [34], is an extension of the principles of the channel vocoder using homomorphic transformations to the *cepstral* (inverse log-Fourier) domain:

$$C_k = \frac{1}{2\pi} \int_{-\pi}^{\pi} \log |S(\omega)| \, e^{j\omega k} d\omega, \tag{1}$$

where C_k are the cepstral coefficients and $S(\omega)$ is the Fourier transform of the voice signal $s(t)$. In this domain, what was multiplication in the frequency domain (and thus convolution in the time domain) becomes a simple addition through the nonlinearity of logarithms. The original voice signal (a convolution of the vocal excitation function and the vocal tract response function – assuming a linear model) is represented as an addition of excitation and source in the cepstral domain making estimation of the separated functions easier. This technique was implemented by Tom Stockham and Neil Miller for the separation of voice from orchestra and subsequent sound restoration in recordings of Enrico Caruso [32].

The cepstral representation and its variants (e.g., MFCCs – Mel-frequency cepstral coefficients) tend to characterize gross features of the spectrum, which makes them useful for tasks requiring generalization, such as speech recognition [38] and speaker (talker) identification [27]. Some of the individual distinctiveness in singing, however, is characterized by finer spectral features that are not captured by low-order MFCCs. Figure 2c demonstrates how the cepstral representation results in a highly smoothed spectral envelope.

4.3 Sinusoidal Voice Coding

Sinusoidal analysis/synthesis for voice coding was first demonstrated by McAulay and Quatieri [29]. The voice signal is modeled as a number of sinusoidal partials:

$$s(t) = \sum_{k=1}^{M} A_k e^{j\omega_k t}. \qquad (2)$$

For voiced sound segments the partials will be largely harmonic, while for unvoiced segments they will be inharmonic. Sinusoidal parameter estimation is typically performed using STFT analysis, and careful peak selection, tracking, and phase matching across analysis frames result in high sound quality at reduced bitrates. Mellody et al. [30] have proposed a sinusoidal analysis technique for singing using a high-resolution time–frequency distribution, called the *modal distribution*. By maintaining high resolution in both time and frequency, precise parameterizations of the individual harmonic partials of the voice can be obtained. The benefit of this analysis was demonstrated in an experiment in which individual soprano singers were automatically identified.

An extension to the sinusoidal representation using a parametric noise model was developed by Serra and Smith [43]. Since this model makes no assumptions about the input signal, it has also been applied to the synthesis of musical instruments [43] and the coding of general audio [36]. The sinusoidal decomposition also lends itself well to modifications, such as pitch- and time-scaling and has been demonstrated to be very effective in modifying speech and singing voice recordings [37] as well as arbitrary audio signals [20].

Primarily grounded in the frequency domain, the sinusoidal model benefits from the intuitions of traditional frequency domain signal processing as related to pitch and timbre. A deficiency of the sinusoidal representation has been the difficulty in relating sinusoidal parameters to the mechanics of voice production. Hybrid models have been built, however, incorporating aspects of the source–filter representation into the sinusoidal framework. At the simplest level, the sinusoidal amplitudes form a spectral envelope roughly corresponding to the vocal tract filter response, implying an excitation signal of equal-amplitude sinusoids. More sophisticated approaches have attempted to capture the amplitude and phase relationships implied by a true glottal excitation with a sinusoidal excitation model. Researchers at the Music Technology Group of the Pompeu Fabra University (UPF) have used such a model, called EpR (excitation-plus-resonance), for singing synthesis and voice morphing in karaoke applications [5]. Recent work by McAulay [28] incorporating a sinusoidal excitation phase model has led to improvements in the quality of pitch, time scale, and spectral envelope modifications of voice signals.

4.4 Linear Predictive Coding

Perhaps the most influential advance in the history of voice coding after Dudley's work was the development of *linear predictive coding* (LPC) of speech, first proposed by Atal and Schroeder in 1970 [1]. LPC estimates a voice signal $s[n]$, using a linear combination of its p past samples. This is illustrated in the following difference equation relating the glottal source $g[n]$ and the voice output:

$$s[n] = \sum_{k=1}^{p} \alpha_k s[n-k] + g[n]. \qquad (3)$$

Taking the z-transform of this difference equation, the transfer function $H(z)$ relating the voice output to the glottal source in the frequency domain can be derived:

$$H(z) = \frac{S(z)}{G(z)} = \frac{1}{1 - \sum_{k=1}^{p} \alpha_k z^{-k}} = \frac{1}{A(z)}. \qquad (4)$$

Equation (4) demonstrates that the transfer function is always an all-pole filter, and for convenience the denominator polynomial is often labeled separately as $A(z)$. This derivation further demonstrates the inherent source–filter formulation of the LPC model where the transfer function $H(z)$ filters the glottal source $G(z)$, resulting in the voice output $S(z)$. For this reason, $H(z)$ is defined as the vocal tract filter. To determine the filter coefficient values *alpha$_k$* the error is minimized between the actual signal $s[n]$ and the predicted values from the previous p samples. The two most common techniques for solving this minimization are the autocorrelation and covariance methods, named for their use of the autocorrelation and covariance matrices, respectively [26], which both satisfy the minimum least squared error criterion. Vocal tract filter estimation via LPC also has the benefit of being a closed-form computation, requiring no heuristics for the determination of parameters. A disadvantage of standard LPC is that all frequencies are weighted equally on a linear scale while the frequency sensitivity of the human ear is close to logarithmic. The LPC coefficients, however, can also be found in a warped frequency domain [13,45,48] that is more faithful to human perception.

When coupled with an appropriate excitation model, LPC can result in very low-bitrate transmission of speech. An extension, *code-excited linear prediction* (CELP), which is the basis of most speech codecs used in today's digital communications devices, employs a closed-loop analysis-by-synthesis to find the best fit within a codebook of excitation vectors. The codebook usually consists of an adaptive component from past excitations along with a fixed component allowing for further modification of the excitation.

By design, LPC finds the best fit of the vocal tract filter to an impulse excitation, which can be a poor rendition of the derivative glottal flow. Closed-phase LPC, using data only from the closed phase of the glottal flow (the time of least source–tract interaction) has been used to calculate glottal flow mode parameter estimates used for talker identification [35]. A recent extension to LPC modeling of singing has been joint optimization with a derivative glottal flow model, minimizing the error across a KLGLOTT88 model and LPC vocal tract filter, accounting for some of the inherent source–tract interaction [24]. After the joint optimization, the glottal flow is re-parameterized using the LF model, which is more intuitive for manipulation and synthesis. The joint source–filter optimization has been further enhanced to use warped LPC [17]. Synthesis quality using the jointly optimized system is improved using a parametric pulsed noise model in [24] and CELP-style residual codebooks in [17].

4.5 Physical Modeling of the Vocal Tract

Kelly and Lochbaum first modeled the vocal tract as a series of cylindrical tube sections represented by a digital ladder filter in 1962 [16]. This model of sound propagation has come to be known as a waveguide digital filter and has been used as the basis for a variety of musical instrument models [44]. Liljencrants first coupled the Kelly–Lochbaum model with the LF glottal excitation model in 1985 [22].

A *singing physical articulatory synthesis model* (SPASM) was created by Perry Cook at the Center for Computer Research in Music and Acoustics (CCRMA) at Stanford University in 1990 [8]. SPASM used a digital waveguide to model the physical acoustics of the vocal tract as well as the nasal cavity and throat radiation, and the system was driven by a frequency domain excitation model. An integrated graphical composition system was provided to control the synthesis parameters. In 1994, Välimäki and Karjalainen extended the vocal tract physical model by using variable length conical sections [52]. A system using coupled physical models of both vocal folds and vocal tract has been developed by Kob [19].

As with the physical models of the vocal folds, it is difficult to estimate model parameters from an audio signal, though several studies at KTH and the University of Arizona have extracted vocal tract area parameters from three-dimensional magnetic resonance images (MRI) [11,47]. Engwall [10] and Story [46] have developed 3D shape models of the vocal tract from measured MRI data. Articulation can be modeled in 3D, but synthesis is performed by calculating area functions from the 3D model. True 3D models, such as *finite element modeling*, are an active area of research.

4.6 PSOLA and WSOLA

Pitch-synchronous overlap-add (PSOLA) [7] and *waveform similarity overlap-add* (WSOLA) [53] are techniques that are commonly used in commercial products for voice processing and modification. Both techniques operate by identifying and manipulating time-domain regions of similarity (by estimating pitch or comparing waveforms). Deletion or replication of segments results in time compression or expansion, respectively. By overlapping and summing the windowed regions, these effects can be achieved while maintaining high sound quality. Pitch manipulation is also possible by altering sampling rates or changing the amount of overlap between windows. The use of devices based on PSOLA/WSOLA is almost ubiquitous in popular music recording, and they are also commonly used to provide pitch correction for live performances as well as karaoke machines. But the separation of parameters using this approach is somewhat limited, e.g., the pitch cannot be varied independently from the vocal tract shape.

4.7 Formant Wave Functions

Formant wave functions (FOFs, from the French) are time-domain functions with a particular resonance characteristic used to model individual formants [41]. Each function is repeated at the desired fundamental period of voicing, and the functions are then summed for synthesis. As with the formant vocoder, parameter estimation for the FOF functions is challenging because of the difficulty in identifying and characterizing formants. Because of this, FOFs have been used primarily for singing synthesis.

Singing voice synthesis with FOFs is implemented in the CHANT system, developed at the Institut de Recherche et Coordination Acoustique/Musique (IRCAM) by Xavier Rodet et al. [42]. It uses five formant wave functions to generate an approximation of the resonance spectrum of the first five formants of a singer. The results of the original system for vowel sounds are impressive (including a famous rendition of Mozart's Queen of the Night aria from *Die Zauberflöte* [2]) but involve a great deal of hand adjustment of parameters. Later work has extended the use of FOFs to consonants and unvoiced phonemes [40]. The CHANT system was primarily designed as a tool for composers.

5 • Synthesis of the Singing Voice

Direct synthesis of singing, as opposed to resynthesis or encoding of an existing signal, adds even more challenges. Most languages can be represented in terms of a limited set of *phonemes* (basic linguistic units of sound), but the rules governing pronunciation and inflection are only guidelines to the actual speaking or singing of that language. The problem of interpreting a musical score in a musically proper and convincing fashion is difficult in the case of any instrument, and adding proper controls to manipulate the parameters of a singing voice model increases the difficulty of the problem.

5.1 Rules-Based Approaches

Several approaches have formulated rules for singing voice performance based on analytical and anecdotal data. The KTH Rule System [3], designed for the MUSSE DIG synthesizer, contains rules for numerous parameters such as pitch change glissando and timing, dynamics, vibrato, and formant transitions. A rule system was also used in the CHANT project to model formant trajectories. Rules-based systems have also been used to enhance other synthesis systems (see below).

5.2 Concatenation Synthesis

An approach favored in speech synthesis uses models of phonemes concatenated together to create words and phrases. The phonemes may be stored simply as short recordings which are smoothly concatenated and pitch-shifted to create a continuous pitch contour. Alternative models can also be used to generate the phonemes. One example is the LYRICOS project [25], which uses a set of phonemes stored as sinusoidal model parameters, incorporated with control data (such as pitch, dynamics, and vibrato) from MIDI. But single-phoneme models generally do not have the flexibility to accommodate the wide variation present in speech or singing, and without dynamic control manipulation, the synthesis is rather limited.

Multi-phoneme models have been used with a greater amount of success. A recent example at UPF uses their EpR sinusoids-plus-noise representation to model a large database of phoneme combinations from an individual singer at a variety of pitches [5]. A rule system was used to enhance the overall quality of the phoneme transitions [6]. This work has been incorporated into a commercial product, the Yamaha Vocaloid [54], a singing voice synthesizer that makes use of "voice libraries" designed to model an individual singer's voice.

5.3 Data-Driven Approaches

Systems have also been designed to use statistical methods to model parameter or phoneme transitions for singing voice models. One project, by Ken Lomax at Oxford University, trained neural networks on input data from famous singers to control the necessary synthesis parameters [23]. This system was limited to mostly vowels, with a few voiced consonants. Yoram Meron at the University of Tokyo used *hidden Markov models* (HMMs) to control a mapping from sinusoidal phoneme models of an individual singer to a target sound, which could be from the same singer or a different singer [31]. This enabled a voice transformation in which one singer's voice model (stored phonemes) is driven by another (target sound). A recent system by Kim [17] uses HMMs to model joint source–filter (LF and warped LPC) model dynamic parameter motion within single phonemes to analyze an individual singer's time-varying expression. This system allows the source–filter model characteristic of one singer to be driven by the expressive characteristics of another, achieving a form of separation of the instrument from the instrumentalist that is not possible with real singers.

6 • Conclusion

This chapter has provided a brief overview of some of the myriad models and techniques used in applications of analysis and synthesis of singing. Some of the signal processing models have been in existence for quite some time, and their utility and robustness are well documented. Appropriate model representations and mappings from control to model parameters remain an active area of research. Dynamic modeling of expressive performance parameters is a relatively new endeavor, but new methods in pattern recognition and machine learning along with the continuing advance of computational power hold great promise for the future.

References

1. B. S. Atal and M. R. Schroeder. Adaptive predictive coding of speech signals. *Bell System Tech. J.*, pp. 1973–1986, 1970.
2. G. Bennett and X. Rodet. Synthesis of the Singing Voice, CD of Sound Examples. *Current Directions in Computer Music Research*, pp. 19–44, MIT Press, Cambridge, MA, 1989.

3. G. Berndtsson. The KTH rule system for singing synthesis. *Computer Music J.*, 20(1):76–91, 1996.
4. G. Berndtsson and J. Sundberg. The MUSSE DIG singing synthesis. *Proc. Stockholm Music Acoustics Conf. (SMAC)*, pp. 279–281, 1993.
5. J. Bonada and A. Loscos. Sample-based singing voice synthesizer by spectral concatenation. *Proc. Stockholm Music Acoustics Conf. 2003*, pp. 439–442, 2003.
6. J. Bonada, O. Celma, A. Loscos, J. Ortolà, and X. Serra. Singing voice synthesis combining excitation plus resonance and sinusoidal plus residual models. *Proc. Int. Computer Music Conf.*, Havana, pp. 139–146, 2001.
7. F. J. Charpentier and M. G. Stella. Diphone synthesis using an overlap-add technique for speech waveforms concatenation. *Proc. Int. Conf. Acoustics, Speech, and Signal Processing*, Tokyo, pp. 2015–2018, 1986.
8. P. R. Cook. *Identification of Control Parameters in an Articulatory Vocal Tract Model, with Applications to the Synthesis of Singing*. PhD thesis, Stanford University, 1990.
9. H. Dudley. Synthesizing speech. *Bell Laboratories Record*, 15: 98–102, 1936.
10. O. Engwall. Synthesising static vowels and dynamic sounds using a 3d vocal tract model. *Proc. Fourth ISCA Tutorial and Research Workshop on Speech Synthesis*, Perthshire, Scotland, pp. 38–41, 2001.
11. O. Engwall and P. Badin. Collecting and analysing two- and three-dimensional MRI data for Swedish. *Speech Transmission Laboratory Quarterly Progress and Status Report (STL-QPSR)*, 3–4:11–38, 1999.
12. G. Fant, J. Liljencrants, and Q.-G. Lin. A four parameter model of glottal flow. *Speech Transmission Laboratory Quarterly Progress and Status Report (STL-QPSR)*, 4:1–13, 1985.
13. A. Härmä. A comparison of warped and conventional linear predictive coding. *IEEE Trans. on Speech and Audio Processing*, 9(5):579–588, 2001.
14. E. J. Hunter, I. R. Titze, and F. Alipour. A three-dimensional model of vocal fold abduction/adduction. *J. Acoust. Soc. Am.*, 115(4): 1747–1759, 2004.
15. K. Ishizaka and J. L. Flanagan. Synthesis of voiced sounds from a two-mass model of the vocal cords. *Bell System Tech. J.*, 51: 1233–1268, 1972.
16. J. L. Kelly and C. C. Lochbaum. Speech synthesis. *Proc. Fourth Int. Congress on Acoustics*, pp. 1–4, 1962.
17. Y. E. Kim. *Singing Voice Analysis/Synthesis*. PhD thesis, Massachusetts Institute of Technology, 2003.
18. D. H. Klatt and L. C. Klatt. Analysis, synthesis, and perception of voice quality variations among female and male talkers. *J. Acoust. Soc. Am.*, 87(2):820–857, 1990.
19. M. Kob. *Physical Modeling of the Singing Voice*. PhD thesis, University of Technology Aachen, 2002.

20. J. Laroche, Time and Pitch Scale Modification of Audio Signals. *Applications of Signal Processing to Audio and Acoustics*. Kluwer Academic Publishers, Dordrecht, 1998.
21. B. Larsson. Music and Singing Synthesis Equipment (MUSSE). *Speech Transmission Laboratory Quarterly Progress and Status Report (STL-QPSR)*, 1:38–40, 1977.
22. J. Liljencrants. *Speech Synthesis with a Reflection-Type Line Analog*. PhD thesis, Royal Institute of Technology, Stockholm, Sweden, 1985.
23. K. Lomax. *The Analysis and Synthesis of the Singing Voice*. PhD thesis, Oxford University, 1997.
24. H.-L. Lu. *Toward a High-Quality Singing Synthesizer with Vocal Texture Control*. PhD thesis, Stanford University, 2002.
25. M. W. Macon, L. Jensen-Link, J. Oliverio, M. A. Clements, and E. B. George. Concatenation-based midi-to-singing voice synthesis. *Audio Eng. Soc. Preprint 4591*, 1997.
26. J. Makhoul. Linear prediction: A tutorial review. *Proceedings of the IEEE*, 63:1973–1986, 1975.
27. R. Mammone, X. Zhang, and R. P. Ramachandran. Robust speaker recognition: A feature-based approach. *IEEE Signal Processing Magazine*, 13(5):58–71, 1996.
28. R. J. McAulay. Voice Modification using the Sinusoidal Analysis/Synthesis Method, *Keynote Presentation at 2003 IEEE Workshop on Applications of Signal Processing to Audio and Acoustics*, New Paltz, NY, October, 2003.
29. R. J. McAulay and T. Quatieri. Speech analysis/synthesis based on a sinusoidal representation. *IEEE Trans. Acoustics, Speech, and Signal Processing*, 34(4):744–754, 1986.
30. M. Mellody. G. H. Wakefield, and F. Herseth, *Model distribution analysis and Synthesis of a Soprano's Sung vowels. J. Voice*, 15(4): 469–482, 2001.
31. Y. Meron. *High Quality Singing Synthesis Using the Selection-Based Synthesis Scheme*. PhD thesis, University of Tokyo, 1999.
32. N. J. Miller. *Filtering of Singing Voice Signal from Noise by Synthesis*. PhD thesis, University of Utah, 1973.
33. J. A. Moorer. The use of the phase vocoder in computer music applications. *J. Audio Eng. Soc.*, 26(1):42–45, 1978.
34. A. V. Oppenheim and R. W. Schafer. *Discrete-Time Signal Processing*. Prentice-Hall, Englewood Cliffs, NJ, 1989.
35. M. D. Plumpe, T. F. Quatieri, and D. A Reynolds. Modeling of the glottal flow derivative waveform with application to speaker identification. *IEEE Trans. Speech and Audio Processing*, 7(5):569–586, September 1999.
36. H. Purnhagen and N. Meine. HILN – the MPEG-4 parametric audio coding tools. *Proc. IEEE Int. Symposium on Circuits and Systems* Geneva, Switzerland, pp. 201–204, 2000.
37. T. F. Quatieri and R. J. McAulay. Speech transformations based on a sinusoidal representation. *IEEE Trans. Acoustics, Speech, and Signal Processing*, 34(6):1449–1464, 1986.

38. L. R. Rabiner and B.-H. Juang. *Fundamentals of Speech Recognition*. Prentice Hall, Englewood Cliffs, NJ, 1993.
39. L. R. Rabiner and R. W. Schafer. *Digital Processing of Speech Signals*. Prentice-Hall, Englewood Cliffs, NJ, 1978.
40. G. Richard, C. d'Alessandro, and S. Grau. Unvoiced speech synthesis using poissonian random formant wave functions. *Signal Processing VI: European Signal Processing Conf.*, pp. 347–350, 1992.
41. X. Rodet. Time-domain formant-wave-function synthesis. *Computer Music J.*, 8(3):9–14, 1984.
42. X. Rodet, Y. Potard, and J.-B. Barrière. The CHANT project: From the synthesis of the singing voice to synthesis in general. *Computer Music J.*, 8(3):15–31, 1984.
43. X. Serra and J. O. Smith. Spectral modeling synthesis: A sound analysis/synthesis system based on a deterministic plus stochastic decomposition. *Computer Music J.*, 14(4):12–24, 1990.
44. J. O. Smith. *Techniques for Digital Filter Design and System Identification with Application to the Violin*. PhD thesis, Stanford University, 1985.
45. K. Steiglitz. A note on variable recursive digital filters. *IEEE Trans. Acoustics, Speech, and Signal Processing*, 28(1):111–112, February 1980.
46. B. H. Story. Using imaging and modeling techniques to understand the relation between vocal tract shape to acoustic characteristics. *Proc. Stockholm Music Acoustics Conf. 2003*, pp. 435–438, 2003.
47. B. H. Story, I. R. Titze, and E. A. Hoffman. The relationship of vocal tract shape to three voice qualities. *J. Acoustical Soc. Am.*, 109(4):1651–1667, 2001.
48. H. W. Strube. Linear prediction on a warped frequency scale. *J. Acoust. Soc. Am.*, 68(4):1071–1076, 1980.
49. J. Sundberg. *The Science of the Singing Voice*. Northern Illinois University Press, Dekalb, IL, 1987.
50. I. R. Titze. Phonetica. *The Human Vocal Cords: A Mathematical Model. Part 1*, 28:129–170, 1973.
51. I. R. Titze and B. H. Story. Voice quality: What is most characteristic about "you" in speech. *Echoes: The Newsletter of The Acoustical Society of America*, 12(4):1, 4, Fall 2002.
52. V. Välimäki and M. Karjalainen. Improving the Kelly–Lochbaum vocal tract model using conical tube sections and fractional delay filtering techniques. *Proc. 1994 Int. Conf. on Spoken Language Processing (ICSLP)*, Yokohama, Japan, pp. 615–618, 1994.
53. W. Verhelst and M. Roelands. An overlap-add technique based on waveform similarity (WSOLA) for high-quality time-scale modifications of speech. *Proc. Int. Conf. on Acoustics, Speech, and Signal Processing*, Minneapolis, Vol. 2, pp. 554–557, 1993.
54. Yamaha Corporation Advanced System Development Center. *New Yamaha VOCALOID Singing Synthesis Software Generates Superb Vocals on a PC*, 2003. http://www.global.yamaha.com/news/20030304b.html.

24

Instrument Modeling and Synthesis

Andrew B. Horner[1] *and James W. Beauchamp*[2]

[1] *Department of Computer Science, Hong Kong University of Science and Technology, Clear Water Bay, Kowloon, Hong Kong, China, e-mail: horner@cs.ust.hk*
[2] *School of Music and Department of Electrical and Computer Engineering, University of Illinois at Urbana-Champaign, Urbana, IL 61801, USA, e-mail: jwbeauch@uiuc.edu*

1 • Introduction

During the 1970s and 1980s, before synthesizers based on direct sampling of musical sounds became popular, replicating musical instruments using frequency modulation (FM) or wavetable synthesis was one of the "holy grails" of music synthesis. Synthesizers such as the Yamaha DX7 allowed users great flexibility in mixing and matching sounds, but were notoriously difficult to coerce into producing sounds like those of a given instrument. Instrument design wizards practiced the mysteries of FM instrument design. Wavetable synthesis was less mysterious, but was limited to rather static organ-like sounds. With a single wavetable, you could easily attain a trumpet-like sound, but a realistic trumpet was still out of reach. Then, sampling came along and soon even cheap synthesizers could sound like realistic pianos and trumpets, as long as the desired articulations happened to match those of the original recorded samples. Sample libraries quickly replaced sound wizards.

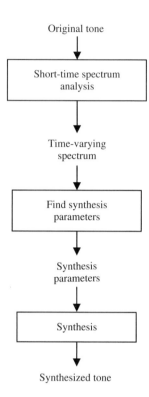

FIGURE 1 Overview of the analysis/matching/resynthesis procedure.

Ironically, at about the time that FM declined, researchers developed algorithms to optimize FM and wavetable parameters. The results were every bit as realistic as sampled sounds, with the added benefits of increased spectral and temporal control.

The basic procedure for FM and wavetable matching is shown in Figure 1. Matching generally begins with a short-time (aka time-varying) spectrum analysis of the original sound. Typically, 500–2,000 consecutive "spectral snapshots" or frames of the tone result from this analysis, each consisting of the instantaneous amplitudes and frequencies of the harmonics of the tone's time-varying fundamental frequency ($f_0(t)$). The next step is to find parameters for the synthesis algorithm that provide the average "best fit" of the spectrum of the tone over the set of these snapshots. The final step is to resynthesize the sound using the matched parameters.

The next section describes the short-time spectral analysis methods the authors have used, the first step of Figure 1. Then, we describe the fundamentals of wavetable synthesis, followed by an overview of wavetable matching procedures. Next, we describe

FM synthesis and FM matching. Finally, we conclude with some practical guidelines for selecting among these methods.

2 • Spectral Analysis

Spectral analysis methods used for wavetable and FM matching synthesis are of necessity pitch-synchronous and are described by Beauchamp (2007). The most often used method is a variation on the *phase vocoder* (Flanagan and Golden 1966; Portnoff 1976). With our method, a digital single-voice signal, which is assumed to have a nearly constant fundamental frequency, is broken into overlapping segments, called windows or frames. For analysis purposes, the fundamental is chosen to be f_a, an approximation to the actual time-varying fundamental frequency $f_0(t)$. The window lengths are taken to be twice the period corresponding to the fundamental (i.e., $T = 2/f_a$), and the overlap time or time-between frames is taken to be a half period (i.e., $\Delta T = 0.5/f_a = 0.25T$). The windowed signal segments are multiplied by a Hamming window function ($w(t) = 0.5 + 0.426\cos(2\pi t/T)$) before applying a fast Fourier transform (FFT) to the signal samples. However, the signal must generally be upsampled to provide an even power-of-two samples per window in order to satisfy the requirements of our version of the FFT. If the original sample rate is f_s, the new sample rate is $f'_s = 2^{\mathrm{ceil}(\log_2(Tf_s))}/T$. After the FFT is taken, only the even components are retained, with the second component corresponding to the first harmonic, the fourth component corresponding to the second harmonic, and so forth. For each spectral snapshot or frame j, the FFT returns the real and imaginary parts of the analysis, and these must be converted to harmonic amplitudes and frequencies. For each harmonic k, the estimated amplitude $A_{k,j}$ is computed as the square root of the sum of the squares of the real and imaginary parts (i.e., the complex magnitude), whereas the estimated instantaneous frequency $f_{k,j}$ is computed using an arctangent formula based on the real and imaginary parts for two successive frames (Beauchamp 2007). The number of harmonics K analyzed and kept is given by the Nyquist frequency divided by f_a; thus, $K = \mathrm{floor}(0.5f_s/f_a)$. The total number of data points kept is twice the number of samples in the sound signal analyzed.

An alternative analysis method is based on frequency tracking of spectral components (McAulay and Quatieri 1986; Smith and Serra 1987). This method works for signals whose fundamental frequency variation exceeds a small percentage of its mean value. A fixed window of 2^N sample length is chosen to be greater than three times the period of the lowest expected frequency.

After a Kaiser-windowed FFT processes the signal, the magnitude spectrum is computed, and spectral peaks above a specified amplitude threshold are retained. Spectral peaks are connected from frame to frame using a nearest neighbor algorithm to form tracks. Next, fundamental frequency vs. time is computed using a "two-way mismatch algorithm" (Beauchamp et al. 1993; Maher and Beauchamp 1994). Finally, harmonics are separated from the total spectrum based on the computed time-varying f_0. For tones with little frequency variation, the result is very similar to that obtained using the phase vocoder, but the frequency-tracking method works much better for sounds with significant frequency deviation.

Although in general these methods result in harmonics which do not track each other perfectly (i.e., $f_{k,j}/k_1 \neq f_{k,j}/k_2$ for $k_1 \neq k_2$), the methods described in this chapter require a common fundamental frequency such that $f_{k,j} = kf_{0,j}$. For the normally used phase-vocoder method, Calculation of $f_{0,j}$ is described by Beauchamp (2007). The result is a signal model given by

$$s[n] = \sum_{k=1}^{K} A_k[n] \cos(k \int f_0[n] \mathrm{d}n/f_s + \phi_k), \qquad (1)$$

where $n =$ sample number, $k =$ harmonic number, and $A_k[n]$ and $f_o[n]$ are given as sample-time harmonic amplitude and fundamental frequency values interpolated from neighboring frame-time values $A_{k,j}$ and $f_{o,j}$ obtained from the spectral analysis. f_s is the sampling frequency in Hz.

3 • Wavetable Synthesis

The music industry currently uses the term *wavetable synthesis* synonymously with *sampling synthesis*. However, in this article, *sampling* means recording an entire note and playing it back, whereas *wavetable synthesis* refers to storing only one period of a waveform in an oscillator table and synthesizing a signal by applying amplitude- and frequency-vs.-time envelopes (Mathews 1969). The waveform is generated by a sum of harmonically related sine waves, whose amplitudes and phases define the wavetable's spectrum. In our methods of wavetable synthesis, several wavetables, each corresponding to a *basis spectrum*, are scaled by time-varying amplitude envelopes (or weight functions) and summed to form the output signal. Figure 2 shows a block diagram for this *multiple-wavetable synthesis* model.

Prior to synthesis, one cycle of each basis waveform is stored in a wavetable. The waveform is generated by a sum of harmonically

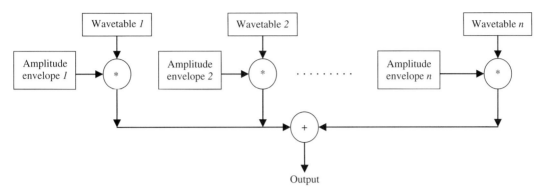

FIGURE 2 A multiple wavetable synthesis block diagram.

related sine waves, i.e., for the mth waveform or table, the table entries are given by

$$\text{table}_m[i] = \sum_{k=1}^{K} A_{k,m} \cos\left(\frac{2\pi k i}{\text{tableLength}} + \phi_{k,m}\right),$$

$$\text{for } 0 \leq i < \text{tableLength}, \quad (2)$$

where $\text{table}_m[i]$ is the ith entry of the mth wavetable, tableLength is the number of entries in the table, and $A_{k,m}$ and $\phi_{k,m}$ are the amplitude and phase of the kth harmonic for the table. The A_ks corresponding to each table form the wavetable's basis spectrum. The phases generally do not usually affect the sound and are often set to zero, but in any case, for a given harmonic, they should be the same for all tables in order to avoid phase cancellations. Then, for M equal-length wavetables, the output signal to be matched with the original of Eq. (1) is given by

$$\hat{s}[n] = \sum_{m=1}^{M} \alpha_m[n] \text{table}_m\left[\left(\int f_o[n] dn / (f_s \text{tableLength})\right)\right], \quad (3)$$

where $\alpha_m[n]$ is the amplitude envelope of the mth wavetable and $f_o[n]$ is the frequency envelope common to all wavetables. Although table entries are specified to be integers in Eq. (2), we assume here that the table values are interpolated according to the integer and fraction values of its argument and that its argument is taken modulo tableLength.

The main advantage of multiple-wavetable synthesis is its efficiency for generating periodic waveforms. A possible disadvantage of this method is that each wavetable only produces a static spectrum, while real sounds produce continuously changing spectra. However, if we carefully select basis spectra and control amplitudes, we can fully reproduce the dynamic spectra of the original tone in wavetable synthesis.

Another possible disadvantage is that it is necessary that harmonics be strictly locked together in frequency so that $f_k[n] = kf_1[n]$. Since phase is the integral of frequency, this implies that relative phases are also locked together. Frequency unlocking (inharmonicity) or phase unlocking certainly occurs in freely vibrating tones, such as struck or plucked string tones (Fletcher et al. 1962), and there is some evidence that phase unlocking (or aperiodicity) also occurs momentarily in sustained tones, especially during vibrato and transients (Beauchamp 1974; Schumacher 1992; Brown 1996; Dubnov and Rodet 2003). Extension of the multiple-wavetable synthesis method for approximately handling inharmonic cases with slightly mistuned wavetables has been explored for the piano (Lee and Horner 1999; Zheng and Beauchamp 1999). Bristow-Johnson (1996) has proposed a wavetable method for changing relative phases as time progresses, which is tantamount to synthesizing unlocked phases. However, the perceptual advantage of unlocked phases in sustained tones has not been fully explored. In any case, synthesis with unlocked phases is beyond the scope of this chapter.

There are two basic types of multiple-wavetable synthesis. *Wavetable interpolation synthesis* is a method whereby tables are interpolated from one (or possibly a few) to the next in succession, usually using straight-line interpolation (Horner and Beauchamp 1996). With *wavetable-index synthesis* all tables are active simultaneously, and the amplitude envelope weights applied to the tables are computed using *least squares* (Horner et al. 1993a). In this case, the amplitude envelopes tend to be quite complex with no restriction on their polarities. Thus, when basis spectra are combined (assuming they have identical phases), their weights may cause them to cancel each other, and, in any case their combination will converge to the original time-varying spectrum as more tables are added.

Basis spectra can be taken directly from the original time-varying spectrum at certain points in time, from spectral averages using a clustering method, or by using a method such as principal components analysis (PCA) to compute basis spectra which decrease in importance as more are added. Optimal matching algorithms, such as genetic algorithms (GA) (Goldberg 1989), can be used to search for the "best" set of basis spectra.

4 • Wavetable Matching

The objective of wavetable matching is to find the best set of parameters to synthesize a musical instrument tone using wavetable synthesis. A number of papers have explored methods for optimizing wavetable basis spectra and their amplitude-vs.-time envelopes (Horner et al. 1993a; Sandell and Martens 1995; Horner and Beauchamp 1996; Horner 2001). Usually a *cost metric* is used to determine the efficacy of the match and to steer the matching process. One metric we have found successful, called the *relative amplitude spectral error*, appears to correspond closely to perceptual similarity (Horner et al. 2006); it is given by

$$\varepsilon = \frac{\sum_{j=0}^{J-1} \sum_{k=1}^{K} \left| A_{k,j} - A'_{k,j} \right|^p}{\sum_{j=0}^{J-1} \sum_{k=1}^{K} \left| A_{k,j} \right|^p}, \qquad (4)$$

where J is the number of analysis frames used for the average. The error ε can be an average over all of the frames of the tone or a suitable subset of the frames. The power p can be either 1 or 2, and good results are obtained for either case.

Maher and Beauchamp (1990) used a wavetable matching synthesis method in their investigation of vocal vibrato. They selected their basis spectra at the low and high points of the vibrato of sung tones and crossfaded the wavetables as a function of the vibrato. Since the spectra at these points virtually repeated in cyclic fashion during the tone, only two wavetables were required for the synthesis.

With wavetable interpolation synthesis (Chamberlin 1980; Serra et al. 1990; Horner and Beauchamp 1996), the signal is represented by a consecutive series of basis spectra, which are generally spectral snapshots take from the tone itself at particular points in time or *breakpoints*, and synthesis proceeds by gradually crossfading from one spectrum (or waveform) to the next. While one basis spectrum ramps down to zero, a new basis spectrum ramps up to take its place.

Serra et al. (1990) give two algorithms for determining spectral interpolation basis spectra. The first draws basis spectra from the original tone. The second uses a least-squares algorithm to compute the basis spectra which minimize the mean-squared error between the original and synthetic spectra. Basis spectra are added until the maximum mean-squared error is brought below a prescribed threshold. Interpolation between basis spectra is done using a linear or nonlinear method. These algorithms cycle through basis spectra between 5 and 20 times per second.

Rather than first specifying a error threshold, Horner and Beauchamp (1996) used a genetic algorithm and a sequential enumeration (aka greedy) method to select a predetermined number of best times (breakpoints) to take basis spectra from the signal. Several questions were explored. It was found that the GA method was slightly better than the greedy method in terms of relative error vs. number of breakpoints, but it was substantially better than random breakpoints or equally spaced breakpoints. While breakpoints optimized independently for each harmonic performed somewhat better than common breakpoints in terms of error vs. the number of breakpoints per harmonic, common breakpoints won in terms of error vs. data storage requirements. Also, independent harmonic breakpoints would not allow wavetable interpolation. Another method tried used quadratic rather than linear curves to interpolate between basis spectra. Like the independent harmonic breakpoint attempt, this resulted in an improvement for the same number of breakpoints, but when total data and computation requirements were taken into account, improvement was negligible or nil.

Genetic algorithms have also been used to optimize wavetable interpolation with multiple (more than two at a time) wavetables (Horner 1996b). In this case, each breakpoint spectrum is the weighted sum of two or more spectra (wavetables) which are crossfaded to the same number of tables at the next breakpoint.

Group additive synthesis (Kleczkowski 1989) is another wavetable variant at the opposite extreme of spectral interpolation. Group additive synthesis divides the spectrum into non-overlapping subsets of harmonics for the various wavetables. As an example, one wavetable might contain only the even harmonics while a second only the odd. Subsequent to Kleczkowski's initial study, researchers have optimized group additive synthesis parameters using an automated clustering scheme (Oates and Eaglestone 1997) and genetic algorithms (Cheung and Horner 1996; Horner and Ayers 1998; Lee and Horner 1999).

For wavetable-index synthesis, researchers have used genetic algorithms to match the multiple-wavetable model shown in Figure 2 (Horner et al. 1993a; Horner 1995). Like the interpolation synthesis method, one approach is to use a genetic algorithm to select the best time values (breakpoints) for spectral snapshots to be taken from the original tone as the basis spectra (Horner et al. 1993a). However, unlike the interpolation method, a least-squares algorithm is used to compute the time-varying basis spectra amplitude weights which minimize the squared error between the original and synthetic spectra. This method generates an exact match at the time points of the selected snapshots and usually results in excellent matches at neighboring points as well. The relative amplitude spectral error (Eq. 4) between the original and matched spectra typically serves as a fitness function to guide the GA's search for the best solution. Most matched instruments require 3–5 wavetables for a good match (less than 5% relative amplitude spectral error), a considerable savings compared to sinusoidal additive synthesis or interpolation synthesis. However, in terms of total computation, interpolation synthesis can be just as efficient as wavetable-index synthesis and is more intuitive in terms of the amplitude weight envelopes. This is because the amplitude weights for wavetable-index synthesis can be either positive or negative and thus the sum of the amplitude-controlled basis spectra only *converges* to the least-error values.

Another approach for selecting basis spectra is based on principal components analysis (PCA). PCA is a statistical procedure for optimizing both weights and basis vectors (Dunteman 1989) and has been applied to the wavetable matching problem (Zahorian and Rothenberg 1981; Sandell and Martens 1992; Horner et al. 1992, 1993a; Sandell and Martens 1995). In our case, PCA first finds a basis spectrum and time-varying amplitude control to minimize the average mean-squared spectral error (rather than the relative amplitude spectral error). It then finds an amplitude and control to further minimize the error. It usually converges to a good approximation with five basis spectra (wavetables). If the number of PCA basis spectra equals the number of harmonics, convergence is perfect. Also, an alternative way to compute the control functions, which we actually use, is to first compute the basis spectra and then employ the same method we used for wavetable-index synthesis, which is the least-squares solution. However, there are a couple of disadvantages to the PCA method in comparison to using spectral snapshots: First, in selecting basis spectra, PCA strongly weighs the high-amplitude steady-state spectra, which is unfortunate because the low-amplitude attack and release spectra are perceptually very important. However, this can be overcome by judicious use of spectra used in computing the basis spectra. Second, except for the first basis spectrum, the PCA basis spectra do not resemble real spectra because they are used to correct the first one (in analogous fashion to Fourier sine waves correcting each other to form a square wave) and are therefore less intuitive to work with than spectral snapshots. Nevertheless, PCA provides an interesting alternative approach to basis spectra generation.

Another method which uses basis spectra and wavetables is *cluster synthesis*. With this method snapshot spectra are sorted into clusters according to some measure of the spectra. One method is to sort the spectra according to their spectral centroids (Beauchamp and Horner 1995; Horner 1995). The spectra within each cluster are amplitude-normalized and then averaged to produce a basis spectrum. For synthesis the control functions are amplitude, centroid, and f_0 vs. time. For each time frame, the centroid value is used to interpolate between two basis spectra whose centroids straddle this value to get an output spectrum which is then synthesized at the appropriate amplitude and f_0. Like wavetable interpolation, it can be shown that interpolation in the frequency domain coupled with additive synthesis is equivalent to interpolation in the time domain with basis-spectra-loaded wavetables.

An alternative method of cluster synthesis comes from the technique of vector quantization (Ehmann and Beauchamp 2002). Here the spectra are clustered using a K-means algorithm (Rabiner and Juang 1993). The basic synthesis method is similar to the centroid-clustering method above. However, with vector quantization, each cluster is represented by an average spectrum which is labeled by a "code-book number" or index rather than by its average spectral centroid. So the basis spectra must be retrieved by index-vs.-time data. For synthesis, the basis spectra can be ordered according to their spectral centroids or corresponding amplitudes in the original signal. Then for each synthesis frame, the best rms basis spectrum match is found, and the index-vs.-time data are generated. To avoid discontinuities, the index-vs.-time data are smoothed, and then interpolation is used to ensure smooth transitions between the basis spectra. With one version of the mehod, the control parameters are time-varying amplitude, index, and f_0. Alternatively, the time-varying spectrum is first normalized in terms of both amplitude and spectral tilt, where tilt is the slope p of a straight line that forms the best least-squares fit to the spectrum plotted as log-amplitude-vs.-log-frequency. The K-means algorithm is then used to cluster the resulting flattened spectra. Synthesis parameters now consist of amplitude, tilt, index, and f_0 vs. time, and synthesis proceeds in a fashion described above, except that now both amplitude and spectral tilt have to be imposed on the retrieved spectra before synthesis. Therefore, synthesis has to be done in the frequency domain before conversion to signal. However, if tilt normalization is omitted, wavetables can be used as in the centroid-clustering case.

Wavetable synthesis is an inherently harmonic synthesis method, so replicating sounds that are nearly harmonic, such as piano and plucked string tones, require some enhancements of the method. By grouping partials with similar frequency stretch factors (Fletcher et al. 1962), genetic algorithms have successfully optimized group additive synthesis parameters to simulate piano tones (Lee and Horner 1999; Zheng and Beauchamp 1999) and string tones (So and Horner 2002).

Sometimes using a complex method can lead to a simpler one. We have found that insights gained from exploring a problem first with the GA method often led to finding better or simpler solutions. Wavetable matching is such an example. Instead of using the GA to approximate the best match to all or a subset of the spectral snapshots of the original tone, an alternative method is to use a combinatorial method to find the best match for a subset of the spectral snapshots (Horner 2001). It turns out that out that this approach is as effective and efficient as the GA method and much simpler.

5 • Frequency Modulation (FM) Synthesis Models

Like wavetable synthesis, FM synthesis can efficiently generate interesting sounds. There are several possible FM configurations, including those with multiple parallel modulators, nested (serial) modulators and feedback (see Figure 3). Several of these "modules" can be combined to form complex FM synthesizers. For example, as with multiple-wavetable synthesis, several carrier oscillators can be combined in parallel to form a single-modulator/multiple-carrier FM synthesizer (see Figure 4). During the height of FM's popularity in the 1980s, synthesizers such as the Yamaha DX7 allowed users great flexibility in mixing and matching models like these.

5.1 Single-Modulator/Single-Carrier Model

The original FM equation John Chowning used for music synthesis consisted of a single sine wave modulating a carrier sine wave in a vibrato-like fashion (Chowning 1973):

$$A\sin(2\pi)\int (f_c + \Delta f_m \cos(2\pi f_m t))dt = A\sin(2\pi f_c t + \alpha_m \sin(2\pi f_m t)), \tag{5}$$

where A is the amplitude, f_c is the carrier frequency, Δf_m is the modulator amplitude, f_m is the modulation frequency, and $\alpha_m = \Delta f_m / f_m$ = modulation index. Vibrato results when the

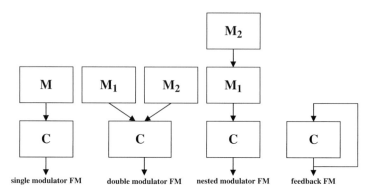

FIGURE 3 Block diagrams for several basic types of FM synthesis.

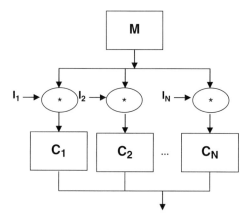

FIGURE 4 Block diagram of single-modulator/multiple-carrier (formant) FM synthesizer.

modulator frequency is low ($< 20\,\text{Hz}$). However, with an audio-rate modulator frequency, a spectrum is heard whose frequencies depend on the carrier and modulator frequencies and whose amplitudes depend on the modulation index. This is made obvious by expanding Eq. (5) in terms of Bessel functions:

$$\sin(2\pi f_c t + \alpha_m \sin(2\pi f_m t)) = \sum_{k=-\infty}^{\infty} J_k(\alpha_m)\sin(2\pi(f_c + kf_m)t). \quad (6)$$

We see that the spectrum frequencies are given by $f_k = |f_c + kf_m|$, for $k = \ldots -3, -2, -1, 0, 1, 2, 3\ldots$, which makes it clear that f_c is the center frequency with amplitude $J_0(\alpha_m)$ and that this component is surrounded by positive and negative side bands of frequency $f_c \pm |k|f_m$ and amplitudes $J_k(\alpha_m)$. As k increases, frequency eventually a point is reached (when $k > f_c/f_m$) where the negative-k side-band frequencies become negative, causing the sine function to flip its sign. Negative frequencies beyond this point are said to "fold over zero". Thus, the frequencies are effectively just $|f_c \pm |k|f_m|$ with amplitudes of $\pm |J_{|k|}(\alpha_m)|$. Because, for the same value of $|k|$, the negative frequency component amplitudes are given by $J_{-k}(\alpha) = (-1)^k J_k(\alpha)$, Eq. (6) can be rewritten as

$$\sin(2\pi f_c t + \alpha_m \sin(2\pi f_m t)) = J_0(\alpha_m) \sin(2\pi f_c t)$$
$$+ \sum_{k=1}^{\infty} J_k(\alpha_m)[\sin(2\pi(f_c + kf_m)t) + (-1)^k \sin(2\pi(f_c - kf_m)t)] \quad , \quad (7)$$

which shows a clear separation between the positive and negative side-bands. The actual sign of each side-band amplitude depends on a combination of (1) whether it is a positive or negative side-band, (2) whether the Bessel function itself is positive or negative at a particular value of α_m, and (3) whether the component has folded-over (i.e., $k > f_c/f_m$). Also, note that the left sides of Eqs. (6) and (7) are actually in the form of phase modulation (PM). FM and PM are closely related because the phase is the integral of the frequency, i.e., $\alpha_m \sin(2\pi f_m t) = \int 2\pi \Delta f_m \cos(2\pi f_m t) dt$. If some other sinusoid phase inside the integral is used (e.g., $\sin(2\pi f_m t)$), the result will be different, although basically the same type of spectra will result. For details see Beauchamp (1992).

The modulation index controls the amount of modulation and the precise shape of the resulting spectrum. Keeping f_m fixed, the spectrum bandwidth generally increases as the modulation index α_m increases. This effect is shown in Figure 5. For large α_m, the $-40\,\text{dB}$ bandwidth approaches $2\Delta f_m$. For small α_m, the bandwidth-to-Δf_m ratio gets larger; for example, at $\alpha_m = 1$, the ratio is approximately 6.2. Thus, a time-varying modulation index produces a dynamically changing spectrum from a single FM carrier–modulator oscillator pair. By contrast, single wavetable synthesis lacks this flexibility. Still, this simple FM model has the unfortunate property that, due to the oscillatory nature of the Bessel functions, as the modulation index changes, individual spectral components fade in and out dramatically, in a way that is not characteristic of typical musical tone spectral evolution.

A special case of Eq. (7) called *formant FM* occurs when the carrier frequency is an integer multiple of the modulator frequency. First, the spectrum is harmonic with fundamental frequency f_m, and second, for limited values of α_m, the spectrum tends to form a symmetrical formant band around the carrier frequency (see Figure 5).

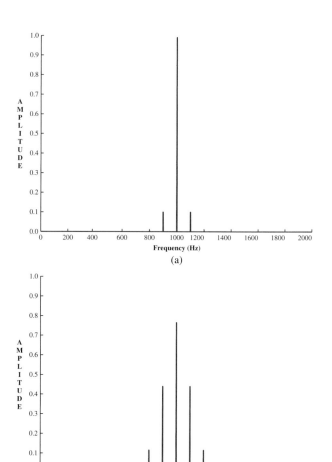
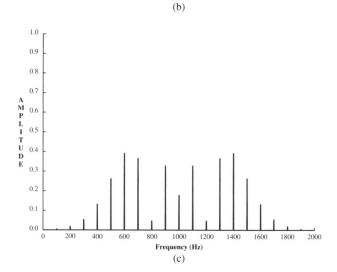

FIGURE 5 FM magnitude spectra for carrier frequency f_c, modulator frequency f_m, and index $\alpha_m =$ (a) 0.2, (b) 1.0, and (c) 5.0.

5.2 Single-Modulator/Multiple-Carrier Synthesis Model

Like wavetable synthesis, FM synthesis is very efficient in terms of computation and storage: A single carrier–modulator FM instrument requires about the same amount of computation as a pair of wavetables. However, only a single sine wavetable is required for all the modulators and carriers of a complex FM synthesizer; thus, FM is more storage-efficient than wavetable synthesis. Also, assuming that modulation indices are not time-varying, each carrier's output has a static spectrum, and so the FM model provides time-varying spectrum control in similar fashion as multiple-wavetable synthesis (see Figure 2), i.e., by means of the time-varying amplitudes of the carriers (see Figure 4). However, unlike wavetable synthesis, the spectrum produced by FM is not arbitrary, but is restricted to the subset of possible FM spectra, so that more modules are generally needed for FM to produce a result of the same quality.

An equation for single-modulator/multiple-carrier FM synthesis, as depicted in Figure 4, is given by

$$s(t) = \sum_{n=1}^{N} A_n(t) \sin(2\pi(r_n f_m t + \alpha_{m_n} \sin(2\pi f_m t))). \quad (8)$$

For each carrier oscillator n, the time-varying amplitude is given by $A_n(t)$, the modulator frequency by f_m, the carrier frequency by $r_n f_m$ (where r_n is an integer), and the modulation index by α_{m_n}. The time-varying harmonic spectrum of $s(t)$ can be calculated by expanding each term of Eq. (8) using Eq. (7) and combining corresponding frequencies.

5.3 More Complex FM Models

A few years after Chowning's original work, Schottstaedt (1977) introduced a double-modulator/single-carrier (DMSC) FM model (see Figure 3), where the outputs of two oscillators are summed to modulate the carrier. Thus,

$$s(t) = A \sin(2\pi f_c t + \alpha_{m_1} \sin(2\pi f_{m_1} t) + \alpha_{m_2} \sin(2\pi f_{m_2} t)). \quad (9)$$

If the carrier and modulator frequencies are all related by integer multiples of the fundamental, a harmonic tone results. As shown by Le Brun (1977), the DMSC FM model with N modulators and a single carrier produces frequencies of $|f_c + k_1 f_{m_1} + \cdots + k_N f_{m_N}|$ with Bessel function product amplitudes $J_{k_1}(\alpha_{m_1}) J_{k_2}(\alpha_{m_2}) \cdots J_{k_N}(\alpha_{m_N})$ for all combinations of (positive, negative, or zero) integers $k_1, \ldots, k_N$. This is a significantly more complicated model, even with just two modulators, than the single-modulator case, as described above, thus making parameter optimization more difficult than for the single-modulator case.

Justice (1979) introduced a two-modulator nested FM model (see Figure 3), where one modulator modulates another, which in turn modulates the carrier:

$$s(t) = \sin(2\pi f_c t + \alpha_{m_1} \sin(2\pi f_{m_1} t + \alpha_{m_2} \sin(2\pi f_{m_2} t))). \quad (10)$$

Like double FM, if the carrier and modulator frequencies are all related by integer multiples of the fundamental, a harmonic tone results. By treating the modulator m_1 as a carrier, we see it can be expanded in terms of Bessel functions as in Eqs. (6) and (7) so that Eq. (10) could be expanded in the form of Eq. (9) with an infinite sum of modulators. Taking this a step further, as for the finite modulator sum case, results in an equation with infinite sums and products. This is a much more complicated relationship than the Bessel function expansion of double FM, where the modulators are summed instead of nested. However, if time samples of Eq. (10) are computed uniformly over the fundamental period (the inverse of the largest common divisor of f_c, f_{m_1}, and f_{m_2}), a discrete Fourier transform can be used to compute the harmonic amplitudes for any combination of the five parameters, f_c, f_{m_1}, f_{m_2}, α_{m_1}, and α_{m_2}.

Another FM variant that proved useful in the 1980s' synthesizers was *feedback FM* (Tomisawa 1981; Mitsuhashi 1982). The output of the carrier at sample n modulates the following sample, scaled by a modulation index:

$$s(n+1) = \sin(2\pi f_c n/f_s + \alpha_m s(n)). \quad (11)$$

Unlike the other FM methods discussed above, this is probably intractable for analytic solution. However, given a starting value, e.g., $s(0) = 0$, it can be computed and converted into a spectrum. It turns out that when the modulation index is less than about 1.5, a monotonically decreasing spectrum results (Tomisawa 1981). Because of this, feedback FM is potentially more easily controlled than the other forms of FM, where the harmonics oscillate in amplitude as the modulation index changes. Another advantage of feedback FM over other FM models is that its harmonic amplitudes are strictly positive when the modulation index is less than 1.5 (other forms of FM produce both positive and negative amplitudes). This avoids cancellation when adding multiple carriers together. Still, the monotonically decreasing spectrum of feedback FM has a disadvantage: Many musical instruments which have strong formants at upper harmonics cannot be directly modeled by the monotonic spectra inherent with feedback FM. Nevertheless, amplitude modulation of the feedback FM carrier signal by a sine wave can shift frequencies upward to overcome this limitation.

6 • Frequency Modulation Matching

One of the factors leading to FM's decline in popularity in the late 1980s was that matching an arbitrary musical instrument tone by FM synthesis is difficult, much more difficult than with wavetable matching. A closed-form analytical solution for determining the best set of FM parameters does not exist, and some form of general-purpose optimization is necessary. Most previous work on FM utilized ad hoc and semi-automated techniques for matching instrument tones. However, hand tuning of multiple carriers quickly exceeds the limits of human ability and endurance. In the early 1990s Horner et al. (1993b) introduced systematic techniques for matching tones to FM models based on genetic algorithm (GA) optimization.

Several researchers have emulated particular instruments by tailoring the time-varying FM indices by hand, starting with Chowning's original FM paper (Chowning 1973). In addition to applying FM to music synthesis, Chowning gave parameters appropriate to various classes of instruments based on simulating various properties of those instruments. For instance, a brass tone's spectral centroid (a strong correlate of perceptual brightness) is usually proportional to its overall amplitude. Chowning simulated this behavior by taking advantage of the fact that the centroid of an FM spectrum generally increases as its modulation index increases. He then varied the modulation index in direct proportion to the amplitude of the carrier to approximate a brass instrument. He produced woodwind-like tones and percussive sounds using similar methods. Chowning also discussed a double-carrier instrument near the end of his 1973 paper. In a later study, Chowning (1980) designed a double-carrier FM instrument to simulate a singing soprano voice. Chowning centered the first carrier at the fundamental frequency and the second at an upper formant, intuitively deriving the parameters of the instrument. He identified vibrato as critically important for achieving a voice-like sound.

Morrill (1977) followed Chowning's lead in trying to determine FM parameters based on detailed knowledge of trumpet sounds. Morrill outlined single- and double-carrier instrument designs for the trumpet and clearly identified the limitations of single-carrier instruments. His double-carrier instrument set carrier frequencies to the fundamental and its sixth harmonic, the latter corresponding to a known upper formant region in the trumpet. Morrill also pointed out the difficulty in predicting the spectral output of the double-carrier instrument.

Schottstaedt (1977) changed the basic FM instrument design by using two modulators to simultaneously frequency-modulate a single carrier. After describing the spectral behavior of the double-modulator FM model, Schottstaedt gave parameters for simulating the piano and string instruments. He used instrument characteristics and trial and error to find the parameters. The technique found small modulation indices most useful.

Justice (1979) introduced the nested modulator FM model and outlined a Hilbert transform procedure to decompose a signal into parameters for a single-carrier FM instrument. The procedure attempted to produce a matched FM signal close to the original, leaving the user to tweak the parameters as desired. However, Justice matched FM-generated signals and not those of acoustic musical instruments. Payne (1987) extended Justice's technique to a pair of carriers with nested modulators. Each carrier contributed to an independent frequency region, giving a more accurate match than with a single carrier. In addition to matching contrived FM signals, Payne matched a cello sound. The result was reportedly string-like, but lacking properties of "liveliness". Payne reported that the matching procedure was computationally very expensive.

More recent work by Delprat and his collaborators used a wavelet analysis and a Gabor transform to find spectral trajectories (Delprat et al. 1990; Delprat 1997). They used these trajectories to estimate modulation parameters. This approach was similar to that used by Justice and Payne except that it broke the frequency range into more component parts. Thus, it was also computationally expensive. They gave examples for a saxophone and trumpet using five carrier–modulator pairs, indicating the growing awareness that precise spectral control requires several carriers.

Meanwhile, Beauchamp (1982) developed a frequency-domain method to find FM parameters as part of a larger study on nonlinear synthesis based on time-varying spectral centroid (aka "brightness") matching. He used a single-modulator/single-carrier model with a centroid-controlled modulation index to match the time-varying spectral centroid of the original signal. Though the level of control was too coarse to provide a convincing synthesis, the technique was notable in its attempt to perform an automated spectral match.

In the early 1990s researchers introduced evolutionary matching techniques for the various FM models, first applying them to the single-modulator/multiple-carrier (formant FM) model (Horner et al. 1992, 1993b). A genetic algorithm (GA) procedure was used to optimize fixed modulation indices and modulator-to-carrier frequency ratios for various numbers of carriers in an effort to best match particular instrument sounds. Fixed rather than time-varying modulation indices were used because allowing them to vary resulted in radically different frame-to-frame indices with their accompanying audible artifacts. Using invariant indices also avoided the considerable extra expense of optimizing time-varying indices. As in wavetable matching, the relative amplitude spectral error between the original and matched spectra served as a fitness function in guiding the GA's search for the best FM parameters. Matching results in terms of relative amplitude spectral error (see Eq. (4)) vary with instrument. Recent measurements on a trumpet tone showed that five formant-FM carriers were needed to achieve an error of 10%, whereas for a Chinese pipa tone an error of 25% required six carriers (Horner 2007). This performance was substantially worse than for the comparable wavetable-index method where only three tables were needed to yield an error of 7.5% for the trumpet and six tables for an error of 10% for the pipa.

Tan et al. (1994) introduced an enumerative procedure for optimizing a steady-state double-modulator FM model. Because this model only produced static spectra, it did not effectively match instruments with dynamic spectra. Since then, genetic algorithms have successfully optimized the double FM problem (Horner 1996a; Tan and Lim 1996; Lim and Tan 1999). GA methods were used to optimize invariant modulation indices in the double FM model, and the modulation indices it found were relatively small. Double FM matches were worse than formant-FM matches when compared with the same number of table lookups. However, double FM matches were better than formant-FM matches for the same number of carriers, an advantage when double FM hardware is available.

Finally, the GA method was applied to nested modulator and feedback FM matching (Horner 1998). Fixed modulation indices were used for nested modulator FM, but time-varying modulation indices were allowed for feedback FM. Like double FM matching, the optimized parameters for nested modulator and feedback FM had relatively small modulation indices. Feedback FM often gave the best matches of all the FM models when compared against the same number of table lookups, indicating feedback FM is a good choice for software synthesis where computation is the main factor. However, if nested modulator FM hardware is already available, then double or triple modulator FM gave the best results of all the FM models for the same number of carriers.

7 • Conclusions

We have reviewed several techniques for instrument modeling/synthesis based on spectral analysis, with a particular focus on wavetable and FM matching. Among the various types of wavetable synthesis, the best method depends on the given situation. For simplicity, group additive synthesis has the advantage of being intuitive, since each harmonic is only in one wavetable. For memory-constrained systems where instruments have to compete for limited wavetable space, wavetable matching is a very good choice. Conversely, for real-time systems where memory is not a problem, wavetable interpolation is a good choice.

Although wavetable matching is simpler and more effective than FM matching in general (Horner 1997), FM synthesis provides real-time flexibility over wavetable synthesis when wavetable memory is limited. Among the various types of FM, the best method again depends on the given situation. For simplicity and ease of control, formant FM is best. For software synthesis where computation is the main factor, feedback FM is best. If FM hardware is available, nested modulator FM is best.

In any case, the various wavetable and FM matching procedures provide an interesting point of departure for instrument designers in applications such as *timbral interpolation* (Grey 1975; Beauchamp and Horner 1998), where the parameters of one spectral match are crossfaded to that of another. The smoothness of the transformation depends on the synthesis technique. Wavetable synthesis gives a smoother interpolation than FM synthesis. Interpolating distantly spaced FM index values will likely produce wildly changing spectral results during the interpolation due to oscillation of the Bessel functions. However, such interpolations may still be interesting and musically useful.

Acknowledgments The Hong Kong Research Grant Council's Projects 613505 and 613806 supported this work.

References

Beauchamp, J.W. (1974), "Time-Variant Spectra of Violin Tones", *J. Acoust. Soc, Am.* **56**(3), 995–1004.

Beauchamp, J. (1982). "Synthesis by Amplitude and 'Brightness' Matching of Analyzed Musical Instrument Tones", *J. Audio Eng. Soc.* **30**(6), pp. 396–406.

Beauchamp, J. (1992). "Will the Real FM Equation Please Stand Up"(letter), *Comput. Music J.* **16**(4), pp. 6–7.

Beauchamp, J.W., R.C. Maher, & R. Brown (1993). "Detection of Musical Pitch from Recorded Solo Performances", *Audio Eng. Soc.* Preprint No. 3541.

Beauchamp, J.W., & A, Horner (1995). "Wave table Interpolation Synthesis Based on Time-Variant Spectral Analysis of Musical Sounds", *Audio Eng. Soc.* Preprint No. 3960.

Beauchamp, J., & A. Horner (1998). "Spectral Modeling and Timbre Hybridization Programs for Computer Music", *Organ. Sound* **2**(3), pp. 253–258.

Beauchamp, J.W. (2007). "Analysis and Synthesis of Musical Instrument Sounds", in Analysis, Synthesis, and Perception of Musical Sounds; J.W. Beauchamp, ed., New York: Springer, pp. 1–89.

Bristow-Johnson, R. (1996). "Wavetable Synthesis 101, A Fundamental Perspective", *Audio Eng. Soc.* Preprint No. 4400.

Brown, J.C. (1996). "Frequency ratios of Spectral Components of Musical Sounds", *J. Acoust. Soc. Am.* **99**(2), 1210–1218.

Chamberlin, H. (1980). "Advanced Real-Timbre Music Synthesis Techniques", *Byte Magazine* April, pp. 70–94, 180–196.

Cheung, N.-M., & A. Horner (1996). "Group Synthesis with Genetic Algorithms", *J. Audio Eng. Soc.* **44**(3), pp. 130–147.

Chowning, J. (1973). "The Synthesis of Complex Audio Spectra by Means of Frequency Modulation", *J. Audio Eng. Soc.* **21**(7), pp. 526–534.

Chowning, J. (1980). "Computer Synthesis of the Singing Voice", *Sound Generation in Wind, Strings, Computers*, Stockholm: The Royal Swedish Academy of Music, pp. 4–13.

Delprat, N., P. Guillemain, & R. Kronland-Martinet (1990). "Parameter Estimation for Non-linear Resynthesis Methods with the Help of a Time–Frequency Analysis of Natural Sounds", *Proc. 1990 Int. Computer Music Conf.*, Glasgow, Scotland, pp. 88–90.

Delprat, N. (1997). "Global Frequency Modulation Law Extraction from the Gabor Transform of a Signal: A First Study of the Interacting Components Case", *IEEE Trans. Speech Audio Process.* **5**(1), pp. 64–71.

Dubnof, S. & X. Rodet (2003), "Investigation of phase Coupling Phenomend in Sustained Portion of Musical Instrument Sound", *J. Acoust, Soc.* **113**(1), 348–359.

Dunteman, G. (1989). *Principal Components Analysis*, Newbury Park, CA: Sage.

Ehmann, A. F., & J. W. Beauchamp (2002) "Musical sound analysis/synthesis using vector-quantized time-varying spectra" (abstract), *J. Acoust. Soc. Am.* **112** (5, pt. 2), p. 2239.

Flanagan, J. L., & R. M. Golden (1966). "Phase Vocoder", *Bell Syst. Tech. J.* **45**, pp. 1493–1509.

Fletcher, H., E. D. Blackham, & R. Stratton (1962). "Quality of Piano Tones", *J. Acoust. Soc. Am.* **34**(6), pp. 749–761.

Goldberg, D. (1989). *Genetic Algorithms in Search, Optimization, and Machine Learning*, Reading, MA: Addison-Wesley.

Grey, J. (1975). "An Exploration of Musical Timbre", Ph.D. Dissertation, Stanford: Dept. of Music, Stanford Univ.

Horner, A., J. Beauchamp, & L. Haken (1992). "Wavetable and FM Matching Synthesis of Musical Instrument Tones", *Proc. 1992 Int. Computer Music Conf.*, San Jose, CA, pp. 18–21.

Horner, A., J. Beauchamp, & L. Haken (1993a). "Methods for Multiple Wavetable Synthesis of Musical Instrument Tones", *J. Audio Eng. Soc.* **41**(5), pp. 336–356.

Horner, A., J. Beauchamp, & L. Haken (1993b). "Machine Tongues XVI: Genetic Algorithms and Their Application to FM Matching Synthesis", *Comput. Music J.* **17**(4), pp. 17–29.

Horner, A. (1995). "Wavetable Matching Synthesis of Dynamic Instruments with Genetic Algorithms", *J. Audio Eng. Soc.* **43**(11), pp. 916–931.

Horner, A. & J. Beauchamp (1995), "Synthesis of Trumpet Tones Using a Wavetable and a Dynamic Filter", *J. Audio Eng. Soc.* **43**(10), 799–812.

Horner, A. (1996a). "Double Modulator FM Matching of Instrument Tones", *Comput. Music J.* **20**(2), pp. 57–71.

Horner, A. (1996b). "Computation and Memory Tradeoffs with Multiple Wavetable Interpolation", *J. Audio Eng. Soc.* **44**(6), pp. 481–496.

Horner, A., & J. Beauchamp (1996). "Piecewise Linear Approximation of Additive Synthesis Envelopes: A Comparison of Various Methods", *Comput. Music J.* **20**(2), pp. 72–95.

Horner, A. (1997). "A Comparison of Wavetable and FM Parameter Spaces", *Comput. Music J.* **21**(4), pp. 55–85.

Horner, A. (1998). "Nested Modulator and Feedback FM Matching of Instrument Tones", *IEEE Trans. Speech Audio Process.* **6**(4), pp. 398–409.

Horner, A., & L. Ayers (1998). "Modeling Acoustic Wind Instruments with Contiguous Group Synthesis", *J. Audio Eng. Soc.* **46**(10), pp. 868–879.

Horner, A. (2001). "A Simplified Wavetable Matching Method Using Combinatorial Basis Spectra Selection", *J. Audio Eng. Soc.* **49**(11), pp. 1060–1066.

Horner, A., J. Beauchamp, & R. So (2006). "A Search for Best Error Metrics to Predict Discrimination of Original and Spectrally Altered Musical Instrument Sounds", *J. Audio Eng. Soc.* **54**(3), 140–156.

Horner, A. (2007). "A Comparison of Wavetable and FM Data Reduction Methods for Resynthesis of Musical Sounds", in *Analysis, Synthesis, and Perception of Musical Sounds*, J. W. Beauchamp, ed., New York: Springer, pp. 228–249.

Justice, J. (1979). "Analytic Signal Processing in Music Computation", *IEEE Trans. Acoust. Speech Signal Process.* **27**(6), pp. 670–684.

Kleczkowski, P. (1989). "Group Additive Synthesis", *Comput. Music J.* **13**(1), pp. 12–20.

Le Brun, M. (1977). "A Derivation of the Spectrum of FM with a Complex Modulating Wave", *Comput. Music J.* **1**(4), pp. 51–52.

Lee, K., & A. Horner (1999). "Modeling Piano Tones with Group Synthesis", *J. Audio Eng. Soc.* **47**(3), pp. 101–111.

Lim, S. M., & B. T. G. Tan (1999). "Performance of the Genetic Annealing Algorithm in DFM Synthesis of Dynamic Musical Sound Samples", *J. Audio Eng. Soc.* **47**(5), pp. 339–354.

Maher, R., & J. Beauchamp (1990). "An Investigation of Vocal Vibrato for Synthesis", *Appl. Acoust.* **30,** pp. 219–245.

Maher, R.C. & J.W. Beauchamp (1994). "Fundamental Frequency Estimation of Musical Signals Using a Two-way Mismatch Procedures", *J. Acoust. Soc. Am.* **95**(4), 2254–2263.

Mathews, M. V. (1969). *The Technology of Computer Music*, Cambridge, MA: MIT Press, p. 56.

McAulay, R. J., & T. F. Quatieri (1986). "Speech Analysis/Synthesis Based on a Sinusoidal Representation", *IEEE Trans. Acoust. Speech Signal Process.* **34**(4), pp. 744–754.

Mitsuhashi, Y. (1982). "Musical Sound Synthesis by Forward Differences", *J. Audio Eng. Soc.* **30**(1/2), pp. 2–9.

Morrill, D. (1977). "Trumpet Algorithms for Computer Composition", *Comput. Music J.* **1**(1), pp. 46–52.

Oates, S., & B. Eaglestone (1997). "Analytic Methods for Group Additive Synthesis", *Comput. Music J.* **21**(2), pp. 21–39.

Payne, R. (1987). "A Microcomputer Based Analysis/Resynthesis Scheme for Processing Sampled Sounds Using FM", *Proc. 1987 Int. Computer Music Conf.*, Urbana, IL, pp. 282–289.

Portnoff, M. R. (1976). "Implementation of the Digital Phase Vocoder Using the Fast Fourier Transform", *IEEE Trans. Acoust. Speech Signal Process.* **ASSP-24**(3), pp. 243–248.

Rabiner, L., & B.-H. Juang (1993). *Fundamentals of Speech Recognition*, New Jersey: Prentice Hall, pp. 122–132.

Sandell, G., & W. Martens (1992). "Prototyping and Interpolation of Multiple Musical Timbres Using Principal Component-Bases Synthesis", *Proc. 1992 Int. Computer Music Conf.*, San Jose, CA, pp. 34–37.

Sandell, G., & W. Martens (1995). "Perceptual Evaluation of Principal Component-Based Synthesis of Musical Timbres", *J. Audio Eng. Soc.* **43**(12), pp. 1013–1028.

Schottstaedt, B. (1977). "The Simulation of Natural Instrument Tones Using Frequency Modulation with a Complex Modulating Wave", *Comput. Music J.* **1**(4), pp. 46–50.

Schumacher, R.T. (1992). "Analysis of Aperiodicities in Nearly Periodic Waveforms", *J. Acoust. Soc. Am.* **91**(1), 438–451.

Serra, M.-H., D. Rubine, & R. Dannenberg (1990). "Analysis and Synthesis of Tones by Spectral Interpolation", *J. Audio Eng. Soc.* **38**(3), pp. 111–128.

Smith, J. O., & X. Serra (1987). "PARSHL: An Analysis/Synthesis Program for Non-harmonic Sounds Based on a Sinusoidal Representation", *Proc. 1987 Int. Computer Music Conf.*, Urbana, IL, pp. 290–297.

So, K. F., & A. Horner (2002). "Wavetable Matching of Inharmonic String Tones", *J. Audio Eng. Soc.* **50**(1/2), pp. 47–57.

Tan, B. T. G., S. L. Gan, S. M. Lim, & S. H. Tang (1994). "Real-Time Implementation of Double Frequency Modulation (DFM) Synthesis", *J. Audio Eng. Soc.* **42**(11), pp. 918–926.

Tan, B. T. G., & S. M. Lim, (1996). "Automated Parameter Optimization for Double Frequency Modulation Synthesis Using the Genetic Annealing Algorithm", *J. Audio Eng. Soc.* **44**(1/2), pp. 3–15.

Tomisawa, N. (1981). "Tone Production Method for an Electronic Music Instrument", U.S. Patent 4,249,447.

Zahorian, S., & M. Rothenberg (1981). "Principal-Components Analysis for Low Redundancy Encoding of Speech Spectra", *J. Acoust. Soc. Am.* **69**(3), pp. 832–845.

Zheng, H., & J. Beauchamp (1999). "Analysis and Critical-Band-Based Group Wavetable Synthesis of Piano Tones", *Proc. 1999 Int. Computer Music Conf.*, Beijing, China, pp. 9–12.

25

Digital Waveguide Architectures for Virtual Musical Instruments

Julius O. Smith

Center for Computer Research in Music and Acoustics (CCRMA), Stanford University, Stanford, CA 94305, USA, Website: http://ccrma.stanford.edu/~jos/

1 • Introduction

Digital sound synthesis has become a standard staple of modern music studios, videogames, personal computers, and hand-held devices. As processing power has increased over the years, sound synthesis implementations have evolved from dedicated chip sets, to single-chip solutions, and ultimately to software implementations within processors used primarily for other tasks (such as for graphics or general purpose computing). With the cost of implementation dropping closer and closer to zero, there is increasing room for higher quality algorithms. A particularly fertile source of natural sound synthesis algorithms is the mathematical models of musical instruments developed within the science of musical acoustics [1–3, 72]. To realize practical instrument voices from these models, it is helpful to develop robust and efficient signal processing algorithms which retain the audible physical behavior while minimizing computational cost [4].

In this chapter, a number of cost-effective synthesis models will be summarized for various musical instrument families, including strings and winds. Emphasis is placed on techniques adapted from the field of digital signal processing [5,6]. Notably absent is any discussion of percussion instruments, which are normally handled via sample-based methods [7–9], although some model-based methods have been proposed based e.g., on the digital waveguide mesh [10].

2 • Vibrating Strings

In a stringed musical instrument, most of the sound energy is stored in the vibrating string at any given time. The expressive range of sound from a stringed instrument arises primarily from an interaction of the vibrating string with the performer. The body of the instrument functions as a passive resonator which is well modeled in most cases, in principle, by a linear, time-invariant filter [11,12].

2.1 Wave Equation

The starting point for a stringed instrument model is typically a *wave equation* for transverse vibrations of the vibrating string [13–16]. For example, a recently proposed [14] partial differential equation (PDE) governing the motion of a *piano string* is given by

$$f(t,x) = \epsilon \ddot{y} - Ky'' + EIy'''' + R_0 \dot{y} + R_2 \ddot{y}' \qquad (1)$$

where symbols are defined as follows:

$y = y(t,x)$ = string displacement at position x and time t
$\dot{y} = \frac{\partial}{\partial t} y(t,x)$, $y' = \frac{\partial}{\partial x} y(t,x)$, etc.
$f(t,x)$ = driving force density (N/m)
ϵ = mass density (kg/m)
K = tension force along the string axis (N)
E = Young's modulus (N/m^2)
I = radius of gyration of the string cross-section (m)

The basic *lossless* wave equation $\epsilon \ddot{y} = Ky''$ is derived in most textbooks on acoustics, e.g., [15].[1] The term $\epsilon \ddot{y}$ is the mass-per-

---

[1] For an online derivation, see, e.g.,
http://ccrma.stanford.edu/~jos/pasp/String_Wave_Equation.html.

unit-length ϵ times the transverse acceleration $\ddot{y}$, and Ky'' approximates the transverse restoring force (per unit length) due to the string tension K.

The term EIy'''' in Eq. (1) models the transverse restoring force exerted by a *stiff string* in response to being bent. In an ideal string, with zero diameter, this force is zero. Stiffness is normally neglected in models for guitars and violins, but modeled in instruments with larger-diameter strings, such as the piano and cello. The test for whether stiffness is needed in the model for plucked or struck strings (any freely vibrating string) is whether the ear can hear the "stretching" of the partial overtones due to stiffness [17]; for bowed strings, dispersion due to stiffness can affect *bow–string dynamics* [18]. In the context of a digital waveguide string model (described in Section 2.3), the dispersion associated with stiff strings is typically modeled indirectly by designing an *allpass filter* for the string model.

The final two terms of Eq. (1) provide *damping*, which is required in any string practical model. Papers including damping models for piano strings include [13,14]. Instead of introducing such terms into the wave equation based on physical considerations, these terms may be determined implicitly by *digital filter design* techniques [11,19,29], as discussed further in Section 2.5.

2.2 Finite Difference Models

The original approach to digitally modeling vibrating strings was by means of *finite difference schemes* (FDS) [13,20–24]. Such models are also called finite difference time domain (FDTD) methods [25,26]. In these models, partial derivatives are replaced by finite differences, e.g.,

$$\ddot{y}(t,x) \approx \frac{y(t+T,x) - 2y(t,x) + y(t-T,x)}{T^2} \quad (2)$$

$$y''(t,x) \approx \frac{y(t,x+X) - 2y(t,x) + y(t,x-X)}{X^2}. \quad (3)$$

These finite differences are substituted into the string wave equation, Eq. (1), from which a time-recursive finite difference scheme is derived. Given boundary conditions characterizing the string endpoints, the finite difference scheme may be used to recursively compute the time evolution of the string state.

2.3 Digital Waveguide Models

More recently, the *digital waveguide* (DW) approach has been developed for modeling vibrating strings [27,28,41,53]. The DW approach is compared quantitatively with FDS approaches in

[14,31]. For strings used in typical musical instruments, the digital waveguide method generally provides a more efficient simulation for a given sound quality level. A combination of digital waveguides and finite differences may be preferred, however, for nonlinear string simulation [26,32,33].

The digital waveguide formulation can be derived by simply *sampling* the *traveling-wave* solution to the ideal wave equation:

$$Ky'' = \epsilon \ddot{y}.$$

It is easily checked that the lossless 1D wave equation is solved by any string shape y which travels to the left or right with speed $c = \sqrt{K/\epsilon}$ [34]. Denote *right-going* traveling waves in general by $y_r(t - x/c)$ and *left-going* traveling waves by $y_l(t + x/c)$, where y_r and y_l are assumed twice-differentiable. Then, as is well known, the general class of solutions to the lossless, 1D, second-order wave equation can be expressed as

$$y(t,x) = y_r\left(t - \frac{x}{c}\right) + y_l\left(t + \frac{x}{c}\right). \tag{4}$$

Sampling these traveling-wave solutions using sampling intervals T and X along the time and position coordinates, respectively, yields

$$\begin{aligned} y(nT, mX) &= y_r(nT - mX/c) + y_l(nT + mX/c) \\ &= y_r[(n-m)T] + y_l[(n+m)T] \\ &= y^+(n-m) + y^-(n+m), \end{aligned} \tag{5}$$

where n and m are integers. Note that a "+" superscript denotes a "right-going" traveling-wave component, while "−" denotes propagation to the "left". This notation is similar to that used for acoustic-tube modeling of speech [5].

Figure 1 shows a signal flow diagram for the computational model of Eq. (5), which is often called a digital waveguide model (for the ideal string in this case) [28,41]. Note that, by the sampling theorem [35, Appendix G],[2] it is an exact model so long as the initial conditions and any ongoing additive excitations are bandlimited to less than half the temporal sampling rate $f_s = 1/T$.

Note also that the position along the string, $x_m = mX = mcT$ meters, is laid out from left to right in the diagram, giving a

---

[2] http://ccrma.stanford.edu/~jos/mdft/Sampling_Theorem.html.

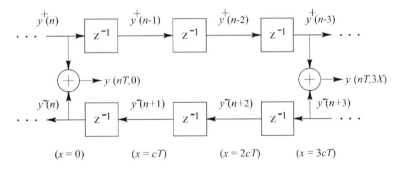

FIGURE 1 Digital simulation of the ideal, lossless, 1D waveguide with observation points at $x=0$ and $x=3X=3cT$. (The symbol "z^{-1}" denotes a one-sample delay.) [Reprinted with permission from [41].]

physical interpretation to the horizontal direction in the diagram, even though explicit spatial samples have been eliminated in the translation of physical variables to traveling-wave components. In Figure 1, "transverse displacement outputs" have been arbitrarily placed at $x=0$ and $x=3X$. The diagram is similar to that of well-known ladder and lattice digital filter structures [5], except for the delays along the upper rail, the absence of "scattering junctions", and the direct physical interpretation. Also closely related is the *Kelly–Lochbaum model* for voice synthesis [5,36–39].

2.4 Bowed Strings

An example DW model for a bowed-string instrument is shown in Figure 2 [40,41]. The main control is bow velocity, but bow force and position also have an effect on the tone produced. The digital waveguide simulates traveling velocity-wave components. The left- and right-going traveling-wave components on the left of the bow are denoted $v_{s,l}^+(n)$ and $v_{s,l}^-(n)$, respectively, where n denotes time in samples. To the right of the bow, the components are $v_{s,r}^+(n)$ and $v_{s,r}^-(n)$. The (abstract) "incoming string velocity" is defined as

$$v_s^+(n) = v_{s,l}^+(n) + v_{s,r}^+(n) \qquad (6)$$

and the "incoming differential velocity" is defined as

$$v_\Delta^+(n) = v_b(n) - v_s^+(n), \qquad (7)$$

where $v_b(n)$ denotes the bow velocity at sample time n. The incoming differential velocity v_Δ^+ can be interpreted physically as

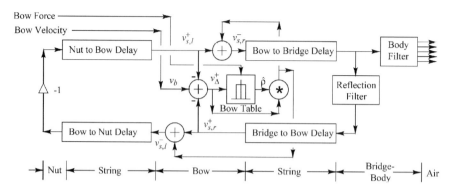

FIGURE 2 Digital waveguide bowed-string model. [From [41].]

the physical differential velocity (bow minus string) that would occur if the bow–string friction were zero (ideal, frictionless "slipping" of the bow along the string). A table lookup (or other nonlinear function implementation) gives the *reflection coefficient* of the bow–string contact point, as seen by traveling waves on the string (labeled "Bow Table" in the figure). This coefficient is then applied to v_Δ^+ and added to the left- and right-going traveling-wave paths. The bow table is derived from the bow–string friction curve characteristic, such as the one shown in Figure 3. Details of this derivation may be found in [41].[3]

The delay lines are drawn in "physical canonical form" for ease of physical interpretation. We see that the string is modeled using two ideal (lossless) digital waveguides, one to the left and one to the right of the bowing point. (A 1D *digital waveguide* is formally defined as a pair of delay lines flowing in opposite directions — a *bidirectional delay line*.) In practice, only two delay lines are normally implemented, one on each side of the bowing point.

Note that delay lines require $\mathcal{O}(1)$ operations per sample, i.e., the number of operations per sample does not increase as the delay-line length is increased.[4] This is the heart of the reason digital waveguide models are more efficient than finite difference models. At present, there is no known $\mathcal{O}(1)$ finite difference model for vibrating strings.

---

[3] Available online at http://ccrma.stanford.edu/~jos/pasp/BSSJ.html.
[4] The notation $\mathcal{O}(K)$ denotes "computational complexity of order K". This means that the computational complexity is bounded by cN^K for some constant c, as $N \to \infty$, where N is the size of the problem (delay-line length in this case).

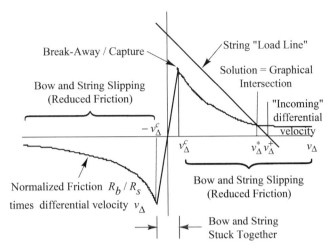

FIGURE 3 Overlay of normalized bow–string friction curve $R_b(v_\Delta)/R_s$ with the string "load line" $v_\Delta^+ - v_\Delta$. The "capture" and "break-away" differential velocity is denoted v_Δ^c. Note that increasing the bow force increases v_Δ^c as well as enlarging the maximum force applied (at the peaks of the curve). [From [41].]

The reflection filter in Figure 2 implements *all losses in one period of oscillation* due to the yielding bridge, absorption by the bow and finger, string losses, etc. Since the string model is linear and time invariant, i.e., Eq. (1) is linear with constant coefficients, superposition applies, and loss/dispersion filtering within the string may be *commuted* to concentrated points. In principle, such filters should appear on either side of the bow, and prior to each output signal extracted. However, because the difference is perceptually moot, normally only one loss/dispersion filter is employed per string loop. For multiple coupled strings, all loss/dispersion filtering may be implemented within the *bridge* at which they share a common termination [41,42].[5]

The bow–string junction is typically implemented as a *memoryless* lookup table ("Bow Table" in Figure 2) or a segmented polynomial. Preferably, however, a *thermodynamic model* should be employed for bow friction, since the bow rosin is known to have a time-varying viscosity due to temperature variations within a period of sound [43]. In [44], thermal models of dynamic friction in bowed strings are discussed, and such models have been incorporated into more recent synthesis models [45–47].

---

[5] http://ccrma.stanford.edu/~jos/pasp/Two_Ideal_Strings_Coupled.html.

A real-time software implementation of a bowed-string model similar to that shown in Figure 2 is available in the Synthesis Tool Kit (STK) distribution [48,49] as `Bowed.cpp`. This prototype can serve as a starting framework for more elaborate models.

2.5 Filter Design Methods for String Models

Methods for designing the string loop filter ("Reflection Filter" in Figure 2) are described in, [11,29,50–53]. For all types of waveguide string models, the loop-filter design software should ideally minimize *audibility of the error*. The loop-filter error naturally divides into two components: the *amplitude response* of the loop filter determines the partial overtone *decay rates*, while the *phase response* determines the precise partial overtone *tuning* [11, pp. 182–184]. These error components are normally addressed separately. That is, string loop filters are normally designed as a series connection of a low-order filter having an inaudible phase response (the *damping filter*) and an allpass filter (the *dispersion filter*), if any. Only stiff strings require an allpass filter component, and only if the stiffness is sufficient to audibly "stretch" the tuning of the partial overtones. An elementary introduction to digital filters and their analyses is given in [12].[6]

In [54], the damping in real piano strings was modeled using a length 17 linear-phase FIR filter for the lowest strings and a length 9 linear-phase FIR filter for the remaining strings. It is not necessary to restrict consideration to linear-phase filters, since a minimum-phase design may sound just as good, and any undesirable aspects of the phase response may be compensated in the dispersion-filter design.

In the case of stiff piano strings, it is possible to well tune the first several tens of partials with total allpass order of 20 or less [55]. Additionally, minimization of the L^∞ norm [56] has been used to calibrate a series of allpass-filter sections [57]. A summary of methods for dispersion-filter design is given in [51].

Filter design methods applicable to acoustic resonators (the "Body Filter" in Figure 2) are discussed in [11,58]. The *commuted synthesis* technique, described in Sect. 2.8 below, replaces a parametric body filter with its nonparametric impulse response (filtered by excitation characteristics). The commuted synthesis technique is applicable to plucked and struck strings, and also to *linearized* bowed-string models [42].

2.6 Electric Guitars

While most musical vibrating strings are well approximated as linear, time-invariant systems, there are special cases in which

---

[6] http://ccrma.stanford.edu/~jos/filters/.

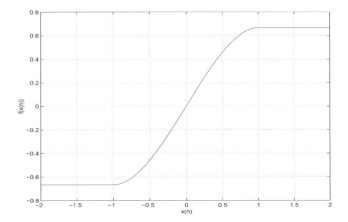

FIGURE 4 Soft clipper defined by Eq. (8). [From [41].]

nonlinear behavior is desired. One example is the distorted electric guitar [30].

A *soft clipper* is similar to a hard clipper (saturation on overflow), but with the "corners" smoothed. A common choice of soft clipper is the *cubic nonlinearity*, e.g., [59],

$$f(x) = \begin{cases} -\frac{2}{3}, & x \leq -1 \\ x - \frac{x^3}{3}, & -1 \leq x \leq 1 \\ \frac{2}{3}, & x \geq 1. \end{cases} \qquad (8)$$

This particular soft-clipping characteristic is diagrammed in Figure 4. An analysis of its spectral characteristics, along with some discussion regarding how to avoid the aliasing it can cause, is given in [41].[7] An input gain may be used to set the desired degree of distortion.

A cubic nonlinearity, as well as *any* odd distortion law,[8] generates only odd-numbered harmonics (like in a square wave). For best results, and in particular for *tube distortion* simulation [60,61], it can be argued that some amount of even-numbered harmonics should also be present. Breaking the odd symmetry in any way will add even-numbered harmonics to the output as well. One simple way to accomplish this is to add an *offset* to the input signal, obtaining

$$y(n) = f[x(n)+c], \qquad (9)$$

---

[7] http://ccrma.stanford.edu/~jos/pasp/Nonlinear_Elements.html.
[8] A function $f(x)$ is said to be *odd* if $f(-x) = -f(x)$.

where c is some small constant. (Signals $x(n)$ in practice are typically constrained to be zero mean by one means or another.)

Another method for breaking the odd symmetry is to add some square-law nonlinearity to obtain

$$f(x) = \alpha x^3 + \beta x^2 + \gamma x + \delta, \qquad (10)$$

where β controls the amount of square law distortion. This is then a more general third-order polynomial. A square law is the gentlest nonlinear distortion, as can be seen by considering the Taylor series expansion of a general nonlinearity transfer characteristic $f(x)$. The constant δ can be chosen to zero the mean, on average; if the input signal $x(n)$ is zero mean with variance 1, then $\delta = -\beta$ compensates the nonzero mean introduced by the squaring term. The term γ can be modified to adjust the "effect mix".

2.7 Amplifier Feedback

A nonlinear feedback effect used with distorted electric guitars is *amplifier feedback*. In this case, the amplified guitar signal couples back into the strings with some gain and delay, as depicted schematically in Figure 5 [59]. The feedback delay can be adjusted to cause different partial overtones to be amplified relative to others.

2.8 Commuted Synthesis

Figure 6 shows a schematic diagram of a plucked or struck stringed instrument, such as a guitar. When the string and resonator may be assumed linear and time invariant, they may be *commuted*, i.e., connected in the opposite series ordering [12], as

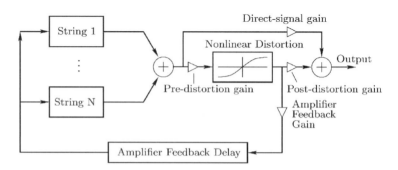

FIGURE 5 Simulation of a basic distorted electric guitar with amplifier feedback. [From [41].]

FIGURE 6 Schematic diagram of a stringed musical instrument. [From [29].]

FIGURE 7 Schematic diagram of commuted synthesis of plucked/struck stringed instruments. [From [29].]

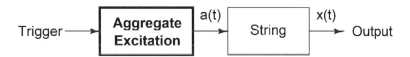

FIGURE 8 Use of an aggregate excitation given by the convolution of original excitation with the resonator impulse response. [From [41].]

shown in Figure 7. Sound synthesis algorithms based on this arrangement are typically called *commuted waveguide synthesis* algorithms [41,42,62,63].

Instead of plucking the string and filtering the string output with a digital filter of extremely high order (to capture the many resonances in the range of human hearing), the "pluck response" of the guitar body (a filtered impulse response) can be fed to the string instead, as shown in Figure 8. In a typical implementation, the guitar-body impulse response (or some filtering of it) is stored in table, just as in sampling synthesis, and a low-order filter is applied to the table playback in order to impart details of the plucking excitation. This simplification exchanges an expensive body filter for an inexpensive "pluck filter". In addition to body resonances, the excitation table may include characteristics of the listening space as well. Commuted synthesis of the piano has also been developed [14,57,64,65].

3 • Wind Instruments

A basic DW model for a single-reed woodwind instrument, such as a clarinet, is shown in Figure 9 [40,41,66].

When the bore is cylindrical (plane waves) or conical (spherical waves), it can be modeled quite simply using a bidirectional delay line [66]. Because the main control variable for the instrument

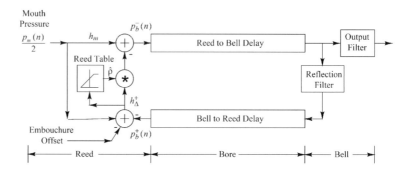

FIGURE 9 Waveguide model of a single-reed, cylindrical-bore woodwind, such as a clarinet. [From [41].]

is air pressure in the mouth at the reed, it is convenient to choose pressure-wave variables. Thus, the delay lines carry left-going and right-going *pressure* samples p_b^+ and p_b^- (respectively) which represent the traveling pressure-wave components within the bore.

To first order, the bell passes high frequencies and reflects low frequencies, where "high" and "low" frequencies may be divided by the wavelength which equals the bell's diameter. Thus, the bell can be regarded as a simple "cross-over" network, as is used to split signal energy between a woofer and tweeter in a loudspeaker cabinet. For a clarinet bore, the nominal "cross-over frequency" is around 1500 Hz [67].

The reflection filter at the right of the figure implements the bell or tone-hole losses as well as the round-trip attenuation losses from traveling back and forth in the bore. The bell output filter is highpass and *power complementary* with respect to the bell reflection filter. This must be true because the incident wave is assumed to either reflect or transmit, and all wave energy must be accounted for. (We assume that the bell itself does not vibrate or otherwise absorb sound.) From continuity of pressure and conservation of flow, we also have that the bell and reflection filters are *amplitude complementary* [41]. Therefore, given the reflection filter $H_r(z)$, the transmission filter is simply $H_t(z) = 1 - H_r(z)$ for velocity waves, or $H_t(z) = 1 + H_r(z)$ for pressure waves. That such filters are also power complementary may be seen by summing the transmitted and reflected power waves:

$$P_t U_t + P_r U_r = (1+H_r)P \cdot (1-H_r)U + H_r P \cdot (-H_r)(-U)$$
$$= [1 - H_r^2 + H_r^2]PU$$
$$= PU,$$

where P and U denote the Fourier transform of the incident pressure and volume-velocity waves, respectively.

The reed may be modeled as a signal- and embouchure-dependent *nonlinear reflection coefficient* terminating the bore. Such a model is possible because the reed mass is neglected.[9] The player's embouchure controls damping of the reed, reed aperture width, and other parameters, and these can be implemented as parameters on the contents of the lookup table or nonlinear function ("Reed Table" in Figure 9).

The controlling mouth pressure is denoted p_m. The *reflection coefficient* of the reed is denoted $\rho(h_\Delta^+)$, where $h_\Delta^+ \triangleq p_m/2 - p_b^+$ ("incoming half-pressure drop"). A simple choice of *embouchure control* is a simple additive offset in the reed-table address. Since the main feature of the reed table is the pressure drop at which the reed begins to open, such a simple offset can implement the effect of biting harder or softer on the reed, or changing the reed stiffness.

The following qualitatively chosen reed table has been used with good results [40]:

$$\hat{\rho}(h_\Delta^+) = \begin{cases} 1 - m(h_\Delta^c - h_\Delta^+), & -1 \leq h_\Delta^+ < h_\Delta^c \\ 1, & h_\Delta^c \leq h_\Delta^+ \leq 1 \end{cases}. \quad (11)$$

This function is shown in Figure 10 for $m = 1/(h_\Delta^c + 1)$. The corner point h_Δ^c is the smallest pressure difference giving reed

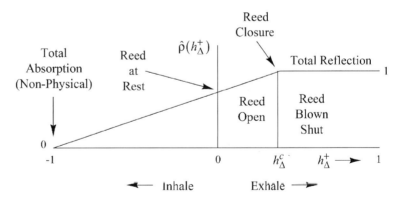

FIGURE 10 Simple, qualitatively chosen reed table for the digital waveguide clarinet. [From [40].]

---

[9]It is valid to neglect the reed mass when the first reed resonance is well above the fundamental frequency of the played note, as is normally the case.

closure. Having the table go all the way to zero at the maximum negative pressure $h_\Delta^+ = -1$ is nonphysical (0.8 would be a better physical estimate), but zero reflectance gives the practical benefit of opening the feedback loop when the lookup-table input signal is too large. Embouchure and reed stiffness correspond to the choice of offset h_Δ^c and slope m. Brighter tones are obtained by increasing the curvature at the point corresponding to the reed opening; for example, one can use $\hat{\rho}^k(h_\Delta^+)$ for increasing $k \geq 1$.

Another variation is to replace the table-lookup contents by a piecewise polynomial approximation. While less general, good results have been obtained in practice with this method [68].

An intermediate approach between table lookups and polynomial approximations is to use interpolated table lookups. Typically, linear interpolation is used, but higher order polynomial interpolation can also be considered [41].[10]

STK software [49] implementing a model as in Figure 9 can be found in the file `Clarinet.cpp`.

4 • Recommended Reading

An excellent musical acoustics textbook is Fletcher and Rossing [1]. For considerable detail on the physics of vibrating strings in musical acoustics, see the chapter by Vallette in **Mechanics of Musical Instruments** [16]. Classic references on acoustics include Pierce [69] and Morse and Ingard [70]. The original monograph **Vibration and Sound** by Morse [15] is still very interesting to read, and contains musical acoustics material that was dropped from Morse and Ingard.

References

1. N. H. Fletcher and T. D. Rossing, *The Physics of Musical Instruments*, Second Edition, Springer Verlag, New York, 1998.
2. A. Hirschberg, J. Kergomand, and G. Weinneich, eds., *Mechanics of Musical Instruments*, Berlin: Springer-Verlag, 1995.
3. T. D. Rossing, F. R. Moore, and P. A. Wheeler, *The Science of Sound*, Third Edition, Addison-Wesley, Reading, MA, 2003.
4. J. O. Smith, "Virtual acoustic musical instruments: Review and update", *J. New Music Research*, 33(3):283–304, 2004.
5. J. D. Markel and A. H. Gray, *Linear Prediction of Speech*, Springer Verlag, New York, 1976.
6. A. V. Oppenheim, R. W. Schafer, and J. R. Buck, *Discrete Time Signal Processing*, Prentice-Hall, Englewood Cliffs, NJ, 1999.

---

[10]http://ccrma.stanford.edu/~jos/pasp/Delay_Line_Interpolation.html.

7. D. Massie, "Wavetable sampling synthesis", in *Applications of DSP to Audio & Acoustics*, M. Kahrs and K. Brandenburg, Eds, pp. 311–341. Kluwer Academic Publishers, Boston/Dordrecht/London, 1998.
8. A. Horner, J. Beauchamp, and L. Haken, "Methods for multiple wavetable synthesis of musical instrument tones", *J. Audio Eng. Soc.*, 41(5):336–356, May 1993.
9. A. Ng and A. Horner, "Iterative combinatorial basis spectra in wavetable matching", *J. Audio Eng. Soc.* 50(12):1054–1063, 2002.
10. F. Fontana and D. Rocchesso, "Physical modeling of membranes for percussion instruments", *Acta Acustica*, 84(3):529–542, May/June 1998.
11. J. O. Smith, *Techniques for Digital Filter Design and System Identification with Application to the Violin*, PhD thesis, Elec. Engineering Dept., Stanford University (CCRMA), June 1983, CCRMA Technical Report STAN-M-14, http://ccrma.stanford.edu/STANM/stanms/stanm14/.
12. J. O. Smith, *Introduction to Digital Filters*, http://ccrma.stanford.edu/~jos/filters/, September 2005.
13. A. Chaigne and A. Askenfelt, "Numerical simulations of piano strings, parts I and II", *J. Acoust. Soc. Am.*, 95(2–3):1112–1118, 1631–1640, Feb.–March 1994.
14. J. Bensa, S. Bilbao, R. Kronland-Martinet, and J. O. Smith, "The simulation of piano string vibration: From physical models to finite difference schemes and digital waveguides", *J. Acoust. Soc. Am.*, 114(2):1095 1107, 2003.
15. P. M. Morse, *Vibration and Sound*, American Institute of Physics, for the Acoustical Society of America, http://asa.aip.org/publications.html, 1948, 1st edition 1936, author's last edition 1948, ASA edition 1981.
16. ibid [2].
17. H. Järveläinen, V. Välimäki, and M. Karjalainen, "Audibility of inharmonicity in string instrument sounds, and implications to digital sound synthesis", in *Proc. 1999 Int. Computer Music Conf.*, Beijing, pp. 359–362, 1999. http://www.acoustics.hut.fi/~hjarvela/publications/.
18. M. E. McIntyre and J. Woodhouse, "On the fundamentals of bowed string dynamics", *Acustica*, 43(2):93–108, Sept. 1979.
19. T. W. Parks and C. S. Burrus, *Digital Filter Design*, Wiley, New York, June 1987, contains FORTRAN software listings.
20. B. Fornberg, *A Practical Guide to Pseudo-spectral Methods*, Cambridge University Press, 1998.
21. P. M. Ruiz, *A Technique for Simulating the Vibrations of Strings with a Digital Computer*, Music Master Thesis, University of Illinois, Urbana, 1970.
22. L. Hiller and P. Ruiz, "Synthesizing musical sounds by solving the wave equation for vibrating objects. Part II", *J. Audio Eng. Soc.*, Part I: 19(6):462–470, June 1971, Part II: 19(7):542–551, July/Aug. 1971.

23. J. C. Strikwerda, *Finite Difference Schemes and Partial Differential Equations*, Wadsworth and Brooks, Pacific Grove, CA, 1989.
24. S. Bilbao, *Wave and Scattering Methods for Numerical Simulation*, Wiley, New York, July 2004.
25. M. Karjalainen, "Mixed physical modeling: DWG + FDTD + WDF", *Proc. 2003 IEEE Workshop on Applications of Signal Processing to Audio and Acoustics, New Paltz, NY*, Oct. 2003, pp. 225–228, IEEE Press.
26. M. Karjalainen and C. Erkut, "Digital waveguides vs. finite difference schemes: Equivalance and mixed modeling", *EURASIP J. Applied Signal Processing*, 2004(7):978–989, June 15, 2004.
27. J. O. Smith, "Music applications of digital waveguides", Tech. Rep. STAN-M-39, CCRMA, Music Department, Stanford University, 1987, CCRMA Technical Report STAN-M-39, line feel http://ccrma.stanford.edu/STANM/stanms/stanm39/.
28. J. O. Smith, "Physical modeling using digital waveguides", *Computer Music J.*, 16(4):74–91, Winter 1992, special issue: Physical Modeling of Musical Instruments, Part I, http://ccrma.stanford.edu/~jos/pmudw/.
29. N. Lee and J.O. Smith, *"Virtual Stringed instruments"*, Dec. 2007, http://ccrma.stanford.edu/realsimple/phys_mod_overview/.
30. J. O. Smith, "Virtual Electric Guitar and Effects Using Faust and Octave", *Proc. 6th Int. Linux Audio Conf. (LAC 2008), paper:* http:// lac.linuxaudio.org/download/papers/zz.pdf supporting website: http://ccrma.stanford.edu/realsimple/faust_strings.
31. M. J. Beeson and D. T. Murphy, "RoomWeaver: A digital waveguide mesh based room acoustics research tool", *Proc. 2004 Conf. Digital Audio Effects (DAFx-04), Naples, Italy*, Oct. 2004, http://www.dafx.de/.
32. A. Krishnaswamy and J. O. Smith, "Methods for simulating string collisions with rigid spatial objects", *Proc. 2003, IEEE Workshop on Applications of Signal Processing to Audio and Acoustics, New Paltz, NY*, Oct. 2003, IEEE Press.
33. R. Pitteroff and J. Woodhouse, "Mechanics of the contact area between a violin bow and a string, part II: Simulating the bowed string", *Acta Acustica*, 84(3):744–757, 1998.
34. J. l. R. d'Alembert, "Investigation of the curve formed by a vibrating string, 1747", in *Acoustics: Historical and Philosophical Development*, R. B. Lindsay, Ed., pp. 119–123. Dowden, Hutchinson & Ross, Stroudsburg, 1973.
35. J. O. Smith, *Mathematics of the Discrete Fourier Transform (DFT) with Audio Applications, 2nd edition*, W3K Publishing, http://w3k.org/books/, 2007, http://ccrma. stanford.edu/~jos/mdft/.
36. P. R. Cook, *Identification of Control Parameters in an Articulatory Vocal Tract Model, with Applications to the Synthesis of Singing*, PhD thesis, Elec. Engineering Dept., Stanford University (CCRMA), Dec. 1990, http://www.cs.princeton.edu/~prc/.
37. J. L. Kelly and C. C. Lochbaum, "Speech synthesis", *Proc. Fourth Int. Congress on Acoustics, Copenhagen*, pp. 1–4, September 1962, Paper G42. Reprinted in [71, pp. 127–130].

38. H.-L. Lu, *Toward a High-Quality Singing Synthesizer with Vocal Texture Control*, PhD thesis, Elec. Engineering Dept., Stanford University (CCRMA), July 2002, http://ccrma.stanford.edu/~vickylu/thesis/.
39. J. O. Smith, "Principles of digital waveguide models of musical instruments", in *Applications of Digital Signal Processing to Audio and Acoustics*, M. Kahrs and K. Brandenburg, Eds, pp. 417–466. Kluwer Academic Publishers, Boston/Dordrecht/London, 1998.
40. J. O. Smith, "Efficient simulation of the reed-bore and bow-string mechanisms", *Proc. 1986 Int. Computer Music Conf., The Hague*, pp. 275–280, 1986, Computer Music Association, also available in [27].
41. J. O. Smith, *Physical Audio Signal Processing*, 2007, http://ccrma.stanford.edu/~jos/pasp/, Online book.
42. J. O. Smith, "Efficient synthesis of stringed musical instruments", *Proc. 1993 Int. Computer Music Conf., Tokyo*, pp. 64–71, 1993, Computer Music Association, also incorporated into [29].
43. J. Woodhouse, "Bowed string simulation using a thermal friction model", *Acta Acustica*, 89(2):355–368, 2003.
44. J. H. Smith and J. Woodhouse, "The tribology of rosin", *J. Mechanics and Physics of Solids*, 48(8):1633–1681, Aug. 2000.
45. S. Serafin, *The Sound of Friction: Real-Time Models, Playability and Musical Applications*, PhD thesis, Music Department, Stanford University, 2004.
46. S. Serafin, P. Huang, S. Ystad, C. Chafe, and J. O. Smith, "Analysis and synthesis of unusual friction-driven musical instruments", *Proc. 2002 Int. Computer Music Conf., Gothemborg, Sweden*, 2002.
47. F. Avanzini, S. Serafin, and D. Rocchesso, "Modeling interactions between rubbed dry surfaces using an elasto-plastic friction model", in *Proc. 2002 COST-G6 Conf. Digital Audio Effects (DAFx-02), Hamburg, Germany* pp. 111–116, September 2002, http://www.dafx.de/.
48. P. R. Cook, "Synthesis toolkit in C++, version 1.0", in *SIGGRAPH Proc.*, May 1996, Assoc. Comp. Mach.
49. P. Cook and G. Scavone, *Synthesis ToolKit in C++, Version 4.2.1*, Feb. 2006, http://ccrma.stanford.edu/CCRMA/Software/STK/.
50. J. Riionheimo and V. Välimäki, "Parameter estimation of a plucked string synthesis model using genetic algorithm with perceptual fitness calculation", *EURASIP J. Applied Signal Processing*, 3(8): 791–805, July 2003, sound examples available at http://www.acoustics.hut.fi/publications/papers/jasp-ga/.
51. B. Bank, F. Avanzini, G. Borin, G. D. Poli, F. Fontana, and D. Rocchesso, "Physically informed signal processing methods for piano sound synthesis: A research overview", *EURASIP J. Applied Signal Processing*, 2003(10):941–952, Sept. 2003, http://www.mit.bme.hu/~bank/publist/jasp03.pdf.
52. V. Välimäki and T. Tolonen, "Development and calibration of a guitar synthesizer", *J. Audio Eng. Soc.*, 46(9):766–778, 1998.
53. M. Karjalainen, V. Välimäki, and T. Tolonen, "Plucked string models: From the Karplus–Strong algorithm to digital waveguides and beyond", *Computer Music Journal*, 22(3):17–32, Fall 1998, http://www.acoustics.hut.fi/~vpv/publications/ cmj98.htm.

54. J. Bensa, S. Bilbao, R. Kronland-Martinet, J. O. Smith, and T. Voinier, "Computational modeling of stiff piano strings using digital waveguides and finite differences", *Acta Acustica*, 91(2): 289–298, 2005.
55. D. Rocchesso and F. Scalcon, "Accurate dispersion simulation for piano strings", in *Proc. Nordic Acoustical Meeting (NAM'96), Helsinki, Finland*, June 12–14 1996, 9 pages.
56. M. Lang, "Allpass filter design and applications", *IEEE Transactions on Signal Processing*, 46(9):2505–2514, 1998.
57. J. Bensa, *Analysis and Synthesis of Piano Sounds Using Physical and Signal Models*, PhD thesis, Université de la Méditérranée, Marseille, France, 2003, http://www.lma.cnrs-mrs.fr/~bensa/.
58. M. Karjalainen and J. O. Smith, "Body modeling techniques for string instrument synthesis", *Proc. 1996 Int. Computer Music Conf., Hong Kong*, pp. 232–239, Aug. 1996, Computer Music Association.
59. C. R. Sullivan, "Extending the Karplus–Strong algorithm to synthesize electric guitar timbres with distortion and feedback", *Computer Music J.*, 14(3):26–37, Fall 1990.
60. E. Barbour, "The cool sound of tubes", *IEEE Spectrum*, pp. 24–35, Aug. 1998.
61. B. Santo, "Volume cranked up in amp debate", *Electronic Engineering Times*, no. 817, pp. 24–35, October 3, 1994, http://www.trueaudio.com/at_eetjlm.htm.
62. M. Karjalainen, V. Välimäki, and Z. Jánosy, "Towards high-quality sound synthesis of the guitar and string instruments", *Proc. 1993 Int. Computer Music Conf., Tokyo*, pp. 56–63, Sept. 10–15 1993, Computer Music Association, http://www.acoustics.hut.fi/~vpv/publications/icmc93-guitar.htm.
63. V. Välimäki, J. Huopaniemi, M. Karjalainen, and Z. Jánosy, "Physical modeling of plucked string instruments with application to real-time sound synthesis", *J. Audio Eng. Soc.*, 44(5):331–353, May 1996.
64. J. O. Smith and S. A. Van Duyne, "Commuted piano synthesis", *Proc. 1995 Int. Computer Music Conf., Banff*, pp. 319–326, 1995, Computer Music Association, http://ccrma.stanford.edu/~jos/pdf/svd95.pdf.
65. S. A. Van Duyne and J. O. Smith, "Developments for the commuted piano", *Proc. 1995 Int. Computer Music Conf., Banff*, pp. 335–343, 1995, Computer Music Association, http://ccrma.stanford.edu/~jos/pdf/vds95.pdf.
66. G. P. Scavone, *An Acoustic Analysis of Single-Reed Woodwind Instruments with an Emphasis on Design and Performance Issues and Digital Waveguide Modeling Techniques*, PhD thesis, CCRMA, Music Dept., Stanford University, Mar. 1997, http://ccrma.stanford.edu/~gary/.
67. A. H. Benade, *Fundamentals of Musical Acoustics*, Dover, New York, 1990.
68. P. R. Cook, "A meta-wind-instrument physical model, and a meta-controller for real time performance control", *Proc. 1992 Int. Computer Music Conf., San Jose*, pp. 273–276, 1992, Computer Music Association.

69. A. D. Pierce, *Acoustics*, American Institute of Physics, for the Acoustical Society of America, 1989, http://asa.aip.org/ publications.html.
70. P. M. Morse and K. U. Ingard, *Theoretical Acoustics*, McGraw-Hill, New York, 1968.
71. J. L. Flanagan and L. R. Rabiner, Eds, *Speech Synthesis*, Dowden, Hutchinson, and Ross, Stroudsburg, PA, 1973.
72. A. Askenfelt, Ed., *Five Lectures on the Acoustics of the Piano*, Royal Swedish Academy of Music, Stockholm, 1990, lectures by H. A. Conklin, Anders Askenfelt and E. Jansson, D. E. Hall, G. Weinreich, and K. Wogram. Sound example CD included. Publication No. 64, http://www.speech.kth.se/music/5_lectures/.

26

Modeling of Musical Instruments

Rolf Bader[1] *and Uwe Hansen*[2]

[1] *Institute of Musicology, University of Hamburg, Hamburg, Germany*
[2] *Department of Physics, Indiana State University, Terre Haute, IN, USA*

1 • Introduction

Signal processing techniques in acoustics address many concerns. Included are such things as wave propagation variables, amplitude considerations, spectral content, wavelength, and phase. Phase is primarily of concern when waves interact with each other, as well as with a medium, and the imposition of boundary conditions leads to normal mode vibrations. Such conditions are prevalent in all musical instruments, and thus relevant signal processing techniques are essential to both understanding and modeling the structure of musical instruments and the sound radiated. A coherent experimental technique for studying such normal mode vibrations is Modal Analysis, where the ratios of a series of FFT response spectra and associated excitations are recorded as transfer functions, and animated normal mode motion is subsequently extracted using appropriate coupled harmonic oscillator modeling. Finite Elements, a complementary theoretical technique is frequently also used to model such normal mode motion. This chapter is primarily concerned with such modeling and associated considerations.

1.1 Timbre

In musical acoustics, instruments can be classified by excitation mechanisms. On the one hand, drive mechanisms have an important influence on the spectra of the steady-state portion of the sound. On the other hand, the initial transients of the tones respond strongly to the nature of the drive mechanism (Bader 2002, Taylor 2001, Krumhansl 1989, Luce 1965). Musicians of all instrument families are aware of the importance of the attack and develop this important part of their artistic individuality. Thus, in terms of drive mechanisms, the following groups can be distinguished:

1. plucked instruments (guitars, lutes...)
2. bowed instruments (violins, double bass...)
3. reed-driven blown instruments (saxophone, oboe...)
4. air jet instruments (flute, flue organ pipes...)
5. lip-driven blown (brass) instruments (trumpet, horn...)
6. struck instruments (percussion, piano...).

Individual instruments of one given family sound different, of course, but the drive mechanisms cluster them into families. Thus, a western oboe is much closer in sound to a Persian zurna or a Thai saranai (all double-reed instruments) than to a western clarinet. Also, a western violin is closer in timbre to a Chinese erhu or an Indonesian rehab, (all bowed instruments) than to a western guitar. Individuals were asked to judge the similarity of pairs of instrument tones presented to them. Such hearing tests confirm instrumental clustering in families, depending on drive mechanisms (Grey 1977). While a true steady state is associated with a harmonic spectrum, departures from harmonicity in transients, such as attack and vibrato, reflect the nature of the mechanics of the transient and are thus instrument specific. Consequently, eliminating the initial transient from the hearing tests resulted in increased difficulty of separating and identifying the instrument type (Grey and Moorer 1977).

These perceptual experiments also found similarity judgments according to the acoustic properties of the sounds. The hierarchy is as follows

1. pitch
2. spectral centroid
3. spectral fluctuations
4. inharmonicities in the initial transient.

For the first parameter, subjects judged the similarity of sounds as far as pitch is concerned (Miller and Carterette 1975). This parameter is prominent in judging instrumental characteristics. If sounds are presented for which the pitch changes, listeners neglect the other parameters when making similarity judgments. Second in importance is the spectral centroid, the "middle" frequency of the spectrum, which from a psychoacoustic standpoint relates to a bright sound for a high spectral centroid, or a dark sound for a low spectral centroid. Indeed, for constant pitch, the identification of musical instruments depends mostly on this spectral centroid. For example, violins have a high spectral centroid because of their resonances. Indeed, listeners will find artificial sounds with high spectral centroid much more similar to violins, than to plucked, or blown instruments.

The parameter third in importance for similarity judgment relates to spectral fluctuations exemplified by a vibrato, or amplitude fluctuations. Violins with their initial transients cut off are much better identified as violins as long as they are played with vibrato. Adding the vibrato, immediately leads to identification. The same holds for blown instruments, though not as strongly. It appears that a typical violin vibrato leads to identification for the instrument.

Finally, inharmonicities in the initial transients contribute to instrument identification. Inharmonicities appear in the attack for several, quite complicated reasons. In the guitar, for example, the resonance frequencies of the guitar body, are more prominent at the beginning of the tone. They are subsequently surpassed in loudness by the eigenfrequencies of the strings. For violin sounds, along with other causes, inharmonicities appear because of changes in bowing pressure, which lead the bow-string system to switch to another bifurcation regime. It has to be mentioned that the secondary role of inharmonicities in the initial attack of musical tones in terms of instrument similarity judgments stands in contrast to the fact that cutting the initial transient makes it hard to identify an instrument. So identification seems to be different from similarity judgments. And, as similarity tasks result in break-down when initial transients are eliminated, the influence on characterization and identification seems to increase in importance in comparison to the similarity feature.

All features mentioned above are associated with detailed structural features of musical instruments, thus the geometrical and physical reasons for the complex behavior of instruments during their initial transient phase need to be understood in order to discuss the character of instruments, including instrument quality, and to make suggestions to instrument builders for improvement or adjustments to the instrument geometry.

Thus modern musical acoustics is concerned with the geometrical features and with the detailed structure of musical instruments. This fine structure is associated with fundamental aspects of instrument quality, to be described on an acoustical level.

1.2 Structure of Musical Instruments

Intricately detailed geometrical structures of musical instruments result in complex fine structures of the sounds produced. When classifying aspects of musical sounds in relation to specific musical instruments, four important features are noted:

1. timbre (sound quality)
2. attack behavior
3. overall loudness
4. degree of possible timbre variation.

Instrument builders and musicians are convinced that it is primarily these features which determine instrument quality and character. The importance of timbre is evident mostly within a given instrument family, rather than in comparison between families. A specific example is the classical guitar, where the timbre difference between Spanish and German instruments is quite small, but it is clearly discernable by experts. It is this difference which determines which type of instrument is to be played for a specific musical style, and furthermore, only that instrument type would be appropriate for that musical style.

As mentioned above, control of the attack behavior is very important. Generally, instruments with fast attack are judged higher in quality than instruments with slow or damped attack. One likely reason is that instruments with rapid response can be played faster and with greater ease. But external influences, such as humidity, may change this feature even for a given instrument. For example, violins played under conditions of high humidity react more slowly and are harder to play than under normal conditions. In the case of the violin, such considerations of individual instruments should, of course, include moisture absorption characteristics of bow hair and rosin. Furthermore, the attack transients of musical sounds contain characteristic instrument information, as well as information unique to the musical expression of the performing artist. This initial transient is short, in the range of about 50 ms, although a high-pitched saxophone sound may finish its attack after about 4 ms and a low pitched organ flue pipe may need 200 ms. In addition, the attack is very complex. Here good results are achieved by using modeling techniques to show the complexity as well as suggesting underlying reasons for the complexity

Builders of musical instruments generally attempt to maximize the overall loudness of the instrument. For example, the sound pressure radiated from a violin is relatively high, considering the size of the instrument. An important element contributing to the radiation efficiency of the violin is the sound post. It forms a rigid connection between the front and back plates of the violin near the foot of the bridge on the treble side (Fletcher and Rossing 1999). In a violin without a sound post, the transverse bowing forces would preferentially excite normal plate modes which exhibit a nodal line along the length of the instrument in the middle. Because of their dipole nature, these modes exhibit rather low radiation efficiencies. The sound post limits the out-of-phase motion of the two sides of the violin thus rendering the overall radiation pattern more spherical, with increased radiation efficiency. Another example is found in some guitar designs, when the guitar body is designed in a fashion resembling a loudspeaker box, with the back plate and ribs very thick and rigid and a very thin top plate. Such a guitar is subjectively judged to be nearly twice as loud as a normal classical guitar, but the guitar timbre is changed and the degree of possible timbre variation by the player is decreased.

This degree of possible timbre variation produced by the player depends on the response of the instrument to loudness variations, as well as to playing position on the string with plucked and bowed instruments, or blowing pressure and mouth position with blown reed and brass instruments. Here, the player expects the instrument to not just behave in a linear fashion. For example, a change of lip pressure with a saxophone also changes the overtone content and thus the brightness of the sound. Moreover, the player may, expect an increase in lip pressure to have just a small influence on the sound in one pressure region and a dramatic effect in another region. This property of musical instruments also shows up in terms of abrupt phase changes. Tone production in reed instruments requires a threshold blowing pressure. Below this threshold pressure, no musical tone is produced, only a blowing noise. At a certain higher blowing pressure, the fundamental tone is no longer excited, and the instrument is overblown sounding in pitches of the harmonic overtone structure of the fundamental. Many instruments, like the horn or the trombone depend partly or completely on the possibility of sudden phase changes. But also small nonlinearities make instruments complex and individual. Some examples are the lip pressure mentioned above with saxophones, additional harmonic generation in strings with large displacement, or additional modes in the initial attack of the piano tone by reflection of the impulse wave at the piano hammer (Fletcher and Rossing 1999). It is this individuality, which among other things is responsible for the richness and quality of instruments.

Again, because many of these characteristics appear in the initial transients of sounds, musical acoustics, being concerned with musical problems, needs to take them into consideration.

1.3 Modeling using the Finite-Element Method (FEM) and the Finite-Difference Method (FDM)

On the one hand, this section is intended to illustrate typical problems and solutions of musical acoustics using geometric modeling of musical instruments. On the other hand, it provides an overview of the theoretical techniques of the Finite-Element Method **(FEM)** (Bathe 1996, Hughes 1987) and the Finite-Difference Method **FDM**. Thus, following the presentation of the FEM and FDM formulations, five basic analytical cases are presented:

Steady-state description: Turbulent $k - \epsilon$ model of a flute.
Eigenfrequency evaluation: Multiphonics in blown instruments as split-phase pressure description.
Time-dependent implementation: Complete geometry of a classical guitar.
Impulse-response technique: Guitar back plate timbre brightening by stress bending.
Contact problem: Church bell clapper time-dependent striking force.

For dynamic structures of complicated geometries, described by higher order differential equation which do not lend themselves to solutions in closed form, numerical methods such as *FDM* and *FEM* are used to obtain solutions. For *FEM*, the structure is subdivided into a mesh or grid, which, depending on the geometry, can consist of one-dimensional line elements, two-dimensional area elements, such as triangles of rectangles, or three-dimensional volume elements, such as regular or irregular tetrahedral or other polyhedra. Node points on the boundaries of these elements are used as discrete points on the structure under investigation. The dependent variables used in the equation systems are defined only at these node points. For a two-dimensional plate, for example, the dependent variables would give the displacement, which would only be known at the node points of the mesh. In addition to the structural and dimensional geometry and the mesh identification, mechanical properties such as density and elastic constants including Young's modulus, Poisson's ratio and shear modulus, as applicable, are needed along with appropriate boundary conditions. In *FDM*, the structure is not divided into a mesh of elements but rather discrete node points are identified on the geometry, and the equation system is solved for these points, considering differences with neighboring points. Thus, the *FEM* formulation is more complicated and involves more computational time. Its advantage is the greater ease of handling boundaries. Furthermore, the choice of element geometry and element density in *FEM* is very flexible, making it very suitable for complicated geometries, in contrast to the usual fixed grid constants in *FDM*.

2 • Finite-Element Formulation

The elements used are most commonly defined as isoparametric elements. Here, an "ideal" element is defined with its own local coordinate system ranging normally from -1 to 1 as convenient integration and differentiation boundaries. Interpolation functions are defined in a way that when using the values of the dependent variable, i.e., the displacement of a plate, a continuous surface is achieved. So, for example, if a tetrahedral element were used, the element functions would be given as

$$N_1 = 1 - \xi - \eta - \zeta$$
$$N_2 = \xi$$
$$N_3 = \eta$$
$$N_4 = \zeta$$

Here, ξ, η, and ζ are local coordinate variables for the x-, y-, and z-direction and the element is defined for

$$-1 \leq \xi, \eta, \zeta \leq 1 \ . \tag{1}$$

This element has the advantage, that it can be precisely integrated analytically once for all elements used. Each element then must be transferred from the local to the global coordinate system using a Jacobian matrix

$$\mathbf{J} = \sum_{i=1}^{n} \mathbf{X}_i \frac{\partial N_i}{\partial \Xi} \ , \tag{2}$$

where $\Xi = \{\xi, \eta, \zeta\}$, $n = 4$ since there are four shape functions, and $\mathbf{X}_i$ are the coordinates in the global coordinate system, which is also called the reference frame. In addition to specifying the differential equation system, only the displacement at the node points needs to be known to obtain a complete interpolation solution for the geometry. This leads to a simple linear equation system which can be solved. For structural mechanics, the differential equation system is often based on a stress–strain formulation like

$$\sigma = E \, \epsilon \tag{3}$$

where σ is the stress, ϵ is the strain, and E is the Young's modulus. With higher dimensional problems, a matrix $\mathbf{C}$ is used which is the material matrix describing Young's modulus, the Poisson ratio, and the shear modulus. When using a three-dimensional model, this matrix is

$$\mathbf{C} = \frac{E}{(1+\nu)(1-2\nu)} \begin{bmatrix} 1 & \frac{\nu}{(1-\nu)^2} & \frac{\nu}{(1-\nu)^2} & 0 & 0 & 0 \\ \frac{\nu}{(1-\nu)^2} & 1 & \frac{\nu}{(1-\nu)^2} & 0 & 0 & 0 \\ \frac{\nu}{(1-\nu)^2} & \frac{\nu}{(1-\nu)^2} & 1 & 0 & 0 & 0 \\ 0 & 0 & 0 & \frac{1-2\nu}{2(1-\nu)^2} & 0 & 0 \\ 0 & 0 & 0 & 0 & \frac{1-2\nu}{2(1-\nu)^2} & 0 \\ 0 & 0 & 0 & 0 & 0 & \frac{1-2\nu}{2(1-\nu)^2} \end{bmatrix}. \quad (4)$$

Here E is the Young's modulus, which can also be formulated as different for all three directions, and ν is the Poisson's ratio. As the dependent variable is the displacement, the stress–strain equation can also be reformulated as

$$\mathbf{D}\,\mathbf{C}\,\mathbf{D}_\epsilon \mathbf{u} = \mathbf{F}, \quad (5)$$

where

$$\mathbf{D} = \mathbf{D}_\epsilon^T = \begin{bmatrix} \frac{\partial}{\partial x} & 0 & 0 & \frac{\partial}{\partial y} & 0 & \frac{\partial}{\partial z} \\ 0 & \frac{\partial}{\partial y} & 0 & \frac{\partial}{\partial x} & \frac{\partial}{\partial z} & 0 \\ 0 & 0 & \frac{\partial}{\partial z} & 0 & \frac{\partial}{\partial y} & \frac{\partial}{\partial x} \end{bmatrix} \quad (6)$$

is the differential matrix, $\mathbf{F}$ is an external force vector on each node point, and $\mathbf{u}$ is the discrete displacement vector

$$\mathbf{u} = \{u_1, v_1, w_1, \ldots u_m, v_m, w_m, u_n, v_n, w_n\}^T. \quad (7)$$

Here, u is the displacement in the x-direction, v in the y-direction and w in the z-direction. If a matrix containing the shape functions of the isoparametric element is defined in the local frame as

$$\mathbf{H} = \begin{bmatrix} N_1 & N_2 & N_3 & N_4 & 0 & 0 & 0 & 0 & 0 & 0 & 0 & 0 \\ 0 & 0 & 0 & 0 & N_1 & N_2 & N_3 & N_4 & 0 & 0 & 0 & 0 \\ 0 & 0 & 0 & 0 & 0 & 0 & 0 & 0 & N_1 & N_2 & N_3 & N_4 \end{bmatrix}, \quad (8)$$

then a matrix of the derivatives can also be formulated as

$$\mathbf{B} = \mathbf{D}^T \mathbf{H} \quad (9)$$

which allows differentiation of the shape functions analytically on a local level. This finally results in the following formulation

$$\sum_e \int_{\Omega_e} \mathbf{B}^T \mathbf{C} \mathbf{B} \, d\Omega_e \, \mathbf{u} = \sum_e \int_{\Omega_e} \mathbf{H}\, \mathbf{f}^{\Omega_e} d\Omega_e + \sum_e \int_{B_e} \mathbf{H}\, \mathbf{f}^{B_e} dB_e + \mathbf{f}^P . \quad (10)$$

Here, integration on the left side is over each element domain Ω_e and the summation is over all elements used in the global domain of the geometry. On the right side is the force vector **F** split up into three parts. There are forces $\mathbf{f}^{\Omega_e}$ on the level of the domain, forces $\mathbf{f}^{B_e}$ on the boundaries of the global domain and $\mathbf{f}^P$, forces on single node points. This can also be written in short like

$$\mathbf{K}\ \mathbf{u} = \mathbf{F}, \tag{11}$$

with dimensions $[n \times n]\ [n] = [n]$,

where **K** is called the stiffness matrix since it contains Young's modulus and the derivative matrices. In the static case, where, for example, an external force is acting on the domain, this linear equation system is solved for **u** when applying a force vector **F**. This is still done most commonly with a Gauss elimination algorithm, although other algorithms have been proposed (Saad 2003). Still the Gauss algorithm seems to be the fastest with problems of structural mechanics. The above equation can also be transferred into a general eigenvalue system like

$$\mathbf{K}\,\mathbf{u} = \lambda\, \mathbf{M}\,\mathbf{u} \tag{12}$$

or in a time-dependent formulation like

$$\mathbf{M}\ddot{\mathbf{u}} + \mathbf{D}\dot{\mathbf{u}} + \mathbf{K}\mathbf{u} = \mathbf{F}\ . \tag{13}$$

Here, the mass matrix **M** is used and the dots denote derivatives with respect to time. Solving these equations can be quite complicated and iterative procedures may be used.

3 • Finite-Difference Formulation

The *FDM* is different from *FEM* as it is most commonly formulated as a time-dependent method. Nevertheless, *FDM* can also be put into a matrix form and solved like the *FEM*. On the other hand, time-dependent *FEM* problems are solved more like *FDM*. When having a time-dependent problem, the differential equation makes use of the definition of the derivative

$$\lim_{\Delta x \to 0} \frac{\Delta y}{\Delta x} = \frac{\partial y}{\partial x}\ . \tag{14}$$

Here Δx and Δy are finite differences in the x- and y-direction of a coordinate system. So for the one-dimensional problem of a string, the differential equation of the string

$$c^2 \frac{\partial^2 y}{\partial x^2} = -\frac{\partial^2 y}{\partial t^2} \tag{15}$$

becomes

$$a_{n,t} = \frac{\partial^2 y}{\partial t^2} = -c^2 \frac{-2y_n + y_{n+1} + y_{n-1}}{\Delta x^2}. \tag{16}$$

This is the discrete version of the string problem with displacements y at node points n. The above equation holds for the node point n and calculates the mass weighted acceleration $a_{n,t}$ of node point n at time point t for a spacial grid constant of Δx. From this acceleration, the velocity of the next time step with width Δt is

$$v_{n,t_+} = v_{n,t} + a_{n,t}\,\Delta t. \tag{17}$$

Finally, the new displacement at this node point is

$$y_{n,t_+} = y_{n,t} + v_{n,t_+}\,\Delta t. \tag{18}$$

This is easy to implement. It is thus a very powerful tool for time stepping algorithms. It can be enlarged to incorporate complicated couplings between differential equations, and nonlinearities are quite easy to implement, too. Nevertheless, this method of solving is an explicit one. So overflow of the system may become a problem when more complicated equation systems are involved.

4 • Turbulent $k - \epsilon$ Model of Flute-Like Instruments

Flute-like instruments, such as the transverse flute, can be looked at in terms of impedance, which represents the resistance of the flute to the blowing of the player (Fletcher et al. 1982). The definition of the impedance Z involves the pressure p required to achieve a flow velocity v into

$$Z = \frac{p}{v}. \tag{19}$$

In experiments (Coltman 1968, 1973), it was shown, that just 3.5 % of the blowing energy of the player gets into the tube[1]. A ratio of energy going into the flute, to the energy radiated into the room outside the flute, can be obtained when modeling the tube of a flute, including the space outside the flute in the region

---

[1] Actually, from this 3.5% only 2.4% are radiated by the flute, so the overall efficiency of the flute is just 0.084%, a reason for its low sound level.

of the sound hole, using an inward flow. If this ratio is in the range of 3.5 %, our model is appropriate.

When modeling the flow entering the flute, it is appropriate to use a Navier–Stokes model of a momentum balance between flow û and pressure p like

$$\partial_t \hat{u}_i + \hat{u}_j \, \partial_j \, \hat{u}_i = -\frac{1}{\rho} \, \partial_i \, p + \nu \nabla^2 \, \hat{u}_i \qquad (20)$$

where ϱ is the density of air, ν is its viscosity and the indexes i and j indicate spacial dimensions. In our two-dimensional model $i, j = 1, 2$, so u_1 and u_2 are the flow in x- and y-directions, respectively.

For a second equation, the incompressibility condition yields

$$\partial_t \, \hat{u}_i = 0 \ . \qquad (21)$$

However, the flow into the flute is turbulent in nature. The oscillation at the blow hole makes the air flow change direction, which is the cause for additional turbulent viscous damping.

When modeling a turbulent field in FEM or FDM there are two ways to proceed. One way is to use a very fine mesh normally consisting of at least one million finite elements. Then all small eddies appearing in the flow are modeled in detail. This is very expensive in terms of memory and, what is more important, in terms of computing time. An alternative method is to use only a fairly rough mesh, but include turbulence in the mathematical equations modeling the system. Using the model with millions of elements would only require the use of the normal Navier–Stokes equation, since it is understood that this equation governs all flows provided it is modeled sufficiently precisely. When using the rough mesh, the most common model in use is Kolmogorov's k-ε model (Kolmogorov, 1941). Here it is assumed, that energy coming from the main flow of blowing is transferred into turbulent eddies. These eddies are no longer modeled in detail. Instead, a turbulent energy k is defined which takes over the energy of the main flow which is assumed to be taken as laminar. Kolmogorov's idea is that these turbulent eddies are getting smaller and smaller up to a smallest eddy size. This size is determined by the damping of the flow which allows no further splitting into ever smaller eddies. The turbulent energy then is thought of as being transported from the larger to the smaller eddies, resulting in turbulent energy dissipation e on the level of smallest eddy size, which then depends on normal viscosity. Since not all structural details of the turbulence are modeled, as mentioned above, a statistical time-averaged flow is used in the Navier–Stokes equation instead of the normal laminar flow.

The flow $\hat{u}$ of the Navier–Stokes model is split here into a mean flow U and a fluctuating flow u:

$$\hat{u} = U + u . \tag{22}$$

Inserting this in the Navier–Stokes equation results in the Reynolds-Averaged Navier–Stokes equation (RANS) (Durbin and Pettersson 2001)

$$\partial_t U_i + U_j \partial_j U_i = -\frac{1}{\rho} \partial_i p + \nu \nabla^2 U_i - \partial_j \overline{u_j u_i} . \tag{23}$$

The result is a normal Navier–Stokes equation with just one additional term on the right side, the turbulent damping term. A problem with this equation is due to averaging over the directions i and j because

$$\overline{u_j u_i} \neq \overline{u_j} \ \overline{u_i} . \tag{24}$$

This results in more variables than equations. Thus, the RANS is a so-called unclosed equation. To close it, several simplifications must be assumed and experimental values must be used (for details see Durbin and Pettersson 2001).

Comparing these two models, i.e., the Navier–Stokes and the Reynolds-Averaged Navier–Stokes model, conclusions can be drawn about the importance of turbulence and turbulent damping in the case of flute-like instruments.

Figures 1 and 2 show the results for the Navier–Stokes (NS) model and the Reynolds-Averaged Navier–Stokes (RANS) model, respectively. The NS model has a split of the blowing flow into the tube and outside the tube with 43% flowing inside the tube. The RANS model on the other hand just shows 0.55 % of the

FIGURE 1 Two-dimensional Navier–Stokes flute model. The streamlines show the blowing flow starting at the embouchure hole, split into a tube part (*lower flow*) and a room part outside the flute tube (*upper part*). The split results in 43% of the flow occurring inside the tube.

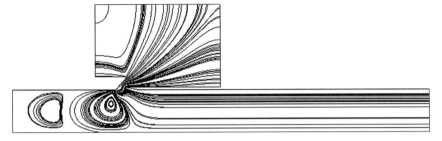

FIGURE 2 Two-dimensional Reynolds-Averaged Navier–Stokes (RANS) flute model. The streamlines again show a split flow as in Figure 1. But here, the split results in just 0.55% of flow inside the tube, which comes much closer to the experimental value of 2.4%.

flow making its way into the tube. This comes much closer to the experimental value of 2.4 %. In the Figures, this behavior can clearly be seen. Around the embouchure hole inside the tube, the NS model in Figure 1 shows less resistance (impedance) than the RANS model in Figure 2, where the flow into the tube is much more difficult. The reason for this is the additional turbulence with associated viscosity concerns, caused by the change of flow direction, which, unlike in the NS model, is not merely in one direction.

Thus, it is noted that turbulent viscous damping not only plays an important role in flute-like blowing behavior but also contributes to the reason for the flute's small amount of radiated sound energy. Additionally, by adjusting the geometry of the computer model, more effective flute impedance adjustments can be suggested to instrument builders.

5 • Multiphonics in Blown Instruments in Terms of a Split-Phase Pressure Description

Multiphonic sounds are mostly known from reed instruments like the saxophone or the clarinet (Gibiat 1988, 2000). Here, two or more tones are produced at the same time, a technique used in modern and free jazz, free improvisation, and contemporary art music. There are four ways of producing multiphonic sounds.

- overblowing
- underblowing
- complex sound hole closing patterns (fingerings)
- inhomogeneity in instrument geometry.

As mentioned above, blown instruments show abrupt phase changes for a linear change of a control parameter. Most prominently, the player's blowing pressure determines whether a "normal" tone is produced or whether the instrument is overblown (increased pressure) or underblown (decreased pressure). Normally, overblowing such an instrument causes it to sound with a tone of its overtone structure. But reed instruments have a second very important parameter of tone production, which is the pressure of the lips of the player on the reed. Since a change of this lip pressure with constant blowing pressure can change the pitch of a tone (up to three semitones) for each sound hole closing pattern (each played note), the lip pressure needs to be adjusted by the player in order not to detune the desired pitch. Now, when overblowing with increased blowing pressure, the lip pressure also has to be increased. If this is not done, and overblowing is done with normal lip pressure, the saxophone or clarinet will sound with more than one pitch. The same thing happens when the instrument is underblown.

Multiphonics may also occur with inhomogeneity of the instrument geometry. An example is the Chinese dizi flute. It has a membrane covering its highest sound hole. This membrane is not too tightly fixed, so it can vibrate to a certain extend. By coupling of this membrane to the air tube of the flute, a perfect bifurcation scenario occurs. The spectrum of the produced time series of radiated sounds by the flute consists of frequencies for which the following relationship holds

$$f = n f_1 \pm m f_2 \text{ with } n = 1, 2, 3 \ldots \text{ and } m = 0, 1, 2, 3 \ldots . \quad (25)$$

This is the formula to which Backus refers for multiphonic sounds (Backus 1978). When m is zero, we have a normal harmonic overtone spectrum. With $m > 0$, two additional frequencies are obtained, differing from f_1 by $\pm m f_2$, which are reduced in amplitude. Since for each $n f_1$, these additional frequencies occur, not only additional inharmonic components are present, but harmonic overtone structures, consisting of frequencies like

$$f(k) = k f_1 \pm k f_2 \text{ with } k = 1, 2, 3 \ldots . \quad (26)$$

For example, if $f_1 = 90$ Hz and $f_2 = 10$ Hz, then in the plus case the frequencies are $f(k) = 100, 200, 300$ Hz....

The formula used by Backus holds perfectly in the case of inhomogeneous geometry, but it is not always appropriate in the case of complex sound hole closing patterns. Here, normally two tones are heard very prominently with additional inharmonic

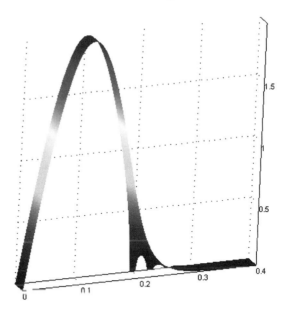

FIGURE 3 Pressure distribution of the first eigenfrequency of a tube with six sound holes open at the right end of the tube.

components around them (Keefe and Laden 1991). Inasmuch as most of the closing patters have many sound holes closed with an additional high-pitched open hole, the modeling of the tube in terms of eigenfrequencies gives insight into the pressure distribution along the tube for such complicated closing patterns.

Figure 3 shows the pressure distribution of the first eigenfrequency of a tube closed at both ends with six sound holes open at the right side of the tube. Note that the sound holes are just open at the upper side of the tube. Because of that, the pressure is not zero at the first open sound hole, but still has a finite value up to about the third open hole. This enlarges the size of the tube up to the sound hole a bit, so compared to the lowest possible frequency of the tube with all sound holes closed, the first overtone shown in Figure 3 is not in a perfect relation of 2:1 to the fundamental frequency.

Figures 4 and 5 now show the cases for tone hole closing patterns typical for multiphonic sounds. Of the six tone holes at the right side of the tube, the lower five are closed and just the highest tone hole is still open. First it is noted that the pressure is distributed over the entire tube. This is because the tone hole is just open at one side of the tube on the upper end. But this distribution can appear in two cases within just one eigenfrequency

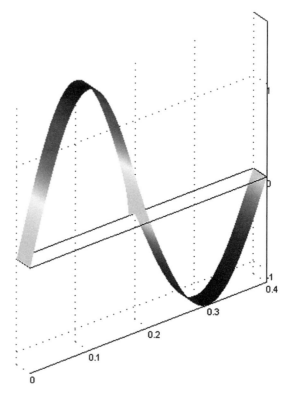

FIGURE 4 Pressure distribution of the anti-phase case of the first eigenfrequency of a tube with all sound holes closed except for one sound hole in the middle of the tube.

pressure distribution: one anti-phase, as shown in Figure 4, where the pressures of the two parts of the tube are of opposite sign; and one in-phase, as shown in Figure 5, where the pressures are of the same sign in the left and right part of the tube. Since this holds for every eigenfrequency of the tube, where at first there was only one frequency with each overtone, there now are two.

Table 1 shows the frequency ratios of the two cases of anti-phase and in-phase pressure distribution of the multiphonic playing technique for each partial. The ratios were calculated in relation to the fundamental frequency of each case. The anti-phase ratios correspond to the case of normal playing without complex sound hole closing patterns. It shows a harmonic overtone structure as expected. But astonishingly, the in-phase ratios of the additional frequencies caused by the multiphonic pattern form within themselves again a harmonic overtone structure. So here, two tones can be heard, which are in an

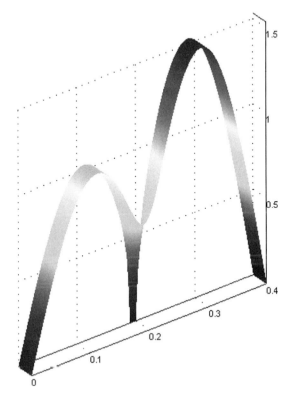

FIGURE 5 Pressure distribution of the in-phase case of the first eigenfrequency for a tube with all sound holes closed except for one sound hole in the middle of the tube.

TABLE 1 Frequency ratios for the anti-phase and in-phase cases, with each eigenfrequency of a tube played with a typical multiphonic sound hole closing pattern. The values are the relations to the fundamental frequency of each case. Not only the anti-phase relations form a harmonic overtone structure as in the normal playing condition but also the additional modes caused by the multiphonic playing technique form for themselves a harmonic structure and so an additional tone.

Partial	Anti-phase frequency ratio	In-phase frequency ratio
1	1.0	1.0
2	1.99966	2.02777
3	2.99984	3.08954
4	3.99972	4.15899

inharmonic relationship to one another. This is the case with multiphonic sounds caused by these complicated tone hole closing patterns, where two tones are heard most prominently.

6 • Time-Dependent Model of a Classical Guitar Body

Time-dependent solutions of complex geometries as shown here with a complete model of a classical guitar (Bader 2003, 2005) lead to a description of the complex development of the initial transient, (For the acoustics of the classical guitar see Fletcher and Rossing 1999, Elejabarrieta et al. 2002, Richardson and Roberts 1985, Meyer 1980). This complexity mainly comes from the inhomogeneous geometry of the guitar body which includes the top plate, the back plate, the ribs, the neck, the enclosed air, and the strings. Detailed modeling must of course include the bracing of the top and back plates, with particular attention to the top plate. However, considerable information about the role of each element can be obtained without that additional resolution. Interaction between these parts cannot be examined experimentally in their entirety. I.e., it is not possible to measure the in-plane waves of the plates. These waves are not radiated, but at discontinuities, like the boundaries of the plates, they are transformed into bending waves which are radiated. So a detailed knowledge of initial tone development leads to a variety of completely new insights into roles of the different guitar parts in the radiated sound. Furthermore, geometrical changes in terms of tone quality improvement can be examined and used by instrument builders.

In Figure 6 a classical guitar is shown as modeled by a Finite-Difference technique. The model distinguishes between bending waves with displacements in the direction normal to the guitar plates (top plate, back plate, ribs...), and so are radiated, and in-plane waves, which are nearly negligible as far as radiation is concerned. Inasmuch as these waves interact within the plates and at the boundaries, most prominently between top and back plate and ribs, and since, in-plane waves propagate about 8-12 times faster than bending waves, the overall development of the initial transient depends on these two coupled waves as described in Section 7. In addition, the ribs are looked at as a curved plate, which is shown in the coordinate system of Figure 6. This again leads to a steady transformation between bending and in-plane waves effecting the eigenfrequencies of the ribs and the temporal development of the sound radiated by the ribs.

Figure 7 shows the time series of the different parts of a classical guitar calculated by the Finite-Difference model. The

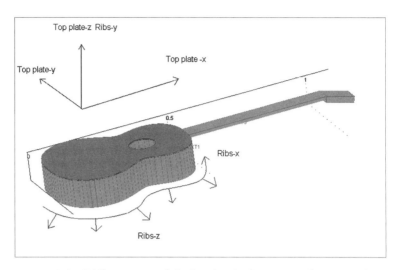

FIGURE 6 Finite-Difference model of a classical guitar with its coordinate system. The modeling of the ribs as a curved plate gives insight into the coupling between radiated bending and non-radiated in plane waves.

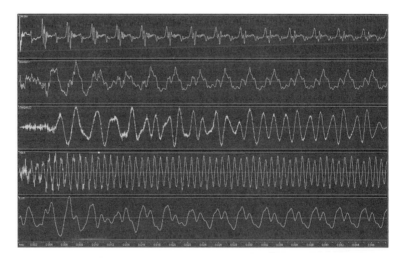

FIGURE 7 Time series calculated by the Finite-Difference model for the different guitar parts within the initial transient as radiated from these parts and recorded 1 m in front of the guitar. From top to *bottom*: top plate, back plate, ribs, neck, enclosed air. The roles of the different guitar parts in the overall sound show up clearly (see text for details), which can also be checked by ear.

open high e-string was virtually plucked (displaced in a trapezoidal shape) and the guitar vibrated "in slow motion" on the computer screen. This visualization aid is important for giving insight into the complex guitar motion since the number of degrees of freedom is too high to find the places and times of interest by pure numerical evaluation. At each iteration time point, the radiation from the different guitar parts was calculated by integrating over these parts with respect to a virtual microphone position 1 m in front of the guitar.

The different roles of the guitar parts in the initial transient show up clearly. The top plate is the only part which reacts nearly instantaneously. When listening to these calculated sounds, the top plate is the only one with a clear, hard attack, and since its sound is very bright, it is perceived as a hard and direct timbre. The back plate and the ribs need some time to come in. In the beginning, they still show some low amplitude vibrations caused by the coupling of the radiated bending waves to the in-plane waves. Although these initial vibrations are low in amplitude, they can still be heard because the frequencies fall in a very sensitive range of the human ear (around 1,000 Hz). The main attack of the back plate and the ribs is caused by the energy transport from the enclosed air, which is much slower in motion and so takes some time to displace the back plate and the ribs in the directions, which can cause radiation. The back plate and the ribs are prominent in a mid-frequency region and therefore responsible for what musicians call the tone color, the spiciness or taste of the sound. Next, the guitar neck is very high pitched as expected. But still this sound is needed in the complete guitar tone. Without it, some brilliance is missing. Finally, the enclosed air is responsible for a kind of boost or pressure impulse in the guitar sound, which can be heard by placing an ear (or microphone) next to the sound hole of the guitar. Some guitar recordings in the sixties or seventies (e.g., by Segovia) use this recording technique to achieve a very "earthy" or pumping sound.

Another aspect of the initial transient of a classical guitar is inharmonicities as discussed in Section 1.1 above about musical timbre. The pluck of a guitar string causes an impulse train which acts on the top plate. The first of these impulses is like a knocking on the plate, since no periodicity is established at this early point in time. This knocking, a kind of impulse, will drive many eigenfrequencies of the guitar body. Only because the impulse train has the constant periodicity of the played tone, the guitar body is forced to take over the string frequencies, and the eigenmodes of the body will disappear. Yet during the initial transient, some eigenfrequencies, namely low ones, appear and interact with the frequencies of the string in an interesting manner.

Figure 8 shows a wavelet transformation (Haase et al. 2002) of a guitar tone (e_4) plucked on the high open E-string calculated by the Finite-Difference model. The temporal development of the frequencies is typical for an initial transient of a classical guitar. The fundamental and the first three overtones can clearly be seen. The typical part of a plucked instrument's initial transient is the disturbance of the string's fundamental frequency component by an eigenmode of the guitar body at the tone beginning. The component's amplitude drops down and comes up again. This is caused by a guitar body resonance frequency absorbing the energy of the string fundamental, as these two frequencies are close to each other. When the guitar body has taken the string mode energy, the guitar mode has to break down as well because it is driven by this string energy. As more than one eigenmode is driven during the initial phase, there is a kind of noise at the beginning of the tone, which can be perceived as "knocking". Indeed, if one knocks on the guitar top plate, the resulting sound

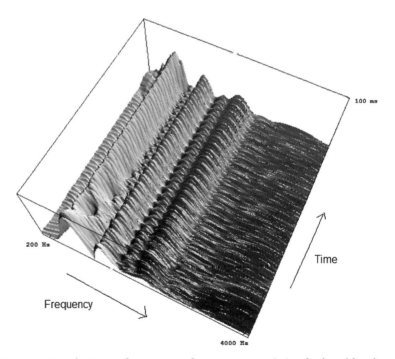

FIGURE 8 Wavelet-Transformation of a guitar tone (e_4) calculated by the Finite-Transient model.

is very similar to the noise during the guitars initial transient. So the eigenfrequencies of the guitar body are not only needed to amplify the string, but they also include a typical aspect of the sound, giving each guitar its own character. And, since we do not expect the frequencies driven by knocking to change, we do find that the development of the initial transient shown in Figure 8 for a particular classical guitar at a given pitch nearly always looks exactly the same, no matter which dynamic or articulation the player chooses.

So, in conclusion, since the initial transient is very important for the quality and character of the guitar, the exact development is of interest here. The complex structure of an instrument with ribs leads to an initial transient where different parts come in, one after another, and so enlarge the whole tone's beginning, causing the guitar to sound more rich and interesting. In addition, the first impulse of the string on the top plate drives many eigenfrequencies of the guitar body during the initial transient. As a rule of thumb, it can be said that the more complex the geometry of an instrument, the more complex the initial transient. This may be a reason for the development of plucked instruments with ribs. Lutes, widely used in ancient times and still in renaissance music, miss this complex initial phase and thus sound more light and free.

7 • Guitar Back-Plate Timbre Brightening by Stress Bending

Another technique for studying musical instruments is the use of impulse response. This corresponds to the knocking technique used by instrument builders, especially with the violin and the guitar, where a knock on the different plates produces their eigenfrequencies. Here, an example of a guitar back plate shall be shown. This flat wooden plate is built a little larger than the geometry of the ribs allow. When added to the ribs, it has to buckle out to fit in and so is under stress. It is claimed by guitar builders that this technique causes an increase of brightness of the guitar sound.

A possibility to model such buckling is the coupling between bending and in-plane waves in the back plate. This was discussed in the previous section. As the plate is too large in the in-plane directions, it accommodates the dimensional discrepancy by deformation in the transverse direction. The stresses applied to the in-plane direction, on the one hand, causes a strain in this in-plane direction, and, on the other hand, a strain in the bending direction.

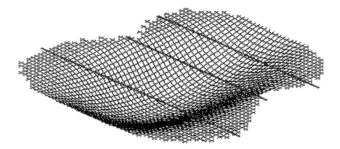

FIGURE 9 Curved back plate of the classical guitar Finite-Difference model. For this geometry, the coupling between bending and in-plane waves leads to an increase of the eigenfrequencies and also to an increased brightness as measured by an impulse response calculation.

If this model is appropriate, an increase of both the eigenfrequencies of the system and its brightness should result.

In Figure 9, the back plate is shown in its equilibrium position as curved inward by an applied force in the in-plane direction at the boundaries. The applied coupling results in the desired curvature. As the model is time-dependent, an impulse can be applied to the back plate. This impulse has a continuous spectrum with all frequencies present. The radiation from the back plate will then produce the eigenfrequencies of the plate. The amplitudes of these frequencies depend on the place of knocking, but the frequencies themselves do not change. However, when comparing the flat back plate to the curved, the place of knocking was held constant.

Table 2 shows the frequencies of the first five modes of the flat and buckled back plate. The buckled values are higher, as expected. Also the maximum spectral centroid value, shortly after the impulse was applied, was increased from 905 Hz for the flat case to 1,430 Hz with the buckled case. This finding was also expected from the experience of instrument builders and the difference in brightness and frequency can clearly be heard when listening to the calculated impulse responses.

TABLE 2 Frequencies of the first five modes of the back plate flat and buckled. The buckled values are higher

Mode	Back plate flat	Back plate buckled
1	229.44	239.04
2	420.48	439.68
3	480.96	500.16
4	652.8	681.6
5	912.0	931.2

8 • Church Bell Clapper Time-Dependent Striking Force

A bell clapper striking a church bell is a contact problem in terms of a Finite-Element Formulation (Wriggers 2002). As this is a time-dependent problem, the contact area between the clapper and the bell needs to be changed for each time step. At first, the clapper is not in contact with the bell. After the first contact, both objects are deformed and so the contact area grows larger. As soon as the velocity of the clapper changes direction and the clapper starts to move from the bell again, the contact area grows smaller until the clapper leaves the bell.

So, for each geometry, there is an external force $F_{external}$, which depends on the deformation and therefore on the distance A between the objects. Thus the contact condition can be formulated as:

$$F_{external} \begin{cases} f(\mathbf{u}), & \text{with contact } A \leq 0; \\ 0, & \text{without contact } A > 0; \end{cases} \qquad (27)$$

Here, during contact, the force generally is a function $f(\mathbf{u})$ of the displacement $\mathbf{u}$.

It has been shown that the contact of a church bell and its clapper is not quite a Hertzian contact and that the contact time is about 1 ms, which is surprisingly small (Fletcher 2002). The resulting impulse function in time of the clapper on the bell is of importance because it determines the amplitudes of the eigenfrequencies of the church bell being driven. An ideal Dirac delta impulse would drive all possible frequencies with the same amplitude. The more the impulse function flattens, the fewer higher frequencies are present.

In Figure 10 a schematic clapper, together with a part of a bell wall is shown. The clapper hits the bell with a speed of 100 m/s moving toward the left. The impulsive time function of the impact, integrated over the contact area of the clapper–bell, is shown in Figure 11, along with its spectrum in Figure 12. It can be seen, that the contact time is just a bit longer than 1 ms as measured in experiments.

In Figure 11, the contact time sequence is divided into two parts, left and right of the maximum. Inserted into the drawing are the values obtained when integrating over time for each segment. This clearly shows a greater value for the approach time than for the separation time. This seems to be because when the impulse function has passed its maximum, the bell rim moves with the clapper causing the impulse to be weaker. Figure 12 shows the spectrum of the impulse function. It has a periodicity of $1/T$ if T is the contact time of about 1 ms. Such a periodic spectrum is known from string instruments, where the periodicity

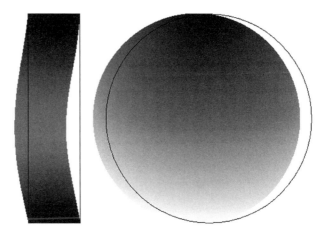

FIGURE 10 Schematic view of a clapper (right) acting on a bell rim (left). The clapper is shown here with a certain distance for clarity. The clapper moves left with an initial velocity and deforms the bell rim until the clapper slows down and moves back again. The overall process takes about 1 ms. The frame shown here corresponds to 0.7 ms after the beginning of the strike.

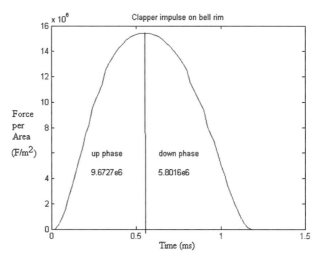

FIGURE 11 Clapper impulse time function on the bell rim integrated over the bell rim height. The impulse length is about 1 ms as expected. The impulse itself is stronger on the left side than on the right side, which can be seen when integrating the force over the area of the two sides as shown in the figure.

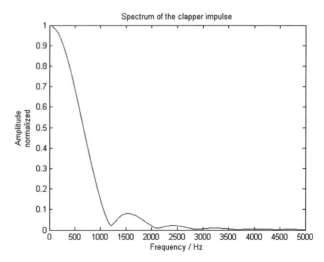

FIGURE 12 FFT spectrum of the clapper–bell rim contact of Figure 11. The spectrum shows a periodicity which comes from the finite length of the impulse forbidding frequencies with positive and negative amplitude parts being distributed equally over the whole impulse function time length. When performing the FFT, these frequency time series are multiplied (or weighted) by the sinusodial. As the impulse function is all positive in amplitude and the sinusodial has equally large positive and negative amplitude parts, the resulting amplitude of the frequency is zero.

depends on the point of plucking (Fletcher and Rossing 1999, p. 42). This seems to be because frequencies having equal positive and negative amplitude values over the impulse time cannot contribute to the final shape of the FFT plot, as their convolution integral over the impulse time, the Fourier integral over the impulse function is zero. As the bell can be thought of as a resonator in terms of impulse response, just the frequencies of the impulse can be excited. But this does not mean that the bell spectrum has to be of the same periodic shape as the clapper impulse FFT function as coupling between modes can occur at places of discontinuities and so can enlarge the spectrum.

References

Backus, J.: Multiphonic tones in the woodwind instruments. J. Acoust. Soc. Am. 63, 591–599, 1978.

Bader, R.: Fractal correlation dimensions and discrete-pseudo-phase-plots of percussion instruments in relation to cultural world view. In: Ingenierias, Octubre-Diciembre 2002, Vol. V. No. 17, 1–11.

Bader, R.: Physical model of a complete classical guitar body. Proc. Stockholm Musical Acoustics Conf. 2003, R. Bresin (ed.) Vol. 1, 121–124, 2003.

Bader, R.: Computational Mechanics of the Classical Guitar. Springer 2005.

Bathe, K.-J.: Finite Element Procedures. Prentice Hall, 1996.

Coltman, J.W.: Acoustics of the flute. Physics Today, 21, 11, 25–32, 1968.

Coltman, J.W.: Mouth resonance effects in the flute. J. Acoust. Soc. Am. 54, 417–420, 1973.

Durbin, P.A. & Pettersson, R.: Statistical Theory and Modeling for Turbulent Flows. Wiley, 2001.

Elejabarrieta, M.J., Ezcurra, A. & Santamaria, C.: Coupled modes of the resonance box of the guitar. J. Acoust. Soc. Am. 111, 2283–2292, 2002.

Fletcher, N.H.: Bell clapper impact dynamics and the voicing of a carillon. J. Acoust. Soc. Am. 111, 1437–1444, 2002.

Fletcher, N.H. & Rossing, Th.D.: The Physics of Musical Instruments. Second Ed. Springer 1999.

Fletcher, N.H., Strong, W.J. & Silk, R.K.: Acoustical characterization of flute head joints. J. Acoust. Soc. Am. 71, 1255–1260, 1982.

Gibiat, V.: Phase space representations of acoustical musical signals. J. Sound and Vibration, 123/3, 529–536, 1988.

Gibiat, V. & Castellengo, M.: Period doubling occurences in wind instruments musical performance. Acustica, 86, 746–754, 2000.

Grey, J.M.: Multidimensional perceptual scaling of musical timbres. J. Acoust. Soc. Am. 61(5), 1977, 1270–1277, 1977

Grey, J.M. & Moorer, J.A.: Perceptual evaluations of synthesized musical tones. J. Acoust. Soc. Am. 62(2), 454–462, 1977.

Haase, M., Widjajakusuma, J. & Bader, R.: Scaling laws and frequency decomposition from wavelet transform maxima lines and ridges. In: Emergent Nature, M.M. Novak, (ed.): World Scientific 2002, 365–374.

Hughes, J.R.: The Finite Element Method. Linear Static and Dynamic Finite Element Analysis. Dover Publications, 1987.

Keefe, D.H. & Laden, B.: Correlation dimension of woodwind multiphonic tones. J. Acoust. Soc. Am. 90(4), 1754–1765, 1991.

Kolmogovov, A.N.: The local structure of turbulence in incompressible viscous fluid for vary large Reynolds number. Doklady Akademi Nauk SSSR, 30, 301–305, 1941.

Krumhansl, C.L.: Why is musical timbre so hard to understand? In: Structure and Perception of Electroacoustic Sound and Music. Nielzn, S. & Olsson, O. (ed.). Amsterdam 1989, 43–53, 1989.

Luce, D. & Clark, M.: Durations of attack transients of nonpercussive orchestral instruments. J. Audio Eng. Soc. 13, 194–199, 1965.

Meyer, J.: Verbesserung der Klangqualitt von Gitarren aufgrund systematischer Untersuchungen ihres Schwingungsverhaltens *Improvement of the sound quality of guitars by systematic investigations of its vibrating behaviour.* Physikalisch-Technische Bundesanstalt Braunschweig, Forschungsvorhaben Nr. 4490, 1980.

Miller, J.R. & Carterette, E.C.: Perceptual space for musical structures. J. Acoust. Soc. Am. 58(3), 711–720, 1975.

Richardson, B.E. & Roberts, G.W.: The adjustment of mode frequencies in guitars: A study by means of holographic interferometry and finite element analysis. In: Proc. Stockholm Musical Acoustics Conf. 1983, 285–302, 1985.

Saad, Yousef: Iterative Methods for Sparse Linear Systems, Cambridge University Press, 2003.

Taylor, Ch.: Sound. In: The New Groove Dictionary of Music and Musicians, S. Sadie (ed.). Second ed. 23, 759–776, 2001.

Wriggers, P.: Computational Contact Mechanics. Wiley, 2002.

PART V
SPEECH

Douglas D. O'Shaughnessy

INRS-Telecommunications, Montreal, Canada

27 Display and Analysis of Speech
Ben Milner 449

Introduction • Speech Production, Sampling and Quantisation • Display of Speech Waveforms • Analysis of Speech Waveforms • Conclusion

28 Estimation of Speech Intelligibility and Quality
Stephen Voran 483

Introduction • Estimators of Speech Intelligibility • Estimators of Speech Quality in Telecommunications • Conclusion

29 Gaussian Models in Automatic Speech Recognition
Jeff Bilmes 521

Introduction • Background • Training Regularized Gaussian Decompositions • Training Shared Regularized Gaussian Decompositions • A Generalized EM Procedure • Conclusions

30 Speech Synthesis
Thierry Dutoit and Baris Bozkurt 557

Concatenative Text-to-Speech Synthesis • Prosody Modification • Smoothing • Estimation of Unit Selection Costs • Voice Quality/Speaker Modification and Other Open Issues • Further Reading and Relevant Resources

31 Speech Coders
Roch Lefebvre and Philippe Gournay 587

Introduction • Speech Coding Concepts • Issues Related to Input Conditions • Issues Related to Channel Conditions • Current Developments and Future Challenges

27

Display and Analysis of Speech

Ben Milner

School of Computing Sciences, University of East Anglia, Norwich, Norfolk, UK

1 • Introduction

The three most important speech processing applications are speech recognition, speech synthesis and speech coding. The display and analysis of speech signals provide knowledge of the important characteristics of speech, facilitating the design of speech processing algorithms for these applications. For example, examination of speech signals allows important properties to be identified which can subsequently be used to discriminate between different speech sounds, which is necessary in both speech recognition and speech synthesis applications. Similarly, structures in speech allow parametric models of speech production to be formulated that are fundamental to recognition, synthesis and coding.

The primary acoustic correlates of speech are the voicing (whether the speech is voiced or unvoiced), the fundamental frequency (for voiced speech) and the frequency and shape of the formants. The extraction of these features has been the subject of research for many years and has led to the development of many forms of speech analysis. Of these, the three most commonly used are linear predictive coding (LPC) analysis, frequency domain analysis and cepstral analysis. These methods all enable voicing, fundamental frequency and formant analysis although it is important to note that none of them are guaranteed to make error-free estimates of these features.

Some speech processing applications make use of these acoustic correlates directly. For example, both speech synthesisers and very low bitrate speech codecs require accurate fundamental frequency estimates in order to produce realistic sounding speech. In contrast to this, many other speech processing applications use derivatives of these fundamental acoustic features. For example, LPC analysis is frequently used in low-bitrate speech coding algorithms to provide vocal tract information rather than specific formant parameters. Similarly, speech recognition systems often extract vocal tract information using cepstral analysis but with a frequency scaling in line with the frequency response of human hearing.

This chapter examines three domains for displaying speech signals: time domain, frequency domain and time–frequency domain. Speech signals are shown in each of these domains and important features identified in each. Three common methods of speech analysis are then discussed. The discussion begins with LPC analysis and its application to both fundamental frequency and formant estimation. Frequency-based analysis is then reviewed and finally cepstral analysis. Finally, this chapter briefly discusses other features that can be extracted from speech signals that are applicable to speech processing applications such as recognition and coding.

However, before the discussion of the display and analysis of speech signals, the next section describes the procedure for converting an acoustic speech signal, as generated by the human speech production mechanism, into a digital signal. This is the necessary first stage of speech processing that a signal must go through to facilitate analysis by computer.

2 • Speech Production, Sampling and Quantisation

A speech signal is generated by the vocal organs in the human speech production mechanism which causes pressure fluctuations in the surrounding air. The speech begins as compressed air produced by the lungs, which in turn passes through the glottis where it can be altered in several different ways. If the vocal cords are under tension, the airflow will cause them to vibrate (at the fundamental frequency) and air will be released in the form of a quasi-periodic sequence of glottal pulses. If the vocal cords are in a relaxed state, this will result in turbulent airflow which generates wideband noise. The distinction between these two modes of operation is not always clear and can result in a mixture of the two forms of signal being produced. A final mode of operation is when the vocal cords close completely and temporarily shut off the airflow. This is followed by a sudden release of air and leads to glottal stop sounds such as the phonemes /g/ and /k/.

The airflow emitted from the glottis is termed the excitation signal and passes through the vocal tract before being expelled as speech from the mouth, nose or both. Cavities in the vocal tract modulate the excitation signal and give rise to resonances, or formants, which define the particular sound being produced. Our ears respond to these pressure fluctuations through the movement of the ear drum. This in turn stimulates hair cells on the basilar membrane of the inner ear which causes neural impulses which we perceive as sound.

For computer-based processing of speech, a microphone is employed to receive the acoustic waveform and convert it to a time-varying voltage. This signal is continuous in both time and amplitude which is suitable for input into analogue devices, such as a tape recorder, but not for processing by computer. To convert this analogue signal into a digital form it must be sampled in time and quantised in amplitude. The remainder of this section discusses the sampling and quantisation processes with application to speech signals.

2.1 Sampling

The sampling operation converts the continuous time analogue signal into a series of regularly spaced amplitude samples. From an analogue signal, $x_a(t)$, the sampling operation can be represented as a multiplication with an impulse train:

$$x_s(t) = \sum_{n=-\infty}^{\infty} x_a(t) \delta(t - nT). \quad (1)$$

The resulting sampled signal, $x_s(t)$, can then be converted into a discrete-time signal, $x(n)$, as

$$x(n) = x_s(nT). \quad (2)$$

An important consideration is the selection of the sampling period, T, which relates to the sampling frequency as $f_s = 1/T$. The sampling frequency places an upper limit on the bandwidth of the sampled, or discrete-time, signal. If a bandwidth of f_b Hz is required of the sampled signal, then the sampling frequency must be at least double that, i.e.

$$f_s \geq 2f_b. \quad (3)$$

The frequency $2f_b$ is also referred to as the Nyquist frequency and indicates the minimum sampling rate necessary to be able to completely reconstruct the signal up to f_b Hz. The choice of sampling frequency is largely governed by the application. For example, telephony speech is sampled at 8 kHz to give a bandwidth of 4 kHz. Higher quality speech applications often increase the sampling frequency to 11 or 16 kHz. For the high-fidelity audio encoded on CD, the sampling rate is 44.1 kHz to give a bandwidth of 22.05 kHz. The downside of a high sampling frequency is the additional storage or processing necessary for the samples.

2.2 Quantisation

After sampling, the continuous-time signal has been converted into a discrete-time signal in the form a series of regularly occurring pulses of continuously valued amplitudes. To complete the digitisation process, these amplitude values must be quantised, which involves dividing the amplitude range of the signal into a set of M discrete levels. Assuming the dynamic range of the signal is from $-V$ to $+V$, then for a linear quantiser, the quantisation step size will be $\Delta = 2V/M$. To perform the quantisation, a set of $M+1$ decision boundaries, $\{b_0, b_1, \ldots, b_M\}$, is needed together with a set of M reconstruction levels, $\{y_1, y_2, \ldots, y_M\}$. For linear quantisation, the lowest decision boundary is set to the minimum amplitude of the signal, $b_0 = -V$. The remaining boundaries are separated by the quantisation step size to give the final boundary at the maximum amplitude level, $b_M = +V$. The reconstruction levels are normally set to the mid-point between two decision boundaries:

$$y_i = \frac{b_i - b_{i-1}}{2} \quad 1 \leq i \leq M. \tag{4}$$

Finally, the quantised amplitude, $x_q(n)$, is obtained:

$$x_q(n) = y_i \quad \text{iff} \quad b_{i-1} < x(n) < b_i. \tag{5}$$

The difference between the quantised and real amplitude values is termed the quantisation error, $q(n)$, and is related to the number of quantisation levels,

$$q(n) = x(n) - x_q(n), \tag{6}$$

and lies in the interval $[-\Delta/2, +\Delta/2]$. If it is assumed that the signal $x(n)$ is uniformly distributed, then the quantisation error is also uniform over this interval with probability density function

$$f_q(q(n)) = \frac{1}{\Delta}, \tag{7}$$

and the quantisation noise power, σ_q^2 can be computed:

$$\sigma_q^2 = \frac{1}{\Delta} \int_{-\Delta/2}^{\Delta/2} q^2(n)\, dq(n) \qquad (8)$$

$$= \frac{\Delta^2}{12}.$$

Given that the signal lies in the interval $[-V, +V]$ and again assuming a uniform PDF, the signal power, σ_x^2, can be computed in a similar way:

$$\sigma_x^2 = \frac{1}{2V} \int_{-V}^{V} x^2(n)\, dx(n) \qquad (9)$$

$$= \frac{V^2}{3}.$$

From the quantisation noise power and signal power, the signal-to-quantisation noise ratio (SQNR) can be calculated:

$$\text{SQNR(dB)} = 10\,\log\left(\frac{\sigma_x^2}{\sigma_q^2}\right) = 10\,\log\left(\frac{V^2}{3}\frac{12}{\Delta^2}\right). \qquad (10)$$

Substituting into this the relationship between the number of bits used in quantisation, b, and quantisation step size, $\Delta = 2V/2^b$, leads to the following expression for the SQNR:

$$\text{SQNR(dB)} = 6.02b\,\text{dB}. \qquad (11)$$

This shows that the SQNR is raised by 6.02 dB for each additional bit used in quantisation. However, it should be noted that in the derivation of this result the assumption of the speech amplitudes having a uniform distribution is far from realistic. This will be demonstrated later in this section. For speech processing applications, 8-bit quantisation is about the lowest acceptable level without introducing too much quantisation noise. For higher quality applications, such as CD encoding, 16-bit quantisation is generally used. An interesting experiment is to quantise a speech signal using only a single bit to give two quantisation levels (set at, for example, $-V/2$ and $+V/2$). Surprisingly, the resulting speech signal is still intelligible although of very poor quality due to the high levels of quantisation noise.

Several improvements can be made to the basic uniform quantiser which exploit the nature of the speech signal to improve

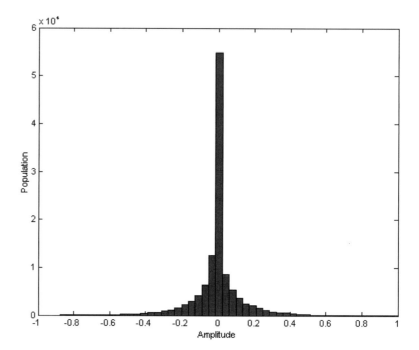

FIGURE 1 Histogram of speech sample amplitudes.

the bitrate to SQNR trade-off. Figure 1 shows a histogram of speech amplitudes taken from several sentences of conversational speech. The distribution of amplitude levels takes a Gaussian-like shape with considerably more small amplitude samples than larger amplitude samples. The large number of samples with close to zero amplitude arises from silence periods in the speech signal.

This implies that a uniform spacing of quantisation levels is not optimal for encoding a speech signal. A better quantiser would therefore use a non-linear spacing of levels which have a closer distribution at low-amplitude regions and coarser distribution at larger amplitude levels. Such quantisers are termed non-linear quantisers and can be designed by methods such as the Lloyd-Max algorithm which iteratively determines the optimal position for the decision boundaries and reconstruction levels based on an estimate of the probability density function (PDF) of the signal.

An alternative to having a set of non-linearly spaced quantisation levels is to apply a logarithmic compression to the amplitude samples. This transforms the amplitude levels into a more uniform distribution which allows standard linear quantisation to be applied without introducing large quantisation errors for larger amplitude signals. Upon de-quantisation a logarithmic expansion is applied to return the amplitude values to the correct range. This is known as *companding* (compression-expanding) and gives a significant increase in SQNR. Several standard companders exist, such as A-law and µ-law, and these are employed in telephony-quality speech to give a SQNR using only 8 bits which is comparable to that obtained using 12-bit linear quantisation.

3 • Display of Speech Waveforms

This section examines three methods of displaying a speech signal, namely time domain, frequency domain and time–frequency domain. In each domain, particular attributes of the speech signal are identified and relationships shown between the different representations.

3.1 Time Domain Representation

The time domain representation is the most straightforward method of displaying a speech signal and shows the amplitude of the signal against time or sample number. Figure 2 illustrates the time domain representation of the speech waveform "*I think that you air your ideas very challengingly*", which has a duration of about 3 s.

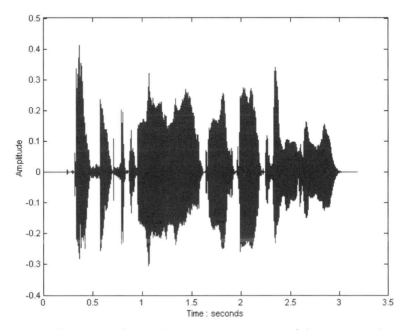

FIGURE 2 Three-second time domain representation of the utterance "*I think that you air your ideas very challengingly*".

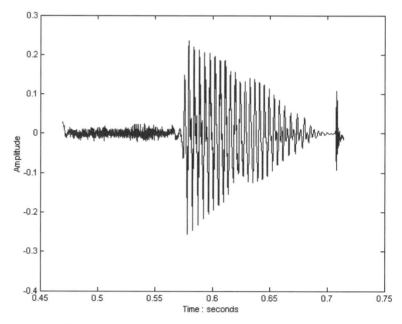

FIGURE 3 0.25-second duration time domain representation of the word *"think"*.

The long time duration of this signal means that fine details of the speech waveform cannot be seen, but instead a good representation of the energy contour is present which provides a useful indication of silence and speech periods. For example at the beginning and end of the utterance can be observed silence periods of about 0.25 s each. Very short silences can also be seen between some of the words in the utterance. To contrast this, Figure 3 shows a much shorter duration segment, taken from the utterance shown in Figure 2, of the word *"think"*. This begins at 0.45 s and lasts for about 0.25 s.

More detail is now visible and in particular the different regions of the speech waveform which correspond to the phonemes which make up the word. The unvoiced phoneme /th/ is recognisable at the start of the word as a noise-like signal. This is followed by the two voiced vowels, /ih/ and /ng/. Their voiced nature is identified by the pitch pulses which dominate that region of the signal. In fact an examination of the time domain waveform in this region would allow measurement of the pitch period. However, the time domain representation does not allow the boundary between the two phonemes to be easily located (this actually occurs at 0.63 s). At the end of the waveform, the structure of the unvoiced stop phoneme /k/ is evident. The beginning of the phoneme /k/ is identified by no speech energy resulting from the closure of the glottis in order to build up sub-glottal pressure. The sudden release of air pressure as the glottis opens is also clearly evident at the end of the phoneme.

In general, the time domain waveform is useful for identifying different classes of speech sound (voiced, unvoiced, silence) but shows insufficient detail to discriminate between phonemes within a particular class. Measurement of the pitch period is also possible from the time domain representation through the location of pitch pulses in the waveform.

3.2 Spectral Domain Representation

The time domain representation of speech enables certain aspects of the signal to be identified, but it is more common to transform the signal into the frequency (or spectral) domain for examination. Historically, this conversion to the frequency domain has been implemented using a series of band-pass filters. The subsequent amplitude outputs of these filters provided a measure of energy in the various frequency regions of the signal. With digital signal processing, an easier method is to transform the signal into the frequency domain using a discrete Fourier transform (DFT). In applying the DFT to a signal, an assumption is made that the signal is stationary. As Figures 2 and 3 have shown, speech is highly non-stationary, and therefore to satisfy this stationarity assumption of the DFT, only a short-time window of the signal should be transformed. The length of time the speech signal remains stationary varies and depends on the particular speech sound under analysis. In general, window widths in the range 10–32 ms are used in speech processing applications.

The shape of the windowing function can take many different forms, and this has an impact on the resulting spectral representation of the signal. The simplest window is the rectangular window, but this introduces severe discontinuities at the beginning and end of the frame of speech under analysis which affects the spectral representation. More commonly used windows are the Hamming and Hanning windows which have much smoother profiles than the rectangular window and are based on cosine functions.

Assuming a time domain signal, $x(n)$, and windowing function, $w(n)$, the complex frequency spectrum, $X(k)$, is computed using the DFT as

$$X(k) = \sum_{n=0}^{N-1} w(n) x(n) e^{-j2\pi k n/N}. \tag{12}$$

Given that the window contains N time domain samples which were sampled at a rate of f_s samples per second, the DFT

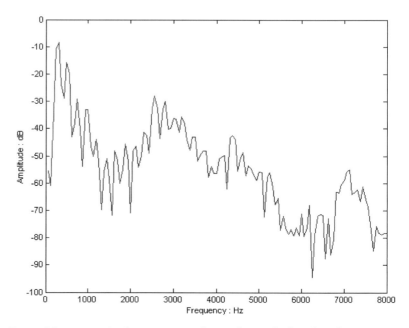

FIGURE 4 Log magnitude spectrum of vowel sound /ih/ taken from a 32-ms window.

coefficients, or DFT bins, $X(k)$, are spaced at f_s/N Hz intervals. For display purposes, it is more useful to convert the complex frequency spectrum into the magnitude spectrum, $|X(f)|$, and phase spectrum, $\angle X(f)$. It is also common to display the log of the magnitude spectrum as this reduces the amplitude of dominant regions of the spectrum which allows examination of lower energy regions. For illustration, Figure 4 shows the log magnitude spectrum of a 32-ms Hamming windowed frame of speech taken from the mid-point of the vowel /ih/ shown in Figure 3.

The figure reveals an important property of speech signals in that the magnitude spectrum can be described in terms of two components: the broad spectral envelope and the finer spectral detail (see Figure 18 for more discussion of this). The spectral envelope shows four spectral peaks (centred at 500, 3,000, 4,200 and 7,000 Hz) which are produced from modulation effects occurring within the vocal tract. These are termed formants and are characterised by their frequency, amplitude and bandwidth. The rapid fluctuations in the magnitude spectrum arise from the fundamental frequency and its harmonics. Changing the fundamental frequency of the speech signal would alter the positions of the harmonics but would leave the spectral envelope broadly unaltered.

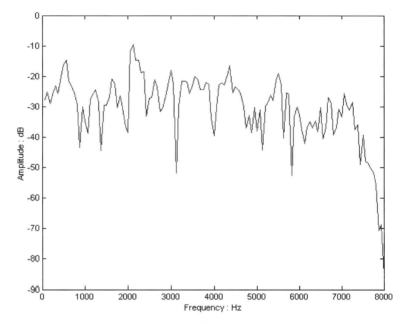

FIGURE 5 Log magnitude spectrum of unvoiced sound /th/ taken from a 32-ms window.

The time domain signal in Figure 3 shows that the unvoiced phoneme /th/ had very different structure to the two voiced phonemes /ih/ and /ng/. These differences can also be examined in the frequency domain, and to illustrate this, Figure 5 shows the log magnitude spectrum of the phoneme /th/ which has been computed using a DFT with a 32-ms window centred at 0.54 s.

The spectrum shows that much less structure is present in the signal. No formants are present and the harmonics no longer exist due to the excitation signal being wideband noise rather than a series of glottal pulses.

In the time domain waveform in Figure 3, it was not obvious that two different voiced phonemes were present and it was certainly difficult to analyse their structure. In the frequency domain, it is much easier to see these differences. For example, Figure 6 shows the log magnitude spectrum of the voiced phoneme /ng/ taken from a 32-ms window centred at time 0.65 s.

Comparing this spectrum with that of the vowel /ih/ shown in Figure 4 shows significant differences in the spectral envelope. The spectral envelope of the vowel /ng/ only contains one formant which has a frequency of 500 Hz in comparison to the four

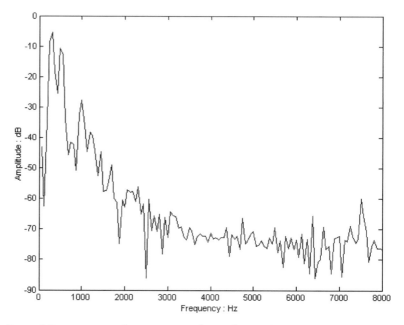

FIGURE 6 Log magnitude spectrum of vowel sound /ng/ taken from a 32-ms window.

formants of the vowel /ih/. In the time domain, these waveforms looked similar, but in the frequency domain very obvious differences in formant structure can be observed. It is these differences in formant structure which give the different sounds to the phonemes.

When using the DFT, it is important to consider the effect of different window widths on the spectral representation of speech. Selecting a large window width of time domain samples gives a very detailed spectral representation of the signal as the frequencies represented by the DFT bins will be very close together. However, the large number of time domain samples may mean that the signal is not stationary over this duration. Certainly, if the speech has changing characteristics within the window, then this will lead to a smearing of the resulting spectrum. This can be avoided by reducing the window width, but the downside to this is a reduction in spectral resolution. For example, halving the window width will double the frequency spacing between the spectral bins. A trade-off therefore exists between the window width and the desired spectral resolution of the signal that the application or analysis requires.

3.3 Time–Frequency Representation

The previous sections have shown that the time domain representation is useful for showing the time-varying nature of a signal, while the frequency domain representation enables the structure of a short-time stationary window of speech to be examined. These two representations can be combined into a time–frequency representation termed a spectrogram. Figures 8 shows an example spectrogram of the waveform "*I think that you air*" which was shown in Figure 2. Time is shown along the x-axis of the spectrogram and frequency along the y-axis. The darkness indicates the energy of the signal at a particular time and frequency. Examination of the spectrogram shows the structure of the speech signal in terms of formants, fundamental frequency and voicing.

The spectrogram is constructed from a series of short-time Fourier transforms which are successively computed along the time domain waveform. As the analysis window slides along the signal, the log energy within each DFT bin is converted into a shade of colour which is then positioned at the appropriate time and frequency. This operation is illustrated in Figure 7.

The resolution of the spectral analysis leads to two different forms of spectrogram. Spectrograms showing the broad spectral envelope are termed wideband spectrograms and those showing the finer harmonic structure are termed narrowband.

3.3.1 Narrowband Spectrogram

Figure 8 shows the narrowband spectrogram of the utterance "*I think that you air*". For the narrowband spectrogram, the Fourier transform is computed from a relatively long-duration

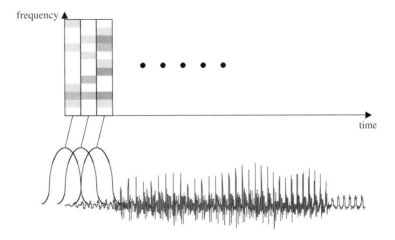

FIGURE 7 Construction of a spectrogram from a time domain waveform.

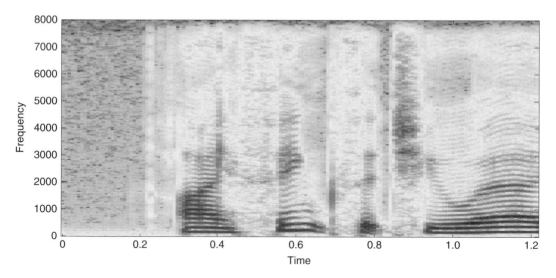

FIGURE 8 Narrowband spectrogram of utterance *"I think that you air"* together with the positions of the phonemes making up the utterance.

time window which gives a detailed spectral representation at the expense of less precise temporal localisation. The spectral bandwidth of the spectrogram shown in Figure 8 is 31.25 Hz. This allows individual frequency harmonics of the glottal pulses to be resolved leading to the characteristic narrow horizontal bands which make up the narrowband spectrogram. Examining a vertical slice of the spectrogram is equivalent to the magnitude spectrum plots shown in Section 3.2.

Voiced regions can be identified by the presence of pitch harmonics such as in the phonemes /ih/ and /ng/ which are shown in Figures 4 and 6. Unvoiced phonemes are characterised by the lack of pitch harmonics such as the phoneme /th/ which exhibits the noise-like structure shown in Figure 5. Silence regions are characterised by a lack of energy as is seen before the speech begins.

3.3.2 Wideband Spectrogram

Figure 9 shows the wideband spectrogram of same utterance shown in Figure 8. The wideband spectrogram is computed in a similar manner to the narrowband spectrogram except that now a much shorter time frame of speech is analysed by the Fourier transform. This results in a smoother spectral representation which shows the spectral envelope but not the individual pitch harmonics. In fact individual glottal pulses can be observed as vertical bands in the spectrogram. The short duration of the analysis window also improves the temporal

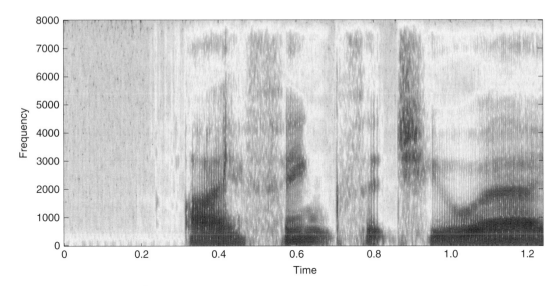

FIGURE 9 Wideband spectrogram of utterance "*I think that you air*" together with the positions of the phonemes making up the utterance.

resolution of the wideband spectrogram in comparison with the narrowband spectrogram. The spectral bandwidth of the wideband spectrogram is around 300 Hz, which means that several harmonics of the glottal pulses fall in each frequency band to give the spectral averaging leading to the spectral envelope.

The use of the wideband spectrogram is preferable when performing phonetic analysis of speech as the effect of the fundamental frequency and its harmonics is removed leaving only formant structure. More accurate temporal localisation is also possible as a result of the shorter time frames. This is demonstrated by the very narrow representation of the glottal stop phoneme /k/ shown in the wideband spectrogram in comparison to the more blurred representation in the narrowband spectrogram.

4 • Analysis of Speech Waveforms

Speech analysis takes many different forms and calls upon a variety of different signal processing methods. The classification of speech as voiced or unvoiced, the estimation of the fundamental frequency and the identification of formants are commonly required parameters which can be obtained through speech analysis. More specific speech features can also be extracted that are related to a particular speech processing application. For example, there are a wide variety of feature extraction algorithms that can be used in the initial stages of speech recognition systems. These extraction methods use various forms of speech analysis to extract features that can discriminate between different speech sounds. In this section, three important methods of speech analysis are discussed and also their application to fundamental frequency and formant estimation. The section is completed with a brief examination of feature extraction methods used in speech recognition, coding and synthesis.

4.1 Linear Predictive Coding (LPC) Analysis

A commonly used model of speech production is the source–filter model which finds applications in low-bitrate speech coding, speech recognition and speech synthesis. The model comprises two parts – source and filter. The source models the excitation signal from the lungs and vocal cords and can take two forms. For voiced speech, the excitation is modelled as a series of impulses, each separated by the pitch period, which represent the glottal pulses produced as the vocal cords open and close. For unvoiced speech, the excitation takes the form of white noise which simulates the airflow when the vocal chords are relaxed leading to turbulent airflow. A separate gain term is used to amplify the excitation signal, and this reflects variations in speech amplitude that can be produced principally by varying the pressure generated in the lungs. The filter component of the model represents the frequency response of the vocal tract. This is usually modelled by an all-pole filter which can reproduce the formant-based resonances in the vocal tract. The all-pole structure is effective for most speech sounds, although to model the anti-resonances found in nasal sounds zeros are also required in the filter. The need for a filter with both poles and zeros increases the complexity of the model and can generally be avoided by increasing the order of the all-pole filter.

The examination of waveforms in the previous section has shown that speech is highly non-stationary. This means that the parameters of the vocal tract model must themselves be time varying. To allow for this the LPC analysis is performed on sequences of short-time frames of speech which are extracted progressively from the utterance. A typical analysis frame rate is 10–30 ms. Figure 10 illustrates the source–filter model of speech production.

The output speech signal, $x(n)$, can be expressed in terms of the excitation signal, $u(n)$, the gain, G, and the vocal tract filter coefficients a_k:

$$x(n) = \sum_{k=1}^{p} a_k x(n-k) + Gu(n). \qquad (13)$$

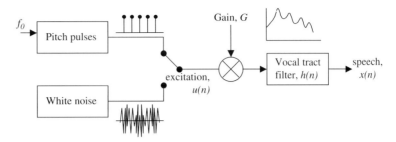

FIGURE 10 Source–filter model of speech production.

The task now remains of how to calculate the parameters of the model, namely the set of filter coefficients, $a = [a_1, a_2, \ldots, a_p]$, the gain, G, the fundamental frequency, f_0, and the voicing decision.

Taking the Z-transform of Eq. (13) followed by re-arrangement leads to

$$H(z) = \frac{X(z)}{U(z)} = \frac{G}{1 - \sum_{k=1}^{p} a_k z^{-k}}, \quad (14)$$

where $H(z)$ is the Z-transform of the vocal tract filter. The similarity between the source–filter model of speech and linear prediction leads to a straightforward method for estimating the model parameters. A linear predictor makes a prediction of a speech sample, $\tilde{x}(n)$, from a weighted summation of the p previous speech samples using a set of predictor coefficients α_k:

$$\tilde{x}(n) = \sum_{k=1}^{p} \alpha_k x(n-k). \quad (15)$$

The prediction error, $e(n)$, can then be defined as

$$e(n) = x(n) - \tilde{x}(n) = x(n) - \sum_{k=1}^{p} \alpha_k x(n-k). \quad (16)$$

Taking the Z-transform of Eq. (16) gives an expression for the filter, $A(z)$, which generates the prediction error samples:

$$A(z) = 1 - \sum_{k=1}^{p} \alpha_k z^{-k}. \quad (17)$$

Comparing Eq. (14) and (17) and setting $\alpha_k = a_k$ and $e(n) = Gu(n)$ allow the vocal tract filter, $H(z)$, to be expressed in terms of the prediction filter $A(z)$:

$$H(z) = \frac{G}{A(z)}. \quad (18)$$

Therefore, the vocal tract filter coefficients of the source–filter model are equal to the linear predictor coefficients of which computation is straightforward. The standard method for computing the predictor coefficients is to minimise the mean squared prediction error over a short-time analysis window. This has been very well documented and is generally solved using either the autocorrelation method or the covariance method. Considering the autocorrelation method of solution results in a set of linear equations:

$$\sum_{k=1}^{p} \alpha_k R(|i-k|) = R(i) \quad 1 \leq i \leq p. \tag{19}$$

$R(k)$ is the kth autocorrelation coefficient of the frame of N speech samples under analysis which is defined as

$$R(k) = \sum_{n=0}^{N-1-k} x(n)\, x(n+k). \tag{20}$$

Equation (19) can be rewritten in matrix form, where the solution for the set of predictor coefficients, **a**, takes the form

$$\mathbf{a} = \mathbf{R}^{-1}\mathbf{p}. \tag{21}$$

The Toeplitz structure of matrix **R** in Eq. (21) allows several efficient methods of solving the equation to be applied such as Durbin's recursive solution. Knowledge of the vocal tract filter coefficients enables estimation of both the fundamental frequency and formants and forms the remainder of this Section.

4.1.1 LPC-Based Formant Analysis

Equation (13) has shown that the frequency response of the vocal tract can be modelled by an all-pole filter with a set of filter coefficients, $\boldsymbol{a} = [a_1, a_2, \ldots, a_p]$. The challenge for formant analysis is to use these filter coefficients to estimate formant parameters such as frequency, amplitude and bandwidth. This can be achieved by factoring the denominator of the vocal tract filter to give the poles. To illustrate this, Figure 11 shows a 32-ms short-time frame of speech taken from the phoneme /ih/ in the word "*think*" illustrated in Figure 3. Figure 12 shows a pole–zero diagram which indicates the poles of the vocal tract filter extracted from the frame of speech using 10th-order LPC analysis.

The pole–zero diagram shows the position of the 10 poles of the vocal tract filter where it can be seen that the poles exist as complex conjugate pairs. Each pair of poles contributes a resonance in the filter frequency response, the frequency of which

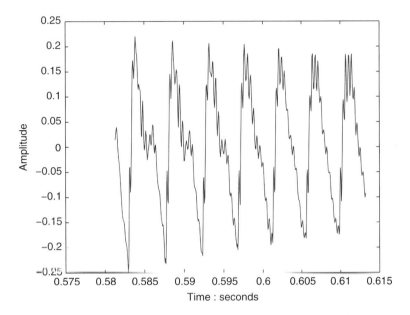

FIGURE 11 32-ms frame of speech from the phoneme /ih/.

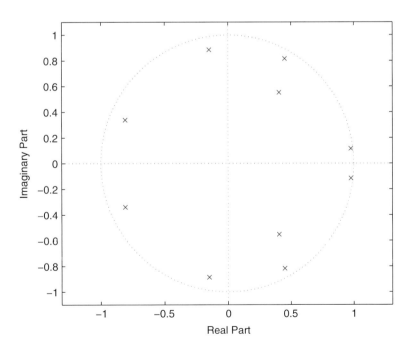

FIGURE 12 Pole–zero diagram showing the poles of the vocal tract filter.

is determined from the angle of the pole and the amplitude by its distance out from the origin. For the frame of speech shown in Figure 11, the pole–zero diagram shows that four poles pairs – 1, 3, 4 and 5 – are very close to the unit circle and will therefore contribute to significant spectral peaks in the vocal tract filter frequency response. The second pole pair has much less amplitude and will have significantly less effect on the frequency response of the vocal tract filter. The effect of these poles can be observed by examining the frequency response of the vocal tract filter. This is illustrated in Figure 13 and has been obtained by taking the Fourier transform of the vocal tract filter's impulse response.

The frequency response shows clearly the four formant resonances which have been modelled by the vocal tract filter. The frequency, F_i, of each formant is determined from the angle of the pole, θ_i, i.e. $F_i = \theta_i f_s / 2\pi$ where f_s is the sampling frequency. The contribution of the second pole pair occurs just before the second formant and its effect is negligible due to its much smaller magnitude. It is interesting to compare the log magnitude spectrum of the vocal tract filter to the log magnitude spectrum, shown in Figure 4, which was obtained from the Fourier transform of the same frame of speech samples. The fine spectral detail shown in Figure 4 arises from the

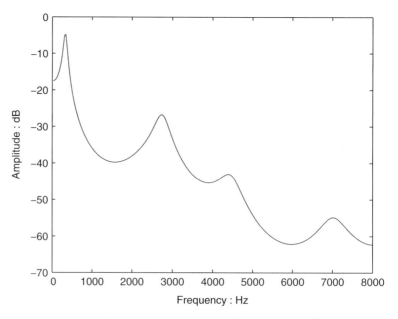

FIGURE 13 Frequency response of the vocal tract filter.

fundamental frequency and its harmonics which have been removed in the LPC analysis leading to the vocal tract filter. Therefore, the LPC analysis provides a useful method of obtaining the spectral envelope of the signal.

4.1.2 LPC-Based Fundamental Frequency Analysis

The fundamental period of the frame of speech shown in Figure 11 can be estimated by measuring the time between the peaks caused by the glottal pulses. Examination of the figure reveals the fundamental period to be approximately 4.5 ms which corresponds to a fundamental frequency of 222 Hz. For the speech shown in Figure 11, pitch estimation is relatively straightforward as the glottal pulses are clearly evident. However, for some voiced speech signals, the glottal pulses are not so clear, which makes their accurate identification and measurement prone to error leading to incorrect fundamental frequency estimates. This situation is made even worse when acoustic noise has contaminated the speech and has the effect of masking the glottal pulses.

A better signal from which to estimate the fundamental frequency is the excitation signal itself. The excitation signal has not been modulated by the vocal tract and has more clearly defined glottal pulses. An estimate of the excitation signal can be obtained by filtering the speech signal with a filter corresponding to the inverse of the vocal tract filter as is illustrated in Figure 14.

Examination of Eq. (18) shows that the inverse of the vocal tract filter, $H(z)$, is the predictor filter, $A(z)$. Passing the speech signal through this inverse filter removes the effect of the vocal tract filter frequency response. This leaves the resulting excitation signal with a flat frequency response, and hence the process is also termed spectral whitening. Figure 15 illustrates the excitation signal which has been generated by passing the frame of time domain samples in Figure 11 through the inverse vocal tract filter obtained from the filter shown in Figure 12.

Comparing the excitation signal to the original time domain signal shows that the glottal pulses are now much more evident.

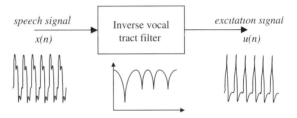

FIGURE 14 Inverse filtering the speech signal to recover the excitation signal.

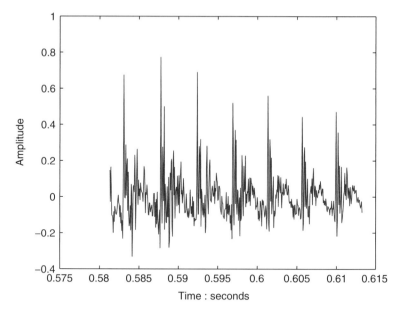

FIGURE 15 Excitation signal produced by inverse filtering.

Identification of the glottal pulses can be made more accurately than from the original speech signal, and measuring the precise time period between pulses is more accurate.

4.1.3 Autocorrelation-Based Fundamental Frequency Analysis

The fundamental frequency measurement in the previous section relies on correctly identifying pitch pulses and accurately measuring the time period between them. A more robust method of estimating the pitch period is to employ the autocorrelation analysis which was used in the calculation of the vocal tract filter coefficients. Autocorrelation analysis applied to a time domain signal is simple to implement and is used to identify patterns and structures in the signal. One such pattern that autocorrelation analysis can be used to identify is the periodic repetition of pulses in a speech waveform which results from the opening and closing of the glottis in voiced speech. The definition for autocorrelation coefficients, $R(k)$, is given in Eq. (20) where the first $p+1$ values are used in LPC analysis to determine the vocal tract filter coefficients. The vocal tract can be estimated from only a small number of autocorrelation coefficients due to its short-term nature. However, to estimate the fundamental frequency, a larger number of autocorrelation coefficients are necessary.

To illustrate the autocorrelation analysis, Figure 16 shows a 32-ms frame of speech taken from the phoneme /eh/ in the word "*air*" shown in Figure 2.

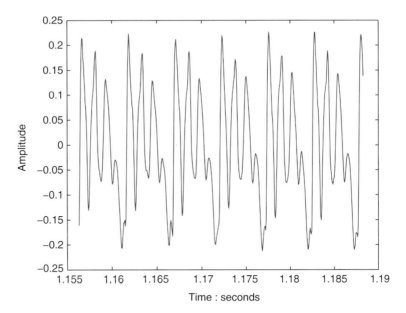

FIGURE 16 32-ms short-time window of speech taken from the phoneme /eh/.

Analysis of the time domain waveform to estimate the pitch period is not as straightforward as for the speech frame shown in Figure 11, and designing robust algorithms to identify glottal pulses would be more challenging. Figure 17 shows the first 512 autocorrelation coefficients, $R(0)$–$R(511)$, computed from the frame of time domain samples using Eq. (20), which corresponds to time lags up to 32 ms. As is normally the case, the autocorrelation coefficients have been normalised by $R(0)$.

The autocorrelation coefficients exhibit similar periodicity to the time domain signal shown in Figure 16 and have an identical period of repetition. Ignoring the peak at time lag zero, the largest peak occurs at a time of $t = 5.4$ ms which corresponds to the pitch period observed in Figure 16. However, the autocorrelation analysis makes identification of the peak corresponding to the pitch period much easier to locate. The smaller peaks found in the autocorrelation analysis occur from resonances in the vocal tract and from the shape of glottal pulses. Applying the autocorrelation-based fundamental frequency estimation to the excitation signal obtained from inverse filtering, discussed in the previous section, leads to even more robust fundamental frequency estimation.

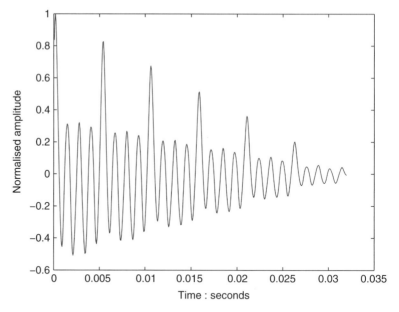

FIGURE 17 Autocorrelation analysis of a frame of speech from phoneme /eh/.

It should be emphasised, however, that robust fundamental frequency estimation across all speech sounds is a very difficult problem to solve and is still the subject of much research even today.

4.2 Frequency Analysis

The frequency domain representation of speech discussed in Section 3.2 shows that the magnitude spectrum comprises a spectral envelope component and a harmonic structure resulting from the glottal pulses. In the time domain, the speech signal is generated by the convolution of the excitation signal with the vocal tract filter. Therefore, in the frequency domain, the spectrum of the speech is given as a multiplication of the excitation signal spectrum and vocal tract frequency response. Figure 18 illustrates this process for a typical voiced vowel segment of speech.

Changing the fundamental frequency will alter the spacing of harmonics but will leave the spectral envelope, or phoneme being spoken, essentially unchanged. In a similar way, changing the spectral envelope will leave the fundamental frequency constant. From the speech spectrum, sufficient information exists to allow estimation of both the fundamental frequency and the formants.

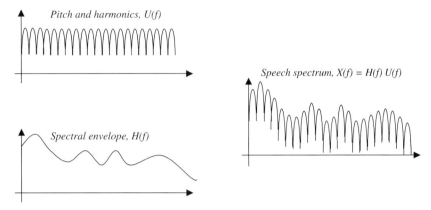

FIGURE 18 Fundamental frequency and spectral envelope components of the speech spectrum.

4.2.1 Frequency-Based Fundamental Frequency Analysis

Examination of the magnitude spectra of voiced speech in Figures 4 and 6 shows that the glottal pulses in the time domain signal appear as closely spaced harmonics in the spectral domain. The first harmonic occurs at the fundamental frequency, f_0, while subsequent harmonics occur at approximately integer multiples of the fundamental frequency, nf_0. The harmonics are in turn modulated by the frequency response of the vocal tract which gives the spectral envelope.

Identification of the fundamental frequency and its harmonics in the frequency domain therefore forms a method of estimating the fundamental frequency. The location of a particular harmonic may not always be reliable as its amplitude will have been modulated by the frequency response of the vocal tract. In fact it is common for the first few harmonics not to be present in situations where the speech has been received across a telephone channel which employs bandlimiting of the signal. To avoid these problems, and to increase robustness, most frequency domain methods of fundamental frequency estimation attempt to identify as many harmonics as possible.

One successful method of frequency domain fundamental frequency estimation is the comb function. This operates by first selecting a set of candidate fundamental frequencies, c_1 to c_N, to compare against the real fundamental frequency of the signal. These candidates may, for example, begin at 100 Hz and end at 400 Hz, incrementing in 10-Hz steps. For each of these candidate frequencies, an estimate of the magnitude spectrum of the glottal pulse train is synthesised by positioning a series of impulses at harmonics of

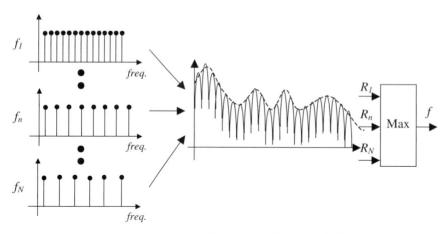

FIGURE 19 Illustration of the comb function for frequency-domain fundamental frequency estimation.

the candidate fundamental frequency. An example of a set of candidate fundamental frequencies and their synthesised magnitude spectra is shown in Figure 19. For each candidate fundamental frequency, the cross-correlation between the synthesised magnitude spectrum and the speech magnitude spectrum is measured. When the candidate fundamental frequency is close to the real fundamental frequency the harmonics will line up leading to high cross-correlation. The estimate of the fundamental frequency is selected as the candidate fundamental frequency which gave the highest cross-correlation with the spectrum of speech. The process is illustrated in Figure 19.

4.2.2 Frequency-Based Formant Analysis

The magnitude spectrum of a frame of speech not only shows the positions of formants but also contains excitation information in the form of harmonics. More reliable estimates of the formants can be obtained from the spectral envelope which does not show the finer spectral details. The spectral envelope can be obtained by applying the Fourier transform to a frame of speech which is sufficiently short so as not to contain a sequence of glottal pulses. However, according to discussions into time-frequency trade-offs in Section 3.2, this means that the frequency resolution will be poor. This can be avoided by zero padding the frame of speech to create a longer duration frame from which a smooth spectral envelope can be computed. Figure 20 shows a smoothed magnitude spectrum which has been computed from a frame of speech comprising just 20 time domain samples taken from the phoneme /ih/. This equates to a duration of 1.25 ms. Before applying the Fourier transform, the frame of speech has been

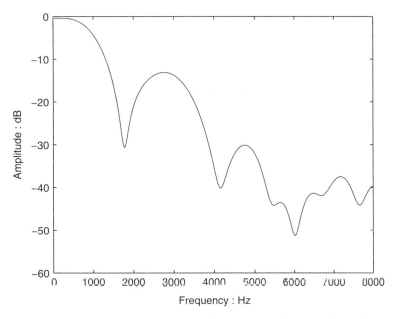

FIGURE 20 Spectral envelope obtained from 1.25-ms frame of speech and zero padding.

zero padded up to a duration of 32 ms through the augmentation of 492 zeros.

The smoothed spectral envelope makes identification of the formants straightforward through peak-picking algorithms. It should be noted that Figure 20 shows the formants very clearly. In practice, when analysing many different frames of speech the formant structure may not be so clear, leading to estimation errors.

4.3 Cepstral Analysis

Analysis of the magnitude spectrum of a speech signal showed that it comprises both spectral envelope information, relating to the vocal tract frequency response, and fine spectral detail from the fundamental frequency and its harmonics. Formant analysis can be applied effectively to the spectral envelope, while fundamental frequency estimation requires the fine spectral detail. Cepstral analysis is a form of homomorphic processing which enables these two components to be separated from each other. The cepstral transform is defined as the inverse discrete Fourier transform of the log power spectrum of a short-time frame of the signal as is illustrated in Figure 21.

The effect of the cepstral transform on a speech signal can be considered as follows. The input speech signal, $x(n)$, comprises

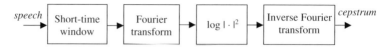

FIGURE 21 Cepstral transformation of an input speech signal.

the convolution of two components: the vocal tract impulse response, $h(n)$, and the excitation signal $u(n)$:

$$x(n) = h(n)^*u(n). \qquad (22)$$

Transforming the signal $x(n)$ into the log power spectrum, $\log |X(f)|^2$, converts the convolutional operation into an addition operation as (ignoring the scaling by 2)

$$\log |X(f)| = \log |H(f)| + \log |U(f)|. \qquad (23)$$

The log power spectrum of the vocal tract component, $\log |H(f)|$, is observed as the spectral envelope which varies slowly along the frequency axis. The component $\log |U(f)|$ gives the fine spectral detail from the fundamental frequency and its harmonics and has much faster variation along frequency than the vocal tract component. To complete the cepstral transform, the inverse discrete Fourier transform is taken of the log magnitude spectrum, $\log |X(f)|$. This serves to separate the vocal tract and excitation components of the signal. The slowly varying vocal tract information is located in the lower quefrency region of the cepstrum (quefrency is the name given to the domain following the cepstral transform and has units measured in seconds). The faster varying excitation component is located in the higher quefrency regions of the cepstrum. Applying a lowpass windowing function to the cepstrum retains the vocal tract information, while applying a highpass window retains the excitation information.

Figure 22 illustrates the log magnitude spectrum of the short-time frame of speech shown in Figure 11 which is taken from the phoneme /ih/.

In this figure, the fast moving spectral detail from the fundamental frequency and its harmonics can be seen together with the slowly varying spectral envelope resulting from the vocal tract. Figure 23 shows the cepstrum of this signal up to a quefrency of 16 ms.

Examining the cepstrum shows significant energy to be present at low quefrencies up to about 1 ms. This corresponds to the slowly time-varying vocal tract information. A peak in the cepstrum can be observed at a quefrency of about 4.5 ms. This

27 • Display and Analysis of Speech   **477**

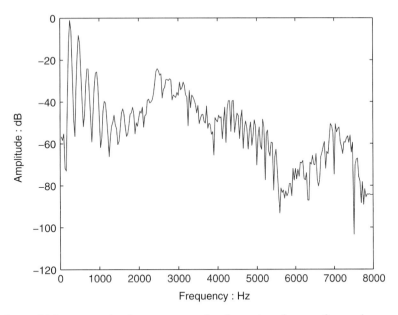

FIGURE 22 Log magnitude spectrum of a short-time frame of speech from phoneme /ih/.

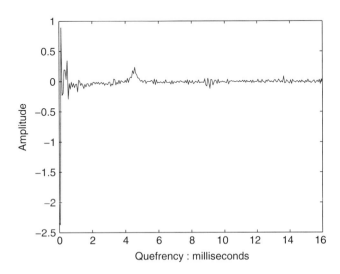

FIGURE 23 Cepstrum of frame of speech shown in Figure 11.

peak corresponds to the pitch period and shows that the cepstrum is an effective method for estimating the fundamental frequency of voiced frames of speech. Above this quefrency the information comprises the faster spectral variations arising from the excitation signal.

Using the fact that the cepstral transform is a homomorphic transform means that the vocal tract and excitation can be separated from one another by either lowpass or highpass filtering the cepstrum (filtering is actually termed liftering in the cepstral domain). For example, Figure 24a shows the resulting log magnitude spectrum when the discrete Fourier transform is applied to a lowpass-filtered cepstrum containing only the first 10 cepstral coefficients (up to 0.625 ms). Similarly Figure 24b shows the resulting log magnitude spectrum when the discrete Fourier transform is applied to a highpass-filtered cepstrum containing all cepstral coefficients except the first 10.

The figure shows the ability of the cepstral transform to separate the vocal tract and excitation components is very effective. From the lowpass-filtered cepstrum, the log magnitude spectrum in Figure 24a shows very clearly the spectral envelope of the vocal tract which makes identification of the formants obvious. The highpass-filtered cepstrum has the vocal tract modulation effects removed as shown by the comparatively flat log magnitude spectrum. This signal comprises mainly excitation signal components.

The position of the cepstral filter cut-off determines the information retained in the two log magnitude spectra. For example,

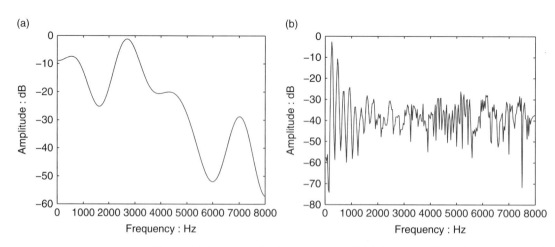

FIGURE 24 Log magnitude spectrum of (a) lowpass-filtered cepstrum and (b) highpass-filtered cepstrum at a quefrency cut-off of 0.625 ms.

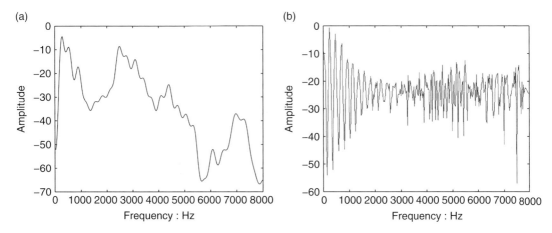

FIGURE 25 Log magnitude spectrum of (**a**) Lowpass-filtered cepstrum and (**b**) highpass-filtered cepstrum at a quefrency cut-off of 3.125 ms.

moving the cut-off from 0.625 to 3.125 ms increases the amount of faster moving information present in the lowpass-filtered cepstrum and removes some of the slower moving information from the highpass-filtered cepstrum. Figures 25a and b shows the resultant log magnitude spectra from lowpass- and highpass-filtered cepstra at a cut-off quefrency of 3.125 ms.

Comparing Figure 25a with Figure 24a shows that the increase in quefrency cut-off allows more faster moving spectral information to be retained in the lowpass-filter cepstrum.

Figures 24 and 25 have shown that the cepstrum is effective at separating vocal tract and excitation information which leads to effective fundamental frequency and formant estimation. To conclude the cepstral analysis of speech signals, Figure 26a shows the time domain representation of a frame of unvoiced speech taken from the phoneme /th/. The log magnitude spectrum of this signal has been shown in Figure 5 where the relatively flat frequency response and lack of harmonics indicate the unvoiced nature of the signal. Figure 26b shows the cepstrum of the signal.

The cepstrum of the unvoiced speech does not contain a peak as there are no harmonics present in the speech spectrum. There is also less energy in the lower quefrency coefficients in comparison to that observed for the voiced speech cepstrum shown in Figures 24 and 25. This is due to the reduction in spectral shape that the relatively flat frequency spectrum of the unvoiced speech has.

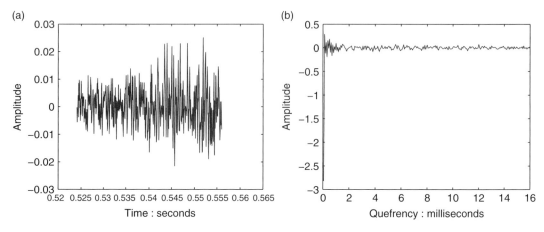

FIGURE 26 (**a**) Time domain representation of a frame of unvoiced speech; (**b**) cepstrum of the frame of speech.

4.4 Analysis for Speech Processing Applications

This section concludes with a brief discussion of feature analysis methods that have been designed specifically for use with particular speech processing applications such as recognition, coding and synthesis. The fundamental acoustic features of fundamental frequency and formants discussed earlier in this section are very important for analysis and understanding purposes but generally do not find direct use in speech processing applications. This is due in part to insufficient understanding of how such features can be successfully integrated into speech processing applications and also because of the difficulty in accurately estimating them across a range of different speech sounds, speakers and noise levels. Certainly, formant and fundamental frequency features have been used in formant-based speech synthesis applications, but for high-quality applications, these have been superseded by concatenative-based systems which are based on very large speech databases.

The majority of speech recognition systems use a variant of cepstral features termed mel-frequency cepstral coefficients (MFCCs). MFCC analysis aims to extract only the vocal tract component of the speech signal but does so using a frequency resolution based on the mel scale. This mimics the frequency response of the human ear by having fine spectral resolution at low frequencies and coarse resolution at high frequencies. Another form of analysis used in speech recognition is perceptual linear predictive (PLP) cepstra. This combines the vocal tract modelling used in LPC analysis with the non-linear frequency response of human hearing using a bark scaled warping. As with the analysis methods described earlier in this section, both MFCC analysis and PLP analysis are based on short-time frames of speech with durations between 10 and 32 ms.

Some very low bitrate speech codecs, such as the vocoder, make use of the fundamental frequency directly but generally have poor quality. Instead, low-bitrate codecs generally use some form of vocal tract filter which is excited by an encoded representation of the excitation signal. Standard LPC analysis provides the vocal tract filter coefficients. In practice, these are converted to features such as line spectral pairs (LSP) which are more robust to the quantisation necessary to reduce the bitrate in low-bitrate speech codecs. The encoding of the excitation signal can take many different forms such as found in code-excited linear prediction (CELP) and mixed excitation linear prediction (MELP).

5 • Conclusion

This chapter has shown that a speech signal must be digitised before it can be displayed and analysed by computers. Various factors need to be considered in the sampling and quantisation processes such as the sampling frequency and number of quantisation levels. The display of speech signals has shown the time domain is useful for showing broad speech classes such as voiced and unvoiced. With sufficient detail in the time domain representation, measurement of the fundamental period is also possible. The frequency domain allows much more structure of the speech signal to be seen. Importantly, this shows how a speech signal can be decomposed into its vocal tract and excitation components. Combining the time and frequency domains in the form of a spectrogram allows the time-varying nature of the fundamental frequency and formants to be seen very clearly.

LPC analysis, frequency analysis and cepstral analysis have been shown as effective methods for recovering vocal tract and excitation information from the speech signal. Demonstrations have shown how these analysis methods can be used to extract the fundamental acoustic correlates of fundamental frequency and formants from a speech signal. It has also been mentioned that these methods are subject to errors, particularly in the presence of acoustic noise, and are still the subject of much research even today.

Finally, some speech analysis methods have been discussed that are applicable to speech processing applications such as recognition, coding and synthesis. The parameters that these systems require are often based on standard analysis methods but with subtle differences.

28

Estimation of Speech Intelligibility and Quality

Stephen Voran

Institute for Telecommunication Sciences, Boulder, CO, USA

1 • Introduction

Speech communication requires a talker and a listener. Acoustical and in some cases electrical representations of the speech are carried from the talker to the listener by some system. This system might consist of the air in a room, or it might involve electro-acoustic transducers and sound reinforcement or telecommunications equipment. Interfering noises (including reverberation of speech) may be present and these may impinge upon and affect the talker, the system, and the listener. A schematic representation of this basic unidirectional speech communication scenario is given in Figure 1.

In general, a talker can be a human or a machine (speech synthesizer) and a listener can also be a human or a machine (automatic speech recognizer). The quality or intelligibility of speech synthesis and the effectiveness of automatic speech recognition are topics unto themselves and are outside the scope of this chapter. Quality and intelligibility issues in bidirectional communications are also outside the scope of this chapter, but these issues are addressed from a telecommunications perspective in [1–3] for example.

Given the situation in Figure 1, there are numerous interesting questions that one might ask. These include: How much of the speech communication is correctly heard and how much effort

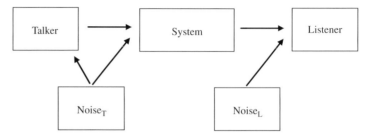

FIGURE 1 A unidirectional speech communication scenario.

does this take? What is the listener's opinion of the quality of the speech that he/she hears? How significant are the impairments to the speech that the listener hears? How do the answers to these questions depend on the talker, the listener, the system, and the noises?

These questions can be considered from multiple perspectives. For example, the question of speech quality could be pursued with respect to acoustic integrity, talker pronunciation and enunciation, or content. These are very different perspectives indeed. Throughout this chapter, our perspective is driven by consideration of how the system and the noises have altered the acoustic integrity of the talker's speech, and what those alterations are likely to mean to a listener with hearing characteristics that fall within normal ranges. Within this context, we use the term "intelligibility" to describe how understandable a speech message is and the term "quality" to describe how pleasing a speech message sounds. Indeed we are only scratching the surface of the myriad of issues in speech communication. One could argue, though, that notions of acoustic integrity are fundamental prerequisites for the more general discussion of the integrity of speech communications.

Each of the questions above ultimately concerns a human listener. Human factors experiments, called subjective tests or subjective experiments, commonly can provide the most direct answers. Further information on subjective speech intelligibility tests and subjective speech quality tests can be found in [4–7]. These tests generally require specialized equipment (e.g., sound-isolated rooms, high-quality transducers, and playback equipment) and they demand significant time associated with careful test design, conduct, and analysis. These costs have led numerous investigators to develop signal processing algorithms that seek to provide approximate answers to some of these questions. In effect, such algorithms need to emulate the behaviors of the human subjects in the experiments that they seek to replace.

In general, the design of a signal processing algorithm that emulates a human behavior can become as complex as one will allow. Some simplification can be gained by assuming that both the talker and the listener are adults with speech production and hearing faculties that fall within the "normal" range. It is generally prudent to focus the algorithm on a single distinct question such as intelligibility, quality, listening effort, or impairment level. One might seek to develop an algorithm that uses speech signals or some specially constructed non-speech test signals. The algorithm may process the system input and output, or it may be driven by the system output alone. Perhaps most challenging is the range of systems that an algorithm seeks to address. These could include rooms with reflections and/or noises, possibly combined with electronic sound reinforcement, telecommunications, or speech playback systems. The electronic portions of the system may be approximately linear and time-invariant (e.g., amplifiers) or they may be highly non-linear, dynamic, signal-adaptive devices (e.g., non-waveform digital speech coders), or they may be mixtures of the two.

Given the many dimensions of the problem, it is understandable that investigators have proposed numerous specific solutions for specific cases. Due to the richness of the field, a comprehensive survey is not possible here. Instead a selection of examples from the range of proposed solutions is provided. These examples were selected to demonstrate the range of signal processing techniques that have proven applicable to this problem.

Section 2 introduces some techniques for estimating speech intelligibility and Section 3 addresses estimation of perceived speech quality in telecommunications. Section 4 offers further observations and references to additional work in the field.

2 • Estimators of Speech Intelligibility

Background noises and reverberation can reduce the intelligibility of the speech that arrives at the listener. Several algorithms have been developed to estimate the relationship between these factors and speech intelligibility. These estimation algorithms are principally applicable to room acoustics problems where any sound reinforcement system that may be present applies only linear processing. They also find application in some basic telecommunications problems where analog or waveform-preserving digital elements are used and no non-linear processing comes into play.

2.1 Articulation Index

About 1921, Harvey Fletcher began a research program at Bell Labs to determine, among other things, how the different portions of the speech band (approximately 200–6,000 Hz) contribute to successful speech communications [8]. His experiments with filtered speech and human subjects led to mathematical models that largely answered this question, at least in the relevant context of the day. Fletcher's work in the 1920s stands today as the fundamental basis for understanding speech intelligibility. His development, called the Articulation Index (AI), was largely held internal to Bell Labs until the publication of [9] and [10] in 1947 and 1953, respectively. (Note that in the case of the AI, articulation refers to effects in speech transmission, not speech production.)

Kryter did additional work to formulate [11], validate [12], and standardize [13] two specific, relatively accessible AI calculations. Both are based on speech and noise levels in multiple frequency bands. One calculation uses 20 bands, each of equal importance to speech intelligibility as determined by Fletcher. The other uses 5 or 6 octave bands or 15 third-octave bands, perhaps making it more practical given the filters available at that time.

The 20 equal importance bands cover the range from 200 to 6,100 Hz. They are specified in [11,14] and are depicted graphically on both the Hertz frequency scale and psychoacoustic Bark (or critical band) frequency scale in Figure 2. This figure also shows the commonly used Hertz to Bark relationship:

$$f_{\text{Bark}} = 6 \cdot \text{arcsinh}\left(\frac{f_{\text{Hz}}}{600}\right). \quad (1)$$

Below 400 Hz this relationship is approximately linear and above 1 Hz it is approximately logarithmic. Critical band frequency scales, such as the Bark scale, have been empirically derived to account for a basic property of human hearing, and additional details are provided in "Bark Spectral Distortion". On the Hertz scale the 20 bands generally increase in width as the center frequency increases, and these widths span about a decade: from 100 to 1,050 Hz. On the Bark scale the bands are wider at the low- and high-frequency ends of the scale and narrower in the center of the scale. The widths span just a factor of 2: from 0.6 Bark near the center of the scale to about 1.2 Bark at either end of the scale.

Thus the widths of the AI bands are significantly more uniform on the Bark scale than on the Hertz scale. Since the Bark scale

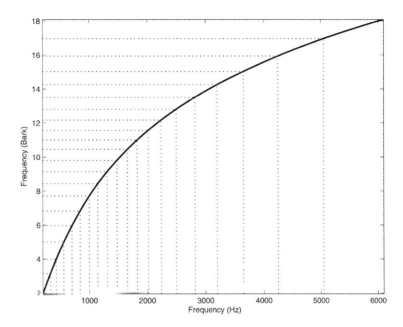

FIGURE 2 The 20 bands used in the calculation of AI, shown on both the Hertz and Bark frequency scales, along with the Hertz to Bark relationship of equation (1).

models an attribute of human hearing, this suggests that human hearing accounts for a portion of the band importances that drive the AI principle. The fact that the AI bands are not completely uniform suggests that human hearing does not account for all of the band importances, and thus the way that information is carried in the human speech spectrum must account for the remainder of the observed effect.

Fletcher determined that overall speech intelligibility can be modeled as the sum of individual contributions from multiple bands. Further, for the cases of band limiting and background noise at least, the contribution of each band is sufficiently described by the long-term signal-to-noise ratio (SNR) in each band. Thus the AI calculations given in [11] include the following steps:

1. Measure or calculate the average power spectrum of the speech and the background noise at the location of interest.
2. Modify the noise spectrum to account for the non-linear growth of masking in the human auditory system for bands where the noise is more than 80 dB above the hearing threshold. Further modify the noise spectrum to account

for the spread of masking across frequencies in the human auditory system.

3. In each band, use the signal power S_j and the noise power N_j to calculate the SNR. Limit that SNR to the range 0–30 dB, then normalize it to the range 0–1 to find the per band contribution to articulation:

$$\text{SNR}_j = 10 \cdot \log_{10}\left(\frac{S_j}{N_j}\right),$$
$$a_j = \frac{\min\left(\max\left(\text{SNR}_j, 0\right), 30\right)}{30}. \quad (2)$$

4. Now combine the contributions of each of the bands to attain AI:

$$\text{AI} = \sum_{j=1}^{N} \alpha_j \cdot a_j, \text{ where } 1 = \sum_{j=1}^{N} \alpha_j. \quad (3)$$

For the $N = 20$ band calculation, each band contributes equally to speech intelligibility and thus $\alpha_j = 0.05$, $j = 1$–20. When the one-octave or third-octave bands are used, the weights α_j reflect the differing contribution of each band. These band definitions and weights can be found in [11]. In either case, since $0 \leq a_j \leq 1$ and $\sum_{j=1}^{N} \alpha_j = 1$, it follows that $0 \leq \text{AI} \leq 1$.

Graphs demonstrating the monotonic relationship between AI and various types of subjective speech intelligibility test results can be found in [11,12,14].

2.2 Speech Transmission Index

In many situations the undesired signals that interfere with speech communication include reverberations of the desired signal. AI was not intended to address this situation. Further, there are practical issues making application of AI in this environment difficult or impossible. Since the desired speech signal produces at least a portion of the undesired interfering signal, separate measurements of the two signals' power spectra may not always be physically possible. The accurate calculation of such power spectra is generally of prohibitive complexity as well.

Steeneken and Houtgast built on the AI principle to develop algorithms that could properly account for effects of reverberations, and possibly some non-linear speech distortions as well. Amplitude-modulated noise signals and modulation transfer functions are key to this work. In early work a noise signal (with power spectrum matched to the long-term speech power spectrum) was modulated by a 3-Hz square wave and this evolved to square waves of multiple frequencies. This approach progressed to a more sophisticated algorithm where the system under test (SUT) is excited with test signals that employ sinusoidal amplitude modulation of the power of seven one-octave bands of noise. Fourteen different modulating frequencies are used. This algorithm is called the Speech Transmission Index (STI) and first appeared in [15]. A recent revision of STI is called STIr. The development of STIr is described in [16], the verification is given in [17], and a complete STIr measurement procedure is standardized in [18]. Various implementations are possible, and one is summarized here:

1. Generate a noise signal and filter it into seven different one-octave-wide bands, centered at 125, 250, 500, 1,000, 2,000, 4,000, and 8,000 Hz. Adjust the relative levels of the seven signals to match the given long-term average speech spectra for female or male talkers. In the case of female talkers, the band centered at 125 Hz is not used. The use of these seven bands allows for the independent characterization and weighting of the SUT behavior at each octave of importance in speech transmission. The time domain samples of these seven signals are $n_j(i)$ where i is the sample time index and $j = 1$–7 is the band index.
2. Amplitude-modulate each of these seven signals at 14 different modulation frequencies f_k, $k = 1$–14, to create 98 modulated noise signals with sample rate f_s:

$$n_j^k(i) = n_j(i)\sqrt{1 + \cos(2\pi f_k i/f_s)}. \qquad (4)$$

The modulation frequencies f_k range from 0.625 to 12.5 Hz and are spaced in third-octave steps. Equation (4) results in a sinusoidal modulation of signal power. Note that these signals are fully modulated (modulation index or modulation depth = 1) in the sense that the modulating function goes to zero on each modulation cycle. The presence of this periodic null in the test signal aids in the evaluation of the noise and reverberation power that may be present in the signal after passing through the SUT. The range of modulation frequencies is consistent with those found in actual human speech signals.
3. Sequentially pass each of these 98 signals $n_j^k(i)$ through the SUT to create 98 new signals $\tilde{n}_j^k(i)$.
4. Filter each of the 98 received signals $\tilde{n}_j^k(i)$ through the jth octave-wide bandpass filter. Square and low-pass filter the result using a cut-off frequency that removes the noise carrier

associated with band j, but that preserves the modulation frequency f_k. Model the resulting extracted envelope as

$$\mathrm{LPF}\left(\tilde{n}_j^k(i)^2\right) \approx g \cdot \left(1 + M_j^k \cos\left(2\pi f_k \cdot (i - i_0)/f_s\right)\right), \quad (5)$$

where M_j^k is the observed modulation index for band j and modulation frequency f_k, g is a gain factor, and i_0 is a time delay associated with the SUT. The observed modulation indices M_j^k can be reduced from their original value of 1 by the presence of noise and reverberations induced by the SUT.

5. Adjust each modulation index M_j^k, to account for the spread of masking from the octave band below as given in [18]. The new modulation indices are designated $\tilde{M}_j^k$. This allows the subsequent calculations to account for how a listener would perceive the test signals that have passed through the SUT.

6. Convert each modulation index to an effective SNR, limit each SNR to the range -15 to $+15$ dB, then normalize to the range 0–1 to find the transmission index for each octave band and modulating frequency:

$$\mathrm{SNR}_j^k = 10 \cdot \log_{10}\left(\frac{\tilde{M}_j^k}{1 - \tilde{M}_j^k}\right)$$

$$\mathrm{TI}_j^k = \frac{15 + \min\left(\max\left(\mathrm{SNR}_j^k, -15\right), 15\right)}{30}. \quad (6)$$

7. Average across the 14 modulating frequencies in each octave band to find the modulation transfer index for each band:

$$\mathrm{MTI}_j = \frac{1}{14} \sum_{k=1}^{14} \mathrm{TI}_j^k. \quad (7)$$

8. Then calculate STIr through a weighted summation of the modulation transfer indices for the seven octave bands along with the corresponding inter-band redundancy correction:

$$\mathrm{STI}_r = \sum_{j=1}^{7} \alpha_j \cdot \mathrm{MTI}_j - \sum_{j=1}^{6} \beta_j \sqrt{\mathrm{MTI}_j \cdot \mathrm{MTI}_{j+1}}, \quad (8)$$

where

$$\sum_{j=1}^{7} \alpha_j - \sum_{j=1}^{6} \beta_j = 1. \quad (9)$$

These weights result from the fitting of experimental data and their values can be found in [16] and [18]. The weights α_j reflect the relative contribution of each octave band to speech transmission. The inter-band redundancy correction (consisting of a weighted sum of the geometric means of adjacent band modulation transfer indices) is the primary difference between STIr and STI. This correction compensates for the apparently redundant contributions made by adjacent octave bands in some cases [16], and their inclusion does improve the agreement between STIr and subjective speech intelligibility tests [17]. Graphs demonstrating the monotonic relationship between STI or STIr and various types of subjective speech intelligibility test results can be found in [15,17,18].

Because the STIr test signals are based on noise, reliable STIr results can require significant measurement intervals for each of the 98 test signals. A typical choice for the measurement interval is 10 s, and this then requires over 16 min to perform the full STIr measurement. For increased measurement efficiency, three abbreviated STI measurement procedures have been developed for specific applications and the associated set of speech impairments: STITEL for telecommunications systems [18], STIPA for room acoustics and public address systems [18], and RASTI [18,19] for room acoustics alone. Each of these uses only a specific subset of the 98 test signals, and in some cases simultaneous rather than sequential excitation of the SUT leads to further efficiency. For example, one implementation of the RASTI algorithm applies only 12 s of test signal to the SUT.

In addition to the modulated noise signal-based measurements, the original STI publication [15] proposed that STI could be made more applicable to non-linear distortions of speech through the use of a more speech-like noise signal. Such a signal would have a pseudo-random modulation (modeled after real speech) in all octave bands other that the current band under test. The noise signal in the band under test would have a sinusoidal modulation. The desired response to non-linear distortions has remained elusive, and several authors have proposed techniques based on actual speech signals. Reference [20] summarizes some of these proposals and offers some new proposals as well.

2.3 Speech Intelligibility Index

The Speech Intelligibility Index (SII) [21] can be described as an updated and expanded version of AI. Expressions for the spread of masking, the notion of standard speech spectrum levels, and the relative importance of the individual bands have been updated. SII places the AI concept into a more general framework that allows one to account for hearing thresholds that deviate from the average and measurements made in a free field or at the eardrum. As a direct descendant of the AI family, SII is driven at its core by per frequency band SNRs. Two main techniques for arriving at the applicable signal and noise level values are described in [21]. One involves direct measurements or calculations, similar to the procedure for AI. The second method extracts the signal and noise levels from modulation transfer functions as STI does, but with different modulation frequencies and signal bands. The first method will generally be faster, but of more limited applicability. The second method is better suited to reverberant environments, but will generally require longer measurement intervals.

3 • Estimators of Speech Quality in Telecommunications

Modern telecommunications and speech storage systems are enabled by digital coding of speech waveforms. Speech coders that operate at higher bit rates can approximately preserve speech waveforms. Lower rate speech coders generally invoke constrained models for speech and extract and transmit model parameters that allow for the approximate reproduction of the speech sound, but with limited or no attempt to preserve the speech waveform. Because of this, these speech coders can exhibit highly non-linear, time-varying, signal-dependent behaviors. The distortions they introduce cannot be modeled as additive noise or reverberations. These speech coders, combined with other elements in telecommunications networks, severely limit the applicability of the speech intelligibility estimators presented above.

Many telecommunications systems incur virtually no impairment to the ability to conduct speech communications. While the intelligibility of the speech delivered by such systems may not be an issue, such systems can still deliver a wide range of perceived speech qualities. For example, one experiment compared the perceived quality of speech constrained to the nominal passband used in narrowband telephony (300–3,400 Hz) with speech constrained to the nominal passband used in wideband telephony (50–7,000 Hz) [22]. A clear difference in perceived speech quality was found and subjects clearly preferred the wideband passband. On the other hand, the AI correctly estimates that speech intelligibility is increased by a miniscule amount when wideband telephony replaces narrowband telephony. Thus speech quality, not speech intelligibility, has been the main recent concern in telecommunications.

It is worth noting, however, that speech intelligibility can be a very real issue in some wireless telephony, Voice over Internet Protocol (VoIP), and land mobile radio systems, yet existing intelligibility estimators generally are not applicable to these situations. One recent effort in this area can be found in [23].

Most estimators of perceived speech quality in telecommunications are focused on the effects of speech coders, transmission channels, and possibly some other network elements. The issues associated with noise at the talker and listener locations are presently under study.

3.1 General Principles

Figure 3 is a high-level block diagram for many speech-based, signal processing-based algorithms that estimate perceived telecommunications speech quality.

These algorithms make use of both the speech signal that goes into the SUT and the speech signal output from the SUT. We denote the time domain samples of these two signals $x(i)$ and

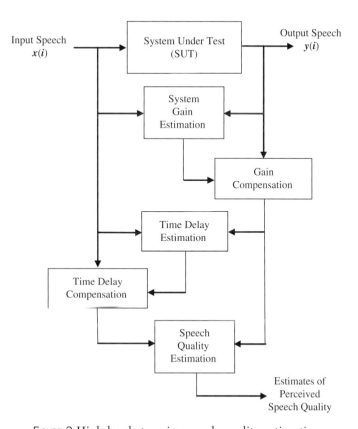

FIGURE 3 High-level steps in speech quality estimation.

$y(i)$, respectively, and they are also known as the reference and test signals. These signals are typically acquired in the analog or digital electrical domains, though it is also possible to acquire one or both of them in the acoustic domain. It is important that speech quality estimation be performed using real or artificial [24,25] speech signals rather than test signals because speech coders and other network elements can respond very differently to different classes of signals.

There is also a signal-free approach to telecommunications speech quality estimation. No signals are required, but complete knowledge of the system elements and operating parameters is required. Previously determined relationships between system elements and speech quality are combined in a mathematical model that produces an estimate of the speech quality that will be delivered by the SUT. This can be very efficient, but is limited to system elements and interactions that have already been completely characterized. Examples of this approach can be found in [1,2,26,27]. We do not consider this approach further in this chapter.

A constant gain in the telecommunications SUT is not typically considered to be a relevant part of the speech quality issue. This gain is often estimated and removed as indicated in Figure 3. The gain estimation can be performed using power ratios or by selecting a gain that minimizes the noise or maximizes the SNR associated with the SUT [28]. It is subsequently assumed that the speech is delivered to the listener at some preferred, standardized listening level.

3.2 Estimation of Constant Delay

As shown in Figure 3, any constant delay in the SUT must be estimated and removed so that the speech quality estimation block receives properly synchronized input signals. If unidirectional speech quality is of interest, this fixed delay is of no consequence. If it is large enough it can affect the quality of a bidirectional conversation. If this is of interest, then the effect of the delay must be combined with the estimated speech quality to fully understand the quality one might experience in the bidirectional case.

Estimation of SUT delay can be a significant problem. If the SUT approximately preserves waveforms, then conventional waveform cross-correlation delay estimation techniques can suffice. However, when waveforms are not preserved, these techniques can fail. In addition, when waveforms are not preserved one must look beyond waveforms to even define what delay means. Since speech envelopes and power spectra are approximately preserved, these are useful representations. The

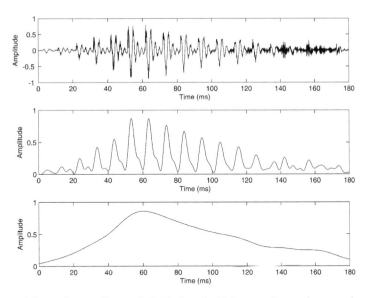

FIGURE 4 Speech waveform, 250-Hz bandwidth speech envelope and 63-Hz bandwidth speech envelope (*top to bottom*), for a male speaking the word "is".

delay estimation technique summarized in [29] and standardized in [30] uses two stages. The first stage performs a wide, low-resolution search to generate an initial delay estimate. This is accomplished by cross-correlating subsampled speech envelopes. In this case, speech envelopes are calculated by rectifying and low-pass filtering speech waveform samples, and examples of such envelopes are given in Figure 4. Subsampling consistent with the envelope bandwidth can then be performed. The peak in the cross-correlation function between the envelopes of the test and reference signals provides the initial estimate of the SUT delay, which is then removed.

The second stage performs cross-correlations between estimated power spectral densities (PSDs). A single reference signal PSD is correlated against a family of test signal PSDs that are calculated from a range of adjacent time windows in the test signal. If these correlations consistently indicate an additional time shift between the test and reference signals, then this second stage result is combined with the first stage result. If the second stage results do not pass the consistency test, then the first stage delay estimate is used alone, and the associated uncertainty in delay is deemed to be an inherent property of the SUT.

3.3 Estimation of Time-Varying Delay

Some SUTs can introduce time-varying delays into speech signals. When speech is transmitted as a series of packets over a shared transmission channel (as in VoIP), transmission errors and congestion can cause the received stream of packets to be non-uniformly spaced in time, and some packets may be lost completely. Packet buffers at the receiving end can mitigate these effects to some extent, but in practice the received speech may still contain temporal discontinuities where the end-to-end delay increases or decreases abruptly. Depending on the duration and location of these temporal discontinuities, they can affect perceived speech quality [31].

Time-varying delays must be estimated and removed so that the quality estimation block receives properly synchronized input signals. Ideally, the perceptual effects of the accompanying temporal discontinuities would be accounted for in the quality estimation algorithm as well. In other words, when time-varying delays are present, the delay estimation and compensation problem and the speech quality estimation problem are no longer truly separable.

3.3.1 A Top-Down Time Delay Estimation Algorithm

One algorithm for estimating time-varying delays is given in [32] and is included as part of a standardized algorithm for estimating perceived speech quality [33]. This algorithm includes a coarse alignment step that performs cross-correlation on logarithmically compressed speech envelopes. The fine alignment step uses 64-ms frames of speech samples. These frames are weighted by the Hann window and are overlapped by 75%. Time domain cross-correlation is applied to these frames and the index and value of the correlation maximum are used to calculate an estimated fine time delay and a confidence in that estimate, respectively.

The coarse alignment step is first applied to the entire available recordings of the reference and test signals to find a single best global time delay. To refine this result each individual contiguous segment of active speech (each utterance) is then considered separately. The coarse and fine steps are applied to each utterance and a measure of confidence is calculated. Multiple two-way splits of each utterance are then explored, and the split that gives the highest confidence is compared with the confidence before splitting. The situation (split or no split) with the higher confidence is considered to be the correct situation. If a split is identified, that utterance is marked as consisting of two different segments, each with its own separate delay value. If an utterance is identified to have two different delays, then the trial splitting is again applied to each segment identified. This recursive trial splitting continues until no segment can be successfully split into a pair of segments. This can be described as a top-down approach, starting with the full recordings and successively working down to smaller and smaller segments of constant delay.

3.3.2 A Bottom-Up Time Delay Estimation Algorithm

A different solution for estimating time-varying delays in speech is described in [34] and is standardized in [35]. This algorithm starts with cross-correlation of subsampled speech envelopes with 63-Hz bandwidth to find the single best global time delay. This delay value is then further refined through additional steps. The next step uses speech envelopes with 250-Hz bandwidth. Unlike the 63-Hz bandwidth, the 250-Hz bandwidth preserves much of the pitch information in the speech signal (see Figure 4), thus allowing for its potential use in delay estimation. A 150-ms window of the test signal speech envelope is cross-correlated with 150-ms windows of the reference signal envelope to find the best delay estimate for each such window in the test signal. This window is advanced in 40-ms steps, and the sequence of results is then further processed and median filtered to generate an initial delay history. For each segment of constant delay identified in this initial delay history, additional cross-correlation is performed using rectified speech. The use of rectified speech makes the algorithm completely robust to phase reversal in the SUT, and it exploits the full bandwidth of the speech signal. If the cross correlation results pass a set of reliability tests, the associated time shift is used to further refine the initial delay history. This algorithm can be described as a bottom-up algorithm, since its core stages move small processing windows over longer speech signals. When a real-time implementation is desired, data-flow issues can make a bottom-up solution more practical than a top-down solution.

In the following summaries of algorithms that estimate telecommunications speech quality, we use $x(i)$ and $y(i)$ to denote the reference and test signals presented to the speech quality estimation algorithm, and we assume that gain compensation and delay compensation have already been performed using one of the methods outlined above.

3.4 Frequency-Domain Estimators

Digital speech coders may exploit the fact that the preservation of speech waveforms is sufficient, but not necessary for the preservation of speech quality. Estimating speech quality directly from the sampled time domain signals $x(i)$ and $y(i)$, or the error signal $y(i) - x(i)$, can be effective for devices that preserve speech waveforms, but it cannot be effective across the broader class of speech coders, network elements, and transmission channel conditions. Since speech power spectra are approximately preserved, estimators that work in the frequency domain provide a good starting place. Two popular frequency-domain estimators are summarized below. Additional details and relationships between them are given in [36].

3.4.1 Log Spectral Error

Perhaps the most basic frequency-domain estimators of perceived speech quality are log spectral error (LSE) based estimators. Here the input and output speech sample streams are typically grouped into frames of $2N$ samples, windowed with a smooth window, and transformed via an FFT to create N complex frequency-domain samples X_j for each frame of the reference signal, and likewise (Y_j) for the test signal. Frame lengths from 8 to 32 ms are typical. Frames shorter than 8 ms typically do not contain enough samples to yield the desired frequency resolution, and frames longer than 32 ms are very unlikely to contain a stationary segment of speech. Frames are often allowed to overlap by 50% so that any portion of a speech signal that is interrupted by a frame boundary will also be processed uninterrupted in the center of a frame as well. When combined with the choice of an appropriate window, overlap can also ensure that every speech sample receives the same total weight in the calculation. For each such frame, the spectral representations can be used to calculate a log spectral error as

$$\text{LSE} = \left(\frac{1}{N} \sum_{j=1}^{N} \left| \alpha_j \cdot 20 \cdot \log_{10} \left(\frac{|Y_j|}{|X_j|} \right) \right|^p \right)^{\frac{1}{p}}. \quad (10)$$

The weights α_j allow for spectral weighting in the LSE calculation. With $p = 1$, the resulting LSE is the mean absolute spectral error in dB. With $p = 2$, it becomes the RMS spectral error in dB. LSE provides a simple scalar indication of how well an SUT preserves speech power spectra on a per-frame basis. It has been commonly used as a figure of merit in the quantization of linear prediction coefficient, reflection coefficient, or line-spectral frequency representations of smoothed speech spectra. It can be averaged or otherwise processed across multiple frames to generate an estimate of perceived speech quality for a speech signal. With the addition of simple normalization steps, LSE can form the basis of a relatively effective estimator of perceived speech quality [37].

3.4.2 Cepstral Distance

An alternate way of comparing log speech power spectra is through cepstral analysis. The discrete time Fourier transform of the logarithm of a sampled speech PSD yields a set of cepstral coefficients. In a typical cepstral coefficient calculation the speech signal is processed in frames of 8–32 ms in length, often with 50% overlap of frames. Each speech frame is multiplied by a smooth window and an FFT is then applied, followed by the magnitude and then the logarithm operators, yielding a sampled estimated log PSD for the given frame of speech. An FFT is then applied to this result to generate a set of cepstral coefficients. Alternately, cepstral coefficients can be calculated from linear prediction coefficients or reflection coefficients. Cepstral coefficients can then be used in the calculation of cepstral distance (CD):

$$\mathrm{CD} = \left(\frac{1}{N} \sum_{i=1}^{N} \left(\alpha_i \cdot |c_x(i) - c_y(i)| \right)^p \right)^{\frac{1}{p}}, \qquad (11)$$

where $c_x(i)$ and $c_y(i)$ represent the cepstral coefficients associated with a given frame of the reference and test speech signals. The cepstral-domain quantity, CD, can then be averaged or otherwise processed across multiple frames to generate an estimate of perceived speech quality for an entire test speech signal. While (11) gives a generalized form of CD, most published works have used $N = 8\text{–}12$, the norm $p = 2$, and weights $\alpha_i = 1$.

Optimization of N, p, and α_i in (11) may yield improved speech quality estimates for specific applications. The cepstral coefficients provide a spectral representation for the log PSD. Thus the lower-order cepstral coefficients describe the amount of smoother, wider variation in reference and test speech frame PSDs with respect to frequency (e.g., spectral tilt) while the higher-order cepstral coefficients indicate the amount of more abrupt, localized variation in the PSDs with respect to frequency (e.g., speech formant structure). The observation that changes in these two types of variation (e.g., fixed SUT frequency response vs. SUT formant distortion) are perceived very differently by listeners motivates a search for an optimal set of weights. More generally though, these and other observations motivate a search for perception-based estimators of perceived speech quality.

3.5 Perception-Based Estimators

When listeners report their perceptions of speech quality, they are necessarily making judgments based on acoustic stimuli that have been processed by their human auditory systems. When seeking to develop a signal processing algorithm that estimates perceived speech quality, it is thus advantageous to account for both the human auditory system and human judgment in that algorithm. Equipped with complete understanding of these processes and accurate mathematical models for them, one might expect to develop a signal processing algorithm that can indicate perceived speech quality in a way that is very close to the indications coming from human listeners. This notion has led to the development of perception-based algorithms that estimate perceived speech quality. These perception-based algorithms often work in the frequency domain, but it is not the Hertz scale frequency domain. Rather, it is a psychoacoustic frequency domain, developed to be consistent with basic properties of human hearing.

Figure 5 shows the common approach to perception-based estimation of perceived speech quality. It can be helpful to think of the perceptual transformation as an emulation of human hearing and the distance measure as an emulation of human judgment. Note, however, that hearing and judgment are complex interdependent processes. Treating them as separable sequential processes is pragmatic simplification. Figure 5 shows a distance measure (judgment) that is fed simultaneously by two perceptually transformed speech signals. This differs from the situation with humans in subjective tests. They are never exposed to the two speech signals simultaneously, and in the most common subjective speech quality tests, called absolute category rating (ACR) tests, they never hear the reference signal at all, yet they reliably judge the quality of the test signal. It may be that human subjects facilitate this task by forming an internal, idealized version of the test signal, against which they can judge the test signal. In the less common degradation category rating (DCR) subjective tests, human subjects hear the reference and test signals sequentially, a situation that would be closer to but different from that described in Figure 5.

It is frequently the case that a perception-based algorithm will contain both a priori models for hearing and judgment and empirically determined components. The a priori models are typically based on hearing and judgment results found in the literature and the empirically determined components are often selected to maximize the agreement between the algorithm

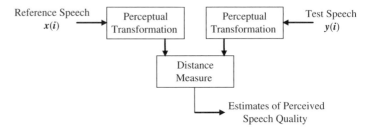

FIGURE 5 Perception-based speech quality estimation.

output and some body of speech quality truth data. Each of the estimators summarized below was designed to estimate speech quality in the 300 to 3,400-Hz telephone bandwidth, but straightforward extensions to wider bandwidths are possible.

3.5.1 Bark Spectral Distortion

Bark spectral distortion (BSD) [38] is an algorithm for estimating perceived speech quality that incorporates three basic properties of human auditory perception. The first of these is simultaneous masking, whereby a spectral component at a given frequency can raise the threshold of hearing at neighboring frequencies, making it harder to detect other spectral components that may be present at that same instant in time. This property of hearing can be conveniently conceptualized and modeled through the notion of critical bands and critical band filters [39]. The shape of a critical band filter serves as a simple model for simultaneous masking at a particular frequency. A set of appropriately spaced critical band filters forms a critical band filterbank, and the filterbank output can exhibit a first-order approximation to the simultaneous masking effects on complex signals. The filterbank output can be described as a critical band power spectrum, or a Bark power spectrum, since Bark is one psychoacoustic unit of measure for frequency. The filters are typically spaced at uniform intervals on the Bark scale and the filterbank output thus provides a power spectrum that is uniformly sampled on the Bark scale.

Figure 6 displays the magnitude response of four example critical band filters on the Hertz scale. These filters are centered at 310, 870, 1,830, and 3,630 Hz, corresponding to 3, 7, 11, and 15 Bark, respectively (see equation (1) and Figure 2). The −3 dB filter bandwidths are 110, 180, 320, and 610 Hz, corresponding to a 1.0-Bark width in all four cases. When viewed on the Bark scale these four filters are uniformly spaced (4.0 Bark), have uniform −3 dB bandwidth (1.0 Bark) and identical shape. Each filter shows asymptotic slopes of +10 dB/Bark below center and −25 dB/Bark above center and these slopes reflect the nominal first-order simultaneous masking properties of human hearing.

A second basic property in human hearing is its frequency-dependent sensitivity. For example, a sinusoidal signal at 3 kHz is perceived as significantly louder than a sinusoidal signal at 100 Hz with the same power. In the BSD algorithm, these first two properties are modeled by a weighted critical band filterbank. Each filter in the bank is weighted to reflect the relative sensitivity of human hearing in the corresponding frequency range. In the BSD algorithm the filters are spaced at 1-Bark intervals and the first filter is centered at 1.5 Bark. This arrangement is a design choice and results in a total of 15 filters across the 1.5–15.5 Bark

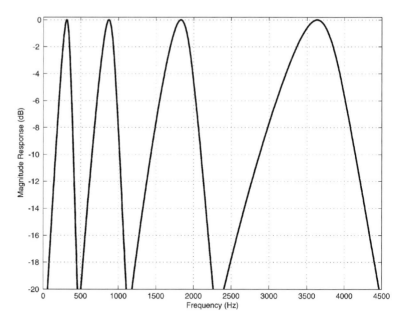

FIGURE 6 Magnitude response of four example critical band filters.

(150–4,000 Hz) band. An example of such a filterbank is given in Figure 7.

In the BSD algorithm, the filtering described above is implemented in the frequency domain. The speech signals are processed in 10-ms frames with 50% overlap. Each frame is then Hamming windowed and an FFT and magnitude function is applied. This results in 40 frequency-domain magnitude samples between 0 and 4,000 Hz for each 10-ms frame of speech that is processed. For a given frame, these 40 samples can be organized into the column vectors x and y for the reference and test signals, respectively. A 15-by-40 matrix, F, can then be used to properly accumulate these samples as the 15 filters (see Figure 7 for example) would, resulting in the length 15 Bark spectrum column vectors $\tilde{x}$ and $\tilde{y}$:

$$\tilde{x} = Fx, \tilde{y} = Fy. \qquad (12)$$

Each row of F contains the power response of one filter.

The third hearing property treated in the BSD algorithm is the perception of loudness. Even at a single fixed frequency, the perception of loudness is not linearly related to signal power. The transformation applied to each sample of $\tilde{x}$ and $\tilde{y}$ to account for this is

$$Lx_j = \alpha 2^{20 \log_{10}(\tilde{x}_j)(\tilde{x}_j)/10} = \alpha \tilde{x}_j^{0.60}, \; Ly_j = \alpha \tilde{y}_j^{0.60}, \qquad (13)$$

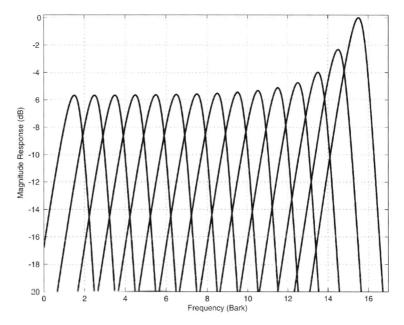

FIGURE 7 Magnitude response of example filterbank with 15 critical band filters centered at 1.5, 2.5,..., 15.5 Bark and weighted to reflect the approximate frequency sensitivity of hearing.

where α is a constant. Reference [38] attributes this transformation to [40]. The samples Lx_j and Ly_j form the Bark loudness spectra of the corresponding frames of the reference and test signals. They reflect the ear's non-uniform integration across frequencies, its non-uniform sensitivity to different frequencies, and the non-linear growth of perceived loudness as a function of signal power. The samples Lx_j and Ly_j are the final output of the perceptual transformations used in BSD.

The distance measure used in the BSD algorithm is squared Euclidean distance:

$$\text{BSD} = \sum_{j=1}^{15} (Lx_j - Ly_j)^2. \qquad (14)$$

The BSD value given by (14) corresponds to one pair of 10-ms frames of reference and test speech. These are then averaged across all frames of interest and the resulting value is normalized by the time average of $\sum_{j=1}^{15} Lx_j^2$, thus creating a final BSD value which might also be described as perceptual noise-loudness-to-signal-loudness ratio.

Other researchers have noted that some smaller loudness differences might be masked by the speech signal and therefore would not be noticed by listeners. Thus they extended the BSD to model this effect [41]. The relationships between BSD values and subjective speech quality test results can be found in [38].

3.5.2 Perceptual Speech Quality Measure

The perceptual speech quality measure (PSQM) algorithm introduced in [42] and standardized in [43] further extends the notion of perception-based speech quality estimation. PSQM transforms signals into a perceptual domain where they are then compared by a distance measure. The PSQM does not use critical band filtering because experiments indicated that modeling simultaneous masking was not advantageous in this application [42]. It does use a critical band or Bark scale spectral representation with 56 samples across the 0–4,000 Hz band, resulting in sample spacing of about 0.31 Bark.

The two speech signals are processed in 32-ms frames with 50% overlap. Each frame is then Hanning windowed and an FFT and magnitude squared function is applied. To transform this Hertz domain representation (with 128 samples of PSD) to a Bark scale representation (with 56 samples), appropriately sized contiguous groups of Hertz domain samples are averaged and then normalized by the local sample density correction factor $\frac{\Delta \text{Hz}(f)}{\Delta \text{Bark}}$. This can be implemented in the same manner as the matrix filterbank implementation used in the BSD, but in this case the filters are rectangular and non-overlapping.

The PSQM algorithm then performs a power normalization step. Next it models the power frequency response of a typical telephone earpiece and the background noise spectrum that is sometimes used in subjective speech quality tests.

The frequency-dependent sensitivity of hearing and the perception of loudness are modeled together as

$$Lx_j = \alpha \left(\frac{T_j}{0.5}\right)^\gamma \cdot \left[\left(0.5 + 0.5 \frac{Px_j}{T_j}\right)^\gamma - 1\right],$$

$$Ly_j = \alpha \left(\frac{T_j}{0.5}\right)^\gamma \cdot \left[\left(0.5 + 0.5 \frac{Py_j}{T_j}\right)^\gamma - 1\right], \qquad (15)$$

where T_j is the threshold of hearing for spectral sample j, Px_j and Py_j are the reference and test signal powers at spectral sample j, and α is a constant. Reference [43] attributes this loudness model to [44]. The loudness growth exponent γ was empirically optimized to maximize agreement between the PSQM output and subjective speech quality test results, resulting in the value

$\gamma = 0.001$. The net effect of this very small exponent is that at each frequency j, the calculated loudness value Lx_j or Ly_j is nearly logarithmically related to the corresponding power Px_j or Py_j. These logarithmic loudness relationships are individually shifted at each frequency, consistent with the hearing threshold at that frequency [45]. A loudness normalization step completes the operations of the perceptual transformation. For each pair of speech frames processed (reference and test), the output of the perceptual transformation is two sets of 56 compressed loudness density values Lx_j and Ly_j.

The distance measure used in PSQM is based on absolute differences of compressed loudness densities $|Ly_j - Lx_j|$. However, it was recognized that added spectral components (which cause $Ly_j > Lx_j$) are likely to be more annoying than missing spectral components (which cause $Ly_j < Lx_j$). Thus the information lost when taking the absolute difference is then restored with additional processing that effectively weights added components about twice as heavily as subtractive components [45].

This is followed by averaging over frequency and time, with a provision to assign different weights to results from frames that contain active speech and results from frames that contain only background noise. The final result is a non-negative value called noise disturbance. Higher values of noise disturbance indicate lower estimated speech quality. Relationships between noise disturbance values and subjective speech quality test results are given in [42].

3.5.3 Measuring Normalizing Block Algorithms

When seeking to emulate human responses, it is logical to use both perceptual transformations and a distance measure, as shown in Figure 5. Further, it is logical that these two components should be equally well developed so that both hearing and judgment receive sufficient consideration. There are an abundance of experimental results available to guide modeling of human hearing and thus guide the development of perceptual transformations. There is significantly less information available to guide the modeling of auditory judgment and thus the development of distance measures. This has understandably resulted in fairly sophisticated perceptual transformations that feed relatively simplistic distance measures. Intuitively, this situation seems less than optimal. Indeed, one study has concluded that distance measures are at least as important as perceptual transformations, and the benefits of sophisticated perceptual transformations tend to be lost when they are followed with simplistic distance measures [46]. Simplistic distance measures may show unrealistic invariance to sign (added vs. missing signal components) or unrealistic invariance to the temporal or spectral scale of impairments (constant vs. rapidly fluctuating gain, low-pass vs. comb filtering).

In the measuring normalizing block (MNB) speech quality estimation algorithms, attention is given to bringing the contributions of the perceptual transformation and the distance measure into balance and eliminating the undesirable invariances cited above. These algorithms are described in [29], verified in [47], and standardized in [48] and [49]. In the MNB algorithms the perceptual transformation models only two basic properties of human hearing. It works in the frequency domain using bands organized according to the critical band concept and uses a simple logarithmic non-linearity to create an approximation of the relationship between signal power and perceived loudness.

The distance measure in the MNB algorithm consists of a hierarchy of processing blocks that can measure and then normalize spectral differences out of the test signal. This can be done on different time and frequency scales, and proceeds from larger scales to smaller scales, following the hypothesis that this would be most likely to emulate listeners' patterns of reaction and adaptation to spectral differences. In addition, the measurements produced by these blocks separate positive and negative spectral differences at the various scales.

To measure spectral differences over a given time-frequency scale, the perceived loudness time-frequency representations $X(i,j)$ and $Y(i,j)$ for the reference and test signals are each individually summed over that time-frequency region. The difference between the two results is the measurement for that scale, and this difference is subtracted from the appropriate region of $Y(i,j)$ to form $\tilde{Y}(i,j)$. Because of this normalization, $X(i,j)$ and $\tilde{Y}(i,j)$ now "agree" at that scale. That is, repetition of this measurement would now yield a measurement of zero. However, further measurements at smaller scales (sub-regions) will continue to reveal additional, smaller scale differences between $X(i,j)$ and $\tilde{Y}(i,j)$. As the measurements and normalizations to the test signal proceed, the normalized test signal becomes more and more similar to the reference signal, first from a large-scale view and then from progressively smaller scale views.

An example of this process is given in Figure 8. For visual simplicity the time scale used is just the time associated with a single 16-ms frame. Thus the figure shows Bark loudness spectra for a single pair of reference (dotted) and test (solid) speech frames using arbitrary loudness units. The frequency scales used

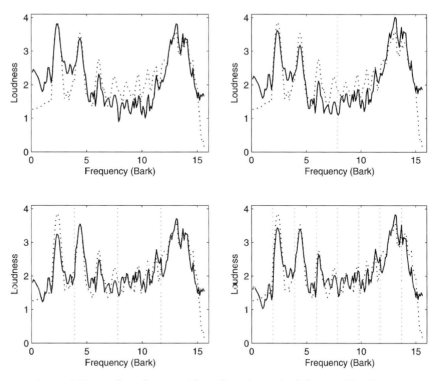

FIGURE 8 Example reference (*dotted*) and test (*solid*) signal loudness spectra for a sequence of MNB steps at decreasing frequency scales (*top left to lower right*). Loudness units are arbitrary.

range from 15.5 to 1.94 Bark and these decrease as one moves through the four panels of the figure. The upper left panel shows the two spectra after the test spectrum has been normalized across the full 15.5-Bark band. Because of this normalization, the areas under the test and reference spectra are the same. The upper right panel shows the two spectra after two additional measurements and normalizations. In this case the measurements and normalizations were conducted at the half-band (7.75 Bark) frequency scale and the consequence is that for both the lower and upper halves of the frequency range, the areas under the reference and test spectra agree. The lower left panel shows the two spectra after four additional measurements and normalizations and these were conducted at the quarter-band (3.88 Bark) frequency scale. Thus for each quarter of frequency range, the area under the reference and test spectra agree. Finally, the lower right panel depicts the result after eight additional measurements and normalizations at the eighth-band (1.94 Bark) scale. As one moves through the four panels in the figure, the test spectrum is evolving toward the reference spectrum and is showing agreement at increasingly finer frequency scales.

In an MNB algorithm a record of the measurements that drove this evolution is preserved and these measurements are organized by sign, time scale, and frequency scale. For example, a measurement associated with a larger frequency scale may track SUT spectral tilt, while a measurement associated with a smaller frequency scale may indicate SUT distortion of speech formants, independent of that spectral tilt.

The MNB distance measure process can also be viewed as a dynamic decomposition of the test signal where the decomposition proceeds in a space that is defined partly by human hearing and judgment (via the MNB structure) and partly by the reference signal. In this view the measurements become the coefficients of the decomposition.

A linear combination of the measurements then forms a single auditory distance (AD) value that describes the perceived distance between the input reference and test signals. Since it is a distance, $0 \leq AD < \infty$ and $AD = 0$ would indicate perfect transmission and thus perfect quality. Subjective speech quality tests generally produce results on a finite scale with increasing values indicating higher speech quality. Thus a logistic function

$$L(AD) = \frac{1}{1 + \exp(\alpha \cdot AD + \beta)} \qquad (16)$$

with $0 < \alpha$ is used to map AD to the interval, $0 < L(AD) < 1$. An auditory distance of zero will give $L(AD) = (1 + e^\beta)^{-1} \approx 1$, while large values of auditory distance will give $L(AD) \to 0$. The linear combination weights used to determine *AD* and the logistic function parameters α and β were selected to maximize correlation between L(AD) and a large set of subjective test results. Thus while the MNB structure can be viewed as an a priori model for human judgment, the details of how that structure operates (the weights) are found empirically. The relationships between L(AD) values and subjective speech quality test results can be found in [47].

While the MNB concept is broad, the specific MNB structures given in [29] operate at a single time scale, effectively measuring and then removing a single time-averaged frequency response of the SUT, and that measurement then figures into the calculation of AD. The MNB structures measure and remove spectral errors at 1–3 different frequency scales. *AD* is then calculated as a linear combination of 11 or 12 resulting linearly independent measurements.

3.5.4 Perceptual Evaluation of Speech Quality

As mentioned in Sect. 3.3, some SUTs can introduce time-varying delays into speech signals. The perceptual evaluation of speech quality (PESQ) algorithm was developed to respond to this situation. It is described in [32] and [50] and standardized in [33] which supercedes [43] and [49]. The algorithm combines the top-down variable delay estimation algorithm described in Sect. 3.2 above and in [32] with an enhanced version of the PSQM algorithm. The PESQ algorithm approximates the delay estimation and quality estimation processes as separable processes. The estimated delay history is removed from the speech signals before quality estimation begins. This delay history does not generally drive the quality estimation algorithm. (The exception is when a speech frame includes a delay increase that exceeds 16 ms. Results from such frames are not included in the final quality estimates produced by PESQ.) If the quality estimation process detects unusually high levels of distortion in specific segments of the speech signals, those segments are further analyzed to see if the delay estimate there can be improved.

The quality estimation stage of the PESQ algorithm is an enhanced version of the PSQM algorithm. One significant difference from the PSQM algorithm follows the creation of the Bark scale power spectral densities for the reference and test signals. The time-averaged frequency response of the SUT is measured and then partially normalized out of the reference signal. This reduces the algorithm's response to fixed filtering effects, consistent with the fact that listeners can often adapt to these effects. In fact, a listener may not even notice or object to mild fixed filtering, especially if no reference signal is available for comparison (as is the case in ACR subjective tests).

Another significant difference from the PSQM algorithm follows immediately. Here any time-varying gain is measured and partially normalized out of the test signal. This is accomplished by calculating the SUT gain for each pair of 32-ms reference and test speech frames, smoothing this gain history (time constant of 16 ms, low-pass corner frequency of 63 Hz), and then using the result to normalize the test signal. Roughly speaking, low-frequency (relative to 63 Hz) gain variations are removed (and presumably are not objectionable to listeners), while high-frequency (relative to 63 Hz) gain variations are retained and their effects are subsequently measured.

The frequency-dependent sensitivity of hearing and the perception of loudness are modeled together as given in (15) for the PSQM algorithm. In the case of the PESQ algorithm, however, the parameter γ takes the value 0.23 for spectral samples at frequencies above 4 Bark (approximately 430 Hz) and a slightly larger value of γ below that frequency to model the slightly faster growth of perceived loudness at those lower frequencies and [50] attributes these γ values to [44]. This step completes the operations of the perceptual transformation. After processing multiple frames of the input speech signals, the perceptual transformations produce compressed loudness densities that are functions of time (i) and perceptual frequency (j), $Lx(i,j)$ and $Ly(i,j)$.

The distance measure in the PESQ algorithm is more sophisticated than that used in PSQM. Both a difference time-frequency distribution

$$D(i,j) = Ly(i,j) - Lx(i,j) \qquad (17)$$

and a corresponding masking time-frequency distribution

$$M(i,j) = 0.25 \cdot \min(Ly(i,j), Lx(i,j)) \qquad (18)$$

are calculated. For each time-frequency location, the bipolar difference time-frequency distribution is then reduced in magnitude according to the positive masking time-frequency distribution:

$$\begin{aligned}\tilde{D}(i,j) &= D(i,j) - M(i,j) \quad \text{when } M(i,j) \leq D(i,j), \\ &= D(i,j) + M(i,j) \quad \text{when } D(i,j) \leq -M(i,j) \qquad (19)\\ &= 0 \qquad\qquad\qquad\quad\ \text{otherwise.}\end{aligned}$$

Equation (19) provides a simple model for masking: a signal at a given time-frequency location can raise the threshold for detecting other signals at or near that time-frequency location.

The resulting masked difference time-frequency distribution $\tilde{D}(i,j)$ is called a disturbance density. It is further processed along two different paths so that the PESQ algorithm can respond differently to positive disturbances (due to added time-frequency components) and negative disturbances (caused by missing-frequency components) consistent with the human response to these two types of disturbances. The symmetric path retains both types of disturbances while the asymmetric path retains and emphasizes only the positive disturbances.

Both paths then use a time- and frequency-weighted L_p absolute value aggregation process across frequency:

$$\tilde{D}(i) = \beta_i \cdot \left[\sum_{j=1}^{N} \left(|\tilde{D}(i,j)| \cdot \alpha_j\right)^p\right]^{\frac{1}{p}}, \qquad (20)$$

where α_j are frequency-specific weights and β_i are temporal weights computed from the power of the ith frame of the reference speech signal in a way that emphasizes disturbances that occur during quieter portions of the speech signal and deemphasizes disturbances that occur during louder portions of the speech signal. Once again, this is an operation that is consistent with masking. The norm $p=3$ is used in the symmetric branch and $p=1$ is used in the asymmetric branch.

The aggregation of $\tilde{D}(i)$ over time is done at two separate time scales in both the symmetric and asymmetric branches. First, groups of $N = 20$ results (spanning approximately 320 ms) are aggregated using

$$\tilde{D} = \left[\frac{1}{N}\sum_{i=1}^{N}\tilde{D}(i)^p\right]^{\frac{1}{p}} \qquad (21)$$

with $p=6$. This models an approximate "max" or "worst case" type of response over this time window. These results are then aggregated over the entire duration of the speech signals (typically lasting 10 s, $N \approx 31$) in a similar manner using $p=2$. This models a more forgiving assessment of the perceived disturbances over the longer term. The two final disturbance values created by the two paths (symmetric and asymmetric) are then combined in a linear combination with empirically determined weights to create a final estimate of perceived speech quality that typically ranges from 1 (worst quality) to 4.5 (best quality). The agreement between these values and subject speech quality test results is described in [50].

3.6 Output-Based Speech Quality Estimation

While Figure 3 describes a common approach to speech-based, signal processing-based speech quality estimation, it is not the only approach. One can also estimate speech quality using only the test signal that is output by the SUT. This is called the single-ended or output-based approach and it has both advantages and limitations.

The approach shown in Figure 3 requires that the input and output of the SUT be available at a common location. This is not a limitation for laboratory testing, but for field testing it requires that a separate, distortion-free, communications channel or storage medium be provided. Alternately the SUT can be removed from service, recorded speech signals can be injected into the SUT, and copies of those signals can be compared to the SUT output. Clearly the output-based approach transcends these limitations. The output-based approach can non-intrusively monitor and estimate speech quality using only the SUT output. An additional advantage is that the issues of gain and delay estimation and compensation disappear. Finally, we note that the output-based approach may appear more pleasing in principle because it provides a closer parallel to the most common subjective testing environment (the ACR environment).

The limitations of the output-based approach stem from the fact that no reference signal is available. While listeners in ACR subjective tests routinely judge speech quality with no reference, the loss of the reference presents a bigger challenge to the signal processing approach. With less information available, it seems inevitable that resulting speech quality estimates must suffer. The perception-based approach described in Figure 5 is no longer possible. The problem moves from finding appropriate measurements of distance in an appropriate perceptual space to the problem of making a perceptually consistent classification of a single signal. Two solutions are summarized below. Examples of other work can be found in [51] and [52].

3.6.1 Auditory Non-intrusive Quality Estimation

The auditory non-intrusive quality estimation (ANIQUE) algorithm [53] incorporates elements of human speech production and human hearing and some of the key steps of are summarized here. The SUT output signal is processed though a bank of 23 critical band filters uniformly spaced by about 0.62 Bark between 125 Hz (1.2 Bark) and 3,500 Hz (14.8 Bark). These filters are implemented in the time domain. Each of the resulting 23 band-limited signals, $s_j(i)$, is then expressed as a carrier or instantaneous phase signal, $\phi_j(i)$, modulated by a temporal envelope, $\gamma_j(i)$, where i is the time sample index and j is the frequency band index. This decomposition follows directly from the introduction of $\hat{s}_j(i)$ which is the Hilbert transform of $s_j(i)$ and the analytic signal $z_j(i) = s_j(i) + \sqrt{-1} \cdot \hat{\mathbf{s}}_{\mathbf{j}}(\mathbf{i})$:

$$s_j(i) = \mathbb{R}\left(z_j(i)\right) = \gamma_j(i) \cdot \cos\left(\phi_j(i)\right), \qquad (22)$$

where

$$\gamma_j(i) = \sqrt{s_j^2(i) + \hat{s}_j^2(i)} \text{ and } \phi_j(i) = \arctan\left(\frac{\hat{s}_j(i)}{s_j(i)}\right). \qquad (23)$$

Based on the known importance of the temporal envelopes and the relative unimportance of the carrier signal or phase information for speech perception, the initial analysis proceeds based on the 23 speech envelopes $\gamma_j(i)$ alone. More specifically, it is the spectral content of these envelopes (or modulation spectra) that drive the initial quality estimation. The modulation spectrum of an envelope is estimated by breaking the envelope into 256-ms processing frames with 75% overlap. A Hamming window and an FFT are applied to each frame and the magnitude of the result is taken. This results in the modulation spectrum $\Gamma_j(k)$, where k is the envelope modulation frequency index.

The next step is to analyze each modulation spectrum and separate the lower frequency components (which can be attributed to natural human speech) from the higher frequency components (which can only be attributed to noise and distortion induced by the SUT) in order to form an articulation-to-nonarticulation ratio (ANR). (In this context, articulation refers to speech production, not speech perception.)

For each frame of the jth envelope, the lower frequency components are those between 2 and 30 Hz (a range consistent with the speed of movement in human articulation) and higher frequency components are those from 30 Hz to $\frac{1}{2}BW_j$ where BW_j is the bandwidth of the jth critical band filter. This upper limit of $\frac{1}{2}BW_j$ is consistent with the highest envelope modulation frequency that is relevant to the perception of speech quality. Thus the modulation power in 2–30 Hz and the modulation power in 30 to $\frac{1}{2}BW_j$ Hz form the basis for the jth ANR, designated Λ_j. These 23 ANRs are then aggregated across the 23 critical band channels according to

$$\nu = \left[\sum_{j=1}^{23} \Lambda_j^{0.80} \right]^{0.38}, \quad (24)$$

where the exponents were determined empirically to maximize agreement between the ANIQUE output and a large set of subjective test scores.

Equation (24) provides a base quality estimate for each 256-ms frame of SUT output speech. Significant additional processing follows, including explicit algorithm branches to account for unnaturally abrupt starts and stops in the SUT output speech (possibly attributable to temporal discontinuities or transmission channel errors). Separate treatments for loud, faint, and inaudible speech frames are also provided. In addition, a normalization process serves to reduce the variance in ν attributable to input speech articulation and speech content, thus making ν a better indicator of the SUT output quality. Ultimately, all results are combined to generate the final output of the ANIQUE algorithm: a scalar variable that increases to indicate higher estimated speech quality. Relationships between this output and subjective test results are given in [53]. An enhanced version of the ANIQUE algorithm is standardized in [54].

3.6.2 ITU-T Recommendation P.563

A different approach is standardized in ITU-T Recommendation P.563 [55]. This algorithm combines the work of multiple research teams and can be viewed as a large and diverse, yet well-organized set of analysis tools, along with a set of signal-driven rules for automatically selecting, using, and combining the results of the various tools. After preprocessing of the SUT output speech, key parameters are extracted to move through a decision tree to classify the dominant impairment: constant noise, temporal impairments, temporally localized noise, speech with artificially high periodicity, unnatural male speech, or unnatural female speech. Once the dominant impairment has been identified, up to approximately 10 additional parameters may be extracted to enable a basic speech quality estimate customized to that type of impairment. This result is then combined linearly with up to approximately 10 additional parameters to form the final estimate of speech quality. The set of all parameters available covers a wide range from pitch-related parameters, to statistics of linear prediction coefficients, to SNR-related parameters, to tube dimensions in a vocal tract model.

One branch through the algorithm uses a linear prediction-based speech enhancer to form a pseudo-reference speech signal. This is then compared with the SUT output signal (the test signal) using an algorithm that has much in common with the PESQ algorithm. Ultimately, all relevant results are combined to generate the final output of the P.563 algorithm: a scalar variable that increases to indicate higher estimated speech quality.

4 • Conclusion

This chapter has presented key concepts associated with the signal processing approaches to the estimation of speech intelligibility and speech quality. Algorithms that estimate speech intelligibility have generally been developed for, and are most applicable to, acoustic transmission combined with noise, reverberation, and possibly some linear sound reinforcement equipment. Algorithms that estimate speech quality have generally been developed for, and are most applicable to, telecommunications equipment and systems. This segmentation of the field may be partially attributed to the inherent attributes of the two situations, but more generalized solutions are very likely possible and are certainly desirable. One rather demanding example is the need to estimate speech intelligibility in adaptive, non-linear digital communications systems with non-stationary channel conditions, possibly compounded with acoustic background noise at the talker or listener ends.

This chapter has focused on a small number of example algorithms that estimate speech intelligibility or quality. It has provided some level of detail on each, especially regarding the various signal processing operations invoked and how they can model or at least emulate various relevant attributes of human hearing, speech production, or speech transmission. Necessarily, much diverse and significant work has gone unmentioned in this chapter. Examples of some of this work can be found in [56–66]. A closely related topic is signal processing algorithms that estimate the quality of transmitted music signals. Examples of work in this area can be found in [67–70].

It is natural to ask how well each of these algorithms work. In general, each algorithm described above has been found to be effective in some application area. Any quantitative answer to this question is encumbered with numerous associated issues which could constitute an entire separate field of study. One issue is truth data. It is generally held that carefully controlled subjective tests conducted by standard procedures provide the truth data that a signal processing algorithm should be judged against. However, multiple, equally valid subjective test methodologies are available. Any of these can only generate estimated mean human responses and some confidence interval about each estimated mean. Further, subjective tests generally tend to provide relative, not absolute, results and the rating of a fixed SUT can be very different between two tests, depending on the other SUTs included in each of those tests. This can present a serious mismatch to the signal processing algorithms which are by nature immune to these context effects.

Different techniques have been employed to bridge this mismatch. These can include the use of reference SUTs in subjective tests and per-test mathematical corrections, transformations, or calibrations of subjective test results and/or algorithm results. These techniques are generally required if useful comparisons are to be made, but they also complicate the situation considerably and lead to a multitude of domains where algorithm performance might be reported.

Another issue stems from the range of SUTs that an algorithm is tested on and the relationships between the classes of SUTs used in algorithm development and those used in algorithm testing. The amount of speech data used, the amount of averaging, and the level of analysis (e.g., per sentence, per talker, per SUT) can also have a significant influence on the results reported.

The references provided in this chapter for each algorithm often provide at least one view of how well the algorithm works. Due to the considerations listed here, these views are often not directly comparable to each other. One blind study of multiple algorithms is available in [71] and the results of other comparative studies are described in the internal working documents of various telecommunications standards bodies.

The field of signal processing-based estimation of speech intelligibility and quality has plenty of room for future advances. There is always room for more effective estimators, where "more effective" can mean better agreement with subjective test results over broader classes of SUTs, transmission channel conditions, and acoustic noise and reverberation environments. This would be a particular challenge for output-based estimators. The continued evolution and deployment of telecommunications services and applications (e.g., mobile, hands-free mobile, conferencing, Internet-based) motivates additional development work, including speech intelligibility estimators for telecommunications and speech quality estimators for telecommunications applications with acoustical or electrical interfaces, supporting narrowband (300–3,400 Hz) or wideband (50–7,000 Hz) transmission, and mixtures of multiple talkers and background noises, reverberations, or music. Finally, there is room for significant progress regarding how subjective tests are done, how their results are processed, and how estimation algorithm results are processed and compared with subjective test results so that the most meaningful conclusions can be drawn.

References

1. N. Johannesson, "The ETSI Computation Model: A Tool for Transmission Planning of Telephone Networks," *IEEE Commun. Mag.*, vol. 35, pp. 70–79, Jan. 1997.
2. ITU-T Recommendation G.107, "The E-Model, A Computational Model for Use in Transmission Planning," Geneva, 2003.
3. S. Möller, *Assessment and Prediction of Speech Quality in Telecommunications*. Boston: Kluwer Academic, 2000.
4. S. Quackenbush, T. Barnwell III, & M. Clements, *Objective Measures of Speech Quality*. Englewood Cliffs, NJ: Prentice-Hall, 1988.
5. ITU-T Recommendation P.800, "Methods for Subjective Determination of Transmission Quality," Geneva, 1996.
6. ANSI S3.2-1989, "Method for Measuring the Intelligibility of Speech over Communications Systems," New York, 1990.
7. ISO TR-4870, "The Construction and Calibration of Speech Intelligibility Tests," Geneva, 1991.

8. H. Fletcher, *The ASA Edition of Speech and Hearing in Communication*. J. Allen, Ed. Woodbury, NY: Acoustical Society of America, 1995, Chap. A1.
9. N. French & J. Steinberg, "Factors Governing the Intelligibility of Speech Sounds," *J. Acoust. Soc. Am.*, vol. 19, pp. 90–119, Jan. 1947.
10. H. Fletcher, *Speech and Hearing in Communication*. New York: Van Nostrand, 1953, Chap. 17.
11. K. Kryter, "Methods for the Calculation and Use of the Articulation Index," *J. Acoust. Soc. Am.*, vol. 34, pp. 1689–1697, Nov. 1962.
12. K. Kryter, "Validation of the Articulation Index," *J. Acoust. Soc. Am.*, vol. 34, pp. 1698–1702, Nov. 1962.
13. ANSI S3.5-1969, "Methods for the Calculation of the Articulation Index," New York, 1969.
14. N. Jayant & P. Noll, *Digital Coding of Waveforms*. Englewood Cliffs, NJ: Prentice-Hall, 1984, Appendix E.
15. H. Steeneken & T. Houtgast, "A Physical Method for Measuring Speech-Transmission Quality," *J. Acoust. Soc. Am.*, vol. 67, pp. 318–326, Jan. 1980.
16. H. Steeneken & T. Houtgast, "Mutual Dependence of the Octave-Band Weights in Predicting Speech Intelligibility," *Speech Commun.*, vol. 28, pp. 109–123, 1999.
17. H. Steeneken & T. Houtgast, "Validation of the Revised STIr Method," *Speech Commun.*, vol. 38, pp. 413–425, 2002.
18. IEC 60268-16, "Sound System Equipment – Part 16: Objective Rating of Speech Intelligibility by Speech Transmission Index," Geneva, 2003.
19. T. Houtgast & H. Steeneken, "A Multi-language Evaluation of the RASTI-Method for Estimating Speech Intelligibility in Auditoria," *Acustica*, vol. 54, pp. 185–199, 1984.
20. R. Goldsworthy & J. Greenberg, "Analysis of Speech-Based Speech Transmission Index Methods with Implications for Nonlinear Operations," *J. Acoust. Soc. Am.*, vol. 116, pp. 3679–3689, Dec. 2004.
21. ANSI S3.5-1997, "Methods for Calculation of the Speech Intelligibility Index," New York, 1998.
22. S. Voran, "Listener Ratings of Speech Passbands," in *Proc. 1997 IEEE Workshop on Speech Coding for Telecommunications*, pp. 81–82, Pocono Manor, PA, 1997.
23. J. Beerends, E. Larsen, N. Iyer, & J. van Vugt, "Measurement of Speech Intelligibility Based on the PESQ Approach," in *Proc. 3rd International Conference on Measurement of Speech and Audio Quality in Networks*, pp. 27–30, Prague, Czech Republic, 2004.
24. ITU-T Recommendation P.50, "Artificial Voices," Geneva, 1999.
25. N. Kitawaki, K. Nagai, & T. Yamada, "Objective Quality Assessment of Wideband Speech Coding Using W-PESQ Measure and Artificial Voice," in *Proc. 3rd International Conference on Measurement of Speech and Audio Quality in Networks*, pp. 31–36, Prague, Czech Republic, 2004.
26. M. Werner, T. Junge, & P. Vary, "Quality Control for AMR Speech Channels in GSM Networks," in *Proc. IEEE Interna-

tional Conference on Acoustics, Speech, and Signal Processing, pp. 1076–1079, Montreal, 2004.
27. A. Karlsson, G. Heikkilä, T. Minde, M. Nordlund, & B. Timus, "Radio Link Parameter Based Speech Quality Index – SQI," in *Proc. 1999 IEEE Workshop on Speech Coding*, pp. 147–149, Porvoo, Finland, 1999.
28. S. Voran, "Compensating for Gain in Objective Quality Estimation Algorithms," in *Proc. IEEE International Conference on Acoustics, Speech, and Signal Processing*, pp. 1068–1071, Montreal, 2004.
29. S. Voran, "Objective Estimation of Perceived Speech Quality, Part I: Development of the Measuring Normalizing Block Technique," *IEEE Trans. Speech Audio Process.*, vol. 7, pp. 371–382, Jul. 1999.
30. ITU-T Recommendation P.931, "Multimedia Communications Delay, Synchronization, and Frame Rate Measurement," Geneva, 1998.
31. S. Voran, "Perception of Temporal Discontinuity Impairments in Coded Speech – A Proposal for Objective Estimators and Some Subjective Test Results," in *Proc. 2nd International Conference on Measurement of Speech and Audio Quality in Networks*, pp. 37–46, Prague, Czech Republic, 2003.
32. A. Rix, M. Hollier, A. Hekstra, & J. Beerends, "Perceptual Evaluation of Speech Quality (PESQ) – The New ITU Standard for End-to-End Speech Quality Assessment, Part I – Time-Delay Compensation," *J. Audio Eng. Soc.*, vol. 50, pp. 755–764, Oct. 2002.
33. ITU-T Recommendation P.862, "Perceptual Evaluation of Speech Quality (PESQ), an Objective Method for End-to-End Speech Quality Assessment of Narrowband Telephone Networks and Speech Codecs," Geneva, 2001.
34. S. Voran, "A Bottom-Up Algorithm for Estimating Time-Varying Delays in Coded Speech," in *Proc. 3rd International Conference on Measurement of Speech and Audio Quality in Networks*, pp. 43–56, Prague, Czech Republic, 2004.
35. ANSI T1-801-04-2005, "Multimedia Communications Delay, Synchronization, and Frame Rate," New York, 2005.
36. A. Gray Jr & J. Markel, "Distance Measures for Speech Processing," *IEEE Trans. Acoust., Speech Signal Process.*, vol. 24, pp. 380–391, Oct. 1976.
37. S. Voran, "Advances in Objective Estimation of Perceived Speech Quality," in *Proc. 1999 IEEE Workshop on Speech Coding*, pp. 138–140, Porvoo, Finland, 1999.
38. S. Wang, A. Sekey, & A. Gersho, "An Objective Measure for Predicting Subjective Quality of Speech Coders," *IEEE J. Sel. Areas Commun.*, vol. 10, pp. 819–829, Jun. 1992.
39. B. Moore, *An Introduction to the Psychology of Hearing*. London: Academic Press, 1989, Chap. 3.
40. R. Bladon, "Modeling the Judgment of Vowel Quality Differences," *J. Acoust. Soc. Am.*, vol. 69, pp. 1414–1422, May 1981.
41. W. Yang & R. Yantorno, "Improvement of MBSD by Scaling Noise Masking Threshold and Correlation Analysis with MOS Difference Instead of MOS," in *Proc. IEEE International Conference on Acoustics, Speech, and Signal Processing*, pp. 673–676, Phoenix, 1999.

42. J. Beerends & J. Stemerdink, "A Perceptual Speech-Quality Measure Based on a Psychoacoustic Sound Representation," *J. Audio Eng. Soc.*, vol. 42, pp. 115–123, Mar. 1994.
43. ITU-T Recommendation P.861, "Objective Quality Measurement of Telephone-Band (300–3400 Hz) Speech Codecs," Geneva, 1996.
44. E. Zwicker & R. Feldtkeller, *Das Ohr als Nachrichtenempfänger*. Stuttgart: S. Hirzel Verlag, 1967.
45. S. Voran, "A Simplified Version of the ITU Algorithm for Objective Measurement of Speech Codec Quality," in *Proc. International Conference on Acoustics, Speech and Signal Processing*, pp. 537–540, Seattle, 1998.
46. S. Voran & C. Sholl, "Perception-Based Objective Estimators of Speech Quality," in *Proc. 1995 IEEE Workshop on Speech Coding for Telecommunications*, pp. 13–14, Annapolis, MD, 1995.
47. S. Voran, "Objective Estimation of Perceived Speech Quality, Part II: Evaluation of the Measuring Normalizing Block Technique," *IEEE Trans. Speech Audio Process.*, vol. 7, pp. 383–390, Jul. 1999.
48. ANSI T1-518-1998, "Objective Measurement of Telephone Band Speech Quality Using Measuring Normalizing Blocks (MNBs)," New York, 1998. Reaffirmed 2008.
49. ITU-T Recommendation P.861 Appendix II, "Objective Quality Measurement of Telephone-Band (300–3400 Hz) Speech Codecs Using Measuring Normalizing Blocks (MNBs)," Geneva, 1998.
50. J. Beerends, A. Hekstra, A. Rix, & M. Hollier, "Perceptual Evaluation of Speech Quality (PESQ) – The New ITU Standard for End-to-End Speech Quality Assessment, Part II – Psychoacoustic Model," *J. Audio Eng. Soc.*, vol. 50, pp. 765–778, Oct. 2002.
51. C. Jin & R. Kubichek, "Vector Quantization Techniques for Output-Based Objective Speech Quality," in *Proc. IEEE International Conference on Acoustics, Speech, and Signal Processing*, pp. 491–494, Atlanta, GA, 1996.
52. W. Li & R. Kubichek, "Output-Based Objective Speech Quality Measurement Using Continuous Hidden Markov Models," in *Proc. 7th International Symposium on Signal Processing and its Applications*, pp. 389–392, Paris, 2003.
53. D. Kim, "ANIQUE: An Auditory Model for Single-Ended Speech Quality Estimation," *IEEE Trans. Speech Audio Process.*, vol. 13, pp. 821–831, Sep. 2005.
54. ANSI ATIS-PP-0100005.2006, "Auditory Non-Intrusive Quality Estimation Plus (ANIQUE+) Perceptual Model for Non-Intrusive Estimation of Narrowband Speech Quality", New York, 2006.
55. ITU-T Recommendation P.563, "Single-Ended Method for Objective Speech Quality Assessment in Narrow-Band Telephony Applications," Geneva, 2004.
56. V. Peutz, "Speech Information and Speech Intelligibility," Preprint, Audio Engineering Society 85th Convention, Los Angeles, 1988.
57. R. Kubichek, "Mel-Cepstral Distance Measure for Objective Speech Quality Assessment," in *Proc. IEEE Pacific Rim Conference on Communications, Computers and Signal Processing*, pp. 125–128, Victoria, British Columbia, 1993.

58. A. De & P. Kabal, "Auditory Distortion Measure for Coded Speech – Discrimination Information Approach," *Speech Commun.*, vol. 14, pp. 205–229, Jun. 1994.
59. A. De & P. Kabal, "Auditory Distortion Measure for Coded Speech – Hidden Markovian Approach," *Speech Commun.*, vol. 17, pp. 39–57, Aug. 1995.
60. M. Hansen & B. Kollmeier, "Using a Quantitative Psychoacoustical Signal Representation for Objective Speech Quality Measurement," in *Proc. IEEE International Conference on Acoustics, Speech, and Signal Processing*, pp. 1387–1390, Munich, 1997.
61. M. Hauenstein, "Application of Meddis' Inner Hair-Cell Model to the Prediction of Subjective Speech Quality," in *Proc. IEEE International Conference on Acoustics, Speech, and Signal Processing*, pp. 545–548, Seattle, 1998.
62. A. Rix & M. Hollier, "The Perceptual Analysis Measurement System for Robust End-to-End Speech Quality Assessment," in *Proc. IEEE International Conference on Acoustics, Speech, and Signal Processing*, pp. 1515–1518, Istanbul, 2000.
63. J. van der Werff & D. de Leeuw, "What You Specify Is What You Get (Parts 1 & 2)," Preprint, Audio Engineering Society 114th Convention, Amsterdam, 2003.
64. J. Holub, M. Street, & R. Smid, "Intrusive Speech Transmission Quality Measurements for Low Bit-Rate Coded Audio Signals," Preprint, Audio Engineering Society 115th Convention, New York, 2003.
65. D. Sen, "Predicting Foreground SH, SL and BNH DAM Scores for Multidimensional Objective Measure of Speech Quality," in *Proc. IEEE International Conference on Acoustics, Speech, and Signal Processing*, pp. 493–496, Montreal, 2004.
66. A. Takahashi & H. Yoshino, "Perceptual QoS Assessment Technologies for VoIP," *IEEE Commun. Mag.*, vol. 24, pp. 28–34, Jul. 2004.
67. T. Thiede, W. Treurniet, R. Bitto, C. Schmidmer, T. Sporer, J. Beerends, C. Colomes, M. Keyhl, G. Stoll, K. Brandenburg, & B. Feiten, "PEAQ – The ITU Standard for Objective Measurement of Perceived Audio Quality," *J. Audio Eng. Soc.*, vol. 48, pp. 3–29, Jan./Feb. 2000.
68. W. Treurniet & G. Soulodre, "Evaluation of the ITU-R Objective Audio Quality Measurement Method," *J. Audio Eng. Soc.*, vol. 48, pp. 164–173, Jan./Feb. 2000.
69. ITU-R Recommendation BS.1387-1, "Method for Objective Measurements of Perceived Audio Quality," Geneva, 2001.
70. B. Moore & C. Tan, "Measuring and Predicting the Perceived Quality of Music and Speech Subjective to Combined Linear and Nonlinear Distortion," *J. Audio Eng. Soc.*, vol. 52, pp. 1228–1244, Dec. 2004.
71. L. Thorpe & W. Yang, "Performance of Current Perceptual Objective Speech Quality Measures," in *Proc. 1999 IEEE Workshop on Speech Coding*, pp. 144–146, Porvoo, Finland, 1999.

29
Gaussian Models in Automatic Speech Recognition

Jeff Bilmes

Department of Electrical Engineering, University of Washington, Seattle, WA, USA

Abstract Most automatic speech recognition (ASR) systems express probability densities over sequences of acoustic feature vectors using Gaussian or Gaussian-mixture hidden Markov models. In this chapter, we explore how graphical models can help describe a variety of tied (i.e., parameter shared) and regularized Gaussian mixture systems. Unlike many previous such tied systems, however, here we allow sub-portions of the Gaussians to be tied in arbitrary ways. The space of such models includes regularized, tied, and adaptive versions of mixture conditional Gaussian models and also a regularized version of maximum likelihood linear regression (MLLR). We derive expectation-maximization (EM) update equations and explore consequences to the training algorithm under relevant variants of the equations. In particular, we find that for certain combinations of regularization and/or tying, it is no longer the case that we may achieve a closed-form analytic solution to the EM update equations. We describe, however, a generalized EM (GEM) procedure that will still increase the likelihood and has the same fixed-points as the standard EM algorithm.

1 • Introduction

In most speech recognition systems, Gaussian and Gaussian-mixture hidden Markov models (HMMs) are used to represent density functions over sequences of acoustic observation vectors. The acoustic observations most often used are Mel-frequency cepstral coefficients (MFCCs) [36], perceptual linear prediction (PLP) [35], or variants thereof—these features are particularly well suited to Gaussian models and have been shown to work well in practice. Moreover, when training such models, the expectation-maximization (EM) update equations are well understood [15,52,6] and widely used.

As the accuracy of automatic speech recognition (ASR) systems has improved over the years, they have naturally become more complex as well. For example, they possess more underlying system parameters, most often in the form of more HMM states and/or more Gaussian components within each Gaussian mixture. The dimensionality of each Gaussian component has also typically increased. If the amount of training data does not increase correspondingly, then a parameter optimization procedure (such as EM) that operates without parameter constraints will produce a poorly trained system. The problem can be explained using a bias-variance argument [31]: the parameter variance increases as the number of parameters grows without an increase in training data. One must thus resort either to optimization under hard constraints or to regularization, or to both.

In speech recognition, optimization with hard constraints most often takes the form of parameter sharing, where a priori fixed decisions are made regarding which Gaussian components should share its parameters with some other Gaussian component. Often these decisions can be made based on phonetic or other high-level knowledge. For example, in a tri-phone system, the center state's Gaussian mixture might share many or all of the same Gaussian components regardless of context if the center phones are similar enough. Often the sharing is decided in a combined knowledge- and data-driven fashion [62]. Other forms of constrained optimization are possible, including partial Gaussian parameter sharing, where a sub-portion of different Gaussian components are shared [23], or sparse Gaussians [8], where Gaussian parameters are constrained to contain a certain set of zeros.

We note that parameter sharing, however, may come at a cost since learning algorithms often assume parameter independence [32,33,34,19], which intuitively means that adjusting one part of a model (either structurally or parameterwise) will not influence another part. In this case, penalized log likelihoods will decompose into separate terms, which may be optimized separately. With parameter sharing, this need not be the case. Indeed, both dynamic Bayesian networks [14,45] and probabilistic relational models [20] can be thought of as network templates and rules, where the rules explain how a template may be expanded into a larger network, and how this larger network should share its parameters among the resulting expanded factors.

Another common procedure to avoid over-training a complex model given a fixed amount of data is regularization, which is generally applicable to regression, classification, and statistical model selection. Regularization is an example of Occam's razor, which roughly states that out of all possible models that can accurately explain an observable phenomenon (i.e., in our case, the data), we should prefer the simplest one possible. There are a variety of ways of measuring simplicity, including Kolmogorov complexity, the VC-dimension, entropy, minimum description length, number of total parameters, Bayesian information criterion, or a general Bayesian prior. From a pattern recognition perspective, we desire a model that achieves a low empirical error on training data while being as simple as possible. The trade-off between these two goals is determined using a "trade-off coefficient" which determines the regularization path [30] or the "accuracy-regularization frontier." This concept is the key behind support-vector and Kernel methods [57], where the function to minimize is a sum of a (hinge) loss function and a regularizer (which is typically the ℓ_2-norm due to its mathematical tractability). Regularization is also commonly used for training neural networks (weight decay [9]), regression (i.e., ridge-regression [31]), and splines [58].

Parameter sharing can also be seen as a form of regularization [31], as the goal of sharing is to optimize the parameters of a model that has been hard constrained to be "simple." In the parameter sharing case, however, simplicity corresponds roughly to the number of free parameters that are available to explain the data, but these free parameters may be pieced together in arbitrarily intricate ways. In other words, more regularization corresponds essentially to more sharing, or to reducing the number of free parameters, but collections of these parameters may combine in a large variety of ways to produce the portion of a model that governs a specific phenomenon needing to be represented (e.g., an individual phone in a speech utterance). There has not, however, been much discussion in the speech recognition literature regarding general forms of regularization as it may be applied to Gaussian mixture modeling other than what has appeared in the Bayesian adaptation area (see, e.g. [26,11]). Recently, the notion of regularization has been applied to produce new forms of speaker adaptation [43,42] for both generative models (such as HMMs) and discriminative models (such as neural networks, support vector machines, and conditional random fields).

In this chapter, we explore how Gaussians, Gaussian mixtures, Gaussian mixture HMMs, and auto-regressive Gaussian mixture HMMs (what we end up calling a "generic Gaussian mixture") can benefit from the simultaneous use of fine-grained sub-Gaussian parameter tying and also parameter regularization. Moreover, we utilize aspects of graphical models to assist in our exposition. Graphical models are a visual mathematical abstraction that are used to describe certain properties of families of probability distributions [40,37,50]. Different types of graphical model (e.g., directed or undirected) can be used to represent factorization properties of such distributions; factorization is the key that allows the distributive law to be used so that algorithms can be developed that are both mathematically sound and algorithmically efficient. Alternatively, many researchers have also ascribed certain high-level meanings (such as causality) to various types of graphs [51]. There have in the past been descriptions of how graphical models can be used to describe certain aspects of speech recognition systems [5]. In this chapter, we use graphical models only to assist us in describing how Gaussian mixture HMMs can be decomposed into partially tied components.

Our main contribution is to derive general forms of EM learning equations in these models. We consider such models "acoustic" since in the field of automatic speech recognition, the Gaussian is where the rubber meets the road—specifically, we consider here only the portion of the statistical model that makes a connection between the acoustic speech signal (or better, the acoustic signal after having undergone some non-probabilistic processing, such as MFCC feature extraction [36]) and the inherent discrete processing that ultimately must correspond to word string (or other high- or meta-level) hypotheses. These models include, as described above, Gaussians, and their various flavors, including sparse Gaussians, auto-regressive Gaussians, and what are known as "trajectory models", where a Gaussian model regresses (or conditions) on other elements in the same or other feature streams. We treat in some detail the various forms of sub-Gaussian parameter sharing (using a tri-part decomposition of a Gaussian's parameters based on Cholesky factorization) that can arise under these scenarios. We also describe the training algorithms using a regularized EM framework. In particular, we find that for certain combinations of regularization and tying, we no longer arrive at a closed-form solution to the EM update equations. We describe, however, a generalized EM (GEM) procedure that will still increase the likelihood and has the same fixed-points as the EM algorithm.

This chapter consists of the following sections. Section 2 gives a brief background on maximum likelihood training, Gaussians, and Gaussian mixtures. Section 3 derives EM update equations for training regularized Gaussians, using a three-way decomposition of the parameters of each Gaussian component. Next, Section 4 derives EM update equations for arbitrary patterns of tying, and these equations immediately lead to adaptation algorithms for mixture conditional Gaussians, and also a regularized maximum-likelihood linear regression [41] procedure. Section 5 describes various methods for training these parameters, and explains how one method is a generalized EM, and lastly Section 6 concludes. Of particular relevance to this chapter (both conceptually and in terms of useful notation) are references [6,5,4].

2 • Background

We begin with notation. Random variables will be depicted using capital letters (e.g., X, Y and Q), and lower case variables will be used to indicate their possible values (x, y, q). Given a set of M variables, the index set $1\ldots M$ will be depicted using the matlab-like notation $1:M$. Given a subset of variable indices $A = \{a_1, a_2, \ldots, a_{|A|}\}$, where $a_i \in 1:M$, then $X_A = \{X_{a_1}, X_{a_2}, \ldots, X_{a_{|A|}}\}$.

Many of the models we describe correspond to temporal sequences of data. We have a size-N sample of training data $\mathcal{D}^{tr} = \{x^{(i)}, q^{(i)}\}_{i=1}^{N}$, where $x^{(i)} = x^{(i)}_{1:T_i}$ is a length T_i sequence of observation vectors. Each observation vector $x_t^{(i)} = \{x_{t,1}^{(i)}, x_{t,2}^{(i)}, \ldots, x_{t,M}^{(i)}\}$ will be of length M, and may correspond to a processed acoustic frame of speech (e.g., it could be a sequence of MFCCs and their deltas). The mth scalar element is denoted $x_{t,m}^{(i)}$, and in general we denote the mth element of vector x as $x_{.,m}$. Thus, we have that $x = x_{.,1:M}$. Also, $q_{1:T_i}^{(i)}$ is a length T_i sequence of integer labels or states, so that $q_t^{(i)}$ is the tth integer of the ith training sequence. When unambiguous, we will often use $\{x, q\}$ as a particular training pair. Each of the training pairs is assumed to be drawn from the same unknown underlying distribution, so that $(x^{(i)}, q^{(i)}) \sim p(x, q)$ $\forall i$, where $p(x, q)$ factors according to the rules that define an HMM. We will also discuss static (non-temporal) data. In this case, we will treat $\{x^{(i)}, q\}$ as a training pair, where $x^{(i)}$ is a fixed length vector. In this case, $p(x, q)$ will be a fixed length distribution. We will express conditional independence relations between random variables using the standard notation $X \perp\!\!\!\perp Y | Z$ [13], which is read "X is independent of Y given Z."

Most speech recognition systems utilize hidden Markov models (HMMs) with mixture density functions for their observation distributions. Specifically, an HMM is a probability distribution over $2T$ variables—there are T state variables $Q_{1:T}$ that are often hidden, and T other variables that are almost always observed $X_{1:T}$. The joint distribution over these $2T$ variables factorizes as follows:

$$p(x_{1:T}, q_{1:T}) = \prod_t p(x_t|q_t) p(q_t|q_{t-1}).$$

Here, $p(q_t|q_{t-1})$ gives the probability of moving from state q_{t-1} at time $t-1$ to state q_t at time t. In this work, we will be discussing the observation distribution $p(x_t|q_t)$ as that is where the acoustic information is typically produced as input into the HMM at time t. For example, with x_t observed, it is often the case that x_t consists of an M-dimensional MFCC feature vector for time t, although many other types feature vectors can be used as well.

In this chapter, x_t will be a continuous valued M-dimensional vector with density $p(x_t|q_t)$. This density will be a mixture model, where components of the mixture consist of Gaussian densities of some form. For example:

$$p(x_t|q_t) = \sum_\ell \alpha_{\ell q_t} \mathcal{N}(x_t; \mu_{\ell q_t}, \Sigma \ell q_t),$$

where $\mathcal{N}(x; \mu, \Sigma)$ indicates that x has an M-dimensional Gaussian density, and where $\alpha_{\ell q_t}$ is the mixture component weight. An M-dimensional Gaussian density has the form

$$p(x) = \mathcal{N}(x; \mu, \Sigma) = |2\pi\Sigma|^{-1/2} e^{-\frac{1}{2}(x-\mu)^T \Sigma^{-1}(x-\mu)},$$

where μ is an M-dimensional mean vector, and Σ is an $M \times M$ covariance matrix. We will use $K = \Sigma^{-1}$ to refer to the inverse covariance (or the concentration) matrix of the density.

It is possible to view a Gaussian density either as a Bayesian network [50] or as a Markov random field—the properties of Gaussians being represented by various forms of graphical model were described in [5] and are very briefly summarized here. First, zeros in the covariance matrix correspond to marginal independence assumptions. For example, given the constraint $\Sigma_{ij} = 0$ for a particular i, j of the concentration matrix, then in the marginal we have $p(x_i, x_j) = p(x_i)p(x_j)$ or that $X_i \perp\!\!\!\perp X_j$. This property of

Gaussians is quite well known. Second, zeros in the concentration matrix correspond to conditional independence assumptions. In particular, if $K_{ij} = 0$ then $X_i \perp\!\!\!\perp X_j | X_{\{1:M\}\setminus\{i,j\}}$. In other words, zeros in the concentration matrix correspond to the pairwise Markov property of undirected graphical models [40]. Since the Gaussian is always positive, the Hammersley–Clifford theorem [40] applies and so also do all the standard Markov properties on undirected graphical models, and we can view a Gaussian as an undirected graph on a set of nodes where there are edges between two nodes i and j if and only if $K_{ij} \neq 0$.

A Gaussian may also be described by a Bayesian network. Assuming a particular ordering of the elements in the variable X, if we Cholesky factorize [27,29] the concentration matrix $K = U'DU$ where U is an upper triangular matrix with unity on the diagonal, and D is a diagonal matrix, then form $U = (I - B)$, we see that B is an upper triangular matrix with zeros on the diagonal, and the upper off-diagonal elements correspond to the edges in a Bayesian network that describes the Gaussian. For example, if $B_{ij} = 0$ with $j > i$, then in the Bayesian network form, there is no edge from node j to node i. With this form of Gaussian (one that we will use extensively in this chapter), the Gaussian distribution becomes [5]:

$$p(x) = (2\pi)^{-M/2} |D|^{1/2} e^{-\frac{1}{2}(x - Bx - \tilde{\mu})^T D (x - Bx - \tilde{\mu})}.$$

We may be interested, for example, in training a Gaussian model where K is constrained to have a certain pattern of zeros. In fact, *Sparse Gaussians* [8] are produced by learning a Gaussian distribution that factors with respect to a particular graph. The ubiquitously utilized and easy to train diagonal covariance matrix, for example, simply corresponds to using a completely disconnected graph. Other forms of constrained Gaussians can be formed simply by utilizing a different graph (i.e., various constraints on either the covariance, concentration, or Cholesky factorized concentration matrices). But as we will see in subsequent sections, regardless of the pattern of zeros in the Cholesky factorized form, the Gaussian will be easy to estimate in a maximum likelihood sense as long as all variables are observed.

A conditional (or autoregressive) Gaussian process is essentially a type of conditional probability density. Suppose that the vector x has been bi-partitioned into two sets U, V such that $U, V \subset 1 : M$ and $U \cap V = \emptyset$. Then $p(x_U | x_V)$ is a conditional Gaussian where the mean vector x_U is a linear function of the parent variable vector x_V, but otherwise X_U is a Gaussian. In other words

x_U has a different Gaussian distribution for each value x_V. This is expressed as

$$p(x_U|x_V) = |2\pi\Sigma_{U|V}|^{-1/2} e^{-\frac{1}{2}(x_U-(B_{U|V}x_V+\mu_{U|V}))^T \Sigma_{U|V}^{-1}(x_U-(B_{U|V}x_V+\mu_{U|V}))},$$

where $\Sigma_{U|V}$ is the conditional covariance matrix, which does not change with varying x_V. The conditional mean, however, does change with x_V as the affine transformation $B_{U|V}x_V + \mu_{U|V}$ is the conditional mean of x_U given x_V. This general form can arise in a number of ways. For example, after factoring a Gaussian using the chain rule of probability

$$p(x_{.,1:M}) = \prod_m p(x_{.,m}|x_{.,1:m-1}),$$

then $p(x_{.,m}|x_{.,1:m-1})$ corresponds to a conditional Gaussian (in this case $U = \{m\}$ and $V = \{1, 2, \ldots, m-1\}$), and in fact the parameters associated with each factor combine together to produce the Cholesky factorization of the concentration matrix K [3].

Using conditional Gaussians, either normal Gaussian distributions or mixtures thereof can be formed over a speech observation at time t, to form $p(x_t)$. Moreover, trajectory type models can be formed over time, where the conditional variables at time t can come from the past, future, present, or all of the above relative to t. The conditional variables might even come from an entirely separate feature stream. For example, linear mixture buried Markov models (BMMs) [7,3] can be formed by employing a mixture of conditional Gaussian distributions, where in this case we equate x_U with x_t, so x_U becomes the feature vector at the current time, and we also equate x_V with z_t, which is an L-dimensional vector of feature elements from previous times, future times, or as mentioned above even different feature streams (that may come from any time). We can denote this using sets, as $z_t \subset \{x_{t-\tau_{11}}, y_{t-\tau_{12}}, \ldots, x_{t-1}, y_{t-1}, y_t, x_{t+1}, y_{t+1}, \ldots, x_{t+\tau_{21}}, y_{t+\tau_{22}}\}$ where y_t might be a different feature stream altogether, and where τ_{ij} are arbitrary fixed integer constants that indicate how much of either the x or y stream are part of z_t. Much has been written on how conditional Gaussians and their mixtures can be used as trajectory models (examples include [53,5]). In the most general case, z_t may also contain a constant element of unity (1) so that both the regression coefficients and the fixed Gaussian mean can all be represented by the single $M \times (L+1)$ matrix B—in other words, if we have that $z_t = [z_t'\ 1]^T$ indicating that the very last element of z_t is unity, then $Bz_t = B'z_t' + \mu$ where μ is the last column of B, and B' is an $M \times L$ matrix consisting of all but the last column of B. This produces a generalized conditional Gaussian of the form:

$$p(x|z) = (2\pi)^{-M/2}|D|^{1/2}e^{-\frac{1}{2}(x-Bz)^T D(x-Bz)},$$

where we assume that everything that might possibly be conditioned on is included in the large length L vector z. We note that the diagonal matrix D corresponds to the inverse conditional variance parameters of the Gaussian (in this chapter, we sometimes refer to D as just the conditional rather than the inverse conditional variances when the meaning is clear).

In general, regardless of how we form a Gaussian (either as a conditional Gaussian $p(x|z)$ or as part of a standard Gaussian that has been factored using the chain rule), we will define a set of parameters of a Gaussian as the following triple: (1) the mean μ; (2) the (conditional) variance D; and (3) the coefficients of B. The matrix B could come from the Cholesky factorized concentration matrix $K = (I - B)^T D(I - B)$, or could also be the regression coefficients on some conditional vector z, or it could come from both; we will not make a distinction between the two since they can be treated identically in the learning equations. Note, however, that in many cases B will be quite sparse—a particular sparsity pattern, then, can be viewed as a form of constrained optimization or itself a form of regularization, where fewer parameters are available in the model to adjust thus reducing parameter estimation variance. This decomposition of the Gaussian's parameters into these three portions (the mean, inverse conditional variance, and regression coefficients) will be the basis for our Gaussian parameter regularization and tying strategies described in Section 4. This treatment of the Gaussian parameters will moreover allow us to represent various training procedures for Gaussians and Gaussian mixtures in a unified way, regardless if the model consists of regular Gaussian components, various sparse Gaussians, or the conditional Gaussians used for temporal trajectory modeling (say in an HMM) or in a multi-stream model.

2.1 Maximum Likelihood Training

Given a set of training data, there are many ways to optimally adjust the parameters to be in best accordance with this data, including maximum likelihood (ML), maximum a-posteriori (MAP), and discriminative parameter training procedures [1,17,16,39,38,31,60,55]. While discriminative parameter training is most appropriate for classification tasks, ML training is still commonly used and can sometimes perform as well or better than discriminative methods, often due to the mathematical or computational simplicity of such techniques, or to statistical estimation issues [48]. Moreover, it is often extremely easy to train a generative model in the ML-framework due to the fact that the ML objective function decomposes exactly as does the generative model (e.g., an HMM). In fact, when all variables in question X and Q are observed, then unconstrained ML training becomes quite easy, namely $p(x,q) = p(x|q)p(q)$ can often be learned simply by counting and averaging.

In the models that we consider, namely $p(x,q)$ we will assume that X is observed and Q hidden when training. The most common form of training in this situation is to use the EM algorithm [15,6,46]. We very briefly outline this procedure in its basic form. Assuming that all of the parameters of the model are described by the variable λ, we start with an initial estimate of the parameters, say λ^p. The goal is to optimize the next set of parameters λ using an augmented function,

$$f(\lambda) = Q(\lambda, \lambda^p) = E_{p(Q|X,\lambda^p)}[\log p(X,Q|\lambda)] = \sum_q p(q|x,\lambda^p) \log p(X,Q|\lambda).$$

This is to be compared to the "incomplete" likelihood function, consisting only of the observed variables, namely $g(\lambda) = \log p(x|\lambda)$. It can be shown that $\partial f(\lambda)/\partial \lambda|_{\lambda=\lambda^p} = \partial g(\lambda)/\partial \lambda|_{\lambda=\lambda^p}$ so the gradients of both the incomplete and the augmented likelihood functions are the same at the location of the current parameters. Moreover, it can be shown that $f(\lambda)$ is a concave lower bound on the true likelihood function at the current point, and that optimizing $f(\lambda)$ is often much easier than optimizing $g(\lambda)$, and therefore is typically preferred. The process is often as simple as finding the unique peak of $f(\lambda)$ by taking a first derivative, setting it to zero, and then solving for λ (possibly with some Lagrangian constraints to ensure we stay within the parameters' probability simplex). This process is repeated. There are also guaranteed convergence properties—once converged, we are guaranteed to have found a local maximum of the likelihood function $g(\lambda)$, which follows from the above since the derivatives are identical.

Details of the derivation of the EM update equations for Gaussian mixtures, and HMMs using Gaussian mixtures are described in [15,46], and notation most similar to what we use in this chapter is described in [6]. In an HMM system, in order to train a Gaussian mixture HMM, one needs the quantities $p(Q_t = i, M_t = j|x_{1:T}, \lambda^p)$, the posterior probability that the Markov chain state at time t is i and the mixture component of state i at time t is j. For a plain Gaussian mixture system, the posterior needed is $p(M_t = j|x_t, \lambda^p)$, which is the posterior of the component for data instance x_t. There is a strong commonality between Gaussian mixtures and Gaussian mixture HMMs, and in fact the only difference in the EM update equations is the form of this posterior. In fact, a Gaussian mixture HMM may be seen as a high dimensional multi-variate Gaussian, specifically $p(x_{1:T}|m_{1:T}, q_{1:T})$ is just an MT-dimensional Gaussian with a block-diagonal covariance matrix. Therefore, in accordance with our unified treatment of Gaussian models as described in the previous section, we will assume what we call a "generic Gaussian mixture", where we use a simple notation p_{mt} to designate the posterior probability under the previous parameters of the current component m for time or sample t, and we will do this using the same notation. As we will see, this will greatly simplify our equations. The general scenario is shown in Figure 1. Lastly, we will let Q be the general auxiliary function to be optimized in EM. For the purposes of this manuscript, it is also sufficient to provide a Q only for a generic Gaussian mixture.

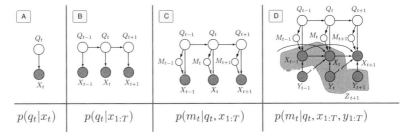

FIGURE 1 A variety of Gaussian-like models utilize the same form of posterior probability of the hidden variables ($q_{1:T}$ and $m_{1:T}$) given the evidence ($x_{1:T}$ or $y_{1:T}$). For each model, the posteriors are needed when training using regularized EM, and the equations for training the different models are identical up to the form of this posterior. In (**A**) (on the *left* of the figure), we have a simple mixture model, with posterior form $p(q_t|x_t)$. In (**B**), we have the basic Gaussian HMM, and its posterior. If we use a Gaussian mixture HMM (**C**), we need the posterior of the component m_t given the current hidden state q_t and the evidence. And last, in (**D**), we have a multi-stream Gaussian mixture auto-regressive (or trajectory) model. The main stream is $X_{1:T}$ with auxiliary stream $Y_{1:T}$, but the form of posterior needed is again still the same. In all cases, a forward–backward procedure [2] can be defined that can compute these posteriors. For our analysis, we will simply use p_{mt} to indicate the posterior of Gaussian component m at data sample (or time) t. Lastly, part D of the figure also shows that Z_t is a vector consisting of elements from X_{t-1}, X_t, Y_t, and Y_{t+1}. Our equations will use the generic notation z_t to indicate this vector, and to produce the other models it is sufficient to realize that z_t may be empty.

3 • Training Regularized Gaussian Decompositions

In this section, the decomposition of a Gaussian component's parameters, as mentioned in the previous section, into the three groups, μ, D, and B, is applied during EM training of a generic Gaussian mixture. Unlike Section 4, in this section we do not yet assume there is any sub-Gaussian parameter sharing.

A desirable property of any EM update equation is that each parameter group at each EM update can be estimated independently without needing any of the other new parameters. In other words, given a collection of previous parameters $\{\mu_m^p, D_m^p, B_m^p\}_m$, then ideally the update equations of each of the new parameters $\{\mu_m, D_m, B_m\}_m$ is expressed entirely in terms of only the data and the previous parameters. Otherwise, either multiple passes through the data might be required thereby making training much more computationally expensive, or even worse there might be cyclic dependencies in the parameter updates (which we ultimately will need to deal with in this chapter as well). A simple example may be demonstrated using the estimation of the mean $\mu \propto \sum_t x_t$ and variance $\sigma^2 \propto \sum_t (x_t - \mu)^2$, where we see that the variance depends on the mean. It is well known that the variance estimate can be reformulated using additional accumulators, namely $\sum_t x_t^2$ and $(\sum_t x_t)^2$, requiring only one pass through the data for both mean and variance, in this latter case accumulating both the sum and sum of squares of x_t. For certain patterns of sharing and regularization of a Gaussian's parameters, however, this type of trick will not always be possible. In any case, the general property as it will apply to our EM updates is depicted in Figure 2.

When there is no parameter sharing, the EM auxiliary function needing to be maximized becomes

$$Q = \sum_m \sum_t \left(\frac{1}{2} \log(|D_m|) - \frac{1}{2}(x_t - B_m z_t)^T D_m (x_t - B_m z_t) \right) p_{mt}$$

$$- \frac{\lambda}{2} \sum_m \|B_m\|_F^2, \tag{1}$$

where $\|A\|_F = \text{tr}(AA^T)$ is the Frobenius norm of matrix A, the last term acts as an ℓ_2 regularizer on the regression coefficients, and λ is an accuracy-regularization trade-off coefficient. This form of regularization is similar to ridge-regression [31] and to weight-decay [9] in the neural-networks literature. In our case, however, we assume that there is only one such global accuracy-regularization λ coefficient for all the Gaussians in the current model, and for all scalar elements within each component. An

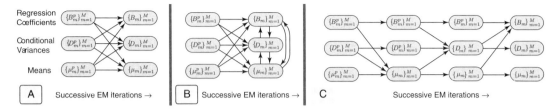

FIGURE 2 EM Gaussian decomposition parameter update dependencies: On the *left* (**A**), we see that each of the new parameters μ_m, B_m and D_m depends only on previous parameters μ_m^p, B_m^p and D_m^p and also on the training data set (not shown in the figure). In this case, it is possible to simultaneously update all new parameters, with only one pass through the training data. In the *middle* (**B**), the new parameters might depend on the other new parameters as well as old parameters, leading potentially to cyclic dependencies. On the *right* (**C**), we break the cycles by only updating one parameter group for each EM iteration, where each EM iteration requires a single pass through the data. This means that to update all the parameters in the model, three passes through the data are required. Still another strategy (*middle* figure, or **B**) updates all parameters simultaneously with one training data pass, but in place of each new parameter that is needed, we use the parameter from the previous EM iteration—this last step is not guaranteed to increase the likelihood of the data, although it does have the same fixed points as the left case, and may work well in practice.

alternative approach would employ one coefficient for each separate Gaussian component, or even one coefficient for each individual regression coefficient. In general, however, the value of such coefficients can only be found using cross-validation [31] on a development or held-out data set and doing this when there are multiple accuracy-regularization coefficients would turn the problem of finding these trade-off coefficients themselves into an intractable search problem.

Equation (1) does not include any constant terms which vanish once derivatives are taken. Also note, for now we have assumed, that the mean parameters are contained within the B matrix—we do this by assuming that one of the elements of z_t is always equal unity. This also means that we are regularizing in the above equation both the regression coefficients (e.g., cross-observation dependences as in a BMM) and the means. We might not wish to regularize the means, however, so in an expansion where the means and regression coefficients are separate (see below), we might not have a regularizer on the means (we utilize separate mean and regression regularization coefficients in Section 4). Also, since there is no sharing, there is a one-to-one correspondence between every value of m and each of the D and B matrices.

Maximizing the regularizer alone $-\frac{\lambda}{2}\sum_m ||B_m||_F^2$ will force the coefficients to converge toward zero. Therefore, it is important to assume that the training data is centered, that is, zero-mean normalized. Also, the coefficients will depend on the scaling of the data as well. Therefore, in practice, it will be important to perform both mean subtraction and variance normalization [44,10] on the observation vectors prior to any training.

Taking derivatives of Eq. (1) with respect to the B_m matrices (again, which includes the mean in the right most column of B_m), gives $\frac{\partial Q}{\partial B_m} = 0$ or that

$$\sum_t D_m(x_t - B_m z_t) z_t^T p_{mt} - \lambda B_m = 0.$$

From this, we must attempt to solve for B_m, and there are several ways to proceed. One simple way next yields

$$\sum_t (x_t - B_m z_t) z_t^T p_{mt} - \lambda D_m^{-1} B_m = 0, \qquad (2)$$

implying that

$$\sum_t x_t z_t^T p_{mt} - B_m \sum_t z_t z_t^T p_{mt} - \lambda D_m^{-1} B_m = 0.$$

We see that we have a matrix equation of the form $A - BC - DB = 0$ where $A = \sum_t x_t z_t^T p_{mt}$, $B = B_m$, $C = \sum_t z_t z_t^T p_{mt}$, $D = \lambda D_m^{-1}$, and the goal is to solve for B. This equation is known as the Sylvester matrix equation [25] (which is a generalization of the more well known Lyapunov equation [21] commonly used in control theory). The Sylvester equation has a unique solution if and only if there is no eigenvalue of C that is the negative of any eigenvalue of D. Therefore, since both C and D are positive definite in our case, this condition will certainly hold. A general (but not efficient) form of solution is given by $\bar{b} = G^{-1}\bar{a}$, where $\bar{b}$ (respectively $\bar{a}$) is a vector consisting of the columns of B (resp. $-A$) stacked on top of each other, and where $G = I \otimes D + C^T \otimes I$ where $\otimes$ is the Kronecker matrix product. Much more efficient solutions are given in [25]. Also, since D is diagonal, significant additional computational efficiencies can be obtained (see Section 3.2 below).

If $\lambda = 0$, however, we have a particularly simple form, which yields:

$$B_m = \sum_t x_t z_t^T p_{mt} \left(\sum_t z_t z_t^T p_{mt} \right)^{-1}. \qquad (3)$$

So if $\lambda = 0$ (no regularization), the above shows that B_m can be estimated quite easily by summing posterior weighted outer products, and that B_m does not depend on any other current parameter. This assumes of course that T is large enough so that the matrix above is invertible, but that is typical for most data sets.

With regularization $\lambda > 0$, however, the new B_m has a much more intricate relationship with the new D_m matrix, as illustrated in the middle of Figure 2.

Next, taking derivatives of Eq. (1) with respect to the diagonal matrix D_m and setting $\frac{\partial Q}{\partial D_m} = 0$ gives

$$\sum_t p_{mt}\left(D_m^{-1} - (x_t - B_m z_t)(x_t - B_m z_t)^T\right) = 0,$$

or that

$$D_m^{-1} = \frac{\sum_t p_{mt}(x_t - B_m z_t)(x_t - B_m z_t)^T}{\sum_t p_{mt}}.$$

In this update, we again have a cross-dependency, namely D_m is dependent on the current estimation of B_m. In this case, therefore, it appears that even if $\lambda = 0$, we have a dependence between the parameters, thus again implying multiple passes through the data on each EM iteration, first to compute B_m and next to compute D_m. The following formula, however, shows that if $\lambda = 0$, we can compute the covariance matrix using the quantities that we have and without running through the data multiple times per iteration.

$$\begin{aligned}D_m^{-1} &\propto \sum_t \left\{p_{mt}\left(x_t x_t^T - x_t z_t^T B_m^T - B_m z_t x_t^T + B_m z_t z_t^T B_m^T\right)\right\} \\ &= \sum_t p_{mt} x_t x_t^T - \left(\sum_t p_{mt} x_t z_t^T\right) B_m^T \\ &\quad - B_m \sum_t p_{mt} z_t x_t^T + B_m \left(\sum_t p_{mt} z_t z_t^T\right) B_m^T,\end{aligned}$$

but when $\lambda = 0$, we have

$$B_m \left(\sum_t p_{mt} z_t z_t^T\right) B_m^T = \left(\sum_t x_t z_t^T p_{mt}\right) B_m^T,$$

and therefore we get

$$D_m^{-1} \propto \sum_t p_{mt} x_t x_t^T - B_m \sum_t p_{mt} z_t x_t^T. \qquad (4)$$

The equations above show that B_m and D_m can indeed be learnt simultaneously when $\lambda = 0$ without needing to make multiple passes over the data. Specifically, with only a single pass over the data, we can accumulate the three quantities $\sum_t x_t z_t^T p_{mt}$, $\sum_t z_t z_t^T p_{mt}^{-1}$, and $\sum_t p_{mt} z_t x_t^T$, which when done are sufficient for both the B_m and D_m updates in Eqs. (3) and (4).

Even though $\lambda = 0$ in this case, it still might appear that we have a problem in that the above equations require outer products of the form $z_t z_t^T$ and $x_t z_t^T$ for every m. Considering that there may be many hundreds or thousands of Gaussian components in a modern ASR system (corresponding to the different HMM states and mixture components therein), and that z_t could easily have many elements, these accumulators could be quite large. For example, in a BMM update equation [3], z_t consists of the collection of *all* context elements grouped together for all individual Gaussian elements, a vector that could be thousands of elements long. When $\lambda > 0$, solving the Sylvester equation would be prohibitive. A solution to this problem is given in Section 3.2, but we first discuss the relationship these models have with ridge regression.

3.1 Ridge Regression

Our procedure may in fact be seen as a multi-variate, multi-class, "soft", and learned-covariance generalization of ridge regression [31]. It is multivariate since we are essentially learning multiple sets of regression coefficients simultaneously. It is multi-class since the regression coefficients may correspond to entirely different Gaussians. It is "soft", since the coefficients are being learned as part of a larger EM procedure, so all data is being essentially weighted by posterior probabilities from previous model parameters. And lastly, unlike conventional ridge regression, we must learn covariance parameters as well. Since the relation with ridge regression is important to understand, we review ridge regression in some detail, and then expand on its generalization.

Ridge regression can be explained from either a frequentist perspective (a Gaussian model with a penalized log-likelihood) or a Bayesian perspective (the maximum a posteriori (MAP) of β under a Gaussian model and a Gaussian prior). In either case, the goal is to learn the parameters of a linear model for predicting the scalar variable x based on a vector variable z and a vector of parameters β of the same length. We are given training samples $\mathcal{D}^{tr} = \{x^{(i)}, z^{(i)}\}_{i=1}^N$ from which to learn β. In the Bayesian case, we find the β that maximizes the posterior $p(\beta|\mathcal{D}^{tr})$ or equivalently that maximizes $p(\mathcal{D}^{tr}|\beta)p(\beta)$. The likelihood $p(\mathcal{D}^{tr}|\beta)$ is composed of the model $p(x|z, \beta) = \mathcal{N}(x; \beta^T z, \sigma^2)$, a conditional Gaussian with mean $\beta^T z$ and variance σ^2. As before, we assume that the last entry of z is always unity, so that we can absorb any mean parameter into the regression parameters β. We also assume a zero-mean and η^2-variance Gaussian Bayesian prior on β, meaning that $\beta \sim \mathcal{N}(0, \eta^2)$ where $p_\eta(\beta) = \mathcal{N}(0, \eta^2) = \frac{1}{\sqrt{2\pi\eta^2}} \exp\left(-\frac{1}{2}\beta^T\beta/\eta^2\right)$. The joint data and β distribution consisting of log-likelihood and prior can be written as

$$L(\beta) = \sum_i \log p(x^{(i)}|z^{(i)}, \beta) + \log p_\eta(\beta) \tag{5}$$

$$= -\frac{N}{2}\log(2\pi\sigma^2) - \frac{1}{2}\sum_i (x^{(i)} - z^{(i)}\beta)^2/\sigma^2$$

$$- \frac{1}{2}\log(2\pi\eta^2) - \frac{1}{2\eta^2}\|\beta\|^2, \tag{6}$$

where $\|\beta\|$ is the ℓ_2-norm. Finding the maximum a posteriori (MAP) value of β reduces to

$$\operatorname*{argmin}_\beta \sum_i \frac{1}{2\sigma^2}(x^{(i)} - z^{(i)}\beta)^2 + \frac{1}{2\eta^2}\|\beta\|^2$$

$$= \operatorname*{argmin}_\beta \sum_i \frac{1}{2}(x^{(i)} - z^{(i)}\beta)^2 + \frac{\sigma^2}{2\eta^2}\|\beta\|^2.$$

If we set $\lambda = \sigma^2/\eta^2$, we get the typical form of ridge regression:

$$\operatorname*{argmin}_\beta \sum_i \frac{1}{2}(x^{(i)} - z^{(i)}\beta)^2 + \frac{\lambda}{2}\|\beta\|^2. \tag{7}$$

This latest form is in fact the standard frequentist starting point for ridge regression, where we have an error term (or a likelihood) along with a complexity penalty that are related by an accuracy-regularization trade-off coefficient λ.

To increase the degree of regularization, we increase λ, and as $\lambda \to \infty$, $\eta^2 \to 0$, which also takes β toward zero for any finite-length data set. λ thus gives us a single normalized parameter to determine the amount of regularization. We note that the variance of the Bayesian prior changes as a function of the data variance since $\eta^2 = \sigma^2/\lambda$. If we wanted to use a fixed Bayesian prior regardless of the data variance, ridge regression would take the form

$$\operatorname*{argmin}_\beta \sum_i \frac{1}{2}(x^{(i)} - z^{(i)}\beta)^2 + \frac{\lambda\sigma^2}{2}\|\beta\|^2, \tag{8}$$

where in this case, we have an interpretation of our regularization coefficient as $\lambda = 1/\eta^2$, the inverse variance of the Bayesian prior.

Of course, the interpretation of λ in Eq. (7) or (8) does not matter here since the regularization coefficient is usually determined using cross-validation for any given data set [31]. Note, however, that the range of values attempted for cross-validation would matter, as in the former case λ is a ratio of variances, while in the latter case λ is an inverse variance.

Suppose now that we wish to solve multiple ridge-regressions simultaneously. We can think of this as multiple separate data sets or, alternatively, we can view this as a single multivariate generalization of ridge regression, where x is now a M-dimensional vector, B is an $M \times L$ matrix of regression coefficients, where each row of B is used to predict the corresponding element of x. In either case, when we wish to use a single λ regression coefficient to control such multiple regressions simultaneously, then the interpretation of λ might start to matter. If we want one global Bayesian prior variance to apply to all regressions simultaneously, then the form of regularization should be as we have defined it in Eq. (2) (for a specific value of m) where each of the variance terms are multiplied by lambda as in λD_m^{-1}. In this case, we interpret λ as the inverse of this globally shared Bayesian prior variance. If, on the other hand, we wish the Bayesian prior for each row of B_m to scale with the variance of the corresponding element in x, then Eq. (2) should take the form

$$\sum_t (x_t - B_m z_t) z_t^T p_{mt} - \lambda I B_m = 0, \tag{9}$$

where I is an appropriately sized identity matrix, in which case B_m has once again a particularly simple update equation, namely Eq. (3) becomes

$$B_m = \sum_t x_t z_t^T p_{mt} \left(\sum_t z_t z_t^T p_{mt} + \lambda I \right)^{-1}. \tag{10}$$

We see that as λ gets larger, the matrix that must be inverted becomes better conditioned, and it moves B toward the zero matrix.

For the remainder of this chapter, we will assume the globally shared Bayesian prior case, but we must keep in mind that some of the equations will change if we take the interpretation shown in Eq. (10).

3.2 Exploiting that D is diagonal

Significant computational simplifications can be achieved by noting that since D_m is a diagonal matrix, only the diagonal portion of the update equations needs to be computed. This requires no loss of generality because of the UDU' factorization mentioned above, whereby the B matrices may contain the factored form of a full covariance matrix, or a generalized regression on z_t, or both.

We let $D_m(r)$ be the rth diagonal scalar element of the diagonal matrix D_m. Also, let $B_m(r)$ be a row-vector consisting of the non-zero (i.e., existing) elements from the rth row of the matrix B_m, and let $z_t(r)$ be a column vector consisting of the elements of the vector z_t that correspond to the non-zero entries in the rth row of B_m. In other words, the scalar quantity $B_m(r)z_t(r)$ equals the rth element of the vector $B_m z_t$ but the computation of $B_m(r)z_t(r)$ does not include the terms that are just zero. This is a slight abuse of notation as the elements contained in $z_t(r)$ depend on the B_m matrix that it is currently being multiplied with, but the meaning should be clear nevertheless. The reason this works at all is the way we formed z_t as being a subset of all possible elements that x_t might condition on lumped together as one large vector—it has been the sparsity pattern of the B_m matrices which has not only included the appropriate regression coefficients, but also acted as a selector for appropriate elements within z_t. Note also that for $r \neq r'$, $z_t(r)$ and $z_t(r')$ might contain some of the same elements from z_t, that is, treating $z_t(r)$ as a set rather than a vector, it is possible to have $z_t(r) \cap z_t(r') \neq \emptyset$. Finally, we let $x_t(r)$ be the rth scalar element of the vector x_t. Using these definitions, we may re-write Eq. (1) as

$$\mathcal{Q} = \sum_m \sum_t \sum_r \left(\frac{1}{2} \log(D_m(r)) - \frac{1}{2}(x_t(r) - B_m(r)z_t(r))^2 D_m(r) \right) p_{mt}$$
$$- \frac{\lambda}{2} \sum_m \sum_r ||B_m(r)||^2. \qquad (11)$$

It follows immediately that the update equations for the diagonal entries of the covariance matrices are

$$\frac{1}{D_m(r)} = \frac{\sum_{t=1}^T p_{mt}(x_t(r) - B_m(r)z_t(r))^2}{\sum_{t=1}^T p_{mt}}$$
$$= \frac{\sum_{t=1}^T p_{mt}[x_t(r)]^2 - B_m(r) \sum_{t=1}^T p_{mt} x_t(r) z_t(r)}{\sum_{t=1}^T p_{mt}}, \qquad (12)$$

where the second equality follows only when $\lambda = 0$, similar to Eq. (4). Note that this is a scalar update equation for each of the diagonal elements of the matrix D_m. Depending on the sparse patterns of the matrices, the cost of the dot products $B_m(r)z_t(r)$ will be significantly smaller than before.

The update equations for the B_m matrices can also be significantly simplified. Since D_m is diagonal, the derivative of Eq. (1)

with respect to the rth row of B_m is uncoupled with the other rows. Also, only the non-zero portions of the row are computed (recall that B_m can be quite sparse so this can be a significant savings). Taking the derivative of Eq. 11 with respect to $B_m(r)$ thus yields:

$$\sum_{t=1}^{T} D_m(r)\big(x_t(r) - B_m(r)z_t(r)\big)z_t(r)^T p_{mt} - \lambda B_m(r) = 0.$$

This leads to the independent update equations for the non-zero portion of each row of the B matrix as follows:

$$B_m(r) = \left(\sum_{t=1}^{T} p_{mt} x_t(r) z_t(r)^T\right)\left(\sum_{t=1}^{T} p_{mt} z_t(r) z_t(r)^T + \frac{\lambda}{D_m(r)} I\right)^{-1},$$

where I is an appropriately sized identity matrix. In this case, only the outer products $z_t(r)z_t(r)^T$ and the SAXPY [27] operations $x_t(r)z_t(r)^T$ are required, yielding a significant reduction in both computational complexity and memory requirements. Also, unlike in the general case above, we see that we get a solution without needing to resort to algorithms that solve the Sylvester equation, regardless of the interpretation or value of λ. We see also again that adding the positive definite diagonal matrix $\lambda/D_m(r)I$ in the above ensures that the matrix is well conditioned (so inversion is possible). And as mentioned earlier, unlike typical ridge regression, we take here the interpretation of one global Bayesian prior variance in order to use a global λ parameter, which means we must appropriately scale by the variance $D_m(r)$.

All is not perfect, however, as when $\lambda > 0$, it appears that we once again obtain cycles in the parameter update dependencies, as in the center of Figure 2. Specifically, the equations show that both $B_m(r)$ depends on $D_m(r)$ and $D_m(r)$ depends on $B_m(r)$. Once again, we can break the cycle by using the previous parameters in place of current parameters and make multiple passes through the data, for each pass updating only one parameter group at a time, as shown on the right in Figure 2. Thus, in each step we are still guaranteed to improve the likelihood, although convergence may not be as fast. Moreover, we clearly have the same fixed points, as when at a particular fixed point, we have that $D_m^p(r) = D_m(r)$. We can also (Figure 2 middle) attempt to update all parameters simultaneously, but such a procedure does not guarantee in all cases an improvement in likelihood on each EM iteration.

4 • Training Shared Regularized Gaussian Decompositions

Normally, and as described in the previous section, each individual Gaussian component has its own unique set of parameters not shared with any other component in the model. In this section, we describe a model where some set of the Gaussian components is constrained to share a portion of their parameters with other components in the model. We wish moreover to analyze the EM update equations under such constraints. This procedure is often called parameter tying, but here the granularity of tying is somewhat finer than what is typically done in a speech recognition system. Specifically, an entire Gaussian component is commonly shared between two mixtures of Gaussians, or alternatively an entire mixture for an HMM state might be shared between two multi-state phone models in an HMM-based system [61]. Here, by contrast, each component Gaussian might share any of its μ, D, and B parameters with other components in the system. Moreover, a given component might share different groups with different other components. For example, a component's μ parameter might be shared differently than the component's D parameter. This leads to an enormous variety of parameter sharing patterns. There have been a number of other models that use partially shared Gaussian parameters, including the ones described in [28,23,49], but this is not done in the same way described herein. In particular, we will see that Cholesky factorized covariance matrices allow us to easily combine sharing and other forms of regularization.

It is crucial to realize that sharing is a form of constrained optimization—we still seek a maximum likelihood solution, but now we are constraining the parameters to live in a space where sub-portions of Gaussian parameters are forced to be identical. It is well known that constrained optimization can be more difficult (either computationally or mathematically) than unconstrained optimization [18], and we will see that our case is no exception. We will find, however, that the update rules are still fairly simple if we use the idea we saw in the previous section of breaking cycles over EM iterations, as it then is possible to find a solution without major modifications to the previously derived equations.

In the most general case of parameter sharing, we have three functions that map from the index m of a general class to the index k of a specific extant mean, diagonal covariance, and regression coefficient used by that index. Those mappings are, respectively, $T_\mu(m)$, $T_D(m)$, and $T_B(m)$. In other words, we have that $T_\mu(m) : \{1, \ldots, M\} \to \{1, \ldots, K_\mu\}$ where M is the total number of components in the system, and K_μ is the total number of unique means in the system. An analogous definition holds for the D and B matrices, where we assume there are a total of K_D unique D matrices, and K_B total unique B matrices in the model. We also define the inverse mappings: $T_\mu^{-1}(k)$, $T_D^{-1}(k)$, and $T_B^{-1}(k)$, which refer to the set of all indices m that lead to a particular index k, that is

$$T_\mu^{-1}(k) \triangleq \{m : T_\mu(m) = k\},$$

and where the analogous definitions hold for $T_D^{-1}(k)$ and $T_B^{-1}(k)$.

Using this representation of tying via the index mapping, the auxiliary function becomes

$$\begin{aligned}
Q = \sum_m \sum_t &\left\{ \frac{1}{2} \log(|D_{T_D(m)}|) \right. \\
&\left. - \frac{1}{2}(x_t - B_{T_B(m)} z_t - \mu_{T_\mu(m)})^T D_{T_D(m)}(x_t - B_{T_B(m)} z_t - \mu_{T_\mu(m)}) \right\} p_{mt} \\
&- \frac{\lambda}{2} \sum_{k=1}^{K_B} \|B_k\|_F^2 - \frac{\eta}{2} \sum_{k=1}^{K_\mu} \|\mu_k\|^2.
\end{aligned} \quad (13)$$

There are several differences between this equation and Eq. (1). First, we have separated out the mean vectors from the B matrices since, in this general case, there could be an entirely different tying pattern for the means and regression coefficients within B. Second, we now have two regularization coefficients, one for the regression coefficients λ and a separate one for the shared mean pool η—note again that this form of regression only makes sense when the data have been centered or zero-mean normalized, as mentioned earlier. Other than that, we still may have a unique posterior probability p_{mt} for each m since various combinations of mean, diagonal covariance, and regression coefficient could still produce a distinct value of p_{mt} for different m and fixed t—the constrained optimization procedure requires that the posterior be filtered down into the appropriate shared component of each Gaussian.

4.1 Covariance Updates

Taking derivatives and setting $\frac{\partial Q}{\partial D_k} = 0$ implies that

$$\sum_t \sum_{m \in T_D^{-1}(k)} p_{mt} \left(D_k^{-1} - (x_t - B_{T_B(m)} z_t - \mu_{T_\mu(m)}) (x_t - B_{T_B(m)} z_t - \mu_{T_\mu(m)})^T \right) = 0,$$

or that

$$D_k^{-1} = \frac{\sum_t \sum_{m \in T_D^{-1}(k)} (x_t - B_{T_B(m)} z_t - \mu_{T_\mu(m)})(x_t - B_{T_B(m)} z_t - \mu_{T_\mu(m)})^T p_{mt}}{\sum_t \sum_{m \in T_D^{-1}(k)} p_{mt}}.$$

In this form, it is not possible to compute the numerator without using multiple passes through the data, so we next expand the numerator in order to see what must be computed. Noting that D_k^{-1} is diagonal, and that any additive contributions to D_k^{-1} in the form of $A + A^T$ can be added as $2A$, we get:

$$D_k^{-1} \propto \sum_{m \in T_D^{-1}(k)} \left\{ \left(\sum_t x_t x_t^T p_{mt} \right) - 2 \left[\left(\sum_t x_t p_{mt} \right) \mu_{T_\mu(m)}^T \right. \right.$$

$$+ \left(\sum_t p_{mt} x_t z_t^T \right) B_{T_B(m)}^T - \mu_{T_\mu(m)} \left(\sum_t z_t^T p_{mt} \right) B_{T_B(m)}^T \right]$$

$$+ B_{T_B(m)} \left(\sum_t z_t z_t^T p_{mt} \right) B_{T_B(m)}^T + \mu_{T_\mu(m)} \mu_{T_\mu(m)}^T \left(\sum_t p_{mt} \right) \right\} \quad (14)$$

From the above, it is particularly easy to see that each major term in the sum does not require any of the other parameters until the very end of the accumulation. For example, computing quantities such as $\sum_t x_t x_t^T p_{mt}$, $\sum_t p_{mt} x_t z_t^T$, and $\sum_t z_t p_{mt}$ can be done in one pass through the data, and at the end combined with the resulting final B matrix or mean vector for the corresponding m in producing the final numerator for the kth covariance matrix.

If there are no B matrices (i.e., all the B entries are zero, which happens when we use purely diagonal covariance matrices), then we have a particularly simple form:

$$D_k^{-1} \propto \sum_{m \in T_D^{-1}(k)} \left\{ \left(\sum_t x_t x_t^T p_{mt} \right) \right.$$

$$\left. - 2 \left(\sum_t x_t p_{mt} \right) \mu_{T_\mu(m)}^T + \mu_{T_\mu(m)} \mu_{T_\mu(m)}^T \left(\sum_t p_{mt} \right) \right\}. \quad (15)$$

4.2 B-Matrix Updates

Next, we find the update equations for the B matrices. Setting $\frac{\partial Q}{\partial B_k} = 0$ yields

$$\sum_{m \in T_B^{-1}(k)} D_{T_D(m)} \sum_t (x_t - B_k z_t - \mu_{T_\mu(m)}) z_t^T p_{mt} - \lambda B_k = 0,$$

or that

$$\sum_{m \in T_B^{-1}(k)} D_{T_D(m)} \left\{ \left(\sum_t x_t z_t^T p_{mt} \right) \right.$$

$$\left. - B_k \left(\sum_t z_t z_t^T p_{mt} \right) - \mu_{T_\mu(m)} \left(\sum_t z_t^T p_{mt} \right) \right\} - \lambda B_k = 0.$$

Again, we have a similar problem in that we are trying to solve for a matrix B in an equation of the form $A - \sum_i D_i B C_i - E - \lambda B = 0$, which is a generalized form of the Sylvester equation. Note that even if we decide not to regularize so that $\lambda = 0$, then Sylvester-like equations still apply since B is still in relation with the other parameters via $A - E = \sum_i D_i B C_i$. In this case, other than the lack of a direct analytical solution for B_k, we see that it depends on other parameters from the current EM iteration, giving us the situation depicted in the center of Figure 2. In particular, B_k depends on all the other mean (μ) and variance (D) parameters that are involved in any of the components $m \in T_B^{-1}(k)$, as described in Figure 3. This means that the dependencies between current parameters as described in the center of Figure 2 will depend entirely on the tying patterns as given by the T functions, each individual parameter having a different set of dependences—in a sense, each parameter is independent of all other parameters given what might be called its variational or parameter Markov blanket.

When $\lambda > 0$ or in order not to solve the general Sylvester equation, we must once again use explicitly the fact that D_k for all k is diagonal—we may write Eq. (13) in a way that is analogous to the way Eq. (1) was written as Eq. (11). Using again $B_k(r)$ to denote the rth row of B_k, and the analogous definitions as used in Section 3.2, we immediately get:

$$B_k(r) = \left(\sum_{m \in T_B^{-1}(k)} D_{T_D(m)}(r) \left\{ \left(\sum_t x_t(r) z_t^T(r) p_{mt} \right) \right.\right.$$
$$\left.\left. - \mu_{T_\mu(m)}(r) \left(\sum_t z_t^T(r) p_{mt} \right) \right\} \right)$$
$$\left(\sum_{m \in T_B^{-1}(k)} D_{T_D(m)}(r) \left(\sum_t z_t(r) z_t^T(r) p_{mt} \right) + \lambda I \right)^{-1}, \quad (16)$$

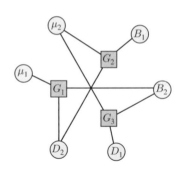

FIGURE 3 Sharing example. There are three Gaussian components, G_1, G_2, and G_3, with components G_2 and G_3 sharing the same mean μ_2, components G_1 and G_3 sharing the same regression matrix B_2, and G_1 and G_2 both sharing D_2. As shown in Eq. (16), the EM update equation for B_2 depends on the other new parameters μ_1, μ_2, D_1, D_2, but B_1's update depends only on μ_2 and D_2.

where I is an appropriately sized identity matrix. We once again see that an effect of the $\lambda > 0$ parameter ensures the invertibility of the matrix in which it is involved.

4.3 Mean Updates

Since the mean parameter can have a separate tying pattern, we must treat it separately from the B matrix. We set $\frac{\partial Q}{\partial \mu_k} = 0$ to get

$$\sum_{m \in T_\mu^{-1}(k)} D_{T_D(m)}^{-1} \sum_t (x_t - B_{T_B(m)} z_t - \mu_k) p_{mt} - \eta \mu_k = 0,$$

leading to

$$\sum_{m \in T_\mu^{-1}(k)} D_{T_D(m)} \left\{ \left(\sum_t x_t p_{mt} \right) - B_{T_B(m)} \left(\sum_t z_t p_{mt} \right) \right\}$$

$$= \sum_{m \in T_\mu^{-1}(k)} D_{T_D(m)} \sum_t \mu_k p_{mt} + \eta \mu_k,$$

or that

$$\mu_k = \left(\sum_{m \in T_\mu^{-1}(k)} D_{T_D(m)} \left(\sum_t p_{mt} \right) + \eta I \right)^{-1} \left\{ \sum_{m \in T_\mu^{-1}(k)} D_{T_D(m)} \left[\left(\sum_t x_t p_{mt} \right) - B_{T_B(m)} \left(\sum_t z_t p_{mt} \right) \right] \right\}.$$

When there is no mean sharing (i.e., the mean is used in only one component), we get that $T_\mu^{-1}(k)$ has only one entry, and if also $\eta = 0$, the inverse covariance matrix cancels out, and we end up with a greatly simplified form for the means:

$$\mu_k = \frac{\sum_t x_t p_{mt} - B_{T_B(m)} \sum_t z_t p_{mt}}{\sum_t p_{mt}}.$$

4.4 When Only the Variances are Tied

In certain cases, it may be that only a particular pattern of tying occurs. For example, when only the variances are tied, a number of simplifications are possible. The auxiliary Q function becomes:

$$Q = \sum_m \sum_t \left\{ \frac{1}{2} \log(|D_{T_D(m)}|) - \frac{1}{2} (x_t - B_m z_t - \mu_m)^T D_{T_D(m)} (x_t - B_m z_t - \mu_m) \right\} p_{mt} - f(B, \mu),$$

where $f(B, \mu)$ indicates only the regularization terms corresponding to $\{B_k : 1 \leq k \leq K_B\}$ and $\{\mu_k : 1 \leq k \leq K_\mu\}$. Because B_k and μ in this case are not tied, we can, as in Section 3, join them together in a single B matrix with the assumption that the final z has a constant of 1, and if we assume that they utilize the same regularization coefficient, we get:

$$Q = \sum_m \sum_t \left\{ \frac{1}{2} \log(|D_{T_D(m)}|) - \frac{1}{2}(x_t - B_m z_t)^T \right.$$
$$\left. D_{T_D(m)}(x_t - B_m z_t) \right\} p_{mt} - \frac{\lambda}{2} \sum_m ||B_m||_F^2.$$

As in Section 3, we set $\frac{\partial Q}{\partial B_m} = 0$ to get that:

$$\sum_t D_{T_D(m)}(x_t - B_m z_t) z_t^T p_{mt} - \lambda B_m = 0,$$

or once again that:

$$\sum_t (x_t - B_m z_t) z_t^T p_{mt} - \lambda D_{T_D(m)}^{-1} B_m = 0,$$

leading to the same update equations for B_m.

To find the update equations for the tied D_k matrix, we set $\frac{\partial Q}{\partial D_k} = 0$ to get:

$$D_k = \frac{\sum_t \sum_{m \in T_D^{-1}(k)} (x_t - B_m z_t)(x_t - B_m z_t)^T p_{mt}}{\sum_t \sum_{m \in T_D^{-1}(k)} p_{mt}} \quad (17)$$

$$= \frac{\sum_{m \in T_D^{-1}(k)} \left(\sum_t p_{mt} x_t x_t^T - B_m \sum_t p_{mt} z_t x_t^T \right)}{\sum_{m \in T_D^{-1}(k)} \sum_t p_{mt}}, \quad (18)$$

where again the second equality holds only when $\lambda = 0$. This equation, therefore, as in Section 3 is relatively easy to compute. Also, note that when $\lambda = 0$, this is an EM rather than a GEM procedure (see Section 5), since in each case we are maximizing rather than just increasing the auxiliary function (since there is no parameter dependence). In general, we see that the form of parameter tying and regularization can interact in complicated ways to determine how we can optimize the parameters, and this is discussed further in Section 5.

4.5 Regularized MLLR

Speaker adaptation [41,26,24,47,59] is one of the most successful parameter adjustment methods in speech recognition. In the adaptation paradigm, the training and test set distributions are not identical [61]. Moreover, there is a large training set so that the training set distribution may be well estimated, but only a small amount of adaptation data from the test distribution is available—this small test distribution data set is used to slightly adjust the parameters of a model to better represent the test set distribution. Of course, if too little restriction is placed on the amount of adjustment made to the model based on the adaptation data, then over-training can occur, and therefore adaptation must be based on a highly constrained model. In the widely used maximum-likelihood linear regression (MLLR) [41,59], a particular form of constraint is used, namely the mean vectors in a Gaussian mixture model are subject to a transformation that consists of no more than a scale, rotation, and shift (an affine transformation).

Given our derivation of shared conditional Gaussians above, we can extend MLLR so that not only the Gaussian means are adapted, but also the covariance matrix, all using non-zero mean Gaussian Bayesian priors—thanks to the Cholesky factorized covariance matrix.

In this framework, we moreover may easily adapt to C the regression coefficients, even if they correspond to a cross-stream conditional Gaussian. For example, the component parameters μ and B can be adapted while leaving the conditional variance parameters D fixed. This form of adaptation is seen essentially as another application of operations similar to the B matrix. In particular, suppose that for each m there is a mapping $T_A(m) : \{1, \ldots, M\} \to \{1, \ldots, K_A\}$ that maps from each component, say m, to a "regression class" [41], which gives an index that identifies the parameters that are to affect that component. The kth adaptation parameters then are to adapt to C all components $\{m : m \in T_A^{-1}(k)\}$ using a similar notation as in Section 4. For each k, there is a general upper triangular (so not symmetric) adaptation coefficient matrix A_k, a shift vector b_k, and a variance scale diagonal matrix C_k. These parameters influence the model as expressed in the following adjusted auxiliary function:

$$Q = \sum_m \sum_t \sum_r \left(\frac{1}{2} \log(C_{T_A(m)}(r)^2 D_m(r)) \right.$$
$$\left. - \frac{1}{2} \left(x_t(r) - A_{T_A(m)}(r) B_m(r) z_t(r) - b_{T_A(m)}(r) \right)^2 C_{T_A(m)}(r)^2 D_m(r) \right) p_{mt}$$
$$+ \sum_k \left(-\frac{\lambda}{2} \|A_k - I\|^2 - \frac{\eta}{2} \|b_k\|^2 - \frac{\nu}{2} \|C_k - I\|^2 \right),$$

where $A_k(r)$ (resp. $b_k(r)$, $C_k(r)$) is the rth row (resp. entry, diagonal element) of matrix A (resp. vector b_k, C_k). Again, any original mean parameters are contained within $B_m(r)z_t(r)$. A_k must be upper triangular so that the form $A_k B$ is still upper diagonal with zeros along the diagonal, and this means that the Gaussian normalization constant (a function of D_m) is still valid. Also, since C_k only occurs in squared form, C_k other than being a diagonal matrix is unconstrained. As can be seen, we adapt parameters of the mth component Gaussian by applying an affine transformation to the conditional parameters as in $A_{T_A(m)}(r)B_m(r)z_t(r) + b_{T_A(m)}(r)$ which becomes the new conditional mean. Also, the rth variance is scaled by $C_k(r)$. Of course, to avoid over-training on the data, we assume that K_A is relatively small, and if $K_A = 1$ we have one global A, C matrix pair and b vector that influences the model by performing one joint shift, scaling, and rotation of the parameters. This generalizes MLLR as well since we are not only allowing the adaptation to occur but we penalize the set of adaptation coefficients depending on its distance from the identity matrix using the Frobenius norm terms $||A_k - I||$ and $||C_k - I||$. This last regression term provides an additional way other than the number of regression classes that accuracy-regularization within MLLR may be controlled. This may be appropriate in cases with extremely small amounts of data, where even one (possibly sparse) A matrix would require too many parameters. Note that the approach outlined above may be seen as applying a Bayesian prior [31] on the adaptation coefficients in MLLR, but this is not the same form as MAP adaptation [26], where a Bayesian prior (in particular, a Dirichlet and Wishart prior) is applied to all of the Gaussian mixture parameters (including the mixture coefficients).

To perform adaptation, then, we can derive an EM update step very similar to the derivation in the shared parameter case in Sections 4.2 and 4.3. First, let $\bar{z}_t^m(r) = B_m(r)z_t(r)$ and $x_t^m(r) = x_t(r) - b_{T_A(m)}(r)$. Let $I(r)$ be a row-vector of zeros but with a one in the rth position (i.e., $I(r)$ is the rth row of the $M \times M$ identity matrix). We find $\frac{\partial Q}{\partial A_k(r)}$ and $\frac{\partial Q}{\partial b_k(r)}$ yielding

$$\frac{\partial Q}{\partial A_k(r)}\left(\sum_{m,t,r}(x_t^m(r) - A_{T_A(m)}(r)\bar{z}_t^m)^2 D_m(r)p_{mt} - \sum_{k',r'}\frac{\lambda}{2}||A_k'(r') - I(r')||^2\right)$$

$$= \sum_t \sum_{m \in T_A^{-1}(k)} D_m(r)(x_t(r) - A_k(r)z_t^m)(z_t^m(r))^T - \lambda(A_k(r) - I(r)).$$

Solving for $A_k(r)$ gives

$$\sum_t \sum_{m \in T_A^{-1}(k)} D_m(r)x_t(r)(z_t^m(r))^T$$

$$- A_k(r)\sum_t \sum_{m \in T_A^{-1}(k)} (D_m(r)I)z_t^m(z_t^m)^T - \lambda A_k(r) + \lambda I(r),$$

yielding

$$A_k(r) = \left(\sum_t \sum_{m \in T_A^{-1}(k)} D_m(r) x_t(r) (z_t^m(r))^T + \lambda I(r)\right)$$

$$\left(\sum_t \sum_{m \in T_A^{-1}(k)} (D_m(r)I) z_t^m (z_t^m)^T + \lambda I\right)^{-1}.$$

We can thus see that as $\lambda \to \infty$, we have that $A_k(r) \to I(r)$, so that $A_k \to I$, thus increasing λ decreases the amount of adaptation for each regression class. Updates for C_k are also relatively straightforward, where for each $C_k(r)$ we get a quadratic form. Each EM update step can be performed one or more times and, like MLLR [41], sets of EM updates can be iteratively interleaved with decodings to potentially further improve results in the unsupervised adaptation case.

It is easy to see that the above corresponds to placing nonzero-mean Gaussian priors on each of the three components of a Gaussian as we have used throughout this chapter, a property enabled by the use of Cholesky factorization. The Cholesky factorization has been used before [24,22]. Moreover, Bayesian priors have also been used for adaptation. For example, a prior may be placed on all parameters in the system [26]. Also, in maximum a posterior linear regression (MAPLR), a matrix Gaussian prior is placed on coefficients governing the affine transformation of the model means [11,56], or on an inner-factor diagonal of the covariance matrix [12]. A kernel ridge regression form of adaptation was also recently developed [54]. The technique presented above is different in that it does not use zero mean Gaussian priors.

5 • A Generalized EM Procedure

A generalized EM (GEM) is a procedure where rather than maximizing the auxiliary function by adjusting the parameters at each EM iteration, we instead only increase the auxiliary function. The above procedures that have been outlined show that, depending on the regularization coefficients and the parameter tying pattern, we may not be able to form a closed form update of each of the parameters as a function only of the previous EM iterations parameters, something as shown on the left of Figure 2. This circularity in the parameter updates cannot be broken even by using multiple passes through the data, as none of the new parameters can be formed without access to the other new parameters.

There have been two strategies for parameter update, however, that break the cycles. The first strategy has suggested that we can still update all parameters simultaneously if we use the previous parameters. In other words, when one parameter type is needing another from the current iteration, we substitute its value with the corresponding parameter at the previous EM iteration. This procedure, again, is guaranteed to have the same fixed points as a standard EM iteration, but we have not provided any guarantees regarding whether it increases the likelihood function, nor if it ever reaches such a fixed point.

An alternate strategy is the one suggested on the right of Figure 2. In this case, we only update one parameter set at a time. For example, we might find the optimal values only for $\{D_k : 1 \leq k \leq K_D\}$, use these new parameters in Eq. (13). Next, we might optimize $\{B_k : 1 \leq k \leq K_B\}$, use these new parameters, and then last optimize $\{\mu_k : 1 \leq k \leq K_\mu\}$. This entire procedure can then be repeated. This alternating optimization can be done without needing to worry about cyclic parameter dependencies since none of the parameters of a given type depend on any of the other parameters of that same type in the same EM iteration (e.g., μ_k for any fixed k does not depend on any $\mu_{k'}$ for any k' in the EM current iteration).

We can easily see that this procedure is a generalized EM since each update is guaranteed not to decrease the original likelihood function—in particular, the auxiliary function, Eq. (13) is concave in each of μ_k, B_k, and D_k separately, and in each case we obtain the same standard EM bounds [15,46]. For example, computing the Hessian matrix of Eq. (13) with respect to $B_k(r)$, we get

$$\frac{\partial^2 Q}{\partial B_k(r)^T \partial B_k(r)} = - \sum_{m \in T_B^{-1}(k)} D_{T_D(m)}(r) \left(\sum_t z_t(r) z_t(r)^T p_{mt} \right) - \lambda I,$$

a matrix that is clearly everywhere negative definite for all $\lambda \geq 0$, implying that Q is concave everywhere. Hessians with respect to the other parameters are also everywhere negative definite.

6 • Conclusions

We have described the EM/GEM update equations for generic Gaussian mixture systems, which include standard Gaussian mixtures, hidden Markov models, auto-regressive HMMs, multi-stream buried Markov models, and regularized MLLR. More specifically, we have described these equations where the

Gaussians are either regularized, partially parameter shared, or both.

We have seen that depending on the regularization parameters or the parameter tying pattern, the standard EM update equations might possess circular dependences, which requires modifications to the normal EM updates to break data dependency cycles. Several schemes were proposed for parameter update in this case, including when only one of each set of parameters is updated at a time. We have seen how this yields a GEM procedure that is still guaranteed to produce a local-maximum of the likelihood function. It would be interesting to relate the parameter tying mechanisms described here with Gaussian generalizations of Probabilistic Relational Models (PRMs) [20], as parameter sharing is a common feature in such models. In their case, however, parameters are typically discrete so the parameter dependency circularities do not arise. If one were to generalize PRMs to include partial Gaussian sharing as we have done above, the same procedures could potentially be adopted. Moreover, all of the above procedures would benefit from the application of ℓ_1-norm regularization [31], which would encourage sparse structure while retaining convexity in the EM update equations.

The above system of equations for tied and regularized Gaussians has been implemented as part of the graphical models toolkit (GMTK) and has been used in a wide variety of contexts. This material is based upon work supported by the National Science Foundation under Grant No. IIS-0093430 and by the Office of Naval Research under Grant No. N000140510388.

References

1. L.R. Bahl, P.F. Brown, P.V. de Souza, and R.L. Mercer. Maximum mutual information estimation of HMM parameters for speech recognition. In *Proc. IEEE Intl. Conf. on Acoustics, Speech, and Signal Processing*, 49–52, Tokyo, Japan, December 1986.
2. L.E. Baum and T. Petrie. Statistical inference for probabilistic functions of finite state Markov chains. *Ann. Math. Statist.*, 37(6):1554–1563, 1966.
3. J. Bilmes. Buried markov models: A graphical modeling approach to automatic speech recognition. *Comput. Speech Lang.*, 17:213–231, April–July 2003.
4. J. Bilmes and C. Bartels. Graphical model architectures for speech recognition. *IEEE Signal Process. Mag.*, 22(5):89–100, September 2005.
5. J. A. Bilmes. Graphical models and automatic speech recognition. In R. Rosenfeld, M. Ostendorf, S. Khudanpur, and M. Johnson, editors,

Mathematical Foundations of Speech and Language Processing. Springer-Verlag, New York, 2003.
6. J.A. Bilmes. A gentle tutorial of the EM algorithm and its application to parameter estimation for Gaussian mixture and hidden Markov models. Technical Report TR-97-021, ICSI, 1997.
7. J.A. Bilmes. Buried Markov models for speech recognition. In *Proc. IEEE Intl. Conf. on Acoustics, Speech, and Signal Process*, Phoenix, AZ, March 1999.
8. J.A. Bilmes. Factored sparse inverse covariance matrices. In *Proc. IEEE Intl. Conf. on Acoustics, Speech, and Signal Processing*, Istanbul, Turkey, 2000.
9. C. Bishop. *Neural Networks for Pattern Recognition*. Clarendon Press, Oxford, 1995.
10. C.-P. Chen, J. Bilmes, and D. Ellis. Speech feature smoothing for robust ASR. In *Proc. IEEE Intl. Conf. on Acoustics, Speech, and Signal Processing*, March 2005.
11. C. Chesta, O. Siohan, and C.-H. Lee. Maximum a posteriori linear regression for hidden markov model adaptation. In *European Conf. on Speech Communication and Technology (Eurospeech)*, 1999.
12. W. Chou and X. He. Maximum a posteriori linear regression (maplr) variance adaptation for continuous density hmms. In *European Conf. on Speech Communication and Technology (Eurospeech)*, 1513–1516, Geneva, Switzerland, 2003.
13. A. P. Dawid. Conditional independence in statistical theory. *J. R. Stat. Soc. B*, 41(1):1–31, 1989.
14. T. Dean and K. Kanazawa. Probabilistic temporal reasoning. *AAAI*, 524–528, 1988.
15. A.P. Dempster, N.M. Laird, and D.B. Rubin. Maximum-likelihood from incomplete data via the EM algorithm. *J. R. Stat. Soc. Ser. B.*, 39, 1977.
16. Y. Ephraim, A. Dembo, and L. Rabiner. A minimum discrimination information approach for HMM. *IEEE Trans. Info. Theory*, 35(5):1001–1013, September 1989.
17. Y. Ephraim and L. Rabiner. On the relations between modeling approaches for information sources. In *Proc. IEEE Intl. Conf. on Acoustics, Speech, and Signal Processing*, 24–27, 1988.
18. R. Fletcher. *Practical Methods of Optimization*. John Wiley & Sons, New York, NY, 1980.
19. N. Friedman. The Bayesian structural EM algorithm. *14th Conf. on Uncertainty in Artificial Intelligence*, 1998.
20. N. Friedman, L. Getoor, D. Koller, and A. Pfeffer. Learning probabilistic relational models. In *IJCAI*, 1300–1309, 1999.
21. Z. Gajic. *Lyapunov Matrix Equation in System Stability and Control*. Academic Press, San Diego, 1995.
22. M.J.F. Gales. Maximum likelihood linear transformations for hmm-based speech recognition. *Comput. Speech Lang.*, 12:75–98, 1998.
23. M.J.F. Gales. Semi-tied covariance matrices for hidden Markov models. *IEEE Trans. Speech Audio Process.*, 7(3):272–281, May 1999.

24. M.J.F. Gales and P.C. Woodland. Mean and variance adaptation within the mllr framework. *Comput. Speech Lang.*, 10, 1996.
25. J.D. Gardiner, A.J. Laub, J.J. Amato, and C.B. Moler. Solution of the sylvester matrix equation $axbt + cxdt = e$. *ACM Trans. Math. Softw.*, 18(2):223–231, 1992.
26. J.L. Gauvain and C.H. Lee. Maximum a-posteriori estimation for multivariate Gaussian mixture observations of markov chains. *IEEE Trans. Speech Audio Process.*, 2:291–298, 1994.
27. G.H. Golub and C.F. Van Loan. *Matrix Computations*. Johns Hopkins, Baltimore, 1996.
28. R. A. Gopinath. Maximum likelihood modeling with Gaussian distributions for classification. In *Proc. IEEE Intl. Conf. on Acoustics, Speech, and Signal Processing*, 1998.
29. D.A. Harville. *Matrix Algebra from a Statistician's Perspective*. Springer, New York, 1997.
30. T. Hastie, S. Rosset, R. Tibshirani, and J. Zhu. The entire regularization path for the support vector machine. *J. Mach. Learn. Res.*, 5:1391–1415, October 2004.
31. T. Hastie, R Tibshirani, and J. Friedman. *The Elements of Statistical Learning: Data Mining, Inference, and Prediction*. Springer Series in Statistics. Springer, New York, 2001.
32. D. Heckerman, D. Geiger, and D.M. Chickering. Learning Bayesian networks: The combination of knowledge and statistical data. Technical Report MSR-TR-94-09, Microsoft, 1994.
33. D. Heckerman and C. Meek. Embedded bayesian network classifiers. Technical Report MSR-TR-97-06, Microsoft Research, Redmond, WA, 1997.
34. D. Heckerman and C. Meek. Models and selection criteria for regression and classification. In *Proc. Thirteenth Conf. on Uncertainty in Artificial Intelligence*, Providence, RI. Morgan Kaufmann, August 1997.
35. H. Hermansky. Perceptual linear predictive (PLP) analysis of speech. *J. Acoust. Soc. Am.*, 87(4):1738–1752, April 1990.
36. M.J. Hunt, M. Lennig, and P. Mermelstein. Experiments in syllable-based recognition of continuous speech. *Proc. IEEE Intl. Conf. on Acoustics, Speech, and Signal Processing*, 1980.
37. F.V. Jensen. *An Introduction to Bayesian Networks*. Springer, New York, 1996.
38. B.-H. Juang, W. Chou, and C.-H. Lee. Minimum classification error rate methods for speech recognition. *IEEE Trans. on Speech and Audio Signal Processing*, 5(3):257–265, May 1997.
39. B.-H. Juang and S. Katagiri. Discriminative learning for minimum error classification. *IEEE Trans. on Signal Process.*, 40(12):3043–3054, December 1992.
40. S.L. Lauritzen. *Graphical Models*. Oxford Science Publications, Oxford, 1996.
41. C.J. Leggetter and P.C. Woodland. Maximum likelihood linear regression for speaker adaptation of continuous density hidden Markov models. *Comput. Speech Lang.*, 9:171–185, 1995.

42. X. Li and J. Bilmes. Regularized adaptation of discriminative classifiers. In *Proc. IEEE Intl. Conf. on Acoustics, Speech, and Signal Processing*, 2006.
43. X. Li, J. Bilmes, and J. Malkin. Maximum margin learning and adaptation of mlp classifers. In *European Conf. on Speech Communication and Technology (Eurospeech)*, 2005.
44. R.J. Mammone, X. Zhang, and R.P. Ramachandran. Robust speaker recognition. *IEEE Signal Process. Mag.*, 13(5):58–71, September 1996.
45. K. Murphy. *Dynamic Bayesian Networks: Representation, Inference and Learning*. PhD thesis, U.C. Berkeley, Dept. of EECS, CS Division, 2002.
46. R.M. Neal and G.E. Hinton. A view of the EM algorithm that justifies incremental, sparse, and other variants. In M.I. Jordan, editor, *Learning in Graphical Models*, 355–368. Kluwer Academic Publishers, Dordrecht, 1998.
47. L. Neumeyer, A. Sankar, and V. Digalakis. A comparative study of speaker adaptation techniques. In *European Conf. on Speech Communication and Technology (Eurospeech)*, 1127–1130, Madrid, Spain, 1995.
48. A. Ng and M. Jordan. On discriminative vs. generative classifiers: A comparison of logistic regression and naive bayes. In *Neural Information Processing Systems (NIPS)*, 14, Vancouver, Canada, December 2002.
49. P. Olsen and R. Gopinath. Modeling inverse covariance matrices by basis expansion. In *Proc. IEEE Intl. Conf. on Acoustics, Speech, and Signal Processing*, 945–948, 2002.
50. J. Pearl. *Probabilistic Reasoning in Intelligent Systems: Networks of Plausible Inference*. Morgan Kaufmann, Los Altos, 2nd printing edition, 1988.
51. J. Pearl. *Causality*. Cambridge University Press, Cambridge, 2000.
52. L.R. Rabiner and B.H. Juang. An introduction to hidden Markov models. *IEEE ASSP Mag.*, 1986.
53. S. Roweis and Z. Ghahramani. A unifying review of linear Gaussian models. *Neural Comput.*, 11:305–345, 1999.
54. G. Saon. A non-linear speaker adaptation technique using kernel ridge regression. In *Proc. IEEE Intl. Conf. on Acoustics, Speech, and Signal Processing*, 2006.
55. F. Sha and L. K. Saul. Large margin Gaussian mixture modeling for phonetic classification and recognition. In *Proc. IEEE Intl. Conf. on Acoustics, Speech, and Signal Processing*, Tolouse, France, 2006.
56. O. Siohan, T. Myrvol, and C. Lee. Structural maximum a posteriori linear regression for fast hmm adaptation. *Comput. Speech Lang.*, 16:5–24, 2002.
57. V. Vapnik. *Statistical Learning Theory*. Wiley, New York, 1998.
58. G. Wahba. *Spline Models for Observational Data*, volume 59 of *CBMS-NSF Regional Conference Series in Applied Mathematics*. SIAM, Philadelphia, 1990.
59. P.C. Woodland. Speaker adaptation: Techniques and challenges. In *Proc. IEEE ASRU*, 1999.

60. P.C. Woodland and D. Povey. Large scale discriminative training for speech recognition. In *ICSA ITRW ASR2000*, 2000.
61. S. Young. A review of large-vocabulary continuous-speech recognition. *IEEE Signal Process. Mag.*, 13(5):45–56, September 1996.
62. S. Young, J. Jansen, J. Odell, D. Ollason, and P. Woodland. *The HTK Book*. Entropic Labs and Cambridge University, 2.1 edition, 1990's.

30

Speech Synthesis

Thierry Dutoit[1] *and Baris Bozkurt*[2]

[1] *Faculte Polytechnique de Mons, Belgium*
[2] *Izmir Institute of Technology (IYTE), Izmir, Turkey*

Text-to-speech (TTS) synthesis is the art of designing talking machines. It is often seen by engineers as an easy task, compared to speech recognition.[1] It is true, indeed, that it is easier to create a bad, first trial text-to-speech (TTS) system than to design a rudimentary speech recognizer. After all, recording numbers up to 60 and a few words ("it is now", "am", "pm") and being able to play them back in a given order provides the basis of a working talking clock, while trying to recognize such simple words as "yes" or "no" immediately implies some more elaborate signal processing. If speech synthesis were really that simple, however, one could only blame the TTS R&D community for not having been able to massively produce a series of talking consumer products as early as the eighties.

The major point is that users are generally much more tolerant to ASR errors than they are willing to listen to unnatural speech. There is magic in a speech recognizer that transcribes continuous radio speech into text with a word accuracy as low as 50%; in contrast, even a perfectly intelligible speech synthesizer is only

---

[1] In an international conference on speech processing, a famous scientist once brought up a tube of toothpaste (whose brand was precisely Signal™) and, pressing it in front of the audience, he coined the words: "This is speech synthesis; speech recognition is the art of pushing the toothpaste back into the tube ... ".

moderately tolerated by users if it delivers nothing else than "robotic voices".[2]

This importance of *naturalness* versus *meaning* is actually very typical of the synthesis of natural signals (as opposed to their recognition). One could thus advantageously compare speech synthesis to the synthesis of human faces: while it is quite easy to sketch a cartoon-like drawing which will be unanimously recognized as a human face, it is much harder to paint a face which will be mistaken for a photograph of a real human being. To a large extent, you can modify the size, position, and orientation of most of the elements of a portrait drawing without breaking the intelligibility barrier (just think of the cubists ...), but even a slight change to a photorealistic painting will immediately make the complete work look like what it actually is: a painting of a face, not a real photograph of it.

Delivering intelligibility *and* naturalness has thus been the holy grail of speech synthesis research for the past 30 years. This chapter outlines the specific problems, encountered when trying to reach these goals, and shows how existing solutions benefit from fine-tuned signal processing techniques.

Section 1 discusses concatenative synthesis and introduces the underlying signal processing requirements, which are then examined with more details: prosody modification (Section 2), smoothing of discontinuities at concatenation points (Section 3), estimation of concatenation costs for corpus-based speech synthesis (Section 4), and the more advanced voice quality control and speaker modification issues (Section 5).

1 • Concatenative Text-to-Speech Synthesis

Figure 1 introduces the functional diagram of a fairly general TTS synthesizer. It consists of a natural language-processing module (NLP, also termed as *front-end*), capable of producing a symbolic linguistic representation of the text to read: a phonetic transcription, together with the desired intonation and rhythm (often termed as prosody). The NLP module is followed by a digital signal processing module (DSP, also termed as *back-end*), which transforms the symbolic information it receives into speech.

---

[2]Strictly speaking, though, there is no such thing as a "robotic voice" : robots have the voice we give them. In practice, the concept of a "robotic voice" is so strongly established that the TTS researchers are now trying to build "robots that do not sound like robots".

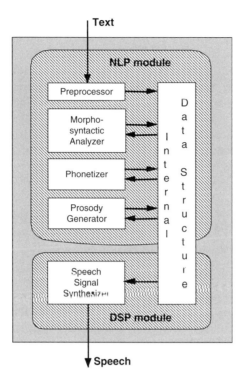

FIGURE 1 The functional diagram of a fairly general text-to-speech conversion system.

A preprocessing (or *text normalization*) module is necessary as a front-end, since TTS systems should in principle be able to read any text, including numbers, abbreviations, acronyms, and idiomatics, in any format. The preprocessor also performs the apparently trivial (but actually intricate) task of finding the end of sentences in the input text. It organizes the input sentences into manageable lists of word-like units and stores them in the internal data structure.[3] The NLP module also includes a morpho-syntactic analyzer, which performs part-of-speech tagging and groups the input sentence into syntactically related sets of words. A phonetizer provides the sequence of phonemes to be pronounced, and a prosody generator computes their duration and intonation (both are *suprasegmental features* and are jointly termed as *prosody*). Once phonemes and prosody have been predicted, the DSP module actually synthesizes them, by

---

[3] In modern TTS systems, all modules exchange information via some *internal data structure* (most often, a multi-level data structure, in which several parallel descriptions of a sentence are stored with cross-level links; sometimes feature structures as used in unification grammars). More on this can be found in Dutoit (1997, chapter 3).

producing speech samples which, when played via a digital-to-analog converter, will hopefully be understood and, if possible, be mistaken for real, human speech.[4]

Producing speech samples automatically, however, does not merely reduce to the playback of a sequence of pre-recorded words or phonemes. Even though we write in terms of isolated words and think speech in terms of discrete phonemes, we produce *continuous* speech, as a result of the coordinated and continuous action of a number of muscles. These articulatory movements are not produced independently of each other; they are frequently altered in a given context to minimize the effort needed to produce a sequence of articulatory movements. This effect is known as *coarticulation*. Coarticulatory phenomena are due to the fact that each articulator moves continuously from the realization of one phoneme to the next. They appear even in the most carefully uttered speech. In fact, they *are* speech. Thus, producing intelligible speech requires the ability to produce continuous, coarticulated speech.

More generally, transients in speech are more important for intelligibility than stable segments (Harris, 1953; Shannon et al., 1995), while modifying stable segments (e.g., the center of vowels) can very easily affect naturalness. Speech, indeed, is never really periodic, or stable. Even sustained vowels exhibit small frequency and amplitude variations (respectively termed as *jitter* and *shimmer*) and have substantial inharmonic components due to non-complete closure of the vocal folds after the so-called glottal closure instant; the presence of these noise-like components is correlated with the actual pitch period. As a result, the intuitive concept of "adding a bit of noise" to intonation curves, to amplitude curves, or to voiced speech waveforms in order to "make it sound more natural" merely leads to ... more noisy voiced speech: inharmonicity in voiced speech is not pure randomness.

Speech signal synthesizers can be roughly classified into two categories: the *expert-based* and the *instance-based* ones.

The former computes speech samples from phonetics either by using acoustic invariants obtained by expert phoneticians (this is often termed as the *rule-based*, or *formant* synthesis approach;

---

[4]In this chapter, we only consider the problems inherent to the design of the speech signal synthesizer in the DSP module; issues related to the NLP module are covered in several general textbooks given in the references, as well as in the Oxford Handbook of Computational Linguistics (Mitkov, 2003). The JASA article by Klatt [87] is still a good reference too, at least for pointing out what problems have to be solved.

see Klatt, 1987; O'Shaughnessy, 1984; Allen et al., 1987) or by describing speech in terms of the movements of the articulators, and perform synthesis using equations derived from mathematical simulations of the vocal tract (this is often termed as the *articulatory* synthesis approach; see Flanagan et al., 1975; Sondhi and Schroeter, 1997; Badin et al., 1998). Expert-based speech synthesis is mostly used for fundamental research on speech, since an improvement of the resulting speech quality necessarily coincides with additional insight into speech and vocal tract dynamics. We do not focus on this approach here.

On the opposite, instance-based (often termed as *concatenative*) speech synthesizers produce synthetic speech by gluing chunks of natural speech (preliminarily recorded and chopped out of human speech) together, in a specific order, so as to deliver a given message. Speech chunks readily embody natural jitter, shimmer, and inharmonicity. They can also be tailored so as to account for coarticulation, by constraining units to contain most coarticulatory effects. As a result, although concatenative synthesis only implicitly refers to the phonetics of speech (it could just as well be used for the synthesis of musical sounds), it is by far the winning paradigm in the industry. As Lindblom (1989) soundly notices: *"After all, planes do not flap their wings!"*

1.1 Diphone-Based Synthesis

Constraining speech chunks to embody coarticulation effects can be achieved to some extent by using *diphones* as basic units. A diphone is a speech segment which starts in the middle of the stable part (if any) of a phoneme and ends in the middle of the stable part of the next phoneme (Dixon and Maxey, 1968). Diphones therefore have the same average duration as phonemes (about 100 ms), but if a language has N phonemes, it typically has about N^2 diphones,[5] which gives a typical diphone database size of 1,500 diphones (about 3 minutes of speech, i.e., about 5Mb for speech sampled at 16 KHz with two bytes per sample). Some diphone-based synthesizers also include a reduced set of multi-phone units of varying length to better represent highly coarticulated speech (such as in /r/ or /l/ contexts).

In order to create such a synthesizer, one needs to set up a list of required diphones; a corresponding list of words is carefully completed, in such a way that each segment appears at least once (twice is better, for security). Unfavorable positions, like inside-stressed syllables or in strongly reduced (over-coarticulated) contexts, are usually excluded. A corpus is then read by a

---

[5] A bit less in practice: not all diphones are encountered in natural languages.

professional speaker (avoiding speaking style variations as much as possible, and even possibly with flat pitch, so as to facilitate speech analysis) and digitally recorded and stored. The elected segments are spotted, either manually with the help of signal visualization tools, or automatically thanks to segmentation algorithms, the decisions of which are checked and corrected interactively. Segment waveforms are then collected into a diphone database. This operation is performed only once.

At run-time, once the synthesizer receives some phonetic input (phonemes, phonemes durations, intonation curves) from the NLP module, it sets up a list of required diphones, together with their required duration and fundamental frequency contour (Figure 2). The available diphones would only match these prosodic requests by chance, since they have generally been extracted from words and sentences which may be completely different from the target synthetic sentence. Some *prosody modification* is therefore necessary. How this is achieved will be examined with more details in Section 2.

Additionally, since segments to be concatenated have generally been extracted from different words—that is, in different phonetic contexts—they often do not match in terms of amplitude and/or timbre. Even in the case of stationary vocalic sounds, for instance, a rough sequencing of waveforms typically leads

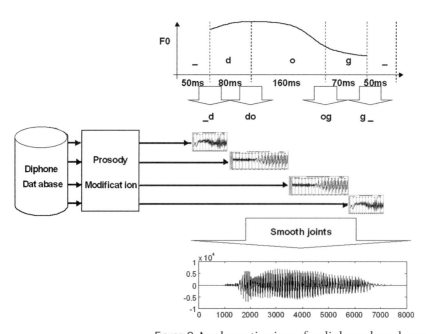

FIGURE 2 A schematic view of a diphone-based speech synthesizer.

to audible discontinuities. Part of this problem can be solved during the construction of the synthesis segments database by *equalization*, in which related endings of diphones are given similar energy levels at the beginning and at the end of segments (by setting the energy of all the phones of a given phoneme to their average value, for instance), in such a way as to eliminate amplitude mismatches. In contrast, timbre (i.e., spectral envelope) conflicts are better tackled at run-time, by *smoothing* (Figure 2) individual couples of segments when necessary rather than equalizing them once for all, so that some of the phonetic variability naturally introduced by co-articulation is still maintained. Smoothing will be discussed in Section 3.

1.2 Unit Selection-Based Synthesis

An extension of the diphone-based approach called *automatic unit selection* (Hunt and Black, 1996) or *corpus-based synthesis* has more recently been introduced, and opened new horizons to speech synthesis. As for diphone-based synthesis, the design of such a system requires the recording of a speech corpus, although much larger: 1–10 h typically, or 150–1,500 Mb. Segmentation in this case can only be achieved semi-automatically, if not completely automatically. The result is a diphone database which comprises much more than one instance of each diphone; all instances differ by their phonetic context and prosodic realization.

At run-time, given a phoneme stream and target prosody for an utterance, this algorithm selects, from this very large speech database, an optimum set of acoustic units (typically isolated diphones[6] or sequences of contiguous diphones, hence the NUU acronym which is sometimes used for this technique: nonuniform units) that best matches the target specifications (Figure 3).

For every *target unit* required (typically, every diphone to be synthesized), the speech database proposes lots of *candidate units*, each in a different context (and in general not exactly in the same context as the target unit). When candidate units cannot be found with the correct pitch and/or duration, prosody modification can be applied. Candidate units usually do not concatenate smoothly (unless a sequence of such candidate units can be found in the speech database, matching the target requirement), some smoothing can be applied. Recent synthesizers, however, tend to avoid prosodic modifications and smoothing, which sometimes create audible artifacts, and keep the speech data as is (Balestri

---

[6]Although sub-phonetic units such as half-diphones have also been proposed as alternatives when complete diphones are not available with the required pitch and duration (Conkie, 1999).

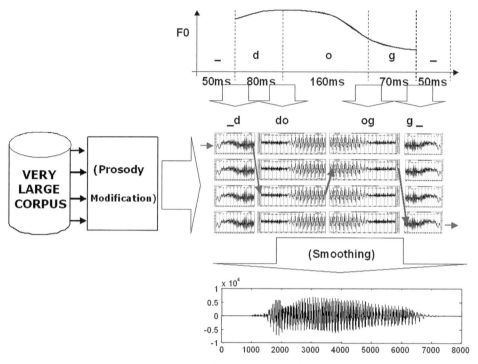

FIGURE 3 A schematic view of a unit selection-based speech synthesizer. The prosody modification and smoothing modules have been mentioned between parentheses, since they are not always implemented.

et al., 1999) (thereby accepting some distortion between the target prosody and the actual prosody produced by the system, and some spectral envelope discontinuities).

The biggest challenge of unit selection synthesis lies in the search for the "best" path in the candidate units lattice. This path must minimize two costs simultaneously: the overall *target cost*, defined as the sum of elementary target costs between each candidate unit chosen and the initial target unit, and the overall *concatenation cost*, defined as the sum of elementary concatenation costs between successive candidate units. The elementary target cost is typically computed as a weighted sum of differences between the linguistic features of the target and candidate units. The elementary concatenation cost is usually computed as a weighted sum of acoustic differences between the end of the left-candidate unit and the beginning of the right-candidate unit. This will be examined in Section 4.

The remarkable naturalness of speech produced by unit selection-based synthesizers comes from the fact that they tend to avoid as many concatenation points as possible, by selecting the largest sequences of consecutive units in the database. This greatly decreases the overall concatenation cost: the elementary concatenation cost between successive candidate units in the database is simply set to zero, so as to force the system to choose consecutive units. Another great advantage of such systems is that they avoid a precise (and often unnatural) prediction of phoneme durations and intonation curves. These are obtained as a by-product of unit selection itself provided the correct linguistic features are used for the estimation of the target cost. Most TTS systems sold today are based on this technique, while more rudimentary diphone-based systems are still used for low cost consumer products (as well as for research: free diphone-based TTS systems are available[7]; this is hardly the case for unit-selection-based systems).

It should be noted, however, that the development of larger and larger segmented speech corpora cost a lot. Hence, it is becoming increasingly difficult for companies to deliver tailored voices on demand. Similarly, the voice quality offered by today's TTS systems is somehow uniform. Producing expressive speech can only be achieved with the parallel recording of several speech corpora, each with an identifiable voice quality; this adds to the cost of a system. An alternative to the development of various speech corpora is precisely the use of speech processing techniques for voice quality modification, or for speaker modification. This problem is the focus of Section 5.

2 • Prosody Modification

The term *prosody* refers to certain properties of the speech signal which are related to audible changes in pitch, loudness, and syllable length. As we have seen, the concatenation of acoustic units taken from natural speech avoids the difficult problem of modeling the way humans generate speech. It also introduces another problem: the possible need for a modification of their prosody, mainly of their intonation and duration.[8]

---

[7]See for instance the synthesizers available from the MBROLA (http://tcts.fpms.ac.be/synthesis/mbrola.html), FESTIVAL (http://www.cstr.ed.ac.uk/projects/festival/), FestVox (http://festvox.org/), or FreeTTS (http://freetts.sourceforge.net/docs/) projects.
[8]In addition to the possible need for smoothing at concatenation points (see Section 3) and for compressing of the unit inventory using a speech

Prosody modification is a requirement for diphone-based synthesizers, in which only one instance of each diphone is available and must be used in all cases (i.e., with various target durations and intonation curves). It tends to be omitted in unit selection-based synthesizers, assuming the speech unit inventory is rich enough for the requested diphone to be available with approximately correct pitch and duration. Prosody modification is also required for *prosody transplantation*, which aims at modifying speech so as to keep the voice of the original speaker (the original speaker is still identifiable) while imposing prosody from another speaker. Prosody transplantation is necessary in industrial applications where synthetic speech must be inserted into fixed, pre-recorded sentences. In order to avoid abrupt voice changes, prerecorded speech is re-synthesized using the synthesizer's voice, but keeping its original, highly natural intonation.

Of the two parameters (pitch and duration), pitch is clearly the more sensitive. Even slightly modifying the shape of a pitch curve indeed readily breaks the naturalness of the speech segment it applies to. This effect is still more audible when the pitch modification ratios on both sides of concatenated units differ too much. Applying a constant pitch modification ratio on a complete segment (as in Bozkurt et al., 2004) does less harm, provided the ratio is not too high (typ. <= 2) or too low (typ. >= 0.5). Similarly, applying duration modification is less sensitive, provided the ratio is reasonable (<= 2 and >= 0.5).

Neither pitch nor duration modification is a straightforward operation. Simply resampling speech data (creating additional speech samples by interpolation of original ones, while keeping the same sampling frequency for playing them) does modify duration, and also the spectral envelope. All formants get compressed/dilated on the frequency axis, which has exactly the same audible effect as that of changing the "pitch" wheel of a tape recorder: human voices turn into doggy or mousy voices.

Two main classes of algorithms are in use for prosody modification, namely *time-domain* and *frequency-domain* algorithms.

2.1 Time-Domain Prosody Modification

The pitch-synchronous overlap-add (PSOLA) algorithms (Charpentier and Stella, 1986; Moulines and Charpentier, 1990) have drawn considerable attention, given their high segmental and suprasegmental efficiency, associated with a virtually unequaled simplicity in the case of their time-domain version:

---

coding technique, concatenative synthesis techniques tend to require large amounts of memory. Speech coding is covered in Chapter 31.

TD-PSOLA (Hamon et al., 1989). They extend the early time-domain harmonic scaling (TDHS) of Malah (1979) and the synchronized overlap-add (SOLA) technique of Roucos and Wilgus (1985), which highlighted the efficiency of simple block operations for time scaling of speech. The distinctive feature of the TD-PSOLA method is the idea that it is possible to perform both the required pitch and duration modifications directly on continuous waveforms.

2.1.1 Pitch Modification

If we restrict ourselves to a purely periodic signal $s(n)$, it is possible to get a perfect pitch-modified version of $s(n)$ by summing *OLA frames* $s_i(n)$, extracting pitch synchronously from $s(n)$ with a weighting window $w(n)$, and changing the time-shift between frames from the original pitch period T_0 to the desired period T:

$$s_i(n) = s(n)w(n - iT_0) \qquad (1)$$

$$\tilde{s}(n) = \sum_{i=-\infty}^{\infty} s_i(n - i(T \quad T_0)) \qquad (2)$$

If $T \neq T_0$, the operation results, according to the Poisson formula (Papoulis, 1962), in a re-harmonization of the spectrum of $s_i(n)$ (which, if we assume perfect periodicity, does not depend on i when the original voiced sound is purely periodic) with the fundamental frequency $1/T$:

$$\text{if } s_i(n) \xleftrightarrow{F} S_i(\omega) \text{ then } \tilde{s}(n) \xleftrightarrow{F} \frac{2\pi}{T} \sum_{i=-\infty}^{\infty} S_i\left(i\frac{2\pi}{T}\right) \delta\left(w - i\frac{2\pi}{T}\right). \qquad (3)$$

Consequently, provided that $w(n)$ can be chosen so that the spectrum of $s_i(n)$ closely matches the spectral envelope of $s(n)$, equation (2) provides a simple and efficient way to change the pitch of a periodic signal (Figure 4).

Notice $w(n)$ also needs to be smoothly zero-ended, so as to avoid clicks at each OLA operation. Typically, Hanning windows are used. Another practical constraint is that F_0 modification produces better audio quality when the size of OLA frames is about two times the local pitch period, and when they are centered on the glottal closure instant, or at least on an instant of important speech excitation (as it is the case on Figure 4). As a result, TD-PSOLA requires the prior computation (or the semi-automatic setting) of *pitch marks* on all speech units. However, computing glottis closure instants (the only reference instant available in voiced speech) has two major

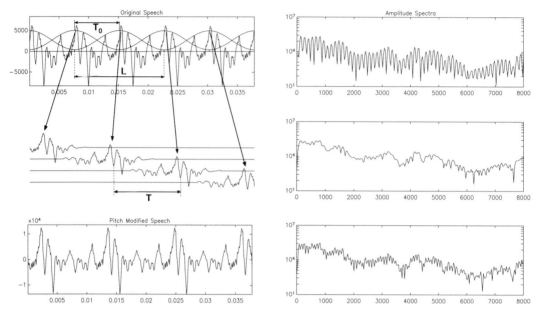

FIGURE 4 The TD-PSOLA re-harmonization process. The pitch-modified waveform (*bottom* plot) has the same spectral envelope as the original waveform (*top* plot). The center plot shows the OLA frames and their Fourier Transform, which is approximately the spectral envelope of the initial signal.

drawbacks: finding glottal closure instants requires expert assistance and it is not very precise if performed automatically (see Cheng and O'Shaughnessy, 1986, for a review), and time consuming if performed manually. A possible alternative to this problem is to measure glottal activity directly, with the help of an electro-laryngograph (Hess and Indefrey, 1987), instead of trying to deduce it from speech. Another solution is to use the local center of gravity of speech as the center of OLA frames (Stylianou, 1998b).

2.1.2 Duration Modification

Duration modification can be achieved in the time-domain by replicating or eliminating some OLA frames, as shown in Figure 5. For duration-only modification, OLA frames need not be pitch-synchronous (see the WSOLA method of Verhelst and Roelands, 1993, and the PICOLA algorithm of Morita and Itakura, 1986, implemented in the new ISO MPEG-4 audio standard). When two noncontiguous frames are overlap-added, a cross-correlation analysis is performed to find the optimal lag or lead between the frame centerings of the source waveforms.

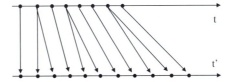

FIGURE 5 OLA-based duration modification process (lengthening in this case, as some frames are duplicated for synthesis). Dots on the top and bottom time lines correspond to centers of frames on the original and modified signals, respectively.

For simultaneous duration and pitch modification, pitch-synchronous frames are used. In this case, in order to prevent the algorithm from introducing artificial short-term autocorrelation in the synthesis signal when frames are duplicated, Charpentier and Moulines (1989) have proposed reversing the time axis of every other duplication of an unvoiced OLA frame. With such an artifice, speech can reasonably be slowed down by a factor of four, even though some tonal effect is encountered in voiced fricatives, which typically combine voiced and unvoiced frequency regions and therefore cannot be inaudibly time reversed.

TD-PSOLA is widely used in concatenative synthesis because of its high synthesis quality and extraordinarily low computational load (typically seven operations per sample; real-time synthesis can be achieved with an Intel 386 processor). Synthesized speech, however, is not perfect: spectral mismatches cannot easily be eliminated at segmental boundaries, and tonal quality appears when large prosodic modifications are applied on the concatenated units.

2.2 Frequency-Domain Prosody Modification

Sinusoidal approaches (e.g., Macon, 1996) and hybrid harmonic/stochastic representations (as Stylianou's HNM, 1998a) have also been proposed for speech synthesis. HNM (Harmonic+Noise Model), in particular, is increasingly used because of its additional smoothing (see Section 3.1) and speech compression capabilities. In this model, the speech spectrum is divided into two bands. A time-varying maximum-voiced frequency determines the limit between the two bands. In the lower band, the signal is represented as a sum of harmonically related sinewaves with slowly varying amplitudes and frequencies. The frequencies, amplitudes, and phases of these sinewaves are estimated from real speech signal by minimizing a weighted time-domain least-squares criterion. The upper band, which contains the noise part, is modeled by a linear prediction (LP) model and it is modulated by a time-domain amplitude envelope. This makes it possible to account for the nonperiodic components of speech, which include frication noise and period-to-period variations of the glottal excitation.

3 • Smoothing

Concatenating speech segments possibly extracted from different words or phonetic contexts is not straightforward. If done without care, it produces audible clicks, which cannot always be removed by simple OLA between the end of the left segment and the beginning of the right segment. This is due to three types of mismatches between successive segments, respectively related to phase, pitch, and spectral envelope mismatches (Dutoit and Leich, 1993), and illustrated in Figure 6 for the concatenation of vowels [a] sampled at 16 kHz.

Phase mismatch originates in overlap-adding frames that are not centered at the same place within the period (Figure 6a): even if the wave shapes to be concatenated are identical, the

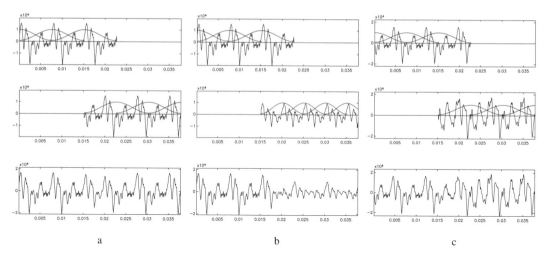

FIGURE 6 *Top* = last frames of the first segment to be concatenated; *center* = first frames of the second one; *bottom* = after OLA; (**a**) *Phase mismatch*: Waveforms are identical but the OLA windows are not centered on the same relative positions within the period. (**b**) *Pitch mismatch*: Both segments have exactly the same spectral envelope but were pronounced with different pitches. OLA windows are positioned coherently. The synthetic signal is obtained after changing the interval between the frames of the right segment, in order to impose a constant pitch. (**c**) *Spectral envelope mismatch*: The left segment is the diphone "ma", the right one is "am". The pitch is constant, and the windows are again positioned coherently. The spectral discontinuity is concentrated in one period.

OLA operation results in an unexpected discontinuity in this case. This can be avoided by adequate pitch-marking of boundary frames (as done in TD-PSOLA for all speech segments). Another solution is to adjust the position of OLA windows on the left and right segments so as to maximize the cross-correlation between windowed frames (as in WSOLA), thereby maximizing the chance that overlap-added frames are synchronous (Meyer et al., 1993). This solution, which corresponds to finding a common relative synchronization point, is easier than the absolute synchronization on glottal closure instants, but it is more time-consuming (it is usually performed online, since synchronization points cannot be pre-computed for all possible pairs of speech segments). It is also possible to use Stylianou's (1998b) center of gravity as an absolute reference point for centering frames.

Pitch mismatch arises when overlapped signals have very different fundamental periods, even if they have identical spectral envelopes and are windowed on similar relative positions. This situation cannot easily be avoided in a database of several thousand segments (Figure 6b). It is nevertheless possible to minimize this factor by recruiting professional speakers to record the segments database and training them to read the corpus with the most constant pitch possible. Another solution is to apply frequency-domain pitch modification before smoothing.

Last but not least, spectral envelope mismatches are due to coarticulation and speaker variability and appear whenever speech segments to be concatenated have been extracted from rather different phonetic contexts (Figure 6c). They can be minimized, by using context-dependent segment boundaries (Conkie and Isard, 1994). It is also possible to smooth them out, by using frequency/parametric domain (3.1) or time-domain (3.2) methods. A recent trend in unit selection-based synthesis is simply to ignore them.

3.1 Frequency/Parametric Domain Smoothing

Speech models can advantageously be used for smoothing discontinuities around concatenation points, provided a parametric form for acoustic units which makes it possible to modify their local spectral envelope without introducing audible artifacts. There has been a considerable amount of research effort directed at the design of adequate speech models for TTS. Linear prediction (LP) has been used first (Markel and Gray, 1976) for that purpose, given its relative simplicity. However, the buzziness inherent in LP degrades perceived voice quality.

The *linear prediction PSOLA* (LP-PSOLA) approach (Moulines and Charpentier, 1988) was introduced as a means of combining the high compression and smoothing capabilities of the LP model with the natural prosodic modifications provided by TD-PSOLA. LP-PSOLA techniques thus associate a parametric LP synthesizer with a TD-PSOLA algorithm. Prosodic modifications are performed with the TD-PSOLA algorithm on the excitation signal of the all-pole synthesis filter rather than on direct speech. The output of the overlap-adder is simply connected to an LPC synthesis filter, on the coefficients of which linear smoothing rules are applied. Several LP-PSOLA techniques can be derived from this principle, depending on the type of excitation signal used, as in the *multipulse linear prediction coding PSOLA* algorithm (MPLPC/PSOLA) of Moulines and Charpentier (1988), or the *residue-excited-PSOLA* (RELP-PSOLA) technique (Macchi et al., 1993).

The *HNM* model (see Section 2.2) also allows to smooth out discontinuities by linearly interpolating harmonic amplitudes and LP coefficients. Special care has to be taken for phase interpolation, due to phase wrapping (Stylianou et al., 1997).

3.2 Time-Domain Smoothing

The standard implementation of TD-PSOLA cannot smooth spectral envelope mismatches. As a result, synthesizing highly fluent speech with TD-PSOLA often requires a painstaking trial-and-error process in which the most probable diphone sequences are systematically tested, and diphones are rejected when an important spectral envelope mismatch is encountered. Other versions of the rejected diphones are then extracted from words with a different phonetic context than the initial one; tests are run again, and so on.

A solution consists of resynthesizing the speech segment database so as to provide segments with interesting properties which can be turned into account at synthesis time. Resynthesis is performed once on the database, off-line. This is the case of the so-called *multiband resynthesis OLA* (*MBROLA*) technique (Dutoit and Leich, 1993), in which a sinusoidal model is used to resynthesize all periods with the same pitch and with the same initial harmonic phases. This maximizes the similarity between periods having the same spectral envelope, and therefore happens to provide an efficient solution to pitch and phase mismatches as well. Additionally, when periods don't have similar spectral envelope (i.e., in case of spectral envelope mismatch), it is possible to perform period-based smoothing in the time domain (Figure 7). This feature compensates the loss of segmental quality due to the resynthesis operation. It is also one of the features of MBROLA which contributed to the success of the MBROLA project (http://tcts.fpms.ac.be/synthesis/mbrola/), by making it possible for nonspeech experts to contribute to the development of new synthetic voices.

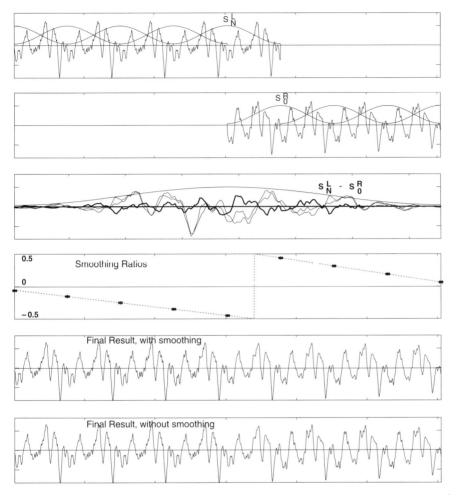

FIGURE 7 The time-domain linear smoothing process. S_N^L is the last frame of the left speech segment; S_0^R is the first frame of the right segment. The difference between waveforms, $S_N^L - S_0^R$, is distributed on several frames around the concatenation point.

For a formal comparison between different smoothing algorithms, see Dutoit (1994) and Syrdal et al. (1998).

4 • Estimation of Unit Selection Costs

The quality of speech produced by corpus-based concatenative synthesis is very sensitive to the definition of *target* and *concatenation* costs (Figure 8). These costs, as we shall see, should not be minimized separately, but rather through a weighted

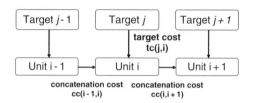

FIGURE 8 Target and concatenation costs.

sum. Target costs should favor the selection of units whose acoustics will match the context in which they will be inserted. Ideally, target costs should thus reflect the acoustic (or even perceptual) difference between target units and candidate units. Since the acoustics of the target are not available (the synthesizer is precisely in charge of producing them), only linguistic features can be used for estimating the segmental difference. We will not focus here on the definition of target costs, as they rely on linguistic descriptions of speech which fall beyond the scope of this chapter. The reader can refer to Möbius (2000).

In contrast, ideal concatenation costs should reflect the perceptual discontinuity (as opposed to the acoustic one) between successive candidate units, so as to favor the selection of units whose "cut-and-paste" concatenation will not be perceived. In particular, if the complete target sentence, or a significant part of it, is available in the corpus, all its consecutive units should obviously be selected. This is easily achieved by setting concatenation cost to zero for originally consecutive units. As a result, the unit selection algorithm tries to use as many consecutive units as possible from the corpus, provided the related target cost is not excessive (which would correspond to a "fluid" pronunciation of the required phonemes, but with the wrong intonation or phrasing, and therefore lead to low intelligibility).

Since Hunt and Black (1996), many proposals have been made for best defining concatenation costs. Costs are systematically based on some parameterization (using some speech model) of the initial and final speech frames of concatenated units, combined to some distance between the values of these parameters. Among the parameters used for this purpose, the most prominent are formant values, LPC spectra, Line Spectrum Frequencies (LSFs), Mel-Frequency Cepstrum Coefficients (MFCCs), various other frequency-modified spectral representations. Distances between the values of these multi-dimensional parameters vary from the Euclidian distance to the Kullback–Leibler divergence, through weighted Euclidian distances or the Mahalanobis distance. Although it is not clear yet which one leads to optimal results in all cases (i.e., for all recording conditions and all languages, if only possible), several studies have been made for comparing combinations of these options on specific corpora, by computing correlations between the values of these parameters/distances combinations and the impression of discontinuity reported by human listeners (see for instance the comparisons performed by Donovan (2001)). In the work of Klabbers and Veldhuis (2001), it was found that using LPC power-normalized spectra and a symmetrical Kullback–Leibler distance function gave the best results. This is close to Stylianou and Syrdal's (2001) choice for a symmetric Kullback–Leibler distance between FFT-based power spectra. In Wouters and Macon (1998), an Euclidian distance on MFCCs was chosen. Vepa and King (2004) report on using the Mahalanobis distance on LSFs. Clearly, this question is still open; its solution will obviously require (and lead to) a better modeling of perceived discontinuity in speech.

5 • Voice Quality/Speaker Modification and Other Open Issues

Voice quality in speech synthesis research usually refers to the perceived degree of characteristics like breathiness, creakiness, and loudness. Voice quality variations are considered to be mainly due to the variations in the phonation (production of the glottal excitation signal) process at the glottis. Voice quality issues have often been studied within the context of formant synthesis in the speech synthesis area since in this type of parametric approach, glottal excitation signals are synthesized parametrically and therefore can rather easily be controlled. Few studies address voice quality issues in concatenative synthesis (Kawai and Tsuzaki, 2004; Stylianou, 1999; Campbell and Marumoto, 2000). But voice quality analysis/synthesis is drawing more and more attention in the domain of concatenative speech synthesis since it is one of the most important features of naturalness in speech. In addition, it is especially important for emotional/expressive speech synthesis (an area which is gaining popularity) since voice quality codes as much information about the state of the speaker as does the prosodic information (Campbell and Marumoto, 2000). There is a strong correlation between voice quality variations and prosodic variations in speech since both are features of the phonation process. Therefore, advances in one of the fields would potentially result in advancement in the other. However, current state of art in voice quality analysis/modification/synthesis of recorded speech is not yet advanced enough to be widely used since tools for automatically estimating and modifying features of the glottal excitation of real speech are necessary; this is a challenging problem due to nonlinear processes in speech production.

From a signal processing point of view, voice quality variations mainly correspond to variations in spectral tilt, in the relative amount of aperiodic components in speech, and in some spectral variations in the low frequency part of the spectrum (like variations in the glottal formant frequency F_g in the bandwidth of first formant, in amplitudes of the first few harmonics, etc.) (Doval and d'Alessandro, 2003). The simplest approach consists of studying how these few features vary in natural speech. In Figure 9, we demonstrate (on synthetic speech signals) how variations in two spectral features (glottal formant frequency F_g and high frequency band energy) correspond to a variation of phonation in the tense-soft dimension. The time-domain signals presented at the *bottom* of Figure 9 include the glottal excitation (glottal flow derivative) signal and the speech signal obtained by filtering this glottal excitation signal with an all-pole vocal tract filter with resonances at 600, 1,200Hz, 2,200, and 3,200Hz. A variation from tense phonation to soft phonation corresponds to a decrease both in glottal formant frequency and high frequency band energy (and vice versa). For a detailed study of acoustic feature variations due to voice quality variations, see Klatt and Klatt (1990).

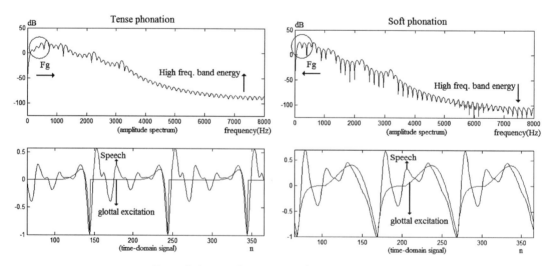

FIGURE 9 Spectral variations due to variations in phonation. *Top* figures show the amplitude spectrum and *bottom* figures show the time domain signals for glottal excitation and speech (Bozkurt, 2005).

5.1 Voice Quality and Concatenative Synthesis

Voice quality studies in concatenative speech synthesis research are concentrated mainly on voice quality labeling of speech corpora. In corpus construction for concatenative synthesis, one of the main difficulties is the need for long recording sessions and the physical difficulties this brings; voice quality changes during sessions (for example, due to the voice fatigue effect at the end of sessions). When long sessions are split into smaller sessions to avoid fatigue, this leads to the problem of matching voice qualities among sessions. Voice quality variations may be induced by many other reasons which are related to physical properties of the vocal folds or even psychological factors (when the speaker gets bored, he/she may start speaking fast-tense to be able to finish the recording quickly). Such variations in speech corpora potentially result in an increased degree and frequency of acoustic discontinuities in synthetic speech. This problem has been addressed in some recent studies (Kawai and Tsuzaki, 2004; Stylianou, 1999; Campbell and Marumoto, 2000) and is an open research area.

Most often, voice quality variations are treated as very slow variations (for example, variations between two recording sessions of several hours) but actually, voice quality differences do not exist only between segments from two different phrases. Voice source characteristics may show quite fast variations and voice quality discontinuities may exist even when two segments of the same sentence are concatenated. One of the reasons why voice quality discontinuities are not seen as a major concern for the current TTS systems is that unit selection is based on phonological structure matching and therefore implicitly selects segments with substantial voice quality continuity. However, voice quality and prosody are not completely interdependent and the realization of various voice quality effects for the same prosodic pattern and phonological segment is possible, and even desirable for expressive speech synthesis.

There are mainly two ways to avoid the discontinuities introduced by voice quality mismatches: using offline signal processing to equalize some voice quality variations, or including voice quality features in concatenation costs for unit selection to guarantee voice quality continuity.

The advantage of offline processing for voice quality equalization is that it does not introduce additional complexity in unit selection or to run-time signal processing. Stylianou (1999) proposes a method to detect some voice quality problems like roughness and breathiness throughout long recording sessions and correct them by using auto-regressive filters. Given a large speech corpus from the same speaker, this technique automatically selects a recording session as reference, sets statistical decision thresholds by using Gaussian mixture models, and corrects speech segments where needed.

The second solution stays rather as an unstudied subject apart from studies investigating which acoustic features are correlated with voice quality variations. One such study is Kawai and Tsuzaki (2004) which investigates the correlation of various measures (MFCC, Spectral tilt, band limited power) to perception scores of voice quality discontinuities constructed by concatenating phrases recorded over a long period of time (up to 202 days) and concludes that band-limited power (8-16Khz) is a good measure for detecting voice quality discontinuities. This area is likely to be studied in the concatenative expressive speech synthesis area in the upcoming years.

One other possible solution to the voice quality discontinuity problem is voice quality smoothing on the fly, which is another unstudied subject. For such an operation, some means of achieving voice quality modification without audible degradation of in the segmental quality of speech are clearly needed. Most of the concatenative synthesis algorithms, which include some spectral smoothing, smooth some of the voice quality discontinuities automatically (for example, spectral envelope smoothing results in smoothing some spectral tilt discontinuity automatically). However, such smoothing is too local for removing all voice quality variations. Higher quality smoothing can be possible once the separate components of discontinuity are understood (for example, the discontinuity in the aperiodic component of speech).

5.2 Voice Quality Modification

Voice quality modification is not only useful for smoothing voice quality variations but also for synthesizing variations on purpose. A potentially very useful and interesting approach in expressive speech synthesis is to use a hybrid synthesis paradigm: concatenative synthesis with more control parameters (like energy, spectral tilt, relative energy of the aperiodic component, glottal formant frequency) than the conventional concatenative synthesizers (which mostly include only duration and intonation information). Such an approach would stretch the limits of what is possible with the available data while keeping naturalness at a higher level compared to formant synthesis.

Natural voice quality modification is one of the unreached goals of speech synthesis research. Defining high-level rules that will drive low-level parameter modifications is too complex with our current understanding of the phenomenon. Signal processing for low-level parameter modification is being addressed in some few studies.

As in many parameter modification schemes, there are mainly two classes of approaches: spectral modification techniques and parametric methods performing some form of decomposition–modification–recomposition. The latter is usually known to introduce some audible artifacts and spectral modifications are preferred when possible. Spectral techniques are specially advantageous for voice quality modification since the perceptual relevance of spectral parameters is high, the representations are rather simplified (various time-domain models can be unified into a single spectral representation, as presented in Doval and d'Alessandro, 1997). They are also less sensitive to phase distortions, while time-domain voice quality modification applied on phase distorted speech signals often results in very low quality speech. D'Alessandro and Doval (2003) draw guidelines for the spectral modification of voice quality in a recent study. Their method includes modification on three dimensions: glottal formant, spectral tilt and PAPR (periodic to aperiodic ratio) the last two being perceptually more relevant for modification. However, the implementation of their ideas in a speech synthesizer is still not tested and is potentially a promising research area.

5.3 Speaker Modification

For commercial speech synthesis systems, customers often ask for the possibility of modifying speaker characteristics on the fly. Voice quality modification tools are potentially very useful for modifying some characteristics of the available voices (for example, making a synthetic voice sound softer to increase pleasantness, or just the converse: making it tenser to draw more attention).

Creating a "new" speaker from an existing one is rather more complex, although some signal processing tricks like changing average pitch frequency and applying spectral envelope modification would provide some speaker variations which are acceptable in some special cases (see Laroche, 2003, for instance).

For concatenation-based systems, creating a completely new voice to the system requires the recording and segmentation of a large speech corpus for the new speaker. An alternative solution is to use a voice conversion algorithm to be able to create a large speech database for a new speaker, given a large speech database from another speaker and a small speech database from the new speaker. Most of the algorithms proposed for this voice conversion task performs frame-by-frame spectral mapping. Long-time suprasegmental variations like intonation, voice quality, and timing are not realistically accounted for. Despite those difficulties, using voice conversion technology for adding new voices to concatenative speech synthesis systems deserves to be listed among future technical goals for speech synthesis research.

5.4 Other Open Issues

The interest of signal processing in speech synthesis by far does not reduce to the topics mentioned in the previous sections. Among other interesting implications, we would like to point out some which deserve special attention:

- *Monitoring tools for speech recordings for TTS database construction*. Efficiency in long recording sessions is essential because database construction is an expensive process. Some of the factors reducing efficiency are variations in general characteristics of the speaker's voice (pitch, loudness, and speech rate), low-quality segments produced due to fatigue or other reasons like dryness of vocal folds. Tools that can monitor pitch, loudness, speech rate, jitter–shimmer, breathiness, etc. and provide warning messages when some limits are exceeded can be potentially useful. We have found no reference to such tools in the literature.
- *Perceptual models for speech features*. Except the considerable effort made on finding the optimum perceptual distance measure for spectral discontinuities for improving quality of unit selection systems, perceptual speech models are still not much used. For example, pitch discontinuities, maybe the most important type of discontinuities in concatenative synthesis but no known measure of perceived pitch discontinuity seems to be available. In the best of all worlds, speech synthesis technology should be equipped with perceptual models and perform smoothing operations in a perceptual domain. Such an approach would perform better smoothing when necessary and reduce the amount of unnecessary smoothing operations which sometimes introduce degradations.
- *High-quality fundamental frequency modification*. F_0 modification still stays as one of important problems in concatenative speech synthesis. This is due to the fact that perceptual relevancy of fundamental frequency is very high, and existing modification algorithms have limited quality. Improvements in the quality of fundamental frequency modification have a great potential for reducing the size and complexity of speech corpora.
- *HMM-based speech synthesis* (Tokuda et al., 2004). Speech recognition and speech synthesis research can be considered to be somehow two independent research areas when we consider the amount of shared tools. HMM-based speech synthesis is a step forward for profiting from existing advances in speech

recognition technology in speech synthesis. One example is the investigation of using speaker adaptation techniques of speech recognition for voice conversion of speech synthesis in such a hybrid system.

- *Advanced articulatory modeling/synthesis.* Although the recent advances in concatenative TTS have resulted in a reduction of popularity and dedicated effort on articulatory speech models, there is still a need for a better insight in speech production phenomena in order to expand our reach in TTS synthesis and speech processing in general. Articulatory speech synthesis should be stated among the future technologies providing plenty of interesting problems to solve and demanding considerable research effort.

For a complementary list of forthcoming speech synthesis research areas, the reader could advantageously refer to Sproat et al.'s (1999) report on the *NSF Workshop for Discussing Research Priorities and Evaluation Strategies in Speech Synthesis*.

6 • Further Reading and Relevant Resources

Most of the information presented in this chapter can be found with a lot more details in Dutoit (1997), Van Santen et al. (1997) or Sproat (1998).

References

Allen, J., Hunnicut, S. and Klatt, D. (1987). *From Text to Speech: The MITTALK System*, Cambridge University Press.

Badin, P., Bailly, G., Raybaudi, M. and Segebarth, C. (1998). "A three-dimensional linear articulatory model based on MRI data". *Proceedings of the International Conference on Speech and Language Processing*, vol. 2, pp. 417–420, Sydney, Australia, November 1998.

Balestri, M., Paechiotti, A., Quazza, S., Salza, P. and L., Sandri, S. (1999). "Choose the best to modify the least: a new generation concatenative synthesis system", Proceedings of Eurospeech, Budapest, Hungary.

Bozkurt, B. (2005). *Zeros of the Z Transform (ZZT) Representation and Chirp Group Delay Processing for the Analysis of Source and Filter Characteristics of Speech Signals*, PhD Dissertation, Faculté Polytechnique de Mons, Belgium.

Bozkurt, B., Dutoit, T., Prudon, R., d'Alessandro, C. and Pagel, V. (2004). "Improving quality of MBROLA synthesis for non-uniform units synthesis". In Narayanan, S. and Alwan, A. (eds.), *Text to Speech Synthesis: New Paradigms and Advances*. Prentice Hall.

Campbell, N. and Marumoto, T. (2000). "Automatic labelling of voice-quality in speech databases for synthesis.", *Proceedings of the International Conference on Spoken Language Processing (ICSLP'00)*, pp. 468–471, Beijing, China.

Charpentier, F. and Stella, M.G. (1986). "Diphone synthesis using an overlap-add technique for speech waveforms concatenation", *Proceedings of the International Conference on Acoustics, Speech, and Signal Processing 86*, pp. 2015–2018.

Charpentier, F. and Moulines, E. (1989), "Pitch-synchronous waveform processing fechniques for text-to-speech synthesis using diphones", *Proceedings of Eurospeech 89*, Paris, vol. 2, pp. 13–19.

Cheng, Y.M. and O'Shaughnessy, D. (1986). "Automatic and reliable estimation of glottal closure instant and period", *IEEE Transactions on Acoustics, Speech, and Signal Processing*, vol. 37, no 12, pp. 1805–1815.

Conkie, A. (1999). "A robust unit selection system for speech synthesis", *Proceedings of the 137th meet. ASA/Forum Acusticum*, Berlin.

Conkie, A. and Isard, S. (1994). "optimal coupling of diphones", *Proceedings of the 2nd ESCA/IEEE Workshop on Speech Synthesis*, Mohonk, Sept. 1994.

D'Alessandro, C. and Doval, B. (2003). "Voice quality modification for emotional speech synthesis." *Proceedings of the European Speech Communication and Technology (Eurospeech'03)*, pp. 1653–1656. Geneva, Switzerland.

Dixon, N.R. and Maxey, H.D. (1968). "Terminal analog synthesis of continuous speech using the diphone method of segment assembly", *IEEE Transactions on ASSP*, AU-16, no. 1, pp. 40–50.

Donovan, R.E. (2001). "A new distance measure for costing spectral discontinuities in concatenative speech synthesisers". *Proceedings of 4th ISCA Tutorial and Research Workshop on Speech Synthesis*, Perthshire, Scotland.

Doval, B. and d'Alessandro, C. (1997). "Spectral correlates of glottal waveform models: an analytic study". *Proceedings of the International Conference on Acoustics, Speech, and Signal Processing (ICASSP'97)*, pp. 1295–1298.

Dutoit, T. (1994). "High quality text-to-speech synthesis : a comparison of four candidate algorithms". *Proceedings of the International Conference on Acoustics, Speech, and Signal Processing (ICASSP'94)*, pp. 565–568. Adelaide, Australia.

Dutoit, T. (1997). *An Introduction to Text-To-Speech Synthesis*, Kluwer Academic Publishers.

Dutoit, T. and Leich, H. (1993). "MBR-PSOLA: text-to-speech synthesis based on an MBE resynthesis of the segments database", *Speech Communication*, 13, pp. 435–440.

Flanagan, J.L., Ishizaka K. and Shipley K.L. (1975). "Synthesis of speech from a dynamic model of the vocal cords and vocal tract". *Bell System Technical Journal*, 54, pp. 485–506.

Hamon, C., Moulines, E. and Charpentier F., (1989). "A diphone system based on time-domain prosodic modifications of speech", *Proceedings of the International Conference on Acoustics, Speech, and Signal Processing 89*, S5.7, pp. 238–241.

Harris, C.M. (1953). "A study of the building blocks in speech", *Journal of the Acoustical Society of America*, 25, pp. 962–969.

Hess, W. and Indefrey, H. (1987). "Accurate time-domain pitch determination of speech signals by means of a laryngograph", *Speech Communication*, 6, pp. 55–68.

Hunt, A.J. and Black A.W. (1996). "Unit selection in a concatenative speech synthesis system using a large speech database". *Proceedings of the International Conference on Acoustics, Speech, and Signal Processing (ICASSP'96)*, vol. 1, pp. 373–376. Atlanta, Georgia.

Kawai H. and Tsuzaki M. (2004). "Voice quality variation in a long-term recording of a single speaker speech corpus". In Narayanan, S. and Alwan, A. (eds.), *Text to Speech Synthesis: New Paradigms and Advances.* Prentice Hall.

Klabbers, E. and Veldhuis, R., (2001). "Reducing audible spectral discontinuities", *IEEE Transactions on Speech and Audio Processing*, 9(1):39–51.

Klatt, D.H. (1987). "Text-to-speech conversion", *Journal of the Acoustical Society of America*, 82(3), 737–793.

Klatt D.H. and Klatt L.C. (1990). "Analysis, synthesis, and perception of voice quality variations among female and male talkers." *Journal of the Acoustical Society of America*, 87(2):820–57.

Laroche, J. (2003). "Frequency-domain techniques for high-quality voice modification", *Proceedings of the International Conference on Digital Audio Effects (DAFx-03)*, London.

Lindblom, B.E.F. (1989), "Phonetic Invariance and the Adaptive Nature of Speech", in B.A.G. Elsendoorn and H. Bouma eds., *Working Models of Human Perception*, Academi Press, New York, pp. 139–173.

Macchi, M., Altom, M.J., Kahn, D., Singhal, S. and Spiegel, M., (1993). "Intelligibility as a 6th Function of speech coding method for template-based speech synthesis", *Proceedings of Eurospeech 93*, Berlin, pp. 893–896.

Macon, M.W. (1996). "*Speech Synthesis Based on Sinusoidal Modeling*", Ph. D Dissertation, Georgia Institute of Technology.

Malah, D. (1979). "Time-domain algorithms for harmonic bandwidth reduction and time-scaling of pitch signals", *IEEE Transactions on Acoustics, Speech, and Signal Processing*, vol. 27, no 2, pp. 121–133.

Markel, J.D. and Gray A.H. (1976). *Linear Prediction of Speech*, Springer.

Meyer, P., Rühl, H.W., Krüger, R., Kluger, M. Vogten, L.L.M., Dirksen, A. and Belhoula, K., (1993). "PHRITTS – A text-to-speech synthesizer for the German language", *Proceedings of Eurospeech 93*, Berlin, pp. 877–890.

Mitkov, R. (2003). Handbook of Computational Linguistics, R. Mitkov, ed., Oxford University Press.

Möbius, B. (2000). "Corpus-based speech synthesis: methods and challenges". *Arbeitspapiere des Instituts für Maschinelle Sprachverarbeitung (Univ. Stuttgart)*, AIMS 6(4), 87–116.

Morita, N. and Itakura, F. (1986). "Time-scale modification algorithm for speech by use of pointer interval control overlap and add (PICOLA) and its evaluation," *Proceedings Of Annual Meeting of Acoustical Society of Japan*.

Moulines, E. and Charpentier, F. (1988). "Diphone synthesis using a multipulse LPC technique", *Proceedings of the FASE International Conference*, Edinburgh, pp. 47–51.

Moulines, E. and Charpentier F. (1990). "Pitch synchronous waveform processing techniques for text-to-speech synthesis using diphones". *Speech Communication*, 9, 5–6.

O'Shaughnessy, D. (1984). "Design of a real-time French text-to-speech system", *Speech Communication*, 3, 233–243.

Papoulis, A. (1962). *The Fourier Integral and Its Applications*, McGraw Hill, p. 47.

Roucos, S. and Wilgus, A. (1985). "High-quality time scale modification of speech", *Proceedings of the International Conference on Acoustics, Speech, and Signal Processing 85*, pp. 236–239.

Shannon, R.V., Zeng, F.G., Kamath, V., Wygonski, J. and Ekelid, M. (1995). "Speech recognition with primarily temporal cues", *Science*, 13;270(5234):303–4.

Sondhi, M.M. and Schroeter, J. (1997). "Speech production models and their digital implementations", *The Digital Signal Processing Handbook*. CRC and IEEE Press.

Stylianou, Y. (1998a). "Concatenative speech synthesis using a Harmonic plus Noise Model". *Proceedings of the 3rd ESCA Speech Synthesis Workshop*, 261–266. Jenolan Caves, Australia.

Stylianou, Y. (1998b), "Removing phase mismatches in concatenative speech synthesis", *Proceedings of the 3rd ESCA Speech Synthesis Workshop*, pp. 267–272.

Stylianou, Y. (1999). "Assessment and correction of voice quality variabilities in large speech databases for concatenative speech synthesis." *Proceedings of the International Conference on Acoustics, Speech, and Signal Processing (ICASSP'99)*, 377–380. Phoenix, A2.

Stylianou, Y. and Syrdal, A.K. (2001). "Perceptual and objective detection of discontinuities in concatenative speech synthesis", Proceedings of ICASSP, Salt Lake City, UT.

Stylianou, Y., Dutoit, T. and Schroeter, J. (1997). "Diphone concatenation using a Harmonic plus Noise Model of speech", *Proceeding of Eurospeech '97*, pp. 613–616.

Sproat, R., ed. (1998). *Multilingual Text-to-Speech Synthesis*. Kluwer Academic Publishers.

Sproat, R., Ostendorf, M. and A. Hunt, eds. (1999). *The Need for Increased Speech Synthesis Research: Report of the 1998 NSF Workshop for Discussing Research Priorities and Evaluation Strategies in Speech Synthesis.*

Syrdal, A., Stylianou, Y., Garisson, L., Conkie A. and Schroeter J. (1998). "TD-PSOLA versus Harmonic plus Noise Model in diphone based speech synthesis". *Proceedings of the International Conference on Acoustics, Speech, and Signal Processing (ICASSP'98)*, 273–276. Seattle, WA.

Tokuda, K., Zen, H. and Black, A. (2004). "An HMM-based approach to multilingual speech synthesis." In Narayanan, S. and Alwan, A. (eds.), *Text to Speech Synthesis: New Paradigms and Advances.* Prentice Hall.

Van Santen, J.P.H., Sproat, R., Olive, J., Hirshberg, J. eds. (1997), *Progress in Speech Synthesis,* Springer.

Vcpa, J. and King, S. (2004). "Join cost for unit selection speech synthesis". In Alwan A. and Narayanan S. (eds.), *Speech Synthesis.* Prentice Hall.

Verhelst, W. and Roelands, M. (1993). "An overlap-add technique based on waveform similarity (WSOLA) for high quality time-scale modification of speech", *Proceedings of ICASSP-93,* Vol. II, pp. 554–557.

Wouters, J. and Macon, M. (1998). "Perceptual evaluation of distance measures for concatenative speech synthesis", *Proceedings of ICSLP,* Vol 6, pp. 2747–2750, Denver Co.

31

Speech Coders

Roch Lefebvre and Philippe Gournay

Université de Sherbrooke, Sherbrooke, QC, Canada

1 • Introduction

Telecommunication systems make intensive use of speech coders. In wireless systems, where bandwidth is limited, speech coders provide one of the enabling technologies to reach more users and furnish better services. In wireline systems, where bandwidth can be less of an issue, speech is also digitized and compressed to a certain extent depending on the system.

In general, a speech encoder takes sampled speech at its input and produces a compressed bitstream for transmission or storage. At the receiver (or at playback from a storage device), a speech decoder takes the bitstream and produces a synthesized speech signal that is desired to sound as close to the original speech as possible. This is illustrated in Figure 1. The compression ratio of the speech encoder and the quality of the synthesized speech depend on the system requirements and application needs. Very often, the application requirement is that synthesized speech should have quality as high as possible with a bit rate as low as possible. Any voice application with limited channel capacity and/or memory space for storage can greatly benefit from existing low-rate speech coders.

The main objective of this chapter is to give the reader an appreciation of the ways signal processing is used inside and around modern speech coders to meet specific application needs and to solve acoustical problems in the communication link. Examples will focus mostly on standards since these are well documented and thoroughly tested. Only the main concepts of speech coders

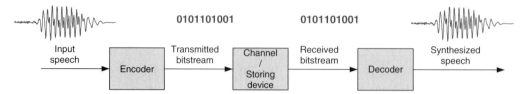

FIGURE 1 Simplified block diagram of a voice transmission or storage system using a speech coder.

will be presented here. For in-depth reviews of speech coders, the reader is referred to the several excellent tutorials written on the subject. For example, see [1,2].

The outline of this chapter is as follows. In Section 2, principles of modern speech coders are briefly reviewed. Section 3 presents methods to address issues related to input conditions. These include the presence of acoustical noise, the effect of audio bandwidth and the variability of the input signal characteristics. In Section 4, problems related to channel conditions, such as packet loss and variable transmission delay, are addressed. Finally, Section 5 discusses current developments and future challenges.

2 • Speech Coding Concepts

Advances in speech coding span several decades. It would be impractical to present here all the advances that have been made in this field. However, looking at most voice communication systems using speech coders today, we observe that the speech coders used in most systems are based on variants of linear predictive (LP) coding [3]. The basic principle in LP coding is to model the speech signal using a time-varying linear filter with a proper excitation signal at the input of the filter. The time-varying linear filter is often called the *synthesis* filter in the sense that it produces the synthetic speech, provided it has an appropriate excitation at its input. Moreover, the excitation signal is very often modeled as the weighted sum of two components: one component selected from an *adaptive codebook* and the other component from the *innovative codebook*. The adaptive codebook is nothing more than the past excitation signal, which is also available at the decoder – or at least approximately if there are channel errors. The adaptive codebook has thus a content which varies in time, as the past excitation is shifted after encoding new speech samples. The innovative codebook is also called the fixed codebook. In the simplest form, its content does not change with time. The innovative codebook essentially models the unpredictable component of the speech samples, that is, the information that is not modeled by the synthesis filter and the adaptive codebook. What distinguishes many modern speech coders is the structure and the search method of the innovative codebook. Figure 2 shows a simplified version of a modern speech decoder using the above-mentioned principles. We see from this figure that the encoded bitstream carries three types of information: (1) the LP, or synthesis filter parameters; (2) the adaptive codebook parameters (typically, a set of delays and gains); and (3) the innovative codebook parameters (typically codevector indexes and the associated gains). From this information, the decoder can reconstruct or synthesize a frame of speech. At the encoder, the input speech will be split into frames, and the parameters in Figure 2 will thus be calculated and encoded for every frame. A typical speech frame will have between 10 and 30 ms in duration.

In Figure 2, only the speech decoder is illustrated. It shows the relative simplicity of a speech decoder, both conceptually and algorithmically. At every frame, the decoder constructs the total excitation signal and then filters this excitation signal through the synthesis filter to produce a frame of synthesized speech. The complexity is more at the encoder, where the parameters actually have to be calculated from the input speech frame, and then encoded for transmission. This is usually carried out in a sequential manner. First, the synthesis filter parameters are calculated to match the input speech. This involves windowing the speech signal, then calculating a correlation matrix for that

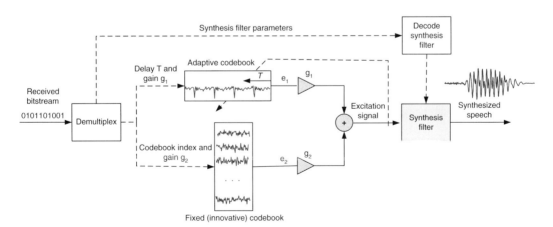

FIGURE 2 Typical speech decoder based on an LP filter model, using two components to encode the excitation: an adaptive codebook and a fixed (innovative) codebook.

speech frame, and inverting the matrix to find the optimal filter parameters. Then, the excitation parameters are calculated in such a way that when the excitation is passed through the LP filter, this produces synthetic speech as close to the input speech as possible according to a given error criterion. This search process is called *analysis-by-synthesis* in the speech coding literature. The adaptive codebook parameters (optimal delays T and gains) are calculated first. This codebook typically models periodic components in speech which are strong when the vocal cords are active. The adaptive codebook thus achieves a significant coding gain with relatively few bits (a delay and a gain). After the adaptive codebook contribution is determined, the innovative codebook is searched for its optimal contribution.

The innovative codebook is shown in Figure 2 as comprising seemingly random components. To make it useful and implementable in real time, this codebook has to be structured. Several approaches have been proposed to build this structure [4–6]. The most commonly used one in speech coders deployed today is referred to as the ACELP model [7]. There are more than 15 international standards today based on ACELP for wireless and wireline communication systems.

The main idea of the ACELP model is to construct the innovative codebook by using an interleaved pulse positions (IPP) structure. Each speech frame is further divided into subframes, typically of 5 ms each. Then, for each subframe, the candidates in the innovative codebook are formed by dividing the samples in the subframe in interleaved *tracks* and by placing few non-zero pulses on each track, each with equal amplitude but individual sign. The number of bits allocated to the ACELP codebook will then depend on the number of tracks and the number of pulses per track. The advantage of this structure is that it requires very little storage (the codebook entries are described by algebraic formulas and not stored in a table). Further more, this pulse structure allows for very fast search algorithms which makes the innovative excitation search implementable in real time even on mobile devices with limited CPU power and battery life. More details on the ACELP structure and search algorithms can be found in [5–7].

In itself, a speech coder introduces perturbations in the speech signal since the encoding process is *lossy*. This means that the decoded, synthesized speech waveform differs from the original speech waveform. These perturbations can however be made perceptually very small. Methods to reduce the perceived distortions in a speech coder include pre- and post-filtering as well as proper weighting of the error signal when selecting the optimal excitation parameters. Modern, state-of-the-art speech coders can produce high-quality speech at compression ratios of 10:1 and often much more. For example, the EFR-GSM standard narrowband speech coder [8], widely used in cellular networks, operates at 12.2 kbits/s. The compression ratio of this coder is 10.5 (128 kbit/s for the uncompressed narrowband signal, with 16 bits per sample and 8 kHz sampling frequency, divided by 12.2 kbit/s). This compression ratio can easily be doubled with the use of voice activity detectors and discontinuous transmission in the case of silence (more than 50% of conversational speech is silence).

Outside the perturbations introduced by the speech coder itself, several factors can influence the speech quality. In the following sections, we group these factors into two general categories:

- Input conditions
- Channel conditions

We will see how signal processing can be used to address these different issues to maintain or improve the speech quality in different system scenarios.

3 • Issues Related to Input Conditions

In Figure 1, we implicitly assume that the input signal is "clean" speech, which is a suitable requirement for the good performance of speech coders based on the model of Figure 1. This is especially important for low-rate speech coders, which take advantage of the speech signal characteristics. We also assume that the input signal is indeed speech and not, for example, loud acoustical noise or even music, as can be the case in telephony applications when a caller is put on hold. The input conditions have significant influence on the synthesized speech quality. In this section, we consider different variants of input conditions and discuss how signal processing can be used to solve the related issues.

3.1 Signal Dynamics

Digitized speech exhibits specific dynamics in the time domain and in the frequency domain. Figure 3 shows about 3/4 of a second of a typical speech signal. Large variations in the time envelope can be observed. In segments with slowly evolving time envelope, the signal typically exhibits local stationarity, for example, as a pseudo-periodic (voiced) signal or as a pseudo-random (unvoiced) signal. This is shown in Figure 4, where the voiced and unvoiced segments of Figure 3 are shown in more detail in both time and frequency domains.

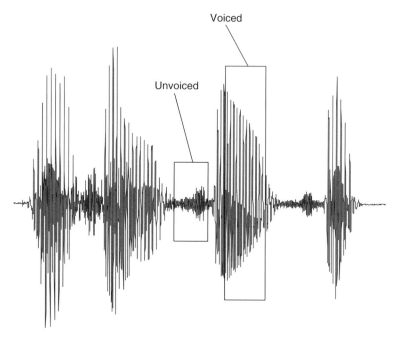

FIGURE 3 Example of a typical speech signal (about 3/4 of a second is shown), with succession of voiced and unvoiced segments.

The speech example in Figures 3 and 4 is sampled at 8 kHz, typical in telephony. The audio bandwidth is thus about 4 kHz (full band). This full-band, non-equalized condition is typically referred to as "flat" speech. Several speech coding algorithms have been optimized for "IRS-filtered" speech. IRS filtering refers to pre-conditioning, or equalizing, the speech signal with a frequency response typical of a telephone handset [9]. Linear-phase FIR filters are typically used to implement the IRS characteristics. Figure 5 shows the gain of the frequency response for the ITU-T IRS8 filter used for narrowband speech signals. Frequencies outside the 300–3,400 Hz telephone band are strongly attenuated, while the telephone band is tilted toward the high frequencies. Low-rate speech coders which have been optimized for IRS-filtered speech can suffer slight quality degradation when encoding flat speech. This is because the quantizers of the speech encoder's parameters do not have enough bits to cover all possible signal characteristics and are thus specialized. Of particular concern is the encoding of the synthesis filter coefficients. It is thus recommended to use a specific speech coding algorithm in the condition for which it was optimized, by properly pre-conditioning the input speech, for example, with

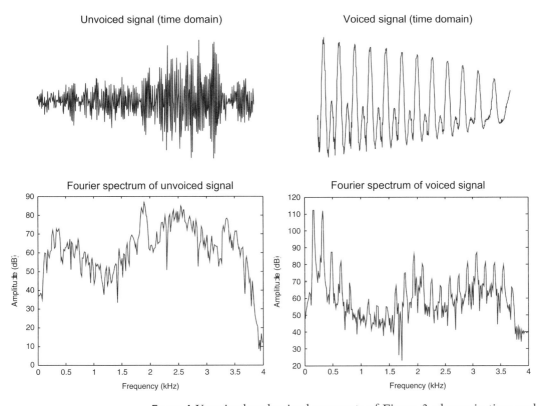

FIGURE 4 Unvoiced and voiced segments of Figure 3, shown in time and frequency domains (time-domain signals not to scale).

an IRS filter characteristic. For example, ITU-T recommendation G.729 was optimized for IRS conditioned signals.

3.2 Acoustical Noise

The input speech of a communication system is rarely exempt from noise. With mobile phones especially, acoustical noise from the speaker's environment is mixed with speech. In extreme cases – talking in a crowded hall for example – the noise level can even surpass the speech level. In most cases, the noise level and characteristics can still perturb the speech signal sufficiently to reduce the performance of the speech coder in the system. Hence, in many systems, different techniques are used to reduce the amount of acoustical noise entering the system. One approach consists of reducing the noise at the transducers, using specific microphones and circuitry. Directional and differential microphones, in particular, can be used to focus on the speech source, which can be seen as unidirectional, while blocking a large part of the acoustical noise, which is generally omnidirectional. With

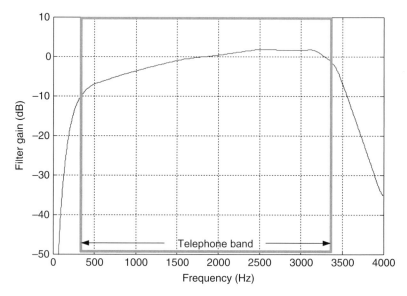

FIGURE 5 Gain of frequency response for IRS filter. Frequencies outside the telephone band are attenuated.

residual noise still present after using directional microphones, noise reduction algorithms can be used to further reduce the level of noise in the speech signal. We consider here the typical case where there is only one microphone, which means that the speech and noise are mixed in a single observation signal.

Several single-microphone noise reduction algorithms have been proposed. A large majority of the ones in use are variants of spectral subtraction methods [10]. The main idea is to subtract an estimate of the noise signal from the "speech-plus-noise" mixture. This subtraction is performed not in the time domain, but in the frequency domain – thus the name spectral subtraction. This is because in the frequency domain, amplitude and phase components can be processed separately. Since in single-microphone systems there is only one observation of the noise and speech mixture, phase information from the noise cannot be extracted in active voice regions – where noise reduction is to be useful. However, if the noise signal is relatively stationary, then an estimate of its amplitude spectrum can be obtained from several successive noise segments where speech is inactive. Hence, a typical spectral subtraction algorithm operates as in Figure 6.

In Figure 6, the noise estimation often relies on a voice activity detector (VAD) to determine whether the current frame has active speech or is only noise. If the current frame is noise only, the noise spectrum is updated. The noise spectrum is hence a smoothed

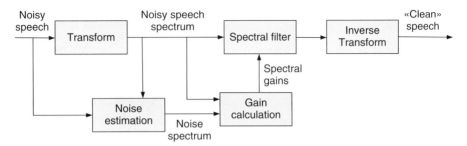

FIGURE 6 Basic blocks of spectral subtraction algorithm for single-microphone noise reduction.

average of the previous frames containing noise only. The update memory length depends on the algorithm. If the current frame is active speech, then it is not included in the noise estimation. Then, in basic spectral subtraction, the noise spectrum (estimate) is subtracted from the noisy speech spectrum (both are amplitude spectra only – phases are left unchanged). The clean speech is recovered by applying the inverse transform. An equivalent, and widely used, variant consists of calculating a gain, between 0 and 1, for each frequency bin of the noisy speech spectrum. When the speech-to-noise ratio is high for a given frequency bin, the gain is set close to 1 (spectrum unchanged). When the speech-to-noise ratio is low, the gain is set closer to 0 (noise attenuation). Using this gain or "spectral filter" approach avoids in particular the problem of negative valued amplitudes as in direct spectral subtraction.

As a specific example of a noise reduction algorithm based on spectral subtraction, consider the VMR-WB wideband speech coder, standardized by the 3GPP2 [11] in 2005. The noise subtraction algorithm of VMR-WB is described in [12].

3.3 Audio Bandwidth

For decades, the audio bandwidth of speech has been limited to roughly 300–3,400 Hz, in particular due to link limitations and handset frequency response. As shown in Figure 7, speech occupies a larger bandwidth compared to the telephone band. Transmitting this larger bandwidth can bring the users much closer to experiencing "face-to-face" quality communications, which strongly benefits applications such as teleconferencing and videoconferencing. The lower band – below 300 Hz – increases presence, while the higher band – above 3,400 Hz – increases intelligibility as several fricatives have significant content in the high frequencies. However, as the spectrum in Figure 7 shows, spectral dynamics and variability increase with bandwidth. In the example given, which is a segment of voiced speech, the telephone

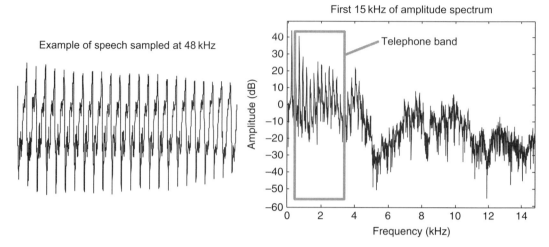

FIGURE 7 Speech example (voiced) sampled at 48 kHz.

band exhibits the harmonic structure discussed above, while the higher band has less structure. This observation suggests that the LP model of Figure 2 is suitable for the low band, but other models may be more applicable to the higher band of speech. Parametric models will be discussed below in the section covering the encoding of non-speech signals.

Depending on the sampling frequency and resulting audio bandwidth, speech can be characterized as

narrowband (or telephone band) for 8 kHz sampling (3 kHz band);
wideband for 16 kHz sampling (7 kHz band);
FM grade for 32 kHz sampling (15 kHz band);
or *full band* for 44.1 or 48 kHz sampling (20 kHz band).

The most significant jump in quality results from using wideband speech compared to narrowband speech. More improvement can be achieved by using a higher bandwidth, but it depends on the quality of the microphones and speakers used in the system. A large fraction of the speech energy lies below 8 kHz. Providing the energy above 8 kHz to the listener requires a proper acoustical environment, including efficient speakers and relatively low environmental noise. Of course, increasing the audio bandwidth usually requires additional bit rate.

Wideband speech can be encoded efficiently using essentially the same coding model as for narrowband speech, namely using the model of Figure 2 at the decoder. There are however some aspects specific to a wider bandwidth. In particular, as shown in Figure 7, the spectrum has larger dynamics (peak to valley ratios) and cannot be considered fully harmonic even in voiced speech. There are also more formants, i.e., spectral envelope peaks, in wideband speech than in narrowband speech. As a consequence, the synthesis filter in Figure 2 has to model a richer spectral envelope in wideband speech. This is accomplished by increasing the filter order, from a typical value of 10 coefficients in narrowband speech coders to a value between 16 and 20 coefficients in wideband speech coders. Of course, increasing the filter order implies more coefficients to encode and transmit, hence more bits to use for transmitting the synthesis filter. Indeed, a 10th-order synthesis filter typically requires 20–25 bits to transmit, while a 16th-order filter, used in a wideband speech system, will require at least 35 bits.

Increasing the synthesis filter order is not sufficient to increase performance of a wideband speech coder. In the analysis-by-synthesis process performed at the encoder to search for the optimum parameters in Figure 2 in a given speech frame, a weighting filter is used to amplify low-energy bands in the speech spectrum. These low-energy bands are typically the frequencies which lie between the speech formants as well as the high frequencies in the spectrum. Since wideband speech has more spectral dynamics than narrowband speech, the weighting filter has to be optimized for this. Specifically, the weighting filter in a narrowband speech coder typically has the form

$$W(z) = \frac{A(z/\gamma_1)}{A(z/\gamma_2)}, \qquad (1)$$

where $1/A(z)$ is the transfer function of the synthesis filter, and γ_1 and γ_2 are adjustable parameters with the constraints $\gamma_1 \leq 1.0$, $\gamma_2 \leq 1.0$, and $\gamma_2 < \gamma_1$. An example is given in Figure 8, using wideband speech (16 kHz sampling) as input. The frequency response of the synthesis filter $1/A(z)$, shown on the left spectrum, follows the envelope of the input speech. Without a weighting filter (i.e., with $W(z) = 1$), the encoder would try to find the best excitation parameters that produce a synthesized frame matching the original speech in the mean squared error sense. In this case, the overall speech encoder would on average introduce white noise in the original speech. Introducing weighting filter $W(z)$ in the excitation parameters search changes things quite a bit. Instead of trying to match the original speech, the encoder now tries to find the best excitation parameters which match a *weighted speech* signal which has the spectral envelope of $W(z)/A(z)$ as shown in dotted line in the left part of Figure 8. It can be easily shown that applying this weighting filter in the excitation parameters search results in shaping the speech coder's

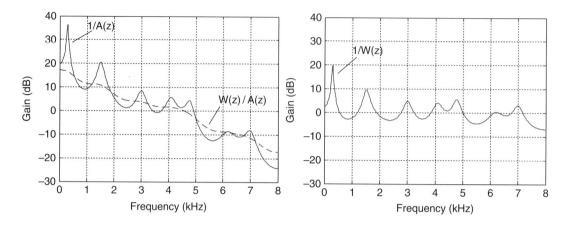

FIGURE 8 Typical example of weighting filter $W(z)$ optimized for a narrowband speech coder. Here, $\gamma_1 = 1.0$ and $\gamma_2 = 0.85$. Left: Frequency response of synthesis filter $1/A(z)$ superimposed with frequency response of *weighted* synthesis filter $W(z)/A(z)$. Right: Frequency response of the inverse weighting filter, $1/W(z)$.

noise with the inverse filter $1/W(z)$, which is shown in the right part of Figure 8. Now, the speech coder's noise, which would otherwise be white, follows more closely the input speech energy distribution – its formants. This reduces the perceived distortion due to the masking properties of the hearing system [13].

Figure 8 shows that this noise weighting strategy works well with narrowband signals, i.e., below 4 kHz. However, for wideband signals having more spectral dynamics, the noise introduced by the speech coder may still be too high in high frequencies due to the large spectral tilt (more than 40 dBs between 500 Hz and 7 kHz in the envelope $1/A(z)$ of Figure 8). Hence, different improvements to the weighting filter have been proposed in the case of wideband speech. A simple modification consists of adding an explicit correction for the spectral tilt in the weighting filter [14]:

$$W(z) = \frac{A(z/\gamma_1)}{A(z/\gamma_2)} T(z), \qquad (2)$$

where $T(z)$ is defined as $T(z) = 1 - \alpha \frac{R(1)}{R(0)} z^{-1}$. Here, $R(k)$ is the autocorrelation of the input speech frame at lag k and $0 < \alpha < 1$ is a parameter controlling the amount of tilt reduction in $W(z)$. Figure 9 shows the effect of adding the tilt factor $T(z)$ in the weighting filter $W(z)$. Here a value of $\alpha = 0.6$ is chosen, with the same values of γ_1 and γ_2 as in Figure 8. Using this tilt factor, the encoder now has to match a signal that is more "flat" (dotted curved in left part of Figure 9) compared to not using the tilt factor as in Figure 8. The effect is that the noise introduced in the synthesized speech by the encoder will not only follow the formant structure (peaks and valleys) but also exhibit a more proper spectral tilt. The noise in the low-energy spectral bands

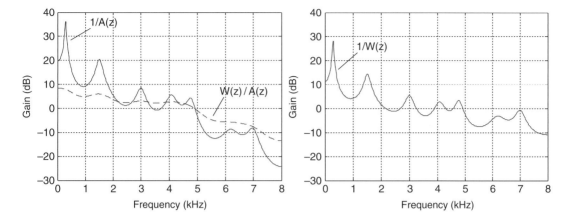

FIGURE 9 Weighting filter with tilt factor as in Equation (2. Here, $\gamma_1 = 1.0$, $\gamma_2 = 0.85$, and $\alpha = 0.6$. *Left*: Frequency response of synthesis filter $1/A(z)$ superimposed with frequency response of *weighted* synthesis filter $W(z)/A(z)$. *Right*: Frequency response of the inverse weighting filter, $1/W(z)$.

(here, in the high frequencies) will therefore be lower and better masked.

A simpler approach to decouple spectral tilt from formant noise shaping is exemplified in the AMR-WB wideband speech coding standard [5]. Here, the input speech is first pre-emphasized using a fixed first-order filter $P(z)$. Then, the synthesis filter $1/A(z)$ is actually computed from the pre-emphasized speech samples. Then, the weighting filter is computed as

$$W(z) = \frac{A(z/\gamma_1)}{P(z)}. \tag{3}$$

This reduces the filtering complexity while achieving essentially the same results as with Equation (2) and noise shaping similar to that in Figure 9.

3.4 Encoding Non-speech Signals

In several applications in communication systems, the audio signal can be different than speech. Examples include call waiting and on-hold music, acoustical noise (covered in a section above), and more recently musical ringtones which give cell phones an audible personality. The growing use of multimedia cell phones also increases the need for low-rate speech and audio coding, where ideally any kind of audio – speech, music, mixed content – can be compressed at low bit rates, transmitted and resynthesized with high quality using a single technology. However, at low rates, speech coders produce annoying artifacts when encoding music. On the other hand, perceptual audio coders [13] cannot be used for speech transmission systems requiring low delay and especially low rates. Perceptual audio coders use more general models than speech coders, but require higher bit rates and delay. In wireless communication systems, in particular, bit rates are critical and must be kept very low.

Seeing the importance of multimedia, and in particular audio, on emerging wireless and cellular systems, the Third Generation Partnership Project (3GPP) initiated in the early 2000s a new standardization activity to seek a new low-rate general audio coding algorithm. This resulted in 2004 in two new recommended standards for low-rate speech and audio transmission on cellular networks. EAAC+ [15], on the one hand, and AMR-WB+ [16] on the other hand were both recommended for complementary applications; EAAC+ for higher rates (32 kbits/s and above) and music signals, and AMR-WB+ for lower rates (24 kbits/s and below) and all types of content from speech to music and including mixed content such as movies soundtracks. The underlying model in these two standards is quite different.

EAAC+ is based on perceptual audio coding, where the input audio is decomposed into frames and each frame is transformed in the frequency domain using a form of Fourier transform [13]. In EAAC+, the Fourier transform used is the modified discrete cosine transform (MDCT). The frequency components are quantized following a bit allocation which depends on a perceptual model estimated from the frequency spectrum of the input signal. The perceptual model essentially calculates a masking curve which determines the maximum allowable noise level per frequency band to keep the noise inaudible. Then, if the bit budget is sufficient for the frame, each frequency component is quantized such that the quantization noise falls below the masking curve. To improve quality at low rates, EAAC+ uses a very low rate parametric approach called spectral band replication (SBR) to encode the high frequencies. Specifically, the signal spectrum is split into two equal and overlapping bands; the lower band is encoded using perceptual audio coding as described above, while the higher band is encoded using SBR. In SBR, the higher band is regenerated at the decoder applying frequency transposition to the lower band and using side information to scale the transposed frequency components. Only a few kbit/s are used as side information, and no bits are used to describe the frequency transposition. Hence, most of the bit rate is used to encode the lower band, where a large fraction of the signal energy usually lies especially in the case of speech signals. A detailed description of the SBR algorithm used in the EAAC+ 3GPP standard can be found in [17]. EAAC+ is a general audio coding algorithm, but it is covered in this chapter since it addresses the problem of encoding speech, music, and mixed signals at low rates.

AMR-WB+, also known as Extended AMR-WB, is another 3GPP standard recommended in 2004 for low-rate general audio transmission on cellular networks [16]. AMR-WB+ uses a hybrid encoding model which alternates between the ACELP model (see Section 2) and a transform-coding model called transform coded eXcitation (TCX) [18]. The TCX model is such that it allows seamless switching with the ACELP model. The ACELP model in AMR-WB+ is actually the AMR-WB speech coding standard from 3GPP, which is a multirate coding standard for speech sampled at 16 kHz (see Section 4.1). The AMR-WB+ encoder selects between the ACELP coding mode (AMR-WB) and the transform-coding mode (TCX) based on the input signal characteristics. It is worthwhile to note that mode selection is not based on speech/music classification; when encoding speech, the TCX model is chosen in a significant proportion, and when encoding music, the ACELP model is also chosen in some frames. Mode selection at the encoder can be made in either closed loop or open loop. In closed-loop mode selection, different coding modes are applied to the input signal, and the best mode is selected based on segmental signal-to-noise measures. In open-loop mode selection, the input signal is classified, and only the selected coding mode is applied based on the classification. Thus, open-loop mode selection results in a reduced-complexity encoder compared to closed-loop mode selection. Different frame lengths can be used and are also determined by the mode selection. As in EAAC+, AMR-WB+ uses a very low rate parametric approach to encode the high frequencies. Instead of frequency transposition as in SBR, AMR-WB+ uses spectral folding from the lower band to synthesize the higher band, along with low-rate side information to properly scale the folded frequency components. Only 800 bits per second per channel are used to encode the high frequencies in AMR-WB+. It can operate from 6 to 36 kbit/s in mono and from 7 to 48 kbit/s in stereo. The sampling frequency can range from 8 to 48 kHz. The algorithmic details of AMR-WB+ can be found in [16].

The EAAC+ and AMR-WB+ standards are both designed for mono and stereo operations, and target general audio at low-rates. For low rate wideband speech and audio communications, another available codec, standardized in 1999, is ITU-T recommendation G.722.1 [19]. The encoder operates on monophonic wideband signals sampled at 16 kHz and supports bit rates of 24 and 32 kbit/s. Lower rates (16 kbit/s) are also supported by the encoder, but significant degradation of speech quality prevented the ITU-T to recommend its use at these rates. Furthermore, this standard is recommended for use in hands-free systems with channels having very low bit error rates. This is suitable for teleconferencing applications, one of the main targets for this standard. G.722.1 uses a transform-based encoder as in perceptual audio coding. However, the frequency components are encoded using very simple scalar quantizers and entropy coding, without the use of sophisticated perceptual models. Hence, one of the main characteristics of G.722.1 is its low complexity, reported at 15 MOPS (million operations per second) for both the encoder and the decoder.

4 • Issues Related to Channel Conditions

The previous sections concentrated on the speech encoder and decoder, without consideration for the transmission channel. In most telecommunications applications, the transmission channel will not provide an error-free service. Bit errors will occur and packets will be delayed or lost. In data communications, this problem is alleviated by adding transport layers above the network layers in order to ensure that all bits are received correctly. This implies retransmission of lost data which adds considerable amounts of delay. This is exemplified by the TCP/IP model for the Internet. For real-time speech and audio communications, delay is critical, so retransmission is unacceptable. Hence, other solutions have emerged to alleviate the channel imperfections and maintain a suitable service quality for speech communications on lossy channels. These different solutions are presented in the following sections.

4.1 Adaptation to Channel Conditions

Adaptive multirate (AMR) speech coders support several coding rates and are capable of switching seamlessly and rapidly (on a frame-by-frame basis) from one rate to another. They were originally developed for wireless communications [20,21], where the quality of the communication channel (noise, interferences) varies greatly over time.

4.1.1 Adaptive Multirate Coders

The motivation for developing AMR solutions was the following. Over communication channels, the available bit rate is shared between source and channel coding. Because of transmission errors, increasing the bit rate dedicated to source coding to the detriment of the bit rate dedicated to channel coding does not necessarily increase the resulting audio quality. Hence, the underlying principle of AMR is to dedicate to source encoding the maximum bit rate allowed by the communication channel. When the communication channel is clean, most of the bit rate is dedicated to source coding. That amount decreases as the

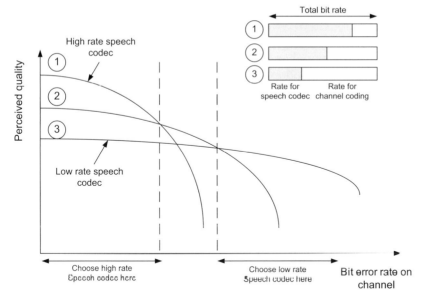

FIGURE 10 Principle of adaptive multirate (AMR) speech coding for channel with varying quality (bit error rate). Constant total bit rate (speech codec + channel codec) is assumed.

quality of the communication channel degrades, leaving more room for error correction. A similar strategy can be used when the communication channel is congested (varying available bit rate). This is illustrated in Figure 10. Case-1 corresponds to good channel conditions (low bit errors), while case-3 corresponds to poor channel conditions (high bit errors).

At the system level, AMR speech coders generally require a link adaptation unit to adjust the channel coding and speech coding schemes according to the quality of the communication channel [22].

There are two examples of standard AMR speech codecs. The "AMR narrowband" codec [23] is an ETSI and 3GPP standard. It uses the ACELP (algebraic code-excited linear prediction) coding algorithm [7]. It operates on narrowband (300–3,400 Hz) speech signals at eight rates in the range of 4.75–12.2 kbits/s. The coding modes at 12.2, 7.9, and 6.7 kbits/s correspond to the European ETSI EFR-GSM, the North American TIA IS-641, and the Japanese ARIB EFR-PDC codecs, respectively. The standard includes algorithms at the decoder for concealing lost frames. In addition to the AMR capability, the standard also includes some source-controlled variable bit rate features: voice activity detection (VAD), discontinuous transmission (DTX), and comfort noise generation (CNG). Finally, the AMR standard describes an associated channel coding, capable of error correction and bad frame detection, for full-rate (22.8 kbits/s) and half-rate

(11.4 kbits/s) channels of GSM. Introduction of AMR as an evolution of GSM systems has improved significantly the quality of service in these systems.

The "AMR wideband" (AMR-WB) codec [5] was the first codec adopted as a standard for both wireless and wireline services. It was standardized by ETSI/3GPP in 2001 and approved by the ITU-T in 2002 under reference G.722.2. It also interoperates with the latest 3GPP2 wideband standard, VMR-WB at mode 3 [11]. Compared to AMR narrowband, the audio bandwidth in AMR-WB is increased from 300–3,400 Hz to 50–7,000 Hz. This increases intelligibility and naturalness of speech and creates a feeling of transparent communication. The AMR-WB codec operates at eight rates in the range of 6.6–23.85 kbits/s and supports VAD/CNG/DTX.

RTP payload and file storage formats for AMR and AMR-WB supporting forward error correction (FEC), interleaving, and VAD/DTX/CNG were defined by the IETF [24].

4.1.2 Embedded Codecs

In communication systems using the AMR approach, whenever a decision is taken to reduce the source bit rate somewhere in the network (for example, to avoid congestion), that decision must be relayed both forward to the decoder and backward to the encoder. Rate switching of the encoder is therefore necessarily delayed, and some data can be lost in the meanwhile. There are also some contexts in which the AMR approach is not appropriate. In point-to-multipoint (or multicast) applications, for example, not all destinations get the same bit rate due to channel limitations or end device capability. In order to accommodate all destinations in a multicast application, a sender using an AMR codec would have to use the bit rate taken by the destination with the smaller bandwidth. This would penalize all the other destinations.

In an embedded codec, the bitstream is split into a core layer and one or several enhancement layers. Encoding is typically performed in stages, where a core encoder is first applied to the input signal. Then, enhancement layers are applied in successive stages to progressively reduce the distortion and improve the quality. Decoding all layers provides the highest quality. The encoder generates all layers at all time. However, at some stage during transmission, some of the upper layers can be stripped off the bitstream. The decoder is still able to decode the layers it has received, with the quality gracefully degrading as the number of received layers decreases. The encoder does not need to know whether some layers have been discarded or not somewhere in the communication chain. The encoder just provides a constant bitstream comprising all layers, and the network nodes or the receiver decides how many layers can be transmitted to the next node or how many layers can be used for decoding. The decision can be based on network congestion or on the ability of the decoder to interpret all layers. For example, an embedded coder could comprise a standard core layer along with improvement layers. Older receivers could decode only the core, standard layer, while newer receivers could decode all layers for improved service quality.

Embedded coding was first studied for sample-based waveform coders. The basic PCM (pulse code modulation) coder, known as ITU-T recommendation G.711, can very easily be embedded, as some of the least significant bits can be stripped off a PCM codeword with gradually increasing degradation of quality. With minor modifications to the encoder and decoder, an ADPCM (adaptive differential pulse code modulation) codec can also be turned into an embedded codec, at the expense of only slight degradation of its performances [25]. This principle is used by the ITU-T G.727 standard and ANSI T1.310 ADPCM codec which operate at 2–5 bits per sample (16–40 kbits/s). As for the ITU standard G.722, it describes a wideband (7 kHz) subband ADPCM codec at 64 kbits/s with three layers at 48, 56, and 64 kbits/s.

The embedded coding principle has also been applied to other coding algorithms. Reference [26], for example, describes a CELP-based embedded coder using one adaptive codebook (ACB) and three cascaded innovative codebooks (ICB), the ACB and first ICB forming the core layer and the two other ICBs forming the upper, refinement layers. Reference [27] describes a two-layer embedded coder, with a core using linear-prediction (ACELP) speech coding and one enhancement layer using transform-based (MDCT) audio coding.

The ITU-T G.729.1 codec [28] is, to date, the most accomplished example of an embedded codec. Its bitstream is organized into 12 layers ranging from 8 to 32 kbits/s. The first two layers (8 and 12 kbits/s) are narrowband and speech specific while the others (14–32 kbits/s with 2 kbits/s steps) are wideband and can deal with more general audio signals. The core layer (8 kbits/s) is compatible with the ITU-T G.729 standard (this was decided to ensure interoperability with already deployed VoIP systems, which widely support G.729). The first two layers (8 and 12 kbits/s) are populated by a cascade CELP codec. The third layer (14 kbits/s) performs a parametric time-domain bandwidth extension to achieve wideband speech quality. The upper layers are populated by a transform codec. The input and output signals can be sampled at either 8 or 16 kHz. Another interesting feature of the ITU G.729.1 embedded codec is that its bitstream includes some frame erasure concealment (FEC) information (namely a voicing classification, a phase, and an energy) for improved concealment of (and recovery after) lost frames similar to that of the 3GPP2 VMR-WB codec.

4.1.3 Variable Bit Rate Codecs

In most communication systems, it is desirable to accommodate as many users as possible while providing each of them with an acceptable quality of service, whatever the condition of the communication channel (traffic load, noise, and interference levels). Channel-controlled AMR codecs offer a means for adapting the source bit rate to the channel conditions. However, AMR codecs do not take into account the fact that some parts of the speech signals can be compressed more efficiently than others. Source-controlled variable bit rate speech coders were designed to minimize the average data rate (ADR) required by each user while continuously maintaining voice quality above a certain level [29].

Significant reduction in ADR is very easily achieved by simply taking into account the fact that, during a conversation, a speaker is silent almost 60% of the time. Most, if not all, modern speech coders use the VAD/DTX/CNG approach to take advantage of that small activity factor typical of conversational speech. Voice activity detection (VAD) partitions the speech signal into active (speech) and inactive (silence and background noise) periods. Discontinuous transmission (DTX) is used during inactive periods, as silence and background noise vary much more slowly than active speech and require very little bit rate. Comfort noise generation (CNG) is used at the decoder to recreate the acoustic background noise. Information about the acoustic background noise is sent within SID (silence descriptor) frames that are encoded at a much lower bit rate than speech (typically, 1 kbits/s) and also sent at a much lower update rate (typically, once every eight frames).

Further reduction in average bit rate can be obtained when the varying nature of speech is also taken into account [30]. During unvoiced segments, for example, the long-term predictor (adaptive codebook in Figure 2) used by most CELP codecs is unnecessary and the corresponding bits can be saved. Also, those noise-like segments are much less sensitive to a coarse quantization of their waveform. Onsets on the other hand are critical events that, though they do not occur very frequently, need to be quantized accurately.

A number of narrowband speech coding standards using the source-controlled variable bit rate approach were designed for wireless telephony over CDMA systems. The first one was the North American TIA/EIA/IS-96 standard (also known as QCELP) created in 1993 [31]. In this standard, each 20-ms input frame is encoded at either 0.8, 2, 4, or 8.5 kbits/s (rate set I) depending on the nature of the input signal. The lowest rate is for background noise. The channel rates corresponding to those source rates are 1.2, 2.4, 4.8, and 9.6 kbits/s. The TIA/EIA/IS-733 standard (also known as Q-13) was established in 1995 and uses another rate set: 1.0, 2.7, 6.2, and 13.3 kbits/s (rate set II). The TIA/EIA/IS-127 (EVRC) standard was created in 1997 to replace IS-96 (same rate set except that the 2-kbits/s quarter rate is not used). EVRC uses the RCELP (relaxed CELP) technique [32] which consists of modifying the input speech signal (specifically, simplifying its pitch contour) so that the pitch information can be coded more efficiently (only one pitch value per frame need to be transmitted instead of four as in conventional CELP). The SMV (selectable mode vocoder) [33] standard was created in 2001 to replace EVRC. It uses the same rate set as the IS-96 standard. It also supports several network-controlled operating modes (premium, standard, economy, and capacity saving) that, by making different usages of the rates, provide different average data rates. Due mainly to its high computational complexity, the SMV standard was never really used however. EVRC-B is an enhancement of EVRC that, mostly by making use of the quarter rate, provides a lower average data rates for a given voice quality. The 4GV codec that should be used in CDMA2000 is based on EVRC-B.

To date, the most accomplished source-controlled variable bit rate codec is the variable rate multimode wideband (VMR-WB). This codec was standardized in 2004 by 3GPP2 for wideband telephony and multimedia streaming services over CDMA2000 [11]. Among its numerous benefits, the VMR-WB codec is compatible with rate sets I and II. It supports several operating modes achieving various ADRs. One of its coding modes (mode 3) is interoperable with the ITU-T and 3GPP standard AMR-WB. It can operate on either narrowband or wideband input signals and deliver either narrowband or wideband decoded signals. This is the first speech coding standard for cellular telephony to include an algorithm for acoustic background noise reduction [12] (only the very low bit rate military standard NATO STANAG 4591 which is based on mixed excitation linear predictive – MELPe – also includes noise reduction). At a given bit rate, different coding strategies are used for different classes of signals. At 6.2 kbits/s, for example, frames are encoded using an unvoiced, a voiced, or a generic coding model. Among other differences, the unvoiced coding model uses a Gaussian innovative codebook while the others use an ACELP innovative codebook. For voiced signals, a signal modification technique equivalent to that of RCELP is applied to improve the coding efficiency [34]. Again, this is the first speech coding standard to include in its bitstream specific FEC (frame erasure concealment) information to improve concealment and recovery in case of frame losses [35]. It is therefore extremely robust to frame erasures (up to 6%). To give an idea of the efficiency of this codec, the quality obtained at 5.5 kbits/s (the codec operating in mode 1 on a wideband speech signal with 60% voice activity) is equal to that of the ITU G.722 standard (subband ADPCM) at 56 kbits/s.

4.2 Robustness to Channel Impairments

The problem of channel impairments is traditionally handled by channel coding. When transmitting general data, two channel-coding approaches are considered: ARQ (automatic repeat request) and FEC (forward error correction). ARQ consists of sending again any data that was not correctly received by the destination. FEC consists of adding redundant information to the source information so that the destination can detect and correct transmission errors. Typical FEC methods include parity checking or cyclic redundancy checking (CRC) to detect bit errors and block convolutional or turbo codes to correct them [36].

Because of the delay it introduces, ARQ is generally not considered when transmitting speech (although it can be used on some high-capacity sections of the network). FEC on the other hand is widely used in voice communication systems. There is a big difference however between data transmission and speech transmission, as speech decoders are generally tolerant, to a certain extent, to bit errors. As a matter of fact, from the perspective of voice quality it is preferable to receive frames containing some bit errors than to receive no frame at all. This is a central paradigm to the way FEC is used with speech coders.

FEC is therefore used in speech communication to minimize the percentage of residual bit errors on the source bitstream. Some techniques that are commonly used in speech coding to minimize the impact of those errors are described in the following subsection. Contingency measures that are taken when a frame is declared totally lost by the system or the channel decoder will then be described in the following subsection (frame erasure concealment).

4.2.1 Improving Robustness to Bit Errors

Channel-Optimized and Robust Vector Quantization

Vector quantization (VQ) [37] is extensively used in speech coding. It consists of replacing an input vector of parameters by another (the "codeword") taken from a finite set of vectors (the "codebook") that is supposed to be representative of the whole input space. In Figure 2, VQ is used to quantize the synthesis filter parameters, the innovative excitation, and the gains of the adaptive and innovative codebooks. In VQ, coding is achieved by only transmitting the position (called the "index") of the selected codeword within the codebook. VQ is known to be sensitive to bit errors, as one single bit error in an index can lead to a codeword that comes from a totally different region of the multidimensional input space.

A number of "joint source-channel coding" techniques have been proposed to improve the robustness of vector quantizers to bit errors. Two main approaches are possible [38]. The first one is referred to as channel-optimized vector quantization (COVQ). In that approach, the distortion measure used when designing the vector quantizer takes into account the possibility of bit errors. The resulting vector quantizer is therefore redundant, as it essentially performs channel coding [39]. The second approach is referred to as "robust vector quantization". In that approach, the impact of bit errors on a given quantizer is minimized by carefully choosing the index assignment (i.e., the pairing between indexes and codewords). Ideally, codewords that are closer in terms of their index (Hamming distance) should also be closer in terms of their value (Euclidian distance, or distortion). Index assignment can be optimized using binary switching [40], for example.

Predictive Quantization

Prediction is another key technique for speech coding. When a parameter is temporally correlated, quantizing the difference between this parameter and a predicted value is more efficient than quantizing the parameter itself. Prediction however causes error propagation in the case of transmission errors. Moving average (MA) predictors are generally preferred to Autoregressive (AR) predictors, as error propagation is limited to a duration on the order of prediction in the MA case while it is potentially infinitely long (depending on the value of the prediction coefficients) in the AR case [41]. Robustness of both AR- and MA-based quantizers can be further improved by using a safety-net approach [42]. Those considerations apply both to bit errors and frame erasures.

Unequal Error Protection

Another common way to improve robustness of speech communication against bit errors is via the use of unequal error protection (UEP). UEP is based on the fact that not all bits within a coded frame are equally sensitive to bit errors. Therefore, not all bits require the same amount of protection from channel coding. In a CELP coder, for example, and specifically in a CELP coder using the ACELP model for the innovative excitation, it is well known that the adaptive codebook (delay and gain) is much more sensitive than the innovation codebook. The sensitivity of the bits that compose a coded frame can be determined automatically, for example, by systematically inverting a given bit and measuring the output degradation using a signal-to-noise ratio or a more perceptually correlated measure such as PESQ (perceptual evaluation of speech quality) [43]. Bits are then generally grouped into three classes of bits: class A, class B, and class C. Class A contains the bits most sensitive to errors, and any error in these bits typically results in a corrupted speech frame which should not be decoded without applying appropriate frame erasure concealment. Classes B and C contain bits where increasing error rates gradually reduce the speech quality, but decoding of an erroneous frame is usually possible without annoying artifacts. Class B bits are more sensitive to errors than class C bits. In the AMR-WB standard codec at 12.65 kbits/s, for example, 72 bits out of the 253 that comprise one coded speech frame are considered very sensitive and belong to class A [44]. They are the VAD bit, some bits from the multistage split LPC quantizer (specifically from the first quantization stage and from low frequencies), the adaptive codebook delay, and the codebook gains.

4.2.2 Improving Robustness to Lost Frames

The techniques used to mitigate the effects of a lost or erased frame can be divided into two main groups. The first group of techniques, generally referred to as "concealment", consists of computing a replacement frame for the lost frame. The second group of techniques deals with the recovery of the decoder when good frames are received again. The main techniques are described in the following. Many other solutions can be found in [45–47].

Frame Loss Concealment

For non-predictive or sample-based codecs such as PCM, frame loss concealment can be done either in the temporal domain using prediction techniques [48] or in the frequency domain using sinusoidal modeling [49]. Combining those two approaches provides a packet loss concealment mechanism which adapts to the signal characteristics and is not restricted to pure speech signals [50].

In CELP-based speech codecs, frame loss concealment is facilitated by the presence of a speech model. An input parameter called the bad frame indicator (BFI) is used to indicate to the decoder that the frame is lost or corrupted. The decoder then generally simply extrapolates the coding parameters of the previous correctly received frame [51]. Interpolation requires a one-frame delay at the decoder but gives better results when a single frame is lost [27].

Frame Loss Recovery

Recovery after a lost frame is not a problem for non-predictive or sample-based codecs. For those codecs, a simple fade-in fade-out between the concealed segment and the following one does the job.

Recovery can be highly problematic however for predictive codecs such as CELP. As a matter of fact, interframe dependencies due, for example, to the use of long-term prediction or of predictive quantizers (for parameters such as the gains or the spectral envelope) can cause the error introduced by one single lost frame to propagate through several frames. This is particularly true when a critical segment of signal (an onset for example) is lost [52]. It has been proposed in [53] to completely eliminate interframe dependencies from the codec. As was shown in [54], however, this approach is not necessarily more efficient than using standard CELP-based codecs together with straightforward bitstream-based redundancy. Another solution to limit error propagation in the decoder is to introduce some constraints in the encoder [55]. Those constraints can be further exploited at the decoder to speed up recovery [56]. Yet another solution used in the VMR-WB and G.729.1 standards [27,35] consists of sending specific frame erasure concealment information (voicing classification, energy, and pitch phase) within the bitstream. That information can be used by the decoder to improve both concealment (using the classification and energy) and recovery (using mainly the phase of the signal). As will be shown later, in some other contexts (voice over packet networks), late frames when they are available can also be used to limit error propagation [57,58].

4.2.3 Adaptation to Packet Networks

Packet switching is a very challenging environment for voice codecs. This section describes the main problems encountered when transmitting voice over packet networks and presents some specific processing technologies and adaptations that are typically brought to the codec to improve voice quality over packet networks (VoIP).

Jitter Buffering

VoIP is characterized by variations in the time that packets take to transit through the network. Jitter buffering is generally used by the receiver to control the effects of the jitter (which can be defined as the difference between the actual arrival time of the packets and a reference clock at the normal packet rate). This buffer works by introducing an additional "playout" delay (which is defined with respect to the reference clock that was, for example, started at the reception of the first packet). The playout delay is chosen so as to provide the best compromise between the latency of the communication and the percentage of late packets. This delay allows transforming the uneven flow of arriving packets into a regular flow of packets as seen at the output of the jitter buffer. In this way, the decoder can provide a sustained flow of speech to the listener, provided the application can sustain the additional delay produced by the jitter buffer.

Time Scaling

Several strategies [59] can be used to adapt the playout delay to the network conditions. In the fixed jitter buffer strategy, the playout delay is set at the beginning of the conversation and maintained afterward. In the adaptive jitter buffering strategy, the playout delay is changed only at the beginning of each silence period ("talk-spurt"-based approach). For quickly varying networks, better results are obtained when the playout delay is also adapted during active speech [60].

Changing the playout delay during periods of active speech requires a means to change the rhythm of the speech signal. As explained in Table 1, playing out shorter frames (i.e., speeding up speech) decreases the playout delay, while playing out shorter frames (i.e., slowing down speech) increases it.

Time scaling can be done on the decoded signal (in the speech samples domain) using techniques such as SOLA [61] or its variants, PSOLA [62], or TDHS [63]. When a predictive codec is used, this operation can also be carried out during decoding, for example, in the excitation domain. As shown in [64], this approach has many advantages in terms of complexity and voice quality. In particular, it does not involve any additional complexity, provided some very reasonable constraints are applied with respect to when and how time scaling is applied.

TABLE 1 The algorithm for playout delay adaptation (the playout delay for packet i is the difference between P_i and the reference clock)

1. Using past jitter values, estimate the "ideal" playout time $\hat{P}_{i+1}$ of frame number i+1;
2. Send frame number i to the time scaling device, requesting it to generate a frame of length $\hat{T}_i = \hat{P}_{i+1} - P_i$ where P_i is the playout time of frame i.
3. The actual playout time of frame i+1 is $P_{i+1} = P_i + T_i$ where T_i is the actual frame length achieved by the time scaling device
4. Play out frame number i when its playout time P_i has arrived and iterate from step 1.

Late Packet Processing

Despite all the efforts done to adapt the playout delay to the jitter, some packets will still arrive too late to be decoded and must be considered as lost. As mentioned before, the concealment procedure used to replace the missing segment of speech is not perfect and introduces an error in the concealed segment. Moreover, in predictive codecs such as CELP, the concealment procedure does not correctly update the internal state of the decoder which becomes desynchronized with the encoder. Thus the error propagates in the following segments. A method to improve the recovery of a speech decoder after the reception of one or several late frames has been proposed in [57,58]. Rather than considering the late frame as "lost", it is used to update the internal state of the decoder. This limits and in some cases stops error propagation.

Other Techniques Used in VoIP Applications

Forward error correction (FEC) is widely used in VoIP applications to minimize the occurrence of lost frames. As mentioned earlier, FEC consists of sending redundant information to the destination. A simple and generic method to get redundant information using parity codes based on the exclusive OR (XOR) is described in [65]. Redundant information can also be obtained by repetition or double encoding [66]. Repetition consists of sending multiple copies of the coded frame in different packets. Double encoding consists of using one primary encoder, and one or several secondary encoders operating at a lower bit rate. FEC can be made channel controlled as proposed in [67,68], or source controlled as proposed in [52]. More elaborate solutions can be found in [69].

Interleaving, which consists of never sending two contiguous frames in the same packet, can be used to avoid bursts of frame erasures. This is useful, as frame erasure concealment procedures perform better for isolated erasures. This technique introduces an additional delay, however, and does not decrease the percentage of lost frames.

Finally, an IP transport protocol called UDP Lite can be used on error-prone networks [70]. UDP normally discards the whole packets when the checksum indicates the presence of bit errors. UPD Lite, by limiting the scope of the checksum, allows the destination to receive frames with bit errors. This is useful as voice codecs, as mentioned above, are rather tolerant to bit errors.

5 • Current Developments and Future Challenges

The last 15 years has seen the emergence of numerous speech and audio coding standards. These standards have followed the evolution of communication systems and application needs. The exponential growth of cellular systems and the Internet as a multimedia environment is largely due to the existence of efficient speech and audio coding algorithms (MP3 is more widely known than speech coding standards, but the latter are no less important for wireless communication systems as we know them today). In this chapter, we have presented several ways that signal processing can be used to improve service and audio quality in various input and channel conditions. Increasingly, these solutions are applied to existing standards, by either combining them or extending their capability while maintaining interoperability. For example, embedded coders typically use a well-deployed standard as the core. Emerging standards often ensure interoperability with other existing standards in a least part of their operating modes.

What has driven recent algorithmic development and standards adoption is in particular the move toward wideband telephony and multimedia handheld devices. Interconnecting the different heterogeneous networks (cellular and Internet, for example) is certainly a driving force to find innovative solutions to allow access to any information, in any format from anywhere. We can expect significant advances in signal processing and information coding to facilitate these interconnections, and to ensure that the large amounts of multimedia data already available in different formats will remain accessible in these evolving networks.

References

1. A.S. Spanias, "Speech coding: a tutorial review", Proceedings of the IEEE, vol. 82, no. 10, pp. 1541–1582, October 1994.
2. B. Kleijn and K. Paliwal, eds., *Speech Coding and Synthesis*, Elsevier, 1995.

3. L.R. Rabiner, R.W. Shafer, *Digital Processing of Speech Signals*, Prentice-Hall Signal Processing Series, 1978.
4. R.A., Salami, L., Hanzo, R., Steele, K.H.J. Wong, and I. Wassell, Speech coding, in R., Steele, eds., *Mobile Radio Communications*, chapter 3, pp. 186–346. IEEE Press – Wiley, 1992.
5. B. Bessette, R. Salami, R. Lefebvre, M. Jelinek, J. Rotola-Pukkila, J. Vainio, H. Mikkola, and K. Jarvinen, "The adaptive multirate wideband speech codec (AMR-WB)", IEEE Transactions on Speech and Audio Processing, vol. 10, no. 8, pp. 620–636, November 2002.
6. R. Salami, C. Laflamme, B. Bessette, and J.-P. Adoul, "ITU-T Recommendation G.729 Annex A: reduced complexity 8 kbit/s CS-ACELP codec for digital simultaneous voice and data", IEEE Communications Magazine, vol. 35, no. 9, pp. 56–63, September 1997.
7. C. Laflamme, J.P. Adoul, H.Y. Su, and S. Morissette, "On reducing computational complexity of codebook search in CELP coder through the use of algebraic codes", IEEE International Conference on Acoustics, Speech, and Signal Processing (ICASSP), pp. 177–180, Albuquerque, New Mexico, USA, April 3–6, 1990.
8. K. Järvinen et al, "GSM enhanced full rate codec", IEEE 1997 International Conference on Acoustics, Speech and Signal Processing, pp. 771–774, Munich, Germany, April 20–24, 1997.
9. ITU-T Recommendation P.48, "Specification for an intermediate reference system, volume V of the Blue Book", pp. 81–86, ITU, Geneva, February 1996.
10. J. Thiemann, *Acoustic Noise Suppression for Speech Signals Using Auditory Masking Effects*, Masters Thesis, McGill University, 2001.
11. S. Ahmadi and M. Jelinek, "On the architecture, operation, and applications of VMR-WB: the new cdma2000 wideband speech coding standard", IEEE Communications Magazine, vol. 44, no. 5, pp. 74–81, May 2006.
12. M. Jelinek and R. Salami, "Noise reduction method for wideband speech coding", 12th European Signal Processing Conference (EUSIPCO 2004), pp. 1959–1962, Vienna, Austria, September 6–10, 2004.
13. A.S. Spanias, "Perceptual coding of digital audio", Proceedings of the IEEE, vol. 88, no. 4, pp. 451–513, April 2000.
14. E. Ordentlich and Y. Shoham, "Low-delay code-excited linear-predictive coding of wideband speech at 32 kbps", IEEE 1991 International Conference on Acoustics, Speech, and Signal Processing (ICASSP'91), pp. 9–12, Toronto, Canada, May 14–17, 1991.
15. 3GPP Technical Specification TS26.401, "General audio codec audio processing functions; Enhanced aacPlus general audio codec; General description", June 2006.
16. 3GPP Technical Specification TS26.290, "Audio codec processing functions; Extended adaptive multi-Rate – wideband (AMR-WB+) codec; Transcoding functions", June 2005.
17. M. Schug, A. Groschel, M. Beer, and F. Henn, "Enhancing audio coding efficiency of MPEG Layer-2 with spectral band replication (SBR) for DigitalRadio (EUREKA 147/DAB) in a backwards

compatible way", 114th Audio Engineering Society Convention, preprint no. 5850, Amsterdam, The Netherlands, March 22–25, 2003.
18. R. Salami, R. Lefebvre, and C. Laflamme, "A wideband codec at 16/24 kbit/s with 10 ms frames", 1997 IEEE Workshop on Speech Coding, pp. 103–104, Pocono Manor, Pennsylvania, USA, September 7–10, 1997.
19. ITU-T Rec. G.722.1, "Coding at 24 and 32 kbit/s for hands-free operation in systems with low frame loss", September 1999.
20. T.B. Minde, S. Bruhn, E. Ekudden, and H. Hermansson, "Requirements on speech coders imposed by speech service solutions in cellular systems", 1997 IEEE Workshop on Speech Coding, pp. 89–90, Pocono Manor, Pennsylvania, USA, September 7–10, 1997.
21. A. Uvliden, S. Bruhn, and R. Hagen, "Adaptive multi-rate. A speech service adapted to cellular radio network quality", 32nd Asilomar Conference on Signals, Systems and Computers, vol. 1, pp. 343–347, Pacific Grove, California, USA, November 1–4, 1998.
22. S. Bruhn, P. Blocher, K. Hellwig, and J. Sjöberg, "Concepts and solutions for link adaptation and inband signalling for the GSM AMR speech coding standard", IEEE Vehicular Technology Conference, pp. 2451–2455, Amsterdam, The Netherlands, September 19–22, 1999.
23. K. Järvinen, "Standardisation of the adaptive multi-rate codec", 10th European Signal Processing Conference (EUSIPCO 2000), pp. 1313–1316, Tampere, Finland, September 4–8, 2000.
24. J. Sjöberg, M. Westerlund, A. Lakaniemi, and Q. Xie, "Real-time transport protocol (RTP) payload format and file storage format for the adaptive multi-rate (AMR) and adaptive multi-rate wideband (AMR-WB) audio codec", IETF RFC 3267, June 2002.
25. D.J. Goodman, "Embedded DPCM for variable bit rate transmission", IEEE Transactions on Communications, vol. 28, no. 7, pp. 1040–1046, July 1980.
26. R.D. De Iacovo and D. Sereno, "Embedded CELP coding for variable bit-rate between 6.4 and 9.6 kbit/s", IEEE 1991 International Conference on Acoustics, Speech, and Signal Processing (ICASSP'91), pp. 681–684, Toronto, Canada, May 14–17, 1991.
27. S.A. Ramprashad, "A two stage hybrid embedded speech/audio coding structure", IEEE 1998 International Conference on Acoustics, Speech, and Signal Processing (ISACCP'98), pp. 337–340, Seattle, Washington, USA, May 12–15, 1998.
28. S. Ragot et al., "ITU-T G.729.1: an 8–32 kbit/s scalable coder interoperable with G.729 for wideband telephony and voice over IP", IEEE 2007 International Conference on Acoustics, Speech and Signal Processing (ICASSP'2007), Honolulu, Hawaii, USA, April 15–20, 2007.
29. A. Gersho, and E. Paksoy, "An overview of variable rate speech coding for cellular networks", International Conference on Selected Topics in Wireless Communications, pp. 172–175, Vancouver, Canada, June 25–26, 1992.

30. E. Paksoy, K. Srinivasan, and A. Gersho, "Variable Bit Rate CELP coding of speech with phonetic classification", European Transactions on Telecommunications and Related Technologies, vol. 5, no. 5, pp. 591–602, September–October 1994.
31. A. DeJaco, W. Gardner, P. Jacobs, and C. Lee, "QCELP: The North American CDMA digital cellular variable rate speech coding standard", 1993 IEEE Workshop on Speech Coding for Telecommunications, pp. 5–6, Sainte-Adèle, Québec, Canada, October 13–15, 1993.
32. W.B. Kleijin, P. Kroon, and D. Nahumi, "The RCELP speech-coding algorithm", European Transactions on Telecommunications and Related Technologies, vol. 5, no. 5, pp. 573–582, September–October, 1994.
33. S.C. Greer, and A. DeJaco, "Standardization of the selectable mode vocoder", IEEE 2001 International Conference on Acoustics, Speech and Signal Processing (ICASSP'01), pp. 953–956, Salt Lake City, Utah, USA, May 7–11, 2001.
34. M. Tammi, M. Jelinek, and V.T. Ruoppila, "A signal modification method for variable bit rate wideband speech coding", IEEE Transactions on Speech and Audio Processing, vol. 13, no. 5, pp. 799–810, September 2005.
35. M. Jelinek, R. Salami, S. Ahmadi, B. Bessette, P. Gournay, C. Laflamme, and R. Lefebvre, "Advances in source-controlled variable bit rate wideband speech coding", Special Workshop in Maui (SWIM), Lectures by Masters in Speech Processing, Maui, Hawaii, January 12–14, 2004.
36. A. Glavieux, *Channel Coding in Communication Networks: From Theory to Turbo Codes*, Iste Publishing Company, 2007.
37. A. Gersho and R.M. Gray, *Vector Quantization and Signal Compression*, Kluwer Academic Publishers, 1991.
38. M. Skoglund, "On channel-constrained vector quantization and index assignment for discrete memoryless channels", IEEE Transactions on Information Theory, vol. 45, no. 6, pp. 2615–2622, November 1999.
39. H. Kumazawa, M. Kasahara, and T. Namekawa. "A construction of vector quantizers for noisy channels", Electronics and Engineering in Japan, vol. 67-B(1), pp. 39–47, January 1984.
40. K. Zeger, and A. Gersho, "Pseudo-gray coding", IEEE Transactions on Communications, vol. 38, no. 12, pp. 2147–2158, December 1990.
41. J. Skoglund, and J. Linden. "Predictive VQ for noisy channel spectrum coding: AR or MA?", IEEE 1997 International Conference on Acoustics, Speech, and Signal Processing (ICASSP'97), pp. 1351–1354, Munich, Germany, April 21–24, 1997.
42. T. Eriksson, J. Linden, and J. Skoglund, "Interframe LSF quantization for noisy channels", IEEE Transactions on Speech and Audio Processing, vol. 7, no. 5, pp. 495–509, September 1999.
43. J.G. Beerends, A.W. Rix, M.P. Hollier, and A.P. Hekstra, "Perceptual evaluation of speech quality (PESQ): The new ITU standard for end-to-end speech quality assessment, Part I – time-delay compensation;

Part II – psychoacoustic model", Journal of the Audio Engineering Society, vol. 50, no. 10, pp. 765–778, October 2002.
44. "AMR Wideband Speech Codec; Frame Structure", 3GPP Technical Specification 3GPP TS 26.201, March 2001.
45. M. Chibani, "Increasing the robustness of CELP speech codecs against packet losses", Ph.D. Thesis, University of Sherbrooke, Canada, January 2007.
46. C. Perkins, O. Hodson, and V. Hardman, "A survey of packet-loss recovery techniques for streaming audio", IEEE Network, pp. 40–48, September–October 1998.
47. B.W. Wah, X. Su, and D. Lin, "A survey of error-concealment schemes for real-time audio and video transmission over the Internet", 2000 International Symposium on Multimedia Software Engineering, pp. 17–24, Taipei, Taiwan, December 11–13, 2000.
48. E. Gündüzhan and K. Momtahan, "A linear prediction based packet loss concealment algorithm for PCM coded speech", IEEE Transaction on Speech and Audio Processing, vol. 9, no. 8, pp. 778–785, November 2001.
49. J. Lindblom and P. Hedelin, "Packet loss concealment based on sinusoidal modeling", 2002 IEEE Speech Coding Workshop Proceedings, pp. 65–67, Ibaraki, Japan, October 6–9, 2002.
50. V. Vilaysouk and R. Lefebvre, "A hybrid concealment algorithm for non-predictive wideband audio coders", 120th Audio Engineering Society Convention, preprint no.6670, Paris, France, May 20–23, 2006.
51. R. Salami, C. Laflamme, J. Adoul, A. Kataoka, S. Hayashi, T. Moriya, C. Lamblin, D. Massaloux, S. Proust, P. Kroon, and Y. Shoham, "Design and description of CS-ACELP: a toll quality 8 kb/s speech coder", IEEE Transactions on Speech Audio Processing, vol. 6, no. 2, pp. 116–130, March 1998.
52. H. Sanneck and N. Le, "Speech property-based FEC for Internet telephony applications", Proceedings of SPIE vol. 3969, pp. 38–51, Multimedia Computing and Networking 2000, San Jose, California, USA, January 24–26, 2000.
53. S.V. Andersen et al., "ILBC – A linear predictive coder with robustness to packet losses", 2002 IEEE Speech Coding Workshop Proceedings, pp. 23–25, Ibaraki, Japan, October 6–9, 2002.
54. R. Lefebvre, P. Gournay, and R. Salami, "A study of design compromises for speech coders in packet networks", IEEE 2004 International Conference on Acoustics, Speech and Signal Processing (ICASSP'2004), pp. 265–268, Montréal, Canada, May 17–21, 2004.
55. M. Chibani, P. Gournay, and R. Lefebvre, "Increasing the robustness of CELP-based coders by constrained optimization", IEEE 2005 International Conference on Acoustics, Speech and Signal Processing (ICASSP'2005), pp. 785–788, Philadelphia, Pennsylvania, USA, March 19–23, 2005.
56. M. Chibani, R. Lefebvre, and P. Gournay, "Resynchronization of the Adaptive codebook in a constrained CELP codec after a frame erasure", IEEE 2006 International Conference on Acoustics, Speech

and Signal Processing (ICASSP'2006), pp. 13–16, Toulouse, France, March 14–19, 2006.
57. P. Gournay, F. Rousseau, and R. Lefebvre, "Improved packet loss recovery using late frames for prediction-based speech coders", IEEE 2003 International Conference on Acoustics, Speech and Signal Processing (ICASSP'2003), pp. 108–111, Hong Kong, April 6–10, 2003.
58. K.D. Anderson, and P. Gournay, "Pitch resynchronization while recovering from a late frame in a predictive decoder", 9th International Conference on Spoken Language Processing (Interspeech 2006 – ICSLP), pp. 245–248, Pittsburgh, Pennsylvania, USA, September 17–21, 2006.
59. A. Ramjee, J. Kurose, D. Towsley, and H. Schulzrinne, "Adaptive playout mechanisms for packetized audio applications in wide-area networks", The Conference on Computer Communications, 13th Annual Joint Conference of the IEEE Computer and Communications Societies, Networking for Global Communications (INFOCOM'94), pp. 680–688, Toronto, Canada, June 12–16, 1994.
60. Y.J. Liang, N. Färber, and B. Girod, "Adaptive playout scheduling using time-scale modification in packet voice communications", IEEE 2001 International Conference on Acoustics, Speech, and Signal Processing (ICASSP'2001), pp. 1445–1448, Salt Lake City, Utah, USA, May 7–11, 2001.
61. S. Roucos and A.M. Wilgus, "High quality time-scale modification for speech", IEEE 1985 International Conference on Acoustics, Speech, and Signal Processing (ICASSP'85), pp. 493–496, Tampa, Florida, USA, March 26–29, 1985.
62. H. Valbret, E. Moulines, and J.-P. Tubach, "Voice transformation using PSOLA technique", Speech Communication, vol. 11, no. 2–3, pp. 175–187, June 1992.
63. D. Malah, "Time-domain algorithms for harmonic bandwidth reduction and time scaling of speech signals", IEEE Transactions on Acoustics, Speech, and Signal Processing, vol. 27, no. 2, pp. 121–133, April 1979.
64. P. Gournay, and K.D. Anderson, "Performance analysis of a decoder-based time scaling algorithm for variable jitter buffering of speech over packet networks", IEEE 2006 International Conference on Acoustics, Speech and Signal Processing (ICASSP'2006), pp. 17–20, Toulouse, France, March 14–19, 2006.
65. J. Rosenberg, and H. Schulzrinne, "An RTP payload format for generic forward Error Correction", IETF RFC 2733, December 1999.
66. C. Perkins, I. Kouvelas, O. Hodson, V. Hardman, M. Handley, J.C. Bolot, A. Vega-Garcia, and S. Fosse-Parisis, "RTP payload for redundant audio data", IETF RFC2198, September 1997.
67. J.C. Bolot, S. Fosse-Parisis, and D. Towsley, "Adaptive FEC-based error control for Internet telephony", Proceedings of IEEE INFOCOM'99, pp. 1453–1460, March 1999.
68. C. Padhye, K. Christensen, and W. Moreno, "A new adaptive FEC loss control algorithm for voice Over IP applications", 19th

IEEE International Performance, Computing and Communication Conference (IPCCC 2000), pp. 307–313, Phoenix, Arizona, USA, February 20 22, 2000.
69. I. Johansson, T. Frankkila, and P. Synnergren, "Bandwidth efficient AMR operation for VoIP", 2002 IEEE Speech Coding Workshop Proceedings, pp. 150–152, Ibaraki, Japan, October 6–9, 2002.
70. L.-A. Larzon, M. Degermark, S. Pink, "The lightweight user datagram protocol (UDP-Lite)", L.-E. Jonsson and G. Fairhurst, eds., IETF RFC 3828, July 2004.

PART VI
AUDIO ENGINEERING

Matti Karjalainen

Helsinki University of Technology, Laboratory of Acoustics and Audio Signal Processing, Espoo, Finland

32 Transducer Models
Juha Reinhold Backman 623

Modelling Fundamentals • Historical Introduction • Current Literature Overview • Overview of the Models • Classical Principles • Fundamentals of Transducers • Dynamic Loudspeakers • Horn Loudspeakers • Electrostatic Loudspeakers • Microphones • Piezoelectric Transducers for Ultrasonic and Underwater Applications

33 Loudspeaker Design and Performance Evaluation
Aki Vihtori Mäkivirta 649

Loudspeaker as a System Design Problem • Crossover Filters • Frequency Response Equalization • Room Response Equalizers • System Delay and Delay Variation • Protection • Understanding Loudspeaker Operation by Modeling • Acoustic Measurement Techniques • Listening Tests

34 PA Systems for Indoor and Outdoor
Jan Felix Krebber 669

Introduction • The PA System • Planning and Setting Up the Loudspeakers • Sound Indoors • Sound Outdoors • Measurement Methods • Special Issues on Signal Processing for PA Systems

35 Beamforming for Speech and Audio Signals
Walter Kellermann 691

The Basic Concept • Data-Independent Beamforming • Differential Microphone Arrays • Statistically Optimum Beamformers • Blind Source Separation as Blind Beamforming • Concluding Remarks

36 Digital Audio Recording Formats and Editing Principles
Francis Rumsey 703

Digital Tape Recording • Digital Recording on Computer Mass Storage Media • Consumer Digital Formats • Digital Audio Editing Using Mass Storage Media

37 Audiovisual Interaction
Riikka Möttönen and Mikko Sams 731
Audiovisual Space • Audiovisual Speech • Neural Basis of Audiovisual Perception

38 Multichannel Sound Reproduction
Ville Pulkki 747
Introduction • Concepts of Spatial Hearing by Humans • Loudspeaker Layouts • Reproduction Methods • Capturing Sound for Multichannel Audio

39 Virtual Acoustics
Tapio Lokki and Lauri Savioja 761
Introduction • Auralization and Signal Processing • Waveguide Meshes in Room Acoustics

40 Audio Restoration
Paulo A.A. Esquef 773
A Brief History of the Recording Technology • Aims and Problems • Overview of Digital Audio Restoration • Future Directions

41 Audio Effects Generation
Udo Zölzer 785
Filters and Delays • Tremolo, Vibrato and Rotary Speaker • Chorus, Flanger and Phaser • Pitch Shifting and Time Stretching • Dynamics Processing • Room Simulation

42 Perceptually Based Audio Coding
Adrianus J.M. Houtsma 797
Introduction • Transparent Coding: General Strategy • Transparent Coding: Specific Examples • Subjective Evaluation of Sound Quality • Objective Evaluation of Sound Quality • Concluding Remarks

32
Transducer Models

Juha Reinhold Backman

Nokia, Espoo, Finland

1 • Modelling Fundamentals

This chapter discusses the basic models with emphasis on audio applications. Loudspeakers are most commonly used as an example of electroacoustic transducers yet, from a modelling point of view, they present the broadest range of challenges to the theoreticians. The fundamental principles are, however, applicable to all transducer problems (microphones, hydrophones, ultrasonics). The reader is assumed to be reasonably familiar with the fundamental concepts of electroacoustics; introductory summaries have been presented by, e.g., Poldy (1994) and Hickson and Busch-Vishniac (1997).

We shall mostly confine ourselves to the classical approach where the description of the full physical system is reduced into a more tractable model. There are different, to some extent independent, means towards achieving this goal:

- Linearization: replacing the system with its small-signal approximation
- Reducing the dimensionality of the problem
- Dividing the system into subsystems with simple coupling

Linearization, the basic step in mathematical treatment of any complex system, renders the solutions independent of amplitude and enables the superposition of solutions at different frequencies. In the discussion below the models are assumed fundamentally linear, except for some aspects of loudspeakers which are discussed separately.

Reducing the dimensionality from a three-dimensional physical description to scalar quantities is a feature more characteristic to the acoustic transducer design. The typical systems are well described through scalar mechanical variables (force, velocity), acoustical variables (pressure, volume velocity), and the electrical quantities.

The ability to divide the system into smaller subsystems better suited for modelling is typical to the audio transducer construction. This is to some extent related to the reduction of the dimensionality, since the well-defined input and output quantities of one-dimensional systems lend themselves quite naturally to the separation (also in the mathematical sense). This is further helped by the fact that often the electromechanical transduction and mechanical subsystems are the only significantly nonlinear parts in a model, and acoustical subsystems can be regarded as linear.

Subsystems common to all electroacoustic systems include

- Electrical, including amplification and driving/loading impedances
- Electromechanical coupling
- Mechanical aspects of the system (moving mass, compliance, damping, resonances, etc.)
- Mechanical–acoustical transformation (sound radiation or reception)
- Enclosure interior sound field
- Exterior sound field, including the effects of immediate transducer housing and acoustical coupling between transducers

In microphones, ultrasonic transducers, underwater transducers, and electrostatic loudspeakers the transducers are used as a single entity, either without an enclosure, etc., or with the enclosure already forming an inseparable part of the total design, whereas in conventional dynamic loudspeakers the same driver is quite often intended to be used in multitude of enclosures, which necessitates independent transducer and enclosure models.

2 • Historical Introduction

The fundamental idea of the correspondence between the electrical impedance and factors affecting the movement of the transducer was discovered during the early stages of audio development (Butterworth 1915; Kennelly and Pierce 1912). The basic ideas grew soon into maturity, and the more complex acoustical systems were discussed by Mason (1942), whose book contains extensive references to early work in equivalent circuits and mechanical and acoustical signal processing. After Mason, Olson (1943) published his less detailed, but more general-purpose and audio-oriented work on the use of equivalent circuits. The theoretical understanding evolved further in the 1950s: Hunt (1954, pp. 92–127) published his treatment of transducer theory, Hueter and Bolt (1955) discussed transducers from ultrasonics point of view, and Fischer (1955) published his very clear introduction to the physical fundamentals of electroacoustic transduction.

For overviews, the author would rather refer the reader for the earliest loudspeaker theory developments towards McLachlan (1935), for overall development of electroacoustics until mid-1950s towards Hunt (1954, pp. 1–91) and Beranek (1954), and for a critical review of the more recently published loudspeaker model related work towards Gander (1998). Groves (1981) has edited an anthology with emphasis on early material.

3 • Current Literature Overview

In the modern loudspeaker literature the most usable materials on modelling are the compilation edited by Borwick (1994) (especially on electrostatic speakers and general aspects of lumped-parameter models) and the monographs by Eargle (1997) (especially on thermal models and magnetic system design) and Geddes and Lee (2002). Literature written for the amateur loudspeaker builders is typically rather weak on modelling methods, but notable exceptions to this are the books by Dickason (2000) and Bullock and White (1991). Most of the microphone literature concentrates on applications, but modelling aspects are discussed in compilations edited by Gayford (1994) and Wong and Embleton (1995). Key publications in loudspeaker, and to some extent microphone, theory have been collected in the Audio Engineering Society Anthology series (1978a, 1978b, 1978c, 1983, 1991a, 1991b, 1996). General transducer theory, especially for piezoelectric transducers and ultrasonic applications, is discussed in the Physical Acoustics series edited by Mason, Thurston, and Pierce. (Mason 1964a, 1964b; Mason 1973, 1979, 1981; Thurston and Pierce 1990).

4 • Overview of the Models

The various aspects of a typical (audio) electroacoustic transducer model are illustrated by the loudspeaker in Figure 1.

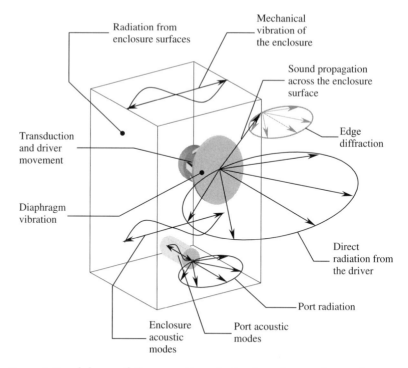

FIGURE 1 Breakdown of the acoustics of a loudspeaker enclosure from modelling point of view.

The classical electroacoustic system models, most often quoted in literature and very widely applicable, are the electrical analogue (or equivalent) circuits represented as lumped-component models. They are well applicable to traditional transducers, where the various subsystems can be clearly identified by physical inspection. The essential assumption for the simplest models is that the system dimensions are smaller than the wavelength, although the lowest drumhead mode of, for example, a microphone diaphragm can be approximated by a lumped-component model. The success of the lumped-component models is largely due to the ability to use the concepts and the tools of electric circuit theory directly in the mechanical–acoustical system modelling. Also, the transducers are always interfaced to an electric circuit, and in some cases the electric circuit has a significant effect on the apparent mechanical behaviour of the transducer (in both loudspeakers and ultrasonic transducers). The components of the equivalent circuit can be frequency-dependent (e.g. flow resistances or radiation impedances).

TABLE 1 A summary of complexity, accuracy, and computational load of some modelling approaches.

Model type	Complexity	Computational load	Accuracy	Remarks
Lumped parameter	Very low	Very low	Usable at low frequencies	Enabler for deriving analytical transducer theory
1-dimensional waveguide	Low	Low	Extended frequency range of validity, horns, etc., included	Strict geometric restrictions, usable also for ultrasonics
Waveguide + exterior diffraction	Moderate	Moderate	Medium to good for conventional loudspeakers	Moderate geometric restrictions
Lossless FEM/FD for enclosure interior or ultrasonic transducers	Moderate to high	High	Good for identifying eigenmodes of low-loss systems	Good for identifying potential problems, not for predicting their magnitude
Lossy (multiphysics) FEM/FD	High	High	Good for identifying realistic modal shapes for lossy systems	Difficulty in formulating the description of material losses
Coupled FEM/FD for acoustics and enclosure vibration	High	Very high	Good	Necessary for thin-walled structures or large enclosures
FEM/BEM/FD for interior (rigid enclosure), FEM/BEM for the exterior	High	Very high	Good, includes also diffraction and coupling through exterior field	Appropriate approximation for typical high-fidelity loudspeakers
Coupled FEM/BEM/FD for interior and enclosure, FEM/BEM/FD for the exterior	High	Very high	Potential for high accuracy	Applicable to all transducer types
Analytical solutions for acoustics	High	High for symbolic computation, low for numerical evaluation	Very high when underlying assumptions are met	Benchmark for numerical calculations, feasible only in limited cases

The electrical equivalent circuit approach can be developed further by introducing one-dimensional waveguide (or transmission line) models. These allow taking longitudinal resonances, propagation delay, reflections from terminations and discontinuities, etc., into account.

The lumped-parameter and one-dimensional waveguide models do not as such address explicitly the sound radiation problem. Sound radiation can be included as a simple source model (monopole or dipole as appropriate) or by some suitable analytical approximation, such as a flat piston in a baffle, at the end of a tube or free, or by more complex shapes, such as a cone or a dome.

Usually the transducer models are represented in the frequency domain. This corresponds to the practices of electric circuit modelling, and if propagation in lossy materials needs to be taken into account, then the theory is commonly formulated in the frequency domain. However, explicit time-domain descriptions can be very useful for more physics-based system description, for instance in describing nonlinear behaviour or in the analysis of an ultrasonic transducer.

The shift in transducer modelling has been towards more common use of direct numerical solution of the partial differential equations using finite element (FE), boundary element (BE), or more recently finite difference (FD) methods. This paradigm shift is quite simply due to increase in computing power and the amount of memory available. The numerical solutions are mostly based on the wave equation, but with the increasing computing capacity the application of the (linearized) Navier–Stokes equation would increase the reliability of the models for cavities, ports, etc. Table 1 summarizes the main properties of different modelling methods.

5 • Classical Principles

An electroacoustic transducer is typically described by two principal transforms: from electrical to mechanical and from mechanical to acoustical. Most types of transducers can be described through a unified description (Woollett 1966), where the nature of the transducer determines the coupling coefficient. The common practice is to use a gyrator to describe the electromechanical (or directly electroacoustic) conversion. A transformer could be equally well applied, with the slight theoretical hitch of having a transformer coefficient that is not dimensionless. The conversion of mechanical vibration to sound is typically represented by a transformer (Figure 2).

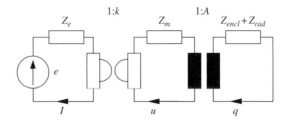

FIGURE 2 Basic equivalent circuit of a simple electroacoustic transducer; Z_e = electrical impedance, Z_m = mechanical impedance, Z_{encl} = enclosure impedance, Z_{rad} = radiation impedance, k = coupling coefficient, A = effective radiating area; e = driving source voltage, I = transducer current, u = transducer velocity, q = transducer volume velocity.

A more widely applicable form of this basic equivalent circuit can be realized by substituting a four-pole for one or more of the components of the circuit above. This allows a broader description of electromechanical transduction, mechano-acoustical conversion, and acoustical loading. Four-pole models are essential for, e.g., nonreciprocal transducers, and on the acoustical side for concise descriptions of any system that can be described as scalar inputs and outputs, such as simple or coupled-cavity loudspeakers, waveguides, or horns. The lumped-parameter and waveguide representations are illustrated by the example presented later for bass reflex loudspeakers.

The generic electroacoustic transducer can be represented by a four-pole illustrated below (Figure 3), where the quantities on the right-hand side can be chosen to be acoustical (volume velocity q and pressure p) or mechanical (velocity u and force F).

The input–output relationships of this four-pole can be described by

$$\begin{cases} e = Z_I I + M_{ue} u \\ F = M_{FI} I + z_l u \end{cases}$$

or alternatively in a matrix form as

$$\begin{bmatrix} e \\ F \end{bmatrix} = \begin{bmatrix} Z_I & M_{ue} \\ M_{FI} & z_l \end{bmatrix} \begin{bmatrix} I \\ u \end{bmatrix}.$$

FIGURE 3 Four-pole representation of transducer.

The matrix coefficients can be assigned a physical interpretation as follows: M_{ue} = transduction from velocity to voltage, M_{FI} = transduction from current to force, Z_I = electrical impedance of a transducer locked immobile, z_l = mechanical impedance of a transducer with open electrical circuit.

This representation can be applied to most simple transducers. Furthermore, if the transducer is passive (i.e. does not contain any energy sources) certain constraints can be placed on the relationship between the coefficients, and if the transducer is reciprocal (i.e. the transduction coefficient from electrical to acoustical/mechanical is the same as from mechanical/acoustical to electrical) the matrix is also symmetrical. It is possible to define multi-port representations for transducers with several inputs or outputs.

The importance of the four-pole representation is that the same formalism can be extended to mechanical–acoustical transform (i.e. sound radiation and reception) and purely acoustic subsystems (cavities, ports, horns, etc.; Figure 4). The matrix representation is comparable to the electromechanical four-poles, and also the physical interpretation of the matrix elements can be understood in the analogous manner.

An important special case of the acoustical four-pole is a homogeneous tube loaded with a known impedance (Figure 5). Besides the obvious simple cases, such a tube can be used as a building block for describing systems with either abrupt or gradual cross-section area or material property changes.

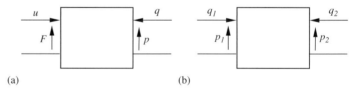

FIGURE 4 Input and output quantities of (**a**) a mechanical-acoustical and (**b**) an acoustical four-pole.

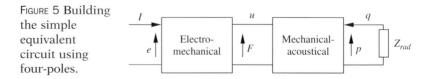

FIGURE 5 Building the simple equivalent circuit using four-poles.

For modelling purposes it is usually more practical to divide the four-pole into two separate entities, one representing the electromechanical transduction (like the voice coil and the magnet assembly of a loudspeaker) and another the mechanical–acoustical transformation (like the cone and the enclosure of a loudspeaker). Furthermore the mechanical and acoustical subsystems can be modelled as combinations of purely mechanical and acoustical four-poles yielding a more manageable detailed description of the system. Several simple systems, like low-frequency models of loudspeakers or microphones, can be quite easily described directly using lumped-component equivalent circuits without reverting to this kind of more generic approach. The four-pole approach, however, allows easy incorporation of both distributed and lumped components.

The choice of the fundamental input and output quantities to be examined depends on the usage situation. In a loudspeaker a suitable input quantity is voltage (although current could be argued to be a more basic input quantity for dynamic loudspeakers), and the suitable output quantity is the volume velocity, from which the far-field sound pressure is readily calculated. In ultrasonic applications the design target is often a certain force (i.e. pressure × area) at the application point. In microphones the usual choice for input quantity is the sound pressure or pressure gradient indirectly specified through pressure, frequency, and the distance from a sound source.

6 • Fundamentals of Transducers

6.1 Dynamic Transducers

The basic dynamic transducer has a conductor moving in a constant magnetic field, usually generated with permanent magnets. This principle appears in loudspeakers, microphones, high-level low-frequency underwater sound sources, and some mechanical actuators. Usually a dynamic transducer is built using a lengthy wire wound into a coil moving in a narrow air gap, but the simplest form, a straight short conductor moving in a transverse magnetic field, is occasionally encountered in ribbon microphones and loudspeakers and in the ultrasonic EMAT (electromagnetic–acoustic) transducers (Thomson 1990), where a wire is mounted directly on the surface of a solid substrate.

The classic description of the use of equivalent circuits for dynamic transducers was presented by Bauer (1976a). Modern dynamic transducer equivalent circuits have been discussed in a concise form by, e.g., Poldy (1994) and, together with other modern modelling tools, by Barnet and Opitz (2001).

6.2 Electromagnetic Transducers

Electromagnetic transducers, as opposed to dynamic transducers, rely on modulating the magnetic flux within a system where the magnetic flux goes through a variable air gap and the coil system is stationary. They were popular in the past in hearing aid microphones (and record player pickups) due to their sensitivity, but now they are found mostly in hearing aid earpieces. Their equivalent circuits resemble those of dynamic transducers; the only difference is in the form of the transduction coefficients.

6.3 Electrostatic Transducers

There are two categories of electrostatic transducers, benefiting from slightly different theoretical approaches. The one-sided transducers, having one fixed electrode and a moving diaphragm (like most condenser microphones, some headphones, and ultrasonic transmitters), rely as sources on the changing force due to the voltage changes and as a receiver on the voltage change of a constant-charge capacitor (carrier wave microphones are deliberately ignored here). The symmetrical transducers, like most electrostatic loudspeakers, where the moving diaphragm is between two fixed perforated electrodes, can be easiest understood as a variable charge moving in a constant electric field. Thus the symmetrical transducer is essentially linear, while the one-sided is essentially nonlinear.

6.4 Piezoelectric Transducers

In piezoelectric transducers both the external electrical field and the deformation are always essentially three-dimensional. A full description of the relationship between the electrical field acting on a piezoelectric device and the deformation or stress of the device therefore requires the use of a tensor description. However, as transducers are manufactured to have one direction of maximum sensitivity, in common applications the models can be simplified towards generic forms used for other transducer types. Characterisation of a piezoelectric transducer needs mechanical, electrical (dielectric), and transduction properties to be taken into account (Mason 1972).

7 • Dynamic Loudspeakers

The conventional dynamic loudspeaker has been one of the great successes of lumped-parameter modelling. Models succeed in predicting the low-frequency response (low frequency implying a frequency not much above system fundamental resonances and with wavelength large compared to dimensions, not a frequency actually regarded as subjectively low), but at higher frequencies they need to be complemented by more precise enclosure acoustics, diaphragm behaviour, and external sound field models.

7.1 Driver Models

The simple description of a dynamic driver consists of the electric circuit (resistance and inductance), linear electromagnetic coupling, and a mass-spring-damper system for the vibrating system. This was the starting point for the classical analytic theory, and this simple model is a surprisingly strong predictor of loudspeaker low-frequency behaviour.

A critical success factor for models is an exact determination of the driver parameters. This cannot, unfortunately, be achieved with high precision from the mechanical dimensions, weight of the moving parts, etc., as there may be frequency dependence in the parameters, the effective radiating area is difficult to determine, etc. The parameters can be better determined by fitting a model to impedance measurements with a driver in free space and sealed enclosure, measurements in ported enclosures, simultaneous current and displacement measurements, or fitting current measurements to a system model (Seidel and Klippel 2001; Clark 1997).

Since the voice coil produces an eddy current with frequency-dependent penetration depth, a simple resistance–inductance circuit suggested by the coil structure is insufficient to model the voice coil impedance (Vanderkooy 1988). Optimal magnet assembly design requires the use of detailed numerical models (Bie 1992).

7.2 Direct-Radiating Loudspeakers, Sealed and Ported Enclosures

In the 1960s and 1970s the Australian scientists J. E. Benson (1996), A. Neville Thiele (1971a, 1971b), and Richard H. Small (1972, 1973a, 1973b, 1973c, 1973d, 1973e) utilized the topological equivalence between electric filters and the loudspeaker-equivalent circuits to take the desired low-frequency transfer function shape as a design starting point. This led quickly to adoption of key parameters used in driver characterisation (Thiele–Small parameters) and to design nomograms, tables, and calculation formulae essentially removing guesswork from the low-frequency design of conventional loudspeakers. Later the low-frequency modelling theory was extended by several authors towards numerical methods and more generally applicable to analytical models (Keele 1973, 1981) and analytical parameter synthesis approach (Rutt 1985). Since all lumped-parameter models of simple direct-radiating enclosures have a similar fundamental structure, they can be described employing a unified mathematical treatment (Bullock 1989).

7.3 Absorption Material Models

The absorption materials can be included in the models as time constants (Putland 1998) or as well suited for wageguide models, through specific impedance and complex propagation coefficient (describing speed of sound and propagation loss) (Delany and Bazley 1970; Miki 1990), utilizing a model for static flow resistance derived from material parameters (Sides et al. 1971). Also description as equivalent circuit in simple lumped-component enclosure models is possible (Leach 1989). The models mentioned here, unfortunately, lead to slightly inconsistent results due to their differing theoretical and experimental starting points.

7.4 Other Enclosure Types

Coupled-cavity loudspeakers (i.e. loudspeakers where the sound is radiated only through ports of resonators or passive radiators on one or both sides of the driver) exhibit typically strong coupling between the acoustical resonances and the driver resonance, thus necessitating the simultaneous modelling of the acoustics on both sides of the driver (Geddes 1989a; Fincham 1979). Although the equivalent circuits employed for deriving the basic theory usually describe ports as lumped components (inductors), it is highly recommended to use distributed (i.e. waveguide) in actual design models for the ports, since the longitudinal port modes have a significant effect on stop-band response even if the cavity modes would be controlled with the application of acoustic absorbents.

The waveguide approach is quite naturally appropriate for modelling of transmission line loudspeakers, where the additional low-frequency load and resonant boost is provided by the longitudinal modes of the enclosure air space (Backman 1992).

The open-baffle loudspeakers are from the modelling point of view much more challenging than their more enclosed counterparts. Describing their response needs a model for the edge diffraction (e.g. geometric diffraction models discussed below are appropriate), and the acoustical asymmetry of the driver needs to be taken into account in order to obtain the correct mid-range frequency response and polar pattern (Backman 1999a). Acoustical resistance enclosures, where the enclosure has a significant well-damped leak, can be approximated using lumped-parameter models (Backman 1999b).

7.5 Thermal Models

A particular feature of loudspeakers is the need for a dynamic thermal model for the voice coil. It is important for system reliability calculations and in predicting the compression and frequency response shape changes as a function of input level

(Eargle 1997, pp. 190–199; Button 1992; Klippel 2003). The parameters describing the thermal circuit need to be determined experimentally (Behler and Bernhard 1998). For other transducer types ambient temperature influence (in microphones, through thermal expansion and air viscosity changes, (Brüel & Kjær 1996); in piezoelectric devices through temperature dependence of coupling coefficients) can be taken into account as a static correction, but loudspeakers, due to their significant input power, have relatively rapid temperature variations affecting the electrical characteristics of the system. The time constants associated with these variations are, however, typically long as compared to the signal period, and thus the thermal effects are, from electroacoustic model point of view, quasi-static parameter changes (mostly change in voice coil resistance).

The thermal behaviour is described through its own equivalent circuit, but as the temperatures are measures of energy, not amplitude, the amplitude quantities of thermal models actually correspond to the temperature and thermal power. In thermal calculations sound radiation in direct-radiating loudspeakers is usually ignored as a minor correction term amounting only at most to a few percent of the total power (Figure 6).

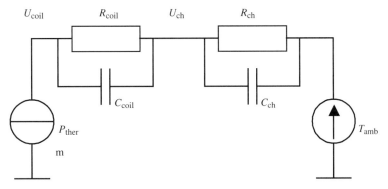

FIGURE 6 Simple thermal model of a loudspeaker describing the effects of the relatively rapid heating of the voice coil, the slower heating of chassis (including the magnet assembly), and the external temperature. P_{therm} = thermal input power (real power to the loudspeaker), R_{coil} = thermal loss of voice coil (radiation, conduction, etc.), C_{coil} = heat capacity of the voice coil, R_{ch} = thermal loss of the chassis (radiation, convection, conduction), C_{ch} = heat capacity of the frame, T_{amb} = ambient temperature inside the enclosure. The U_{coil} represents the voice coil temperature, and the voltage U_{ch} represents the chassis temperature as seen by the voice coil.

7.6 Nonlinear Models

Loudspeakers, especially woofers, are linear for only a very narrow part of their operating range. Thus nonlinear models are essential for both identifying the appropriate targets for transducer development and assessing the full system performance.

Nonlinear models have been developed for all types of transducers (Sherman and Butler 1995). Besides the nonlinearity in the transduction between the input signal and the exciting force the specific feature of loudspeakers is the nonlinearity of the passive mechanical structure also, which can be modelled through a direct numerical solution for the nonlinear differential equation (Olsen and Christiensen 1996). The direct solution is rather tedious, and thus there has been interest in developing computationally more efficient methods, such as the use of Volterra series (Kaizer 1987), which unfortunately suffer from the great difficulty in obtaining the needed kernel functions experimentally. In general, the determination of nonlinearity parameters is the critical factor in model selection, as all the computational methods provide essentially correct descriptions of nonlinearity within their range of validity (in terms of displacement and frequency). When the nonlinearity is described in terms of displacement-dependent coupling coefficient, compliance, and inductance, each of these can be approximated using a low-order polynomial (Geddes and Phillips 1991; Clark and Michelich 1999; Klippel 1999).

The enclosure itself also exhibits some nonlinearity. As the change in the internal volume is not negligible, the compliance (as a simple approximation) changes inversely proportional to the total volume. The phase shift due to damping material losses (and nonlinearity mechanisms of the damping material itself) complicates the practical situation somehow. This nonlinear compliance can be incorporated easily into the nonlinear suspension model. A more problematic aspect, both from modelling and practical engineering point of view, is the behaviour of the reflex port. There are some quasi-empirical formulae and empirical results, but predicting both the onset level and seriousness of the nonlinearity is difficult (Roozen et al. 1998).

7.7 Radiation Models

Although the analysis of sound radiation could be fundamentally regarded as a single computational effort, it is usually beneficial, from both application and computation point of view, to divide the problem into the following subtopics:

- Radiation load on the transducer (seen as Z_{rad} in the equivalent circuits); this can be extended to describe the coupling between multiple sound sources
- Sound radiation from the transducer itself

- Diffraction effects on the transducer housing/enclosure
- Radiation from, and through, enclosure surfaces, necessitating a numerical vibroacoustics model

7.7.1 Radiation Impedance

Analytical descriptions for a simple piston in an infinite baffle have been well discussed in the classic theoretical acoustics literature. Also a circular piston at the end of an infinite tube (Levine and Schwinger 1948) and a rectangular piston in an infinite baffle (Bank and Wright 1990) can be described analytically. The radiation load can often be approximated by a simple two-component equivalent circuit (Bauer 1976b). For direct-radiating loudspeakers and reflex ports it is usually sufficient to consider only the reactive component ("added mass"). Relatively simple approximations can be derived for the interaction of two pistons in an infinite baffle (Jacobsen 1976). A more precise model of the radiation impedance needs also taking into account the backscattering of the sound from the transducer housing, and thus diffraction models must be employed.

7.7.2 Diaphragm Radiation

A simple, but useful, approximation is to describe an actual sound source as a rigid piston. This model yields an estimate for the displacement needed for a certain sound pressure at low frequencies and defines a lower limit for the frequency at which the source (or receiver) becomes directive. The rigid-piston model is an excellent model for longitudinal ultrasonic transducers coupled to a fluid, as the transducer is excited very homogenously; so even if the dimensions of the transducer would allow for higher modes, the piston mode remains dominant.

Due to the finite bending wave propagation speed the rigid-piston model will fail for loudspeaker diaphragms at higher frequencies. Typically, the model inaccuracy will take effect around the same frequency range the driver in general becomes directive. Thus, rather than regarding the loudspeaker diaphragm as a rigid structure, a better conceptual model is to consider the sound radiation to be due to bending waves propagating on the diaphragm surface (Frankfort 1978). The propagating wave concept has been applied with reasonable success to driver and system modelling (Murphy 1993), but for development purposes the use of element models is recommended, as the pseudo-analytical expressions for local bending stiffness needed for the travelling wave model can be derived only for very simple geometries (flat, straight cone), and radial modes are difficult to include.

7.7.3 Enclosure Diffraction

In modern high-quality loudspeakers the major cause for mid-range and high-frequency colouration can be enclosure diffraction. For accuracy and computational efficiency the best general-purpose approach is the time-domain formulation of geometric diffraction theory (Svensson et al. 1999). This approach, however, has geometric limitations, e.g. curved surfaces necessitate the use of more generic numerical models.

7.8 Distributed-Mode Loudspeakers

As opposed to conventional loudspeakers, where higher vibration modes are avoided, distributed-mode loudspeakers use a rigid plate that is excited to produce a high number of vibration modes, which are designed to have reasonably even distribution as a function of frequency. Despite this complexity, a good qualitative description of their function can be obtained using a combination of statistical and lumped-parameter models (Harris and Hawksford 2000; Panzer and Harris 1998; Panzer and Kavanagh 1999).

7.9 Full Numerical Enclosure Acoustic and Structural Models

Although the models based on analytic theory have the advantage of being computationally exceptionally fast, all of them contain built-in constraints and approximations for the geometry of the system. Comparisons of numerical models, analytical models, and measured results indicate that although simple geometries are described with reasonable accuracy by analytical models (Ih 1991), more realistic systems need numerical approach (Sakai et al. 1984), especially when fluid–structure interaction needs to be taken into account (Karjalainen et al. 2001). The need of analysing the fluid–structure interaction stems from the unfortunate coincidence that the speed of the bending waves in typical enclosures has the same order of magnitude as the sound.

Microphones and miniature loudspeakers (by which in this context we mean audio frequency loudspeakers with maximum dimensions of a few centimetres) would appear excellent candidates for a lumped-parameter description, as they are much smaller than the wavelength, but also their analysis can benefit from finite-element approach. One reason for partial failure of lumped-parameter models is that in small devices distinguishing between clearly defined cavities, ports, etc., is often difficult, and describing losses call for precise geometrical details.

8 • Horn Loudspeakers

Horn models are equally well needed for audio and the solid horns in ultrasonic applications. Classic plane-wave horn theory is quite similar for both fluid-filled (audio) horns and solid horns, but the differences arise in solid transverse waveguides, like those applied in some ultrasonic tools. They need a fundamentally different approach due to the dispersive propagation, since there is no general approach corresponding to the horn equation for the fourth-order partial differential equation describing the propagation of the bending waves.

8.1 Classic Theory

It can be shown that the wave propagation in any system where the system can be described with one space variable obeys the horn equation (Putland 1993). The classic horn theory stems from the fortunate fact that there are families of closed-form analytical solutions for the horn equation for some of the horn shapes with practical importance (Salmon 1998). The limits of the classic theory have been discussed extensively by Geddes (1989b, 1993), who introduced curvilinear coordinate techniques for a more general class of horns and also proved that a fully generalized analytical solution is not feasible. However, Holland et al. (1991) have experimentally shown that with reasonable limitations a horn with a straight axis can be modelled as consecutive simple blocks. A very useful four-pole representation of the Salmon family of solutions was developed by Leach (1996). Arbitrary profiles not included in the analytical solutions are commonly constructed with connected straight (or conical) sections approximating the actual horn profile. The classic theory, however, is unable to predict horn polar patterns and high-frequency behaviour, for which numerical models (FEM, BEM) are needed.

8.2 Nonlinear Models

Horn drivers share the electromechanical transduction nonlinearity mechanisms with direct-radiating drivers. The nonlinear aspects of horn and driver acoustics, on the other hand, have a more direct link to the fundamental physical concepts. The fundamental theory of horn driver nonlinearity has been discussed in great detail by Voishvillo (2002, 2003), who is to be credited for discussing clearly the fundamental distinction between nonlinear effects (e.g. the inverse relationship between air pressure and density) and parametric effects (e.g. modulation of acoustical parameters due to changes in system dimensions with the movement of loudspeaker diaphragm).

In large audio systems at high power levels also the propagation of sound over large distances contributes to the overall distortion. This is mainly due to the nonlinear asymmetric steepening of the propagating wave (Czerwinski et al. 2000). (The nonlinear propagation can also be utilized to produce very directional audio frequency sound from modulated ultrasound in both air and underwater applications.)

9 • Electrostatic Loudspeakers

The analysis of electrostatic loudspeaker behaviour needs attention to fundamentally different aspects as compared to dynamic loudspeakers. Besides the transduction principle the fundamental differences are

- Diaphragm moving mass is close to negligible as compared to the radiation load.
- Listener is often in the near field.
- Diaphragm is in an open baffle.

Due to the fundamentally linear nature of transduction and the small mechanical impedance it is relatively easy to come up with a quasi-analytical description covering the basic behaviour even in greater detail than is possible for dynamic loudspeakers (Baxandall 1994). The near-field radiation aspect can be analysed using generally applicable approaches derived for conventional transducers.

10 • Microphones

10.1 General Features of Microphone Models

Modelling of small-scale acoustics needs closer attention to viscosity and heat conduction than the relatively large structures found in loudspeakers (Maisano 1994; Stinson 1991). The critical parameter determining the smallness of the structure is not here the ratio of the dimensions to the wavelength, but to the thickness of the boundary layer close to the surface, within which both viscosity and heat conduction play significant part.

Microphone models need an accurate description of acoustical resistance. The acoustical resistance can be in the form of a thin air layer (as in condenser microphones) or separate acoustical resistance elements (as often in dynamic microphones). These structures are the greatest challenge for accurate microphone theory. The difficulty of small-scale acoustics comes partially from the coupled nature of the problem and partially from the difficulty of characterizing the acoustical resistance elements in a satisfying manner. The geometry of even the simplest woven meshes is very complex for an accurate model of the flow, let alone more random porous materials, especially if the structural compliance of the materials needs to be taken into account. Thus, acoustical resistance materials are usually best described through quasi-experimental numerical models, although simplified theoretical models lend very usable insight into their design. The model of a thin flat air layer between the diaphragm and the condenser microphone back plate is more feasible due to the precisely definable geometry. Even then the solution needs some simplifying approximations, but the accuracy can already be very high.

The noise sources characteristic to microphones include thermal acoustical noise in the air and the mechanical damping of the diaphragm structure itself, thermal electrical noise in the resistive parts of the microphone electrical circuit, and the noise of the possible semiconductor devices. The acoustically and mechanically generated noise does not need any addition to the model, but it is calculated simply by evaluating the thermal noise in the resistances of the equivalent circuit (Olson 1972). The effect of external air load can be taken into account using the simple equivalent circuit mentioned earlier for loudspeaker radiation or a more detailed model (Leach 1990).

10.2 Condenser Microphones

The relatively accurate theory that can be derived for condenser microphones has enabled their development into trustworthy precision measurement instruments. The main features of condenser microphone operation can be described by a relatively simple model found in most electroacoustics textbooks, but describing the behaviour of the microphone diaphragm with the accuracy needed for development purposes requires several factors to be taken into account: diaphragm tension, electrostatic forces acting on the diaphragm, compliance of the air space behind the diaphragm, and the damping caused by the air layer between the diaphragm and the back electrode, and in the holes of the back electrode. The computational models have been summarized (with extensive references) by Hawley et al. (1995) and Zuckerwar (1995).

It should be remembered that one of the key achievements of transducer theory was the early correct model for condenser microphones (Wente 1922). This had very profound implication on the overall development of experimental acoustics, since it enabled the independent determination of the microphone performance parameters from basic physical quantities, in turn enabling the early development of precision acoustical measurements. Another fruitful relationship between the fundamental transducer theory and precision measurements led to the theory of reciprocal calibration, which still is the primary precision transducer calibration method for microphones and hydrophones.

10.3 Dynamic Microphones

Conventional dynamic microphones appear to some extent similar to dynamic loudspeakers. The essential difference is that the models are intended to cover the full operating frequency range instead of just describing the low-frequency behaviour. Thus relatively complex acoustical structures are needed to maintain a flat frequency response across the entire operating range (Rossi 1988). The problem of modelling dynamic miniature loudspeakers and earpieces is to some extent similar to the microphones.

11 • Piezoelectric Transducers for Ultrasonic and Underwater Applications

The feature that is common to ultrasonic and underwater transducers, and which separates them from conventional loudspeakers and microphones, is the focus on impedance matching between the transducer and the acoustical load. As opposed to audio transducers, for which there are a few well-defined typical equivalent circuit descriptions, the models for the ultrasonic transducers need to be selected according to the application. The transducers (especially transmitters) are quite often employed as narrow-band resonant devices, which is reflected in the choice of the equivalent circuit representation and allows a significant simplification of the description around the principal (usually the lowest) resonance, but wide-band models require the thickness of the transducers to be taken into account (Hickson and Busch-Vishniac 1997).

References

AES Anthology: Loudspeakers vol. 1, Audio Engineering Society, New York, 1978a, 448 pp.
AES Anthology: Sound Reinforcement, Audio Engineering Society, New York, 1978b, 339 pp.
AES Anthology: Microphones, Audio Engineering Society, New York, 1978c, 392 pp.
AES Anthology: Loudspeakers vol. 2, Audio Engineering Society, New York, 1983, 464 pp.
AES Anthology: Loudspeakers vol. 3, Audio Engineering Society, New York, 1991a, 456 pp.
AES Anthology: Loudspeakers vol. 4, Audio Engineering Society, New York, 1991b, 496 pp.
AES Anthology: Sound Reinforcement, vol. 2, Audio Engineering Society, New York, 1996, 496 pp.
Backman, J.: A Computational Model of Transmission Line Loudspeakers. 92nd Convention of the Audio Eng. Soc., March 1992, Preprint 3326.
Backman, J.: A Model of Open-Baffle Loudspeakers. 107th Convention of the Audio Eng. Soc., September 1999a, Preprint 5025.

Backman, J.: Theory of Acoustical Resistance Enclosures. 106th Convention of the Audio Eng. Soc., May 1999b, Preprint 4979.

Bank, G.; Wright, J. R.: Radiation Impedance Calculations for a Rectangular Piston. J. Audio Eng. Soc., vol. 38, no. 5, May 1990, pp. 350–354.

Barnet, R.; Opitz, M.: Modern Development Tools for Dynamic Transducers. 111th Convention of the Audio Eng. Soc., December 2001, Preprint 5438.

Bauer, B. B.: Equivalent Circuit Analysis of Mechano-acoustic Structures (Reprint). J. Audio Eng. Soc., vol. 24, no. 8, October 1976a, pp. 643–652 (originally in Transactions of IRE, vol. AU-2, July–August 1954, pp. 112–120).

Bauer, B. B.: On the Equivalent Circuit of a Plane Wave Confronting an Acoustical Device. J. Audio Eng. Soc., vol. 24, no. 8, October 1976b, pp. 653–655.

Baxandall, P. J.: Electrostatic Loudspeakers, in Borwick, J. (ed.): Loudspeaker and Headphone Handbook, 2nd ed., Focal Press, Oxford, 1994.

Behler, G. K.; Bernhard, A.: Measuring Methods to Derive the Lumped Elements of the Loudspeaker Thermal Equivalent Circuit. 104th Convention of the Audio Eng. Soc., May 1998, Preprint 4744.

Benson, J. E.: Theory & Design of Loudspeaker Enclosures, Prompt Publications (Howard W. Sams & Company), Indianapolis, 1996, 244 pp. (reprints of Benson's articles from 1960s).

Beranek, L. L.: Loudspeakers and Microphones. J. Acoust. Soc. Am., vol. 26, no. 5, September 1954 (Twenty-Fifth Anniversary Celebration Issue), pp. 618–629.

Bie, D.: Design and Theory of a New Midrange Horn Driver. 93rd Convention of the Audio Eng. Soc., October 1992, Preprint 3429.

Borwick, J. (ed.): Loudspeaker and Headphone Handbook, 2nd ed., Focal Press, Oxford, 1994, 601 pp.

Brüel & Kjær: Microphone Handbook, Brüel & Kjær, Naerum, Denmark, 1996, pp. 3-18–3-19.

Bullock III, R. M.: A Unified Model for Closed-Boxes, Vented-Boxes, and Passive-Radiators with Losses. 87th Convention of the Audio Eng. Soc., October 1989, Preprint 2841.

Bullock, R. M.; White, R.: Bullock on Boxes. Old Colony Sound Laboratory, 1991, 72 p.

Butterworth, S.: On Electrically Maintained Vibrations. Proc. Phys. Soc. (London), vol. 26, 1915, p. 264; vol. 27, 1915, p. 410.

Button, D. J.: Heat Dissipation and Power Compression in Loudspeakers. J. Audio Eng. Soc., vol. 40, no. 1/2, January 1992, pp. 32–41.

Clark, D.: Precision Measurement of Loudspeaker Parameters. J. Audio Eng. Soc., vol. 45, no. 3, March 1997, pp. 129–141.

Clark, D.; Michelich, R. J.: Modeling and Controlling Excursion-Related Distortion in Loudspeakers. 106th Convention of the Audio Eng. Soc., May 1999, Preprint 4862.

Czerwinski, E.; Voishvillo, A.; Alexandrov, S.; Terekhov, A.: Propagation Distortion in Sound Systems – Can We Avoid It? J. Audio Eng. Soc., vol. 48, no. 1/2, January/February 2000, pp. 30–48.

Delany, M. E.; Bazley, E. N.: Acoustical Properties of Fibrous Absorbent Materials. Appl. Acoust., vol. 3, 1970, pp. 105–116.

Dickason, V.: The Loudspeaker Design Cookbook, 6th ed., Audio Amateur Press, Peterborough, 2000, 233 pp.

Earlgle, J. M.: Loudspeaker Handbook, Chapman & Hall, New York, 1997, 325 pp.

Fincham, L. R.: A Bandpass Loudspeaker Enclosure. 63rd Convention of the Audio Eng. Soc., May 1979, Preprint 1512.

Fischer, F. A.: Fundamentals of Electroacoustics, Interscience, New York, 1955, 186 pp. (Originally Grundzüge der Elektroakustik, Fachverlag Scheile & Schön, Berlin, 1950).

Frankfort, F. J. M.: Vibration Patterns and Radiation Behavior of Loudspeaker Cones. J. Audio Eng. Soc., vol. 26, no. 9, September 1978, pp. 609–622.

Gander, M. R.: Fifty Years of Loudspeaker Developments as Viewed Through the Perspective of the Audio Engineering Society. J. Audio Eng. Soc., vol. 46, no. 1/2, January 1998, pp. 43–58.

Gayford, M. (ed.): Microphone Engineering Handbook, Focal Press, Oxford, 1994, 449 pp.

Geddes, E. R.: An Introduction to Band-Pass Loudspeaker Systems. J. Audio Eng. Soc., vol. 37, no. 5, May 1989a, pp. 308–342.

Geddes, E. R.: Acoustic Waveguide Theory. J. Audio Eng. Soc., vol. 37, no. 7/8, July 1989b, pp. 554–569.

Geddes, E. R.: Acoustic Waveguide Theory Revisited. J. Audio Eng. Soc., vol. 41, no. 6, June 1993, pp. 452–461.

Geddes, E.; Lee, L.: Audio Transducers, GedLee Associates, 2002, 304 pp.

Geddes, E.; Phillips, A.: Efficient Loudspeaker Linear and Nonlinear Parameter Estimation. 91st Convention of the Audio Eng. Soc., October 1991, Preprint 3164.

Groves Jr., I. D.: Acoustic Transducers, Benchmark Papers in Acoustics, vol. 14, Hutchinsorn Ross Publishing Company, Stoudsburg, 1981, 391 pp.

Harris, N. J.; Hawksford, M. O.: Introduction to Distributed Mode Loudspeakers (DML) with First-Order Behavioural Modelling. IEE Proc. Circuits Devices Syst., vol. 147, no. 3, June 2000, pp. 153–157.

Hawley, M. S.; Romanov, F. F.; Warren, J. E.: The Western Electric 640AA Capacitance Microphone: Its History and Theory of Operation, in Wong, G. S. K.; Embleton, T. F. W. (eds.): AIP Handbook of Condenser Microphones, American Institute of Physics, New York, 1995, pp. 8–34.

Hickson, E. L.; Busch-Vishniac, I. J.: Transducer Principles, in Crocker, M. J. (ed.): Encyclopedia of Acoustics, vol. 4, Wiley, New York, 1997, pp. 1889–1902.

Holland, K. R.; Fahy, F. J.; Morfey, C. L.: Prediction and Measurement of the One-Parameter Behavior of Horns. J. Audio Eng. Soc., vol. 39, no. 5, May 1991, pp. 315–337.

Hueter, T. F.; Bolt, R. H.: Sonics, Wiley, New York, 1955, pp. 109–118, 180–186.

Hunt, F. V: Electroacoustics, 1954 (reprinted by the Acoustical Society of America, 1982).

Ih, J.-G.: Acoustic Wave Action Inside Rectangular Loudspeaker Cabinets. J. Audio Eng. Soc., vol. 39, no. 12, December 1991, pp. 945–955.

Jacobsen, O.: Some Aspects of the Self and Mutual Radiation Impedance Concept with Respect to Loudspeakers. J. Audio Eng. Soc., vol. 24, no. 2, March 1976, pp. 82–92.

Kaizer, A. J. M.: Modeling of the Nonlinear Response of an Electrodynamic Loudspeaker by a Volterra Series Expansion. J. Audio Eng. Soc., vol. 35, no. 6, June 1987, pp. 421–433.

Karjalainen, M.; Ikonen, V.; Antsalo, P.; Maijala, P.; Savioja, L.; Suutala, A.; Pohjolainen, S.: Comparison of Numerical Simulation Models and Measured Low-Frequency Behavior of Loudspeaker Enclosures. J. Audio Eng. Soc., vol. 49, no. 12, December 2001, pp. 1148–1166.

Keele Jr., D. B.: Sensitivity of Thiele's Vented Loudspeaker Enclosure Alignments to Parameter Variations. J. Audio Eng. Soc., vol. 21, no. 4, May 1973, pp. 246–255.

Keele Jr., D. B.: Direct Low-Frequency Driver Synthesis from System Specifications. 69th Convention of the Audio Eng. Soc., May 1981, Preprint 1797.

Kennelly, A. E.; Pierce, G. W.: The Impedance of Telephone Receivers as Affected by the Motion of Their Diaphragms. Proc. Am. Acad. Arts Sci., vol. 48, 1912, pp. 113–151. Reprinted in Groves, 1981, pp. 328–352.

Klippel, W.: Measurement of Large-Signal Parameters of Electrodynamic Transducer. 107th Convention of the Audio Eng. Soc., September 1999, Preprint 5008.

Klippel, W.: Nonlinear Modeling of the Heat Transfer in Loudspeakers. 114th Convention of the Audio Eng. Soc., March 2003, Preprint 5733.

Leach, W. M.: Electroacoustic-Analogous Circuit Models for Filled Enclosures. J. Audio Eng. Soc., vol. 37, no. 7/8, July 1989, pp. 586–592.

Leach, M. W.: On the Electroacoustic-Analogous Circuit for a Plane Wave Incident on the Diaphragm of a Free-Field Pressure Microphone. J. Audio Eng. Soc., vol. 38, no. 7/8, July 1990, pp. 566–568.

Leach Jr., M. W.: A Two-Port Analogous Circuit and SPICE Model for Salmon's Family of Acoustic Horns. J. Acoust. Soc. Am., vol. 99, no. 3, March 1996, pp. 1459–1464.

Levine, H.; Schwinger, J.: On the Radiation of Sound from an Unflanged Circular Pipe. Phys. Rev., vol. 73, no. 4, February 1948, pp. 383–406.

Maisano, J.: An Improved Transmisson Line Model for Visco-thermal Lossy Sound Propagation. 96th Convention of the Audio Eng. Soc., February 1994, Preprint 3828,

Mason, W. P.: Electromechanical Transducers and Wave Filters, D. Van Nostrand Company, New York, 1942, 333 pp.

Mason, W. P. (ed.): Physical Acoustics, vol. IA, Academic Press, New York, 1964a, 515 pp.

Mason, W. P. (ed): Physical Acoustics, vol. IB, Academic Press, New York, 1964b, 376 pp.

Mason, W. P.: Properties of Transducer Materials, in American Institute of Physics Handbook, 3rd ed., McGraw-Hill, New York, 1972, pp. 3-118–3-129.

Mason, W. P.; Thurston, R. N. (eds.): Physical Acoustics, vol. X, Academic Press, New York, 1973, 403 pp.

Mason, W. P.; Thurston, R. N. (eds.): Physical Acoustics, vol. XIV, Academic Press, New York, 1979, 561 pp.

Mason, W. P.; Thurston, R. N. (eds.): Physical Acoustics, vol. XV, Academic Press, New York, 1981, 382 pp.

McLachlan, N. W.: Elements of Loud Speaker Practice, Oxford University Press, London, 1935, 160 pp.

Miki, Y.: Acoustical Properties of Porous Materials – Modifications of Delany-Bazley Models. J. Acoust. Soc. Jpn (E), vol. 11, no. 1, 1990, pp. 19–24.

Murphy, D. J.: Axisymmetric Model of a Moving-Coil Loudspeaker. J. Audio Eng. Soc., vol. 41, no. 9, September 1993, pp. 679–690.

Olsen, E. S.; Christiensen, K. B.: Nonlinear Modeling of Low-Frequency Loudspeakers: A More Complete Model. 100th Convention of the Audio Eng. Soc., May 1996, Preprint 4205.

Olson, H. F.: Dynamic Analogies, D. Van Nostrand Company, New York, 1943 (2nd ed., 1958, 278 pp.).

Olson, H. F.: Microphone Thermal Agitation Noise. J. Acoust. Soc. Am., vol. 51, no. 2 (pt 1), 1972, pp. 425–432.

Panzer, J. W.; Harris, N.: Distributed-Mode Loudspeaker Simulation Model. 104th Convention of the Audio Eng. Soc., May 1998, Preprint 4739.

Panzer, J.; Kavanagh, S.: Modal Network Solver for the Simulation of Complex Mechanoacoustical Systems. 107th Convention of the Audio Eng. Soc., September 1999, Preprint 5022.

Poldy, C. A.: Headphones, in Borwick, J. (ed.): Loudspeaker and Headphone Handbook, 2nd ed., Focal Press, Oxford, 1994, pp. 495–500.

Putland, G. R.: Every One-Parameter Acoustic Field Obeys Webster's Horn Equation. J. Audio Eng. Soc., vol. 41, no. 6, June 1993, pp. 435–451.

Putland, G. R.: Thermal Time Constants and Dynamic Compressibility of Air in Fiber-Filled Loudspeaker Enclosures. J. Audio Eng. Soc., vol. 46, no. 3, March 1998, pp. 139–151.

Roozen, N. B.; Bockhots, M.; van Eck, P.; Hirschberg, A.: Vortex Sound in Bass-Reflex Ports of Loudspeakers. J. Acoust. Soc. Am., vol. 104, no. 4, October 1998, pp. 1914–1918 (Part I), 1919–1928 (Part II).

Rossi, M.: Acoustics and Electroacoustics, Artech House, Norwood, 1988, pp. 518–536.

Rutt, T. E.: Root-Locus Technique for Vented-Box Loudspeaker Design. J. Audio Eng. Soc., vol. 33, September 1985, pp. 659–668.

Sakai, S.; Kagawa, Y.; Yamabuchi, T.: Acoustic Field in an Enclosure and Its Effect on Sound-Pressure Responses of a Loudspeaker. J. Audio Eng. Soc., vol. 32, no. 4, April 1984, pp. 218–227.

Salmon, V.: Horns, in Crocker, M. J. (ed.): Encyclopedia of Acoustics, vol. 4 Wiley Interscience, 1998, pp. 1925–1931.

Seidel, U.; Klippel, W.: Fast and Accurate Measurement of the Linear Transducer Parameters. 110th Convention of the Audio Eng. Soc., May 2001, Preprint 5308.

Sherman, C. H.; Butler, J. L.: Analysis of Harmonic Distortion in Electroacoustic Transducers. J. Acoust. Soc. Am., vol. 98, no. 3, September 1995, pp. 1596–1611.

Sides, D. J.; Attenborough, K.; Mulholland, K. A.: Application of a Generalized Acoustic Propagation Theory to Fibrous Absorbents. J. Sound Vibr., vol. 19, no. 1, 1971, pp. 49–64.

Small, R. H.: Closed-Box Loudspeaker Systems Part I: Analysis. J. Audio Eng. Soc., vol. 20, no. 10, December 1972, pp. 798–808.

Small, R. H.: Closed-Box Loudspeaker Systems Part II: Synthesis. J. Audio Eng. Soc., vol. 21, no. 1, January/February 1973a, pp. 11–18.

Small, R. H.: Vented-Box Loudspeaker Systems, Part I: Small-Signal Analysis. J. Audio Eng. Soc., vol. 21, no. 5, June 1973b, pp. 363–372.

Small, R. H.: Vented-Box Loudspeaker Systems, Part II: Large-Signal Analysis. J. Audio Eng. Soc., vol. 21, no. 6, July/August 1973c, pp. 438–444.

Small, R. H.: Vented-Box Loudspeaker Systems, Part III: Synthesis. J. Audio Eng. Soc., vol. 21, no. 7, September 1973d, pp. 549–554.

Small, R. H.: Vented-Box Loudspeaker Systems, Part IV: Appendices. J. Audio Eng. Soc., vol. 21, no. 8, October 1973e, pp. 635–639.

Stinson, M. R.: The Propagation of Plane Sound Waves in Narrow and Wide Circular Tubes, and Generalization to Uniform Tubes of Arbitrary Cross-Sectional Shape. J. Acoust. Soc. Am., vol. 89, no. 2, 1991, pp. 550–558.

Svensson, U. P.; Fred, R. I.; Vanderkooy, J.: An analytic secondary source model of edge diffraction impulse responses. J. Acoust. Soc. Am., vol. 106, no. 5, November 1999, pp. 2331–2344.

Thiele, A. N.: Loudspeakers in Vented Boxes, Part I. J. Audio Eng. Soc., vol. 19, no. 5, May 1971a, pp. 382–392.

Thiele, A. N.: Loudspeakers in Vented Boxes, Part II. J. Audio Eng. Soc., vol. 19, no. 6, June 1971b, pp. 471–483.

Thomson, R. B.: Physical Principles of Measurements with EMAT Transducers, in Thurston and Pierce 1990, pp. 157–200.

Thurston, R. N., Pierce, A. D. (eds.): Physical Acoustics, vol. XIX, Ultrasonic Measurement Methods, Academic Press, New York, 1990.

Vanderkooy, J.: A Model of Loudspeaker Driver Impedance Incorporating Eddy Currents in the Pole Structure. 84th Convention of the Audio Eng. Soc., Paris, March 1988, Preprint 2619.

Voishvillo, A.: Nonlinearity in Horn Drivers – Where the Distortion Comes From? 113th Convention of the Audio Eng. Soc., October 2002, Preprint 5641.

Voishvillo, A.: Nonlinear Versus Parametric Effects in Compression Drivers. 115th Convention of the Audio Eng. Soc., October 2003, Preprint 5912.

Wente, E. C.: The Sensitivity and Precision of the Electrostatic Transmitter for Measuring Sound Intensities. Phys. Rev., vol. 19, 1922, pp. 498–503.

Wong, G. S. K.; Embleton, T. F. W. (eds.): AIP Handbook of Condenser Microphones, American Institute of Physics, New York, 1995, 321 pp.

Woollett, R. S.: Effective Coupling Factor of Single-Degree-of-Freedom Transducer. J. Acoust. Soc. Am., vol. 40. 1966, pp. 1112–1123.

Zuckerwar, A. J.: Principles of Operation of Condenser Microphones, in Wong, G. S. K.; Embleton, T. F. W. (eds.): AIP Handbook of Condenser Microphones, American Institute of Physics, New York, 1995, pp. 37–69.

33

Loudspeaker Design and Performance Evaluation

Aki Vihtori Mäkivirta

Genelec Oy, Iisalmi, Finland

1 • Loudspeaker as a System Design Problem

A loudspeaker comprises transducers converting an electrical driving signal into sound pressure, an enclosure working as a holder for transducers, front baffle and box to contain and eliminate the rear-radiating audio signal, and electronic components. Modeling of transducers as well as enclosures is treated in Chap. 32 of this handbook. The purpose of the present chapter is to shed light on the design choices and options for the electronic circuits conditioning the electrical signal fed into loudspeaker transducers in order to optimize the acoustic performance of the loudspeaker.

Loudspeakers with wide bandwidth, flat and smooth magnitude response, and uniformly wide directivity are generally rated high by experienced listeners (Olive, 2003). This type of response has become the widely accepted design target in the industry, although there are also examples of highly directive designs mainly for consumer use. To achieve such high quality, the audio signal is split in two, three, or more frequency bands by *crossover filters*, and separate transducers are used for low frequencies (called "woofer"), high frequencies ("tweeter"), and possibly for mid-frequencies ("midrange").

Loudspeakers come in two flavors – passive and active. A *passive loudspeaker* contains transducers, an enclosure, and crossover filters. Crossover filters are built by passive electrical components to divide the signal frequency components from a power amplifier to transducers (see Figure 1a).

In professional applications the passive principle is used less and less because the active loudspeaker concept offers many significant benefits. The response of a passive loudspeaker is influenced by the output impedance of the driving amplifier, modified by the feeding cable. These factors are not a concern with active loudspeakers.

An *active loudspeaker* contains power amplifiers directly connected to transducers with filtering and signal conditioning electronics placed before the power amplifiers (see Figure 1b). The signal conditioning electronics include crossover filtering, system equalization, and protection. Modern active

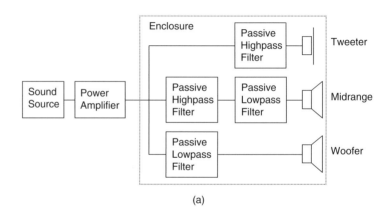

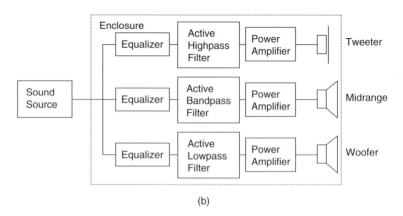

FIGURE 1 Fundamental functional blocks of a loudspeaker: (**a**) passive loudspeaker and (**b**) active loudspeaker.

loudspeakers also provide adjustable room equalization to optimize the frequency response in the final room installation. Active loudspeakers are built using both analogue and digital signal processing techniques. Because power amplifiers determine the source impedances seen by transducers, the source impedances can be carefully designed and maintained in an active loudspeaker. Precise circuitry to protect transducers can be implemented. Active loudspeakers can use high-order crossover filtering and precise equalization without impacting the source impedance seen by transducers.

2 • Crossover Filters

The basic crossover filter consists of a complementary filter pair of a lowpass filter $H_{\text{LP}}(s)$ and a highpass filter $H_{\text{HP}}(s)$. The frequency band where the crossing over between the two outputs occurs can be loosely defined as the frequency range where the two outputs are both contributing to form the acoustic output signal. Both outputs are contributing at equal level at the *crossover frequency*. In a modern crossover filter, the complex sum $H(s)$ of the filter outputs results in a flat magnitude response on the *acoustic axis* of the loudspeaker. The actual complex summation occurs in the air, at the position of the listener's ear. Other types of crossover alignments are used when the power output of a loudspeaker is optimized to be flat (see, e.g., Garde, 1980 for an overview of traditional alignments).

The requirements for crossover filter are characterized by (a) a defined flatness of the response (complex sum output) across the audio band, (b) complementary magnitude responses across the crossover transition band and (c) matching phase responses for the two filter outputs:

$$|H(s)| = |H_{\text{LP}}(s) + H_{\text{HP}}(s)| = 1.0 \quad \text{(requirements } a \text{ and } b\text{)}, \quad (1)$$

$$\arg(H_{\text{LP}}(s)) = \arg(H_{\text{HP}}(s)) \quad \text{(requirement } c\text{)}. \quad (2)$$

Our basic two-output crossover filter could be feeding two transducers. If requirements b or c are violated, output phases are no longer the same, and the magnitude of the complex sum of outputs is no longer exactly unity. Gain variations across the crossover region result.

Regarding requirement c, there are several factors that can contribute to produce a change in acoustic output phases. The transducers are mechanical bandpass devices with several resonant mechanisms, resulting in variations in the rate of phase angle change. The phase for each transducer must be determined at the listening position. Excluding the listening room effects (assuming measurement in an anechoic chamber), the differences in transducer characteristics, the mechanical layout of transducers in a loudspeaker enclosure, and diffractions created by the loudspeaker mechanics all contribute to change the phases of audio signals from those of the electronic signals fed into transducers. If the relative phase between transducers does not remain constant across the crossover, the acoustic axis can shift in direction. When optimizing a crossover filter, all these factors must be accounted for. It is generally not a good idea to design a crossover filter in isolation as an academic exercise, without measurements of transducers in the actual loudspeaker enclosure.

The now ubiquitous *Linkwitz–Riley (LR) crossover filter* (Linkwitz, 1976, 1978; Lipshitz and Vanderkooy, 1986) was an improvement over earlier crossover networks that showed directional shifts and an increase in the main lobe magnitude. Cascading two identical Butterworth filters forms an even-order LR crossover network. The highpass and lowpass LR filters have identical phase characteristics except for a 180-degree phase difference for certain filter orders. The LR filter has a flat passband and monotonously increasing stopband attenuation. The combined magnitude response of LR high and lowpass filters is flat. The attenuation of the low and highpass filters separately is 6.0 dB at the crossover point (see Figure 2), corresponding to the 3.0-dB passband edges of the prototype Butterworth filters used as building blocks of an even-order LR filter.

As other conventional crossover filter types, the LR filter is a minimum-phase filter showing peaking of the throughput delay near the crossover frequency. Already at the time of its introduction, Linkwitz was concerned about this local change in delay in the LR filter because certain signals (he gives the example of a square wave) can show strong changes in the waveform shape due to shifting in time of harmonics constituting a signal waveform. Linkwitz did some listening tests to conclude that the increase of delay did not cause audible changes to such signals for a then-realistic fourth-order LR filter (Linkwitz, 1976).

A faster crossover transition can be obtained by increasing the filter order for designs using filters with monotonously increasing stopband attenuation, such as LR filters. When the order of a crossover filter is increased, an analysis of coefficient sensitivity is needed to understand how realizable the filter is with the chosen technology. It is possible to reduce the sensitivity of the design by selecting favorable circuit topologies (Cochenour et al., 2003).

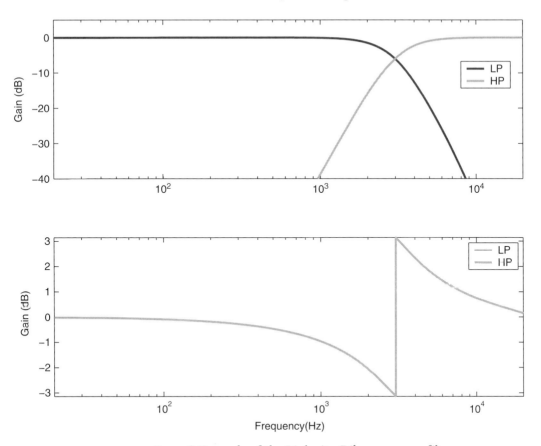

FIGURE 2 Example of the Linkwitz–Riley crossover filter response: complementary magnitude responses (*top*) and phase responses (*bottom*). LP = lowpass filter, HP = highpass filter.

A faster transition, and hence a narrower transition bandwidth, can also be obtained with filters having transfer function zeros in their stopband. Thiele (2000) describes a class of crossover filters having a null in their highpass and lowpass transfer functions at frequencies close to the crossover transition. The initial attenuation rate is higher than with LR filters, while the sum of the high and lowpass outputs shows an allpass response. When nulls are moved to very high and low frequencies, these transfer functions degenerate into Butterworth functions for the odd order and into Linkwitz–Riley functions for the even order. Thiele presents both active and passive realizations. Compared to LR filters, with monotonously increasing stopband attenuation, the faster transition of these filters results in a larger delay variation close to the crossover frequency.

Certain minimum-phase filter types such as Bessel and Gaussian filters show constant group delay at low frequencies. A crossover filter with reduced local increase of delay at the crossover point can be constructed using these filters (e.g., Fink, 1980). The main drawback is that these filters do not show a gain of −6 dB at the crossover point for designs that have equal phase response for both the high and lowpass filters. The complex sum of outputs will show level deviations at the crossover point.

By subtracting the filter output from the delayed input signal we can conveniently obtain two complementary outputs:

$$H_{HP}(s) = 1 - H_{LP}(s) \tag{3}$$

This principle has been applied to obtain *high-slope analogue* crossover filters (Lipshitz and Vanderkooy, 1983; Wilson et al., 1989). In the digital domain it is possible to obtain an exactly linear phase filter using a *finite impulse response* (FIR) filter. Such a filter can be designed to have a delay of an exact multiple of the sampling intervals, making the delay needed by the complementary filter design very easy to implement in the digital domain. *Infinite impulse response* (IIR) filters with similar behavior have also been suggested (Reviriego et al., 1998).

The transition bandwidth of crossover filters can in principle be made arbitrarily narrow, but narrowing the transition band also results in the increase of filter delay, particularly for linear phase FIR filters. Depending on the application, there will be a practical limit to how large the real-time delay of a loudspeaker can be. Another potential problem with linear phase filters is related to the fact that the impulse response of such a filter is symmetric so that the amount of energy in a linear phase system impulse response is equal before and after the maximum energy peak. FIR systems with fast transition bands also show long impulse responses before the main lobe, also known as pre-echo. Moving zeros in the FIR response inside the unit circle makes an FIR response increasingly minimum phase, and this can circumvent the problem of pre-echo. As a consequence, the FIR begins to show increasing local delay close to the crossover point, very much in the same way as with traditional designs.

In a conventional crossover design, where the phase is changing systematically and fairly slowly across the crossover region, one can always find a listening angle at a given distance where the two radiators are either in phase or in antiphase. An interesting idea due to Hawksford (1997) suggests a high-order crossover filter with random phase variation in both the highpass and lowpass outputs across the crossover frequency. This *stochastic interleave* crossover could in theory minimize angle-dependent co-radiation effects.

3 • Frequency Response Equalization

Equalization techniques are needed to produce a flat and smooth frequency response. Rudimentary equalization is possible by modifying passive crossover filters based on anechoic acoustic measurements, which is a standard practice in loudspeaker design. Precise implementation of system equalization becomes easier with the active loudspeaker concept where component values and circuit properties can be chosen more freely than in passive designs.

Equalization is typically implemented as modifications of standard crossover filter responses. Optimization to a flat overall target response can be performed manually when the number of variables or electronic components to optimize is fairly small. There is also the possibility for automated optimization using nonlinear optimization methods. Optimization of circuitry and component values can be done off-line based on anechoic response measurements using standard system simulation and modeling tools, such as Spice (Spice, 2005) and Matlab (Matlab, 2005), or software intended for loudspeaker design such as Calsod (Audiosoft, 2005), or Leap (LinearX, 2005). A particular challenge in loudspeaker design is to combine the equalization with the mechanical design of the loudspeaker so that the frequency response, when moving away from the acoustic axis, remains flat (for an example, see Figure 3).

Digital signal processing offers further advantages in system equalization. Digital filters can have negligible transfer function deviation from the ideal target because coefficients can be presented very accurately, and coefficient sensitivity can be made small compared to their analogue counterparts. Once designed, an equalizer transfer function can therefore be implemented very precisely. Infinite impulse response filters can be used in a similar fashion to analogue systems (White, 1986; Orfanidis, 1997). Finite impulse response (FIR) filters offer new possibilities to use easily designable and unconditionally stable high-order equalizers in practical loudspeaker systems (Hawksford, 1999). FIR filters can also be used to equalize frequency-dependent delay (Clarkson et al., 1985; Murray, 1994), a feature seldom practical in most analogue implementations. Special filter implementations such as warped FIR filters and Kautz filters allow filter resolution to be frequency specific, for example with increasing zero-pole

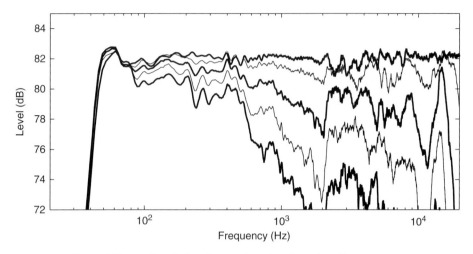

FIGURE 3 Example of the change in a loudspeaker frequency response when moving horizontally off-axis; *traces from top to bottom*: on-axis, 15 degrees, 30 degrees, 45 degrees and 60 degrees off-axis on the horizontal plane.

density toward low frequencies (Karjalainen et al., 1999; Wang et al., 2000; Paatero and Karjalainen, 2003).

Precise equalization can be effected by filtering the system H_S feed signal with the transfer function inverse H_E:

$$H_E(s) = 1/H_S(s), \qquad (4a)$$

$$H(s) = H_S(s)\, H_E(s) = H_S(s)/H_S(s) = 1.0. \qquad (4b)$$

Direct inverse of a loudspeaker response is usually not desirable. Theoretically, an inverse exists if the magnitude response at all frequencies of interest is larger than zero. Straightforward application of an inverse response, resulting in spectral whitening, leads to widening of the loudspeaker passband (for example, Clarkson et al., 1985). Typically there are certain frequencies, such as passband edges or transition bands, where equalization should not change the inherent response of the loudspeaker. Normally only the range of frequencies considered as the passband should be equalized. This leads to special requirements for the equalizing filter design. Several solutions for this problem have been proposed (Greenfield and Hawksford, 1991; Karachalios et al., 1995; Potchinkov 1998a, b; Kirkeby and Nelson, 1999; Karjalainen et al., 1999; Paatero and Karjalainen, 2003).

The ultimate precision of equalization depends on three factors – the capability of the equalizing filter to correct the acoustic response, the precision of the acoustic measurements used as the data on which the design is based, and the capacity of the human hearing system. The practical limit to how flat a loudspeaker must be rendered by equalization is shifting from being a filter design and implementation problem to being a problem of precision and relevance of measurements used as the basis for equalizer design as well as the lack of capacity in the human hearing system to recognize small imprecision in the acoustic response flatness.

A loudspeaker can in principle be equalized to be perfectly flat in anechoic conditions for all practical purposes (e.g., Wilson et al., 1989; Karjalainen et al., 1999), but flatness only exists relative to the acoustic measurements used as the basis for equalizer design. Acoustic measurements show *variability* and *measurement uncertainty*, limiting how precisely the true acoustic response can be known. Variability is related to measurement noise. Measurement uncertainty is related to the instrumentation used, notably the microphone and anechoic chamber. These two should be calibrated relative to some (traceable) reference. With all this done, the true response is only known to a certain tolerance. Equalizer design can only be as good as the measurement upon which its design is based.

Also the human auditory system has limits beyond which further improvement in audio system flatness is no longer readily recognized. Broadband deviations from a flat response are heard more sensitively than narrow-band ($Q > 10$) peaks or notches, and narrow-band peaks are more readily heard than notches (Toole and Olive, 1988; Olive et al., 1997). Broadband continuous signals, such as pink noise, are the best to reveal wideband flatness deviations because the stationary spectral content in pink noise allows the spectral balance to be recognized. Narrow-band resonances and antiresonances become audible with transient signals (such as clicks). These work as short duration exciters for narrow-band resonances eliciting their natural impulse responses that become audible after the masking related to the exciting signal has ceased. We do not yet know in full detail the ability of the human auditory system to hear small impairments in the loudspeaker frequency response, yet practical equalizer design must strike a balance between complexity and subjective benefit.

4 • Room Response Equalizers

A loudspeaker with flat response in anechoic conditions will not appear flat when taken into a listening room. This is because of *early reflections,* or reflections producing a large number of delayed replicas of the output signal, arriving at the listener very shortly after the direct sound, and because the *late reverberant field* in a room may be colored due to the sound absorption varying as a function of frequency. The aim of a loudspeaker installation is to set the loudspeaker up such that the frequency response at the primary listening area is sufficiently wide, with flat and smooth frequency response, so that there is a uniform directivity of sound over the whole intended listening area.

Equalization of the loudspeaker so that the frequency response at the listening area is flat can greatly improve perceived audio quality. Modern active loudspeakers typically provide special *room response adjustments* for this purpose. Low-order controls correcting the room response have a gentle shaping effect on the loudspeaker response. Typically tilt and roll-off controls are provided for the low frequencies of the woofer and for the high frequencies of the tweeter. In multi-way loudspeakers also transducer signal levels may be adjustable.

A loudspeaker is usually designed to produce a flat response in the free field. At low frequencies a loudspeaker is omnidirectional (radiating to 4π space), but changes to forward directing (radiating to 2π space) at a frequency determined approximately by the size of its front baffle. The woofer response controls are needed to optimize the low-frequency response for the 2π–4π transition when the loudspeaker is installed near or at a wall. The wall loading will limit the low-frequency radiation space, and boost the sound level below the 4π–2π transition frequency. When the loudspeaker is used behind a screen or if the acoustic axis of the speaker is not properly aligned, the tweeter response adjustment can be used to calibrate the very high-frequency response at the listening area.

Digital signal processing allows for easy implementation of more sophisticated and higher-order room response corrections. While it is possible to greatly improve reproduction quality, careless use of sophisticated equalizations can also result in audio quality worse than had no room equalizers been available. Typically a room response is optimized not at a single point in the listening space but within a listening area, typically with a diameter of 0.5–0.8 m. The response measured in a room can show strong *comb filtering effects* (transfer function zeros at a series of frequencies) due to strong reflections. To alleviate this problem, the response may be averaged in frequency either before filter design or indirectly as a part of it (Hatziantoniou and Mourjopoulos, 2000). When multiple responses are measured within the listening area, a weighted average of them may be calculated as a basis for equalizer design (Wilson, 1991;

Mourjopoulos, 1994; Bharitkar et al., 2002). This weight may also be frequency dependent. Particularly at low frequencies it is more likely that there are certain frequencies where a single microphone location is inadvertently placed in a room mode null. Moving the microphone around will enable this to be detected and accounted for (Mäkivirta et al., 2003).

It is also possible to reduce the perceived reverberation of a listening space to some degree by careful design of room equalization, such as by *complex smoothing* of the measured response before equalizer filter design (Hatziantoniou and Mourjopoulos, 2004). The so-called *modal equalization* can be applied to shorten the too long decay time of low-frequency modes of a room (Mäkivirta et al., 2003).

5 • System Delay and Delay Variation

Traditional loudspeaker designs have used fairly low-order minimum-phase filtering, and the system throughput delay or its variation as a function of frequency has not been an issue (e.g., Linkwitz, 1976: 7–8). Steep minimum-phase crossover filters, such as the Linkwitz–Riley crossovers, can produce excessive amounts of delay variation in frequency. As mechanical systems, loudspeakers are bandpass systems, and therefore will not have uniform delay characteristics. In fact, the system delay may show marked increase at a mechanical resonance of the transducer-enclosure system or the transducer itself.

The woofer displacement increases rapidly toward the low-frequency cut-off in a closed box design. Bass reflex and passive radiator designs can limit this increase, and are therefore frequently used to maintain low-frequency distortion at a low level, particularly in small enclosure designs. Both bass reflexes and passive radiators are mechanical resonators and will therefore increase loudspeaker delay close to their mechanical resonance. Despite a long-standing debate over the subjective importance of this added system delay, and demands in the popular press to make the phase response linear also at very low audio frequencies, there does not seem to be strong evidence suggesting that the additional delay generated in such designs would be detrimental to audio fidelity.

Frequency-dependent *delay variation* can shift in time the frequency components constituting a sound waveform, and can therefore change time-domain waveforms. In order to understand whether this has audible consequences we must know how much delay variation there can be before the human auditory system can notice a change. Audibility of delay variation is frequency dependent. Early investigations report subtle audible differences for delay variation caused by high-frequency roll-off (Meyer, 1984), mid-frequencies (4 kHz Preis and Bloom, 1984; 2 kHz Fincham, 1985), and low-frequency roll-off (Preis and Bloom, 1984; Deer et al., 1985). At very low frequencies and at very high frequencies significant increases in the delay can remain inaudible. Variation of several milliseconds appears to be inaudible at frequencies below a few hundred Hertz while in the mid-frequencies (1–5 kHz) variation of more than 1 ms may be audible (Preis, 1982).

Linear phase systems are interesting because they do not cause any frequency-dependent delay variation. However, they can produce large system *latency*. Latency produces a problem similar to a large physical distance – the visual event and sound event happen at different times. Tolerance to latency depends on the type of use of an audio system. If performance depends on audiovisual coordination, for example in dubbing or ensemble performance mediated over loudspeakers, a large latency deteriorates performance. In the case of audiovisual monitoring, for example in film and television, the sound cannot be delayed much in relation to the picture. For these types of applications the loudspeaker designer must balance between minimizing the latency while keeping the frequency-dependent delay variation in the loudspeaker acceptably small, and linear phase filtering can be used only with limitations. For audio-only playback applications much larger system delay is tolerable, which allows more efficient use of linear phase filters.

6 • Protection

A loudspeaker can be destroyed by several mechanisms. The most obvious is *thermal overload*. The other central destruction mechanism is *displacement overload*. Most of the input power fed into a dynamic transducer is turned into heat. There are also several mechanisms of voice coil instability and diaphragm movement that can potentially lead to destruction of a transducer (Klippel, 1990, 1992).

In large woofers, excess heat is conducted away from the voice coil efficiently. In tweeters protection levels and time constants must be set more stringent than for woofers. A typical overheat protection uses a double integral of the output voltage (voltage integrated into power, power integrated into energy). When the heat energy estimated in this way exceeds a design limit, the output level is reduced to maintain safe long-term power level and to allow cooling. Short-term instantaneous input power can be made very large compared to continuous power, particularly for tweeters. A more precise protection is possible if the voice coil and magnet structure temperatures can be known. In most practical systems such methods have not been deemed necessary because the incremental benefit in terms of additional output capacity usually is rather modest.

Several methods have been devised for displacement protection. The basic idea is to measure the voice coil location and to use this information to activate protection. It is also possible to calculate the voice coil position by using the physical force–mass relations. By double integrating the transducer driving current, related in electro-dynamic transducers to the force exerted by the voice coil, we can obtain an estimate of the voice coil location. This principle works when the nonlinearities of the suspension and variation of the magnetic flux in the air gap are also considered in calculations (Klippel, 1998, 2003).

7 • Understanding Loudspeaker Operation by Modeling

Lumped parameter modeling can be considered the principal method for developing understanding about the interaction of quantities governing loudspeaker system behavior (see Section 33.3). It may be beneficial to describe the whole electro-mechanic-acoustic system as a lumped parameter model. Computer-aided design techniques then allow various system scenarios to be examined early in the design phase without expensive and time-consuming prototype building. Different transducer parameters can also be inserted into simulations to determine how overall performance of a system might be affected. This enables the designer to identify important system design variables and to try different signal processing options to see their effects on the loudspeaker response.

Lumped parameter modeling using the Thiele–Small parameters was initially used to enable designers to predict the performance of a woofer in an enclosure and to optimize the low-frequency performance of the loudspeaker with a given transducer. Today, computer programs automate the use of lumped parameter calculation. Extensions to the original Thiele–Small model for large displacement conditions have been developed. Commercial measurement equipment exists for determining nonlinear changes to the lumped model parameters under large signal conditions (e.g., Klippel, 2005).

Directionality of a sound radiator depends primarily on its size (Kinsler et al., 1982). Design of a multi-way loudspeaker seeks a good compromise between the transducer size and displacement-related distortion at the desired maximum sound level. The uniformity of energy dispersion across the listening area can be improved by mechanical design of the front baffle where the transducers are mounted. Shallow horns (Holland et al., 1996), pioneered by the Genelec DCW (Genelec, 2005), have become popular as a means to tailor directivity and to match transducer directivities across the crossover transition. Advanced modeling techniques based on *finite and boundary element methods* can be used to extend modeling to these structures, allowing simulations starting from fundamental physical properties, and therefore able to yield deeper understanding of mechanical phenomena of coupling of mechanical vibration of the transducer to air and propagation of the sound in air; and interaction with mechanical boundaries of this propagating sound wave suffers while it is traveling from the transducer to the listener.

A loudspeaker is typically designed as a stiff structure to eliminate secondary or parasitic vibrations. The intention is that only the transducers radiate sound. A notable exception from this principle is the distributed mode loudspeaker (DML, Prokofieva et al., 2002) where the radiation is produced by shaking a large surface area, evoking randomly distributed modal vibrations, and breaking up any pistonic operation. Such a radiator has fewer tendencies to be directive than a similar size pistonic radiator, but cannot be easily designed to produce a well-defined on-axis frequency response in the same way a conventional radiator can. Element modeling methods and *statistical modeling* have been used to study the properties of such radiators.

8 • Acoustic Measurement Techniques

Acoustic measurements reveal the actual behavior of a loudspeaker. Acoustic measurements can be categorized in free field and in situ measurements. *Free field measurements*, typically performed in anechoic rooms, are central to loudspeaker development work and in principle enable the study of a loudspeaker in isolation. Methods have been developed to simulate free field conditions in a real room by signal windowing (Struck and Temme, 1994).

In situ measurements study the combination of a loudspeaker and a listening room. The measurement methodology is in principle similar to that used in anechoic measurements. Additionally, smoothing in frequency is often used to obtain average response level in the presence of dense notching produced by the complex acoustic reflection field present in a listening room. Time windowing of the impulse response data is also frequently employed to dissociate direct sound from reverberation.

Multitone techniques, where a signal is a composite of multiple suitably phased sinusoids, allow very fast approximate measurement of system magnitude and phase responses at all constituent frequencies (Czerwinski et al., 2001). Contemporary methods to estimate system impulse responses include the use of *maximum length sequences* (MLS; see, e.g., Rife and Vanderkooy, 1989; Vanderkooy, 1994) and *log-swept sinusoidal* signals (e.g. Müller and Massarani, 2001; Stan et al., 2002). The MLS measurement appears robust against environmental noise and can produce accurate results in noisy conditions (Stan et al., 2002). Also log-frequency sinusoidal sweep techniques can capture impulse responses with high signal-to-noise ratio (SNR) due to the possibility of rejecting simultaneous harmonic distortions and allowing analysis of linear responses and harmonic distortions simultaneously (Müller and Massarani, 2001). Measurement quality can degrade in the presence of *time variance*, for example swaying of the microphone stand (Svensson and Nielsen, 1999) or when a system is strongly nonlinear (Vanderkooy, 1994).

The loudspeaker response is optimized on the *acoustic axis*, whereby the acoustic axis defines the assumed direction of the listener. The *directivity* characteristics of a loudspeaker are usually studied by measuring frequency responses in certain angles (Seidel and Staffeldt, 1996). A typical selection of angles is from zero (on the acoustic axis) to 60 degrees with 15-degree increments. Rotating the loudspeaker around its horizontal and vertical planes on a turntable and measuring the frequency responses can obtain a more accurate understanding of directivity. It is important to rotate the loudspeaker around its acoustic and not the mechanical center. These two may not coincide. If the loudspeaker is rotated around its mechanical center, there will be a direction-related error in the acoustic distance and hence on the recorded sound level.

The list of commercially available acoustic measurement systems is rather volatile, but some examples include (primary technologies are given in parenthesis) MLSSA (2005, MLS sequence), WinMLS (Morset, 2005, log-sine sweep and MLS), Clio (Audiomatica, 2005, log-sine sweep and MLS), and LMS (LinearX, 2005, sine sweep). This list could be a lot longer and may not be understood to be an endorsement for any of the products mentioned. There are several other quite capable measurement systems available, and the reader should use the list merely as a starting point for evaluating products for his purposes.

9 • Listening Tests

Modern measurement and modeling methods produce detailed data on loudspeaker performance. Yet we lack a method to map this information to the listener's experience. The subjective quality judgments relate to the main categories of flatness of the frequency response, quality and accuracy of virtual imaging, and lack of distortion or extraneous noises during operation. A loudspeaker is expected not to color the sound even when used in a reverberant room. Virtual sound images constructed between loudspeakers are expected to be precise and well defined (not stretched or blurred). The loudspeaker is expected to be able to reproduce the whole dynamic range in the sound recording without detectable increase in distortion at high output levels and without producing extraneous sounds. Although discussion about listening tests goes beyond the scope of this article, the importance of first-hand listening testing can never be overemphasized.

References

Audiomatica. Clio Measurement System. Audiomatica Srl. http://www.mclink.it/com /audiomatica/home.htm, 2005

Audiosoft. http://members.optusnet.com.au/~ audiosoft/, 2005

Bharitkar S., Hilmes P., Kyriakakis C., "Robustness of Spatial Averaging Equalization Methods: A Statistical Approach." IEEE 36th Asilomar C. Signals, Systems & Computers, Pacific Grove, CA, Nov. 3–6, 2002

Clarkson P. M., Mourjopoulos J., Hammond J. K., "Spectral, Phase, and Transient Equalization for Audio Systems", J. Audio Eng. Soc., Volume 33, Number 3, pp. 127–132. March 1985

Cochenour B., Chai C., Rich D. A., "Sensitivity of High-Order Loudspeaker Crossover Networks with All-Pass Response", J. Audio Eng. Soc., Volume 51, Number 10, pp. 898–911. October 2003

Czerwinski E., Voishvillo A., Alexandrov S., Terekhov, A., "Multitone Testing of Sound System Components, Some Results and Conclusions, Part 2: Modeling and Application", J. Audio Eng. Soc., Volume 49, Number 12, pp. 1181–1192. December 2001

Deer J. A., Bloom P. J., Preis, D., "Perception of Phase Distortion in All-Pass Filters", J. Audio Eng. Soc., Volume 33, Number 10, pp. 782–786. October 1985

Fincham L. R., "The Subjective Importance of Uniform Group Delay at Low Frequencies", J. Audio Eng. Soc., Volume 33, Number 6, pp. 436–439. June 1985

Fink D. G., "Time Offset and Crossover Design", J. Audio Eng. Soc., Volume 28, Number 9, pp. 601–611. September 1980

Garde P., "All-Pass Crossover Systems", J. Audio Eng. Soc., Volume 28 Number 9, pp. 575–584. 1980

Genelec. Genelec DCW Technology. http://www.genelec.com/support/dcw_tech.php, 2005

Greenfield R., Hawksford M. O., "Efficient Filter Design for Loudspeaker Equalization", J. Audio Eng. Soc., Volume 39, Number 10, pp. 739–751. October 1991

Hatziantoniou P. D., Mourjopoulos J. N., "Generalized Fractional-Octave Smoothing of Audio and Acoustic Responses", J. Audio Eng. Soc., Volume 48, Number 4, pp. 259–280. April 2000

Hatziantoniou P. D., Mourjopoulos J. N., "Errors in Real-Time Room Acoustics Dereverberation", J. Audio Eng. Soc., Volume 52, Number 9, pp. 883–899. 2004

Hawksford M. O., "Digital Signal Processing Tools for Loudspeaker Evaluation and Discrete-Time Crossover Design", J. Audio Eng. Soc., Volume 45, Number 1/2, pp. 37–62. January/February 1997

Hawksford M. O., "MATLAB Program for Loudspeaker Equalization and Crossover Design", J. Audio Eng. Soc., Volume 47, Number 9, pp. 706–719. September 1999

Holland K. R., Fahy F. J., Newell P. R., "The Sound of Midrange Horns for Studio Monitors", J. Audio Eng. Soc., Volume 44, Number 1/2, pp. 23–36. January/February 1996

Karachalios G., Tsoukalas D., Mourjopoulos J., "Multiband Analysis and Equalisation of Loudspeaker Responses", AES 98th Convention, 1995

Karjalainen M., Piirilä E., Järvinen A., Huopaniemi, J., "Comparison of Loudspeaker Equalization Methods Based on DSP Techniques", J. Audio Eng. Soc., Volume 47, Number 1/2, pp. 14–31. January/February 1999

Kinsler L., Frey A., Coppens A., Sanders J., Fundamentals of Acoustics. Third edition. Wiley, New York. 1982

Kirkeby O., Nelson P. A., "Digital Filter Design for Inversion Problems in Sound Reproduction", J. Audio Eng. Soc., Volume 47, Number 7/8, pp. 583–595. July/August 1999

Klippel W., "Dynamic Measurement and Interpretation of the Nonlinear Parameters of Electrodynamic Loudspeakers", J. Audio Eng. Soc., Volume 38, Number 12, pp. 944–955. December 1990

Klippel W., "Nonlinear Large-Signal Behavior of Electrodynamic Loudspeakers at Low Frequencies", J. Audio Eng. Soc., Volume 40, Number 6, pp. 483–496. June 1992

Klippel W., "Direct Feedback Linearization of Nonlinear Loudspeaker Systems", J. Audio Eng. Soc., Volume 46, Number 6, pp. 499–507. June 1998

Klippel W., "Assessment of Voice-Coil Peak Displacement Xmax", J. Audio Eng. Soc., Volume 51, Number 5, pp. 307–324. May 2003

Klippel W., http://www.klippel.de/pubs/default.asp, 2005

LinearX. http://www.linearx.com/, 2005

Linkwitz S. H., "Active Crossover Networks for Noncoincident Drivers", J. Audio Eng. Soc., Volume 24, Number 1, pp. 2–8. January/February 1976

Linkwitz S. H., "Passive Crossover Networks for Noncoincident Drivers", J. Audio Eng. Soc., Volume 26, Number 3, pp. 149–150. March 1978

Lipshitz S. P., Vanderkooy J., "A Family of Linear-Phase Crossover Networks of High Slope Derived by Time Delay", J. Audio Eng. Soc., Volume 31, Number 1/2, pp. 2–20. January/February 1983

Lipshitz S. P., Vanderkooy, J., "In-Phase Crossover Network Design", J. Audio Eng. Soc., Volume 34, Number 11, pp. 889–894. November 1986

Mäkivirta A., Antsalo P., Karjalainen M., Välimäki, V., "Modal Equalization of Loudspeaker – Room Responses at Low Frequencies", J. Audio Eng. Soc., Volume 51, Number 5, pp. 324–343. May 2003

Matlab. http://www.mathworks.com/, 2005

Meyer J., "Time Correction of Anti-aliasing Filters Used in Digital Audio Systems", J. Audio Eng. Soc., Volume 32, Number 3, pp. 132–137. March 1984

MLSSA. MLSSA Measurement Software. DRA Laboratories. http://www.mlssa.com/, 2005

Morset. WinMLS Software. Morset Sound Development. http://www.winmls.com/, 2005

Mourjopoulos J., "Digital Equalization of Room Acoustics", J. Audio Eng. Soc., Volume 42, Number 11, pp. 884–900, 1994

Müller S., Massarani, P., "Transfer-Function Measurement with Sweeps", J. Audio Eng. Soc., Volume 49, Number 6, pp. 443–471. June 2001

Murray J. A., "Microalignment of Drivers via Digital Technology", J. Audio Eng. Soc., Volume 42, Number 4, pp. 254–264. April 1994

Olive S. E., "Differences in Performance and Preference of Trained Versus Untrained Listeners in Loudspeaker Tests: A Case Study", J. Audio Eng. Soc., Volume 51, Number 9, pp. 806–825. September 2003

Olive S. E., Schuck P. L., Ryan, J. G., Sally, S. L., Bonneville, M. E., "The Detection Thresholds of Resonances at Low Frequencies", J. Audio Eng. Soc., Volume 45, Number 3, pp. 116–128. March 1997

Orfanidis S. J., "Digital Parametric Equalizer Design with Prescribed Nyquist-Frequency Gain", J. Audio Eng. Soc., Volume 45, Number 6, pp. 444–455. June 1997

Paatero T., Karjalainen, M., "Kautz Filters and Generalized Frequency Resolution: Theory and Audio Applications", J. Audio Eng. Soc., Volume 51, Number 1/2, pp. 27–44. January/February 2003

Potchinkov A., "Frequency-Domain Equalization of Audio Systems Using Digital Filters, Part 1: Basics of Filter Design", J. Audio Eng. Soc., Volume 46, Number 11, pp. 977–987. November 1998a

Potchinkov A., "Frequency-Domain Equalization of Audio Systems Using Digital Filters, Part 2: Examples of Equalization", J. Audio Eng. Soc., Volume 46, Number 12, pp. 1092–1108. December 1998b

Preis D., "Phase Distortion and Phase Equalization in Audio Signal Processing-A Tutorial Review", J. Audio Eng. Soc., Volume 30, Number 11, pp. 774–794. November 1982

Preis D., Bloom P. J., "Perception of Phase Distortion in Anti-alias Filters", J. Audio Eng. Soc., Volume 32, Number 11, pp. 842–848. November 1984

Prokofieva E., Horoshenkov K. V., Harris N., "Intensity Measurements of the Acoustic Emission from a DML Panel", Preprint Number 5609, Proc. 112 Convention of AES, April 2002

Reviriego P., Parera J., García R. "Linear-Phase Crossover Design Using Digital IIR Filters", J. Audio Eng. Soc., Volume 46, Number 5, pp. 406–411. May 1998

Rife D., Vanderkooy, J. "Transfer-Function Measurement with Maximum-Length Sequences", J. Audio Eng. Soc., Volume 37, Number 6, pp. 419–444. June 1989

Seidel F., Staffeldt, H., "Frequency and Angular Resolution for Measuring, Presenting, and Predicting Loudspeaker Polar Data", J. Audio Eng. Soc., Volume 44, Number 7/8, pp. 555–568. July/August 1996

Spice. http://bwrc.eecs.berkeley.edu/Classes/IcBook/SPICE/, 2005

Stan G.-B., Embrechts J.-J., Archambeau, D., "Comparison of Different Impulse Response Measurement Techniques", J. Audio Eng. Soc., Volume 50, Number 4, pp. 249–262. April 2002

Struck C. J., Temme, S. F., "Simulated Free Field Measurements", J. Audio Eng. Soc., Volume 42, Number 6, pp. 467–482. June 1994

Svensson U. P., Nielsen, J. L., "Errors in MLS Measurements Caused by Time Variance in Acoustic Systems", J. Audio Eng. Soc., Volume 47, Number 11, pp. 907–927. November 1999

Thiele N., "Loudspeaker Crossovers with Notched Responses", J. Audio Eng. Soc., Volume 48, Number 9, pp. 786–799. September 2000

Toole F. E., Olive, S. E., "The Modification of Timbre by Resonances: Perception and Measurement", J. Audio Eng. Soc., Volume 36, Number 3, pp. 122–142. March 1988

Vanderkooy J., "Aspects of MLS Measuring Systems", J. Audio Eng. Soc., Volume 42, Number 4, pp. 219–231. April 1994

Wang P., Ser W., Zhang, M., "A Dual-Band Equalizer for Loudspeakers", J. Audio Eng. Soc., Volume 48, Number 10, pp. 917–921. October 2000

White S. A., "Design of a Digital Biquadratic Peaking or Notch Filter for Digital Audio Equalization", J. Audio Eng. Soc., Volume 34, Number 6, pp. 479–483. June 1986

Wilson R., "Equalization of Loudspeaker Drive Units Considering Both On- and Off-Axis Response", J. Audio Eng. Soc., Volume 39, Number 10, pp. 127–139, 1991

Wilson R., Adams, G., Scott, J., "Application of Digital Filters to Loudspeaker Crossover Networks", J. Audio Eng. Soc., Volume 37, Number 6, pp. 455–464. June 1989

34

PA Systems for Indoor and Outdoor

Jan Felix Krebber

Institute of Communication Acoustics, Ruhr-University Bochum, Bochum, Germany, e-mail: jan.krebber@rub.de

1 • Introduction

Sound reinforcement (SR) of the voice has allowed to build bigger halls and to address large gatherings. Between 1915 and 1960, public address (PA) systems were mainly used to address gatherings. The first recorded use of an outside PA system or SR system goes back to December 30th 1915 (Yaxley, 2003). In San Francisco, Governor Hiram Jones was intended to deliver a speech in the new Civic Auditorium. However, unfortunately he fell ill and engineers Jensen and Pridham thought of a solution of how to transmit Hiram Jones' voice to the public. The auditorium was equipped with "Magnavox" equipment to reinforce the talker's voice. Jensen and Pridham installed a special line between the Governor's home and the sound reinforcement system in the auditorium. Eventually, Hiram addressed the public by means of a microphone while sitting in front of his fireplace.

SR systems were capable of distributing sound for more than 90,000 people (e.g., opening ceremony of the Wembley Exhibition 1925 by King George). In the 1960s, Rock'n' Roll music started to need amplification, especially for electric guitars and vocals. A new phenomenon occurred: The audience was not silent anymore during the performance, and PA systems had to be louder than the audience. In 1965, the Beatles were the first rock band to play in a stadium of more than 55,000 visitors (Shea Stadium, New York). People screamed at such a volume, that even the musicians on stage could not hear themselves anymore. This was one of the reasons, why the Beatles stopped performing live concerts in 1966 (Cross, 2005). The band "The Who" was the first band known to the author using a stage monitor in 1968, which partly transmitted the vocals back to the stage. In the early years of monitoring, sidefill stage loudspeakers provided the same signal as the PA system.

Over the years, the monitoring strategy changed from onstage self-monitoring by the musicians' own amplifiers to individual monitor mixes and monitor wedges for each member of the band. The most important step in monitoring was to move from ear-damaging sound pressure levels (SPL) on stage to ear-friendly in-ear monitoring.

The latest deep impact on PA systems was made by L-ACOUSTICS by introducing line arrays in the early 1990s and the research results of "Wavefront Sculpture Technology"[1] (Urban et al., 2003). Line arrays deliver a higher and more directive sound pressure level than any other system known so far. Nowadays, it is a must for every PA system manufacturer to offer at least one line array system from their own product range.

Another interesting "back to the roots" PA solution is introduced by Bose's "Personalized Amplification System"[2] for smaller venues. The system provides again own loudspeakers for every musician. These loudspeakers serve as monitor and as PA at the same time.

This chapter presents a first insight into PA systems. It will shortly discuss all components of a PA system and give some ideas how to plan and how to set up the sound system regarding indoor and outdoor effects and measurements. After all, there will be a short discussion about digital vs. analog equipment and about things to be taken care besides signal processing. However, the design of sound systems is some kind of art, and it requires much experience, gained from practical and theoretical knowledge.

For further reading, Davis and Jones (1989), Davis and Davis (1997), and Stark (1989) will discuss the field of sound reinforcement from a practical point of view, whereas Ahnert and Steffen (1999) and Ballou (2002) provide very detailed

---

[1] Wavefront Sculpture Technology and WST are trademarks of L-ACOUSTICS.
[2] Personal Amplification System is a trademark of Bose.

and theoretical information on sound systems. For the first steps using mixing desks, the reader may find helpful advice in Trubit (2000), and for microphone techniques, Pawera (2003), Huber and Williams (1922) or Alldrin (1997) will give advice.

2 • The PA System

The core of a PA system consists of one or more input transducers, one or more signal processing stages, and one or more output transducers. A conceptual model of a sound system is shown in Figure 1.

Input transducers convert either sound waves or mechanical, optical, or magnetic changes over time from storage media into electrical signals. The types of input transducers normally found in sound reinforcement systems are as follows:

Air pressure and velocity microphones are used to convert sound waves traveling through air into electrical signals. Usually dynamic, condenser, or electret microphones are used. Other types of microphones like ribbon microphones or carbon microphones are not of interest for sound reinforcement applications due to the lack of suitable frequency range (carbon microphone) or stability (ribbon and piezoelectric microphone) for hard hiring company environment.

Contact pickups convert sound waves traveling through a solid medium into electrical signals.

Magnetic pickups convert sound waves of induced magnetism, e.g., metal strings into electrical signals.

Phono pickups convert mechanical movement of a stylus into electrical signals. There exist two different systems: MM = moving magnet and MC = moving coil.

Tape heads, optical pickups, and *laser pickups* convert recorded material from a certain medium (magnetic tape, photographic film, and imprinted patterns on a CD via digital-analog conversion) into electrical signals. The signal shape of all the electrical signals after conversion (and equalizing frequencies if

FIGURE 1 Conceptual model of a sound system.

necessary, e.g., RIAA[3]) look very similar (regarding the use of high-quality equipment), but the overall level of the converted electrical signals may be very different depending on the type of conversion.

The signal processing is divided into several functional units. The same units may occur several times at different positions in the signal paths. These units are pre-amplification, delay, equalization, dynamic level processing, room and delay effects, mixing and distributing, and power amplification.

Pre-amplification brings the weak microphone signal to line level. The microphone pre-amplification stage may allow gains up to 70 dB.

Delays allow a simple phase reverse or the time alignment for different microphone positions targeting the same source. This technique requires much knowledge as it might lead to severe comb-filter effects. In the later signal processing path, delays may also be used for the time alignment of loudspeakers.

Equalization or *filtering* allows to adjust the tonal balance of the audio signal. There are several types of filters: highpass and lowpass filters, shelving filters, and peaking filters. Highpass and lowpass filters may allow a variable cutoff frequency. Shelving filters can be controlled in cut or boost and depending on the implementation also in cutoff frequency. Fixed peaking filters or graphic EQs can cut or boost the audio signal at a fixed frequency. The semi-parametric or quasi-parametric EQs can cut or boost the audio signal with filters at a variable frequency, whereas the fully parametric EQ allows also controlling of the filter width. A special type of filter is the notch filter. It cancels a certain frequency with a very narrow bandwidth. Variable notch filters are usually not found in analog equipment because circuit design for proper control of the frequency is too complicated. Instead, peaking EQs with a fixed gain cut and very narrow filter width (1/10 octave) are implemented. All the relevant filter types are shown in Figure 2.

Dynamic Level Processing units are compressors, limiters, de-essers, and noise. These units reduce the dynamic range. For linear dynamic level processing, the ratio output level/input level is a constant of value 1. A *compressor* allows to change this ratio (compression ratio) from a certain input level (threshold) upward.

A special case of a compressor is the *limiter*. Above a set threshold, the ratio is $\infty : 1$, which means that the input signal

---

[3]Record Industry Association of America. The RIAA response curve is used for records to overcome technical problems while cutting and playing records.

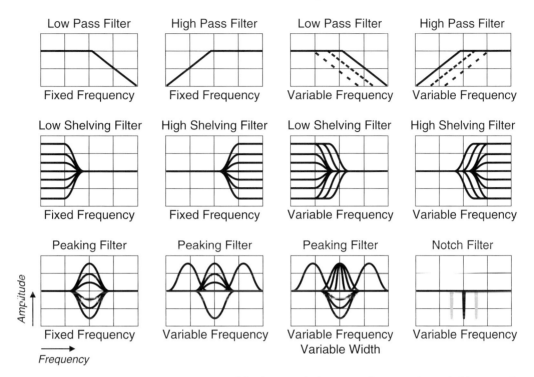

FIGURE 2 Lowpass and highpass, shelving, peaking, and notch filters with fixed or variable frequency.

could theoretically reach infinity, but the output signal would stay at the threshold level.

Another member of the compressor family is the *de-esser*. There is an additional bandpass filter in the control path of the compressor, which allows for frequency-dependent compression of the audio signal. De-essers are mainly used to reduce the loudness of "s"-sounds (which may occur to be louder than the rest of the vocal signal) leaving the rest of the material untouched. Without a de-esser, one might reduce the high-mid frequencies to avoid too loud "s"-sounds, but this might result in a dull sound. The "s"-phoneme frequencies are mainly found around 4.5 and 6.7 kHz and the "sh"-phoneme frequencies around 3.4 and 5.1 kHz for male and female voices, respectively.[4]

Gates cut off a signal for which power is below a certain threshold. Usually, not all microphones are used at the same time and they can be "turned off" for the duration of their source

---

[4]Fricative resonance frequencies vary according to vowel context.

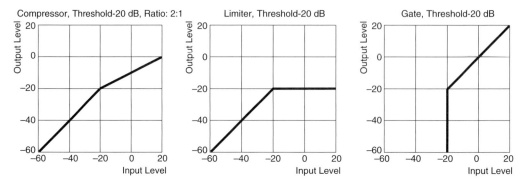

FIGURE 3 Level dependency of input and output signals for different dynamic level processors.

being "silent". In this way, gates allow to reduce the amount of microphones "being open" and finally increase the gain-before-feedback.[5] The level dependency of input and output signals for dynamic level processors is depicted in Figure 3.

Delay and reverb effects are based on simulating the idea of reflecting a sound source when it hits a boundary. The delay repeats the sound source several times with constant time interval and more and more reduced intensity. Usually, the delay effect is implemented as a single delay stage with an intensity adjustable feedback loop. The delay effect is more artificial than realistic, and it is mainly used to thicken the sound with echoes in a controlled, rhythmic way. The reverb effect aims at modeling a room with some early reflections and subsequent dense reverberation. The parameter settings allow either more or less realistic simulations of certain rooms, but also some extreme and artificial room effects which would be impossible in the real world. The reverb effect is mainly used to add liveliness and spaciousness to a dry signal.

Mixing and distributing the sound from several sources to different destinations is the main task of a mixer, although in the first step all sources are processed individually (e.g., by the preamplifier, delay, dynamic processing, and EQ). After that, sources are mixed for different purposes, e.g., monitoring, effects for a group of sources, and for the main PA system. A practical guidance how to use mixers can be found in Trubit (2000). Besides the mixing capability, mixing desks also provide EQs and sometimes dynamic effects like gates or compressors.

---

[5] Gain-before-feedback is the maximum gain allowed between microphone and loudspeaker just before the sound system starts to ring. Doubling the number of microphone being "open" decreases the gain-before-feedback theoretically by 3 dB.

Power amplification is the last stage of the signal processing chain before the electrical signal is sent to the output transducers. There are different topologies of amplifiers available on the market. Nowadays, switch-mode technology can be found in the power supply, or in the amplification stage itself. The advantages are reduced weight and less loss of energy due to a better degree of efficiency. The limitation of power being available from an amplifier is usually caused by the limits of the power supply (overload or clipping) or heatsink capabilities (most times resulting in the minimum load impedance allowed at the output[6]).

So far, only headphones are capable of propagating nearly the full audio spectrum by a single driver unit. However, single output transducers are not able to convert the electrical signal in the full audio frequency spectrum between 20 Hz and 20 kHz (4 decades or 10 octaves) with high sound pressure level. Therefore, transducers are optimized for several frequency ranges. In PA systems, transducers are mainly loudspeakers from different frequency ranges which are combined to cover the full spectrum. The types of loudspeakers most commonly encountered in SR systems are listed in Table 1. The types of enclosures found in PA systems are discussed in the following:

TABLE 1 Speakers, their frequency range, and the possible types of transducers

Speaker	Frequency range	Type of transducer (diameter)
Subwoofer	20–300 Hz, usually 30–100 Hz	15″–21″ cone-type drivers
Woofer	Usually < 500 Hz for LF,[a] < 1.5 kHz for LF + MF[b]	10″–18″ cone-type drivers
Mid-range	Typically > 500 Hz, < 6 kHz	5″–12″ cone-type drivers, 2″–4″ compression drivers
Tweeter	> 1.5 kHz, usually > 6 kHz	2″–5″ cone-type drivers, 1″–4″ compression drivers, electrostatic drivers
Super tweeter	Above 10 kHz	compression drivers, piezoelectric drivers

[a] Low frequencies.
[b] Mid-frequencies.

[6] Often amplifiers are capable of driving lower impedance loads than specified, but they might switch off after a couple of minutes as they are getting too hot.

Sub-lows are usually used for frequencies below 100 Hz down to subsonic.

Lows do not reach as low a frequency as sub-lows, but may go up to 250 Hz.

Full-range enclosures contain loudspeakers which cover the frequency range from 100 Hz or below and up to 13 kHz or above. In case full-range enclosures are used with lows, their low cut-off will shift to higher frequencies to gain more power for the low-mid frequency range.

Monitor Wedges are full-range loudspeakers that are built in a special shape to be laid on the floor and to point to the performer. They are used to return a "monitoring signal" to the performer to help him or her to stay in tune or to keep the timing by listening some certain instruments from the band.

3 • Planning and Setting Up the Loudspeakers

The typical area for sound distribution range is between 10 m for close distances up to 200 m for far distances from a single source. For a rough estimation of the sound pressure level required for a PA system, one may take the following items into consideration:

- The *noise floor* of the venue. A babbling crowd will be around 86 dB.
- The *sound level* should be remarkably higher than the noise floor, e.g., perceived with double loudness than the noise floor. The perceived loudness doubles with 10 dB SPL.
- The *longest distance* of your audience area. The loss will be 6 dB SPL by doubling the distance, e.g., in the far distance of 200 m, the SPL will be 46 dB less than at 1 m distance.
- Allow the system a *headroom*; 10 dB is a reasonable value for normal applications.

The speaker should provide $86 + 10 + 46 + 10 \text{ dB} = 152 \text{ dB}$ if the atmospheric effects are not taken into account. This high SPL cannot be distributed by a single speaker and would cause too high SPL in the near field. Therefore, large area sound distribution requires different types of loudspeakers for short- (up to 20 m), intermediate- (up to 100 m), and long-throw (up to 200 m) sound distribution. However, correctly installed line arrays can handle all distances with one type of loudspeaker except very close distance to the stage (less than 10 m).

Large venues or fixed installations are planned by the help of computer programs. Usually these programs require a computer-aided design (CAD) drawing of the room with some additional parameters concerning the boundaries, as there is the degree of scattering, the degree of transmission, and the degree of absorption. Loudspeaker parameters are provided by the loudspeaker vendors and are available in databases. Good programs should deliver parameters like SPL,[7] RT,[8] STI,[9] RASTI,[10] C_{80},[11] D_{50},[12] echograms, and auralization.[13] These parameters are necessary for estimating the quality of sound related to the installed PA system.

However, most venues do not need, or the budget does not allow, a computer-aided planning of the loudspeaker positions. In the following, some common setups are discussed and depicted in Figure 4.

1. Central public address: The sound is propagated from a central point. The sound waves from the PA system cover mainly the back area of the listeners area, where the direct sound from the podium would be too weak. To avoid additional reverberation, boundaries of the first reflections

---

[7]Calculation of Sound Pressure Level can be found in IEC61672-1-1:2002.

[8]Reverberation Time is the time the reverberation takes to die away. For example, RT_{60} is the time it takes for a sound's reverberation field to attenuate by 60 dB.

[9]Further information about the Speech Transmission Index can be found in IEC60268-16:1998-16:1998.

[10]The RApid Speech Transmission Index is a faster version of the STI. Further information can be found in IEC60268-16:1998-16:1998.

[11]C_{80} is usually named clarity and is related to music performances. It is an objective value which estimates the perceptibility of tones spread in time or differentiation of concurrent sounds made by different instruments. C_{80} is the ratio of energy within the first 80 ms vs. the energy of the subsequent sound.

$$C_{80} = \frac{\int_{0\,\text{ms}}^{80\,\text{ms}} p^2(t)\text{dt}}{\int_{80\,\text{ms}}^{\infty} p^2(t)\text{dt}}. \tag{1}$$

Further information can be found in ISO3382:1997.

[12]D_{50} "Definition" or "Deutlichkeit" is sometimes used for speech conditions. It is the ratio of energy within the first 50 ms vs. overall energy of the sound. Together with C_{80}, it gives the balance between early and late arriving energy.

$$D_{50} = \frac{\int_{0\,\text{ms}}^{50\,\text{ms}} p^2(t)\text{dt}}{\int_{0\,\text{ms}}^{\infty} p^2(t)\text{dt}}. \tag{2}$$

Further information can be found in ISO3382:1997.

[13]Auralization is the convolution of the estimated room impulse response with a source file at predefined places and predefined directions for playback by headphones.

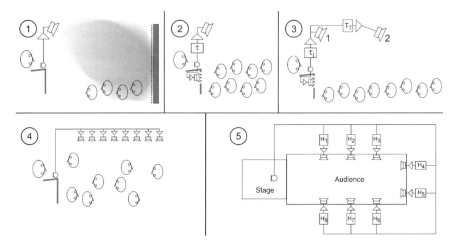

FIGURE 4 Different setups of loudspeakers, adopted from Blauert (2000).

are built in a sound absorbing manner. Usually this is the rear panel and the audience area.

2. The next approach makes use of the law of the first wave front. In case the amplified sound wave is delayed 10–20 ms (related to the original sound wave from the talker), the auditory event is localized at the position of the talker. The amplified sound wave may be up to 10 dB louder at the listener's position without being localized as such, as loudness and intelligibility are related to the amplified sound. For enhanced PA systems for medium-sized events, a loudspeaker is placed in or next to the podium, in addition to main loudspeakers. For live concerts, a short delay of the main PA (5–20 ms) may bring sound from stage more directly to the audience.
3. For large rooms or places or rooms with niches or protrusions covering the listening area, a column setup will be applied. The delay unit τ_2 will compensate the delay of sound between the first and second column. It will cancel out echoes from the first column at the place of the second row.
4. In shopping malls or train stations, distributed PA systems can be found. They include many synchronously driven full-range enclosures which emit a diffuse sound field. The talker will not be localized at his original position, but this does not disturb the listener. The propagated SPL of every loudspeaker has to be set very carefully.
5. Artificial reverberation is added to originally dry rooms which show a short reverberation time, e.g., multipurpose venue places or which should sound larger than they are,

e.g., theaters in cruiser ships. The signal processing is either a set of parallel delay units or an impulse response. These systems are uniquely tuned for rooms they are installed in.

4 • Sound Indoors

Important physical phenomena for indoor sound reinforcement are refraction, reflection, diffusion, diffraction, and absorption.

Refraction bends the sound wave due to changes in velocity of sound through a medium. This will be explained as temperature gradient in Section 5.

Reflection is the ability of the surface material to mirror an incoming sound wave with a certain amount of loss. It is the counterpart of absorption. The surface material acts as a filter to the incoming sound; some energy is turned into heat but other fractions of the energy are reflected back into the room.

Diffusion is a special form of reflection. Diffusor materials have desired surface irregularities which should spread the incoming sound wave equally in all directions.

Diffraction is the bending of a sound wave around an obstacle, in case the obstacle is small with respect to the wavelength. Otherwise the obstacle shades the sound, which means that diffraction and shading are frequency dependent as shown in Figure 5. For further information, Ballou (2002) may provide helpful information.

Absorption is the loss of energy when a sound wave strikes a given material. It is specified by the absorption coefficient (examples given in Table 2), developed by Sabine (1922)

$$n \text{ dB} = 10 \log \frac{1}{1-\bar{a}}, \qquad (3)$$

where

n dB is the reduction in sound pressure level and
a the absorption coefficient.

FIGURE 5 Diffraction and shading of sound waves.

TABLE 2 Absorption coefficient chart

Material	125 Hz	250 Hz	500 Hz	1 kHz	2 kHz	4 kHz
Carpet	0.01	0.02	0.06	0.15	0.25	0.45
Concrete (sealed or painted)	0.01	0.01	0.02	0.02	0.02	0.02
Theater seats (wooden, empty)	0.03	0.04	0.05	0.07	0.08	0.08
Theater seats (wooden, fully occupied)	0.50	0.30	0.40	0.76	0.80	0.76
Glass (6 mm plate, large pane)	0.18	0.06	0.04	0.03	0.02	0.02
Fiberglass board (25 mm thick)	0.06	0.20	0.65	0.90	0.95	0.98
Fiberglass board (100 mm thick)	0.99	0.99	0.99	0.99	0.99	0.99
Plasterboard (12 mm in suspended ceiling grill)	0.15	0.11	0.04	0.04	0.07	0.08
Metal deck (perforated channels, 75 mm batts)	0.73	0.99	0.99	0.89	0.52	0.31
People, adults (per 1/10 person)	0.25	0.35	0.42	0.46	0.50	0.50
Ventilating grilles	0.30	0.40	0.50	0.50	0.50	0.40
Water or ice surface	0.01	0.01	0.01	0.02	0.02	0.03

In an area fully occupied by audience on wooden theater seats, the absorption coefficient is around 0.76 for high and mid frequencies, whereas an area with wooden and empty theater seats has an absorption coefficient below 0.1. The attendance of audience may dramatically change the absorption coefficient for at least one boundary (floor or seating areas) and therefore the impulse response of the venue will vary. The absorption coefficient of the various portions of boundaries are averaged arithmetically by using the respective weighting factors.

$$\bar{\alpha} = \frac{1}{S}\sum S_i \alpha_i, \qquad (4)$$

where

S_i is the area of the portion of the boundary, and
α_i is the absorption coefficient of the boundary portion.

All boundary reflections of a room will result in reverberation. There are several equations for calculating the reverb time based on the absorptiveness of a given room. The Sabine formula (5) gives the best correspondence with $\bar{\alpha} < 0.2$, the Eyring formula (6) (Eyring, 1930) should be used for well-behaved rooms having $\bar{\alpha} > 0.2$, and the Millington–Sette formula (7) (Millington, 1932; Sette, 1932) should be used for rooms with a wide variety of absorption coefficients.

$$T = \frac{0.16V}{S\bar{\alpha} + 4mV}, \qquad (5)$$

$$T = \frac{0.16V}{-S\ln(1-\bar{\alpha}) + 4mV}, \quad (6)$$

$$T = \frac{0.16V}{\sum -S_i \ln(1-\bar{\alpha}_i) + 4mV}, \quad (7)$$

where

T is the decay time in seconds for 60 dB level reduction,
S_i are the surface areas in m²,
V is the volume in m³,
$\bar{\alpha}_i$ are the average absorption coefficients of the individual surfaces,
m is the attenuation constant of air which is responsible for the attenuation of sound traveling through air. For small rooms, the term $4mV$ related to air absorption can be neglected

When a continuous sound hits a reflective boundary (hard boundary surface) the reflected sound wave combines with subsequent incoming waves. In case the wave crests, i.e., maximum pressure coincide, they combine and reinforce each other. The troughs, i.e., minimum pressure also combine. This results in a stationary pattern in air. It will consist of nodes (low pressure, but high velocity of air) and antinodes (high pressure, but low velocity of air), resulting in very soft and very loud sound areas, respectively. The same effect will occur for integral multiples of the frequency. However, the effect of standing waves is most prominent for low frequencies. Absorption materials have to be placed at the nodes (lambda 1/4 distance from the boundary) for best absorption results. Further ideas for room design for minimizing stationary patterns can be found in Cox et al. 2004.

A good sound system shall deliver a sound as clear as possible and it should allow as much control over the sound as possible. Therefore, the critical distance has to be maximized. It is the distance from the loudspeaker, where the sound pressure level of the direct sound and the sound pressure level of the reverberant sound are equal. When moving further away from the source toward the respective boundary, the reverberant sound field will predominate. The critical distance can be enlarged when using directional loudspeakers instead of omnidirectional ones. Further techniques to maximize the critical distance are discussed in Davis and Jones 1989.

5 • Sound Outdoors

Atmospheric factors of interest especially for outdoor venues are wind, temperature, and humidity. Wind effects are divided into two classes: velocity effects and gradient effects.

Wind velocity: Crosswind can change the direction of sound as it adds a velocity vector to the propagated sound wave. Sound will appear to come from a different location. Tail or head wind (related to the PA) will propagate greater or less distances respectively.

Wind velocity gradient: This effect occurs in case there are two different layers of air next to each other and air layers are moving with different speeds. This might happen in case the audience is shielded by a wall or a barrier of trees. In case wind passes through a velocity gradient, velocity will add a vector to the propagated sound wave. As shown in Figure 6, let's assume that the upper layer is with higher velocity wind than the lower layer. Then sound which is propagated against the wind will be refracted upward, whereas sound propagated with the wind, will be refracted downward. The actual effect of wind is minimal as the wind speed is negligible compared to the speed of sound.

Temperature gradients: The speed of sound is dependent on the temperature of the air[14]

$$c = (331.4 + 0.607T), \qquad (8)$$

where

c is the sound speed in meters per second,
T is the temperature in degrees Celsius.

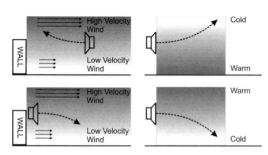
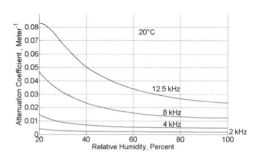

FIGURE 6 Effect of wind and temperature gradient, and absorption of sound in air, values taken from Harris (1966).

---

[14]Linearized formula shows an error of ±0.2% within +40° and −20°C.

When an acoustic wave passes through a temperature gradient, it will bend toward the cooler air, as shown in Figure 6.

This phenomenon occurs mainly during the hot daytime (sound will bend upward) and in the evening hours (sound will bend downward). This fact has to be noticed when tuning a PA during the afternoon for an evening concert.

Humidity: The medium which sound waves travel through will absorb the energy from the sound and attenuate it. The attenuation of sound in air is affected by the relative humidity. However, the effect will occur only for frequencies above 2 kHz.

6 • Measurement Methods

The measurement of a PA system should always be an essential part of the initial operation, whether it is just a simple check with well trained ears or a set of measured impulse responses. The goal of these measurements is always the same:

- Provide a homogenous sound field for the whole listening area
- Optimize the frequency response at all positions in the listening area
- Reach the maximum gain-before-feedback ratio
- Compensate delays
- Avoid reflections

Some of the commonly used measurement software are MLSSA, WINMLS, SMAART, ETF5, EASERA, Aurora, or the WaveCapture series. Besides measuring impulse responses, these softwares are capable of estimating most of the following parameters: SPL, RT, RASTI, STI, C80, and D50. The measurement systems use special measurement signals like maximum length sequences (MLS), swept sines, pink noise, impulses, or wobbles. It has to be made sure that the PA system and the ears of persons attending the measurement procedure can handle these signals without taking any damage.

An optimal measurement and tuning procedure may look as follows:

- Measure and compensate the delays
- Estimate the mean frequency response out of many measurement positions
- Transfer the filter function gained from the inverse mean frequency response into the sound system
- Measure again the frequency response, especially at former measurement positions with unwanted results, and tune the sound system again if necessary
- Measure the feedback path of the relevant microphones

- Tune the PA system with the results obtained by the feedback measurement
- In case of commercial applications, document the complete procedure, measurement results and steps taken to optimize the sound system

The whole idea of technical measurement has a major drawback: Ears do not correlate with the specs. Measurements may help finding faults or bad frequency shapes, but it is essential to do a final check of the sound system with an expert listener with trained ears.

7 • Special Issues on Signal Processing for PA Systems

There are some signal processing topics which are very closely related to PA systems and which are rarely found within other fields of acoustics:

- Loudspeaker time alignment
- Feedback reduction or gain-before-feedback boost
- Automatic maintenance of room equalization

The *loudspeaker time alignment* is necessary to avoid comb-filter effect sound. As shown in Figure 7, in case there are two sources with same signal and same amplitude but different delay time, the reinforcement and cancelation frequencies depend on the delay time between the arrival of the sound of the first source and the second source.

The cancelation occurs at the frequencies $f = 2x + 1/2t$, where $x = 0, 1, 2, 3, ..$ and t is the delay time in seconds. The reinforced frequencies occur at the frequencies $f = x/t$, where $x = 0, 1, 2, 3, ...$ and t is the delay time in seconds. The reinforced frequencies

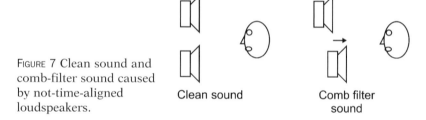

FIGURE 7 Clean sound and comb-filter sound caused by not-time-aligned loudspeakers.

Clean sound   Comb filter sound

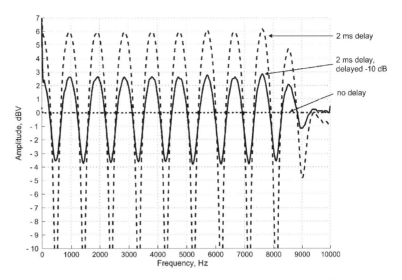

FIGURE 8 Frequency response of two loudspeakers without delay, with a delay of 2 ms and equal SPL, and with a delay of 2 ms, when the delayed loudspeaker is attenuated by 10 dB SPL.

increase in amplitude by 6 dB, while the out-of-phase frequencies cancel to $-\infty$ dB. In case the delayed loudspeaker is reduced in SPL compared to first loudspeaker, e.g., 10 dB as shown in Figure 8, the comb-filter effect will be reduced.

Comb-filter effects are unavoidable to some degree in most sound systems, and they cannot be corrected with equalization. Fortunately, most comb-filter problems can be minimized by carefully synchronizing the signals and reducing the amplitude of the delayed signal. Comb-filter effects occur mostly in speaker clusters, when different types of speakers are combined, e.g., longthrow and downfill or in case the speakers are not fixed properly within the cluster.

The delay time between main PA and delay line speaker as shown in Figure 9 will always be a compromise as the geometry will not allow an exact synchronization of signals for every location behind the delay line. For spoken word programs, the signal from the delay line speaker should arrive within 10 ms of the signal from the center cluster. For programs that are mostly music, one can allow a little more reverberation which means a little bit higher τ. Finally, comb-filter effects have to be reduced by time and level alignment.

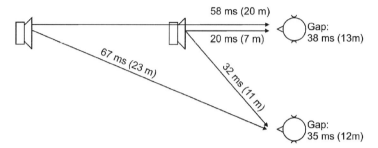

FIGURE 9 Problem of different delay times of two distributed loudspeakers related to the seating position.

For a first estimation one may use

$$\text{delay [ms]} = 1,000 \frac{\text{distance [m]}}{344[\frac{m}{s}]}. \tag{9}$$

The problem of *feedback reduction or gain-before-feedback boost* occurs for the house systems as well as for the monitoring systems. In case the gain of the feedback loop is ≥ 1 and the phasing of the system matches, the system gets instable. The positive feedback of the sound system limits the maximum gain. It is the intention to raise the point of self-excitation to higher gains without colorizing the sound too much. There are several techniques to battle this problem.

- The use of *directional sound sources* and directional microphones placed in geometric favorable positions;
- The use of *boundaries with high absorption coefficient* which will be hit directly by the amplified sound;
- A *slow-phase modulation* of the audio signal sent to the amplifier avoids a match of phases while the feedback loop is ≈ 1. This technique is only suitable for speech, not for music;
- The *frequency shifting* goes back to Schroeder (1961). After each run through the feedback loop, the frequency will be shifted by 1–3 Hz. Here as well, the phases do not match while the feedback loop is ≈ 1. The peaks of the feedback loop transfer function are canceled out by averaging. Also this approach is not suitable for musical signals as the instruments sound detuned for each run through the feedback loop;
- For onstage personal monitoring applications, one may *invert the signals* (constant phase shift of 180°) between two monitor wedges which point to a single performer. The ear is not able to detect a frequency shift between left and right ear, but the sound field distributed from the two monitor wedges around

the microphone cancels itself up to a certain degree. The comb-filter effect is negligible, as long as the actor stays within the sweet spot;
- Another approach for onstage feedback reduction is the *insertion of short delays* (e.g., 5–20 ms) for each monitor-send path. This reduces the danger of "hot spots" on stage but may cause timing problems for professional musicians, especially within the rhythm section. Here as well, the effect of comb filters is negligible within most setups.
- By *equalizing,* one may cut the peaks of the feedback loop transfer function. This can be done by manual setting of graphic or parametric peaking EQs, however it will colorize the sound. The automatic feedback reduction algorithms usually use notch filters between 1/5 and 1/80 octave. The general problems of equalizing the signal to increase the gain before feedback are discussed by McPeek (2000). There are two main approaches for feedback detection:

  a. *Level comparison.* If a certain frequency band (1/10–1/100 octave bandwidth) raises in volume to a more-or-less consistent level or disproportionate level, the algorithm assumes that there is a feedback within this frequency band. This algorithm fails for long constant pure tones, like a constant string sound.
  b. *Waveform analysis.* This approach is based on the fact that the waveform signature of a feedback in almost all instances differs quite much from the rest of the audio signal. The algorithm compares the signal with different feedback signatures trying to find correlations. In case the correlation is above a certain threshold, the algorithm assumes that a feedback related to the waveform is stored in the feedback signature database. This algorithm requires more processing power than the level comparison algorithm.

A good feedback reduction algorithm can provide an additional 6–9 dB of overall system gain before feedback.

The *automatic maintenance of room equalization* may be necessary for venues which change their frequency response remarkably with the attendance of audience or during the performance due to the change of humidity. Therefore, the automated EQ is connected to a measurement microphone, trying to keep the actual frequency response close to the predefined frequency response between the PA system and the measurement microphone. Therefore, the signal is bandpass-filtered related to the frequency to be analyzed. The EQ inserts an inaudible measurement signal related to the frequency to be measured, which is typically 60 dB below the average program level. Meanwhile, a reciprocal filter is placed in the measurement microphone path, removing all program material, so that the reference signal is remaining (Sabine 2001).

References

Ahnert, W. and Steffen, F. (1999). *Sound reinforcement engineering, fundamentals and practice*. E & FN Spon, GB-London.

Alldrin, L. (1997). *The home studio guide to microphones*. Mix Books, CA-Emeryville.

Ballou, G. M. (2002). *Handbook for sound engineers, Third Edition*. Focal Press, MA-Boston.

Blauert, J. (2000). *Lecture notes, communication acoustics, only in German language, (original title: "Akustik 2, Kommunikationsakustik")*. Ruhr-University Bochum, D-Bochum.

Cox, T. J., Antonio, P. D., and Avis. M. R. (2004). Room sizing optimization at low frequencies. *Journal of the Audio Engineering Society*, 52(6):640–651.

Cross, C. (2005). www.beatles-discography.com.

Davis, D. and Davis, C. (1997). *Sound system engineering, Second Edition*. Focal Press, MA-Boston.

Davis, G. and Jones, R. (1989). *The sound reinforcement handbook*. Hal Leonard Corporation, WI-Milwaukee.

Eyring, C. F. (1930). Reverberation time in dead rooms. *Journal of the Acoustical Society in America*, 1:217–249.

Harris, C. M. (1966). Absorption of sound in air versus humidity and temperature. *Journal of the Acoustical Society in America*, 40(1): 148–159.

Huber, D. M. and Williams, P. (1922). *Professional microphone techniques*. Primedia Intertec Publishing Corporation, CA-Emeryville.

IEC60268-16:1998 (1998). Sound system equipment – part 16: Objective rating of speech intelligibility.

IEC61672-1:2002 (2002). Electroacoustic – sound level meters, part 1.

ISO3382:1997 (1997). Acoustics – measurement of the reverberation time of rooms with reference to other acoustical parameters.

McPeek, R. (2000). www.svconline.com.

Millington, G. (1932). A modified formula for reverberation. *Journal of the Acoustical Society in America*, 4:69–82.

Pawera, N. (2003). *Microphones in practical use, only in German language (original title: "Mikrofonpraxis, Tipps und Tricks für Bühne und Studio")*. PPV Medien GmbH, D-Bergkirchen.

Sabine (2001). *Real Q2 operating guide, Version 7.5*. Sabine, Inc., FL-Alachua.

Sabine, W. C. (1922). *Collected papers in acoustics*. Harvard University Press, MA-Cambridge.

Schroeder, M. R. (1961). Improvement of feedback stability of public address systems by frequency shifting. *In audio engineering society preprint, 13th annual meeting, New York*.

Sette, W. J. (1932). A new reverberation time formula. *Journal of the Acoustical Society in America*, 4:193–210.

Stark, S. H. (1989). *The sound reinforcement handbook, Second Edition*. Mix Books, CA-Vallejo.

Trubit, R. (2000). *Compact mixers, Second Edition*. Hal Leonard Corporation, WI-Milwaukee.

Urban, M. Heil, C. and Bauman, P. (2003). Wavefront sculpture technology. *Journal of the Acoustical Society in America*, 51(10): 912–932.

Yaxley, K. (2003). www.historyofpa.co.uk.

35
Beamforming for Speech and Audio Signals

Walter Kellermann

Multimedia Communications and Signal Processing, University Erlangen-Nuremberg, Erlangen, Germany, e-mail: wk@lnt.de

If microphone arrays instead of a single microphone are employed for sampling acoustic wavefields, signal processing of the sensor data can exploit the spatial diversity to better detect or extract desired source signals and to suppress unwanted interference. Beamforming represents a class of such multichannel signal processing algorithms and suggests a spatial filtering which points a beam of increased sensitivity to desired source locations while suppressing signals originating from all other locations. While beamforming techniques are also extensively used in other areas, e.g. in underwater acoustics, ultrasound diagnostics, and radio communications [1–3], the treatment is concentrating here on wideband acoustic signals in the audio frequency range.

In the following, first the basic concept is introduced and then several classes of beamforming techniques are outlined: Common data-independent beamformers form the most popular class for all areas of beamforming applications, whereas differential microphone arrays are especially attractive for wideband audio signals. Statistically optimum beamformers are able to better match given noise and interference scenarios and to follow its changes over time. Finally, blind source separation of convolutive mixtures is linked to beamforming. Please note that in view of the long tradition and the wealth of literature, relatively few original references are cited, and often survey papers are referenced instead, which provide the links to the original sources.

1 • The Basic Concept

In a general scenario as shown in Figure 1, an acoustic wavefield is sensed by M microphones at positions $\mathbf{r}_\mu$, $\mu = 1,\ldots,M$. The continuous-time microphone signals $x_{c,\mu}(\mathbf{r}_\mu,t)$ will in general capture components from a point source of interest, $s(t)$, at position $\mathbf{r}_S$ radiating a wavefield $s(\mathbf{r},t)$, and additive noise components $n_\mu(t)$, which may represent coherent interferers, a diffuse noise field, or sensor noise. After A/D conversion, M sampled microphone output signals $x_\mu(k)$ are available as inputs for the beamformer, with k as the discrete time in integer multiples of the sampling interval $T_s = 1/f_s$ (sampling frequency f_s). Considering a generic beamformer for real-valued acoustic signals, each digitized microphone signal $x_\mu(k)$ is filtered by a possibly time-variant linear filter with real-valued impulse response $w_\mu(i,k)$:

$$y(k) = \sum_{\mu=1}^{M} \sum_{i=-\infty}^{\infty} w_\mu(i,k) x_\mu(k-i). \qquad (1)$$

As a typical characterization, the response of the beamformer to a harmonic plane wave [4,5]

$$s(\mathbf{r},t) = S_0 \cdot e^{j(\omega t - \mathbf{k}\cdot\mathbf{r})} \qquad (2)$$

is considered, where S_0 is the scalar pressure amplitude, ω is the temporal frequency, and the wave vector $\mathbf{k}$ describes the direction of propagation of the wave and its spatial frequency $|\mathbf{k}| = 2\pi/\lambda = \omega/c$ (with λ representing the wavelength and c the sound velocity,

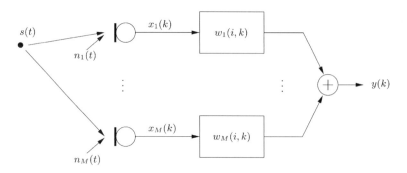

FIGURE 1 Scenario for acoustic beamforming.

respectively). Inserting Eq. (2) for the microphone positions $\mathbf{r}_\mu$, the array output reads

$$y(k) = S_0 \cdot e^{j\omega k T_s} \cdot \sum_{\mu=1}^{M} \sum_{i=-\infty}^{\infty} w_\mu(i,k) e^{-j(\omega i T_s + \mathbf{k} \cdot \mathbf{r}_\mu)}$$

$$:= S_0 \cdot e^{j\omega k T_s} \cdot W(\mathbf{k}, \omega, k), \qquad (3)$$

with $W(\mathbf{k}, \omega, k)$ representing the – possibly time-varying – *array response* to a harmonic plane wave with wave vector $\mathbf{k}$.

For the wavefield of a point source S arriving at the microphone array, the plane wave assumption (*far-field assumption*) is said to be valid if the aperture width, i.e. the geometric extension of the array as seen by the impinging wavefront, R, and the distance from the source to the array centre, r_S, fulfill $|r_S| > 2R^2/\lambda$.

Essentially, most beamforming algorithms can be viewed as techniques that try to extract a sampled and delayed but undistorted version of the signal of interest $s(t)$ from $y(k)$ while at the same time suppressing the undesired noise contributions $n_\mu(t)$.

2 • Data-Independent Beamforming

The most intuitive and simple approach to beamforming is to coherently add the desired components in the observations expecting that thereby the undesired noise components cancel out to a certain extent. Assuming free-field conditions and that only the desired source S located at $\mathbf{r}_S$ is active, the continuous-time sensor signals can be written as $x_{c,\mu}(\mathbf{r}_\mu, t) = a_\mu s(t - \tau_\mu)$, where a_μ is a constant gain and τ_μ is a delay, both depending on the Euclidean distance between the source and the μth microphone $\|\mathbf{r}_\mu - \mathbf{r}_S\|$. For coherent superposition of the desired source components, the beamforming filters are time-invariant and only have to equalize the delays $\tau_\mu = \|\mathbf{r}_\mu - \mathbf{r}_S\|/c$ (*delay-and-sum beamformer*). If all the relative delays between the various sensors are integer multiples of the sampling interval T_s, then all filter impulse responses reduce to time-shifted unit impulses $w_\mu(i) = \delta(i - i_\mu)$. Noninteger delays, however, require interpolation sequences as impulse responses, which are typically approximations of $w_\mu(i) = \text{sinc}\left[\pi(i + (\tau_\mu - \tau')/T_s)\right]$ [6], where $\text{sinc}(x) = (\sin x)/x$ and $\tau' \geq \max_\mu\{\tau_\mu\}$ is a necessary common delay if causality of all $w_\mu(i)$ must be assured.

If all the desired source components are aligned perfectly, if the gains a_μ are equal, and if the noise components $n_\mu(t)$ have the same level at each microphone and are mutually orthogonal, then the *array gain*, i.e. the ratio of desired source power relative to the noise power compared to the ratio at a single microphone, of the *delay-and-sum beamformer* is equal to the number of microphones M. If it is assumed that the noise is due to a spherically isotropic noise field, then the logarithmic array gain, $10\log_{10} M$, is termed *directivity index* and corresponds to the logarithmic ratio of the source power from the steering direction relative to the noise power integrated over all directions [1–3].

As a widely-used special case, a linear equispaced array is considered as shown in Figure 2, where the microphone positions are given by multiples of the spacing d along the x axis. For a planar wave according to Eq. (2), the $M = 2K+1$ sampled microphone signals $x_\kappa(k)$, $\kappa = -K,\ldots,K$, are given by

$$x_\kappa(k) = s(\mathbf{r}_\kappa, kT_s) = S_0 \cdot \exp\left[j\omega(kT_s + \kappa\frac{d}{c}\cos\Theta)\right]. \quad (4)$$

Assuming now that the time-invariant beamforming filters realize constant gains w_κ and causal delays $\tau_\kappa = \tau' + \kappa\frac{d}{c}\cos\Theta_0$, the output of the beamformer reads

$$y(k) = S_0 \cdot e^{j\omega(kT_s-\tau')} \sum_{\kappa=-K}^{K} w_\kappa \exp\left[j\kappa d\frac{\omega}{c}(\cos\Theta - \cos\Theta_0)\right]. \quad (5)$$

If for simplicity, $\tau' = 0$ (allowing noncausal delays), the *array response* reads with $u = \cos\Theta$

$$W(u,\omega) = \sum_{\kappa=-K}^{K} w_\kappa \exp\left[j\kappa d\frac{\omega}{c}(u - u_0)\right] \quad (6)$$

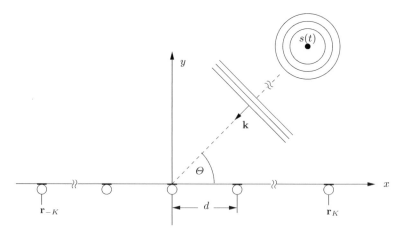

FIGURE 2 Point source in the far field and a linear equispaced array with $M = 2K+1$ sensors.

and shows that frequency ω and spacing d can be interchanged: Doubling the frequency ω and simultaneously halving the spacing d yield the same array response. In Figure 3, the frequency-independent gain can be seen for the mainlobe at $u = 0$ ($\Theta_0 = \pi/2$), but periodic repetitions of the mainlobe appear at $u = \pm 1$ for frequencies above 5,000 Hz. This is due to the periodicity of the array response in u with $W(u, \omega) = W(u + \nu \cdot \frac{2\pi c}{\omega d}, \omega)$ ($\nu \in \mathbb{Z}$) and is called *spatial aliasing*. To avoid it for all steering angles Θ_0, $u \pm \frac{2\pi c}{\omega d}$ must be outside the *visible range* of u, $|u| \leq 1$, which requires $\frac{2\pi c}{\omega d} > 2$, i.e. $d < \lambda/2$.

With regard to the choice of w_κ, it is interesting to note that the array response Eq. (6) can be equated to an FIR filter with frequency response

$$H(\omega) = \sum_{\kappa=-K}^{K} h_\kappa e^{-j\omega\kappa T_s}. \qquad (7)$$

While Eq. (7) implies a filtering along a uniformly sampled time axis (at multiples of T_s), Eq. (6) implies FIR filtering along a uniformly sampled spatial coordinate axis (at multiples of d). This means that for a given frequency, the magnitude of the array response (*beampattern*) as a function of u directly corresponds to the magnitude response of the corresponding FIR filter. As a consequence, besides the popular choices of a rectangular

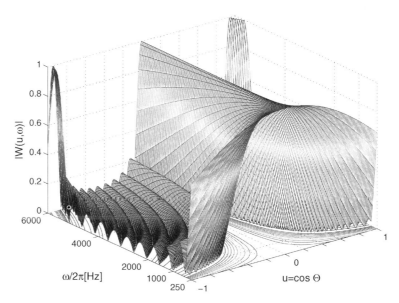

FIGURE 3 Array response (*beam pattern*) for a linear equispaced array with $M = 11$ sensors, $d = 6$ cm, $w_\kappa = 1/M$.

window or Dolph-Chebyshev windows [7], many other window types known from FIR filtering or spectrum analysis [8] are directly applicable to beampattern design. For beamforming, the various windows define different tradeoffs between mainlobe width and sidelobe suppression in the angular domain, just as they do in the frequency domain for spectrum analysis.

Obviously, the filters w_κ are not limited to realizing constant gains and delays only, but may be general linear filters (*filter-and-sum beamforming*), so that the response of a time-invariant beamformer for a linear equispaced array to a plane wave reads in accordance with Eq. (1):

$$W(u,\omega) = \sum_{\kappa=-K}^{K} \sum_{i=-\infty}^{\infty} w_\kappa(i) \exp\left[j(\omega\kappa\frac{d}{c}(u-u_0) - \omega i T_s)\right]. \quad (8)$$

From this it is obvious that the design of a filter-and-sum beamformer for plane waves impinging on linear equispaced arrays is equivalent to a 2D-FIR filter design. As a typical design goal with wideband acoustic signals, filter-and-sum beamformers will aim at a frequency-independent beamwidth for the mainlobe steered towards the desired source. The according design methods [9–11] can well overcome the narrowing of the mainlobe at high frequencies (cf. Figure 3). However, regarding low frequencies, the minimum beamwidth for any given frequency, as, e.g. measured by the first zero in the beampattern off the mainlobe, is always obtained if, at this frequency, the gain for all microphone signals is equal. Therefore, the widening of the beam towards low frequencies can only be overcome if the aperture size is increased, that is, the spacing d is increased, or more microphones are added. For audio and speech applications, this led to the concept of nested arrays, which foresee one microphone array per octave while doubling the spacing d from one octave to the next [4,12].

Although linear arrays are still prevalent for practical applications, the concept of *delay-and-sum* beamforming has been extended from linear arrays to two-dimensional (2D) planar arrays very early [4], and multiple 2D arrays were proposed to realize 3D beamforming [4,13].

3 • Differential Microphone Arrays

The idea of differential microphone arrays is to approximately measure spatial derivatives of the sound pressure by pressure differences between closely spaced locations in the sound field instead of the pressure itself, and thereby obtain better directivity with small aperture arrays even for low frequencies, which is not possible for ordinary *filter-and-sum* beamformers with larger microphone spacings. For illustration of the principle, the linear array as given by Eq. (5) is considered for only two sensors, $\kappa = 0, -1$, and opposite weights $w_0 = 1, w_{-1} = -1$. Then, the output for a plane wave is given by

$$y(k) = S_0 \cdot e^{j\omega(kT_s - \tau')} \left(1 - \exp\left[-jd\frac{\omega}{c}(\cos\Theta - \cos\Theta_0)\right]\right). \quad (9)$$

If the spacing d is now much smaller than the wavelength, $d \ll \lambda$, the truncated Taylor series $e^x \approx 1 + x$ can be used and the array response reads for $\tau' = 0$:

$$W(u, \omega) = j\omega\frac{d}{c}(\cos\Theta - \cos\Theta_0) = j\omega\frac{d}{c}(u - u_0). \quad (10)$$

Obviously, this *first-order differential array* exhibits the same first-order highpass characteristic for all angles of arrival Θ, and has frequency-independent zeroes at $\Theta = \pm\Theta_0$. For any given frequency, the magnitude of the array response is maximum for $\Theta = 0$ or $\Theta = \pi$ depending on the sign of $\cos\Theta_0$, that is, the maximum sensitivity coincides always with the array axis (*endfire array*). The directivity index assumes its maximum of 6.02 dB for $u_0 = \cos 109^0 \approx 1/3$ (hypercardioid, see [14]) and is greater than for a delay-and-sum beamformer for all values of u_0 in the visible range, which leads to the notion of *superdirective beamforming*. The highpass characteristic can be easily equalized by a matching lowpass filter applied to $y(k)$. However, it should be noted that, thereby, spatially uncorrelated low-frequency noise is amplified accordingly. Higher-order differential arrays using $M > 2$ microphones for approximating higher-order derivatives have been investigated in [5,14] where it is shown that the maximum possible directivity index is $10 \log_{10} M^2$ for a differential array of order $M - 1$. However, the order of the equalizing lowpass filters has to increase accordingly, which aggravates the *noise susceptibility* of the array and thereby limits the practical use of this concept currently to orders less than four, especially as any mismatch between the individual spatio-temporal transfer functions of the microphones has an equivalent effect as additive noise [14]

4 • Statistically Optimum Beamformers

In order to optimally suppress all interferers and noise sources, it is highly desirable that, especially for broadband signals and complex noise scenarios as typical for audio and speech applications, beamformers match the statistics of the involved signals. Two classes are in common use: One is based on minimizing a squared error criterion based on the difference $e(k)$ between the beamformer output $y(k)$ and a reference signal $y_{\text{ref}}(k)$. Minimizing the expectation of the mean squared error $\mathcal{E}\{e^2(k)\}$ produces a *minimum mean squared error* (MMSE) estimate for the optimum beamforming filters [1,3,15,16]. In its deterministic form, it is often called *least-squares error (LSE) beamforming* if a weighted sum of squared error samples is minimized instead of an estimated mean squared error. This concept can be viewed as a multichannel Wiener Filter which not only accounts for the second-order statistics within each channel but also optimizes with regard to the coherence between the channels. In practice, the main problem in this context is to obtain a useful reference signal $y_{\text{ref}}(k)$ which may represent either the desired signal or the noise component in the beamformer output [15–17].

The second class minimizes the squared output of the beamformer subject to a linear constraint which essentially tries to preserve the integrity of the desired signal. This approach is called *linearly constrained minimum variance (LCMV) beamforming* [18]. A special case is the *minimum variance distortionless response* (MVDR) *beamformer* which assures a distortionless response for a predefined steering direction Θ_0 of the array and is especially important in acoustic signal processing [15,17]. In Figure 4, an example for an MVDR beamformer is shown: A broadband interferer impinging from $\Theta = 0.37\pi$ is strongly rejected except for the very low frequencies, whereas the desired source assumed at $\Theta = 0.5\pi$ will be left undistorted.

For implementing the MVDR beamformer, the *generalized sidelobe canceller (GSC)* [19] represents an efficient choice, as here, the constraint is assured by a generally simple data-independent structure (a fixed beamformer and a so-called blocking matrix), so that the data-dependent statistical optimization reduces to an unconstrained interference canceller. This also gives way to very efficient adaptive implementations that are able to continuously follow nonstationary signals and time-varying acoustic environments. As, however, the GSC is very vulnerable to movements of the desired sources, robust versions with adaptive blocking matrices have been developed for its application to speech and audio [17,20].

The filter characteristics $w_\mu(i,k)$ of the above statistically optimum beamformers are determined by the power spectral densities of the desired source and the noise sources, respectively, and by the spatial and temporal coherence of the noise. Typically, the performance in terms of noise and interference suppression will exceed that of a fixed filter-and-sum beamformer especially for coherent interferers, as its flexibility allows for

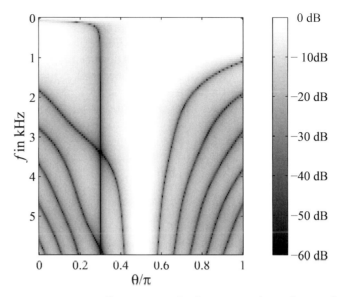

FIGURE 4 Array response (*beam pattern*) of an MVDR beamformer for a linear equispaced array with $M = 9$ sensors, $d = 4$ cm to an interferer emitting white noise from $\Theta = 0.3\pi$ (with permission from [17]).

signal-dependent suppression of certain spatial regions. For low frequencies, where the sensor spacing is relatively small compared to the wavelength, the optimum solutions will often be similar to those for differential arrays, so that constraints on the noise susceptibility may be advisable, whereas for high frequencies the filters will approach *filter-and-sum* characteristics [15,17].

5 • Blind Source Separation as Blind Beamforming

As a relatively recent development, blind source separation (BSS) has been introduced to acoustic applications, where it can be understood as a kind of blind beamforming [21,22]. As opposed to the techniques described in the previous paragraphs, it does not presume any knowledge about the spatial positions of desired sources and interferers (*blindness*), but only assumes statistical independence or at least uncorrelatedness between several sources. In a typical quadratic configuration, with M acoustic sources $s_\nu(t)$ and M microphones, a so-called demixing system will produce M outputs $y_\nu(k)$, and each output should contain only one of the source signals. For each output, the demixing is a linear time-variant multichannel filtering operation as described by Eq. (1). This filtering can be interpreted as an adaptive beamformer, which treats one of the sources as desired source and tries to suppress all other competing sources as unwanted interferers. For identifying the optimum demixing filters, cost functions are minimized, which capture statistical dependencies between all M outputs of the separation system, so that the optimum demixing matrix produces M mutually independent or at least uncorrelated outputs [23–25].

6 • Concluding Remarks

In this brief treatment of beamforming techniques for acoustic wideband signals such as speech and audio, fundamental concepts have been emphasized and recent advanced developments have only been outlined. The proportions are meant to reflect current practical importance; however, the author is convinced that statistically optimum adaptive beamformers and blind beamforming algorithms will gain more importance in the near future.

References

1. B.D. Van Veen and K.M. Buckley. Beamforming: A versatile approach to spatial filtering. *IEEE ASSP Magazine*, 5(2):4–24, April 1988.
2. D.H. Johnson and D.E. Dudgeon. *Array Signal Processing: Concepts and Techniques*. Prentice Hall, Englewood Cliffs, NJ, 1993.
3. H.L. van Trees. *Optimum Array Processing*. Wiley, New York, NY, 2002.
4. J.L. Flanagan, J.D. Johnston, R. Zahn, and G.W. Elko. Computer-steered microphone arrays for sound transduction in large rooms. *Journal of Acoustical Society of America*, 78(5):1508–1518, November 1985.
5. G. Elko. Microphone array systems for hands-free telecommunication. *Speech Communication*, 20:229–240, 1996.
6. T. Laakso, V. Välimäki, M. Karjalainen, and U.K. Laine. Splitting the unit delay. *IEEE Signal Processing Magazine*, 13(1):34–60, January 1996.
7. C.L. Dolph. A current distribution for broadside arrays which optimizes the relationship between beamwidth and sidelobe level. *Proceedings of the IRE*, 34:335–348, June 1946.
8. F.J. Harris. Use of windows for harmonic analysis. *Proceedings of the IEEE*, 66(1):51–83, January 1978.
9. M. Goodwin. Constant beamwidth beamforming. *Proceedings of the International Conference on Acoustics, Speech, and Signal Processing*, I-169 to I-172, Minneapolis, MN. IEEE, April 1993.

10. T. Chou. Frequency-independent beamformer with low response error. *Proceedings of the International Conference on Acoustics, Speech, and Signal Processing*, 2995–2998, Detroit, MI. IEEE, May 1995.
11. D.B. Ward, R.A. Kennedy, and R.C. Williamson. Constant directivity beamforming. In M.S. Brandstein and D. Ward, editors, *Microphone Arrays: Signal Processing Techniques and Applications*, chapter 13, pp. 3–18. Springer, Berlin, May 2001.
12. J.L. Flanagan, D.A. Berkley, G.W. Elko, J.E. West, and M.M. Sondhi. Autodirective microphone systems. *Acustica*, 73:58–71, 1991.
13. H. Silverman, W.R. Patterson, J.L. Flanagan, and D. Rabinkin. A digital processing system for source location and sound capture by large microphone arrays. *Proceedings of the International Conference on Acoustics, Speech, and Signal Processing*, 251–254, Munich, Germany. IEEE, April 1997.
14. G. Elko. Superdirectional microphone arrays. In S. Gay and J. Benesty, editors, *Acoustic Signal Processing for Telecommunication*, pp. 181–238. Kluwer, Dordrecht, 2000.
15. J. Bitzer and K.U. Simmer. Superdirective microphone arrays. In M.S. Brandstein and D. Ward, editors, *Microphone Arrays: Signal Processing Techniques and Applications*, chapter 13, pp. 19–38. Springer, Berlin, May 2001.
16. S. Doclo and M. Moonen. GSVD-based optimal filtering for multi-microphone speech enhancement. In M.S. Brandstein and D. Ward, editors, *Microphone Arrays: Signal Processing Techniques and Applications*, chapter 6, pp. 111–132. Springer, Berlin, May 2001.
17. W. Herbordt and W. Kellermann. Adaptive beamforming for audio signal acquisition. In J. Benesty and Y. Huang, editors, *Adaptive Signal Processing: Application to Real-World Problems*, chapter 6, pp. 155–194. Springer, Berlin, January 2003.
18. O.L. Frost. An algorithm for linearly constrained adaptive array processing. *Proceedings of the IEEE*, 60(8):926–935, August 1972.
19. L.J. Griffiths and C.W. Jim. An alternative approach to linear constrained adaptive beamforming. *IEEE Transactions on Antennas and Propagation*, 30(1):27–34, January 1982.
20. O. Hoshuyama and A. Sugiyama. A robust adaptive beamformer with a blocking matrix using constrained adaptive filters. *Proceedings of the International Conference on Acoustics, Speech, and Signal Processing*, pp. 925–928. IEEE, 1996.
21. J.-F. Cardoso and A. Souloumiac. Blind beamforming for non-gaussian signals. *IEE Proceedings-F*, 140(6):362–370, December 1993.
22. L. Parra and C. Fancourt. An adaptive beamforming perspective on convolutive blind source separation. In G. Davis, editor, *Noise Reduction in Speech Applications*. CRC Press LLC, 2002.
23. J.-F. Cardoso. Blind signal separation: Statistical principles. *Proceedings of the IEEE*, 86(10):2009–2025, October 1998.
24. L. Parra and C. Spence. Separation of non-stationary natural signals. In S. Roberts and R. Everson, editors, *Independent Components*

Analysis, Principles and Practice, pp. 135–157. Cambridge University Press, 2001.
25. H. Buchner, R. Aichner, and W. Kellermann. Blind source separation for convolutive mixtures: A unified treatment. In Y. Huang and J. Benesty, editors, *Audio Signal Processing for Next-Generation Multimedia Communication Systems*, chapter 10, pp. 255–296. Kluwer, Dordrecht, 2004.

36

Digital Audio Recording Formats and Editing Principles

Francis Rumsey

Institute of Sound Recording, University of Surrey, Guildford, UK

This chapter summarises digital audio recording formats and the principles of digital audio editing. A brief overview of the most common storage formats is given followed by an introduction to audio editing technology and techniques. Digital audio conversion and encoding principles are covered elsewhere in this book and will not be described here.

1 • Digital Tape Recording

There are still a number of dedicated digital tape recording formats in existence, although they are being superseded by computer-based products that use removable disks or other mass storage media. Tape has a relatively slow access time because it is a linear storage medium. However, a dedicated tape format can easily be interchanged between recorders, provided another machine operating to the same standard can be found. Disks, on the other hand, come in a very wide variety of sizes and formats. Even if the disk fits a particular drive it may not be possible to access the audio files thereon, owing to the multiplicity of levels at which compatibility must exist between systems before interchange can take place.

1.1 Background to Digital Tape Recording

When commercial digital audio recording systems were first introduced in the 1970s and early 1980s, it was necessary to employ recorders with sufficient bandwidth for the high data rates involved (a machine capable of handling bandwidths of a few megahertz was required). Analogue audio tape recorders were out of the question because their bandwidths extended only up to around 35 kHz at best, so video tape recorders (VTRs) were often utilised because of their wide recording bandwidth. PCM (pulse code modulation) adaptors (Doi et al. 1978) converted digital audio data into a waveform which resembled a television waveform, suitable for recording on to a VTR. The Denon Company of Japan developed such a system in partnership with the NHK broadcasting organisation, and they released the world's first PCM recording on to LP (long-playing vinyl disk) in 1971. In the early 1980s, devices such as Sony's PCM-F1 became available at modest prices, allowing 16-bit, 44.1-kHz digital audio to be recorded on to a consumer VTR, resulting in widespread proliferation of stereo digital recording.

Dedicated open-reel digital recorders using stationary heads were also developed. High-density tape formulations were then manufactured for digital use, and this, combined with new channel codes, improvements in error correction and better head design, led to the use of a relatively low number of tracks per channel, or even single-track recording of a given digital signal, combined with playing speeds of 15 or 30 in. per second. Dedicated rotary-head systems, not based on a VTR, were also developed – the R-DAT format being the most well known.

Digital recording tape is thinner (27.5 μm) than that used for analogue recordings; long-playing times can be accommodated on a reel, but also thin tape contacts the machine's heads more intimately than does standard 50-μm thickness tape which tends to be stiffer. Intimate contact is essential for reliable recording and replay of such a densely packed and high-bandwidth signal.

1.2 Overview of Digital Tape Formats

There have been a number of commercial recording formats over the last 20 years, and only a brief summary will be given here of the most common ones.

Sony's PCM-1610 and PCM-1630 adaptors [11] dominated the CD-mastering market for a number of years, although by today's standards they use a fairly basic recording format and rely on 60 Hz/525 line U-matic cassette VTRs. The system operated at a sampling rate of 44.1 kHz and used 16-bit quantisation, being designed specifically for the making of tapes to be turned into CDs. Recordings made in this format could be electronically edited using the Sony DAE3000 editing system, and the playing time of tapes ran up to 75 min using a tape specially developed for digital audio use.

The R-DAT or DAT format (Ishida et al. 1985) is a small stereo, rotary-head, cassette-based format offering a range of sampling rates and recording times, including the professional rates of 44.1 and 48 kHz. Originally, consumer machines operated at 48 kHz to avoid the possibility for digital copying of CDs, but professional versions became available which would record at either 44.1 or 48 kHz. Consumer machines will record at 44.1 kHz, but usually only via the digital inputs. DAT is a 16-bit format, but has a non-linearly encoded long-play mode as well, sampled at 32 kHz. Truly professional designs offering editing facilities, external sync and IEC-standard timecode have also been developed. The format became exceptionally popular with professionals owing to its low cost, high performance, portability and convenience. Various non-standard modifications were introduced, including a 96-kHz sampling rate machine and adaptors enabling the storage of 20-bit audio on such a high sampling rate machine (sacrificing the high sampling rate for more bits). The IEC timecode standard for R-DAT was devised in 1990. It allows for SMPTE/EBU timecode of any frame rate to be converted into the internal DAT "running-time" code and then converted back into any SMPTE/EBU frame rate on replay.

The Nagra-D recorder was designed as a digital replacement for the world-famous Nagra analogue recorders and as such was intended for professional use in field recording and studios. The format was designed to have considerable commonality with the audio format used in D1- and D2-format digital VTRs, having rotary heads, although it used open reels for operational convenience (Kudelski et al. 1987). Allowing for 20–24 bits of audio resolution, the Nagra-D format was appropriate for use with high-resolution convertors. The error correction and recording density used in this format were designed to make recordings exceptionally robust, and recording time could be up to 6 h on a 7-in. (18 cm) reel, in two-track mode. The format is also designed for operation in a four-track mode at twice the stereo tape speed, such that in stereo the tape travells at $4.75\,\text{cm}\,\text{s}^{-1}$ and in four track at $9.525\,\text{cm}\,\text{s}^{-1}$.

The DASH (Digital Audio Stationary Head) format (Kogure et al. 1983) consisted of a whole family of open-reel stationary-head recording formats from two tracks up to 48 tracks. DASH-format machines operated at 44.1 or 48 kHz rates (and sometimes optionally at 44.056 kHz), and they allowed varispeed ±12.5%.

They were designed to allow gapless punch-in and punch-out, splice editing, electronic editing and easy synchronisation. Multitrack DASH machines gained wide acceptance in studios, but the stereo machines did not. Later developments resulted in DASH multitracks capable of storing 24-bit audio instead of the original 16 bits.

In more recent years, budget modular multitrack formats were introduced. Most of these were based on 8-track cassettes using rotary-head transports borrowed from consumer video technology. The most widely used were the DA-88 format (based on Hi-8 cassettes) and the ADAT format (based on VHS cassettes). These offered most of the features of open-reel machines and a number of them could be synchronised to expand the channel capacity.

1.3 Editing Digital Tape Recordings

Razor blade cut-and-splice editing was possible but risky on open-reel digital formats, and the analogue cue tracks could be monitored during these operations. The thin tape could easily be damaged during the cut-and-splice edit procedure making electronic editing far more desirable. A butt joint was used for the editing of digital tape. The discontinuity in the data stream caused by the splice would cause complete momentary dropout of the digital signal if no further action were taken; so circuits were incorporated which sensed the splice and performed an electronic crossfade from one side of the splice to the other, with error concealment to minimise the audibility of the splice.

Electronic editing normally required the use of two machines plus a control unit. A technique was employed whereby a finished master tape could be assembled from source tapes on player machines. This was a relatively slow process, as it involved real-time copying of audio from one machine to another, making modifications to the finished master difficult. The digital editor often stored several seconds of programme in its memory and this could be replayed at normal speeds or under the control of a search knob enabling very slow to-and-fro searches to be performed in the manner of rock-and-roll editing on an analogue machine. Edits could be rehearsed prior to execution. When satisfactory edit points had been determined, the two machines were synchronised using timecode, and the record machine switched to drop in the new section of the recording from the replay machine at the chosen moment. A crossfade was introduced between old and new material to smooth the join. The original source tape was left unaltered.

2 • Digital Recording on Computer Mass Storage Media

The increasing availability, performance and cheapness of computer mass storage media has led to their increasing use for digital audio storage. The majority of digital audio recording and editing is now carried out on computer-based workstations, and the days of dedicated audio formats are waning. It was most common in the early days of hard disk audio systems for manufacturers to develop dedicated systems with fairly high prices. This was mainly because mass-produced desktop computers were insufficiently equipped for the purpose, and because large-capacity mass storage media were less widely available than they are now, having a variety of different interfaces and requiring proprietary file storage strategies. Furthermore, the size of the market was relatively small to begin with, and considerable R&D investment had to be recouped. Nowadays, the average desktop computer is more than capable of handling multiple channels of digital audio, with storage capacity in abundance, even carrying out much of the audio signal processing on the host CPU if required. Alternatively, third-party audio DSP hardware can be employed for greater independence of desktop CPU power.

2.1 Storage Requirements of Digital Audio

Table 1 shows the data rates required to support a single channel of digital audio at various resolutions. Media to be used as primary storage would need to be able to sustain data transfer at a number of times these rates to be useful for audio purposes. The table also shows the number of megabytes of storage required per minute of audio. Storage requirements increase pro rata with the number of audio channels to be handled. Storage systems may use removable media but many have fixed media. It is advantageous to have removable media for audio purposes because it allows different jobs to be kept on different media and exchanged

TABLE 1 Data rates and capacities required for linear PCM

Sampling rate (kHz)	Resolution (bits)	Bit rate (kbit/s)	Capacity/min (Mbytes/min)	Capacity/h (Mbytes/h)
96	16	1,536	11.0	659
48	20	960	6.9	412
48	16	768	5.5	330
44.1	16	706	5.0	303
44.1	8	353	2.5	151

at will, but the use of removable media drives can compromise performance. An alternative is the use of removable drives where the whole mechanism including the storage surfaces can be unplugged and exchanged.

2.2 Disk Drives

Disk drives are probably the most common form of mass storage. They have the advantage of being random access systems – in other words, any data can be accessed at random and with only a short delay. This may be contrasted with tape drives, which only allow linear access, resulting in a considerable delay. Disk drives come in all shapes and sizes from the floppy disk at the bottom end to high-performance hard drives at the top end. The means by which data are stored is usually either magnetic or optical, but some use a combination of the two, as described below.

The general structure of a disk drive is shown in Figure 1. It consists of a motor connected to a drive mechanism, which causes one or more disk surfaces to rotate at anything from a few hundred to many thousands of revolutions per minute. This rotation may either remain constant or may stop and start, and it may either be at a constant rate or a variable rate, depending on the drive. One or more heads are mounted on a positioning mechanism which can move the head across the surface of the disk to access particular points under the control of hardware and software called a disk controller. The heads read data from and write data to the disk surface by whatever means the drive employs. It should be noted that certain disk types are read-only,

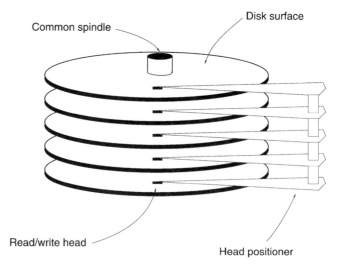

FIGURE 1

some are write-once–read-many (WORM), and some are fully erasable and rewritable.

The disk surface is normally divided up into tracks and sectors, not physically but by means of "soft" formatting (see Figure 2). Formatting places logical markers which indicate block boundaries, among other processes. On most hard disks, the tracks are arranged as concentric rings, but with some optical disks there is a continuous spiral track.

Disk drives look after their own channel coding, error detection and correction so there is no need for system designers to devise dedicated audio processes for disk-based recording systems. The formatted capacity of a disk drive is all available for the storage of "raw" audio data, with no additional overhead required for redundancy and error checking codes. "Bad blocks" are mapped out during the formatting of a disk, and not used for data storage. If a disk drive detects an error when reading a block of data, it will attempt to read it again. If this fails, then an error is normally generated and the file cannot be accessed, requiring the user to resort to one of the many file recovery packages on the market. Disk-based audio systems do not resort to error interpolation or sample hold operations, unlike tape recorders. Replay is normally either correct or not possible.

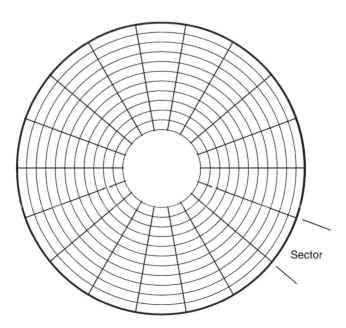

Figure 2

2.3 Recording Audio on to Disks

The discontinuous "bursty" nature of recording on to disk drives requires the use of a buffer RAM (random access memory) during replay, which accepts this interrupted data stream and stores it for a short time before releasing it as a continuous stream. It performs the opposite function during recording, as shown in Figure 3. Several things cause a delay in the retrieval of information: the time it takes for the head positioner to move across a disk, the time it takes for the required data in a particular track to come around to the pickup head and the transfer of the data from the disk via the buffer RAM to the outside world, as shown in Figure 4. Total delay, or data access time, is in practice several milliseconds. The instantaneous rate at which the system can accept or give out data is called the transfer rate and it varies with the storage device.

Sound is stored in named data files on the disk, the files consisting of a number of blocks of data stored either separately or together. A directory stored on the disk keeps track of where the blocks of each file are stored so that they can be retrieved in correct sequence. Each file normally corresponds to a single recording of a single channel of audio, although some stereo file formats exist (see below). Multiple channels are handled by accessing multiple files from the disk in a time-shared manner, with synchronisation between the tracks being performed subsequently in RAM. The storage capacity of a disk can be divided between channels in whatever proportion is appropriate, and it is not necessary to pre-allocate storage space to particular audio channels.

The number of audio channels that can be recorded or replayed simultaneously depends on the performance of the storage device and the host computer. Slow systems may only be capable of handling a few channels, whereas faster systems with multiple disk drives may be capable of expansion up to a virtually unlimited number of channels.

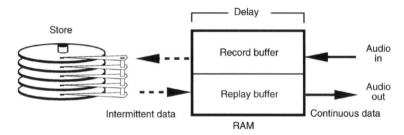

FIGURE 3

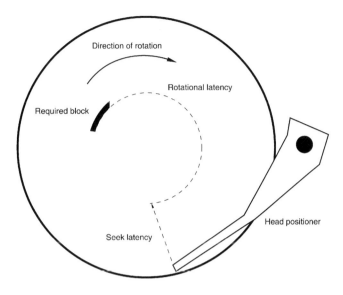

FIGURE 4

2.4 Sound File Formats

As the use of networked workstations grows, the need for files to be transferred between systems also grows, and either by international standardisation or by sheer force of market dominance, certain file formats are becoming the accepted means by which data are exchanged. The recent growth in the importance of metadata (data about data), and the representation of audio, video and metadata as "objects", has led to the development of interchange methods that are based on object-oriented concepts and project "packages" as opposed to using simple text files and separate media files. There is increasing integration between audio and other media in multimedia authoring, and some of the file formats mentioned below are closely related to international efforts in multimedia file exchange. The following is a relatively brief summary of the most commonly encountered file formats. Further details may be found in Rumsey (2004).

2.4.1 Sound Designer Formats

Sound Designer files originate from the Californian company Digidesign. Many systems handle Sound Designer files because they were used widely for such purposes as the distribution of sound effects on CD-ROM and for other short music sample files.

The Sound Designer I format (SD I, file extension ".sd") is for mono sounds and it is recommended principally for use in storing short sounds. It originated from the Apple Macintosh (Mac) computer, so numerical data are stored in big-endian byte order but it has no resource fork. The data fork contains a header of 1,336 bytes, followed by the audio data bytes. The header contains information about how the sample should be displayed in Sound Designer editing software, including data describing vertical and horizontal scaling. It also contains details of "loop points" for the file (these are principally for use with audio/MIDI sampling packages where portions of the sound are repeatedly cycled through while a key is held down, in order to sustain a note). The header contains information on the sample rate, sample period, number of bits per sample, quantisation method (e.g. "linear", expressed as an ASCII string describing the method) and size of RAM buffer to be used. The audio data are normally either 8 or 16 bits, and always MSbyte (most significant byte) followed by LSbyte (least significant byte) of each sample.

Sound Designer II (file extension ".sd2") has been one of the most commonly used formats for audio workstations and has greater flexibility than SD I. Again it originated from a Mac file and unlike SD I it has a separate resource fork which contains the file's "vital statistics". The data fork contains only the audio data bytes in twos complement form, either 8 or 16 bits per sample. SD II files can contain audio samples for more than one channel, in which case the samples are interleaved, as shown in Figure 5, on a sample-by-sample basis (i.e. all the bytes for one channel sample followed by all the bytes for the next, etc.). It is unusual to find more than stereo data contained in SD II files, and it is recommended that multichannel recordings are made using separate files for each channel.

2.4.2 AIFF and AIFF-C Formats

The AIFF (Audio Interchange File Format) format is widely used as an audio interchange standard because it conforms to the EA IFF 85 (Electronic Arts Interchange File Format) standard for interchange format files used for various other types of information such as graphical images. AIFF (file extension ".aiff" is an Apple standard format for audio data and is encountered widely on Mac-based audio workstations and some Silicon Graphics

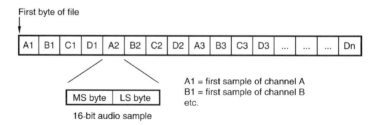

FIGURE 5

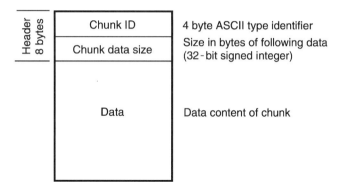

FIGURE 6

systems. Audio information can be stored at a number of resolutions and for any number of channels if required, and the related AIFF-C (file extension ".aifc") format allows also for compressed audio data. A file consists only of a data fork, with no resource fork, making it easier to transport to other platforms.

All IFF-type files are made up of "chunks" of data which are typically made up as shown in Figure 6. A chunk consists of a header and a number of data bytes to follow. The simplest AIFF files contain a "common chunk", which is equivalent to the header data in other audio files, and a "sound data" chunk containing the audio sample data. These are contained overall by a "form" chunk as shown in Figure 7. AIFC files must also contain a "Version Chunk" before the common chunk to allow for future changes to AIFC.

2.4.3 RIFF WAVE Format

The RIFF (Resource Interchange File Format) WAVE (often called WAV) format is the Microsoft equivalent of Apple's AIFF. It has a similar structure, again conforming to the IFF pattern, but with numbers stored in little-endian rather than big-endian form. It is used widely for sound file storage and interchange on PC workstations and for multimedia applications involving sound. Within WAVE files, it is possible to include information about a number of cue points and a playlist to indicate the order in which the cues are to be replayed. WAVE files use the file extension ".wav".

A basic WAV file consists of three principal chunks, as shown in Figure 8, the RIFF chunk, the FORMAT chunk and the DATA chunk. The RIFF chunk contains 12 bytes, the first four of which

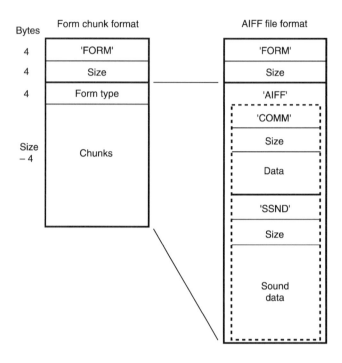

FIGURE 7

are the ASCII characters "RIFF", the next four indicating the number of bytes in the remainder of the file (after the first eight) and the last four of which are the ASCII characters "WAVE". The format chunk contains information about the format of the sound file, including the number of audio channels, sampling rate and bits per sample, as shown in Table 2.

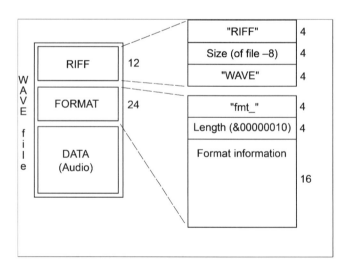

FIGURE 8

TABLE 2 Contents of FORMAT chunk in a basic WAVE PCM file

Byte	ID	Contents
0–3	ckID	"fmt_" (ASCII characters)
4–7	nChunkSize	Length of FORMAT chunk (binary, hex value: &00000010)
8–9	wFormatTag	Audio data format (e.g. &0001 = WAVE format PCM) Other formats are allowed, for example, IEEE floating point and MPEG format (&0050 = MPEG 1)
10–11	nChannels	Number of channels (e.g. &0001=mono, &0002=stereo)
12–15	nSamplesPerSec	Sample rate (binary, in Hz)
16–19	nAvgBytesPerSec	Bytes per second
20–21	nBlockAlign	Bytes per sample: e.g. &0001=8 bit mono; &0002=8 bit stereo or 16 bit mono; &0004=16 bit stereo
22–23	nBitsPerSample	Bits per sample

The audio data chunk contains a sequence of bytes of audio sample data, divided as shown in the FORMAT chunk. Unusually, if there are only 8 bits per sample or fewer, each value is unsigned and ranges between 0 and 255 (decimal), whereas if the resolution is higher than this, the data are signed and range both positively and negatively around zero. Audio samples are interleaved by channel in time order so that if the file contains two channels, a sample for the left channel is followed immediately by the associated sample for the right channel. The same is true of multiple channels (one sample for time-coincident sample periods on each channel is inserted at a time, starting with the lowest numbered channel), although basic WAV files were nearly always just mono or two-channel.

The RIFF WAVE format is extensible and can have additional chunks to define enhanced functionality such as surround sound and other forms of coding. This is known as "WAVE-format extensible" (see http://www.microsoft.com/hwdev/tech/audio/multichaud.asp). Chunks can include data relating to cue points, labels and associated data, for example.

2.4.4 Broadcast WAVE Format

The Broadcast WAVE format, described in EBU Tech. 3285 (EBU 2001), was standardised by the European Broadcasting Union (EBU) because of a need to ensure compatibility of sound files and accompanying information when transferred between workstations. It is based on the RIFF WAVE format described above, but contains an additional chunk that is specific to the format (the "broadcast_audio_extension" chunk, ID = "bext") and also limits some aspects of the WAVE format. version 0 was published in 1997 and version 1 in 2001, the only difference being the addition of a SMPTE (Society of Motion Picture and Television Engineers) UMID (Unique Material Identifier) in version 1 (this is a form of metadata). Such files currently only contain either PCM or MPEG-format audio data.

A multichannel extension chunk has recently been proposed for Broadcast WAVE files that define the channel ordering, surround format, downmix coefficients for creating a two-channel mix and some descriptive information. There are also chunks defined for metadata describing the audio contained within the file, such as the "quality chunk" (ckID = "qlty"), which together with the coding history contained in the "bext" chunk make up the so-called capturing report. These are described in Supplement 2 to EBU Tech. 3285. Finally, there is a chunk describing the peak audio level within a file, which can aid automatic programme level setting and programme interchange.

2.4.5 MPEG Audio File Formats

It is possible to store MPEG-compressed audio in AIFF-C or WAVE files, with the compression type noted in the appropriate header field. (MPEG is the Motion Pictures Expert Group that defines standards for data-reduced audio.) There are also older MS-DOS file extensions used to denote MPEG audio files, notably ".mpa" (MPEG Audio) or ".abs" (Audio Bit Stream). However, owing to the ubiquity of the so-called MP3 format (MPEG 1, Layer 3) for audio distribution on the Internet, MPEG audio files are increasingly denoted with the extension ".mp3". Such files are relatively simple, being really no more than MPEG audio frame data in sequence, each frame being preceded by a frame header.

2.4.6 DSD-IFF File Format

The DSD-IFF file format is based on a similar structure to other IFF-type files, described above, except that it is modified slightly to allow for the large file sizes that may be encountered with the high-resolution DSD (Direct Stream Digital) format used, for SuperAudio CD. Specifically the container FORM chunk is labelled "FRM8" and this identifies all local chunks that follow as having "length" indications that are 8 bytes long rather than the normal four. In other words, rather than a 4-byte chunk ID followed by a 4-byte length indication, these files have a 4-byte ID followed by an 8-byte length indication. This allows for the definition of chunks with a length greater than 2 Gbytes, which may be needed for mastering SuperAudio CDs. There are also various optional chunks that can be used for exchanging more detailed information and comments such as might be used in project interchange. Further details of this file format, and an excellent guide to the use of DSD-IFF in project applications, can be found in the DSD-IFF specification (Philips 2002a and b).

2.4.7 Edit Decision List Files

EDL (edit decision list) formats have usually been unique to the workstation on which they are used, but the need for open interchange is increasing the pressure to make EDLs transportable between packages. There is an old and widely used format for EDLs in the video world that is known as the CMX-compatible form. CMX is a well-known manufacturer of video editing equipment, and most editing systems will read CMX EDLs for the sake of compatibility. These can be used for basic audio purposes, and indeed a number of workstations can read CMX EDL files for the purpose of auto-conforming audio edits to video edits performed on a separate system. The CMX list defines the cut points between source material and the various transition effects at joins, and it can be translated reasonably well for the purpose of defining audio cut points and their timecode locations, using SMPTE/EBU form, provided video frame accuracy is adequate.

Software can be obtained for audio and video workstations that translate EDLs between a number of different standards to make interchange easier, although it is clear that this process is not always problem-free and good planning of in-house processes is vital. The Open Media Framework Interchange (OMFI) structure also contains a format for interchanging edit list data. The AES 31 standard (AES 1999, 2001) is now gaining considerable popularity among workstation software manufacturers as a simple means of exchanging audio editing projects between systems. The Advanced Authoring Format (AAF) is becoming increasingly relevant to the exchange of media project data between systems and is likely to take over from OMFI as time progresses.

2.4.8 MXF

MXF (Media Exchange Format) was developed by the Pro-MPEG forum as a means of exchanging audio, video and metadata between devices, primarily in television operations. It is based on the modern concept of media objects that are split into "essence" and "metadata". Essence files are the raw material (i.e. audio and video) and the metadata describes things about the essence (such as where to put it, where it came from and how to process it).

MXF files attempt to present the material in a "streaming" format, that is one that can be played out in real time, but they can also be exchanged in conventional file transfer operations. As such they are normally considered to be finished programme material, rather than material that is to be processed somewhere downstream, designed for playout in broadcasting environments. The bit stream is also designed to be compatible with recording on digital videotape devices.

2.4.9 AAF

AAF (Advanced Authoring Format) is an authoring format for multimedia data that is supported by numerous vendors, including Avid which has adopted it as a migration path from OMFI (Gilmer 2002). Parts of OMFI 2.0 form the basis for parts of AAF and there are also close similarities between AAF and MXF (described in the previous section). Like the formats to which it has similarities, AAF is an object-oriented format that combines essence and metadata within a container structure. Unlike MXF it is designed for project interchange such that elements within the project can be modified, post-processed and resynchronised. It is not, therefore, directly suitable as a streaming format but can easily be converted to MXF for streaming if necessary.

Rather like OMFI, AAF is designed to enable complex relationships to be described between content elements, to map these elements onto a timeline, to describe the processing of effects, synchronise streams of essence, retain historical metadata and refer to external essence (essence not contained within the AAF package itself). It has three essential parts: the AAF Object Specification (which defines a container for essence and metadata, the logical contents of objects and rules for relationships between them); the AAF Low-Level Container Specification (which defines a disk filing structure for the data, based on Microsoft's Structured Storage); and the AAF SDK Reference Implementation (which is a software development kit that enables applications to deal with AAF files). The Object Specification is extensible in that it allows new object classes to be defined for future development purposes.

3 • Consumer Digital Formats

3.1 Compact Discs and Drives

Compact discs (CDs) are not immediately suitable for real-time audio editing and production, partly because of their relatively slow access time compared with hard disks, but can be seen to have considerable value for the storage and transfer of sound material that does not require real-time editing. Broadcasters use them for sound effects libraries and studios and mastering facilities use them for providing customers and record companies with "acetates" or test pressings of a new recording. They have also become quite popular as a means of transferring finished masters to a CD pressing plant in the form of the PMCD (pre-master CD). They are ideal as a means of "proofing" CD-ROMs and other CD formats and can be used as low-cost backup storage for computer data. Detailed coverage of CD formats and "books" may be found in Pohlmann (1992).

CDs are familiar to most people as a consumer read-only optical disk for audio (CD-DA) or data (CD-ROM) storage. Standard audio CDs (CD-DA – DA for digital audio) conform to the Red Book standard published by Philips. The CD-ROM standard (Yellow Book) divides the CD into a structure with 2,048 byte sectors, adds an extra layer of error protection and makes it useful for general-purpose data storage including the distribution of sound and video in the form of computer data files. It is possible to find disks with mixed modes, containing sections in CD-ROM format and sections in CD-Audio format. The CD Plus is one such example.

CD-R is the recordable CD and may be used for recording CD-Audio format or other CD formats using a suitable drive and software. The Orange Book, Part 2, contains information on the additional features of CD-R, such as the area in the centre of the disk where data specific to CD-R recordings is stored. Audio CDs recorded according to the Orange Book standard can be "fixed" to give them a standard Red Book table of contents (TOC), allowing them to be replayed on any conventional CD player. Once fixed into this form, the CD-R may not subsequently be added to or changed, but prior to this there is a certain amount of flexibility, as discussed below. CD-RW (RW for read–write) disks are erasable and work on phase-change principles, requiring a drive compatible with this technology, being described in the Orange Book, Part 3.

The degree of reflectivity of CD-RW disks is much lower than that of typical CD-R and CD-ROM. This means that some early drives and players may have difficulties reading them. However, the "multi-read" specification developed by the OSTA (Optical Storage Technology Association) describes a drive that should read all types of CD, so recent drives should have no difficulties here.

3.2 Minidisc

CD has been available for some years now as a 16-bit, 44.1-kHz digital playback medium; it was joined by CD-ROM, CD-R (recordable) and CD-RW (recordable and rewritable). The minidisc (MD) is now an established consumer recording and playback format, and it is of the M-O (magneto-optical) type. Sampling frequency is fixed at 44.1 kHz, and resolution is nominally 16 bits. A coding system similar to those originally developed for digital audio broadcasting (DAB) known as

Adaptive Transform Acoustic Coding (ATRAC) is used (Tsutsui et al. 1992). A data rate of about one-fifth of that required for CD is adequate to encode the resulting signal (CD's data stream is 1.4 Mb/s, MD's is 292 kb/s), and this allows an equivalent playing time to be obtained from a disc which is somewhat smaller than a CD at 64 mm in diameter. Since the format involves considerable data compression (a slight misnomer for data reduction), it is not suitable for professional master recording or archiving, but is used quite widely in applications where the highest sound quality is not required such as broadcast journalism.

3.3 DVD

DVD (Digital Versatile Disk) is the natural successor to CD, being a higher-density optical disk format aimed at the consumer market, having the same diameter as CD and many similar physical features (De Bont et al. 1996). It uses a different laser wavelength to CD (635–650 nm as opposed to 780 nm) so multi-standard drives need to be able to accommodate both. Data storage capacity depends on the number of sides and layers on the disk but ranges from 4.7 Gbytes (single-layer, single-sided) up to about 18 Gbytes (double-layer, double-sided). The data transfer rate at "one times" speed is just over 11 Mbit/s.

DVD can be used as a general-purpose data storage medium. Like CD, there are numerous different variants on the recordable DVDs, partly owing to competition between the numerous different "factions" in the DVD consortium. These include DVD-R, DVD-RAM, DVD-RW and DVD+RW, all of which are based on similar principles but have slightly different features, leading to a compatibility minefield that is only gradually being addressed (see Table 3). The "DVD Multi" guidelines produced by the DVD Forum are an attempt to foster greater compatibility between DVD drives and disks, but this does not really solve the problem of the formats that are currently outside the DVD Forum.

Writeable DVDs are a useful option for backup of large projects, particularly DVD-RAM because of its many-times overwriting capacity and its hard disk-like behaviour. It is possible that a format like DVD-RAM could be used as primary storage in a multitrack recording/editing system, as it has sufficient performance for a limited number of channels and it has the great advantage of being removable. However, it is likely that hard disks will retain the performance edge for the foreseeable future.

DVD-Video is the format originally defined for consumer distribution of movies with surround sound, typically incorporating MPEG-2 video encoding and Dolby Digital surround sound encoding. It also allows for up to eight channels of 48 or 96 kHz linear PCM audio, at up to 24-bit resolution. DVD-

TABLE 3 Recordable DVD formats

Recordable DVD type	Description
DVD-R (A and G)	DVD equivalent of CD-R. One-time recordable in sequential manner, replayable on virtually any DVD-ROM drive Supports "incremental writing" or "disk at once" recording. Capacity either 3.95 (early disks) or 4.7 Gbytes per side. "Authoring" (A) version (recording laser wavelength = 635 nm) can be used for pre-mastering DVDs for pressing, including DDP data for disk mastering "General" (G) version (recording laser wavelength = 650 nm) intended for consumer use, having various "content protection" features that prevent encrypted commercial releases from being cloned
DVD-RAM	Sectored format, rather more like a hard disk in data structure when compared with DVD-R. Uses phase-change (PD-type) principles allowing direct over-write. Version 2 disks allow 4.7 Gbyte per side (reduced to about 4.2 Gbytes after formatting). Type 1 cartridges are sealed and Type 2 allow the disk to be removed. Double-sided disks only come in sealed cartridges. Can be rewritten about 100,000 times. The recent Type 3 is a bare disk that can be placed in an open cartridge for recording
DVD-RW	Pioneer development, similar to CD-RW in structure, involving sequential writing. Does not involve a cartridge Can be re-written about 1,000 times. 4.7 Gbytes per side
DVD+RW	Non-DVD-Forum alternative to DVD-RAM (and not compatible), allowing direct over-write. No cartridge. Data can be written in either CLV (for video recording) or CAV (for random access storage) modes. There is also a write-once version known as DVD+R

Audio is intended for very high quality multichannel audio reproduction and allows for linear PCM sampling rates up to 192 kHz, with numerous configurations of audio channels for different surround modes, and optional lossless data reduction (MLP – see below).

DVD-Audio (DVD-A) has a number of options for choosing the sampling frequencies and resolutions of different channel groups, it being possible to use a different resolution on the front channels from that used on the rear, for example. The format is more versatile in respect of sampling frequency than DVD-Video, having also accommodated multiples of the CD sample frequency of 44.1 kHz as options (the DVD-Video format allows only for multiples of 48 kHz). Consequently, the allowed sample frequencies for DVD-Audio are 44.1, 48, 88.2, 96, 176.4 and 192 kHz. The sample frequencies are split into two groups – multiples of 44.1 and multiples of 48 kHz. While it is possible to split frequencies from one group among the audio channels on a DVD-A, one cannot combine frequencies across the groups for reasons of simple clock rate division. Bit resolution can be 16, 20 or 24 bits per channel, and again this can be divided unequally between the channels, according to the channel group split described below.

Meridian Lossless Packing (MLP) is licensed through Dolby Laboratories and is a lossless coding technique designed to reduce the data rate of audio signals without compromising sound quality (Gerzon et al. 2004). It has both a variable bit rate mode and a fixed bit rate mode. The variable mode delivers the optimum compression for storing audio in computer data files, but the fixed mode is important for DVD applications where one must be able to guarantee a certain reduction in peak bit rate.

3.4 Super Audio CD

Version 1.0 of the Super Audio CD (SACD) specification (Verbakel et al. 1998) is described in the "Scarlet Book", available from Philips licensing department. SACD uses DSD (Direct Stream Digital) as a means of representing audio signals, so it requires audio to be sourced in or converted to this form. SACD aims to provide a playing time of at least 74 min for both two-channel and six-channel balances. The disc is divided into two regions, one for two-channel audio, the other for multichannel, as shown in Figure 9. A lossless data packing method known as Direct Stream Transfer (DST) can be used to achieve roughly 2:1 data reduction of the signal stored on disc so as to enable high-quality multichannel audio on the same disc as the two-channel mix.

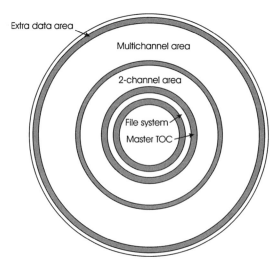

FIGURE 9

SACDs can be manufactured as single- or dual-layer discs, with the option of the second layer being a Red Book CD layer (the so-called hybrid disc). SACDs, not being a formal part of the DVD hierarchy of standards (although using some of the optical disc technology), do not have the same options for DVD-Video objects as DVD-Audio. The disc is designed first and foremost as a super-high-quality audio medium. Nonetheless, there is provision for additional data in a separate area of the disc. The content and capacity of this is not specified but could be video clips, text or graphics, for example.

4 • Digital Audio Editing Using Mass Storage Media

The random access nature of disk-based mass storage media led to the coining of the term non-linear editing for the process of audio editing. With non-linear editing, the editor may preview a number of possible masters in their entirety before deciding which should be the final one. Even after this, it is a simple matter to modify the edit list to update the master. Edits may also be previewed and experimented with in order to determine the most appropriate location and processing. Crossfades may be modified and adjustments made to equalisation and levels, all in the digital domain. Non-linear editing has also come to feature very widely in post-production for video and film.

Non-linear editing is truly non-destructive in that the edited master only exists as a series of instructions to replay certain parts of sound files at certain times, with specified signal processing overlaid, as shown in Figure 10. The original sound files remain intact at all times, and a single sound file can be used as many times as desired in different locations and on different tracks without the need for copying the audio data. Editing may involve the simple joining of sections, or it may involve more complex operations such as long crossfades between one album track and the next, or gain offsets between one section and another. All these things are possible without affecting the original source material.

4.1 Sound Files and Sound Segments

In the case of music editing sound files might be session takes, anything from a few bars to a whole movement, while in picture dubbing they might contain a phrase of dialogue or a sound effect. Specific segments of these sound files can be defined while editing, in order to get rid of unwanted material or to select useful extracts. The terminology varies but such identified parts of sound files are usually termed either "clips" or "segments". Rather than creating a copy of the segment or clip and storing

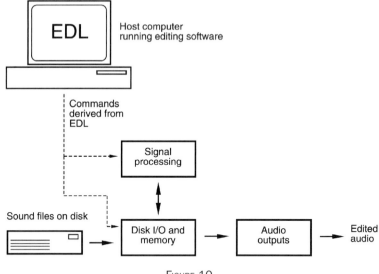

FIGURE 10

it as a separate sound file, it is normal simply to store it as a "soft" entity – in other words as simply commands in an edit list or project file that identify the start and end addresses of the segment concerned and the sound file to which it relates. It may be given a name by the operator and subsequently used as if it were a sound file in its own right. An almost unlimited number of these segments can be created from original sound files, without the need for any additional audio storage space.

4.2 Edit Point Handling

Edit points can be simple butt joins or crossfades. A butt join is very simple because it involves straightforward switching from the replay of one sound segment to another. Since replay involves temporary storage of the sound file blocks in RAM (see above), it is a relatively simple matter to ensure that both outgoing and incoming files in the region of the edit are available in RAM simultaneously (in different address areas). Up until the edit, blocks of the outgoing file are read from the disk into RAM and thence to the audio outputs. As the edit point is reached, a switch occurs between outgoing and incoming material by instituting a jump in the memory read address corresponding to the start of the incoming material. Replay then continues by reading subsequent blocks from the incoming sound file. It is normally possible to position edits to single sample accuracy, making the timing resolution as fine as a number of tens of microseconds if required.

The problem with butt joins is that they are quite unsubtle. Audible clicks and bumps may result because of a discontinuity in the waveform. It is normal, therefore, to use at least a short crossfade at edit points to hide the effect of the join. This is what happens when analogue tape is spliced because the traditional angled cut has the same effect as a short crossfade (of between 5 and 20 ms depending on the tape speed and angle of cut). Most workstations have considerable flexibility with crossfades and they are not limited to short durations. It is now common to use crossfades of many shapes and durations (e.g. linear, root cosine, equal power) for different creative purposes. This, coupled with the ability to preview edits and fine-tune their locations, has made it possible to put edits in places previously considered impossible.

The locations of edit points are kept in an edit decision list (EDL) which contains information about the segments and files to be replayed at each time, the in and the out points of each section and details of the crossfade time and shape at each edit point. It may also contain additional information such as signal processing operations to be performed (gain changes, EQ, etc.)

4.3 Crossfading

Crossfading is similar to butt joining, except that it requires access to data from both incoming and outgoing files for the duration of the crossfade. The crossfade calculation involves simple signal processing, during which the values of outgoing samples are multiplied by gradually decreasing coefficients while the values of incoming samples are multiplied by gradually increasing coefficients. Time-coincident samples of the two files are then added together to produce output samples, as described in Chapter 35. The duration and shape of the crossfade can be adjusted by altering the coefficients involved and the rate at which the process is executed.

Crossfades are either performed in real time, as the edit point passes, or pre-calculated and written to disk as a file. Real-time crossfades can be varied at any time and are simply stored as commands in the EDL, indicating the nature of the fade to be executed. The process is similar to that for the butt edit, except that as the edit point approaches samples from both incoming and outgoing segments are loaded into RAM in order that there is an overlap in time. During the crossfade, it is necessary to continue to load samples from both incoming and outgoing segments into their respective areas of RAM, and for these to be routed to a crossfade processor. The resulting samples are then available for routing to the output. Alternatively, the crossfade can be calculated in non-real time. This incurs a short delay while the system works out the sums, after which a new sound file is stored which contains only the crossfade. Replay of the edit then involves playing the outgoing segment up to the beginning of the crossfade, then the crossfade file and then the incoming segment from after the crossfade. Load on the disk drive is no higher than normal in this case.

The shape of the crossfade can usually be changed to suit different operational purposes. Standard linear fades (those where the gain changes uniformly with time) are not always the most suitable for music editing, especially when the crossfade is longer than about 10 ms. The result may be a momentary drop in the resulting level in the centre of the crossfade that is due to the way in which the sound levels from the two files add together. If there is a random phase difference between the signals, as there will often be in music, the rise in level resulting from adding the two signals will normally be around 3 dB, but the linear crossfade is 6 dB down in its centre resulting in an overall level drop of around 3 dB (see Figure 11). Exponential crossfades and other such shapes may be more suitable for these purposes because they have a smaller level drop in the centre. It may even be possible to design customised crossfade laws. It is often possible to alter the offset of the start and end of the fade from the actual edit point and to have a faster fade up than fade down. Many systems also allow automated gain changes to be introduced as well as fades, so that level differences across edit points may be corrected. A lot of the difficulties that editors encounter in making edits work can be solved using a combination of these facilities.

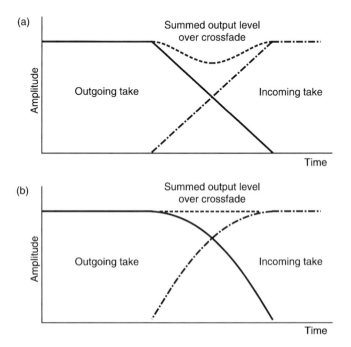

FIGURE 11

4.4 Editing Modes

During the editing process, the operator will load appropriate sound files and audition them, both on their own and in a sequence with other files. The exact method of assembling the edited sequence depends very much on the user interface, but it is common to present the user with a visual analogy of moving tape, allowing files to be "cut-and-spliced" or "copied and pasted" into appropriate locations along the virtual tape. These files, or edited clips of them, are then played out at the timecode locations corresponding to their positions on this "virtual tape". It is also quite common to display a representation of the audio waveform that allows the editor to see as well as hear the signal around the edit point (see Figure 12).

In non-linear systems the tape-based approach is often simulated, allowing the user to roughly locate an edit point while playing the virtual tape followed by a fine trim using simulated reel-rocking or a detailed view of the waveform. Some software presents source and destination streams as well, in further simulation of the tape approach. It is also possible to insert or change sections in the middle of a finished master, provided the EDL and source files are still available. To take an example, assume that an edited opera has been completed and that the producer now wishes to change a take somewhere in the middle. The replacement take is unlikely to be exactly the same length, but it is possible simply to shuffle all of the following material along or back slightly to accommodate it, this being only a matter of changing the EDL rather than modifying the stored music in any way. The files are then simply played out at slightly different times than in the first version of the edit.

It is also normal to allow edited segments to be fixed in time if desired, so that they are not shuffled forwards or backwards

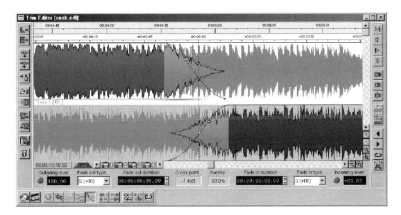

FIGURE 12

when other segments are inserted. This "anchoring" of segments is often used in picture dubbing when certain sound effects and dialogue have to remain locked to the picture.

4.5 Simulation of "Reel-Rocking"

It is common to simulate the effect of analogue tape "reel-rocking" in non-linear editors, providing the user with the sonic impression that reels of analogue tape are being "rocked" back and forth as they are in analogue tape editing when fine-searching edit points. Editors are used to the sound of tape moving in this way and are skilled at locating edit points when listening to such a sound.

The simulation of variable speed replay in both directions (forwards and backwards) is usually controlled by a wheel or sideways movement of a mouse which moves the "tape" in either direction around the current play location. This magnitude and direction of this movement is used to control the rate at which samples are read from the disk file, via the buffer, and this replaces the fixed sampling rate clock as the controller of the replay rate. Systems differ very greatly as to the sound quality achieved in this mode because it is in fact quite a difficult task to provide convincing simulation. So poor have been many attempts that many editors do not use the feature, preferring to judge edit points accurately "on the fly", followed by trimming or nudging them either way if they are not successful the first time. Good simulation requires very fast, responsive action and an ergonomically suitable control. A mouse is very unsuitable for the purpose. It also requires a certain amount of DSP to filter the signal correctly in order to avoid the aliasing that can be caused by varying the sampling rate.

References

AES (1999) *AES 31-3-1999: AES 31-3-1999: AES standard for network and file transfer of audio. Audio file transfer and exchange: Part 3 – Simple project interchange.*

AES (2001) *AES 31-1-2001: AES standard for network and file transfer of audio. Audio file transfer and exchange: Part 1 – Disk format.*

De Bont, F. et al. (1996) The high density DVD. *Proceedings of the AES UK Conference on Audio for New Media.* Audio Engineering Society.

Doi, T., Tsuchiya, Y. and Iga, A. (1978) On several standards for converting PCM signals into video signals. *J. Audio Eng. Soc.* 26, 9, pp. 641–649.

EBU (2001) *Tech. 3285. BWF – a format for audio data files in broadcasting. Version 1.* (See also Supplements 1, 2 and 3.). European Broadcasting Union.

Gerzon, M. et al. (2004) The MLP lossless compression system for PCM audio. *J. Audio Eng. Soc.* 52, 3, pp. 243–260.

Gilmer, B. (2002) AAF – the advanced authoring format. *EBU Tech. Rev.*, July.

Ishida, Y. et al. (1985) On the development of a rotary head digital audio tape recorder. *Presented at 79th AES Convention*, Preprint 2271. Audio Engineering Society.

Kogure, T., Doi, T. and Lagadec, R. (1983) The DASH format: an overview. *Presented at 74th AES Convention*, Preprint 2038. Audio Engineering Society.

Kudelski, S. et al. (1987) Digital audio recording format offering extensive editing capabilities. *Presented at 82nd AES Convention*, Preprint 2481. Audio Engineering Society.

Nakajima, H., Doi, T., Tsuchiya, Y. and Iga, A. (1978) A new PCM audio system as an adapter of video tape recorders. *Presented at 60th AES Convention*, Preprint 1352. Audio Engineering Society.

Philips (2002a) *Direct Stream Digital Interchange File Format: DSD-IFF version 1.4, revision 2.* Available from www.superaudiocd.philips.com.

Philips (2002b) *Recommended usage of DSD-IFF, version 1.4.* Available from www.superaudiocd.philips.com

Pohlmann, K. (1992) *The Compact Disc Handbook.* A-R Editions.

Rumsey, F. (2004) *Desktop Audio Technology.* Oxford and Boston: Focal Press.

Tsutsui, K. et al. (1992) ATRAC: Adapative transform acoustic coding for minidisc. *Presented at 93rd AES Convention*, Preprint 3456. Audio Engineering Society.

Verbakel, J. et al. (1998) Super Audio CD format. *Presented at 104th AES Convention*, Preprint 4705. Audio Engineering Society.

37

Audiovisual Interaction

Riikka Möttönen and Mikko Sams

Department of Biomedical Engineering and Computational Science,
Helsinki University of Technology, Helsinki, Finland

Information about the objects and events in the external world is received via multiple sense organs, especially via eyes and ears. For example, a singing bird can be heard and seen. Typically, audiovisual objects are detected, localized and identified more rapidly and accurately than objects which are perceived via only one sensory system (see, e.g. Welch and Warren, 1986; Stein and Meredith, 1993; de Gelder and Bertelson, 2003; Calvert et al., 2004). The ability of the central nervous system to utilize sensory inputs mediated by different sense organs is called multisensory processing.

In this chapter we will concentrate on reviewing how perception of acoustic information is affected by concurrent visual information. We first discuss briefly how visual information can affect sound localization. Then we focus on audiovisual speech perception, i.e. how seeing a talker's articulatory gestures affects perception of acoustic speech. We finish by discussing the brain mechanisms of audiovisual perception.

1 • Audiovisual Space

Typically, acoustic and visual inputs coming from the same location at the same time originate from the same audiovisual object. Therefore, spatially coherent acoustic and visual inputs are integrated and perceived as a single audiovisual object. As a result of integration, orientation to new objects in the environment becomes more accurate than orientation to unimodal objects alone (Stein et al., 1989). In contrast, orientation to spatially disparate, but simultaneous, objects becomes worse than that to unimodal objects alone (Stein et al., 1989).

1.1 Ventriloquism Illusion

When a visual stimulus is presented simultaneously with an acoustic stimulus but in a slightly different location, both inputs are perceived to arrive from the same location. The common location is mainly determined by the location of the visual stimulus. This illusion is known as the ventriloquism illusion (Howard and Templeton, 1966), because an illusory change of sound location is experienced when ventriloquist's speech is perceived to come from the moving mouth of a puppet. In modern everyday life, a similar illusion is experienced when visual events (e.g. actor's articulatory gestures) on a TV or movie screen determine the perceptual locations of sounds (e.g. actor's voice), which actually come from loudspeakers.

The ventriloquism illusion has been extensively studied in laboratories (for a review, see Bertelson and Gelder, 2004). In psychophysical experiments, participants are instructed to point toward locations of sounds when they are presented alone and when they are presented together with simultaneous, but slightly disparate, flashes of light. Perceived sound location is typically affected by flashes of light. The ventriloquism effect is the strongest when the distance between acoustic and visual inputs is small and they are perceived as a unified object (e.g. Wallace et al., 2004). When the distance of acoustic and visual inputs increases, the inputs are more often perceived to be disparate and the ventriloquism effect becomes weaker (e.g. Wallace et al., 2004). Furthermore, temporal asynchrony between acoustic and visual inputs weakens the influence of vision on perceived sound location (e.g. Slutsky and Recanzone, 2001; Wallace et al., 2004).

2 • Audiovisual Speech

Speech perception usually refers to perception of acoustic speech. However, speech is also visual. Viewing a talker's articulatory gestures allows the observer to understand speech to some extent (for reviews, see Dodd and Campbell, 1987; Campbell et al., 1998). This ability to *speechread* varies a lot across observers. Hearing impaired and deaf people are often very skilful speechreaders, but normal-hearing people are able to speechread as well. In everyday conversations, we typically hear talkers' voice and see their articulatory gestures simultaneously, thus we perceive speech audio-visually.

2.1 Audiovisual Perception of Degraded Acoustic Speech

Sumby and Pollack (1954) published a landmark paper on integration of acoustic and visual speech. They presented bisyllabic words at different signal-to-noise ratios (SNR) to subjects who either only heard the stimuli or also saw the talker to articulate them. SNR was varied from zero to −30 dB. In auditory-only condition, the identification of the words deteriorated progressively as SNR decreased. When the same stimuli could also be seen, the overall performance improved greatly. For example, at −18 SNR in the auditory-only condition the subjects correctly identified about 28% of the words, but in the audiovisual condition 78% of them.

In another pioneering study, O'Neill (1954) studied audiovisual perception of vowels, consonants, words and sentences. Seeing the speech improved identification of each stimulus class. MacLeod and Summerfield (1987) used meaningful sentences to study whether the lipreading skill in normal subjects correlates with the ability to utilize visible speech in audiovisual speech perception. Visual alone performance correlated significantly with the audiovisual benefit. The mean benefit of seeing speech was equivalent to 11-dB increase in SNR, the range being 6–15 dB. In addition, those sentences which were easier to lipread were also recognized at a lower SNR in the audiovisual condition.

Figure 1 shows results from a speech recognition experiment in 20 normal-hearing subjects (Möttönen, 1999). Meaningless acoustic vowel–consonant–vowel stimuli were presented both

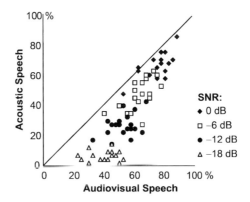

FIGURE 1 Proportions of correctly identified acoustic and audiovisual speech items at four signal-to-noise-ratio (SNR) levels (0, −6, −12, −18 dB) in 20 Finnish subjects. The stimuli were 39 vowel–consonant–vowel words produced by a native Finnish speaker. Each acoustic stimulus was presented alone and together with a matching visual articulation. Modified from Möttönen (1999).

alone and with matching visual speech. Noise was added to acoustic stimuli to acquire four SNR levels: 0, −6, −12 and −18 dB. In each subject, the proportion of correctly recognized audiovisual speech stimuli was greater than that of acoustic speech stimuli at all SNR levels. The lower the SNR of acoustic speech the greater the benefit of visual information.

Voicing and nasality are the articulatory features which are the least affected by noise and the place of articulation is the most difficult one to hear (Binnie et al., 1974; Miller and Nicely, 1955). On the other hand, the place of articulation is relatively easy to identify visually. Therefore, acoustic and visual speech complement each other in an important way. What is hard to hear is easy to see. What is easy to hear needs not to be seen. Binnie et al. (1974) separated five visually distinct consonant clusters, i.e. *visemes*: bilabials, labiodentals, interdentals, rounded labials and linguals. Also other viseme classifications have been suggested (for a review, see Jackson, 1988). In Finnish, Pesonen (1968) demonstrated five consonantal visemes: labials (b, m, p), labiodentals (v, f), incisives (s, t), alveolar dentals (d, l, n, r) and velars (g, k, h). The number of distinct visemes is dependent on the individual talkers (Kricos, 1996). Auer and Bernstein (1997) suggest a measure (phoneme equivalence class – PEC) that reflects both individual speech perceiver skills and the visibility of the visual signal. Twelve PECs suffice to identify most American English words.

2.2 The McGurk Effect

Using videoclips of a single speaker, Harry McGurk and John MacDonald (MacDonald and McGurk, 1978; McGurk and MacDonald, 1976) dubbed the auditory token /ba/ onto the visual articulation of /ga/, and something interesting happened. Most perceivers did not recognize the conflict between the stimuli. Moreover, they typically *heard* a new syllable /da/, or even /ga/. Very few perceivers were sensitive to the conflict and reported the acoustic syllable correctly. This change in auditory percept due to conflicting visual stimulus is called the "McGurk effect". An important feature of the McGurk effect is that it occurs when the acoustic stimulus is clear and can be easily recognized, when presented alone.

A strong fusion of acoustic and visual speech stimuli to a new percept does not occur for all stimulus combinations. For example, when Finnish acoustic /o/ (as in the English "dog") is dubbed onto visual /i/ (as in the English "lip"), it is easy to recognize that the two stimuli are discordant and they do not fuse (Klucharev et al., 2003). Strong fusions are obtained with some consonants. When the visual non-labial consonant

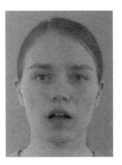 

Visual	/ka/	/pa/
	+	+
Acoustic	/pa/	/ka/
	=	=
Auditory percept	/ta/ or /ka/	/pka/ or /kpa/

FIGURE 2 Examples of the McGurk effect. Photos show a talker who is articulating bilabial /p/ and velar /k/ sounds. When the articulatory gestures are presented simultaneously with conflicting acoustic stimuli, the auditory perception is modified, i.e. the McGurk effect is produced.

/ka/ is presented with the bilabial acoustic /pa/, people generally perceive either /ta/ or /ka/ (see Figure 2). That is, they *hear* the visual feature "not labial" in this combination. When visual /pa/ is combined with acoustic /ka/, a "combination" such as /pka/ or /kpa/ is often *heard* (see Figure 2). The strength and quality of the McGurk effect depends on the vowel context in which the consonant is presented. For example, in American English the effect is stronger in /i/ and /a/ context than in /u/ context (Green, 1996). The nature of the fusion percept also depends on the vowel context. Visual /g/ combined with acoustic /b/ is often perceived as "th" in /a/ vowel context but "d" in /i/ context (Green et al., 1991). A conflicting visual consonant at the beginning of a word may even cause perceptual omission of a consonant. For example, the Finnish visual word /kuola/ combined with acoustic word /puola/ was often perceived as /uola/ (Sams et al., 1998).

2.3 Development of Audiovisual Speech Perception

Babbling is an important stage in the development of speech production and perception. Parents "baby talk" to their infants and also reinforce their utterances by repeating them. Some features of this preliminary speech are hard to hear but they are readily visible, such as bilabials. The syllables /ba/ and /ga/ are acoustically rather similar and difficult to differentiate by hearing, but the corresponding articulations are highly visible and can be easily discriminated. Both deaf and blind children babble. However, relatively soon important differences can be observed in their babbling. The proportion of easy-to-see bilabials increases in deaf children, but in blind children it decreases. Blind children can face problems in learning such contrasts that are easy to see but hard to hear (Mills, 1987).

Children start to prefer audiovisual speech where the channels are congruent at an early age. Dodd (1979) showed that infants aged 10–20 weeks preferred to watch a face where lip movements were in rather that out of synchrony with the acoustic speech. In this study, the synchrony of the two stimulus sources could have been the critical cue for their common origin. When such a cue was not available and audiovisual vowels were presented in synchrony, 4-month-old infants were able to detect a correct combination of acoustically and visually presented vowels /i/ and /a/ (Kuhl and Meltzoff, 1982). Moreover, infants only 20 weeks old can match the face and voice in their native language, but not in a foreign language (Dodd and Burnham, 1988). Burnham (1998) concludes that the children develop the ability for matching acoustic and visible speech by the age of 3 months, and this ability is the clearest in the native tongue.

Strong evidence that children are really integrating acoustic and visual stimulation would be that they show the McGurk effect. This is the case in 4- to 5-month-old infants (Burnham and Dodd, 1996, 2004). McGurk and MacDonald (1976) compared the audiovisual integration of consonant–vowel syllables in pre-school children (3–5 years old), school children (7–8 years old) and adults (18–40 years old). The McGurk effect was stronger in adults than in the child groups, which did not differ from each other. This suggests that even though infants start to integrate at an early age, the effect of the visible speech on speech perception becomes greater with audiovisual speech experience.

2.4 Effects of Attention

The original study (McGurk and MacDonald, 1976) together with several other subsequent reports suggests that integration of auditory and visual information is strongly automatic since the McGurk effect occurs even when the observer is aware of how the stimuli are constructed. Furthermore, in a study on audio-visual speech perception in children, Massaro (1984) found that audiovisual percepts were very similar both when a child just reported what the talker said and when he also had to report whether the talker's mouth moved. The second task was used to ascertain that the child was paying attention to the visual information. Supposing that visual attention was stronger in the latter case, this result suggests that it does not influence perception of audiovisual speech.

Integration of heard and seen speech is often considered to occur in an obligatory and automatic manner. Soto-Faraco et al. (2004) recently studied a syllabic interference task, where two syllables are presented in rapid consecution. The second syllable only interferes with the classification of the first if it varies; not if it is constant. This happens even when the subjects are instructed to concentrate on the first syllable and ignore the irrelevant second syllable. This is interpreted as auditory attention failing to select the relevant stimulus so that the second syllable is obligatorily processed. The authors made a perceptual change to the second syllable by a conflicting visual syllable, causing the McGurk effect. Under these conditions, the illusory percept rather than the actual acoustic stimulus determined whether interference occurred. Soto-Faraco et al. (2004) concluded that the visual influence on the auditory speech percept must have occurred before attentional selection.

However, the issue of automaticity is not as clear-cut as it may seem. Massaro (1998) has shown that by instructing subjects to respond according to either auditory or visual information only, responses to audiovisual consonants are biased toward the instructed modality. Furthermore, the influence of incongruent visual speech on vowel recognition is greater when subjects are asked to watch and listen to and report what was spoken, than when they are told about the dubbing and asked to report only what they heard (Summerfield and McGrath, 1984).

In order to study the effect of endogenous visual attention on audiovisual speech perception, Tiippana et al. (2004) conducted an experiment in which visual attention was modulated using a distractor stimulus presented together with the talking face, and the McGurk effect was measured in two conditions: with full attention at the talking face and with attention directed towards the distractor. The McGurk effect was found to be weaker in the latter condition for many different combinations of auditory and visual tokens, indicating that visual attention modulated audiovisual speech perception.

Additional evidence of the strong influence of attention was obtained by Andersen et al. (2002). The visual stimulus was a movie of two faces displaced laterally on each side of a central fixation point. Each of the two faces articulated a different utterance. Subjects were instructed to always maintain fixation on the central fixation cross but to covertly attend to the face indicated by the cueing arrow. In audiovisual trials, where a voice was dubbed onto the movie, subjects responded according to what they heard. The voice and the attended face could be matching or conflicting, the latter combination producing a McGurk effect. The results were very clear. The direction of visual attention strongly determined what the subjects heard. Moreover, Alsius et al. (2005) showed that a difficult concurrent visual or auditory discrimination task impairs audiovisual integration, as indexed by reduced McGurk effect. While audiovisual speech processing may not require extensive attentional resources, nevertheless it is sensitive to attentional contingencies and can be disrupted under specific attentional demand conditions.

2.5 Effects of Spatial and Temporal Asynchronies

A good example of the robustness of the McGurk effect was provided by Green et al. (1991). They dubbed a male voice onto a female talker and vice versa. Although the different genders were easy to recognize, this failed to reduce the strength of the McGurk effect. Interestingly, the McGurk effect is generated even when acoustic and visual speech inputs come from clearly different directions (Jones & Munhall, 1997). This finding is surprising, because a common spatial origin has been thought to be one of the main features of a multisensory event.

The acoustic and visual speech tokens do not need to be in perfect temporal synchrony for the McGurk effect to occur. The time window for the integration can be astonishingly wide. Green (1996) varied the asynchrony of acoustic and visual stimuli from 0 to 225 ms and found no effect on the strength of the McGurk effect. A wide integration window was also found by Munhall et al. (1996). The integration, as indicated by the strength of the McGurk effect, was the strongest with smaller asynchronies but a significant effect was obtained even when the auditory stimulus lagged the visual one by 240 ms. An important finding was that the integration window is strongly asymmetric. Much greater asynchrony is tolerated when the visual stimulus is leading.

3 • Neural Basis of Audiovisual Perception

As the behavioural findings reviewed above show the human brain is able to combine acoustic and visual inputs in order to create unified percepts. Modern non-invasive neuroimaging methods, e.g. electroencephalography (EEG), magnetoencephalography (MEG), functional magnetic resonance imaging (fMRI) and positron emission tomography (PET), provide tools to study how the human brain accomplishes this task (for reviews, see Calvert, 2001; Calvert and Thesen, 2004). These methods provide complementary information about the human brain activity underlying audiovisual processing. EEG and MEG directly measure electric potentials and neuromagnetic fields generated by neural currents, providing information about the brain activity with millisecond accuracy. The haemodynamic responses measured by fMRI and PET do not provide accurate information about the timing of brain activity. However, fMRI and PET are superior to both EEG and MEG in terms of spatial resolution.

3.1 Multisensory Neurons

Single-cell recording studies in non-human mammals have found multisensory neurons which are activated by inputs mediated by multiple sense organs. These kinds of neurons have been found at various levels of central nervous system in: (1) the subcortical structures (e.g. the superior colliculus (SC), Stein and Meredith, 1993; Stein et al., 2004), (2) the sensory-specific cortices (e.g. auditory belt and parabelt, Schroeder and Foxe, 2002; Schroeder et al., 2003) and (3) the heteromodal cortices (e.g. anterior and posterior regions of superior temporal sulcus (STS), for a review, see Cusick, 1997). These findings support the view that convergence of different sensory inputs to the same neurons enables interaction between sensory modalities.

The most thoroughly studied multisensory neurons are located in the mammalian SC (Stein and Meredith, 1993; Stein et al., 2004), which is a subcortical structure thought to be involved in orientation and attentive behaviours. Some of the multisensory neurons in the SC are able to integrate inputs from different sensory systems: two stimuli presented in the same location at the same time produce *response enhancement* in these neurons. The response to two simultaneous stimuli typically exceeds the sum of responses to the same stimuli presented separately. The weaker the unimodal stimuli the stronger the enhancements; this principle is called *inverse effectiveness*. In contrast, two stimuli presented in different locations (or at different times) produce *response suppression* in these neurons. There are parallels between the responses of the SC multisensory neurons and behaviour of the animal. For example, a cat detects and localizes visual targets more accurately when they are presented together with a spatially coherent sound (Stein et al., 1989). On the other hand, presenting a sound in a different location impairs detection of a visual target (Stein et al., 1989).

3.2 Audiovisual Interaction in the Human Auditory Cortex

A widely used method to study auditory processing in humans is to record the so-called "mismatch responses" to acoustic changes in sound sequences with EEG or MEG (for a review, see Näätänen, 1992). The mismatch responses are considered to be generated by pre-attentive change-detection mechanisms in the auditory cortices. An obvious question which arises is whether these mechanisms are activated when there is no acoustic change in the sound sequence, but a person perceives an illusory auditory change due to McGurk or ventriloquism illusion.

Sams et al. (1991) presented incongruent audiovisual stimuli (acoustic/pa/and visual/ka/), which were perceived as/ta/or/ka/, among frequent congruent syllables (acoustic/pa/and visual/pa/), and measured neuromagnetic responses over the left hemisphere. Perceptually deviant stimuli elicited mismatch responses peaking at 180 ms after stimulus onset in the left auditory cortex. A number of studies have replicated this finding (Möttönen et al., 2002; Colin et al., 2002a, 2004). Furthermore, it has been shown that illusory change in sound location due to ventriloquism illusion is detected in the auditory cortices (Stekelenburg et al., 2004, see also Colin et al., 2002b). These findings show that "illusory" changes in sound sequences are processed in the auditory cortices in a similar manner as "real" acoustic changes.

There is also evidence that the early auditory cortex responses to speech sounds are modulated during presentation of a visual speech stimulus (Klucharev et al., 2003; Besle et al., 2004; Möttönen et al., 2004; van Wassenhove et al., 2005). These findings, together with above-mentioned auditory change-detection studies, suggest that acoustic and visual speech inputs are integrated at an early level of sensory processing, within 200 ms after sound onset.

Which sub-regions of the auditory cortex visual speech has access to can be best studied using fMRI. Calvert et al. (1997) found that auditory regions of the human brain were robustly activated during silent speechreading. Activity was observed in the superior temporal cortex and in medial parts of Heschl's gyrus, where the primary auditory cortex is located. Several later studies have replicated the finding that visual speech has access to the auditory regions in the posterior superior temporal cortex (MacSweeney et al., 2000; Campbell et al., 2001; Calvert and Campbell, 2003; Santi et al., 2003; Pekkola et al., 2005). Some of these studies have found activity also in Heschl's gyrus (MacSweeney et al., 2000; Calvert and Campbell, 2003; Pekkola et al., 2005); however, some others have failed to see such activity during speechreading (Bernstein et al., 2002; Paulesu et al., 2003). Moreover, it has been demonstrated that speechreading can modify processing in the auditory brain stem (Musacchia et al., 2006). A brain stem response component peaking at around 11 ms was of longer latency and smaller amplitude when the acoustic speech stimulus was presented together with visual speech rather than alone.

In sum, the human auditory system is capable of processing visual input, not just acoustic input. Access of visual input into the early levels of the auditory processing stream might explain why vision can influence on auditory perception as demonstrated by McGurk and ventriloquism illusions. It should however be noted that heteromodal regions in the human brain, such as STS, also participate in the integration of audiovisual speech and other audiovisual objects (Calvert et al., 2000; Raij et al., 2000; Sekiyama et al., 2003; Wright et al., 2003; Callan et al., 2004; Beauchamp et al., 2004; Möttönen et al., 2004; van Atteveldt et al., 2004). Furthermore, there is evidence that motor regions in the frontal lobe are involved especially in audiovisual speech processing (Ojanen et al., 2005; Skipper et al., 2005). Finding out how the different brain regions contribute to integration of inputs from different senses is a challenge for future research.

References

Alsius A, Navarra J, Campbell R, Soto-Faraco S (2005) Audiovisual integration of speech falters under high attention demands. Curr Biol 15:839–843.

Andersen TS, Tiippana K, Sams M (2002) Endogenous visual spatial attention affects audiovisual speech perception. International Multisensory Research Forum, 3rd Annual Meeeting, Geneva, Switzerland, May, 24th–26th.

Auer ET, Jr., Bernstein LE (1997) Speechreading and the structure of the lexicon: computationally modeling the effects of reduced phonetic distinctiveness on lexical uniqueness. J Acoust Soc Am 102:3704–3710.

Beauchamp MS, Lee KE, Argall BD, Martin A (2004) Integration of auditory and visual information about objects in superior temporal sulcus. Neuron 41:809–823.

Bernstein LE, Auer ET, Jr., Moore JK, Ponton CW, Don M, Singh M (2002) Visual speech perception without primary auditory cortex activation. Neuroreport 13:311–315.

Bertelson P, de Gelder B (2004) The psychology of multimodal perception. In: Crossmodal Space and Crossmodal Attention (Spence C, Driver J, eds), pp. 141–177. Oxford: Oxford University Press.

Besle J, Fort A, Delpuech C, Giard MH (2004) Bimodal speech: early suppressive visual effects in human auditory cortex. Eur J Neurosci 20:2225–2234.

Binnie CA, Montgomery AA, Jackson PL (1974) Auditory and visual contributions to the perception of consonants. J Speech Hear Res 17:619–630.

Burnham D (1998) Language specificity in the development of auditory-visual speech perception. In: Hearing by Eye 2: Advances in the Psychology Speechreading and Auditory-Visual Speech (Campbell R, Dodd B, Burnham D, eds). Hove, East Sussex, UK: Psychology Press Ltd.

Burnham D, Dodd B (1996) Auditory-visual speech perception as a direct process: the McGurk effect in infants and across languages. In: Speechreading by Humans and Machines: Models, Systems and Applications (Stork DG, Hennecke ME, eds), pp. 103–114. Berlin: Springer-Verlag.

Burnham D, Dodd B (2004) Auditory-visual speech integration by prelinguistic infants: perception of an emergent consonant in the McGurk effect. Dev Psychobiol 45:204–220.

Callan DE, Jones JA, Munhall K, Kroos C, Callan AM, Vatikiotis-Bateson E (2004) Multisensory integration sites identified by perception of spatial wavelet filtered visual speech gesture information. J Cogn Neurosci 16:805–816.

Calvert GA (2001) Crossmodal processing in the human brain: insights from functional neuroimaging studies. Cereb Cortex 11:1110–1123.

Calvert G, Campbell R (2003) Reading speech from still and moving faces: the neural substrates of visible speech. J Cogn Neurosci 15:57–70.

Calvert GA, Thesen T (2004) Multisensory integration: methodological approaches and emerging principles in the human brain. J Physiol Paris 98:191–205.

Calvert GA, Bullmore ET, Brammer MJ, Campbell R, Williams SC, McGuire PK, Woodruff PW, Iversen SD, David AS (1997) Activation of auditory cortex during silent lipreading. Science 276:593–596.

Calvert GA, Campbell R, Brammer MJ (2000) Evidence from functional magnetic resonance imaging of crossmodal binding in the human heteromodal cortex. Curr Opin Biol 10:649–657.

Calvert G, Spence C, Stein BE (2004) The Handbook of Multisensory Processes. Cambridge, MA: The MIT Press.

Campbell R, Dodd B, Burnham D (1998) Hearing by Eye II: Advances in the Psychology of Speech-Reading and Audio-visual Speech. Hove: Psychology Press.

Campbell R, MacSweeney M, Surguladze S, Calvert G, McGuire P, Suckling J, Brammer MJ, David AS (2001) Cortical substrates for the perception of face actions: an fMRI study of the specificity of activation for seen speech and for meaningless lower-face acts (gurning). Brain Res Cogn Brain Res 12:233–243.

Colin C, Radeau M, Soquet A, Dachy B, Deltenre P (2002a) Electrophysiology of spatial scene analysis: the mismatch negativity (MMN) is sensitive to the ventriloquism illusion. Clin Neurophysiol 113:507–518.

Colin C, Radeau M, Soquet A, Demolin D, Colin F, Deltenre P (2002b) Mismatch negativity evoked by the McGurk–MacDonald effect: a phonetic representation within short-term memory. Clin Neurophysiol 113:495–506.

Colin C, Radeau M, Soquet A, Deltenre P (2004) Generalization of the generation of an MMN by illusory McGurk percepts: voiceless consonants. Clin Neurophysiol 115:1989–2000.

Cusick CG (1997) The superior temporal polysensory region in monkeys. In: Cerebral Cortex (Rockland KS, Kaas JH, Peters A, eds), pp. 435–468. New York: Plenum.

de Gelder B, Bertelson P (2003) Multisensory integration, perception and ecological validity. Trends Cogn Sci 7:460–467.

Dodd B (1979) Lip reading in infants: attention to speech presented in- and out-of- synchrony. Cognit Psychol 11:478–484.

Dodd B, Burnham D (1988) Processing speechread information. Volta Rev 90:45–60.

Dodd B, Campbell R (1987) Hearing by Eye: The Psychology of Lip-Reading. Hove: Lawrence Erlbaum.

Green KP (1996) The use of auditory and visual information in phonetic perception. In: Speechreading by Humans and Machines: Models, Systems, and Applications (Stork DG, Hennecke ME, eds), pp. 55–77. Berlin: Springer.

Green KP, Kuhl PK, Meltzoff AN, Stevens EB (1991) Integrating speech information across talkers, gender, and sensory modality: female faces and male voices in the McGurk effect. Percept Psychophys 50:524–536.

Howard IP, Templeton WB (1966) Human Spatial Orientation. London: Wiley.

Jackson PL (1988) The theoretical minimal unit for visual speech perception: visemes and coarticulation. Volta Rev 90:99–115.

Jones JA, Munhall KG (1997) The effects of separating auditory and visual sources on audiovisual integration of speech. Can Acoust 25: 13–19.

Klucharev V, Möttönen R, Sams M (2003) Electrophysiological indicators of phonetic and non-phonetic multisensory interactions during audiovisual speech perception. Brain Res Cogn Brain Res 18:65–75.

Kricos PB (1996) Differences in visual intelligibility across talkers. In: Speechreading by Humans and Machines: Models, Systems, and Applications (Stork DG, Hennecke ME, eds), pp. 43–53. Berlin: Springer.

Kuhl PK, Meltzoff AN (1982) The bimodal perception of speech in infancy. Science 218:1138–1141.

MacDonald J, McGurk H (1978) Visual influences on speech perception processes. Percept Psychophys 24:253–257.

MacLeod A, Summerfield Q (1987) Quantifying the contribution of vision to speech perception in noise. Br J Audiol 21:131–141.

MacSweeney M, Amaro E, Calvert GA, Campbell R, David AS, McGuire P, Williams SC, Woll B, Brammer MJ (2000) Silent speechreading in the absence of scanner noise: an event-related fMRI study. Neuroreport 11:1729–1733.

Massaro DW (1984) Children's perception of visual and auditory speech. Child Dev 55:1777–1788.

Massaro DW (1998) Perceiving talking faces. Cambridge, MA: MIT Press.

McGurk H, MacDonald J (1976) Hearing lips and seeing voices. Nature 264:746–748.

Miller GA, Nicely PE (1955) An analysis of perceptual confusions among some English consonants. J Acoust Soc Am 27:338–352.

Mills AE (1987) The development of phonology in the blind child. In: Hearing by Eye: The Psychology of Lipreading (Dodd B, Campbell R, eds), pp. 145–161. London: Lawrence Erlbaum.

Möttönen R (1999) Perception of Natural and Synthetic Audiovisual Finnish Speech. Master's Thesis, Department of Psychology, University of Helsinki.

Möttönen R, Krause CM, Tiippana K, Sams M (2002) Processing of changes in visual speech in the human auditory cortex. Brain Res Cogn Brain Res 13:417–425.

Möttönen R, Schürmann M, Sams M (2004) Time course of multisensory interactions during audiovisual speech perception in humans: a magnetoencephalographic study. Neurosci Lett 363:112–115.

Munhall KG, Gribble P, Sacco L, Ward M (1996) Temporal constraints on the McGurk effect. Percept Psychophys 58:351–362.

Musacchia G, Sams M, Nicol T, Kraus N (2006) Seeing speech affects acoustic information processing in the human brainstem. Exp Brain Res 168:1–10.

Näätänen R (1992) Attention and Brain Function. Hillsdale, NJ: Lawrence Erlbaum.

Ojanen V, Möttönen R, Pekkola J, Jääskeläinen IP, Joensuu R, Autti T, Sams M (2005) Processing of audiovisual speech in Broca's area. Neuroimage 25:333–338.

O'Neill JJ (1954) Contributions of the visual components of oral symbols to speech comprehension. J Speech Hear Disord 19:429–439.

Paulesu E, Perani D, Blasi V, Silani G, Borghese NA, De Giovanni U, Sensolo S, Fazio F (2003) A functional–anatomical model for lipreading. J Neurophysiol 90:2005–2013.

Pekkola J, Ojanen V, Autti T, Jääskeläinen IP, Möttönen R, Tarkiainen A, Sams M (2005) Primary auditory cortex activation by visual speech: an fMRI study at 3 T. Neuroreport 16:125–128.

Pesonen J (1968) Phoneme Communication of the Deaf. Helsinki: Suomalaisen kirjallisuuden kirjapaino Oy.

Raij T, Uutela K, Hari R (2000) Audiovisual integration of letters in the human brain. Neuron 28:617–625.

Sams M, Aulanko R, Hamalainen M, Hari R, Lounasmaa OV, Lu ST, Simola J (1991) Seeing speech: visual information from lip movements modifies activity in the human auditory cortex. Neurosci Lett 127:141–145.

Sams M, Manninen P, Surakka V, Helin P, Kättö R (1998) McGurk effect in Finnish syllables, isolated words, and words in sentences: effects of word meaning and sentence context. Speech Commun 26:75–87.

Santi A, Servos P, Vatikiotis-Bateson E, Kuratate T, Munhall K (2003) Perceiving biological motion: dissociating visible speech from walking. J Cogn Neurosci 15:800–809.

Schroeder CE, Foxe JJ (2002) The timing and laminar profile of converging inputs to multisensory areas of the macaque neocortex. Brain Res Cogn Brain Res 14:187–198.

Schroeder CE, Smiley J, Fu KG, McGinnis T, O'Connell MN, Hackett TA (2003) Anatomical mechanisms and functional implications of multisensory convergence in early cortical processing. Int J Psychophysiol 50:5–17.

Sekiyama K, Kanno I, Miura S, Sugita Y (2003) Audio-visual speech perception examined by fMRI and PET. Neurosci Res 47:277–287.

Skipper JI, Nusbaum HC, Small SL (2005) Listening to talking faces: motor cortical activation during speech perception. Neuroimage 25:76–89.

Slutsky DA, Recanzone GH (2001) Temporal and spatial dependency of the ventriloquism effect. Neuroreport 12:7–10.

Soto-Faraco S, Navarra J, Alsius A (2004) Assessing automaticity in audiovisual speech integration: evidence from the speeded classification task. Cognition 92:B13–B23.

Stein BE (1998) Neural mechanisms for synthesizing sensory information and producing adaptive behaviors. Exp Brain Res 123:124–135.

Stein BE, Meredith MA (1993) Merging of the Senses. Cambridge, MA: The MIT Press.

Stein BE, Meredith MA, Huneycutt WS, McDale L (1989) Behavioral indices of multisensory integration: orientation to visual cues is affected by auditory stimuli. J Cogn Neurosci 1:12–24.

Stein BE, Jiang H, Stanford TE (2004) Multisensory integration in single neurons of the midbrain. In: The Handbook of Multisensory Processes (Calvert G, Spence C, Stein BE, eds), pp. 243–264. Cambridge, MA: The MIT Press.

Stekelenburg JJ, Vroomen J, de Gelder B (2004) Illusory sound shifts induced by the ventriloquist illusion evoke the mismatch negativity. Neurosci Lett 357:163–166.

Sumby WH, Pollack I (1954) Visual contribution to speech intelligibility in noise. J Acoust Soc Am 26:212–215.

Summerfield Q, McGrath M (1984) Detection and resolution of audiovisual incompatibility in the perception of vowels. Q J Exp Psychol 36A:51–74.

Tiippana K, Andersen TS, Sams M (2004) Visual attention modulates audiovisual speech perception. Eur J Cogn Psychol 16:457–472.

van Atteveldt N, Formisano E, Goebel R, Blomert L (2004) Integration of letters and speech sounds in the human brain. Neuron 43:271–282.

van Wassenhove V, Grant KW, Poeppel D (2005) Visual speech speeds up the neural processing of auditory speech. Proc Natl Acad Sci U S A 102:1181–1186.

Wallace MT, Roberson GE, Hairston WD, Stein BE, Vaughan JW, Schirillo JA (2004) Unifying multisensory signals across time and space. Exp Brain Res 158:252–258.

Welch RB, Warren DH (1986) Intersensory interactions. In: Handbook of Perception and Human Performance: Vol. 1. Sensory Processes and Perception (Boff KR, Kaufman L, Thomas JP, eds), pp. 25.21–25.36. New York: Wiley

Wright TM, Pelphrey KA, Allison T, McKeown MJ, McCarthy G (2003) Polysensory interactions along lateral temporal regions evoked by audiovisual speech. Cereb Cortex 13:1034–1043.

38
Multichannel Sound Reproduction

Ville Pulkki

Department of Signal Processing and Acoustics, Helsinki University of Technology, Helsinki, Finland, e-mail: Ville.Pulkki@tkk.fi

1 • Introduction

Spatial reproduction of sound is a field in which the spatial attributes of a real recording room or a virtual space are reproduced to the listener. Spatial attributes include for example directions of sound sources, directions of reflections and envelopment by reverberation. Many such systems employ more than two loudspeakers to create virtual sources. This is called multichannel sound or spatial sound reproduction. Binaural techniques also try to recreate sound faithfully but only in listeners' ear canals. These techniques employ in most cases two audio channels through headphone or loudspeaker reproduction. In this chapter, the main methods for spatial reproduction of sound are reviewed, and the quality that can be achieved with them is discussed.

2 • Concepts of Spatial Hearing by Humans

As a sound signal travels from a sound source to the ear canals of a listener, the signals in both ear canals will be different from the original sound signal and from each other. The transfer functions from sound source to the ear canals are called the head-related transfer functions (HRTF) [1]. They are dependent on the direction of a sound source, and they yield temporal and spectral differences between left and right ear canals. Humans decode these differences and use them to localize sound sources. These differences are called binaural directional cues. Temporal difference is called the interaural time difference (ITD), and spectral difference is called the interaural level difference (ILD). ITD and ILD provide information on where a sound source is in the left vs. right direction dimension. The angle between the sound source direction and the median plane can thus be decoded. The median plane is the vertical plane which divides the space related to a listener to left and right parts. To decode the direction more accurately, monaural spectral cues are applied, and the effect of head rotation on binaural cues is taken into account [1,2]. The precedence effect [1,3] is a suppression of early delayed versions of the direct sound in source direction perception. This helps in reverberant rooms to localize sound sources.

3 • Loudspeaker Layouts

In the history of multichannel audio [4–6], multiple different loudspeaker layouts have been specified. The most frequently used layouts are presented in this chapter. The most common loudspeaker layout is the two-channel setup, called the standard stereophonic setup. It was widely taken into use after the development of single-groove 45°/45° two-channel record in the late 1950s. Two loudspeakers are positioned in front of the listener, separated by 60° from the listener's viewpoint, as presented in Figure 1. In the 1970s, the quadraphonic setup was proposed, in which four loudspeakers are placed evenly around the listener in azimuth angles ±45° and ±135°. This layout was never successful because of the problems related to reproduction techniques of that time and because the layout itself has too few loudspeakers to provide good spatial quality to all directions around the listener [7].

For cinema sound, a system was evolved in which the frontal image stability of standard stereophonic setup was enhanced by one extra center channel, and two surround channels are used to create atmospheric effects and room perception. This kind of a setup was first used in Dolby's surround sound system for cinemas from 1976 [5]. Later, the layout was investigated [8], and ITU gave a recommendation about the layout in 1992 [9]. In the late 1990s, this layout became common also in domestic use. It is widely referred to as the 5.1 system, where 5 stands for the number of loudspeakers, and .1 stands for the low-frequency

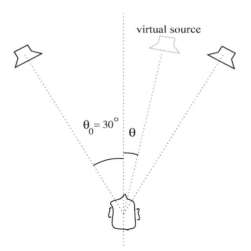

FIGURE 1 Standard stereophonic listening configuration.

channel. In the recommendation, three frontal loudspeakers are in directions 0° and ±30°, and two surround channels in ±110 ± 10°, as shown in Figure 2. The system has been criticized that it can not deliver good directional quality elsewhere than in the front [7]. To achieve better quality, it can be extended by adding loudspeakers. Layouts with 6–12 loudspeakers have been proposed and are presented in [7].

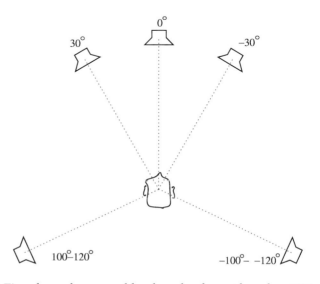

FIGURE 2 Five-channel surround loudspeaker layout based on ITU recommendation BS775.1.

In computer music, media installations and academic projects, loudspeaker setups, in which the loudspeakers are in equal spacing, have been used. In horizontal arrays, the number of loudspeakers can be, for example, six (hexagonal array) or eight (octagonal array). In wave field synthesis, see Sect. 4.3, the number of loudspeakers is typically between 20 and 200. In theaters and in virtual environment systems, there exist systems in which loudspeakers are placed also above and/or below the listener.

4 • Reproduction Methods

In this chapter, reproduction methods which use the loudspeaker layouts presented in the previous section are reviewed. The methods aim to position monophonic sound tracks in arbitrary directions. The directions do not necessarily coincide with any loudspeaker, as shown in Figure 1. Such perceived sound sources are called virtual sources.

4.1 Amplitude Panning

Amplitude panning is the most frequently used virtual source positioning technique. In it, a sound signal is applied to loudspeakers with different amplitudes, which can be formulated as

$$x_i(t) = g_i x(t), \quad i = 1, \ldots, N, \tag{1}$$

where $x_i(t)$ is the signal to be applied to loudspeaker i, g_i is the gain factor of the corresponding channel, N is the number of loudspeakers, and t is the time variable. The listener perceives a virtual source, the direction of which is dependent on the gain factors.

4.1.1 Two-Channel Stereophony

In stereophony, two loudspeakers are placed in front of a listener, (Figure 1). Variable θ denotes the perceived azimuth of the virtual source. A panning law estimates θ from the gain factors of loudspeakers. The estimated direction is called the panning direction or panning angle. In [10], it has been found that amplitude panning provides consistent ITD cues up to 1.1 kHz, and roughly consistent ILD cues above 2 kHz for a listener in the best listening position. Outside the best listening position, the virtual source is localized toward the nearmost loudspeaker which emanates sound due to the precedence effect.

In the sine law [11] of perceived direction, the wave propagation path from loudspeakers to ears was modeled with straight lines. The path from the contralateral loudspeaker therefore penetrates the listener's head, which is highly unnatural. It is presented as

$$\frac{\sin\theta_S}{\sin\theta_0} = \frac{g_1 - g_2}{g_1 + g_2}, \quad (2)$$

where θ_S is an estimate of perceived azimuth angle (panning angle) of a virtual source. θ_0 is the loudspeaker base angle, as in Figure 1. The equation does not set limitations to θ_S, but in most cases, its value is set to satisfy $|\theta_S| \leq \theta_0$. If $|\theta_S| > \theta_0$, the amplitude panning will produce antiphase loudspeaker signals which may distort the virtual source [1].

Bennett et al. [12] derived the tangent law by approximating the propagation path from contralateral loudspeaker to ear with a curved line around the head

$$\frac{\tan\theta_T}{\tan\theta_0} = \frac{g_1 - g_2}{g_1 + g_2}, \quad (3)$$

which has been found to estimate perceived direction best in listening tests in anechoic listening [10]. There are also other panning laws, reviewed in [10].

The panning laws set only the ratio between the gain factors. To prevent undesired change in loudness of virtual source depending on panning direction, the sum of squares of gain factors should be normalized:

$$\sqrt{\sum_{n=1}^{N} g_n^2} = 1. \quad (4)$$

This normalization equation is used in real rooms with some reverberation. Depending on listening room acoustics, different normalization rules may be used [13].

4.1.2 2-D Loudspeaker Setups

In 2-D loudspeaker setups, all loudspeakers are on the horizontal plane. Pair-wise amplitude panning [14] methods can be used in such reproduction systems. The sound signal is applied only to two adjacent loudspeakers of the loudspeaker setup at one time. The pair between which the panning direction lies is selected. Different formulations for pair-wise panning are Chowning's law [14], which is not based on any psychoacoustic criteria, or 2-D vector base amplitude panning (VBAP) [15], which is a generalization of the tangent law (Eq. 3) for stereophonic panning.

In VBAP, a loudspeaker pair is specified with two vectors. The unit-length vectors $\mathbf{l}_m$ and $\mathbf{l}_n$ point from the listening position to the loudspeakers. The intended direction of the virtual source (panning direction) is presented with a unit-length vector **p**. Vector **p** is expressed as a linear weighted sum of the loudspeaker vectors

$$\mathbf{p} = g_m \mathbf{l}_m + g_n \mathbf{l}_n. \tag{5}$$

Here g_m and g_n are called gain factors of respective loudspeakers. The gain factors can be solved as

$$\mathbf{g} = \mathbf{p}^T \mathbf{L}_{mn}^{-1}, \tag{6}$$

where $\mathbf{g} = [g_m \; g_n]^T$ and $\mathbf{L}_{mn} = [\mathbf{l}_m \; \mathbf{l}_n]$. The calculated factors are used in amplitude panning as gain factors of the signals applied to respective loudspeakers after suitable normalization, e.g., $\|\mathbf{g}\| = 1$.

The directional quality achieved with pair-wise panning was studied in [10]. When the loudspeakers are symmetrically placed on the left and right of the listener, VBAP estimates the perceived angle accurately. When the loudspeaker pair is not symmetrical with the median plane, the perceived direction is biased toward the median plane [10], which can be more or less compensated [16].

When there is a loudspeaker in the panning direction, the virtual source is sharp, but when panned between loudspeakers, the binaural cues are unnatural in some degree. This means that the directional perception of the virtual source varies with panning direction, which can be compensated by applying sound always to more than one loudspeaker [17,18]. Since the perceived direction collapses to the nearmost loudspeaker producing a specific sound, the maximal directional error is of the same order of magnitude with the angular separation of loudspeakers from the listener's viewpoint in pair-wise panning. The curvature of the wavefront generated with pair-wise panning cannot be controlled.

4.1.3 3-D Loudspeaker Setups

A 3-D loudspeaker setup denotes here a setup in which all loudspeakers are not in the same plane with the listener. Typically, this means that there are some elevated and/or lowered loudspeakers added to a horizontal loudspeaker setup. Triplet-wise panning can be used in such setups [15]. In it, a sound signal is applied to a maximum of three loudspeakers at one time that form a triangle from the listener's viewpoint. If more than three loudspeakers are available, the setup is divided into triangles, one of which is used in the panning of a single virtual source at one time, as shown in Figure 3. Three-dimensional vector base amplitude panning (3-D VBAP) is a method to formulate such setups [15]. It is formulated in an equivalent way as with pair-wise panning

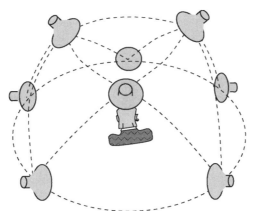

FIGURE 3 A 3-D triangulated loudspeaker system for triplet-wise panning.

in the previous section. However, the number of gain factors and loudspeakers is naturally three in the equations. The angle between the median plane and virtual source is estimated correctly with VBAP in most cases, as in pair-wise panning. However, the elevation of a sound source is individual to each subject [10].

4.1.4 Ambisonics

Ambisonics is basically a microphone technique [19]. However, it can also be simulated to perform synthesis of spatial audio [20] to 2- or 3-D loudspeaker setups. In this case, it is an amplitude panning method in which a sound signal is applied to all loudspeakers placed evenly around the listener with gain factors

$$g_i = \frac{1}{N} \sum_{m=1}^{M} \{1 + 2p_m \cos(m\alpha_i)\}, \qquad (7)$$

where g_i is the gain of ith speaker, N is the number of loudspeakers, α is the angle between loudspeaker and panning direction, $\cos(m\alpha_i)$ represents a single spherical harmonic with order of m, M is the order of Ambisonics and p_m are the gains for each spherical harmonic [21,22]. When the order is low, sound signal emanates from all loudspeakers, which causes some spatial artifacts due to unnatural behavior of binaural cues [23]. In such cases also, when listening outside the best listening position, the sound is perceived at the nearmost loudspeaker which produces the sound. This effect is more prominent with first-order Ambisonics than with pair- or triplet-wise panning, since in Ambisonics, virtually all loudspeakers produce the same sound signal. The sound is also colored, for the same reason, i.e., multiple propagation paths of the same signal to the ears produce comb-filter effects. Conventional microphones can be used to realize first-order Ambisonics, as shortly reviewed in Sect. 5.

When the order is increased, the cross talk between loudspeakers can be minimized by optimizing gains of spherical harmonics for each loudspeaker in a listening setup [24]. Using higher order spatial harmonics increases both directional and timbral virtual source quality, since there is less cross talk between loudspeakers. The physical wave field reconstruction is then more accurate, and different curvatures of wavefronts, as well as planar wavefronts can be produced [21], if a large enough number of loudspeakers is in use.

4.2 Time Panning

When a constant delay is applied to one loudspeaker in stereophonic listening, the virtual source is perceived to migrate toward the loudspeaker that radiates the earlier sound signal [1]. Maximal effect is achieved asymptotically when the delay is approximately 1.0 ms or more.

In tests with different signals, it has been found, however, that the perceived direction of virtual sources is dependent on frequency [25,26]. The produced binaural cues vary with frequency, and different cues suggest different directions for virtual sources [27]. It may thus generate a "spread" perception of direction of sound, which is desirable in some cases. The effect is dependent on listening position. For example, if the sound signal is delayed by 1 ms in one loudspeaker, the listener can compensate the delay by moving 30 cm toward the delayed loudspeaker.

4.3 Wave Field Synthesis

Wave field synthesis is a technique that requires a large number of carefully equalized loudspeakers [28,29]. It targets to reconstruct the whole sound field in a listening room. When a virtual source is reproduced, the sound for each loudspeaker is delayed and amplified in a manner that a desired circular or planar sound wave occurs as a superposition of sounds from each loudspeaker. The virtual source can be positioned far behind the loudspeakers, or in some cases, even in the space inside the loudspeaker array, as shown in Figure 4. The loudspeaker signals have to be equalized depending on virtual source position [30].

Theoretically, the wave field synthesis is superior as a technique, the perceived position of the sound source is correct within a very large listening area. Unfortunately, to create a desired wave field in the total area inside the array, it demands that the loudspeakers are at a distance of maximally a half wavelength from each other. The area where a perfect sound field synthesis is achieved shrinks with increasing frequency [21]. In practice, due to the perception mechanisms of humans, more error can be allowed above approximately 1.5 kHz. Arrays for wave field synthesis have been built for room acoustics control

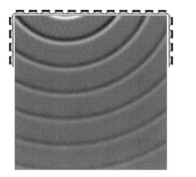 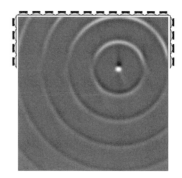

FIGURE 4 Wave field synthesis concept. A desired 2-D sound field can be constructed using a large loudspeaker array. Figure reprinted from [29] with permission.

and enhancement to be used in theatres and multipurpose auditoria [29].

4.4 Binaural Reproduction

Binaural reproduction is mentioned here for completeness, although it is reviewed in more detail in Chapter 11.1, part II. In headphone listening, a monophonic sound signal can be positioned virtually in any direction, if HRTFs for both ears are available for the desired virtual source direction [31,32]. A sound signal is filtered with a digital filter modeling the measured HRTFs. The method simulates the ear canal signals that would have been produced if a sound source existed in a desired direction. If a listener moves his/her head during listening, then the movements of his/her should also be taken into account in processing, otherwise the sound stage will be moving along when the listener rotates his/her head and that may cause inside-head localization.

If the cross talk between a loudspeaker and the contralateral ear in stereophonic setup is canceled by feeding an antiphase signal to the other loudspeaker in a carefully controlled manner, it is possible to control precisely the sound signals in a listener's ear canals [32–35]. However, the best effect is audible only in a small listening area, and the listener should keep his ears in constant distance from the loudspeakers with about 1 cm accuracy. The listening room should also be anechoic, since acoustic reflections deteriorate cross talk canceling.

5 • Capturing Sound for Multichannel Audio

Techniques, in which a microphone is positioned close to a sound source, capture the direct sound with prominently higher amplitude than the reverberation of the room. In mixing phase, the recorded sound tracks of different sound sources are positioned with a multichannel loudspeaker system using a desired panning technique. However, often it is desired to also introduce reverberation of the room into the mixed signals. It can be captured with distant microphones, and it can be introduced to the sound stage in the mixing. Alternatively, a desired room effect can be introduced using digital reverberators, which either mimic room response with a digital filter structure or use measured room impulse responses in convolution [36,37].

The microphones which have flat frequency response over audible frequency range have first-order directional patterns, such as figure-of-eight, cardioid or hypercardioid, or they have zeroth-order pattern, which means the omnidirectional microphone. Figure-of-eight microphones and hypercardioids have two lobes in their directional patterns. One of the lobes reproduces the sound in correct phase and is called the positive lobe, and the other lobe reproduces the sound phase reversed, which is called the negative lobe. Higher order microphones suffer from coloration in sound spectrum.

When the microphones are relatively far from sound sources, the acoustical influence of the recording room is prominent. There are different approaches to record for a multichannel loudspeaker system in such cases. In the coincident microphone techniques, two or more directive microphones are positioned as close to each other as possible but pointing to different directions [26,38,39]. The microphone signals thus differ generally only in amplitude, in a fashion similar to amplitude panning presented in Sect. 4.1. Only in cases where dipole microphones are used, the negative lobes may introduce 180° phase shifts between microphone channel signals.

In spaced microphone techniques, there are considerable distances between microphones [7,39]. These techniques produce both amplitude and phase differences between microphone channels. They can be treated as a combination of amplitude panning and time panning (Sects. 4.1 and 4.2).

5.1 Microphone Techniques for Stereophonic Setup

The most significant microphone techniques for the two-channel stereophonic setup are presented here briefly. An analysis of the presented techniques computed with a binaural auditory model of the presented techniques has been published in [40]

Coincident techniques are often denoted XY techniques. If cardioids or hypercardioids are used, the angle between microphones is typically 60°–120°, although angles upto 180° may be used [39]. It produces good directional accuracy [26]. However, these techniques have been characterized to lack "sense of space" [39].

Two figure-of-eight microphones in XY positioning was the first XY technique presented by Blumlein [38]. In it, the positive lobes of dipole microphones are directed 45° and −45°, respectively. It reproduces equally the sound energy from all directions. However, the sound arriving from side produces 180° phase reversal between the microphone signals. The sound is thus reproduced in loudspeakers out of phase, which may produce coloration and directional artifacts.

MS stereo was also presented by Blumlein [38]. In it, a directive microphone (M) points to the front and a figure of eight microphones (S) to the side. Later, weighted sums are performed from the signals, which create two virtual microphone signals. The directional patterns of these microphones can be adjusted as needed in mixing phase by changing the weights in summing.

In spaced microphone techniques, the distance between microphones is typically between 60 and 100 cm. The microphones can be either omnidirectional or directional microphones. The spacing between the microphones gives better feeling of "ambience" or "air" [26], however it deteriorates the directional accuracy and may produce some comb-filter effects.

In near coincident microphone techniques, the distance between microphones is selected in a way that the microphone signals are effectively in same phase at low frequencies, and in different phase at high frequencies. This can be thought to compromise coincident and spaced techniques: they provide moderately good directional quality combined with better "sense of depth" [39]. Probably the most used near coincident technique is the ORTF technique, in which two cardioid microphones are spaced by 17 cm in left–right dimension and face to ±110°.

5.2 Microphone Techniques for 5.1 Loudspeaker Setup

A fundamental problem with microphone techniques for the 5.1 loudspeaker setup is that good-quality microphones are available only with directional patterns of zeroth or first order. When such microphones are used with a 5.1 setup, there will be a considerable amount of cross talk between loudspeaker channels, which results in coloration and directional artifacts. To overcome this, a microphone has been recently proposed which consists of eight omnidirectional microphones and postprocessing which composes virtual microphones with moderately narrow directional patterns [41]. The too high coherence produced by first-order directional patterns can be enhanced using information of directional properties of sound field in frequency bands [46], or by cardioid microphones pointing to opposite direction combined with digital signal processing [47].

In practice, the recording is typically performed by positioning omnidirectional or directional microphones with distances of

10 cm to 1.5 m from each other and by directing the microphones in a desired manner. There is no generally accepted theory for an optimal setup, although some heuristic models to formalize common practices have been presented [42]. A large number of spaced microphone techniques have been tested, which are reviewed by Rumsey [7].

5.3 Microphone Techniques for Loudspeaker Arrays

Both higher order Ambisonics and wave field synthesis utilize large horizontal arrays of loudspeakers. Higher order Ambisonics is based on measuring the sound field with a set of highly directional microphones in a single point. Unfortunately such microphones do not exist as single units. It has been however proven that microphone arrays can be used for recording for high-order Ambisonics [43].

Wave field synthesis requires capturing the sound field with a large number of microphones [28]. Circular microphone arrays have been found superior to linear arrays [44]. The measured impulse responses with such arrays have been used in the analysis of room acoustics, which is called wave field analysis [45].

References

1. J. Blauert. *Spatial Hearing, Revised edition*. MIT Press, Cambridge, MA, 1997.
2. R. H. Gilkey and T. R. Anderson, editors. *Binaural and Spatial Hearing in Real and Virtual Environments*. Lawrence Erlbaum, Mahwah, NJ, 1997.
3. P. M. Zurek. The precedence effect. In W. A. Yost and G. Gourewitch, editors, *Directional Hearing*, p. 3–25. Springer, New York, 1987.
4. G. Steinke. Surround sound — the new phase. an overview. In *Audio Engineering Society 100th Convention Preprint # 4286*, Copenhagen, Denmark, 1996.
5. M. F. Davis. History of spatial coding. *J. Audio Eng. Soc.*, 51(6): 554–569, June 2003.
6. E. Torick. Highlights in the history of multichannel sound. *J. Audio Eng. Soc.*, 46(1/2):27–31, January/February 1998.
7. F. Rumsey. *Spatial Audio*. Focal Press, Oxford, England, 2001.
8. G. Theile. HDTV sound systems: How many channels? In *Proceedings of AES 9th International Conference "Television Sound Today and Tomorrow"*, p. 217–232, Detroit, MI, Feb. 1–2 1991.
9. ITU-R Recommendation BS.775-1. Multichannel stereophonic sound system with and without accompanying picture. Technical report, International Telecommunication Union, Geneva, Switzerland, 1992–1994.
10. V. Pulkki. *Spatial Sound Generation and Perception by Amplitude Panning Techniques*. Doctoral thesis, Helsinki University of Technology, Laboratory of Acoustics and Audio Signal Processing, Espoo, Finland, August 2001.

11. B. B. Bauer. Phasor analysis of some stereophonic phenomena. *J. Acoust. Soc. Am.*, 33(11):1536–1539, November 1961.
12. J. C. Bennett, K. Barker, and F. O. Edeko. A new approach to the assessment of stereophonic sound system performance. *J. Audio Eng. Soc.*, 33(5):314–321, May 1985.
13. F. R. Moore. *Elements of Computer Music*. Prentice Hall, Englewood Cliffs, NJ, 07632, 1990.
14. J. Chowning. The simulation of moving sound sources. *J. Audio Eng. Soc.*, 19(1):2–6, 1971.
15. V. Pulkki. Virtual source positioning using vector base amplitude panning. *J. Audio Eng. Soc.*, 45(6):456–466, June 1997.
16. V. Pulkki. Compensating displacement of amplitude-panned virtual sources. In *the AES 22nd International Conference on Virtual, Synthetic and Entertainment Audio*, p. 186–195, Espoo, Finland, June 15–17 2002.
17. V. Pulkki. Uniform spreading of amplitude panned virtual sources. In *Proceedings of the 1999 IEEE Workshop on Applications of Signal Processing to Audio and Acoustics*, Mohonk Mountain House, New Paltz, NY, 1999.
18. R. Sadek and C. Kyriakakis. A novel multichannel panning method for standard and arbitrary loudspeaker configurations. In *AES 117th Convention paper # 6263*, San Francisco, CA, 2004.
19. M. A. Gerzon. Panpot laws for multispeaker stereo. In *The 92nd Convention 1992 March 24–27 Vienna*. Audio Engineering Society, Preprint No. 3309, 1992.
20. D. G. Malham and A. Myatt. 3-d sound spatialization using ambisonic techniques. *Comput. Music J.*, 19(4):58–70, 1995.
21. J. Daniel, R. Nicol, and S. Moreau. Further investigations of high order ambisonics and wavefield synthesis for holophonic sound imaging. In *Proceedings of 114th Audio Engineering Society Convention*, 2003. Paper # 5788.
22. G. Monro. In-phase corrections for ambisonics. In *Proceedings of Internation Computer Music Conference*, p. 292–295, Berlin, Germany, 2000.
23. V. Pulkki and T. Hirvonen. Localization of virtual sources in multichannel audio reproduction. *IEEE Trans Speech Audio Proc.*, 2005.
24. P. G. Craven. Continuous surround panning for 5-speaker reproduction. In *AES 24th Int. Conf. Multichannel Audio*, 2003.
25. D. H. Cooper. Problems with shadowless stereo theory: Asymptotic spectral status. *J. Audio Eng. Soc.*, 35(9):629–642, September 1987.
26. S. P. Lipshitz. Stereophonic microphone techniques. Are the purists wrong? *J. Audio Eng. Soc.*, 34(9):716–744, 1986.
27. V. Pulkki, M. Karjalainen, and J. Huopaniemi. Analyzing virtual sound source attributes using a binaural auditory model. *J. Audio Eng. Soc.*, 47(4):203–217, April 1999.
28. A. J. Berkhout, D. de Vries, and P. Vogel. Acoustic control by wave field synthesis. *J. Acoust. Soc. Am.*, 93(5), May 1993.
29. D. Vries and M. Boone. Wave field synthesis and analysis using array technology. In *Proceedings of the 1999 IEEE Workshop on Applications of Signal Processing to Audio and Acoustics*, p. 15–18, Mohonk Mountain House, New Paltz, NY 1999.

30. A. Apel, T. Röder, and S. Brix. Equalization of wave field synthesis systems. In *Audio Eng. Soc. 116th Convention Paper # 6121*, 2004.
31. H. Møller, M. F. Sørensen, D. Hammershøi, and C. B. Jensen. Head-related transfer functions of human subjects. *J. Audio Eng. Soc.*, 43(5):300–321, May 1995.
32. D. R. Begault. *3-D Sound for Virtual Reality and Multimedia*. AP Professional, Cambridge, MA, 1994.
33. D. H. Cooper and J. L. Bauck. Prospects for transaural recording. *J. Audio Eng. Soc.*, 37(1/2):3–39, January/February 1989.
34. B. Gardner. *3-D Audio Using Loudspeakers*. PhD thesis, Massachusetts Institute of Technology, Massachusetts, USA, 1997.
35. O. Kirkeby, P. A. Nelson, and H. Hamada. Local sound field reproduction using two closely spaced loudspeakers. *J. Acoust. Soc. Am.*, 104:1973–1981, October 1998.
36. B. Gardner. Reverberation algorithms. In M. Kahrs and K. Brandenburg, editors, *Applications of Digital Signal Processing to Audio and Acoustics*. Kluwer Academic Publishers, Norvell, MA, 1998.
37. J. Merimaa and V. Pulkki. Spatial impulse response rendering I: Analysis and synthesis. *J. Audio Eng. Soc.*, 53(12):1115–1127, 2005.
38. A. D. Blumlein. U.K. Patent 394,325, 1931. Reprinted in Stereophonic Techniques, Audio Eng. Soc., NY, 1986.
39. R. Streicher and W. Dooley. Basic stereo microphone perspectives – a review. *J. Audio Eng. Soc.*, 33(7/8), 1985.
40. V. Pulkki. Microphone techniques and directional quality of sound reproduction. In *Proceedings of 112th AES Convention, Preprint # 5500*, Munich, Germany, 2002. Audio Engineering Society.
41. A. Laborie, R. Bruno, and S. Montoya. Designing high spatial resolution microphones. In *Audio Eng. Soc. 117th Convention Paper # 6231*, 2003.
42. M. Williams. Multichannel sound recording using 3, 4 and 5 channel arrays for front sound stage coverage. In *Proceedings of 117th AES Convention, Preprint # 6230*, San Francisco, CA, 2004.
43. R. Nicol and M. Emerit. 3d-sound reproduction over an extensive listening area: a hybrid method derived from holophony and ambisonic. In *Proceedings of AES 16th International Conference*, p. 436–453, 1999.
44. E. Hulsebos, D. de Vries, and E. Bourdillat. Improved microphone array configurations for auralization of sound fields by wave-field synthesis. *J. Audio Eng. Soc.*, 50(10):779–790, October 2002.
45. A. Berkhout, D. de Vries, and J. Sonke. Array technology for acoustic wave field analysis in enclosures. *J. Acoust. Soc. Am.*, 102(5): 2757–2770, 1997.
46. V. Pulkki. Spatial sound reproduction with directional audio coding. *J. Audio Eng. Soc.*, 55(6):503–516, 2007.
47. C. Faller. A highly directive 2-capsule based microphone system. In *Proceedings of 123th AES Convention*, Preprint #7313, New York, 2007.

39
Virtual Acoustics

Tapio Lokki and Lauri Savioja

*Department of Media Technology,
Helsinki University of Technology, Helsinki, Finland,
e-mail: Tapio.Lokki@tkk.fi, Lauri.Savioja@tkk.fi*

1 • Introduction

The term virtual acoustics is often applied when sound signal is processed to contain features of a simulated acoustical space and sound is spatially reproduced either with binaural or with multichannel techniques. Therefore, virtual acoustics consists of spatial sound reproduction and room acoustics modeling. A synonym for virtual acoustics is auralization–rendering audible– which was first defined by Kleiner et al. [1].

Virtual acoustics can be created with perceptually or physically based modeling techniques. The physically based approach seeks to simulate exactly the propagation of sound from the source to the listener for a given room. Conceptually, this approach is straightforward since by modeling physical principles of sound propagation in air and reflections from boundaries as correctly as possible a high-quality virtual acoustics should be achieved. However, for many applications, e.g., in computer music and professional audio, perceptually based approach produces relevant and accurate enough results. Rendering in perceptually based virtual acoustics is often performed without any knowledge of the room geometry, and perceptually optimized efficient algorithms are used [2]. In other words, perceptually based virtual acoustics endeavors to reproduce only the perceptually salient characteristics of reverberation (Section 75.7) [3].

Room acoustics modeling, required for physically based virtual acoustics, can be realized with several different ways. In principle, the most accurate results are obtained with wave-based techniques such as finite element method (FEM), boundary element method (BEM) [4], and finite-difference time domain (FDTD) methods [5–7]. They all try to numerically solve the wave equation in a modeled space. The more practical techniques are ray based, such as ray-tracing [8,9], beam-tracing [10], and the image source method [11,12]. None of the room acoustics modeling techniques is optimal in all geometries and over the whole audible frequency range, but each of them can be applied in the areas they are best at. A survey of room acoustics modeling has been presented by Svensson and Kristiansen [13].

In this chapter, signal processing methods applied in physically based virtual acoustics are presented. The presentation follows mainly the design of the DIVA auralization system [14,15] which has been developed at the Helsinki University of Technology since 1994. In addition to auralization, one signal processing-oriented room acoustics modeling method, namely the waveguide mesh, is presented in Section 3.

2 • Auralization and Signal Processing

In ray-based methods, the wavelength of sound is assumed to be zero and sound propagation is modeled with sound rays. With them reflection paths inside an enclosure can be efficiently found, but all frequency-dependent acoustical phenomena caused by the wave nature of sound are neglected. An energy response, needed to compute room acoustical parameters of a space, can be obtained with ray-based methods. Such parameters are often computed in octave bands. If path tracing is performed separately on each band, frequency-dependent attenuations can be simulated with gain factors. However, in auralization it is practical to perform room acoustics modeling with sound pressure signals. In addition, the whole audible frequency range is needed to be processed at once. These requirements are fulfilled if room acoustics modeling is performed with the image source method and the frequency-dependent phenomena are added by processing the sound emitted by each image source with digital filters. In other words, with image sources the propagation paths are found and with digital filters each impulse, corresponding to a ray path, is filtered to add wave nature features and correct reproduction direction. Figure 1 presents two reflection paths and the filters applied on both paths.

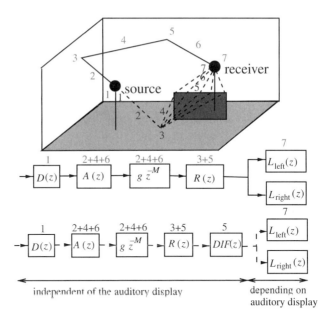

FIGURE 1 Two sound propagation paths with applied filters to perform physically based rendering. The filters simulate sound source directivity ($D(z)$), air absorption ($A(z)$), delay and gain based on distance (gz^{-M}), material absorption ($R(z)$), diffraction (DIF(z)), and binaural filtering ($L_{\text{left,right}}(z)$). Illustration adapted from Strauss [16].

The filtering approach in auralization enables interactive applications in which the listener or the sound source can move in a virtual space. In such dynamic rendering, the filter parameters are updated, e.g., 20 Hz update rate [17] while the sound processing is performed at 48 kHz frequency. The interactivity makes filter design challenging since as low-order filters as possible have to be used. Filter design is thus often optimized from a perception point of view. In addition, filters have to be designed so that interpolation of filter coefficients between updates is possible. In the following, different solutions to design and apply filters, presented in Figure 1, are discussed.

2.1 Sound Source Directivity

Sound sources have radiation characteristics that are dependent on frequency and direction of radiation. In auralization, a sound source is typically modeled as a point source and radiation to different directions is modeled with directional filtering [18–21]. Such simplification overrules the fact that in reality sound sources have a finite volume and they emit sound from several locations of their body. For example, most musical instruments radiate sound from their whole body, and in addition the radiation characteristics vary according to the played note.

The modeling of sound source directivity is a problem similar to HRTF modeling (Section 11.1). For filter design, the directivity data of each sound source has to be measured or derived with mathematical models. The sparse measurement data is usually completed by interpolating measurements to form a dense grid of directivity data. Then, digital filters are fitted to the frequency response of each direction and these filters are applied in rendering. The filter fitting can be performed in a warped frequency domain [22,23] to optimize low-order filters for human auditory perception [15].

In auralization, one directional filter ($D(z)$ in Figure 1) has to be applied for every single reflection. One way to enhance the performance in real-time rendering is to separate the directivity filter into two parts, namely diffuse field part and directional part. Since the diffuse field filter is common to all directions, it has to be applied only once, while the directional filter is applied to every reflection. The diffuse field power spectrum for diffuse field filter can be derived from all measured responses ($H_{\mathrm{dir},i}(\omega, \theta, \phi)$) by power-averaging:

$$S_{\mathrm{d}}(\omega, \theta, \phi) = \sqrt{\frac{1}{k}\sum_{i=1}^{k} |H_{\mathrm{dir},i}(\omega, \theta, \phi)|^2 \cos\phi}, \quad (1)$$

where k is the number of responses, θ is the azimuth angle, ϕ is the elevation angle, and ω is the angular frequency.

The use of a diffuse field filter has many advantages. For example, it makes the design of the primary directivity filters much easier, by flattening the response at low and high frequencies. This is also called diffuse field equalization and it allows the use of computationally efficient low-order filters for the primary directivity filter.

2.2 Air Absorption

The absorption of sound in air depends mainly on distance, temperature, and humidity. The equations for calculation of air absorption are standardized [24], and they can be applied to calculate the target responses for filter design [14,25]. The filters ($A(z)$ in Figure 1) can be implemented as low-order IIR filters which are fitted, e.g., to minimum-phase complex frequency responses. In dynamic rendering, the coefficients of air absorption filters need not be interpolated, despite the recursive filters. Instead, filter coefficients can be changed without any audible effects if the grid of the precalculated filters is dense enough, e.g., in 1-m steps [15].

2.3 Distance Delay and Attenuation

The distance of an image source to the listener, i.e., the length of the ray path, defines the delay and attenuation of sound. The frequency independent attenuation according to $1/r$-law, caused by the spherical propagation of sound waves (see, e.g., [26]), is trivial to compute. In dynamic rendering, in which sound source or listener can move, the distance delay has to be implemented with fractional delays [27] to obtain smooth and continuous output signal. The first-order FIR fractional delay filter has been found accurate enough for this purpose. Correspondingly, interpolation of the distance attenuation coefficient between updates has to be applied [14].

2.4 Material Absorption

The effect of boundary material to a sound wave should be modeled somehow to obtain realistic auralization. The temporal and spectral behavior of a reflected sound is a function of incident angle. In addition, scattering and diffraction make it impossible to develop numerical models that are accurate in all aspects. For these reasons, in room acoustics modeling usually simplified material absorption models are applied. Typically, only angle-independent material absorption that can be measured in reverberation room or with the impedance tube technique is applied.

In auralization, the material absorption is modeled by utilizing reflection filters ($R(z)$ in Figure 1). The input data for filter design is measured absorption coefficients, available in octave bands. For filter design, absorption coefficients are transformed into reflectance data. The resulting amplitudes are transformed into the frequency domain, and filter fitting is performed in a warped frequency domain as in the case of sound source directivity filters. When sound is reflected from two or more surfaces, the absorption data can be cascaded so that only one filter is needed in rendering [25].

2.5 Diffraction

Modeling of sound propagation paths with rays totally neglects diffraction. However, Svensson et al. [28] have derived a mathematical solution with which edge diffraction can be added to the image source method. In fact, edge diffraction modeling compensates also the low-frequency error of the image source method. Such error is an inherent problem of the image source method due to the finite size of surfaces. In edge diffraction modeling, multiple rays are traced via all points of the edge and the sum of all these rays forms an impulse response for that edge which can be of a finite length. Although sound passes the edge through all points along the edge, most of the energy is concentrated to the shortest path point of the edge. Based on this, it can be assumed that diffraction can be modeled with a point-like secondary source positioned to the shortest path point. So, diffraction can be approximated with one edge image source, filtered with a diffraction filter DIF(z) in Figure 1. For efficient auralization a low-order filter can be fitted to a computed edge impulse response [29].

2.6 Directional Filtering

Each modeled reflection reaches the listening position from a defined direction, and in auralization, the correct reproduction direction is obtained with directional filtering, filters $L_{\{left,right\}}(z)$ in Figure 1. The filtering required depends of the applied auditory display. For binaural reproduction HRTF filters [30] (Section 11.1) are applied, and for multichannel reproduction many different techniques are available (Chapter 38).

For dynamic auralization, HRTF filters have to be designed so that interpolation between them is possible. In the DIVA system [14], the HRTF filters were realized by decomposing the measured HRTF functions to a frequency-independent delay (ITD) and a minimum-phase part. Then, FIR filters were fitted to the minimum-phase responses. Such minimum-phase FIR filtering enables direct interpolation of filter coefficients in a dynamic auralization.

2.7 Late Reverberation

In real-time auralization, only early reflections can be modeled with image sources and digital filtering. As reflection order grows, more and more reflections reach the listening position and reflections build a nearly diffuse reverberation field which is close to exponentially decaying random noise [31]. Under these assumptions, the late reverberation does not have to be modeled as individual reflections with certain wave features and directions. Therefore, to optimize computation in late reverberation modeling, recursive digital filter structures have been designed [3,32], whose responses model the characteristics of real room responses, such as the frequency-dependent reverberation time (Section 75.7).

3 • Waveguide Meshes in Room Acoustics

The digital waveguide mesh provides a tool for simulating wave propagation in a multidimensional system. The most practical applications lie in physical modeling of musical instruments and room acoustics.

A digital waveguide mesh [33] is formed of a regular array of discretely spaced 1D digital waveguides arranged along each perpendicular dimension, interconnected at their intersections

(Chapter 25). The resulting mesh of a 3D space is a regular rectangular grid in which each node is connected to its six neighbors by unit delays [33,34].

Two conditions must be satisfied at a lossless junction connecting lines of equal impedance [33]:

1. Sum of inputs equals the sum of outputs (volume velocity flows add to zero).
2. Sound pressures at each intersecting waveguide are equal at the junction (continuity of pressure).

Based on these conditions, a difference equation for the nodes of an N-dimensional rectangular mesh [34] can be derived:

$$p_k(n) = \frac{1}{N}\left[\sum_{l=1}^{2N} p_l(n-1)\right] - p_k(n-2), \tag{2}$$

where p represents the sound pressure at a junction at time step n, k is the position of the junction to be calculated, and l represents all the neighbors of k. This equation is equivalent to a difference equation derived from the Helmholtz equation by discretizing time and space.

3.1 Dispersion Error

An inherent problem with the rectangular digital waveguide mesh method is the direction- and frequency-dependent dispersion of wavefronts [33]. High-frequency signals parallel to the coordinate axes are delayed, whereas diagonally the waves propagate undistorted.

The dispersion error of digital waveguide mesh structures can be investigated with Von Neumann analysis [33,35]. It is based on the two-dimensional discrete-time Fourier transform of the difference scheme. After the Fourier transform, the scheme can be represented by means of the spectral amplification factor, and the stability and the dispersion characteristics of the scheme may be analyzed with it.

3.2 Mesh Topologies

The two-dimensional rectangular mesh is the most simple mesh structure as shown in Figure 2a. Several mesh topologies have been experimented to reduce the direction-dependent dispersion in that structure, but all of them suffer from some kind of dispersion. The two main approaches are triangular [36,37] and interpolated rectangular ones [38] as illustrated in Figure 2b and c, respectively. The 3D counterpart for the triangular mesh is the tetrahedral structure [37], while the interpolated mesh can be applied in both 2D and 3D [39].

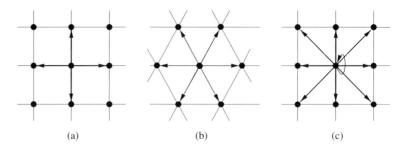

FIGURE 2 Various 2D digital waveguide mesh structures: (**a**) in the rectangular mesh each node is connected to four neighbors with unit delays, (**b**) triangular mesh has six connections per neighbor, and (**c**) in the interpolated mesh the signal propagates to all the eight neighbors and to the node itself.

3.3 Reducing Dispersion with Frequency Warping

The triangular and interpolated rectangular mesh structures render the dispersion practically independent of propagation direction. In addition, the frequency error curves are smooth. Due to these reasons, the dispersion error can be reduced remarkably by applying a frequency warping technique [38,39].

The frequency warping is applied to the input and output signals of the mesh using a warped FIR filter [22,23]. This filter shifts frequencies to compensate for the dispersion error. For this purpose a first-order allpass warping is suitable. The applied amount of warping determines the extent of warping, and there are various criteria to find the optimal value to be used with digital waveguide meshes [38].

3.4 Boundary Conditions

The boundary conditions of a mesh dictate the reflection characteristics of the surfaces surrounding the space under study, and thus are fundamental for acoustic simulation. In general, the boundary conditions can be represented by digital filters [25]. However, in practice setting the boundary conditions accurately are difficult, and the topic is still under research at the time of writing [40]. In general, it is required that users can set arbitrary boundaries for modeling reflection of sound from various materials. Challenging topic, especially, is the modeling of an anechoic boundary that enables simulation of semi-open spaces.

References

1. M. Kleiner, B.-I. Dalenbäck, and U.P. Svensson. Auralization – an overview. *J. Audio Eng. Soc.*, 41(11):861–875, Nov. 1993.
2. J.-M. Jot. Real-time spatial processing of sounds for music, multimedia and interactive human–computer interfaces. *Multimedia Syst., Special Issue Audio Multimedia*, 7(1):55–69, 1999.

3. W.G. Gardner. Reverberation algorithms. In M. Kahrs and K. Brandenburg, editors, *Applications of Digital Signal Processing to Audio and Acoustics*, pp. 85–131. Kluwer Academic Publishers, Norwell, MA, 1997.
4. A. Pietrzyk. Computer modeling of the sound field in small rooms. *Proceedings of the AES 15th International Conference on Audio, Acoustics & Small Spaces*, pp. 24–31, Copenhagen, Denmark, Oct. 31–Nov. 2, 1998.
5. D. Botteldooren. Finite-difference time-domain simulation of low-frequency room acoustic problems. *J. Acoust. Soc. Am.*, 98(6): 3302–3308, 1995.
6. L. Savioja, J. Backman, A. Järvinen, and T. Takala. Waveguide mesh method for low-frequency simulation of room acoustics. *Proceedings of the 15th International Congress Acoustics on (ICA '95)*, volume 2, pp. 637–640, Trondheim, Norway, June 1995.
7. L. Savioja. *Modeling Techniques for Virtual Acoustics*. PhD thesis, Helsinki University of Technology, Telecommunications Software and Multimedia Laboratory, report TML-A3, 1999.
8. A. Krokstad, S. Strom, and S. Sorsdal. Calculating the acoustical room response by the use of a ray tracing technique. *J. Sound Vib.*, 8(1):118–125, 1968.
9. A. Kulowski. Algorithmic representation of the ray tracing technique. *Appl. Acoust.*, 18(6):449–469, 1985.
10. T.A. Funkhouser, N. Tsingos, I. Carlbom, G. Elko, M. Sondhi, J. West, G. Pingali, P. Min, and A. Ngan. A beam tracing method for interactive architectural acoustics. *J. Acoust. Soc. Am.*, 115(2): 739–756, February 2004.
11. J.B. Allen and D.A. Berkley. Image method for efficiently simulating small-room acoustics. *J. Acoust. Soc. Am.*, 65(4):943–950, 1979.
12. J. Borish. Extension of the image model to arbitrary polyhedra. *J. Acoust. Soc. Am.*, 75(6):1827–1836, 1984.
13. U.P. Svensson and U.R. Kristiansen. Computational modeling and simulation of acoustic spaces. *AES 22nd International Conference on Virtual, Synthetic and Entertainment Audio*, pp. 11–30, Espoo, Finland, June 15–17, 2002.
14. L. Savioja, J. Huopaniemi, T. Lokki, and R. Väänänen. Creating interactive virtual acoustic environments. *J. Audio Eng. Soc.*, 47(9):675–705, 1999.
15. T. Lokki. *Physically-Based Auralization – Design, Implementation, and Evaluation*. PhD thesis, Helsinki University of Technology, Telecommunications Software and Multimedia Laboratory, report TML-A5, 2002. Available at http://lib.hut.fi/Diss/2002/isbn9512261588/.
16. H. Strauss. Implementing doppler shifts for virtual auditory environments. *The 104th Audio Engineering Society (AES) Convention*, Amsterdam, the Netherlands, May 16–19, 1998. Preprint no. 4687.
17. T. Lokki, L. Savioja, J. Huopaniemi, R. Väänänen, and T. Takala. Creating interactive virtual auditory environments. *IEEE Comput. Graph. Appl.*, 22(4):49–57, July/Aug. 2002.

18. M. Kompis and N. Dillier. Simulating transfer functions in a reverberant room including source directivity and head-shadow effects. *J. Acoust. Soc. Am.*, 93(5):2779–2787, May 1993.
19. J. Huopaniemi, K. Kettunen, and J. Rahkonen. Measurement and modeling techniques for directional sound radiation from the mouth. *Proceedings of the IEEE Workshop on Applications of Signal Processing to Audio and Acoustics (WASPAA '99)*, pp. 183–186, Mohonk Mountain House, New Paltz, New York, Oct. 17–20, 1999.
20. J.-M. Jot, V. Larcher, and O. Warusfel. Digital signal processing issues in the context of binaural and transaural stereophony. *The 98th Audio Engineering Society (AES) Convention*, Paris, France, 1995. Preprint no. 3980.
21. M. Karjalainen, J. Huopaniemi, and V. Välimäki. Direction-dependent physical modeling of musical instruments. *Proceedings of the 15th International Congress on Acoustics (ICA '95)*, pp. 451–454, Trondheim, Norway, June 1995.
22. U.K. Laine, M. Karjalainen, and T. Altosaar. Warped linear prediction (WLP) in speech and audio processing. *Proceedings of the International Conference on Acoustics, Speech, Signal Processing (ICASSP'94)*, volume 3, pp. 349–352, Adelaide, Australia, April 19–22, 1994.
23. A. Härmä, M. Karjalainen, L. Savioja, V. Välimäki, U.K. Laine, and J. Huopaniemi. Frequency-warped signal processing for audio applications. *J. Audio Eng. Soc.*, 48(11):1011–1031, Nov. 2000.
24. ISO Standard 9613-1. *Acoustics – Attenuation of Sound During Propagation Outdoors – Part 1: Calculation of the Absorption of Sound by the Atmosphere*. 1993.
25. J. Huopaniemi, L. Savioja, and M. Karjalainen. Modeling of reflections and air absorption in acoustical spaces – a digital filter design approach. *Proceedings of the IEEE Workshop on Applications of Signal Processing to Audio and Acoustics (WASPAA'97)*, Mohonk, New Paltz, New York, Oct. 19–22, 1997.
26. A.D. Pierce. *Acoustics – An Introduction to Its Physical Principles and Applications*. The Acoustical Society of America, 2nd ed., 1994.
27. T.I. Laakso, V. Välimäki, M. Karjalainen, and U.K. Laine. Splitting the unit delay – tools for fractional delay filter design. *IEEE Signal Process. Mag.*, 13(1):30–60, Jan. 1996.
28. U.P. Svensson, R.I. Fred, and J. Vanderkooy. Analytic secondary source model of edge diffraction impulse responses. *J. Acoust. Soc. Am.*, 106(5):2331–2344, 1999.
29. T. Lokki, U.P. Svensson, and L. Savioja. An efficient auralization of edge diffraction. *AES 21st International Conference on Architectural Acoustics and Sound Reinforcement*, pp. 166–172, St. Petersburg, Russia, June 1–3, 2002.
30. D.R. Begault *3-D Sound for Virtual Reality and Multimedia*. AP Professional, Cambridge, MA, 1994.
31. M.R. Schroeder. Natural-sounding artificial reverberation. *J. Audio Eng. Soc.*, 10(3):219–223, 1962.

32. J.-M. Jot. *Etude et réalisation d'un spatialisateur de sons par modéles physique et perceptifs*. PhD thesis, l'Ecole Nationale Superieure des Telecommunications, Télécom Paris 92 E 019, Sept. 1992.
33. S. Van Duyne and J.O. Smith. Physical modeling with the 2-D digital waveguide mesh. *Proceedings of the International Computer Music Conference*, pp. 40–47, Tokyo, Japan, Sept. 1993.
34. L. Savioja, T. Rinne, and T. Takala. Simulation of room acoustics with a 3-D finite difference mesh. *Proceedings of the International Computer Music Conference*, pp. 463–466, Aarhus, Denmark, Sept. 1994.
35. J. Strikwerda. *Finite Difference Schemes and Partial Differential Equations*. Chapman & Hall, New York, NY, 1989.
36. F. Fontana and D. Rocchesso. A new formulation of the 2D-waveguide mesh for percussion instruments. *Proceedings of the XI Colloquium on Musical Informatics*, pp. 27–30, Bologna, Italy, Nov. 1995.
37. S. Van Duyne and J.O. Smith. The 3D tetrahedral digital waveguide mesh with musical applications. *Proceedings of the International Computer Music Conference*, pp. 9–16, Hong Kong, Aug. 1996.
38. L. Savioja and V. Välimäki. Reducing the dispersion error in the digital waveguide mesh using interpolation and frequency-warping techniques. *IEEE Trans. Speech Audio Process.*, 8(2):184–194, March 2000.
39. L. Savioja and V. Välimäki. Interpolated rectangular 3-D digital waveguide mesh algorithms with frequency warping. *IEEE Trans. Speech Audio Process.*, 11(6):783–790, Nov. 2003.
40. A. Kelloniemi, L. Savioja, and V. Välimäki. Spatial filter-based absorbing boundary for the 2-d digital waveguide mesh. *IEEE Signal Process. Lett.*, 12(2):126–129, 2005.

40
Audio Restoration

Paulo A.A. Esquef

Nokia Institute of Technology, Rod. Torquato Tapajós, 7200, Tarumã
69048-660 Manaus-AM, Brazil, e-mail: paulo.esquef@indt.org.br

1 • A Brief History of the Recording Technology

The first reproducible recording of human voice was made in 1877 on a tinfoil cylinder phonograph devised by Thomas A. Edison. Since then, much effort has been expended to find better ways to record and reproduce sounds. By the mid-1920s, the first electrical recordings appeared and gradually took over purely acoustic recordings. The development of electronic computers, in conjunction with the ability to record data onto magnetic or optical media, culminated in the standardization of compact disc format in 1980. Nowadays, digital technology is applied to several audio applications, not only to improve the quality of modern and old recording/reproduction techniques, but also to trade off sound quality for less storage space and less taxing transmission capacity requirements. For a comprehensive time-line and description of the most prominent events regarding the recording technology history see [1,2].

2 • Aims and Problems

Audio restoration basically aims at improving the sound of old recordings. The primary goal is to reduce spurious artifacts, which are usually introduced by the recording/playback mechanisms, while preserving, as much as possible, the original recorded sound.

The first step in a typical audio restoration procedure consists of transferring the sound from old matrices to a digital form. This task involves practical matters such as locating the original matrices or the best sounding copies; finding the best way or equipment to play back a given matrix; and dealing with the usual lack of standardization associated with obsolescent recording/playback systems.

Sound transferring from old media to more modern ones, for instance from 78 RPM to LP disks,[1] and attempts to improve the sonic quality of old recordings within the sound transferring process were a common practice even before the digital audio era. Thanks to the progressive increases in the computational power of digital processors, more sophisticated and powerful processing of digitized data has become feasible in practice. Nowadays, audio restoration is accomplished via DSP algorithms devoted to reducing or suppressing from the recorded material spurious noises and distortions introduced by the old recording/playback systems.

The most commonly found types of degradation associated with old recordings can be roughly classified into localized and global disturbances [3]. For example, short impulsive noises (clicks, crackles, and pops) as well as low-frequency long pulses (thumps) belong to the former class. Continuous background disturbances or interferences, such as broadband noise (hiss), buzz, and hum, are usually classified as global disturbances. Other examples of audio degradations include slow and fast frequency modulations, which are known as wow and flutter, respectively; surface noise; and non-linear distortions, e.g., amplitude clipping.

3 • Overview of Digital Audio Restoration

Digital audio restoration takes place after the sound transfer process from analog sources to digital domain. A review of the main techniques available for treating disturbances, such as thumps, clicks, and hiss, is given in the ensuing sections. The presentation follows the order of precedence in which the problems are usually tackled. Methods and algorithms devised for dealing with other types of degradations can be found elsewhere [3].

---

[1]The abbreviations RPM and LP stand for revolutions per minute and long-playing, respectively.

3.1 De-thumping

Thumps are produced by long pulses of low-frequency content that additively corrupt an audio signal. These pulses are related to the mechanical response of the playback mechanisms to an abnormal excitation, e.g., the response of the stylus-arm apparatus to large discontinuities on the groove walls of a disk.

3.1.1 Template Matching

Apart from crude techniques, such as highpass filtering of the signal, a few methods, with various degrees of sophistication, exist for treating long pulses. The template-matching method introduced by [4] figures among one of the first propositions for treating long pulses. Figure 1 shows a block diagram that illustrates the functional stages of the template-matching method for de-thumping. The basic assumption behind this method is that long pulses are identical in shape, varying only in amplitude.

In practical terms, a corrupting pulse is considered to be a scaled version of a representative template of the pulse $\mathbf{d}$ added to the clean signal $\mathbf{x}_n$, leading to $\mathbf{y}_n = \mathbf{x}_n + \mu \mathbf{d}$, with $\mathbf{y}_n = [y_n, y_{n+1}, \ldots, y_{n+N-1}]^T$, $\mathbf{x}_n = [x_n, x_{n+1}, \ldots, x_{n+N-1}]^T$, $\mathbf{d} = [d_0, d_1, \ldots, d_{N-1}]^T$, and μ a scale factor. Moreover, the matched filter, whose coefficients $\mathbf{h}_{mf}$ consist of the time-reversed samples of $\mathbf{d}$ appropriately scaled, works as an estimator of the cross-correlation coefficient between $\mathbf{d}$ and $\mathbf{y}_n$. Therefore, the presence of pulses is indicated by values of the filter's output that are near unity in magnitude. Thus, after synchronizing the template with a given pulse occurrence, the estimated de-thumped signal is obtained by $\hat{\mathbf{x}}_n = \mathbf{y}_n - \mu \mathbf{d}$, where the scaling factor can be shown to be given by $\mu = (\mathbf{h}_{mf}^T \mathbf{y}_n)/(\mathbf{h}_{mf}^T \mathbf{h}_{mf})$.

The high-amplitude click that usually appears at the beginning of a long pulse is to be removed by a de-clicking algorithm (to be discussed in Sect. 3.2). The main limitation of the template-matching method is the lack of robustness in detecting and suppressing pulses whose shape varies over time or pulses that are superimposed [3].

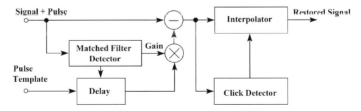

FIGURE 1 Processing stages of the template-matching method for de-thumping. Adapted from [4].

3.1.2 Other Methods

A model-based approach to long pulse removal has been proposed in [5] and further developed in [6]. In this method the corrupted audio signal is modeled as a mixture of two distinct AR processes. Thus, signal restoration is accomplished by separating these two processes. The AR-separation method can produce excellent signal restorations even for superimposed pulses. Moreover, it incorporates the suppression of the initial pulse clicks within the de-thumping procedure. The main drawback of the AR-separation method is its high computational cost.

An alternative audio de-thumping procedure, which is based upon non-linear [7] and polynomial smoothing filters [8], has been proposed in [9]. It is comparable to the AR-separation method in restoration performance, despite being less demanding computationally.

3.2 De-clicking

Impulsive-like disturbances contaminating audio signals are heard as clicks. Such disturbances are usually caused by the intrinsic porosity of the disk material, solid particles adhering to the groove walls, and the presence of superficial scratches on the disk surface [10].

3.2.1 Click Detection

Techniques based on autoregressive (AR) models have been demonstrated to be successful for de-clicking purposes. In these methods the degradation mechanism is assumed to be additive, with the corrupted signal given by $y_n = x_n + i_n d_n$, where x_n is the clean signal and the term $i_n d_n$ corresponds to the noise component. The sequence i_n is binary and indicates the presence of noise at instants n where $i_n = 1$.

The goal of click detection algorithms is to estimate the sequence i_n that minimizes the percentage of both miss detection and false alarm of clicks. Figure 2 depicts a block diagram with the functional stages of a typical AR-based de-clicking scheme.

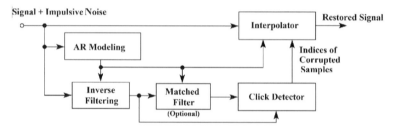

FIGURE 2 Model-based de-clicking method. Adapted from [4].

Given a frame of the corrupted signal $\mathbf{y}_l = [y_l, y_{l+1}, \ldots, y_{l+N-1}]^T$ and a pth-order estimate $\hat{\mathbf{a}}$ of its AR coefficient vector,[2] the associated sequence i_n can be straightforwardly obtained via

$$i_n = \begin{cases} 1, & |e_n| \geq T \\ 0, & \text{otherwise} \end{cases}, \text{ for } n = l+p, l+p+1, \ldots, l+N-1, \quad (1)$$

where $e_n = \hat{\mathbf{a}}^T \mathbf{y}'$, with $\mathbf{y}' = [y_n, y_{n-1}, \ldots, y_{n-p}]$, is the output of the inverse filter and $T = g \operatorname{median}([|e_{l+p}|, |e_{l+p+1}|, \ldots, |e_{l+N-1}|])$ is the value of a variable threshold. The parameter g is a scaling factor. Considering a sample rate of 44.1 kHz, the values of the processing parameters are typically set within the following ranges: $500 \leq N \leq 2{,}000$, $10 \leq p \leq 40$, and $5 \leq g \leq 7$.

Further improvements in click detection performance can be achieved by using matched filtering [4], two-sided linear prediction [11], and more elaborated threshold-based schemes [12].

3.2.2 Signal Reconstruction

Click suppression is attained by replacing the corrupted samples with others that resemble the underlying signal. Since, in general, audio signals exhibit resonant characteristics, AR-based interpolation methods constitute a suitable tool for recovery of missing samples in audio signals. In AR-based interpolators, signal reconstruction is usually attained through the minimization of the variance of the model error w.r.t. the unknown samples [13]. However, due to constraints related to the non-stationarity of audio signals, meaningful data reconstruction is restricted to relatively short-duration (up to about 5 ms) portions of audio material.

Interpolation across longer gaps tends to produce reconstructed signals that are smoother and weaker in energy than the surroundings of the gap. The simplest way to circumvent these shortcomings is to increase the order of the AR model. Other solutions include the use of: extended AR models [11]; interpolators combining AR models and with cosine basis functions [3]; interpolators with constant excitation energy [14,15]; interpolators based on random sampling [16,17]; and interpolators that employ two different AR models, one to model the segment that immediately precedes a gap and another for the fragment that succeeds the gap [18,19].

---

[2] Any standard AR estimation algorithm, such as Yule–Walker and Burg's methods, can be used to obtain this estimate.

3.2.3 Other Methods

Various kinds of approaches have been attempted to tackle the reconstruction of discrete-time signals across gaps of missing samples [20]. In addition to AR-based techniques, audio signal recovery can be attained via: band-limitation criterion [21]; sinusoidal modeling interpolation [22]; waveform substitution [23,24]; multirate signal decomposition and synthesis [25,26]; Bayesian methods [3,5,27]; neural-network algorithms [28]; and model-based adaptive filtering techniques [29,30].

3.3 De-hissing

Signal contamination with additive broadband noise is perceived as hiss. The noise component is usually due to thermic measurement noise, circuitry noise, and intrinsic characteristics of the storage medium, such as in analog tape recordings.

3.3.1 Short-Time Spectral Attenuation Methods

Short-time spectral attenuation (STSA) [31,32] is one of the most common means of audio and speech de-hissing. The basic assumption behind STSA methods is that the samples of the corrupted signal are given by $y_n = x_n + d_n$, where d_n is an additive zero-mean white Gaussian process, uncorrelated to the underlying clean signal x_n.

Due to the non-stationarity of general audio signals, STSA methods employ a block-based overlap-and-add processing scheme [33]. Moreover, the core processing is performed in the frequency domain. In that regard, the transformations from time into frequency and vice versa are carried out through the discrete Fourier transform (DFT). Figure 3 illustrates the typical processing stages of STSA de-hissing methods.

In STSA methods, the frequency components present in a given signal frame are attenuated in magnitude based on a suppression rule, which takes into account the signal-to-noise ratio (SNR) estimated at each frequency bin. For that, estimates of power spectral density of both the disturbance and the corrupted signal

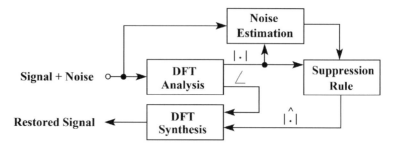

FIGURE 3 STSA-based hiss reduction system. Adapted from [34].

are needed. In practice, those densities can be approximated by the power spectra $|D_k|^2$ and $|Y_k|^2$ of the respective signals, computed via the DFT.

The most common suppression rules can be formulated as

$$|\hat{X}_k| = H_k |Y_k|, \qquad (2)$$

where H_k is a frame-varying frequency-domain filter and $|\hat{X}_k|$ is the magnitude spectrum of the estimated clean signal, with k being the DFT bin index. Typically, the filter assumes the form $H_k = [1 - (|D_k|/|Y_k|)^b]^{1/b}$. For example, the so-called magnitude and power spectrum subtraction rules [32] are attained by choosing $b = 1$ and $b = 2$, respectively. The Wiener filter rule is simply given by H_k^2 for $b = 2$. For comparisons among different suppression rules see [33,35].

In general, noise reduction via STSA methods comes at the cost of losses of valuable signal content, especially on the high-frequency range. Another typical side effect affecting the processed signals is the presence of a phenomenon known as musical noise [35]. There exist several means to reduce this artifact. The simplest way consists in overestimating the power of the noise spectrum. Alternatively, musical noise can be reduced by: averaging the spectral information over successive signal frames [32]; applying heuristic rules to the spectral attenuation factor [4,32]; and leaving a minimum noise floor to mask the musical noise components [3,36,37].

The suppression rule proposed by Ephraim and Malah (EMSR) [38,39] is known to be one of the most effective in reducing musical noise. Simpler alternatives to the EMSR have also been proposed in [34]. Recently, some interest has been aroused in incorporating psychoacoustic-related phenomena into the suppression rules used in STSA methods [40–44].

3.3.2 Other Methods

Within the class of non-parametric de-hissing techniques, wavelet-based shrinkage has received substantial attention in the pertinent literature [45–48]. In this method the noisy signal is mapped into a multi resolution time frequency representation via a discrete wavelet transform. In the wavelet domain, the underlying signal is supposed to be compactly represented by a few high-amplitude coefficients. Conversely, the noise process is spread over a large number of low-level coefficients. Thus, denoising can be accomplished, in the simplest way, by discarding these coefficients (hard threshold) before transforming the signal back to the original domain. More involved wavelet shrinkage methods exist [49].

As for parametric audio de-hissing, techniques that make use of AR modeling of degraded speech and audio signals have been proposed in [50]. Statistical-based methods that employ AR and ARMA modeling have also been developed for joint treatment of broadband and impulsive noise [3,29]. Adaptive filtering schemes for audio de-hissing are presented in [4,29,30]. Hybrid methods that combine either different non-parametric schemes [51] or mix non-parametric and parametric strategies [52] can also be found in the relevant literature.

4 • Future Directions

Currently available DSP tools for audio restoration offer a certain degree of automation for some de-noising tasks. However, most restoration methods do not integrate a priori knowledge on the contents of the signal to be restored. Therefore, the sonic quality of a restored sound strongly depends on how an expert listener chooses the processing parameters as well as personally judges the attained results.

It seems reasonable to believe that, in the future, audio restoration techniques can benefit from the incorporation of high-level information on the contents of the signals under processing. However, as pointed out in [53], such strategy implies working in an object-based audio representation [54], which is less general than processing approaches based on conventional signal modeling techniques. Nevertheless, a priori knowledge on the contents of the audio material at hand, for example, genre of the program and types of generative sound sources, could be used to guide the choice of the most suitable restoration technique to employ as well as to better tune its processing parameters. A far more audacious and involved step would consist of first extracting high-level content information from old recordings and then realizing restoration via sound resynthesis within an object-based audio representation framework.

References

1. D. L. Morton, *Sound Recording: The Life Story of a Technology*, Greenwood Technologies, Greenwood Press, Oxford, UK Sept. 2004.
2. S. Schoenherr, "Recording Technology History," http://history.acusd.edu/gen/recording/notes.html, Retrieved Dec. 2003.
3. S. J. Godsill and P. J. W. Rayner, *Digital Audio Restoration – A Statistical Model Based Approach*, Springer, London, 1998, http://www-sigproc.eng.cam.ac.uk/~sjg/springer/.
4. S. V. Vaseghi and R. Frayling-Cork, "Restoration of Old Gramophone Recordings," *J. Audio Eng. Soc.*, vol. 40, no. 10, pp. 791–801, Oct. 1992.

5. S. J. Godsill, *The Restoration of Degraded Audio Signals*, Ph.D. thesis, Engineering Department, Cambridge University, Cambridge, England, Dec. 1993, http://www-sigproc.eng.cam.ac.uk/~sjg/thesis/.
6. S. J. Godsill and C. H. Tan, "Removal of Low Frequency Transient Noise from Old Recordings Using Model-Based Signal Separation Techniques," *Proc. IEEE ASSP Workshop on Applications of Signal Processing to Audio and Acoustics*, New Paltz, New York, Oct. 1997.
7. W. A. Struzinski and E. D. Lowe, "A Performance Comparison of Four Noise Background Normalization Schemes Proposed for Signal Detection Systems," *J. Acoust. Soc. Am.*, vol. 76, no. 6, pp. 1738–1742, Dec. 1984.
8. W. H. Press and S. A. Teukolsky, "Savitzky-Golay Smoothing Filters," *Comput. Phys.*, vol. 4, no. 6, pp. 669–672, 1990.
9. P. A. A. Esquef, L. W. P. Biscainho, and V. Välimäki, "An Efficient Algorithm for the Restoration of Audio Signals Corrupted with Low-Frequency Pulses," *J. Audio Eng. Soc.*, vol. 51, no. 6, pp. 502–517, June 2003.
10. P. Wilson, "Record Contamination: Causes and Cure," *J. Audio Eng. Soc.*, vol. 13, no. 2, pp. 166–176, Apr. 1965.
11. S. V. Vaseghi and P. J. W. Rayner, "Detection and Suppression of Impulsive Noise in Speech Communication Systems," *IEE Proc.*, vol. 137, no. 1, pp. 38–46, Feb. 1990.
12. P. A. A. Esquef, L. W. P. Biscainho, P. S. R. Diniz, and F. P. Freeland, "A Double-Threshold-Based Approach to Impulsive Noise Detection in Audio Signals," *Proc. X European Signal Processing Conf.*, Tampere, Finland, Sept. 2000, vol. 4, pp. 2041–2044.
13. A. J. E. M. Janssen, R. N. J. Veldhuis, and L. B. Vries, "Adaptive Interpolation of Discrete-Time Signals that Can Be Modeled as Autoregressive Processes," *IEEE Trans. Acoust. Speech Signal Process.*, vol. ASSP-34, no. 2, pp. 317–330, Apr. 1986.
14. P. J. W. Rayner and S. J. Godsill, "The Detection and Correction of Artefacts in Degraded Gramophone Recordings," *Proc. IEEE ASSP Workshop on Applications of Signal Processing to Audio and Acoustics*, Oct. 1991, pp. 151–152.
15. M. Niedźwiecki, "Statistical Reconstruction of Multivariate Time Series," *IEEE Trans. Signal Process.*, vol. 41, no. 1, pp. 451–457, Jan. 1993.
16. J. J. Ó Ruanaidh and W. J. Fitzgerald, "Interpolation of Missing Samples for Audio Restoration," *Electron. Lett.*, vol. 30, no. 8, pp. 622–623, Apr. 1994.
17. S. J. Godsill and P. J. W. Rayner, "Statistical Reconstruction and Analysis of Autoregressive Signals in Impulsive Noise Using the Gibbs Sampler," *IEEE Trans. Speech Audio Process.*, vol. 6, no. 4, pp. 352–372, July 1998.
18. W. Etter, "Restoration of a Discrete-Time Signal Segment by Interpolation Based on the Left-Sided and Right-Sided Autoregressive Parameters," *IEEE Trans. Signal Process.*, vol. 44, no. 5, pp. 1124–1135, May 1996.

19. I. Kauppinen, J. Kauppinen, and P. Saarinen, "A Method for Long Extrapolation of Audio Signals," *J. Audio Eng. Soc.*, vol. 49, no. 12, pp. 1167–1180, Dec. 2001.
20. R. Veldhuis, *Restoration of Lost Samples in Digital Signals*, Prentice-Hall, New Jersey, 1990.
21. P. J. S. G. Ferreira, "Iterative and Noniterative Recovery of Missing Samples for 1-D Band-Limited Signals," in *Nonuniform Sampling – Theory and Practice*, F. Marvasti, Ed., Chap. 5, pp. 235–281. Kluwer Academic/Plenum, New York, 2001.
22. R. C. Maher, "A Method for Extrapolation of Missing Digital Audio Data," *J. Audio Eng. Soc.*, vol. 42, no. 5, pp. 350–357, May 1994.
23. D. J. Goodman, G. B. Lockhart, O. J. Waen, and W. C. Wong, "Waveform Substitution Techniques for Recovering Missing Speech Segments in Packet Voice Communications," *IEEE Trans. Acoust. Speech Signal Process.*, vol. ASSP-34, no. 6, pp. 1440–1448, Dec. 1986.
24. M. Niedźwiecki and K. Cisowski, "Smart Copying – A New Approach to Reconstruction of Audio Signals," *IEEE Trans. Signal Process.*, vol. 49, no. 10, pp. 2272–2282, Oct. 2001.
25. S. Montresor, J. C. Valiere, J. F. Allard, and M. Baudry, "Evaluation of Two Interpolation Methods Applied to Old Recordings Restoration," Presented at the 90th Convention of the AES, Paris, France, Feb. 1991, Preprint 3022.
26. B.-S. Chen and Y.-L. Chen, "Multirate Modeling of AR/ARMA Stocastic Signals and Its Application to the Combined Estimation-Interpolation Problem," *IEEE Trans. Signal Process.*, vol. 43, no. 10, pp. 2302–2312, Oct. 1995.
27. S. J. Godsill and P. J. W. Rayner, "A Bayesian Approach to the Restoration of Degraded Audio Signals," *IEEE Trans. Speech Audio Process.*, vol. 3, no. 4, pp. 276–278, July 1995.
28. A. Czyzewski, "Learning Algorithms for Audio Signal Enhancement, Part I: Neural Network Implementation for the Removal of Impulsive Distortions," *J. Audio Eng. Soc.*, vol. 45, no. 10, pp. 815–831, Oct. 1997.
29. M. Niedźwiecki and K. Cisowski, "Adaptive Scheme for Elimination of Broadband Noise and Impulsive Disturbances from AR and ARMA Signals," *IEEE Trans. Signal Process.*, vol. 44, no. 3, pp. 528–537, Mar. 1996.
30. A. Bari, S. Canazza, G. De Poli, and G. A. Mian, "Improving the Extended Kalman Filter Method for the Restoration of Electroacoustic Music," *Proc. Int. Computer Music Conf.*, Berlin, Germany, Aug. 2000, pp. 233–236.
31. J. Lim, "Evaluation of a Correlation Subtraction Method for Enhancing Speech Degraded by Additive White Noise," *IEEE Trans. Acoust. Speech Signal Process.*, vol. 26, no. 5, pp. 471–472, Oct. 1978.
32. S. F. Boll, "Suppression of Acoustic Noise in Speech Using Spectral Subtraction," *IEEE Trans. Acoust. Speech Signal Process.*, vol. ASSP-27, no. 2, pp. 113–120, Apr. 1979.

33. S. J. Godsill, P. J. W. Rayner, and O. Cappé, "Digital Audio Restoration," in *Applications of Digital Signal Processing to Audio and Acoustics*, M. Kahrs and K. Brandenburg, Eds., Chap. 4, pp. 133–194. Kluwer Academic, Boston, USA, 1998.
34. P. J. Wolfe and S. J. Godsill, "Efficient Alternatives to the Ephraim and Malah Suppression Rule for Audio Signal Enhancement," *EURASIP J. Appl. Signal Process. – Special Issue on Digital Audio for Multimedia Communications*, vol. 2003, no. 10, pp. 1043–1051, Sept. 2003.
35. O. Cappé and J. Laroche, "Evaluation of Short-Time Spectral Attenuation Techniques for the Restoration of Musical Recordings," *IEEE Trans. Speech Audio Process.*, vol. 3, no. 1, pp. 84–93, Jan. 1995.
36. M. Berouti, R. Schwartz, and J. Makhoul, "Enhancement of Speech Corrupted by Acoustic Noise," *Proc. IEEE Int. Conf. Acoustics, Speech, and Signal Processing*, Apr. 1979, vol. 4, pp. 208–211.
37. J. A. Moorer and M. Berger, "Linear-Phase Bandsplitting: Theory and Applications," *J. Audio Eng. Soc.*, vol. 34, no. 3, pp. 143–152, Mar. 1986.
38. Y. Ephraim and D. Malah, "Speech Enhancement Using a Minimum Mean-Square Error Short-Time Spectral Amplitude Estimator," *IEEE Trans. Acoust. Speech Signal Process.*, vol. 32, no. 6, pp. 1109–1121, Dec. 1984.
39. O. Cappé, "Elimination of the Musical Noise Phenomenon with the Ephraim and Malah Noise Suppressor," *IEEE Trans. Speech Audio Process.*, vol. 2, no. 2, pp. 345–349, Apr. 1994.
40. D. Tsoukalas, J. Mourjopoulos, and G. Kokkinakis, "Perceptual Filters for Audio Signal Enhancement," *J. Audio Eng. Soc.*, vol. 45, no. 1/2, pp. 22–36, Jan./Feb. 1997.
41. M. Lorber and R. Hoeldrich, "A Combined Approach for Broadband Noise Reduction," *Proc. IEEE ASSP Workshop on Applications of Signal Processing to Audio and Acoustics*, Oct. 1997.
42. P. J. Wolfe and S. J. Godsill, "Perceptually Motivated Approaches to Music Restoration," *J. New Music Res. – Special Issue: Music and Mathematics*, vol. 30, no. 1, pp. 83–92, Mar. 2001.
43. A. Bari, S. Canazza, G. De Poli, and G. A. Mian, "Toward a Methodology for the Restoration of Electroacoustic Music," *J. New Music Res. – Special Issue: Preserving Electroacoustic Music*, vol. 30, no. 4, Dec. 2001.
44. I. Kauppinen and K. Roth, "Adaptive Psychoacoustic Filter for Broadband Noise Reduction in Audio Signals," *Proc. 14th Int. Conf. Digital Signal Processing*, Santorini, Greece, July 2002, vol. 2, pp. 962–966.
45. J. Berger, R. R. Coifman, and M. J. Goldberg, "Removing Noise from Music Using Local Trigonometric Bases and Wavelet Packets," *J. Audio Eng. Soc.*, vol. 42, no. 10, pp. 808–818, Oct. 1994.
46. D. L. Donoho, "De-noising by Soft-Thresholding," *IEEE Trans. Inf. Theory*, vol. 41, no. 3, pp. 613–627, May 1995.
47. H. K. S. Mallat, D. L. Donoho, and A. S. Willsky, "Best Basis Algorithm for Signal Enhancement," *Proc. IEEE Int. Conf. Acoustics, Speech, and Signal Processing*, May 1995, vol. 3, pp. 1561–1564.

48. P. K. Ramarapu and R. C. Maher, "Methods for Reducing Audible Artifacts in a Wavelet-Based Broad-Band Denoising System," *J. Audio Eng. Soc.*, vol. 46, no. 3, pp. 178–190, Mar. 1998.
49. B. Vidakovic, "Nonlinear Wavelet Shrinkage with Bayes Rules and Bayes Factors," *J. Am. Stat. Assoc.*, vol. 93, pp. 173–179, 1998.
50. J. Lim and A. V. Oppenheim, "All-Pole Modeling of Degraded Speech," *IEEE Trans. Acoust. Speech Signal Process.*, vol. 26, no. 3, pp. 197–210, June 1978.
51. S. B. Jebara, A. Benazza-Benyahia, and A. B. Khelifa, "Reduction of Musical Noise Generated by Spectral Subtraction by Combining Wavelet Packet Transform and Wiener Filtering," *Proc. X European Signal Processing Conf.*, Tampere, Finland, Sept. 2000, vol. 2.
52. I. Kauppinen and K. Roth, "Audio Signal Extrapolation – Theory and Applications," *Proc. 5th Int. Conf. on Digital Audio Effects*, Hamburg, Germany, Sept. 2002, pp. 105–110, `http://www.unibw-hamburg.de/EWEB/ANT/dafx2002/papers.html`.
53. P. A. A. Esquef, *Model-Based Analysis of Noisy Musical Recordings with Application to Audio Restoration*, Ph.D. thesis, Helsinki University of Technology, Espoo, Finland, Apr. 2004, `http://lib.hut.fi/Diss/2004/isbn9512269503/`.
54. B. L. Vercoe, W. G. Gardner, and E. D. Scheirer, "Structured Audio: Creation, Transmission, and Rendering of Parametric Sound Representations," *Proc. IEEE*, vol. 86, no. 5, pp. 922–940, May 1998.

41
Audio Effects Generation

Udo Zölzer

Helmut Schmidt University, Hamburg, Germany

An **audio effect** is the modification of an audio signal to make it sound more interesting or to change its sound characteristic. The use of effects for sound modification has changed over the centuries and has been applied with different means. All effects simulate physical sound production and propagation by acoustical, mechanical or electronic means. The human perception of these sound effects leads to a classification according to the psychoacoustic terminology:

- Spectral envelope modification by filtering
- Time-domain amplitude envelope modification by dynamics processing or amplitude modulation
- Pitch shifting (affects the pitch by phase modulation)
- Time stretching (affects the duration without pitch modification)
- Room simulation (affects the room impression)

In the following we will introduce the effect topics:

- **Filters and delays** – like low-/high-/band-pass, shelving, peak and comb filters based on feedforward/feedback delays.
- **Tremolo, vibrato and rotary speaker** – loudness and pitch modification.
- **Chorus, Flanger and Phaser** - doubling, jet airplane effect.
- **Pitch shifting and time stretching**.

- **Dynamics processing** – limiter, compressor, expander, noise gate and distortion.
- **Room simulation** – infuses a room impulse response on the sound.

These effects will be introduced by their musical or physical meaning and then described by their time-domain operations and corresponding frequency-domain representations, where $x(t)$ is the input signal and $y(t) = T\{x(t)\}$ is the output signal with their corresponding Fourier transforms $X(\omega)$ and $Y(\omega)$ with $\omega = 2\pi f$.

1 • Filters and Delays

Filters are applied for simple input spectrum modifications. The standard filter operations for effects generation can be divided into first-order lowpass filters

$$y(t) = x(t) * h_{LP1}(t) \rightarrow H_{LP1}(\omega) = \omega_c/[\omega_c + j\omega], \quad (1)$$

first-order highpass filters

$$y(t) = x(t) * h_{HP1}(t) \rightarrow H_{HP1}(\omega) = j\omega/[\omega_c + j\omega], \quad (2)$$

and allpass filters (first and second order) given by

$$y(t) = x(t) * h_{AP1}(t) \rightarrow H_{AP1}(\omega) = [\omega_c - j\omega]/[\omega_c + j\omega]$$
$$= H_{LP1}(\omega) - H_{HP1}(\omega) \quad (3)$$

$$y(t) = x(t) * h_{AP2}(t) \rightarrow H_{AP2}(\omega) = [\omega_c - j\omega\omega_c/Q - \omega^2]/$$
$$[\omega_c + j\omega\omega_c/Q - \omega^2]. \quad (4)$$

These standard filters can easily be controlled by the cutoff frequency ω_c and the Q factor, which is the cutoff frequency divided by the $-3\,\text{dB}$ bandwidth. These basic filters can also be implemented by special allpass decompositions which lead to a lowpass filter

$$y(t) = 0.5[x(t) + x(t) * h_{AP1}(t)] \rightarrow H_{LP}(\omega) = 0.5[1 + H_{AP1}(\omega)], \quad (5)$$

highpass filter

$$y(t) = 0.5[x(t) - x(t) * h_{AP1}(t)] \rightarrow H_{HP}(\omega) = 0.5[1 - H_{AP1}(\omega)], \quad (6)$$

and bandpass filter

$$y(t) = 0.5[x(t) - x(t) * h_{AP2}(t)] \rightarrow H_{BP}(\omega) = 0.5[1 - H_{AP2}(\omega)]. \quad (7)$$

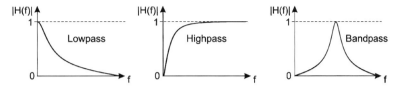

FIGURE 1 Frequency responses of simple filters.

These filters can easily be tuned or adjusted by changing the first- and second-order allpass coefficients to control the cutoff or center frequency of the corresponding filter. The frequency responses are shown in Figure 1.

Based on the allpass decompositions of lowpass, highpass and bandpass filters special weighting filters, such as a shelving filter given by

$$y(t) = x(t) + v \ x(t) * h_{\text{LP/HP}}(t) \rightarrow H_{\text{SF}}(\omega) = 1 + v \ H_{\text{LP/HP}}(\omega) \quad (8)$$

for low-frequency and high-frequency boost and cut and the peak filter for the mid-frequencies given by

$$y(t) = x(t) + v \ x(t) * h_{\text{BP}}(t) \rightarrow H_{\text{PF}}(\omega) = 1 + v \ H_{\text{BP}}(\omega), \quad (9)$$

can be implemented [7,13,21,28]. These allpass decompositions allow independent control of boost and cut, cutoff or center frequency and bandwidth for shelving and peak filters (see Figure 2).

Delays occur in sound propagation and can be described for a single delay by

$$y(t) = x(t - \tau) \rightarrow H(\omega) = 1 \exp(-j\omega\tau). \quad (10)$$

In a variety of applications we have a sum of several delayed replica of the input signal $x(t)$. If the direct signal $x(t)$ and the

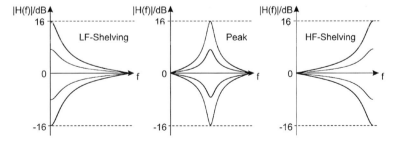

FIGURE 2 Frequency responses of shelving and peak filters.

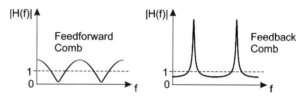

FIGURE 3 Frequency responses of feedforward and feedback comb filters.

delayed signal $x(t-\tau)$ are summed up the so-called feedforward comb filter given by

$$y(t) = x(t) + b\, x(t-\tau) \rightarrow H(\omega) = 1 + b\, \exp(-j\omega\tau) \quad (11)$$

is performed, which leads to a frequency response with repetitive notches controlled by coefficient b (see Figure 3). If the output is delayed the feedback comb filter given by

$$y(t) = x(t) - a\, y(t-\tau) \rightarrow H(\omega) = 1/(1 + a\, \exp(-j\omega\tau)) \quad (12)$$

is implemented, which leads to repetitive peaks in the frequency response controlled by coefficient a (see Figure 3).

The combination of weighted and delayed input and output signals leads to an allpass comb filter given by

$$y(t) = ax(t) + x(t-\tau) - a\, y(t-\tau) \rightarrow H(\omega)$$
$$= [a + \exp(-j\omega\tau)]/[1 + a\, \exp(-j\omega\tau)]. \quad (13)$$

The allpass comb filter has a unity gain frequency response. All three types of comb filters have been intensively used in room simulation devices [19,24,25].

2 • Tremolo, Vibrato and Rotary Speaker

Tremolo is the rapid repetition of one note in music terminology or a rapid alternation between two or more notes (string instruments, mandolin, flute, piano, harpsichord, spinet, guitar). In technical terminology tremolo is a rapid repetitive increase and decrease in volume. The effect is based on modifying the time-domain envelope (amplitude) of a signal according to

$$y(t) = m(t)\, x(t) \text{ with } \mathrm{FT}\{y(t)\} = Y(\omega) = M(\omega) * X(\omega)/2\pi \quad (14)$$

with the modulating signal $m(t) = 1 + a\sin(\omega_0 t)$. The repetition rate is controlled by $\omega_0 = 2\pi f_0$. The depth of the amplitude modulation is adjusted by a.

Vibrato is the rapid and repetitive variation in pitch for the duration of a note (string instruments, guitar, singing voice, yodeling). In technical terminology the effect is based on modifying the pitch of a signal by delay line modulation (phase modulation) according to

$$y(t) = x(t - m(t)) = x(t) * \delta(t - m(t)) \text{ with}$$
$$\text{FT}\{y(t)\} = Y(t, \omega) = X(\omega) \exp(-j\omega m(t)) \qquad (15)$$

with the time-varying delay $m(t) = t_1 + a \sin(\omega_0 t)$ with the fixed delay t_1 and the sinusoidal term controlled by $\omega_0 = 2\pi f_0$. The depth of the phase modulation is adjusted by a. The pitch modification by delay line modulation has its physical correspondence in the Doppler effect. Methods for varying the length of fractional delay lines are discussed in [4,14].

A **rotary speaker** produces a Doppler effect and an amplitude modulation. The Doppler effect leads to a pitch modulation and the amplitude modulation affects a loudness modulation. This effect can be described by a delay line modulation and an amplitude modulation according to

$$y(t) = a_1(t) x(t - m_1(t)) \qquad (16)$$

with parameters $m_1(t) = t_1 + a \sin(\omega_0 t)$ and $a_1(t) = 0.5[1 + \sin(\omega_0 t)]$ [5,6,27].

3 • Chorus, Flanger and Phaser

Chorus simulates the result of an ensemble of musicians (singers, violins) playing together with slight pitch and amplitude differences. The chorus effect performs a loudness increase. These amplitude and pitch differences are simulated by amplitude modulation and delay line modulation (phase modulation) and all signals are summed together to form the output signal

$$y(t) = \Sigma_i x(t - m_i(t)) \qquad (17)$$

with $m_i(t) = t_i + r_i(t) * h_{\text{LP}}(t)$, where t_i is a fixed delay, $r_i(t)$ is a random process which is bandlimited by a lowpass filter with impulse response $h_{\text{LP}}(t)$.

A **flanger** simulates a special filter effect which occurs if a slowly time-varying single reflection is added to the direct signal (jet airplane effect). This effect was achieved by playing two copies of a single signal from two tape machines while slowing

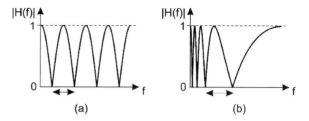

FIGURE 4 Frequency responses of (**a**) flanger and (**b**) phaser.

down the replay speed of one tape machine by placing a finger to the flange of one tape reel. The filter effect is described by

$$y(t) = 0.5x(t) + 0.5x(t - m_1(t)) \text{ and} \tag{18}$$

$$\text{FT}\{y(t)\} = Y(\omega) = X(\omega)H(\omega) = 2X(\omega)|\cos(\omega m_1(t))|\exp(-j\omega m_1(t)), \tag{19}$$

where the slowly time-varying delay is $m_1(t) = t_1 + a\sin(\omega_0 t)$ with $t_1 = 1/f_1$. The frequency f_1 controls the frequency spacing of the zeros in the transfer function. This effect produces equally spaced attenuations to zero across the frequency spectrum, where the zeros move slowly up and down the frequency axis (see Figure 4) by choosing a time-varying delay time $m_1(t)$.

A **phaser** performs a similar effect as a flanger, but instead of the time-varying delay line, which performs a frequency-independent delay, a series of frequency-dependent allpass filters is used which leads to the output signal

$$y(t) = 0.5x(t) + 0.5x(t) * [h_{\text{AP1}}(t) * h_{\text{AP2}}(t) * \cdots * h_{\text{APN}}(t)]. \tag{20}$$

Single allpass sections $h_{\text{AP}i}(t)$ can be first-order allpass filters which have a phase response which ranges from 0 to $-\pi$, in the series connection the allpass phase response ranges from 0 to $-N\pi$. The phase crosses multiples of $-\pi$ $N/2$ times, so notches will be produced at those frequencies. Due to the non-linear phase response the multiples of π are spaced non-uniformly across the frequency spectrum (see Figure 4), where the zeros move slowly up and down the frequency axis when the allpass filter coefficients are varied with a slow speed. Further details on chorus, flanger and phaser can be found in [4,5,10,11,26,29].

4 • Pitch Shifting and Time Stretching

Pitch shifting modifies the pitch of a signal by increasing the playback speed of a recorded signal and thus rising the pitch or decreasing the playback speed of the signal and thus lowering

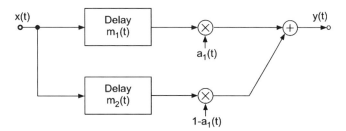

FIGURE 5 Pitch shifting by delay line modulation and overlap and add.

the pitch. The fundamental correspondence is given by the time scaling theorem of the Fourier transform $FT\{x(at)\} = X(\omega/a)/|a|$, which shows that time-domain scaling leads to a frequency-domain scaling. First pitch shifting devices have been introduced in the late 1930s with the help of tape machines and rotary replay heads for secure audio transmission applications. The basic operations of pitch shifting can also be achieved by two or several parallel modulated delay lines (see Figure 5), where the output signals are cross-faded as given by

$$y(t) = a_1(t)x(t - m_1(t)) + [1 - a_1(t)]x(t - m_2(t)). \qquad (21)$$

The time-varying delays are $m_1(t) = t_0 + (t \bmod t_1)$ and $m_2(t) = m_1(t - t_1/2)$, which represent sawtooth signals with a special slope $1/t_1$ for controlling the Doppler effect for pitch shifting each delay line output. The cross-fade functions are given by

$$a_1(t) = 0.5[1 + \sin(\omega_1 t)] \text{ and } \omega_1 = 2\pi/t_1, \qquad (22)$$

which allows the cross-fading when the read pointer for delay $m_1(t)$ jumps back to fade over to delay $m_2(t)$ and vice versa [29, Chap. 7]. This time-domain approach for pitch shifting modifies the spectral envelope of the spectrum according to the scaling theorem of the Fourier transform. A method to perform pitch shifting with preservation of the spectral envelope is explained below by using a frequency-domain approach [1,15–17].

Time stretching can be achieved by segmentation of the input signal $x(t)$ into over-lapping blocks with an analysis hop size S_a, taking the unmodified input blocks $x_i(t)$, and performing an overlap and add operation with synthesis hop size $S_s = \alpha S_a$ to form the output signal $y(t)$ (see Figure 6). The coefficient α controls the increase ($\alpha > 1$) or decrease ($\alpha < 1$) of the stretching operation. In the overlap region a suitable window function (cross-fading between block overlap) is used [29, Chap. 7].

The same analysis and synthesis approach can be used to perform a frequency-domain processing (see Figure 7) of the individual

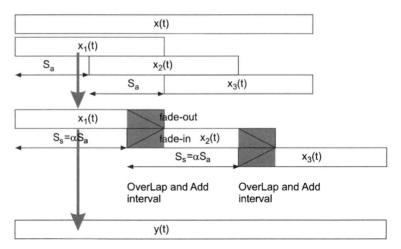

FIGURE 6 Time stretching by input segmentation and modified overlap and add operation.

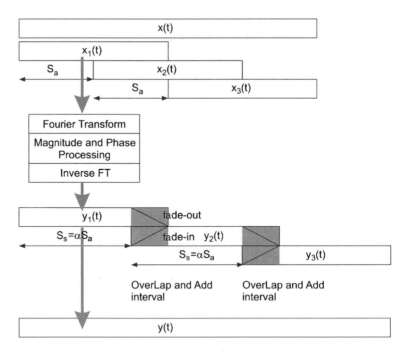

FIGURE 7 Frequency-domain processing for pitch shifting and time stretching.

input blocks $x_i(t)$, modifying the magnitude and the phase of the corresponding Fourier transforms $Y_i(\omega) = f\{X_i(\omega)\}$, and perform an inverse Fourier transform yielding modified blocks $x_i(t)$. The weighted overlap and add operation reconstructs the modified output signal $y(t)$. This frequency-domain approach allows several audio effects (denoising, stable/transients separation, robotization, whisperization, sinusoidal modeling, source separation) besides time stretching and pitch shifting which can preserve the spectral envelope of the input spectrum thus avoiding the so-called Mickey Mouse effect [1,15–17].

5 • Dynamics Processing

Dynamics processing is necessary to reduce or expand the dynamics of the audio signal. It is used in recording or transmission applications and fits the dynamic range to the following signal processing steps. Dynamics processing (see Figure 8) is an operation which performs a modification of the time-domain amplitude envelope according to

$$y(t) = m(t)\,x(t) \text{ with } \mathrm{FT}\{y(t)\} = Y(\omega) = M(\omega)*X(\omega)/2\pi. \quad (23)$$

The time-varying gain function $m(t) = f\{x(t)\}$ is derived from the input signal and represents the instantaneous root mean square or absolute value of the input, which is measured with appropriate time constants. Typical operation modes are limiting, compression, expansion and noise gating [18,28]. The amplitude modification by multiplication shows the convolution of the corresponding spectra. Therefore a bandlimited gain function is necessary, otherwise a spreading of the input spectrum is performed. In several applications a non-linear behavior of the dynamics system is of special interest, for example

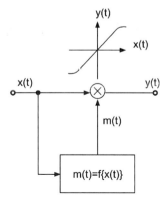

FIGURE 8 Dynamic range processing by non-linear processing.

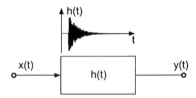

FIGURE 9 Room simulation by fast convolution.

when non-linear transfer characteristics like triode, pentode or tape saturation characteristics are preferred for sound shaping. The spreading of the input spectrum leads to a higher signal bandwidth of the output signal [29].

6 • Room Simulation

Room simulation is the generation of artificial reverberation [2,3,9,12,19,22–25,28] and infuses a room impulse response on the signal. There have been several simplified approximations of room impulse responses over the past 40 years which are based on feedforward and feedback comb and allpass filters in a variety of different configurations. If a room impulse response $h(t)$ (see Figure 9) can be derived by measurements (pseudo-random signals or swept-sine techniques) of a real room or model-based computations (ray-tracing or image model methods), the room simulation applied to the input signal is given by the convolution operation

$$y(t) = x(t)*h(t) \text{ with } \text{FT}\{y(t)\} = Y(\omega) = X(\omega)H(\omega). \quad (24)$$

Digital signal processing techniques allow the segmentation of the input signal and the impulse response thus leading to a parametric convolution operation given by

$$y(t) = x(t)*[a_1 h_1(t) + a_2 h_2(t) + a_3 h_3(t) + a_4 h_4(t)]. \quad (25)$$

Performing the partial convolutions in the frequency domain allows the fast convolution of the entire impulse response [8,20] with the input signal with nearly perfect psychoacoustical properties (see Figure 9).

References

1. D. Arfib, "Time-Frequency Processing" and "Source-Filter Processing", in U. Zölzer (ed.) DAFX – Digital Audio Effects, Wiley, 2002.
2. B. Blesser, "An Interdisciplinary Synthesis of Reverberation Viewpoints", J. Audio Eng. Soc., Vol. 49, No. 10, pp. 867–903, Oct. 2001.

3. J. Dattorro, "Effect Design: Part 1 Reverberator and Other Filters", J. Audio Eng. Soc., Vol. 45, No. 9, pp. 660–684, Sep. 1997.
4. J. Dattorro, "Effect Design: Part 2 Delay-Line Modulation and Chorus", J. Audio Eng. Soc., Vol. 45, No. 10, pp. 764–788, Oct. 1997.
5. J. Dattorro, "Effect Design, Part 3 Oscillators: Sinusoidal and Pseudonoise", J. Audio Eng. Soc., Vol. 50, No. 3, p. 115, 2002.
6. S. Disch, U. Zölzer, "Modulation and Delay Line Based Digital Audio Effects, Proc. DAFx-99 Workshop on Digital Audio Effects, pp. 5–8, Trondheim, 1999.
7. F. Fontana, M. Karjalainen, "A Digital Bandpass/Bandstop Complementary Equalization Filter with Independent Tuning Characteristics", IEEE Signal Process. Lett., Vol. 10, No. 4, pp. 119–122, Apr. 2003.
8. W. G. Gardner, "Efficient Convolution Without Input–output Delay," J. Audio Eng. Soc., Vol. 43, No. 3, pp. 127–136,1995.
9. W. G. Gardner, "Reverberation Algorithms," in M. Kahrs and K. Brandenburg (eds), Applications of Digital Signal Processing to Audio and Acoustics, Kluwer, pp. 85–131, 1998.
10. W. M. Hartmann, Signals, Sound, and Sensation, AIP Press, 1997.
11. W. M. Hartmann, "Flanging and Phasers", J. Audio Eng. Soc., Vol. 26, No. 6, pp. 439–443, June 1978.
12. J. M. Jot, A. Chaigne, "Digital Delay Networks for Designing Artificial Reverberators", Proc. 94th AES Convention, Preprint No. 3030, 1991.
13. F. Keiler, U. Zölzer, "Parametric Second- and Fourth-Order Shelving Filters for Audio Applications", Proc. of IEEE 6th Workshop on Multimedia Signal Processing, Siena, Italy, September 29–October 1 , 2004.
14. T. I. Laakso, V. Välimäki, M. Karjalainen, U. K. Laine, "Splitting the Unit Delay—Tools for Fractional Delay Filter Design", IEEE Signal Process. Mag., Vol. 13, No. 1, pp. 30–60, Jan. 1996
15. J. Laroche, "Time and Pitch Scale Modifications", in M. Kahrs and K.-H. Brandenburg (eds), Applications of Digital Signal Processing to Audio and Acoustics, Kluwer 1998.
16. J. Laroche and M. Dolson, "Improved Phase Vocoder Time-Scale Modification of Audio", IEEE Trans. Speech Audio Process., Vol. 7, No. 3, pp. 323–332, 1999.
17. J. Laroche and M. Dolson, "New Phase-Vocoder Techniques for Pitch Shifting, Chorusing, Harmonizing, and Other Exotic Audio Modifications", J. Audio Eng. Soc., Vol. 47, No. 11, pp. 928–936, 1999.
18. G. McNally, "Dynamic Range Control of Digital Audio Signals", J. Audio Eng. Soc., Vol. 32, No. 5, pp. 316–327, 1984.
19. J. A. Moorer, "About this Reverberation Business", Comput. Music J., Vol. 3, No. 2, pp. 13–28, 1978.
20. A. V. Oppenheim, R. W. Schafer, Digital Signal Processing, Prentice-Hall, Englewood Cliffs, NJ, 1975.
21. S. J. Orfanidis, Introduction to Signal Processing, Prentice-Hall, Englewood Cliffs, NJ, 1996.

22. D. Rocchesso, "Spatial Effects", in U. Zölzer (ed.) DAFX – Digital Audio Effects, Wiley, 2002.
23. D. Rocchesso and J. O. Smith, "Circulant and Elliptic Feedback Delay Networks for Artificial Reverberation", IEEE Trans. Speech Audio Process., Vol. 5, No. 1, pp. 51–63, Jan. 1997.
24. M. R. Schroeder, "Natural Sounding Artificial Reverberation", J. Audio Eng. Soc., Vol. 10, No. 3, pp. 219–223, 1962.
25. M. R. Schroeder, B. F. Logan, "Colorless Artificial Reverberation," J. Audio Eng. Soc., Vol. 9, No. 3, pp. 192–197, 1961.
26. J. O. Smith, "An Allpass Approach to Digital Phasing and Flanging", Proceedings of the 1984 International Computer Music Conference, pp. 103–108, 1984.
27. J. O. Smith, S. Serafin, J. Abel, D. Berners, "Doppler Simulation and the Leslie", Proc. of the 5th Int. Conference on Digital Audio Effects (DAFx-02), Hamburg, Germany, September 26–28, 2002.
28. Udo Zölzer, Digital Audio Signal Processing, 2e, Wiley, 2008.
29. U. Zölzer (ed), DAFX - Digital Audio Effects, Wiley, 2002.

42
Perceptually Based Audio Coding

Adrianus J.M. Houtsma

*Aircrew Protection Division, U.S. Army Aeromedical Research Laboratory,
Fort Rucker, AL 36362-0577, USA,
e-mail: adrian.houtsma@amedd.army.mil*

1 • Introduction

High-quality audio is a concept that is not exactly defined and not always properly understood. To some, it refers directly to the physical similarity between a real sound field and its electro-acoustical reproduction. In this viewpoint, acoustical knowledge and electronic technology are the only limiting factors preventing audio quality from being perfect. To others, however, audio quality refers to the audible similarity between a real life sound event and an electronic reproduction. Given this viewpoint, the human auditory system with all its limitations becomes an essential factor determining audio quality.

From a physical viewpoint, audio quality is to a large extent determined by the dynamic frequency profile of an acoustic signal as a function of time. The frequency profile at any moment of time is a quasi-instantaneous spectrum measured over a very short time interval. For the human ear, this interval is in the order of 10 ms. A convenient way of displaying a dynamic frequency profile in a two-dimensional plot is the spectrogram, where time is mapped along the abscissa, frequency along the ordinate, and intensity (of each frequency element) as degree of saturation on a black/white or color scale. Examples of spectrogram displays can be found in Chapter 3 Frequency is usually mapped on a logarithmic scale (octaves, decades, semitones), similar to the spatial layout of frequencies along a piano keyboard or the natural frequency-space mapping along the basilar membrane in the human inner ear (cochlea). Intensity, whose physical units are Watt/m², is often represented on the decibel (dB) scale. This is a relative logarithmic scale, where intensity level, L_I, is defined as:

$$L_I = 10\log_{10} I/I_0, \tag{1}$$

and where $I_0 = 10^{-12}$ W/m². For alternative and equivalent formulations of the decibel scale, based on sound pressure or sound power, the reader is referred to Chapter 51.

If one measures the locus of all pure tone frequency–intensity combinations that a young and healthy listener is able to perceive without experiencing discomfort, one typically finds a frequency range extending from 20 Hz to 20 kHz and an intensity range from about 10 (average threshold) to 110 dB (discomfort level). Frequency spectra of musical instruments and the human voice cover a large portion of this frequency range. The dynamic range of a full symphony orchestra is, depending on size and concert hall acoustics, somewhere around 90 dB, although dynamic ranges of individual instruments are considerably smaller [1]. Figure 1 shows the typical frequency–intensity area for human hearing and, inside that area, a conservative estimate of the area covered by conventional music.

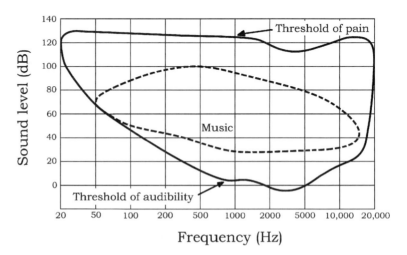

FIGURE 1 The auditory frequency and dynamic range. The "music" range is represented by the *dashed* area (from Rossing [47]).

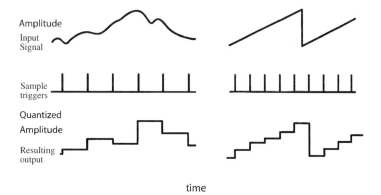

FIGURE 2 Illustration of pulse code modulation sampling, with horizontal steps reflecting the sampling period, and vertical steps varying inversely with sample size (from Rossing [47]).

If an audio signal is to be represented digitally, it first must be sampled, which is schematically illustrated in Figure 2. At regular time intervals, called the sampling period, the amplitude of the analog signal is measured and held constant until the next sampling event. The inverse of the sampling period is known as the sampling frequency or sampling rate. Each sampled amplitude is, to the nearest approximation, represented by an N bit digital word called a sample. If N is small, there are few (2^N) discrete values to represent signal amplitude, resulting in a very inaccurate signal representation. The perceptual effect of such inaccuracy is audible noise, know as quantization noise. The dynamic range of a digital N-bit system, DR, being the signal to noise ratio between the maximum signal output and the minimum unit digital step and expressed in decibels, is given by the expression:

$$\mathrm{DR} = 20 \log_{10} 2^N. \qquad (2)$$

Moreover, according to the sampling theory, the bandwidth of an audio signal to be digitized is limited to one-half of the sampling frequency. Consequently, for digital storage and reproduction of sound with a bandwidth of 20 kHz, one needs a sampling frequency of at least 40 kHz. To reproduce sounds with a dynamic range of 96 dB without audible quantization noise, one needs a sample word size of at least 16 bits. These considerations led to the Compact Disc format proposed by Philips and Sony in the early 1980s, and standardized in 1987 [2]. The CD uses a 16-bit sample size and a 44.1-kHz sampling rate. The total bitrate in two-channel CD audio is therefore a little over 1.4 megabits per second (Mb/s). The CD replaced the "vinyl" record in a few years and quickly became the new standard for audio quality across the world.

In subsequent years, storage and transmission of audio bits became progressively cheaper through advances in digital technology and, consequently, a mostly technology-driven push toward an even higher digital audio quality standard ensued. In some studies, evidence was found that sound components well above 20 kHz can cause measurable effects on α-EEG activity [3], difference-tone perception [4], and other perceptual features [5,6]. The UK-based Acoustic Renaissance for Audio Foundation has repeatedly argued for the establishment of stricter standards for high-quality audio on high-density CD carriers [7]. An example of an industrial response to such an argument is the Philips/Sony Super Audio CD [8] which is capable of audio bandwidths up to 100 kHz and a dynamic range of 120 dB. Critical evaluations of claims that sound components above 20 kHz are perceivable, however, have often uncovered artifacts (e.g., equipment distortion) as the cause of the observed effects [9]. Altogether, it seems that, so far, there is no compelling psychophysical evidence of a need for a stricter and more bit-intensive CD standard. For purposes of documenting and archiving sound recordings for later analysis, however, a more broadband format than the present CD standard could have its merits.

In many applications of digital audio other than compact disc or digital audio tape, there is an inherent bandwidth limitation that impedes or prevents straightforward application of the CD standard. Examples are digital audio broadcast (DAB), digital compact cassette (DCC), cinema sound (Sony Dynamic Digital Sound, Dolby ProLogic, Dolby Digital), high-definition television (HDTV), and sounds transmitted over the Internet. In some cases, the required audio bandwidth cannot be reached because of hardware limitations (DCC), and in other cases, there is bandwidth competition with other desirable features of the communication such as video quality or the number of available radio stations. In all of these cases, one ideally wants to reduce the audio bitrate of 1.4 Mb/s to something considerably less without affecting perceived sound quality. A straightforward approach of lowering the sampling rate would immediately be audible as a reduction of brightness due to the absence of high frequencies, whereas a reduction of the word size would result in audible and annoying sound effects caused by quantization noise.

Bitrate reduction without audible sound degradation can be achieved by (a) reduction of statistical redundancy and (b) minimization of perceptual irrelevance. Both processes are independent and cumulative. The first process removes redundancies in the code, is a strictly statistical mathematical operation, and results in a lossless code (i.e., a code from which the original sound waveform is completely reconstructible). An example of such a code is the open-source Free Lossless Audio Codec (FLAC), readily available on the Internet. The second process exploits known limitations of our hearing system in order to remove inaudible sound elements and to selectively allow quantization noise that cannot be perceived. This results in a lossy, but at the same time, perceptually transparent code. From such a code, the original sound waveform cannot be exactly reconstructed any more, but a listener will not be able to tell the difference between the original and the bitrate-reduced versions. Finally, if bitrates are reduced to a level where audible sound elements are removed or audible quantization noise is allowed to enter, the result will be a lossy and perceptually degraded code.

It is important to realize that, in this development of audio signal processing, the science of sound perception (psychoacoustics) is not merely used to evaluate performance of an end product, but has become a critical and indispensable ingredient of the various coding systems themselves.

2 • Transparent Coding: General Strategy

A generalized scheme for a perceptually transparent audio coding algorithm is shown in Figure 3. The analog acoustic waveform is first subjected to a time-frequency analysis, shown in the top-left part of the figure. This stage can take many different forms, depending on coding strategy. Short time segments, comparable to the temporal resolution of our hearing system, can be passed through a set of uniform or nonuniform bandpass filters, as is done in subband coding [10]. The output parameters of this time-frequency analysis block in this case are the time responses of each subband to each input segment. Transform coders [11]

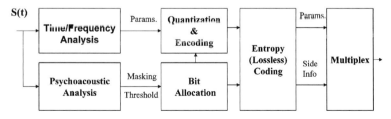

FIGURE 3 General algorithm for perceptual entropy coding. See text for functions of various building blocks (after Painter and Spanias [32]).

typically use some unitary transformation into the frequency domain, for instance a cosine transform.[1] In this case, the output parameters are represented by a set of transform coefficients for each transformed time segment, similar to a Fourier transform. Both subband and transform coders are purely waveform-based and do not assume any knowledge of how the waveform was produced. Another time/frequency analysis strategy is to assume that the sound waveform is, in principle, made up of simple building blocks that can be modeled and adequately described by a limited set of parameters. Parametric coders try to describe the input sound segments in terms of sinusoidal parameters [12], combinations of sinusoids and noise [13], or frequency-modulated sinusoids [14]. Output parameters then take the form of sinusoidal amplitudes and frequencies, noise intensities and bandwidths, and modulation indices. Another example is the linear predictive coder (LPC) that has become very successful in speech coding [15]. It is based on the idea of a source-filter model of sound, where the source is either a periodic pulse or a noise and the filter is an n-pole linear filter. Such a sound model seems a reasonable representation for the human voice and, to some extent, for harmonic musical instruments. It appears less appropriate as a model for inharmonic tonal instruments such as bells, gongs, or other percussion instruments. There are, however,

---

[1]The Modified Discrete Cosine Transform (MDCT) is a modification of the Discrete Cosine Transform (DCT) which, for a sequence of time samples $[x_n]$ of length N, is defined by the following pair of transforms:

(a) The set of transform coefficients, $[X_k]$, is given by the formula

$$[X_k] = \sqrt{1/N} x_0 + \sqrt{2/N} \sum_{n=1}^{N-1} x_n \cos \pi (2n+1)k/2N$$

$$(k = 0, 1, 2, \ldots, N-1);$$

(b) The inverse transform, reconstructing the sequence of time samples from the set of transform coefficients, is

$$[X_n] = \sqrt{1/N} X_0 + \sqrt{2/N} \sum_{k=1}^{N-1} X_k \cos \pi (2n+1)k/2N$$

$$(n = 0, 1, 2, \ldots, N-1).$$

The modification in the MDCT consists primarily of a weighted overlap and add algorithm (instead of a straight sequencing of data blocks as in DCT), resulting in less artifacts at block boundaries and better time-domain aliasing cancellation. A comprehensive and informative description of the MDCT algorithm is found in reference [32].

excellent musical instrument models available these days that could successfully be used as alternatives for the initial time-frequency decomposition stage of the coder. To some extent, the same could be said about the Spectral Band Replacement (SBR) technique [16], which has recently become very popular and is an integral part of the MPEG-4 standard [17]. This technique uses the fact that, at least for certain classes of sound signals, there is a relationship between information in the low and in the high frequencies. Therefore, if very low bitrates must be used that normally would produce audible coding artifacts, the sound quality can be improved by coding and transmitting only low-frequency information and reconstructing the high-frequency information at the receiver end using the known (or assumed) relationships. Success of such a coding strategy will, of course, depend on the existence of assumed relationships in the sound to be coded, which may not always be the case.

The next general operation is the quantization and encoding of the parameters yielded by the time–frequency analysis. The quantization process is typically adaptive, where the number of bits available for encoding parameters of each time frame is determined by the results of a psychoacoustic analysis shown in the *lower* portion of Figure 3. This analysis assesses (1) which parameters can safely be dropped because their effect is inaudible and (2) how much quantization noise can be allowed for signal parameters within that time frame. The principal psychoacoustic effect exploited is simultaneous masking, where the perception threshold for quantization noise in any frequency band is raised by the simultaneous presence of a sound signal [18]. Forward and backward masking [19], in which a sound signal masks quantization noise in succeeding or preceding time frames, can, in principle, also be included. In two-channel stereo sound, masked thresholds for quantization noise may be lowered by binaural masking release [20], resulting in a somewhat higher bit allocation in comparison with mono sound. On the other hand, signal information in the different sound channels can, to some extent, be combined, lowering the overall bitrate [21,22]. For multichannel 3D sound presentation, sound channels may also be partially combined [22], and even transformed into a single monaural channel plus a stream of binaural parameters, as is done in Binaural Cue Coding [23]. This leads to very low bitrates with retention of transparency and, in addition, provides more flexible control over source positioning at the output in comparison with conventional multichannel coding and transmission techniques.

In the final stage of the general encoder of Figure 3, all mathematical and statistical redundancies that are left in the signal code are removed, which is known as entropy coding. This stage of the processor is lossless, whereas the previous stages of the coder yielded a lossy code. Hence, the end result is a lossy code. Whether or not this code is perceptually transparent depends primarily on the available bitrate (i.e., the average number of bits that can be allocated in each time frame). It also depends, however, on details of the design of the time/frequency analysis block and on the finesse and correctness of the psychoacoustical model that is used.

If available bitrates are too low in comparison with the degree of sophistication of signal preprocessing and psychoacoustic analysis, coding artifacts will occur that are likely to affect the perceived sound quality in a negative way. Subjectively, such artifacts may sound like brief irregular bursts of audible noise, timbre distortion, or disturbances in perceived location of sound source images. A particularly well-known and frequently occurring artifact is pre-echo. This happens when time frames are chosen too long with respect to aural time resolution, and a strong transient (i.e., a castanet or bell sound) falls in the later portion of a time frame. Because allowable quantization noise estimated by the psychoacoustic model is spread over the entire time frame, it is likely to be audible just before the onset. To estimate obtainable limits for transparent coding, Johnston [24] developed a general computation scheme for Perceptual Entropy based on quantitative psychoacoustic models for simultaneous and nonsimultaneous masking. He predicted that, for a variety of coders, bitrates could be typically reduced to about 2.1 bits/sample (at sampling rates above 40 kHz) before perceptual transparency is lost. Early listening tests with an MPEG-1 layer I coder (Section 3) and simple fixed-difference or adaptive forced choice procedures between full 16-bit/sample and bit-compressed signals yielded difference detection thresholds between 2 and 3 bits/sample, dependent on music fragment [25]. Subsequent technical developments of combining the best elements of different coding strategies, however, have yielded hybrid coders that are able to produce perceptually transparent codes at rates less than 1 bit/sample, which, at a CD sampling rate, is equivalent to less than 44 kilobits per second per channel (kb/s/ch) [26].

3 • Transparent Coding: Specific Examples

3.1 MPEG-1, Layers I and II

The first example is a specific subband coding scheme known as MUSICAM [27] that became the basis for the ISO/IEC MPEG-1 layers I and II standard [28]. The acronym MPEG stands

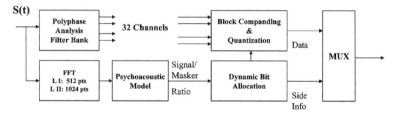

FIGURE 4 Block algorithm of ISO/IEC MPEG-1 layers I and II encoder (after Painter and Spanias [32]).

for Motion Picture Expert Group, indicating that the sound coding standard is part of a larger package covering cinema and TV. The standard is schematically illustrated in Figure 4. The time/frequency analysis is realized by a 32-channel polyphase filter bank that splits up an input signal into 32 band-limited signals, each with a bandwidth of 690 Hz, after the signal has been segmented into short (8-ms) time frames. The filter is a perfect reconstruction filter so that, if these band-limited signals were added, the input signal would be obtained again. Consequently, total bitrates at the filter's input and output are the same, since each of the 32 output channels can be represented at a sampling rate that is 1/32nd of the input signal's sampling frequency. The real bit savings occur in the next stage, where frame amplitudes are normalized to full-scale, indicated as Block Companding, where scale factors are being stored as side information, and where each frame in each frequency channel is quantized using an externally assigned word size. This word size is adaptively determined by a parallel analysis process shown in the lower half of the figure. A Fast Fourier Transform (FFT) of signal segments allows computation of an expected auditory masking pattern, yielding a set of minimum word sizes for each frequency band necessary to keep quantization noise below masked threshold. Quantization of audio data and scale factors is done in a slightly more sophisticated and computationally more intensive manner in layer II than it is in layer I, resulting in a greater bit efficiency. Finally, signal data and side information are multiplexed into a single bitstream.

To illustrate how this coder works, a schematic representation of a three-tone music signal against a quantization noise background typically produced by pulse-code modulation (PCM), as is done in the CD, is shown in Figure 5. It shows the masked threshold of our auditory system, given three masking tones at 250, 1,000, and 4,000 Hz and at a sound pressure level of approximately 70 dB. One can see that at the three tone frequencies, auditory threshold has been raised by about 50 dB compared with the threshold in quietness. Quantization noise for some arbitrary

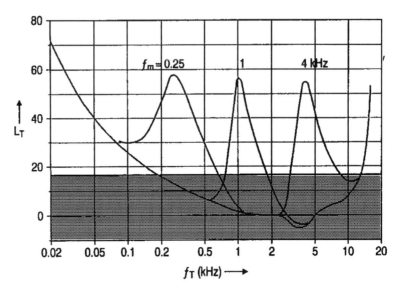

FIGURE 5 Threshold (L_T) of a test tone in quiet (lower contour) and in the presence of three simultaneous masking tones (upper contour). *Shaded area* represents quantization noise of arbitrary level.

bitrate has been represented by a spectrally flat band of noise. Actually, this representation is a slight oversimplification because quantization noise, although spectrally broadband, is not really spectrally flat nor is it stationary. From the figure, it is clear that, for this situation, the quantization noise is audible in the valleys of the masking profile, at approximately 800, 2,500, and 10,000 Hz. Figure 6 illustrates how one should ideally shape this quantization noise so that it is always below the masked threshold created by the signal components. A comparison reveals that the total quantization noise power in Figure 6 is larger than that in Figure 5. Nevertheless, the noise shown in Figure 6 will be inaudible. It is also clear that, because a music signal continuously changes with time, the quantization noise shaping should be adaptive and be updated frequently.

Quantization noise shaping was achieved because the input signal was segmented into short time frames, and each time frame was decomposed into a set of responses of a bank of constant-bandwidth filters, as explained earlier. Parallel to this signal decomposition, an FFT of the time frame and a masking profile were computed, based on elementary psychoacoustic rules, as was shown in Figure 4. From the total number of bits available for that time frame, bits are allocated to each response signal in a manner that varies inversely with the amount

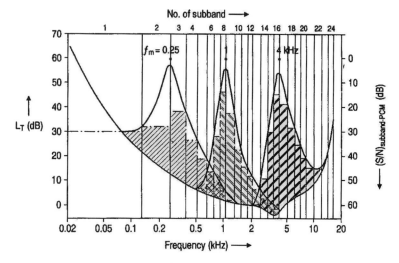

FIGURE 6 Same as Figure 3, with maximum inaudible quantization noise divided over 24 subbands (after Wiese and Stoll [27]).

of masked threshold elevation at the center frequency of the passband for that signal. Since allocation of fewer bits implies generation of more quantization noise, the overall result will be that the spectral profile of the noise will generally conform to the masking profile of the signal, as is shown in Figure 6. Moreover, since each frequency channel is quantized separately, the quantization noise that is generated will remain within the bandwidth of each particular channel.

Figure 7 illustrates the actual performance of the coder for a vowel segment of a female-voice, displaying sound pressure level (SPL) as a function of frequency. Shown are (a) 20 spectral components of the vowel sound, (b) the masking profile of this vowel segment computed by the psychoacoustic analyzer, and (c) the spectrum of quantization noise allowed by the coder. One can observe that bitrate saving is achieved in two ways: (a) by allocating fewer bits in subbands where the masking profile is high and (b) by allocating no bits at all where all signal components within a subband are inaudible, as is the case in subband 5. It is also clear that, if one wants to maintain perceptual transparency, any uncertainty about the correctness of the psychoacoustic model will require a greater safety margin between computed masked threshold and quantization noise levels. Therefore, the quality of the psychoacoustic model has a direct influence on the transparent bitrates that a coder can ultimately achieve. Continued efforts are made to further improve psychoacoustic masking models [29]. The larger FFT resolution of layer II in

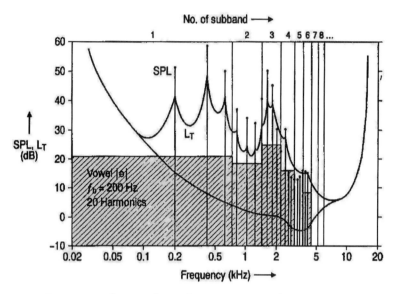

FIGURE 7 Spectrum of a vowel, masking pattern, and quantization noise as allocated in an MPEG-1 layer II coder (after Wiese and Stoll [27]).

comparison with layer I, and also some temporal feature differences that allow exploitation of forward and backward masking between successive time frames, cause layer II to achieve a better average sound quality at a fixed bitrate or, equivalently, a lower bitrate for a given sound quality. Layer I was used in the Philips/Matshusita digital compact cassette recorder/player (DCC), a portable consumer product that was backwards compatible with the popular analog tape cassette recorder. It was found to yield a perceptually transparent code at a bitrate of 192 kb/s/ch, roughly a factor 4 reduction compared with CD code. Layer II was developed for use in Europe's digital audio FM radio broadcast system. It was found to be perceptually transparent at bitrates as low as 96 kb/s/ch, a factor 8 more efficient than the CD.

3.2 MPEG-1 Layer III

The second example, schematically illustrated in Figure 8 and formally known as ISO/IEC MPEG-1 layer III [28], is better known under the popular name "MP3", and often erroneously referred to as MPEG-3 (which does not exist). It has become the de facto standard for sound coding on the Internet, as well as for many handheld devices. MP3 decoders can presently be found as an extra feature on audio-CD players, allowing them to play computer-made CR-ROMs written in MP3 format. MP3 files are considerably shorter than CD audio files, so that much more music can be put on a single CD.

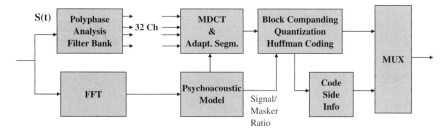

FIGURE 8 Block algorithm of ISO/IEC MPEG-1 layer III encoder (after Painter and Spanias [32]).

There are three major differences with the layers I and II versions of the MPEG-1 standard. First, the initial filter is a hybrid combination of a polyphase filter bank and a modified discrete cosine transform (MDCT).[1] Thus, the layer III version represents a combination of the subband coding [10] and the transform coding [11] strategies. The main advantage of such a hybrid filter is that, through its better frequency resolution, it allows a closer approximation to critical band behavior of the human ear. Secondly, an adaptive instead of a fixed signal segmentation in the cosine transform allows better control of pre-echo artifacts preceding high-intensity signal transients. Thirdly, the quantization stage is different since, its input being a Fourier transform, it quantizes spectral rather than temporal information. Because of layer III's closer approximation to human cochlear frequency and temporal resolution, it will render a better subjective audio quality than layers I and II at some fixed low bitrate or, equivalently, yield a same audio quality at a lower bitrate. Because of the higher complexity of the layer III encoder, this layer was originally not intended for use in portable or handheld devices. With its increasing popularity for exchanging sound files on the Internet, however, commercial sound editing software programs have added MP3 encoding features to their packages, and almost all computer-based sound reproduction tools contain an MP3 decoder. At the moment this chapter is being written, the first small portable sound recording devices containing an MP3 encoder have appeared on the consumer market. With MP3, transparent audio can be achieved at bitrates as low as 64 kb/s/ch, a factor 11 reduction compared with CD code.

It is important to realize that the MPEG-1 standard, just as all later standards, merely regulates the format of the encoder output. This is to make sure that an MPEG-1 decoder, designed in compliance with the standard, can handle every MPEG-1 signal, regardless of the layer of its encoding [30]. This leaves considerable freedom to designers of encoding algorithms to fill in technical details of the building blocks shown in Figure 4 and 8. Consequently, there are significant design differences between various commercial MP3 encoders, with often quite audible performance differences.

3.3 MPEG-2

The final example is the MPEG-2 NBC/AAC (non-backwards compatible advanced audio coder) [31]. Because of growing demands by the movie industry to transmit more sound channels at a finite bitrate, new coding tools had to be adopted that made this code intrinsically incompatible with previous MPEG codes. The coder, schematically illustrated in Figure 9 represents the best that is presently available. It formally met the strictest ITU-R BS.1116 criteria (see Section 5) at an average bitrate of 64 kb/s/ch, but can actually produce perceptually transparent code at bitrates as low as 32 kb/s/ch. The complete algorithm has been designed as a set of coding tools organized in three different levels of complexity. The one shown here represents the main complexity profile (the others are low complexity and scalable sample rate profiles).

Comparing the algorithm of MPEG-2 shown in Figure 9 with the MPEG-1 algorithms (Figures 4 and 8), one sees some general common features. There are parallel signal processing and psychoacoustic analysis chains, leading to the formation of a bitstream of coded audio data and side information. In MPEG-2, however, the output code obtains input from just about every signal processing stage, as can be seen in the top half of Figure 9. Moreover, the quantization, scale factor extraction, and entropy

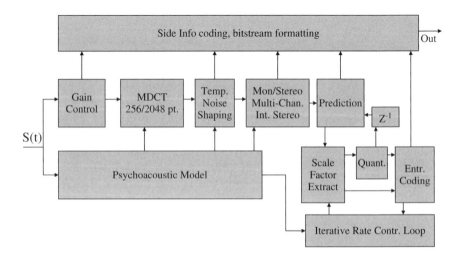

FIGURE 9 Block algorithm of ISO/IEC MPEG-2 NBC AAC encoder (after Painter and Spanias [32]).

coding processes are not performed in a single pass, as in MPEG-1, but in an iterative loop procedure shown in the bottom-right part of the figure. The purpose of this is to achieve an optimal masking effect of quantization noise by the music or speech signal. In the signal processing chain, filtering is done entirely in the frequency domain by an MDCT, using two window sizes (256 points for transients, 2,048 points for stationary signal parts), and two different window shapes (sine shape for best frequency selectivity, Kaiser–Bessel–Dolby shape for best stopband suppression). Pre-echo control for frames that contain strong and sharp signal transients is being achieved by the gain modification control (a time-domain technique) stage, the temporal noise shaping (a frequency-domain technique) stage, and by the block-size switching capability of the MDCT filter. The coder also uses a prediction algorithm that efficiently makes use of the fact that signal coefficients for successive time frames are not statistically independent.

The MPEG-2 NBC/AAC standard is presently used in Japan throughout its digital broadcast system, including SDTV, HDTV, digital radio, and multimedia services. Encoder/decoder chips are available from most of the world's major electronics manufacturers.

To conclude this section, we should point out that there are many other coding standards, e.g., Lucent Technologies' PAC code, Sony's ATRAC, and Dolby's AC-2 that have not been treated here. For a more elaborate technical review, the reader is referred to a comprehensive review paper by Painter and Spanias [32].

4 • Subjective Evaluation of Sound Quality

Because perceptual entropy coders exploit limitations inherent to the human auditory system, classical measures of sound quality such as signal/noise ratio or total harmonic distortion are inadequate and irrelevant. Therefore, the performance of coding schemes is typically evaluated with the use of human listeners.

If one only wants to evaluate whether a sound from a particular coder at a certain bitrate is perceptually transparent, the simplest and most relevant method is a forced-choice discrimination test. The underlying thought is that, if listeners cannot aurally detect the difference between a sound signal from an accepted quality standard such as a CD and the same sound signal from a lower bitrate coder, then the two signals are perceptually equivalent in all respects and therefore the lower-bitrate code is perceptually transparent. The fastest form of such a discrimination test is an adaptive 2-down 1-up comparison method [33] in which each trial consists of a presentation of a reference and a coded signal in random order. Listeners will have to respond on each trial what the presentation order was, and typically will receive response feedback. A run will typically start with a low bitrate for the coded signal, so that differences between reference and coded signals are not difficult to perceive. After two successive correct responses, the bitrate of the coded signal is increased by a certain increment, making the discrimination task more difficult. This procedure is continued until an incorrect response is given, whereupon the bitrate of the coded signal is decreased by one increment. Somewhere, the adaptive trace will end up oscillating between increments and decrements, at which point the adaptive run is ended. The bitrate at which oscillation of the trace occurs is considered the perceptual discrimination threshold and corresponds to a 71% correct discrimination level if fixed-difference runs were used.

Although forced-choice discrimination methods are straightforward in classical psychophysics, its application in the present case is not without problems. First, meaningful comparison of music fragments requires some minimum length of these fragments, which interferes with short-term memory and often causes listener fatigue. Secondly, audible coding artifacts appear in different perceptual forms and can occur at random times, so that the subject is in constant uncertainty about when to listen, what to listen for and, in cases of spatial sound, where to listen. To make matters worse discrimination cues will not only change quantitatively when the bitrate is changed, but they also change qualitatively. This makes adaptive discrimination runs difficult to perform, and often causes unstable non-converging traces and significant learning effects.

A more popular test method is the double-blind A–B–C hidden reference paradigm, standardized in ITU-R BS.1116 [34], which is used for the evaluation of both perceptually transparent and degraded codes. On each trial, expert listeners hear three identical music fragments A, B, and C, where A is always the high-quality reference and B and C are, in random order, the reference and the coded signal. Listeners must (a) identify which of B and C is the hidden reference and (b) rate the difference between B and C on a 41-point scale. A variant on this method is ITU-T Recommendation P.800/P.830, which uses a 7-point scale and is intended for more heavily degraded signals. Incorrect identification of the hidden reference automatically assigns the highest rating ("perceptually equivalent") to the coded signal for that trial. Average scale values assigned to perceived differences provide an indication of the amount of perceptual degradation of coded signals. This method of rating is important for situations where the highest possible quality of audio is desired, but perceptual transparency cannot be achieved because of technical limitations.

5 • Objective Evaluation of Sound Quality

Given the many problems encountered with subjective quality testing, including the finding that "expert" listeners tend to be systematically biased in their sensitivity to particular types of perceptual degradation [35], it is not surprising that several attempts have been made to develop objective measures for perceived sound quality [36–38]. A fundamental problem underlying all these attempts is the fact that, since perceptual entropy coding is based on imperfections of the human ear, an accurate and complete model of human hearing is required to build a reliable and objective evaluation algorithm. It is obvious that a complete model of human hearing does not exist and will not exist for some time to come. Moreover, there is always a hidden danger that an "objective" evaluation algorithm adopts the same or a similar hearing model as was used in a particular coding algorithm, since all perceptual entropy coders use some kind of psychoacoustic model. If this is the case, the evaluation algorithm is likely to favor the coder that uses the same hearing model, and will therefore be everything but "objective".

Despite inherent pitfalls and potential shortcomings of objective evaluation methods, an international standard for objective sound quality testing has been developed and agreed upon by the International Telecommunications Union as Recommendation ITU-R BS.1387 [39]. This standard contains a psychoacoustic model yielding quantitative measures for the internal representation of signal features, and a cognitive model for describing feature extraction and combination. It incorporates six earlier proposed perception and cognition models, specifically the Disturbance IndeX (DIX) model [40], the Noise-to-Masked-Ratio (NMR) model [35], the Objective Audio Signal Evaluation (OASE) model [41], the Perceptual Audio Quality Measure (PAQM) model [42], the PERCeptual EVALuation (PERCEVAL) model [43], and the Perceptual Objective Measurement (POM) model [44]. The essence of the evaluation model is that it always compares a degraded signal with a high-quality reference and estimates specific loudness differences in each critical band following a classical method outlined by Zwicker and Feldtkeller [18]. These quantities are then weighted and combined and ultimately yield a single quantitative measure representing the perceptual distance between the reference and the coded signal.

6 • Concluding Remarks

One should always keep in mind that lossy, perceptually transparent codes are meant to be listened to only. They should never be used for documentation purposes in which sounds are recorded and stored for physical analysis at a later time. The fact that such codes are transparent to our ears does not imply that they are transparent to acoustical or signal processing analysis procedures.

The original waveform-based coding techniques basically allow only playback of a music signal as it was recorded. Modern model-based parametric and binaural cue coding techniques allow not only considerable additional bit savings but also a great deal of flexibility and control of sound features at the playback site [45,46]. This last feature is very useful for music entertainment where, for instance, spatial features of a prerecorded sound track can be manipulated in the spot to fit the needs of a particular concert setting. It also could be used to control spatial properties of auditory icons or warning signals during playback in motor vehicles, airplanes, and helicopters, when the spatial information to be conveyed is only available at the playback site.

Several of the coding strategies, such as LPC, parametric coding, or SBR, are strictly speaking not perceptually based, but rather physics based, and could therefore have been excluded from this chapter. They were included here to illustrate that there are several other bases of knowledge, besides psychoacoustics, that can contribute to significant bit savings without loss of sound quality.

Acknowledgement and Disclaimer. The author is indebted to Armin Kohlrausch, Steven van de Par, Jeroen Breebaart, and an anonymous reviewer for helpful suggestions during the preparation of this chapter. Opinions, interpretations, and conclusions contained in this chapter are those of the author and are not necessarily endorsed by the U.S. Army and/or the U.S. Department of Defense.

References

1. Patterson, B., "Musical dynamics", Sci. Am. 231 (5):78–95, 1974.
2. International Electrotechnical Commission IEC 908, "Compact Disc Digital Audio System", Geneva, Switzerland, 1987.
3. Oohashi, T., Nishina, E., Kawai, N. and Fuwamoto, Y., "Hypersonic effect by high frequency sound above the audible range", Proc. Int. Symp. Mus. Acoust., Tokyo, 119–122, 1992.

4. Nakamura, N., Toraichi, K. and Yamaura, I., "Audible tone on a human tympanic membrane evoked by ultrasonic components", Proc. Int. Symp. Mus. Acoust., Tokyo, 123–126, 1992.
5. Nakamura, N., Toraichi, K., Kamada, M. and Iwaki, M., "Contribution of ultrasound to timbre", Proc. Int. Symp. Mus. Acoust., Tokyo, 127–130, 1992.
6. Sato, H., Yoshida, M. and Yasushi, M., "Evaluation of sensitivity of the frequency range higher than the audible frequency range", Pioneer R&D, 7, 10–16, 1997.
7. Stuart, J.R., "A proposal for high-quality application of high-density CD carriers", Stereophile, August 1995.
8. Aarts, R.M., Reefman, D. and Janssen, E. "Super Audio CD – An Overview", Proc. Int. Congress of Acoustics (ICA-2004), Vol. 3, 2179–2182, 2004.
9. Ashihara, K. and Kiryu, S., "Audibility of components above 22 kHz in a harmonic complex tone", Acta Acustica, 89, 504–546, 2003.
10. Stoll, G., Link, M. and Theile, G., "Masking-pattern adapted subband coding: Use of the dynamic bit-rate margin", Proc. 84th Conv. Aud. Eng. Soc., 1988, preprint 2585.
11. Brandenburg, K., "High-quality sound coding at 2.5 bits/sample", Proc. 84th Conv. Aud. Eng. Soc., 1988, preprint 2582.
12. Serra, X. and Smith, J.O. III, "Spectral modeling and synthesis: A sound analysis/synthesis system based on a deterministic plus stochastic decomposition", Comput. Mus. J., 12–24, 1990.
13. Edler, B. "Technical description of the MPEG-4 audio coding proposal from University of Hannover and Deutsche Bundespost Telekom", ISO/IEC MPEG96/MO632 1996.
14. Chowning, J. "The synthesis of complex audio spectra by means of frequency modulation", J. Aud. Eng. Soc., 526–529, 1973.
15. Makhoul, J. "Linear prediction: a tutorial review", Proc. IEEE, 63, 561–580, 1975.
16. Liljeryd, L.G., Ekstrand, P.R.A., Henn, L.F. and Kjorling, H.M.K. "Sound encoding enhancement using spectral-band replication", Patent 6680972, U.S. Patent Bureau, 2004.
17. ISO/IEC 14496-3, "Coding audio-visual objects", International Standard Organization, 2001.
18. Zwicker, E. and Feldtkeller, R. Das Ohr als Nachrichtenempfänger, Hirzel Verlag, Stuttgart, 1967.
19. Elliott, L.L. "Backward and forward masking of probe tones of different frequencies", J. Acoust. Soc. Am. 34, 1116–1117, 1962.
20. Houtsma, A.J.M., Trahiotis, C., Veldhuis, R.N.J. and van der Waal, R. "Bit rate reduction and binaural masking release in digital coding of stereo sound", Acustica, 82, 908–909, 1996.
21. Houtsma, A.J.M., Trahiotis, C., Veldhuis, R.N.J. and van der Waal, R. "Further bit rate reduction through binaural processing", Acustica, 82, 909, 1996.
22. Van de Par, S.L.J.D.E., ten Kate, W.R.T., Kohrausch, A. and Houtsma, A.J.M. "Bit-rate saving in multichannel sound: Using a

band-limited channel to transmit the center signal", J. Aud. Eng. Soc. 42, 555–564, 1994.
23. Faller, C. and Baumgarte, F. "Binaural cue coding: A novel and efficient representation of spatial sound", Proc. ICASSP, 1841–1844, 2002.
24. Johnston, J. "Estimation of perceptual entropy using noise masking criteria", Proc. ICASSP, 2524–2527, 1988.
25. Houtsma, A.J.M. "Psychophysics and modern digital audio technology", Philips J. Res., 47, 3–14, 1992.
26. Sinha, D. and Johnston, J. "Audio compression at low bit rates using a signal adaptive switched filterbank", Proc. ICASSP, 1053–1056, 1996.
27. Wiese, D. and Stoll, G. "Bitrate reduction of high quality audio signals by modeling the ear's masking thresholds", Proc. 89th Conv. Aud. Eng. Soc., 1990, preprint 2970.
28. "Information technology – Coding of moving pictures and associated audio for digital storage media at up to about 1.5 Mbit/s- IS 11 172-3 (audio)", ISO/IEC, JTC1/SC29, 1992.
29. Van de Par, S., Kohlrausch, A., Charestan, G. and Heusdens, R. "A new psychoacoustical masking model for audio coding applications", Proc. ICASSP, II, 1805–1808, 2002.
30. Noll, P. "MPEG digital audio coding", IEEE Sign. Process. Mag. 14, 59–81, 1997.
31. ISO/IEC, 13818-7, "Information technology – Generic coding of moving pictures and associated audio-Part 7: Advanced audio coding", 1997.
32. Painter, T. and Spanias, A. "Perceptual coding of digital audio", Proc. IEEE, 88, 451–513, 2000.
33. Levitt, H. "Transformed up-down methods in psychoacoustics", J. Acoust. Soc. Am. 49, 467–477, 1971.
34. "Methods for subjective assessment of small impairments in audio systems including multichannel sound systems", ITU-R Rec. BS 1116, 1994.
35. Milne, A. "New test methods for digital audio data compression algorithms", Proc. 11th Conf. Aud. Eng. Soc., 210–215, 1992.
36. Karjalainen, M. "A new auditory model for the evaluation of sound quality of audio systems", Proc. ICASSP, 608–611, 1985.
37. Brandenburg, K. "Evaluation of quality for audio encoding at low bit rates", Proc. 82nd Conv. Aud. Eng. Soc., 1987, preprint 2433.
38. Beerends, J. and Stemerdink, J. "Measuring the quality of speech and music codecs: An integrated psychoacoustic approach", Proc. 98th Conv. Aud. Eng. Soc., 1995, preprint 3945.
39. International Telecommunications Union, "Method for objective measurements of perceived audio quality", ITU-R. BS 1387, 2001.
40. Thiede, T. and Kabot, E. "A new perceptual quality measure for bit rate reduced audio", Proc. 100th Conv. Aud. Eng. Soc., 1996, preprint 4280.
41. Sporer, T. "Objective signal evaluation – applied psychoacoustics for modeling the perceived quality of digital audio", Proc. 103rd Conv. Aud. Eng. Soc., 1997, preprint 4512.

42. Beerends, J. and Stemerdink, J. "A perceptual audio quality measure based on a psychoacoustic sound representation", J. Aud. Eng. Soc., 40, 963–978, 1992.
43. Paillard, B., Mabilleau, P., Morisette, S. and Soumagne, J. "Perceval: Perceptual evaluation of the quality of audio signals", J. Aud. Eng. Soc., 40, 21–31, 1992.
44. Colomes, L., Lever, M., Rault, J.B. and Dehery, Y.F. "A perceptual model applied to audio bit-rate reduction", J. Aud. Eng. Soc., 43, 233–240, 1995.
45. Levine, S. Audio representations for data compression and compressed domain processing, PhD Thesis, Dept. of EE, Stanford University, 1998.
46. Purnhagen, H. "Advances in parametric audio coding", Proc. IEEE Workshop on Applications of Signal Processing to Audio and Acoustics, 17–20, 1999.
47. Rossing, T.D. "The Science of Sound," Addison Wesley, Reading, MA, 1990.

PART VII
TELECOMMUNICATIONS

H.W. Gierlich

Telecom Division, HEAD acoustics GmbH, Herzogenrath, Germany

43 Speech Communication and Telephone Networks
H.W. Gierlich 821
General Aspects of Speech Communication • Modern Telephone Networks

44 Methods of Determining the Communicational Quality of Speech Transmission Systems
H.W. Gierlich 831
Assessment Methods • Subjective Assessment • Objective Methodologies • Network Planning

45 Efficient Speech Coding and Transmission Over Noisy Channels
Ulrich Heute and Norbert Goertz 853
Introduction and Fundamentals • Speech Coding • Speech Transmission over Noisy Channels • Further Work, Results and Literature

46 Echo Cancellation
Walter Kellermann 883
Echo Cancellation and Suppression

47 Noise Reduction and Interference Cancellation
Walter Kellermann 897
Single-Channel Techniques • Multichannel Techniques • Blind Signal Separation of Convolutive Mixtures

48 Terminals and Their Influence on Communication Quality
Walter Kellermann 909
Acoustic Interfaces

49 Networks and Their Influence on Communication Quality
H.W. Gierlich 915
PSTN (Public Switched Telephone Networks) • Packet-Based Transmission • Mobile Networks

50 Interaction of Terminals, Networks and Network Configurations
H.W. Gierlich 921
Tandeming of Speech Coders • Delay • Tandeming of Speech Echo Cancellers • Tandeming VAD and NLP • Other Signal Processing

List of Important Abbreviations 927

43

Speech Communication and Telephone Networks

H.W. Gierlich

HEAD accoustics GmbH, Herzogenrath, Germany

Speech communication over telephone networks has one major constraint: The communication has to be "real time". The basic principle since the beginning of all telephone networks has been to provide a communication system capable of substituting the air path between two persons having a conversation at 1-m distance. This is the so-called orthotelephonic reference position [7]. Although many technical compromises must be made to enable worldwide communication over telephone networks, it is still the goal to achieve speech quality performance which is close to this reference.

This chapter gives an overview of terminals, networks, their effects on speech quality, and the digital signal processing involved in keeping the quality as high as possible.

1 • General Aspects of Speech Communication

Speech communication is influenced by a variety of parameters. In contrast to broadcast services, the parameters contributing to the quality of communication service perceived by the user – for the service as a whole – are more complex than the factors contributing to speech quality only. As described by Möller [13], various quality entities are relevant. The main quality entities and their relationship are shown in Figure 1.

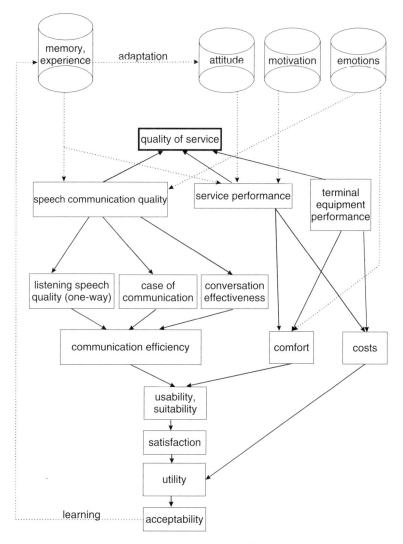

FIGURE 1 Parameters influencing the quality of service (QoS) for a telecommunication service (from [13]).

When focusing on the speech communication aspect, the listening situation, the talking situation, the conversational situation (including double talk), and the background noise situation have to be taken into account and need to be regarded separately. An overview of the parameters influencing speech quality is shown in Figure 2.

The major criterion used in early times for the optimization of traditional telephone communication systems is "speech

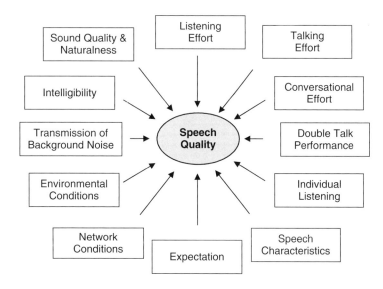

FIGURE 2 Factors influencing the speech quality perceived by the user.

intelligibility" [6,14]. This parameter purely focuses on the transport of information. Based on different kinds of intelligibility tests [6], the requirements for bandwidth, signal-to-noise ratio, distortion, and loudness requirements of the traditional telephone system (defined in ITU-T as 3.1 kHz telephony) were introduced from the beginning of telecommunication system development. Although the technical restrictions are no longer in place for many of the modern communication systems, narrowband telephony is still used in the majority of the connections today. Most speech quality parameters were derived for narrowband (3.4 kHz) telecommunication systems. Wideband communication systems are available, but not widely used yet. This may change with the use of IP-based transmission where a much more flexible bandwidth allocation for the individual call is possible.

When judging the listening situation, the quality of the transmitted speech signal is of major importance besides the intelligibility.

In the *listening situation* the speech signal may be degraded by a variety of parameters.

Loudness. The loudness perceived in the listening situation mainly contributes to the perceived speech quality. The speech signal loudness should be comparable to the loudness perceived when having a conversation between two persons at 1-m distance, the "orthotelephonic reference position" [7]. The loudness in today's telecommunication networks is considered the main parameter used for the planning of telephone networks and the main parameter which has to be determined for all devices involved in a communication chain. The loudness calculation in telephonometry is different from the loudness calculation used, for example, in psychoacoustics and is described in detail in Chapter 44.3.

Frequency response, linearity, distortion, and noise. In modern telephone systems, the frequency response characteristics are mainly determined by the telephone terminal. Except for older analog transmission systems where the telephone line may strongly influence the overall connection quality due to frequency-dependent attenuation and group delay distortion, it is assumed that the network components typically do not influence the frequency response characteristics significantly. Requirements for distortion, linearity, and noise can be found in many relevant standards.

Speech sound quality. The speech sound quality in classical telephones can be described most comprehensively by the loudness, frequency response, distortion, and the noise introduced by the transmission system. In modern telecommunication, however, speech coding and other types of non-linear and time-variant signal processing are widely used. The speech quality of these systems must be determined subjectively. For known transmission systems, the speech quality can be described using objective speech quality measures like PESQ or TOSQA [4,12]. However, whenever new non-linear or time-variant transmission systems are to be investigated, subjective evaluation is required.

In the *talking situation* the speech quality is influenced by impairments typically related to the user's own voice, which is referred to as *sidetone*, coupled either acoustically or electrically (by the handset receiver) to the user's ear and echo. The echo signal is produced either by acoustical coupling between the receiver and the microphone of the far-end terminal, or may be due to a hybrid echo, e.g. resulting from a hybrid in an analog telephone or a hybrid somewhere else in the connection. The perception of echo depends mainly on two parameters: the echo delay and the echo attenuation. The longer the delay the more echo attenuation is required (see [9]). For typical telephones, an echo loss of 45 dB is required as a minimum.

In the *conversational situation*, speech quality is influenced by many parameters.

Delay. Delay is introduced in every communication system. Besides delay introduced by the acoustical paths between mouth and microphone of a telephone and between the telephone receiver and the listener's ear, there are other sources of delay. Algorithmic delay may be introduced by the signal processing in the terminal such as background noise reduction (noise cancellation), echo cancellation, speech coding, voice-controlled attenuation, and others. Packetization, for example, as used in VoIP (voice over Internet protocol) systems may further contribute to the delay introduced in a connection. The main effect of delay is the impairment of the conversational dynamic. For highly interactive conversational tasks, a delay of not more than 150 ms is recommended (see [8]). Another effect of delay is the increased audibility of echo components for increasing delay times (see [9]).

Double talk capability. The double talk capability of a telephone system is mainly determined by its echo and switching performance. The longer the delay in a connection the more echo attenuation is required. Since echo cancellation is never perfect, additional means have to be introduced in order to avoid echo components. Typically, additional gain switching is introduced in the terminals as well as in speech echo cancellers in the network. The switching may be introduced in sending and receiving directions. Requirements for switching and echo during double talk and their relationship to the perceived quality of a telephone system can be found in [11].

A parameter not directly related to speech but highly influencing the overall quality perceived by the user is the *background noise transmission performance*. Background noise is present during many telephone calls and may influence the perceived overall quality of a connection quite significantly. Background noise may influence the transmission quality during periods where no speech signal is present, during periods with near-end speech and with far-end speech. Furthermore, the speech signal may be degraded by noise reduction systems in the presence of background noise. The subjective (auditory) relevance of the background noise situation in comparison to the other conversational situation is still under investigation.

2 • Modern Telephone Networks

2.1 PSTN (Public-Switched Telephone Networks)

In general, all parameters determining the speech quality in circuit-switched networks are well controlled. PSTN networks are typically analog-switched networks or TDM (time division multiplex)-type networks. Proper transmission planning generally can ensure high speech quality in these networks. Digital PSTN-type networks work isochronously. The loss of data is fairly unlikely. The delays in a connection are constant and within the control of the network operator. The network interfaces in such networks are well defined. In analog networks, the transmission line characteristics (length and type of lines) have to be taken into account since it may have impact on signal level, frequency response characteristics, and noise. The delay inserted by traditional national networks is well controlled; proper transmission planning guarantees low delay ($< 15\,\text{ms}$) and echo-free connection for national connections. Network speech echo cancellation which is under the control of the network operator ensures mostly echo-free connections in case of interconnection to international networks or satellite communication. Echoes which might result from insufficient hybrid loss of analog terminals connected to the PSTN network are cancelled.

2.2 Packet-Based Transmission

Packet-based networks can be divided into ATM types (asynchronous transfer mode) of networks and IP networks. IP networks differ considerably from circuit-switched networks but in ATM-type networks, the same rules can be applied as for TDM-type networks, except for the packetization process itself. Control of the parameters relevant for achieving a sufficient or good communication quality is impossible, or achievable only within limits. Therefore, transmission planning only can be made based on assumptions, statistical knowledge, and best-worst-case estimations within the constraints given by a specific network and its interconnection to other types of networks. Any sort of guaranteed speech quality can be given on a statistical basis only.

In packet-based networks, the input signal (speech signal) is typically segmented into packets of fixed length. In IP networks, all users connected to the specific IP-network share the bandwidth given by the IP network; no guarantee for a required bandwidth is given. The packets of different sources (e.g. terminals or subscribers) have to be sequenced for transmission. Neither delay nor jitter is predictable in IP networks. Both can be described only by statistical parameters such as average delay (including delay distribution) and packet loss distribution.

IP networks do not provide any means to control the speech signals, so the sender and receiver have to provide the appropriate signal processing mechanisms. Speech coding, packetization, echo cancellation, noise reduction, jitter buffer, packet loss concealment, and other types of signal processing have to be integrated either into the terminals or the gateways interconnecting the IP network to other types of networks.

2.3 Mobile Networks

In mobile networks, the speech quality is highly dependent on the transmission network. A wide variety of network conditions can be expected due to the time-variant characteristics of the radio channel. In all mobile networks, speech coding is used in order to reduce the bandwidth required for the individual speech channel and to increase the number of channels which may be transmitted. The upper limit for the listening speech quality, therefore, is set by the codecs (see [1–3]). While the speech codec quality is predictable as long as clean speech is encoded, the overall speech quality may be degraded by different impairments introduced by the network or the mobile terminals.

The mobile terminal itself should provide all means to ensure a good conversational quality within the constraints given by the mobile transmission itself. Background noise cancellation, speech coding, acoustical echo cancellation, frame loss concealment, and other types of signal processing are integrated in the mobile terminals. The mobile network itself is responsible for routing and switching of the speech signal. The general configuration of a mobile network is shown in Figure 3 for the example of a GSM network.

Disturbances introduced by the radio channel between the mobile station and the base transceiver station result in bit errors. A high bit error rate (BER) contributes to degraded speech. In addition, transmission errors may occur between BTS and MSC. While the PCM-coded speech used between TRC and MSC is less sensitive to the effects of high BER, the GSM-

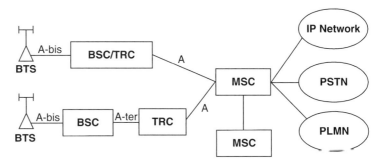

FIGURE 3 Network configuration (see [5]). BTS, base transceiver station; BSC, base station controller; TRC, transcoder controller; MSC, mobile switching centre; PLMN, public land mobile network; PSTN, public-switched telephone network; A, interface between the BSS and the MSC, manages the allocation of suitable radio resources to the MSs and mobility management; A-bis, BSS internal interface linking the BSC and a BTS; A-ter, BSS internal interface linking the BSC and a TRC.

coded speech transmitted between the BTS and the TRC is more sensitive to bit errors. The audible effects on speech quality due to increasing or decreasing BER, on the A-bis and A-ter interfaces, are unnoticeable, provided the transmission fulfills the performance objectives stated in [10].

Varying speech levels in the network affect the loudspeaker level, distortion, and echo canceller performance. Poor network echo cancelling results in disturbing echo for calls to the PSTN and PLMN. The network echo cancellation is provided by the PSTN. If a proper echo control cannot be guaranteed by the PSTN, echo cancellation has to be provided by the MSC (see Figure 3). In the mobile transmission network, a high bit error rate causes serious degradation of the listening speech quality. The most sensitive part of the connection is the connection between the A-bis and A-ter interfaces. The radio network environment is affected by the following factors (see [5]):

- Co-channel interference (C/I),
- Noise limitations (C/N),
- Mobile speed (fading frequency),
- Time dispersion
- Adjacent channel interference (C/A)

During a call, these factors together contribute to the bit error rate (BER – the average amount of bit errors in a speech frame) and frame erasure rate (FER – the percentage of erased frames). In a mobile network, good cell planning is of critical importance to minimize these types of interferences. The quality of the radio environment can be further improved by using the radio network features such as DTX, power control, and frequency hopping.

Discontinuous transmission (DTX) reduces the total interference in the network since signal is only transmitted during speech activity. DTX and comfort noise generation are parts of the speech codec and can therefore differ for different speech codecs.

Frequency hopping can reduce the effect on speech quality caused by multipath fading and interference. The signal strength variations, or interference, are broken up into pieces of duration short enough for the de-interleaving and speech-decoding process to correct the errors.

Dynamic power control may result in a further increase in the protection against fading dips and interference. The output power is controlled with respect to signal strength of the received radio signal as well as with respect to the BER.

The speech quality is always impaired during *handover* (when the call is passed from one BSC to another). During handover, the speech is interrupted shortly; speech frames are used to transmit signalling information.

References

1. 3GPP: TS 46.010: Full Rate Speech Encoding, Third Generation Partnership Project (3GPP), 2002.
2. 3GPP: TS 46.051: GSM Enhanced Full Rate Speech Processing Functions: General Description, Third Generation Partnership Project (3GPP), 2002.
3. 3GPP: TS 46.090: AMR Speech Codec; Transcoding Functions, Third Generation Partnership Project (3GPP), 2002.
4. Berger, J.: Instrumentelle Verfahren zur Sprachqualitätsschätzung – Modelle auditiver Tests, PHD thesis, Kiel, 1998.
5. ETSI EG 201 377-2: Specification and Measurement of Speech Transmission Quality; Part 2: Mouth-to-Ear Speech Transmission Quality Including Terminals, 2004.
6. Fletcher, H. and Galt, R.H.: The Perception of Speech and Its Relation to Telephony. J. Acoust. Soc. Am., 1950, Vol. 22, No. 2, pp. 90–119.
7. ITU-T Handbook on Telephonometry, ITU-T, Geneva, 1992.
8. ITU-T Recommendation G. 114: One-Way Transmission Time, International Telecommunication Union, Geneva, 2000.
9. ITU-T Recommendation G. 131: Control of Talker Echo, International Telecommunication Union, Geneva, 2000.
10. ITU-T Recommendation G. 821: Error Performance of an International Digital Connection Operating at a Bit Rate Below the Primary Rate and Forming Part of an Integrated Services Digital Network, International Telecommunication Union, Geneva, 2002.
11. ITU-T Recommendation P. 340: Transmission Characteristics and Speech Quality Parameters of Hands-Free Telephones, International Telecommunication Union, Geneva, 2000.
12. ITU-T Recommendation P.862: Perceptual Evaluation of Speech Quality (PESQ): An Objective Method for End-to-End Speech Quality Assessment of Narrow-Band Telephone Networks and Speech Codecs, International Telecommunication Union, Geneva, 2001.
13. Möller, S.: Assessment and Prediction of Speech Quality in Telecommunications, Kluwer Academic Press, 2000, ISBN 0-7923-7894-6.
14. Richards, D.L.: Telecommunication by Speech, Wiley New York, 1973, ISBN 0-470-71949 4.

44

Methods of Determining the Communicational Quality of Speech Transmission Systems

H.W. Gierlich

HEAD accoustics GmbH, Herzogenrath, Germany

Telecommunication systems generally are designed for bi-directional exchange of information. Therefore, quality assessment of telecommunication systems has to reflect all aspects of realistic conversational situations. Speech quality assessment is part of the more global assessment of the service quality [34]. Since all quality parameters are based on sensations perceived by the subjects using the communication system, the basis of all objective assessment methods are subjective tests. They have to be defined carefully in order to reflect the real use situation as closely as possible and, on the other hand, provide reproducible and reliable results under laboratory conditions.

1 • Assessment Methods

Modern telecommunication networks provide a wide range of voice services. The rapid deployment of different types of new technologies leads to an increased need for evaluating the transmission characteristics of new transmission equipment. In many circumstances, it is necessary to determine the subjective effects of new transmission equipment or modification to the transmission characteristics of a telephone network. Subjective assessment techniques are required to (a) investigate the performance of new and unknown services and systems where no objective evaluation procedure is known yet and (b) form the basis of objective evaluation and measurement procedures. Every objective evaluation method has to be validated based on results achieved from subjective assessment of the impairment the objective method is developed for. Therefore new services and systems typically have to be evaluated subjectively first before an objective testing procedure can be developed and applied.

2 • Subjective Assessment

The basic description of the subjective assessment of speech quality in telecommunication is ITU-T Recommendation P.800 [24]. The methods described here are intended to be generally applicable whatever forms of degradation factors are present. Further recommendations [25–28] describe more specialized tests for specific processing techniques and devices used in modern networks such as speech codecs, speech echo cancellers, hands-free equipment and noise reduction techniques.

2.1 Opinion Scales Used for Subjective Performance Evaluation

Subjective testing requires exactly defined scaling of the scores derived from the subjects participating in subjective experiments. The scale needs to be easily understandable by the subjects, unambiguous and widely accepted. The layout and wording of opinion scales, as seen by subjects in experiments, is very important. Different ways of scaling and different types of scales may be used [32,34]; their individual benefits and drawbacks are not discussed here, only those most often used in telecommunications are described.

The following opinion scale is the most frequently used one for telecommunication applications, and equivalent wording is to be used in different languages. This is a category rating obtained from each subject at the end of each subjective experiment [24,25]:

Opinion of (the overall quality, the listening speech quality, etc.) of the connection you have just been using:

1 – Excellent
2 – Good
3 – Fair
4 – Poor
5 – Bad

If the subjects' opinion is asked for in conversational tests, the result is termed *mean opinion score, conversational* (MOSc). If the scale is used for speech quality rating in listening tests then the result is called *mean listening quality-opinion score* (MOS). Such tests are named *absolute category rating* (ACR).

Often, especially in conversational tests, a binary response is obtained from each subject at the end of each conversation, asking for difficulty in talking or hearing over the connection used. The experimenter allocates the score "Yes" or "No". If a subject experienced difficulties, the experimenter may carefully ask for further information about the type of difficulties experienced by the subject.

If a degradation introduced to a transmission system is to be evaluated, degradation category rating (DCR) tests are required. The scale used is [24]

5 – Degradation is inaudible.
4 – Degradation is audible but not annoying.
3 – Degradation is slightly annoying.
2 – Degradation is annoying
1 – Degradation is very annoying.

The quantity evaluated from the scores is termed *degradation mean opinion score* (DMOS).

Sometimes, when evaluating higher quality systems, comparison rating is used. The scale used for *comparison category rating* (CCR) tests is given as follows [24]:

"The quality of the second system compared to the quality of the first one is –"

3 – Much better
2 – Better
1 – Slightly better
0 – About the same
−1 – Slightly worse
−2 – Worse
−3 – Much worse

Other scales are also used; additional information can be found in [24–27].

2.2 Conversation Tests

Conversation tests best represent typical communication: both conversational partners exchange their information bi-directionally, acting as both talker and listener. The conversation test is the most natural test used in telecommunication and allows the subjects to behave very similarly to a real conversation. The naturalness which can be achieved in a conversation test depends highly on the task chosen within the test. The task should

- Be easy to perform for each subject
- Avoid emotional engagement of the test subjects
- Be mostly symmetrical (the contribution of each test subject to the conversation should be similar)
- Be independent of the individual personal temperament (the task must stimulate people with low interest in talking and must reduce the engagement of people who always like to talk)
- Reflect a typical conversation

Typical tasks used in telecommunication are the so-called *"Kandinsky tests"* [26,27] or *"short conversational tests"* [27,33].

In the "Kandinsky test", the subjects are asked to describe to their partner the position of a set of numbers on a picture. Both subjects have similar pictures, but with some of the numbers in different positions. The pictures used should have simple figures and objects which are easy to describe. This can be achieved by using pictures consisting of colored, geometrical figures (e.g., paintings by Kandinsky or others, see Figure 1).

In the "short conversational tests", the test subjects are given a task to be conducted at the telephone similar to a daily-life

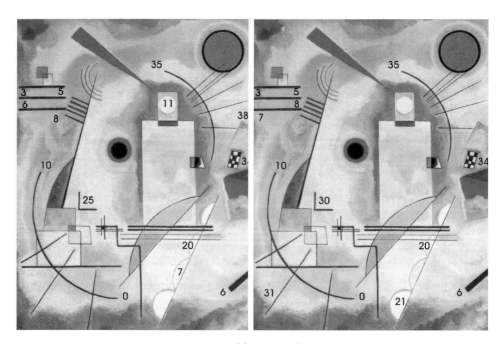

FIGURE 1 Two pictures used by two subjects for conversational tests, "Kandinsky" pictures with different numbering of the different geometrical figures.

situation. Ordering a specific pizza at a pizza service and finding a specific railway connection are examples of typical tasks.

Many conversational tests have been carried out successfully with observers (operators) present in the test rooms together with the subjects. It is their task to register and list all comments which the subjects mention during or after test. This can be useful for further analysis.

In a conversation test, a sufficient number of speakers (voices) should be used in order to minimize talker/speaker-dependent effects.

Subjects mostly rate their opinion about the overall quality of a connection based on the ACR scale. Furthermore, they are asked about difficulties in talking or listening during the conversation. Careful investigation of the nature of these difficulties may require more specialized tests than described below.

2.3 Double Talk Tests

Due to the interactive nature of a conversation, the ability to interact in all conversational situations and especially to interact during double talk with no or only minor impairments is of vital importance. Since many technical implementations may impair the speech quality during double talk, specific double talk tests may help to evaluate the system performance under such conditions. The benefit of this double talk testing method [27] is that it is designed especially for quality assessment during double talk periods and that the test duration is very short. Therefore, double talk tests are very efficient in evaluating this very important quality aspect. Because of the short duration for single tests, double talk tests efficiently determine the influence of the relevant parameters perceived subjectively in detail.

Comparable to the conversational test, the double talk test involves two parties. The purpose of double talk testing tends to be different depending on whether experienced or untrained subjects are used. Untrained subjects are used when it is important to get an indication of how the general telephone-using population would rate the double talk performance of a hands-free telephone, for example. The test procedure is sensitive enough that untrained subjects can assess the relevant parameters even during sophisticated double talk situations. Experienced subjects are used in situations where it is necessary to obtain information about the subjective effects of individual degradations.

During double talk tests, two subjects have a written text in front of them. The texts differ slightly. Subject 1 (talking continuously) starts reading the text. It consists of simple, short and meaningful sentences. Subject 2 (double talk) follows the text (without reading aloud) and starts reading his text simultaneously at a clearly defined point.

Typical questions within a double talk test are about the dialog capability, the completeness of the speech transmission during double talk, echo and clipping during double talk.

The scales used are typically ACR or DCR scales, but other scales can be used as well. More information can be found in [47].

A drawback of double talk tests is that the testing method is more artificial compared to conversational tests. Even if the text is very simple, the subjects have to concentrate in a different way compared to a free conversation.

2.4 Talking and Listening Tests

In all connections where delay is inserted (e.g., network delay, delay due to signal processing or delay due to packetization) talking-related disturbances like echo or background noise modulation are of importance. In order to investigate such types of impairments in more detail, talking and listening tests can be used. Such tests are mainly used to investigate the performance of speech echo cancellers (EC). All aspects of EC function that influence the transmission quality for subscribers while they are either talking and listening or only listening (without having a conversational partner on the other end of the connection) are covered by this procedure.

A typical test setup is shown in Figure 2.

This measurement setup shows an echo canceller on one side of the connection which is the one under test. From the subject's point of view, this is the far-end echo canceller. The far-end subscriber is simulated by an artificial head. If double talk sequences are required, then the artificial mouth is used to produce exactly defined double talk sequences. Environmental conditions will depend on the characteristics of the test room used, the background noise and other factors. Generally the test procedure is separated into two parts. The first part is designed to examine initial convergence of an echo canceller and the second part has to measure the performance during steady-state conditions. Which of these procedures is used will depend on the purpose of the test. Subjects are first instructed on how to conduct the test.

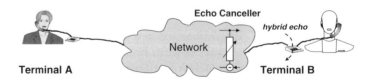

FIGURE 2 Test setup for a talking and listening test.

When testing the initial convergence of a speech echo canceller, subjects answer an incoming telephone call with the same greeting, for example, "[company], [name], [greeting]". After the greeting, the call is terminated, and subjects give their rating.

When testing "steady-state conditions", the echo canceller should first be fully converged. Subjects are asked to perform a task, such as to describe the position of given numbers in a picture similar to the "Kandinsky" test procedure described for the conversational tests. The artificial head can be used to generate double talk at defined points in time in order to introduce disturbing signal components for the speech echo canceller and test the canceller's ability to handle double talk. After the termination of the call, the subjects are asked to give a rating.

The scales used are typically ACR or DCR scales, but other scales can be used as well. More information can be found in [25].

2.5 Listening-Only Tests and Third Party Listening Tests

Listening-only tests are most commonly used in telecommunication. The main purpose of listening-only tests (LOT) is the evaluation of impairments under well-defined and reproducible conditions. When used to evaluate parameters influencing the speech quality in telecommunications, it should be noted that listening tests are very artificial. Listening-only tests are typically used to evaluate the speech sound quality or other listening-related impairments of speech. They are strongly influenced by the selection of the speech material used in the tests; the influence of the test stimuli is much stronger than in conversation tests. Therefore, a very careful test design is required. The test sequences used must represent the typical phoneme distribution of the language used [24]. Furthermore, at least two male and two female speakers should be used for each condition in order to avoid speaker-dependent effects. A sufficient number of presentations must be integrated into a test, ranging from the best to the worst-case condition of the impairment investigated in the test. Reference conditions (simulated, defined impairments) can be integrated in order to check the validity of the test. Additional rules on how to conduct listening-only tests for various purposes are found in [25–28].

In general, listening-only tests are used by presenting pre-recorded, processed speech to the subjects, using reference handset terminals which are operated by the subjects. In third party listening tests [26,27], the speech material is recorded by an artificial head which is used to record the complete acoustical situation including the background. By this procedure, all types of handset, headset and hands-free configurations can be included into a listening-only test including the environmental conditions at the terminal location. For playback, equalized headphones are used. The equalization must guarantee that during playback the same ear signals are reproduced which were measured during recording. By this procedure, binaural reproduction (for details see [4]) is possible which leads to a close-to-original presentation of the acoustical situation during recording. The third party listening setup allows use of this type of test for investigating conversational situations by third parties. Therefore, a complete conversation is recorded and presented to the listeners. Although the listeners are not talking themselves but listening to other persons' voices, these tests have proven their usefulness in investigating conversational impairments.

The ACR or DCR scales are typically used but other scales, like loudness preference, can be used as well. More information can be found in [26,27].

Instead of ACR or DCR tests, the *comparison category rating* (CCR) can be used. CCR tests are based on paired comparisons of samples. They offer a higher sensitivity and may be used for the quality evaluation of high-quality systems, typically to compare processed and unprocessed speech material.

3 • Objective Methodologies

Besides the typical evaluation procedures used for electroacoustical systems, such as frequency responses, distortions, dynamic ranges and others, a variety of analyses exist for telecommunications. While the well-known procedures for electroacoustical systems are not described here, an overview is given of the objective evaluation procedures specific to telecommunication systems. Table 1 gives an overview of the relationship between impairments perceived subjectively and the correlating objective parameters.

3.1 Loudness

Speech is a time-varying process. Speech sounds are short-term stationary within a time interval of approximately 20 ms. The speech spectrum is continuous. In contrast to loudness calculations (such as are described in ISO R 532 B [33], which is based on the work of Zwicker [38] and others), in telephonometry the concept of *loudness ratings* is used. Although the speech signal is not continuous, useful conclusions can be drawn from experiments with speech and referring them to a continuous spectrum-based model, the loudness ratings concept. This concept is based on the speech signal energy calculated in fixed (frequency) bands and a subjective comparison of these values to the signal energy

TABLE 1 Impairments perceived subjectively and their correlating objective parameters

Impairments perceived subjectively	Description	Correlating objective parameters
Loudness of the speech signal	In send and receive direction, perceived in the listening situation	Loudness ratings in send and receive
Speech sound quality	In send and receive direction, perceived in the listening situation	Frequency response Distortions Objective speech quality measures based on hearing models
Delay	Impairment discovered in conversational situations, especially with high interaction between the conversational partners leading to reduced conversational dynamics and unintended double talk situations	Delay (Time-variant) echo loss TCL (terminal coupling loss) Switching Occurrence of double talk impairments
Echo disturbances (under single talk conditions)	Talking-related disturbance correlated to the user's voice, perceived in the presence of transmission delay	Echo attenuation Echo level fluctuation over time Spectral echo attenuation
Double talk capability	In send and receive direction: Loudness variation between single and double talk periods Loudness variation during double talk Echo disturbances Occurrence of speech gaps Occurrence of artifacts uncorrelated to the transmitted speech signal	Attenuation range Attenuation in send and receive direction during double talk Switching characteristics Minimum activation level to switch over from receive to send direction and from send to receive direction Echo attenuation during double talk Spectral and time-dependent echo characteristics
Quality of background noise transmission	Impairments introduced to the background noise signal in the send direction during Idle mode Far-end speech active in the presence of background noise Near end speech active in the presence of background noise	Minimum activation level in send direction Comfort noise spectral and level adaptation Level variation due to noise reduction systems Attenuation in send direction Sensitivity of background noise detection (activation level, absolute level, level fluctuations)

(Continued)

TABLE 1 (Continued)

Impairments perceived subjectively	Description	Correlating objective parameters
(System) noise	In send and receive direction, perceived during silent intervals	Attenuation introducing noise modulation Switching characteristics D-value (sensitivity for background noise compared to the speech signal sensitivity) Noise level Level fluctuations Spectral characteristics

of a reference system, the IRS [11,21,35], when adjusting the perceived loudness of the unknown system to the reference system. The different frequency-dependent contributions to the overall loudness impression are reflected by a weighting factor B':

$$B' = \frac{(\Delta l)_c}{(\Delta f)_c} \cdot \frac{1}{l_{max}} = \frac{0.0156}{10^{K_c/10}},$$

where K_c is the critical bandwidth expressed in dB, $(\Delta l)_c$ is the loudness contribution in the frequency band $(\Delta f)_c$.

The total loudness impression is proportional to the integral over $Q(Z) \times B'$ where $Q(Z)$ is the loudness growth function depending on the speech level Z measured in dB [35]. $Q(Z)$ can be approximated by

$$Q(Z) = \text{const.} \cdot (10^{Z/10})^m,$$

where a single mean value for m can be determined by auditory experiments depending on the interest on the range of Z. Z is defined as

$$Z = \beta'_S - L_{ME} - (\beta_0 - K_C),$$

β'_S	long-term power density spectrum of speech referred to the mouth reference point MRP [20]
β_0	pure-tone threshold of hearing
K_χ	width of the critical band frequencies
$(\beta_0 - K_\chi)$	can be treated as the perception threshold of sounds with a continuous spectrum
L_{ME}	mouth-to-ear transmission loss (between the reference points β'_S and β_0)

From these terms, with

$$Z_{R0} = \beta'_S - (\beta_0 - K_C)$$

the loudness for an unknown object (λ_U) can be determined as

$$\lambda_U = \text{const.} \int_0^\infty 10^{(-m/10) \cdot L_{UME}} \cdot 10^{(m/10) \cdot Z_{R0}} B' df$$

and can be compared to the loudness of the reference system (λ_R):

$$\lambda_R = \text{const.} \int_0^\infty 10^{(-m/10) \cdot L_{RME}} \cdot 10^{(m/10) \cdot Z_{R0}} B' df.$$

The *loudness rating* is defined as the loudness loss Δx, removed from the speech path of the unknown system to achieve $\lambda_U = \lambda_R$, independent of frequency:

$$\textit{Loudness Rating} = \Delta x = \overline{L_{UME}} - \overline{L_{RME}}.$$

For calculation, the frequency band of interest is divided into N bands (20 bands approximately spaced in one-third octaves from 100 Hz to 8 kHz).
 Using the substitution

$$G = [10^{(m/10)Z_{R0}} B']$$

and using the values G for each frequency band based on the subjective loudness experiment conducted in the CCITT laboratory, the tabulated data for the frequency weights (w-weights) are derived. Based on this, the weighted average mouth-to-ear loss of the speech path for an unknown object can be written as

$$\overline{L_{UME}} = -\frac{10}{m} \log_{10} \sum_{i=1}^N 10^{(m/10)(L_{UME,i}+w_i)}$$

with the values for w_i found in ITU-T Recommendation P.79 [21]. The loudness rating is then defined as

$$\textit{Loudness Rating} = -\frac{10}{m} \log_{10} \sum_{i=1}^N 10^{(-m/10)(L_{UME,i} - \overline{L_{RME}} + w_i)}.$$

For a given telephone or transmission system, the values of L_{UME} can be derived from the measurement of different sensitivities $S_{M\vartheta}$ (mouth-to-junction) for calculation of the SLR (sending loudness rating), from $S_{\vartheta E}$ (junction-to-ear) for the calculation of the RLR (receiving loudness rating) or from S_{ME} (mouth-to-ear) for the overall loudness rating, OLR.

In a similar manner, the sidetone paths can be described: STMR is the sidetone masking rating describing the perceived loudness of the user's own voice and LSTR is the listener sidetone rating describing the perceived loudness of room noise coupled to the user's ear.

3.2 Speech Sound Quality – Listening Speech Quality

Speech sound quality is influenced by many parameters: frequency response, loudness rating, non-linear distortions, switching, noise and others. Since any degradation of the speech sound quality is perceived mostly in the listening situation, the main focus of speech sound quality is during one-way transmission (single talk condition).

Due to the use of different types of speech coders and the impairments introduced by the transmission network (packet loss, jitter, interference), the listening speech quality may be degraded. The type of distortion added to the signal is typically uncorrelated with the speech signal. This effect cannot be assessed by loudness ratings, frequency response or traditional distortion measurements. The assessment methods used under these conditions are hearing model-based comparison techniques. The principle of these assessment methods is shown in Figure 3.

All methods [2,3,36] compare an original signal $x(t)$ with a degraded signal $y(t)$ that is the result of passing $x(t)$ through a communication system. The output is a prediction of the perceived quality that would be given to $y(t)$ by subjects in a

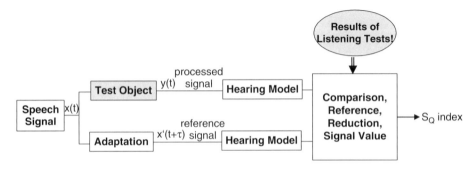

FIGURE 3 Principle of reference signal-based objective speech quality assessment procedures.

subjective listening test. In a first step the signal $x(t)$ is adapted to the delay and the bandwidth limitation introduced by the transmission system. The time alignment between original signal and transmitted signal is made in blocks given by the objective method in order to take care of time-varying delays in transmission systems. It is assumed that the delay within a block is constant.

Based on the time-aligned signals $x'(t+\tau)$ and $y(t)$, the comparison is made. The key to this process is transformation of both the original and the degraded signals to an internal representation that is comparable to the psychophysical representation of audio signals in the human auditory system, taking into account the frequency resolution and the loudness sensation of the human ear.

Almost all of the procedures provide level alignment to a calibrated listening level, time–frequency mapping, frequency warping and loudness scaling. Since not all differences found in the representations of $x'(t)$ and $y(t)$ contribute to the perceived degradation of the processed signals minor steady-state differences between original and degraded signals are compensated and not taken into account for the comparison. More severe effects, like rapid (level) variations, are typically only partially compensated so that a residual effect remains and contributes to the overall perceptual disturbance. Segments with missing signal energy in the processed signal are weighted less than any components added to the processed signal $y(t)$ (asymmetry weighting). By this procedure a small number of quality indicators are used to model the most important subjective effects.

The differences at the hearing model outputs are computed in a cognitive model and combined to an internal objective listening quality score. By appropriate mapping, this internal score is adjusted to the MOS results of the subjective listening experiments (MOS-LQS, mean opinion score-listening quality subjective [23]) which are conducted using the same types of impairments which the objective speech quality assessment method is intended to be used for. In order to distinguish this objective result from the subjective MOS-LQS results, this score is named MOS-LQO (mean opinion score-listening quality objective [23]).

The ITU-T recommended objective speech quality evaluation procedure is PESQ, as described in P.862 [29]. As with many other methods, this procedure is suitable for the electrical access of speech signals, such as using interfaces provided at the network termination points. The influence of terminals, such as degradations of the speech signals introduced by non-ideal frequency responses, transducer distortions, etc., is not modeled. TOSQA2001 [3,6] allows the acoustical access and takes into account the impairments introduced by the acoustical components. In combination with a proper acoustical coupling, taking into account the behavior of a terminal when used by a human operator, a realistic speech quality assessment can be made. In order to achieve realistic conditions, an artificial head (HATS – head and torso simulator, [17,18]) equipped with an artificial mouth and a type 3.4 artificial ear is used for assessing the acoustical input and output signals of a terminal.

A general drawback of all hearing model-based approaches is the assumption of the same listening level in a subjective listening test under all conditions. The standard listening level assumed is $79\,dB_{SPL}$ (dB re $20\,\mu Pa$). For higher- or lower-level acoustical output signals, deviating from the standard listening level, the predicted MOS-LQO value may be incorrect.

3.3 Delay and Echo Loss

Delay is introduced by any communication system. Algorithmic delay may be introduced by signal processing in the terminal, such as background noise reduction (noise cancellation), echo cancellation, speech coding, voice-controlled attenuation and others. Packetization, as used in VoIP systems, may further contribute to the delay introduced in a connection. The main effect of delay is the impairment of the conversational dynamics. For highly interactive conversational tasks a delay of not more than 150 ms is recommended [14], although more recent investigations [31] indicate that in typical conversations higher delays may be acceptable. Another effect of delay is the increased sensitivity of the human ear for echo components with increasing delay [15]. The more delay in a connection, the higher the importance of the echo loss between the acoustical receiver and the terminal microphone. Any signal component received by the terminal and transmitted back to the far-end user may be perceived as echo by the far-end user. A further source of echo may be insufficient trans-hybrid loss in analog networks, which is not properly compensated by the network speech echo cancellers. From the perception point of view the actual echo source is of minor importance. The requirement for the echo attenuation depending on the delay inserted in a connection is described in [15]. The echo loss has to be provided during the entire call without any temporal change. Testing techniques useful for evaluating the convergence behavior of echo cancellers in the networks can be found in [16], those for hands-free terminals in [22].

3.4 Double Talk Performance

The double talk performance of a hands-free terminal and the resulting requirements have been investigated in great detail (see, e.g., [33]). These investigations led to a classification system which allows to assess the performance of a system during double talk objectively by a set of parameters [15,22]. Although the subjective tests were mainly based on the evaluation of hands-free systems, the results can be used for other terminals and network equipment as well, since the impairments are mostly perceived by the far-end user. From the perception point, the source of the impairments (terminal or network component) is not relevant. Table 2 gives an overview of the required performance parameters and their impact on the performance perceived subjectively.

The double talk performance is described by a set of parameters:

- *Attenuation range in send and receive directions*: The attenuation range is the difference between minimum and maximum amplification in one transmission direction. For the measurement, typically a speech-like, time-multiplexed test signal is inserted simultaneously in send and receive direction. The test signal consists of a series of partially overlapping composite source signals [8,19].
- *Talker echo loudness rating/echo attenuation*: The talker echo contributes significantly to the perceived speech quality in the double talk situation. The talker echo assessment during double talk can be made by inserting a speech-like frequency-multiplexed test signal consisting of two voiced sounds with non-overlapping spectra simultaneously in send and receive directions [19]. By applying an appropriate comb filter technique, the echo signal can be extracted from the double talk signal and the echo loss during double talk can be determined.

TABLE 2 Definition of double talk performance according to [22]

	Type 1 (full duplex)	Type 2 (partial duplex)			Type 3 (non-duplex)
		Type 2a	Type 2b	Type 2c	
$TELR_{DT}$ (dB)	≥ 37	≥ 33	≥ 27	≥ 21	< 21
a_{Hsdt} (dB)	≤ 3	≤ 6	≤ 9	≤ 12	> 12
a_{Hrdt} (dB)	≤ 3	≤ 5	≤ 8	≤ 10	> 10

$TELR_{DT}$, talker echo loudness rating during double talk; a_{Hsdt}, attenuation range in send direction during double talk; a_{Hrdt}, attenuation range in receive direction during double talk

- *Switching characteristics, Switching times*: The measurement and optimization of system's gain switching and switching times are required for all systems which are not fully duplex-operable. Such systems typically show a switching behavior, sometimes a certain (speech) level is required to activate the system. The switching characteristics can be evaluated basically in a similar manner as the evaluation of the attenuation range. The definitions of switching times, etc., can be found in, e.g., [22].

Further characterization may be useful depending on the type of duplex capability of a communication system:

- Minimum activation level to switch over from receive to send direction and from send to receive direction
- Frequency responses in send and receive directions during double talk
- Loudness ratings in send and receive directions during double talk

Details can be found in [9,22].

3.5 Background Noise Performance

In background noise situations, the perceived speech quality is influenced by a variety of parameters. The human listener focuses on two signals: the speech signal and the background noise signal. The quality judgment is made based on the perception of the transmitted background noise both during periods where no speech signal is present and during periods with a speech signal. Furthermore, the perception is based on the quality of the transmitted speech signal. Therefore, an objective method describing the perceived quality must take into account these different quality dimensions. Initial subjective experiments conducted with expert listeners indicate that at least two groups of "typical listeners" seem to exist: The first group is very sensitive to any impairments introduced to the speech signal but is not sensitive against the background noise level, the other group is less sensitive against any impairments added to the speech signal but prefers a maximum of noise reduction. A complete background noise evaluation method does not exist yet, but for the different situations instrumental procedures can be used.

When focusing purely on the background noise transmission without speech signal present simultaneously, a level variation analysis can be used to analyze the transmitted signal. Typically the measured output signal level is compared to the input signal level and the level difference is analyzed. This method can only be applied if the reference (input signal) can be assessed. Preliminary auditory investigations indicate that a level variation of

±3 dB should not be exceeded. More information about the transmission characteristics of background noise, however, can be found when comparing output and input signals based on a spectral analysis. The output/input signals are analyzed spectrographically and displayed as a spectral difference between input and output. Again, a reference signal is needed to perform the analysis. Once more, there is some indication that variations in time and/or frequency should not exceed ±3 dB. The disadvantage of such simple spectral difference methods is the non-existent relationship to the human ear signal processing. There is a high probability of getting misleading results due to poorly adapted frequency resolution, not taking into account the masking effects in time and frequency, disregarding the non-linearity of the human ear signal processing and others.

Another possibility is the use of the psychoacoustically motivated "relative approach" [7] which seems to be more promising. The basis for the analysis is a hearing model according to [37]. In contrast to all other methods, the "relative approach" does not use any reference signal in its present form. The nonlinear relationship between sound pressure level and auditory perceived loudness is taken into account by time/frequency warping in a Bark filter bank and proper integration of the individual outputs. The filter bank is realized in the time domain. The output signals of the filter bank are rectified and integrated, thus generating the envelope. The three-dimensional output of the hearing model is the basis for the relative approach. In each critical band the long-term level (integration time 2–4 s) is compared to the short-term level (2 ms).

An overall value can be derived for example by applying the following equation [7]:

$$Q = f(N, S) + f\left(\sum_{i=1}^{24}\left[|F_G(i-1) - F_G(i)| \cdot w_1(i, F_G(i)) + \sum_{n=1}^{T} |F_G(i,n) - F_G(i, n+1)| \cdot w_2(i, F_G(i))\right]\right),$$

where $F_G(i)$ is a mean value of the critical band level over a period T of 2–4 s, $F_G(0) = F_G(1)$, $F_G(i, n)$ is a mean value of the critical band level over a much shorter period (approx. 2 ms), n is the current (time-dependent) value. The weighting factors $w_1(i, F_G(i))$, $w_2(i, F_G(i))$ depend on the critical band level $F_G(i)$. In addition, the overall value is influenced by the function $f(N,S)$ which describes an auditory factor, dependent on loudness N and sharpness S. Figure 4 shows the block diagram including the

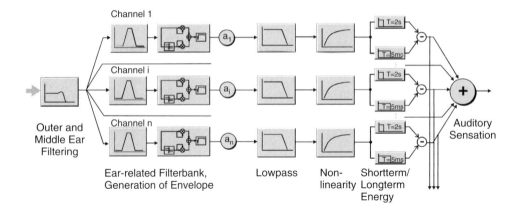

FIGURE 4 Model of the "relative approach": hearing model and calculation of the energy differences in critical bands.

hearing model [37] and the calculation of the relative approach output signals in the Bark domain.

The method can be applied for short-term stationary background noise signals, such as car noise during constant driving conditions, or office noise with low dynamic variations.

When focusing on the quality of the transmitted speech signal, all objective methods proposed so far have some disadvantages. S/N-based methods, as described in [1], do not take into account the perceptual effects of the human ear and are limited in their application to specific types of noise reduction algorithms. More advanced methods like PESQ [29] or TOSQA [3] take into account the human ear signal processing; however, it is known [3,5] that such technologies do not yield sufficient results in conditions with S/N ratios of less than about 20 dB.

4 • Network Planning

The state-of-the-art planning model in today's communication is the *"E-model"* [12]. The E-model is intended for assessing the combined effects of variations in several transmission parameters that determine conversational quality (except double talk and background noise transmission) of 3.1 kHz handset telephony. The E-model is based on the OPINE model [30] which first introduced the concept of impairment factors. The typical application of the E-model is to help planners to ensure that users will be satisfied with end-to-end transmission performance.

The primary output of the model is the *"rating factor"* R; but this can be transformed to give estimates of customer opinion. They can be used for transmission planning purposes, but are not suitable for a customer opinion prediction, since the E-model has not been fully verified. Nevertheless, for many combinations important to transmission planners, the E-model can be used with confidence. Time-variant behavior of transmission systems, however, is not modeled.

The E-model splits a connection into a send side and a receive side. It estimates the conversational quality from mouth to ear as perceived by the user at the receive side, both as listener and talker. The basic assumption of the E-model is the additivity theorem: all impairments are additive on a psychological scale. Based on this assumption a rating factor (R) is calculated:

$$R = R_o - I_s - I_d - I_{e,\text{eff}} + A.$$

R_o is the basic signal-to-noise ratio which all impairments are subtracted from. I_s (simultaneous impairment factor) describes all impairments simultaneous to the speech signal. I_d (delayed impairment factor) summarizes all impairments caused by delays in a connection. $I_{e,\text{eff}}$ describes the impairments introduced by low-bit-rate codecs. For the different codecs used in telecommunication, the I_e values can be derived from tables found in [13]. The perceived listening speech quality in the case of packet loss for the different types of speech coders can be modeled by the packet-loss robustness factor Bpl in combination with the packet-loss probability Ppl [12,13]. In this situation the following formula is applicable:

$$I_{e,\text{eff}} = I_e + (95 - I_e) \cdot \frac{Ppl}{Ppl + Bpl}.$$

The advantage factor A allows for compensation of impairment factors when there are other advantages of access to the user which are not covered by the E-model.

The calculation of R_o can be found in [12].

I_s is defined as

$$I_s = I_{\text{olr}} + I_{\text{st}} + I_q.$$

I_{olr} represents the decrease in quality caused by too-low values of OLR (overall loudness rating), I_{st} represents the impairment caused by non-optimum sidetone and I_q represents impairment caused by quantizing distortion. The detailed description is found in [12].

I_d is defined as

$$I_d = I_{\text{dte}} + I_{\text{dle}} + I_{\text{dd}}.$$

I_{dte} represents the impairments due to talker echo, I_{dle} represents impairments due to listener echo and I_{dd} represents the impairment caused by too-long absolute transmission delay. The detailed description is found in [12].

The advantage factor A may be used to model the advantage of a transmission system in special conditions or situations (for example, a system used for communication in emergency situations where the reliable function is more important than excellent speech quality). There is no relation to all other transmission parameters and typically the factor is not used in transmission planning.

The transmission rating for a narrowband transmission system with very high quality is 93.2 (close to a maximum of 100).

If required, an MOS value can be calculated from the R value in order to better interpret the output of the E-model as an estimated average subjective rating. The following equations are used for mapping:

For $R < 0$: $\quad$ MOS $= 1$
For $0 < R < 100$: $\quad$ MOS $= 1 + 0.035R + R(R-60)(100-R)7 \cdot 10^{-6}$
For $R > 100$: $\quad$ MOS $= 4.5$

References

1. 3GPP TS 26.077: 3rd Generation Partnership Project; Technical Specification Group Services and System Aspects; Minimum Performance Requirements for Noise Suppresser; Application to the Adaptive Multi-Rate (AMR) Speech Encoder, 3GPP, 2003
2. Beerends, J.G., Stemerdink J.A.: A Perceptual Speech-Quality Measure Based on a Psychoacoustic Sound Representation, J. Audio Eng. Soc., Vol. 42, No. 3, pp. 115–123, March 1994
3. Berger, J.: Instrumentelle Verfahren zur Sprachqualitätsschätzung – Modelle auditiver Tests, PhD thesis, Kiel, 1998
4. Blauert, J.: Spatial Hearing: The Psychophysics of Human Sound Localization, MIT Press, Cambridge, MA, 1997
5. Dreiseitel, P.: Quality Measures for Single Channel Speech Enhancement Algorithms. 7th International Workshop on Acoustic Echo and Noise Control, 10–13 Sept. 2001, Darmstadt
6. ETSI EG 201 377-2: Specification and Measurement of Speech Transmission Quality; Part 2: Mouth-to-Ear Speech Transmission Quality including Terminals, 2004
7. Genuit, K.: Objective Evaluation of Acoustic Quality Based on a Relative Approach. Internoise '96, 1996, Liverpool, UK
8. Gierlich, H.W.: A Measurement Technique to Determine the Transfer Characteristics of Hands-Free Telephones, Signal Process., Vol. 27, No. 3, pp. 281–300, 1992
9. Gierlich, H.W.: The Auditory Perceived Quality of Hands-Free Telephones: Auditory Judgements, Instrumental Measurements

and Their Relationship, Speech Commun., Vol. 20, p. 241–254, October 1996
10. ISO R 532 B: Acoustics – Method for Calculating Loudness Level, International Organization for Standardization, Geneva, 1975
11. ITU-T: ITU-T Handbook on Telephonometry, ITU-T, Geneva, 1992
12. ITU-T Recommendation G.107: The E-Model, a Computational Model for Use in Transmission Planning, International Telecommunication Union, Geneva, 2003
13. ITU-T Recommendation G.113: Transmission Impairments due to Speech Processing, International Telecommunication Union, Geneva, 2001
14. ITU-T Recommendation G.114: One-Way Transmission Time, International Telecommunication Union, Geneva, 2000
15. ITU-T Recommendation G.131: Control of Talker Echo, International Telecommunication Union, Geneva, 2000
16. ITU-T Recommendation G.168: Digital Network Echo Cancellers, International Telecommunication Union, Geneva, 2004
17. ITU-T Recommendation P.57: Artificial ears, International Telecommunication Union, Geneva, 2002
18. ITU-T Recommendation P.58: Head and Torso Simulator for Telephonometry, International Telecommunication Union, Geneva, 1996
19. ITU-T Recommendation P.501: Test Signals for Use in Telephonometry, International Telecommunication Union, Geneva, 2000
20. ITU-T Recommendation P.64: Determination of the Sensitivity/Frequency Responses of a Local Telephone Systems, International Telecommunication Union, Geneva, 1999
21. ITU-T Recommendation P.79: Calculation of Loudness Ratings for Telephone Sets, International Telecommunication Union, Geneva, 2000
22. ITU-T Recommendation P.340: Transmission Characteristics and Speech Quality Parameters of Hands-Free Telephones, International Telecommunication Union, Geneva, 2000
23. ITU-T Recommendation P.800.1: Mean Opinion Score (MOS Terminology), International Telecommunication Union, Geneva, 2003
24. ITU-T Recommendation P.800: Methods for Subjective Determination of Speech Quality, International Telecommunication Union, Geneva, 2003
25. ITU-T Recommendation P.830: Subjective Performance Assessment of Telephone-Band and Wideband Digital Codes, International Telecommunication Union, Geneva, 1996
26. ITU-T Recommendation P.831: Subjective Performance Evaluation of Network Echo Cancellers, International Telecommunication Union, Geneva, 1998
27. ITU-T Recommendation P.832: Subjective Performance Evaluation of Hands-Free Terminals, International Telecommunication Union, Geneva, 2000

28. ITU-T Recommendation P.835: Subjective Performance of Noise Suppression Algorithms, International Telecommunication Union, Geneva, 2003
29. ITU-T Recommendation P.862: Perceptual Evaluation of Speech Quality (PESQ): An Objective Method for End-to-End Speech Quality Assessment of Narrow-Band Telephone Networks and Speech Codecs, International Telecommunication Union, Geneva, 2001
30. ITU-T Supplement 3 to ITU-T Series P Recommendations: Models for Predicting Transmission Quality from Objective Measurements, International Telecommunication Union, Geneva, 1988
31. ITU-T SG12: Delayed Contribution D.45: Report on a New Subjective Test on the Relationships Between Listening, Talking and Conversational Qualities when Facing Delay and Echo, France Telecom, Geneva, 18–27 January 2005
32. Jekosch, U.: Voice and Speech Quality Perception, Springer, Berlin, 2005, ISBN3-540-24095-0
33. Kettler, F., Gierlich, H.W., Diedrich, E.: Echo and Speech Level Variations During Double Talk Influencing Hands-Free Telephones Transmission Quality. IWAENC '99, 27–30 Sept. 1999, Pocono Manor, USA
34. Möller, S.: Assessment and Prediction of Speech Quality in Telecommunications, Kluwer Academic, New York, 2000, ISBN 0-7923-7894-6
35. Richards, D.L.: Telecommunication by Speech, Wiley, New York, 1973, ISBN 0-470-71949-4
36. Rix, A.W., Reynolds R., Hollier M.P.: Perceptual Measurement of End-to-End Speech Quality over Audio and Packet-Based Networks. 106th AES Convention, pre-print no. 4873, May 1999
37. Sottek, R.: Modelle zur Signalverarbeitung im menschlichen Gehör, PhD thesis, RWTH Aachen, 1993
38. VDA: VDA-Specification for Car Hands-Free Terminals, Version 1.5, VDA, 2005

45

Efficient Speech Coding and Transmission Over Noisy Channels

Ulrich Heute[1] and Norbert Goertz[2]

[1] Inst. for Circuit and System Theory, Faculty of Engineering, Christian-Albrecht University, Kaiserstr. 2, D-24143 Kiel, Germany
[2] Inst. for Digital Communications, School of Engineering and Electronics, University of Edinburgh, King's Buildings, Mayfield Road, Edinburgh EH9 3JL, UK

1 • Introduction and Fundamentals

1.1 Goals

"Efficient speech coding" addresses the digital representation of discrete-time acoustic signals by few bits. "Transmission over noisy channels" may corrupt these bits. Error correction and, if imperfect, concealment are needed.

1.2 Signal Digitization and Efficient Coding

Bandwidth and sampling: The continuous speech signal $u_0(t)$ is sampled at instances $t_k = kT_s, k \in \mathbb{Z}$. The sampling frequency $f_s = 1/T_s$ is chosen according to the sampling theorem:

$$f_s \geq 2f_c. \tag{1}$$

"Narrow-band telephone speech" is usually band-limited to

$$f_c < 4\,\text{kHz} \quad (\text{e.g. } 3.4\,\text{kHz}); \tag{2}$$

"wide-band speech" has an improved naturalness due to a larger value

$$f_c < 8\,\text{kHz} \quad (\text{e.g. } 7.0\,\text{kHz}). \tag{3}$$

Quantization and coding: A sample $u_0(kT_s) \doteq u(k)$ is then sorted into the ith of 2^w possible quantization intervals of size Q_i and described by its w-bit binary index i. Globally, the sizes Q_i are proportional to the signal-amplitude range D; usually, D is chosen symmetrically as $\pm D/2$ around $u = 0$ and proportionally to the signals' root-mean-square value σ_u:

$$D = \lambda \cdot \sigma_u \text{ (with, e.g. } \lambda \approx 4\text{)}. \tag{4}$$

So, halving D allows to have the same interval sizes with one bit less.

In the simplest case of a linear quantizer (or "linear ADC", or "linear *pulse-code modulation*, PCM"), the size of all intervals is constant:

$$Q_i \equiv Q = D \cdot 2^{-w}.$$

At a DAC output as well as for further digital processing, all signal values $u(k)$ within an interval indexed by $i = i(k)$ are then represented by the same "representative" $[u(k)]_Q$. The quantization error $[u(k)]_Q - u(k)$ is usually modelled as an additive random signal. For linear PCM, with the representatives chosen as the interval centres, signal-independent, white, uniformly distributed quantization noise is a good model, with zero mean and variance

$$\sigma_Q^2 = Q^2/12. \tag{5}$$

The description of $[u(k)]_Q$ by an index $i(k)$ is a digital coding step. "Coding" is also often used, however, to address error protection of $i(k)$. This will be addressed separately (see Section 1.4). At this point, "efficient coding" denotes the replacement of the w-bit indices $i(k)$ by other, linearly or non-linearly "transformed" values $x(k)$ such that the bit rate f_B needed to reconstruct $\tilde{u}(k) \approx u(k)$ at a prescribed quality is much smaller than that needed for a linear PCM. So, "coding" here refers to *data compression*.

A common (though not generally "valid") quality measure is the *s*ignal-to-*n*oise (power) *r*atio, defined logarithmically as

$$\text{SNR}/\text{dB} = 10 \cdot \lg[\sigma_u^2/\sigma_n^2] = 20 \cdot \lg[\sigma_u/\sigma_n], \tag{6}$$

with σ_n^2 denoting the general distortion variance. We got used to

$$\text{SNR} \approx 35,\ldots,40\,\text{dB} \qquad (7)$$

in the "classical" telephone network. Equations (1), (2), and (4)–(7), with $\sigma_n \doteq \sigma_Q$, tell us, after some considerations about the speaker-dependent, varying dynamic range D, that $w = 11$ bits are necessary for linear PCM; thereby,

$$f_{B_{\text{PCM}}} \doteq w \cdot f_s = 11\,\text{bit} \cdot 8000/\text{s} = 88\,\text{kbit/s}. \qquad (8)$$

Speech production and perception modelling: Understanding how some information in a speaker's brain is transferred into a speech sound and how this sound is processed by a listener is more than background knowledge: models for both operations are practical tools in efficient speech coding.

Figure 1 shows a simplified production model [62,77]: the "vocal tract" or "acoustic tube" between larynx and lips can be described, under certain assumptions, by an all-pole-filter transfer function of order $n (\approx 10)$:

$$H^{-1}(z) = z^n / \sum_{v=0}^{n} \beta_v \cdot z^v = 1 / \sum_{v=0}^{n} \beta_{n-v} \cdot z^{-v}. \qquad (9)$$

The frequency response $H^{-1}(e^{j\Omega})$, with the normalized frequency $\Omega \doteq 2\pi \cdot f/f_s$, is time constant for some $5,\ldots,10,\ldots,20,\ldots$ ms. It defines the uttered sound for this time. Especially, the maxima of $|H^{-1}(e^{j\Omega})|$, closely related to the poles of $H^{-1}(z)$, are sound-characteristic and therefore termed "formants".

Some information about the sound stems also from the excitation signal $d(k)$ in Figure 1. It can be white noise with

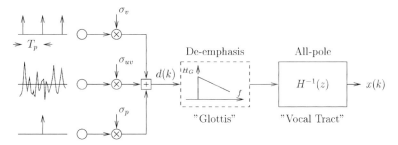

FIGURE 1 Simplified speech-production model with three generators exciting the vocal-tract filter (including a possible "de-emphasis" glottal pre-filter).

a variance σ_{uv}^2 for *un*voiced sounds, a pulse of size σ_{pl} for *p*losives, a periodic pulse sequence with period T_p and size σ_v for *v*oiced sounds which are periodic with a fundamental (or "*p*itch") frequency $f_p = 1/T_p$, or a combination of these. So, $d(k)$ has, ideally, a constant spectral envelope in all cases. The actually slightly decaying spectra can be taken into account by inserting a fixed de-emphasis-type "glottis pre-filter" $H_G(e^{j\Omega})$. Refinements of this model are discussed in [53].

When a sound reaches a listener, it is first filtered in the outer and middle ear. This causes a spectral shaping and a band-limitation to an upper value $f_c \leq 20\,\text{kHz}$ and a lower boundary $f_L \geq 20\,\text{Hz}$. However, not all signals within these limits are perceived: a frequency-dependent minimum level is needed.

In the inner ear, the signal's frequency components are separated by a bank of (...24...) band-pass filters; the bandwidths grow with a warped frequency axis termed "Bark scale" [77,79]. Non-linear processing follows: depending on both their absolute and relative amplitudes, one of two neighbouring components may be "masked" by the other, i.e. become inaudible. So, beyond the absolute hearing threshold, a signal-dependent masking threshold can be computed. Besides, wide-band but time-dependent maskings of weaker succeeding or preceding sounds are possible. Finally, the resulting excitation of the inner-ear sensors is non-linearly transformed into a perceived loudness.

1.3 Speech Coding and the "Coding Gap"

In the "classical" speech-coding literature, "high-rate" and "low-rate" codings were discriminated: while digital speech transmission or storage was possible with "telephone quality" as in Eqs. (2) and (7) via various schemes between log. PCM at $f_B = 64\,\text{kbit/s}$ (ITU G.711, [43]) and ADPCM at $f_B = 32\,\text{kbit/s}$ (ITU G.726, [45]), really low rates like $f_B = 2.4\,\text{kbit/s}$ were possible, for the same narrow-band speech, only with "synthetic" quality via VOCODER techniques like LPC-10 [75]. Here, no waveform reproduction is aimed at and, therefore, no SNR description can be used appropriately. Early attempts to code wide-band speech according to Eq. (3) somehow "naturally" worked at "high rates", like a split-band ADPCM at $f_B = 64\,\text{kbit/s}$ (ITU G.722, [44], see also [61]). So, there is a "gap" – topic of "lab research" mainly in the 1970s and beginning 1980s, with results invading realization in the 1990s [38].

1.4 Robust Transmission

Bit errors: Up to now, we assumed that the source decoder would receive the bitstream as sent out by the optimized coder. In practice, however, the transmission can be corrupted by noise and interference, perhaps strongly, like in mobile radio. Since the quality of the speech output is severely deteriorated by bit errors (the lower the data rate, the higher the sensitivity), error protection is needed.

Real channels may vary widely. Errors may happen randomly or can become "bursty". We assume error probabilities which may be very low sometimes, but may go up to, say, $p_e = 0.2$. Only if required for analysis, we will assume errors and noise to be uncorrelated and independent of other signals.

Channel coding: A channel coder computes "channel codewords" from the source-coder output, enlarging the bitstream by redundant bits which are to be used to correct errors. The ratio $R = f_B/f_{\text{total}}$ of the source coder and the total bit rates is termed "code rate". As detailed later, error-correcting codes are often concatenated by error-detecting codes, enabling at least the awareness of (perhaps "bursty") disturbances which are too strong to be corrected. This means that, in fact, no ideal error protection is assumed. This is due to realizability restrictions violating the conditions of ideal channel coding, especially that of infinite-length data.

Source and channel coder combination: Realizability also prevents the source coder from being ideal – its output bits are not at all free of redundancy or of equal importance. So, the conditions for applying Shannon's separability theorem are violated altogether, and optimizing a source coder to some best possible average quality at a small bit rate which is then fed into a separately optimized channel coder does not lead to the actual aim: the system should yield the highest possible similarity between input and output signals at a possible *total* bit rate after source *and* channel coding.

2 • Speech Coding

2.1 Origin of the Gap

The rate reduction from that of a simple linear PCM with $f_{B_{\text{PCM}}} = 88$ kbit/s as in Eq. (8) over 64 kbit/s down to 32 kbit/s at the same quality uses three essential steps, namely, non-linear quantization, adaptivity and, last but not least, the concept of linear prediction (LP): a non-recursive filter with a transfer function $H_0(z) = \sum_{i=1}^{n} a_i \cdot z^{-i}$ computes a linear combination of past signal values as an estimation $\hat{u}(k) = \sum_{i=1}^{n} a_i \cdot u(k-i)$ of the next sample. The subtraction of the actual sample $u(k)$ yields the difference or residual signal $d(k) = \hat{u}(k) - u(k)$ as the output of a prediction-error filter with the transfer function

$$H(z) = 1 - H_0(z) = 1 - \sum_{i=1}^{n} a_i \cdot z^{-i}. \tag{10}$$

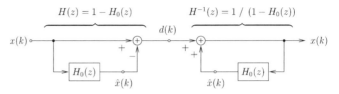

FIGURE 2 Typical structure of a predictive system.

Excitation of $1/H(z)$ by $d(k)$ in the receiver regenerates $u(k)$ (see Figure 2).

The minimization of the residual variance σ_d^2 leads to the optimal predictor-coefficient vector $\mathbf{a} = \mathbf{R}^{-1} \cdot \mathbf{r}_0$, found from the correlation matrix $\mathbf{R}$ and correlation vector $\mathbf{r}_0$. Due to this minimization, $d(k)$ is "smaller" than $u(k)$, and as concluded after Eq. (4), it can be quantized with fewer bits, even considerably less if both prediction and quantizer are adapted with time. This is no heavy algorithmic load, since the above matrix operations can be computed very efficiently, e.g. by the "Levinson–Durbin" recursion [62,77]. But, if the SNR quality is to be kept, $w \geq 4$ is found to be necessary. Together with $f_s = 8\,\text{kHz}$, this limits the rate reduction to $f_B = 32\,\text{kbit/s}$ as mentioned above. On the other hand, large parts of the signal information *do now* reside in the predictor and quantizer parameters, either adapted backwards (ADPCM) in both transmitter and receiver or transmitted as side-information with an additional rate $f_B = \ldots 2 \ldots \text{kbit/s}$ (APC). This, however, is the data rate of the (LPC-) VOCODER [75], synthesizing speech from parameters – with synthetic quality: following the speech-production model of Figure 1, the filter with the transfer function $1/H(z)$ is used to model the vocal tract. This is possible, since it has the same all-pole form as found in Eq. (9); we just have to identify the coefficients in Eqs. (9) and (10) as

$$\beta_{n-i} = -a_i, i \in \{1, \ldots, n\}, \text{ and } \beta_n = 1.$$

Beyond, little information is needed to replace the large residual data volume and control the possible excitations – their type, size, and periodicity.

2.2 Approaches to Close the Gap

RELP coding: If "parameters-only" representations reduce quality, and residual rates cannot be reduced by smaller values of w, f_s should be diminished. This is not possible for speech, because of Eqs. (1) and (2), but for the residual, which is whitened by the (necessarily) decorrelating filter $H(z)$: since $H(z)$ *exploits* the speech correlation to reduce σ_d^2, the correlation remaining in $d(k)$ must be smaller, ideally, a pulse-like autocorrelation; thus a constant power spectrum should remain – the residuum should become white. This can be interpreted as an inversion of the vocal-tract filtering – $H(z)$ denotes the inverse of the vocal-tract filter, which for this reason was already termed $H^{-1}(z)$ before.

After band-limiting $d(k)$ to $f_c/r, r \in\ <001>$, the remaining (constant!) base band needs only a sampling rate f_s/r; in the receiver, a spectrally widened signal can be applied to $1/H(z)$ in a "baseband residual"-excited LP system (BB-RELP). However, some unnatural, metallic distortions are unavoidable. Two reasons cause these artefacts: an *unflexible down-sampling*, first, which especially does not take care of the *spectral fine structure* second, in $d(k)$, namely its lines at harmonics of the fundamental frequency f_p.

Further redundancy reduction – LTP: Equispaced spectral lines are due to periodicity. They carry redundancy like the spectral shape, now (more or less) removed by $H(z)$. So, they can be removed in a similar way by a second, "pitch predictor", which uses correlation not at distances of $n \cdot T_s (n \approx 10)$, but of $m_p \cdot T_s \approx T_p$ ($m_p \in \{32, 160\}$ for normal voices). This long-time range gives rise to the notation "long-term predictor" (LTP; see Figure 3). The parameter β_p may be found by the same principle used before (see Section 2.1) for determining the coefficients a_i in Eq. (10): If m_p is known, e.g. from some pitch-detection algorithm [37], the variance σ_e^2 of the second-difference signal $e(k)$ is minimized with respect to β_p. However, the search for m_p can also be included into this minimization. Then, m_p may be not a "good" pitch estimate, but it is most suitable in terms of the criterion σ_e^2.

Flexibilities in down-sampling – RPE and MPE: Down- and upsampling of the baseband signal results, for voiced sounds, in spectral lines which are harmonic, i.e. lie at multiples of f_p within the baseband, but are then shifted according to f_s/r, irrespective of f_p.

Instead of taking every rth sample of the residual $d(k)$ in a naive manner, one may try to dissolve the misfit between sampling and pitch frequencies by, e.g. a decimation switching between various factors r with a reasonable mean value $\bar{r}$. However, this is less

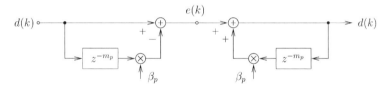

FIGURE 3 Long-term predictor ("1-tap LTP") and its inversion.

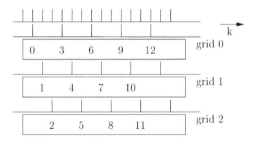

FIGURE 4 The RPE principle with $r = 3$ possible sub-sampling phases.

successful than a simple switch between the possible r phases of the down-sampling grid, according to some optimality criterion (see Figure 4). What remains is a decimated residual, applied in the receiver as a "regular-pulse excitation" (RPE) of $1/H(z)$. As always, an inverse LTP has to precede $1/H(z)$, if an LTP is included in the transmitter.

In a generalized RPE, optimized amplitudes may be applied instead of just samples. Furthermore, the N/r remaining values after down-sampling a block of N data could as well be placed on any optimal positions rather than a grid, leading to the concept of multi-pulse excitation (MPE). Both RPE and MPE can, on the other hand, be simplified by allowing only simple amplitudes like $\{-1, 0, 1\}$, with a common gain factor per block (or for suitable sub-blocks), thus separating a simplified shape and a gain of the excitation.

Vector quantization (VQ): After quantization, samples with a limited number of allowed values remain. So, only a certain set of possible excitation blocks is available at the receiver. Another, "direct" way to such excitations applies a table ("codebook") with, say, $L = 2^w$ length N (decimated or non-decimated) vectors from which the best one is chosen and coded by w bits per N samples; for, e.g. $L = 1024 = 2^{10}$ and $N = 40$, this means that $w = 10/40 = 1/4$ bits/sample are used in such a "vector-quantization" (VQ) scheme. The same principle can then also be applied to parameters like the predictor-coefficient vector **a** or any equivalent, transformed coefficient set, like the so-called reflection-coefficient vector **k**, the log-area-ratio vector **lar** [67], or the line-spectral-frequencies' vector **lsf** [42,77].

2.3 Closing the Gap

A first important practical realization helping to bridge the coding gap was the GSM full-rate (FR) codec (see Figure 5), standardized at the end of the 1980s by ETSI, applied world-wide since 1992 [12], based on those of the above ideas which were

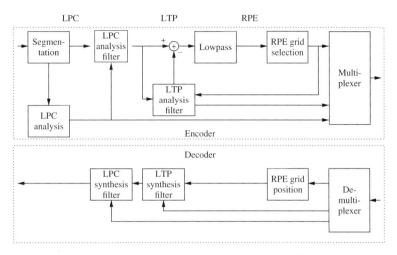

FIGURE 5 GSM-FR system (RPE-LTP-RELP with $r = 3$).

realizable with then available hardware, and with the best quality achievable thereby–reckoned as "not too good", nowadays.

This still quite limited quality as well as the tremendous growth of user numbers caused a wave of research towards improvements. They relied, firstly, on an enormous increase in computational power: nowadays, a 32-fold complexity and more can be envisaged. Secondly, the enhancements were based on some of the above-named tools, but above all, thirdly, on three *decisive* ideas described below.

2.4 The Breakthrough Basis

Vector quantization: Firstly, the VQ principle [31] turned out to be a very powerful concept. Work both on the theory of an N-dimensional extension of the pdf-based scalar optimum quantizer [58] and its limitations [7] and on practical codebook design [57] was helpful. VQ is applied to both receiver-excitation and parameters. Especially, VQ of the vector **lsf** achieves very good quality at only $20\ldots24$ bits per 20 ms-frame, i.e. a rate of $1\ldots1.2$ kbit/s [63].

Improved codebooks via efficient training, especially for $d(k)$ or $e(k)$, enhanced quality. Simple entries (0 and ± 1 only) or sparse vectors (adopting the RPE and MPE principles of Section 2.2) reduced computations and led to realizability. The simple (shape) vectors were combined with separate gains ("shape-gain VQ") and also efficiently transformed into more complex varieties by linear operations ("vector-sum (VS) VQ", "algebraic codebooks").
Analysis by synthesis: Secondly, and above all, however, a better optimization criterion was decisive: instead of minimizing the

MSE of the residual approximation, the distortion of the receiver-output signal (synthesized in the transmitter!) was reduced to a minimum. Furthermore, this minimization enclosed also the LTP into its closed loop, viewed as a second, (pitch-) adaptive codebook. Again, not an optimal pitch estimation is then found, but the most suitable period according to the criterion.

The underlying idea is (a) as simple as many good ideas, and (b) more than 40 years old ([4,35], see also [67]), but became realizable only with today's electronics. Actually, it means that the computation of the inverse filter $H(z)$ (plus the pitch predictor, if applicable) and the scalar or vector quantization of the residual signal cannot be regarded as independent steps. Rather, the best combination of all quantized (!) parameters and excitations has to be searched for - "best" with respect to the reconstructed, full-length signal. In practice, however, finite-length blocks are considered (with only some inclusion of the past), and the (LPC-) filter analysis is made "off-loop" as in Section 2.1 for a given input-signal block, after some pre-emphasis filtering by $H_G^{-1}(e^{j\Omega})$. But (within realizability:) "all" possible excitations are indeed applied to the full synthesis process, including all delayed versions of preceding excitations, i.e. possible (pitch-) periodic components.

Perceptual weighting: Furthermore, this "analysis-by-synthesis" technique included a perceptual weighting (as a crude perception model). For the above "best-match", the simple MSE is not an appropriate criterion: it maximizes the SNR only, which does not reflect a listener's quality impression for low-rate coders. Therefore, the difference between the synthesized and the original speech signals is spectrally weighted prior to the mean-square calculation and minimization. The weighting is carried out by a transfer function

$$W(z) = H(z)/H(z/\gamma), \gamma \approx 0.9. \tag{11}$$

$H(z)$ is the "inverse-vocal-tract" filter of Eq. (10). With $\gamma < 1$, $H(z/\gamma)$ corresponds to a "damped" version of $H(e^{j\Omega})$. Thus, $|W(e^{j\Omega})|$ varies with minima near the formant frequencies F_ν and maxima between them. Thereby, distortions are less important close to strong spectral components, and emphasized otherwise: masking effects are coarsely modelled.

2.5 CELP Coders

Analysis by synthesis as well as perceptual weighting techniques can be applied in various systems, including LP-based RPE and MPE coders. Most successful, however, was the combination of analysis by synthesis, perceptual weighting by $W(z)$ (see Eq. (11)), vector or *codebook* *e*xcitation, and LP: such coders are termed

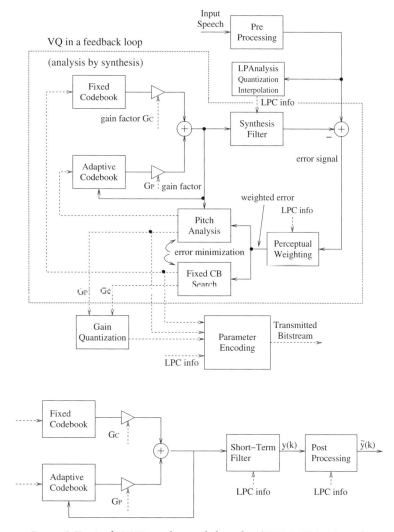

FIGURE 6 Typical CELP coder and decoder (ITU G.729 ACELP).

CELP systems [69]. Their numerous variants can be understood from the above considerations. Figure 6 shows, as a typical codec, one of the ITU standards [47]. It uses the approach of a simple-entry codebook with matrix manipulation, giving rise to its denotation as "algebraic CELP" (ACELP). This is one of many popular CELP systems in use nowadays. Some of them are listed below [39,71,77].

ITU G.729:	Algebraic CELP coder (ACELP); $f_B = 8\,\text{kbit/s}$; high quality; high complexity (18 MOps) [47]
GSM-HR:	ETSI half-rate coder; vector-sum excitation (VSELP); $f_B = 5.6\,\text{kbit/s}$; quality $\leq$ GSM-FR; high complexity (20 MOps) [13]
GSM-EFR:	ETSI-enhanced full-rate coder; ACELP similar to [47]; $f_B = 12.2\,\text{kbit/s}$; enhanced quality and error protection compared to GSM-FR; high complexity (18 MOps) [14]
ITU proposal:	ITU coder for high quality at low rates; "eX-CELP"; $f_B = 4\,\text{kbit/s}$; very high complexity (not yet fixed) [73]

ITU G.728:	ITU codec with low delay (2 ms, "LD-CELP");
	 high-order prediction ($n = 50!$), backward adaptation;
	 no LTP; very short codewords ($N = 5$);
	 $f_B = 16\,\text{kbit/s}$;
	 high quality; high complexity (20 MOps) [46]

MPEG-4 CELP:	MPEG-4 system with scalable data rate; ACELP with MPE/RPE variants; $f_B = 4\ldots24\,\text{kbit/s}$; variable quality; high to very high complexity (simple narrow-band coding: ≈ 20 MOps, enhanced narrow-band version: ≈ 40 MOps, enhanced wide-band coding: $\approx 50\ldots70\ldots$ MOps) [32]
GSM-AMR:	ETSI adaptive multi-rate system; ACELP; variable rates $f_B = 4.75\ldots12.2\,\text{kbit/s}$, variable quality; up to high complexity (17 MOps) [15]
GSM WB-AMR:	ETSI wide-band AMR system: ACELP; low to high rates $f_B = 6.6\ldots23.85\,\text{kbit/s}$, up to very high quality; up to very high complexity (35 MOps) [76]

The two last ones adapt to changing transmission channels and/or user requirements, with 8 and 9 different rates, respectively; the latter is to be applied also in the enhanced GSM system and in UMTS. The notation "eX-CELP" refers to an "extension" of the CELP principles by combining it with other techniques and, above all, by varying sub-codebooks and even parts of the algorithm depending on a signal-segment classification [72]. The quality of these systems has to be compared to that of the common ISDN telephone and that of the wide-spread GSM-FR codec, i.e. to a relatively "high" and a "somewhat reduced" quality. A good CELP system may indeed yield "high quality", much better than RPE-LTP-RELP. But also the complexities have to be taken into account: while the ITU standards G.711 (PCM) and G.726 (ADPCM) need almost no computation, ETSI's GSM needs some 3.5 MOps (= Mega-Operations per second) already, while a good CELP coder may need 5–10 times as much or even more; 3.5 MOps present no problem for a simple digital signal processor (DSP), more advanced DSPs can handle 20 MOps easily, and cost start to really increase at 100 MOps – although hardware performance is still growing quickly.

2.6 Frequency-Domain Coding

In the early GSM preparation phase, most proposals aimed at "medium rates" with a different approach [38], namely, by means of "continuous" sub-band or "block-oriented" transform coders (SBC, TC). In fact, both are systems with analysis-synthesis filterbanks, whether realized by parallel band-pass filters or by filterbank-equivalent DFTs or DCTs; the true differences were their down-sampling factors r and corresponding numbers $M = r$ of frequency channels/components: $M = \ldots 8 \ldots 16 \ldots$ in SBC, $M = \ldots 128 \ldots$ in TC.

Figure 7 displays a typical TC system with adaptive quantization and bit allocation following the spectral envelope

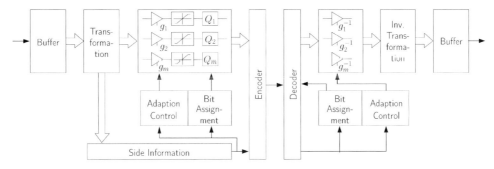

FIGURE 7 Adaptive transform codec (ATC).

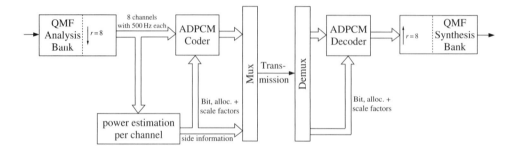

FIGURE 8 Sub-band codec (SBC) with adaptive bit allocation and ADPCM.

(ATC). This was realized in [24] with $f_B = 14\,\text{kbit/s}$, a quality close to GSM-FR, and a similar complexity. Figure 8 shows an SBC system with lower quality, but also lower complexity [55].

Of course, ideas applied in CELP systems could be transferred to frequency-domain coders: VQ is readily used for both main and side information – to spectral-value sets and to a spectral-envelope description for the bit allocation, based on a vector of averaged amplitudes as in [24,60] or on LPC results like **a** or **lsf**. Perceptual weighting could lead to a "damped envelope" in bit allocation, modelling (partial) masking again. Many such ideas were examined, but (for speech in contrast to audio coding!) frequency-domain coders were much less successful than the predictive systems discussed above, although there is no simple, generally accepted reason.

2.7 Other Ways Towards Closing the Gap

Describing speech segments as a finite sum of sinusoids with suitable amplitudes (coded directly or via an envelope, e.g. by means of LPC coefficients), frequencies and phases was intensively investigated in the late 1980s [74]. A trade-off between quality, complexity and rate was possible by choosing more or less refined descriptions especially of the frequencies and phases. A simplified version of this "sinusoidal modelling" (SM) applies a (pitch) frequency grid; this "harmonic-coding" (HC) scheme can be augmented by allowing narrow-band noise in unvoiced sections. SM and HC in their original forms can be seen as frequency-domain approaches; together with their "relatives" from Section 2.6, they have been of less interest in the past 10 years. From here, however, the way is not far to interesting hybrid time–frequency codecs with mixed or multi-band excitation (MELP, MBE, IMBE, EMBE), aiming at rates near 4 kbit/s and below [77].

2.8 Wide-Band Speech Coding

There has been some work on ATC or SBC for wide-band speech, especially when combined with music. At present, however, CELP systems, with an enlarged prediction order $n > 10$, dominate the scene. A variant – ACELP, similar to that in Figure 6 – is included in the list of Section 2.5, as a part of the ETSI WB-AMR standard.

A detail in this coder [6], already seen in some earlier development [11,64], is worth mentioning: the "split-band" principle of [44] reappears. But instead of separating frequencies below and above 4 kHz, an asymmetrical splitting takes place: $0\ldots 6\,\text{kHz}$ (6.4 kHz in [6]) are directly (ACELP-) coded. The remaining band is replaced by noise (or aliased parts of the lower band, similar as in BB-RELP, see Section 2.2), which is filtered by $H^{-1}(z)$; only a gain factor needs to be transmitted. The gain is quite sensitive, but its word length can be reduced by linear prediction. The idea behind this was already used in an approach to "wide-band speech coding without *any* upper-band information": in [8,9], the components above 4 kHz were not transmitted, but substituted in the receiver as above, with a spectral weighting and gain term estimated from the lower band, using CELP techniques.

3 • Speech Transmission over Noisy Channels

3.1 Preliminary Remarks

Speech transmission is usually applied in conversational applications (telephony), which means that the tolerable delay is rather low. This limits the number of source samples that may be encoded commonly to, e.g. 160 in the European mobile radio standards [12–15].

As stated above, speech encoding is carried out in two steps: first, each block of source samples is decomposed into a set of parameter-vectors $x^1, x^2\ldots$ which consist, e.g. of the LPC coefficients or the residual signal. In the second step the vectors x^j are quantized and the output bit vectors i^j are generated. This scenario is depicted in Figure 9. The two steps may be carried out jointly if the "analysis-by-synthesis" coding principle (as in CELP) is used. This, however, does *not* change the principle of

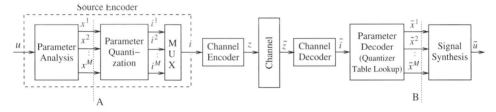

FIGURE 9 Basic model of a speech transmission system.

decomposing the actual source signal into parameter descriptions which are quantized and transmitted.

A channel code is applied to protect the source bits i against channel noise. Very often convolutional channel codes [56] are used, mainly because soft-information (unquantized matched-filter outputs) from the channel can be easily exploited by moderate-complexity decoding algorithms (Viterbi algorithm or BCJR algorithm [3]), which leads to strong gains over hard-decision decoding [56].

The actual goal of the system design is to obtain a decoder output signal $\tilde{u}$ that is as similar as possible to the input speech signal u. If the "similarity" shall be obtained in terms of *auditory* speech quality, it is very hard (if not impossible) to find a mathematically tractable measure for this, especially if it must have moderate complexity and has to operate on relatively short blocks of speech samples. But even if we used the simple mean-squared reconstruction error of the speech signal as a distortion measure, it would still be impossible to analytically optimize the system by variational techniques, because source codecs are complex algorithms that are hard to describe analytically (in the sense of a "transfer function").

There are two concepts to circumvent this problem. Firstly, one can use listening tests to get rid of the problem that we need an analytically tractable quality measure. The problem with this approach is that one must implement an idea which is mainly based on a "guess" and then it must be proven by listening tests that the scheme works. Clearly, this way of system design is extremely time-consuming and only a very small subset of potential options can be checked. Nevertheless, these methods have led to significant improvements in speech quality.

Another way is to optimize the speech quality *indirectly* by minimizing the bit error rate after channel decoding or by minimizing the reconstruction error of the speech codec *parameters* (i.e. minimization of the mean-squared error between the points "A" and "B" in Figure 9). Although there is no guarantee that the best possible quality is obtained, the concept is reasonable and it opens the way for mathematically tractable optimizations without the need to conduct listening tests.

Although we have a number of different parameters in Figure 9 we will, for brevity of notation, only consider one representative parameter signal below, and therefore we will omit superscripts. In what follows we will present a collection of methods to improve the quality for noisy transmission of coded speech signals.

3.2 Some Approaches for Quality Improvements

Comments on the separation principle: The foundation for the structure of today's communication systems is information theory: it guarantees that (possibly by use of infinitely large block lengths) one may realize the transmitter by a cascade of a source and a channel encoder with a binary interface in between (the same holds for the corresponding receiver), at which independent and uniformly distributed bits are exchanged – without any loss in performance compared with the best possible "one-step" encoder. The basic notion is to apply channel coding to achieve error-free bits at a bit rate that equals the channel capacity. The latter describes the amount of information that can be reliably transmitted over the channel. A source code is applied to reduce the bit rate required to represent the source signal down to the channel capacity. Clearly, this reduction is only possible at the price of some distortion when the source alphabet is continuous as in the case of speech and audio signals.[1]

If the separation principle is applied in a practical system with *limited block lengths* the following problems occur:

- It is impossible to realize channel codes that have a residual bit error rate of zero. The reason is that, as long as the block length is limited, in some blocks more random channel errors may occur than can be corrected.
- Practical source encoders do not produce independent and uniformly distributed bits. This is again due to the limited block length. If, for instance, we encode a correlated signal, there is always correlation left between adjacent blocks and, hence, between the quantizer indices.
- The distortion at the source decoder output depends on which bit is in error, because any practical quantizer for a limited block length has more or less "significant" bits. In case of parametric coding such as CELP, even the codec parameters we extract from the signal are not equally important for the (auditory) quality: a bit error in the quantizer index for the signal power leads to severe quality degradations while a flipped sign-bit of the excitation signal is sometimes not even noticeable.

In spite of these "non-idealities," source and channel coding are separate processing steps in all practical systems (with some important modifications discussed below). Besides the

---

[1] It is common practice to consider the amplitudes of speech and audio signals as continuous, although in an implementation the samples are typically represented by 16 bits (linear PCM) and, thus, they are pre-quantized before the actual source compression takes place.

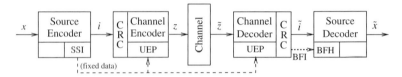

FIGURE 10 Unequal error protection (UEP) for which source significance information (SSI) is used. Error concealment with a cyclic redundancy check (CRC) that at the receiver sets the bad frame indicator (BFI) initiating bad frame handling (BFH).

separate history of source and channel coding motivated by information theory, the practical reason is that the separation allows to construct "partial" encoders/decoders that, compared with asymptotic results, sometimes may have only moderate performances[2] but their complexities are tolerable. Moreover, the system design is greatly simplified.

Unequal error protection and error concealment: Figure 10 shows some system modifications that account for the drawbacks of the separation principle stated above. Implicitly, a channel with discrete input alphabet (application of digital modulation) is assumed, because the channel codeword z, a bit vector, is directly fed into the channel. Some of the source-encoder output bits cause stronger quality degradations than others if they are in error but are used for decoding anyhow. Source significance information (SSI) can be derived from that and be used by the channel-coding scheme to apply stronger error protection to the more sensitive bits.

The residual redundancies in the source-encoder output bits can be exploited to conceal bit errors, e.g. by repetition of the old parameters from the previous block (bad frame handling (BFH)). Clearly, frame repetitions only make sense, if the codec parameters are correlated in time. This is the case, e.g. for the signal power, the LPC coefficients or the "pitch-lag". For the initiation of the error concealment, an error detection (bad frame indication, BFI) is required, which is usually realized by an error-detecting channel code (cyclic redundancy check (CRC)). The strongly sensitive bits of the source encoder are CRC-encoded (Class-1-bits) and afterwards the CRC code word is channel-encoded together with the remaining data bits. Thus, in combination with UEP, the bits of the source encoder are divided into

---

[2]It should be mentioned that in the "waterfall-region" Turbo Codes [5] and the developments based thereon brought the performance of channel coding schemes with fairly long block lengths very close to the information theoretical bounds, if a residual bit error rate of, say, 10^{-6} is accepted.

several classes which are protected by adjustable amounts of channel-code redundancy. Fortunately, the more sensitive bits are typically the stronger correlated ones at the same time, i.e. UEP and error concealment can be combined very efficiently. Hence, both schemes are frequently used, e.g. in mobile radio standards [17–20].

Source-controlled channel decoding: The idea of source-controlled channel decoding is to exploit the residual redundancies systematically as a-priori information in the decision procedure for the data bits in channel decoding to reduce the bit error rate [1,33]; the basic setup is depicted in Figure 11. Source-controlled channel decoding can be implemented very efficiently for the decoding of binary convolutional channel codes and it can be extended to soft-output channel decoders (APRI-SOVA, [33]), which can be combined with estimation-based source decoders described below. For a correct weighting of the a-priori information and the soft channel outputs, a channel state information (CSI, e.g. noise variance) must be available.

A drawback is that the redundancies are usually exploited on bit basis³ (when, as in many cases, binary channel codes are used), although the source encoder often emits correlated *indices* that consist of several bits. Thus, a large part of the redundancies is removed if only the marginal distributions for the individual bits are used.

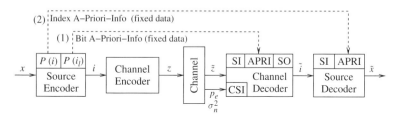

FIGURE 11 (**1**) Source-controlled channel decoding with soft inputs (SI) and outputs (SO) using channel state information (CSI) and source-bit a-priori information (APRI); (**2**) estimation-based source decoding using index a-priori information.

---

³In [36] a channel-decoding scheme for binary convolutional codes is given that is able to optimally exploit index-based a-priori information, if all the bits of an index are adjacently mapped into the input of the channel encoder. However, the method may not be applicable because usually bit-reordering/interleaving is employed at the channel-encoder input to account for different sensitivities of the bits or to improve the distance properties of the concatenated code.

Estimation-based source decoding: In conventional systems, simple procedures, e.g. quantizer table-lookups, are frequently applied for source decoding. If, however, source encoding is not perfect and residual redundancies – e.g. correlations in time – are left in the data bits, this a-priori knowledge can be exploited for better source decoding by performing estimations of the source-encoder input [23,54,59,66,68,70]; such a system is also depicted in Figure 11. In contrast to source-controlled channel decoding, it is now possible to exploit the full correlations on *index* basis and the soft-in/soft-out (SISO) channel decoder merely acts as a device that improves the reliability of the virtual channel "seen" by the source decoder.

Frequently, we want to minimize the squared error in the reproduction $\tilde{x}_k$ of the parameter values x_k. For this, the estimator

$$\tilde{x}_k = \sum_{\forall i} \underbrace{\Pr(I_k = i | \tilde{i}_k, \tilde{i}_{k-1}, \ldots)}_{\text{index APP}} \cdot x_q(i) \tag{12}$$

is a suitable decoding rule [23,70]. The decoder output $\tilde{x}_k$ is computed as a weighted average of all quantizer reproducer values (or vectors) $x_q(i)$. The weighting factors are the a-posteriori probabilities (APPs) of the quantizer indices (for which we use the random variable I_k at time k and the realizations i). The APPs are conditioned on the current and all previous channel decoder outputs $\tilde{i}_k, \tilde{i}_{k-1}, \ldots$, in order to exploit the memory of the source. Hence, the main problem is how to compute these APPs. If we assume a first-order Markov chain to model the autocorrelations of the quantizer indices we obtain [23,66]

$$\underbrace{\Pr(I_k = i | \tilde{i}_k, \tilde{i}_{k-1}, \ldots)}_{\text{index APP}} = A_k \cdot \underbrace{p(\tilde{i}_k | I_k = i)}_{\text{channel term}} \cdot$$

$$\sum_{\forall i'} \underbrace{\Pr(I_k = i | I_{k-1} = i')}_{\text{first-order Markov model}} \cdot \underbrace{\Pr(I_{k-1} = i' | \tilde{i}_{k-1}, \tilde{i}_{k-2}, \ldots)}_{\text{"old" index APP}}, \tag{13}$$

where A_k is a normalizing constant that ensures that $\Pr(I_k = i | \tilde{i}_k, \tilde{i}_{k-1}, \ldots)$ sums up to one over all possible indices i. Equation (13) is a recursion that allows to compute the index APPs from the "old" APPs at the previous time instant,[4] from the Markov transition probabilities $\Pr(I_k = i | I_{k-1} = i')$ of the statistical model for the quantizer indices and from the channel term. The latter weights the statistical knowledge about the source with

---

[4]At time $k = 0$ we start with the unconditional probability distribution of the quantizer indices.

the probability density function of the channel-decoded currently received channel information $\tilde{i}_k$. Hence, a soft-output channel decoder is needed [3] that provides reliability information about the channel decoded data bits and not only hard decisions. As in the previous section, bit-based a-priori information may be used to aid channel decoding, but it is not clear how the a-priori information shall be optimally divided between the channel decoder and the source decoder. This problem gave rise to the idea of "true" joint source-channel decoding.

Joint source-channel decoding: Many schemes in the area of "joint source-channel decoding" use residual redundancies to improve *either* channel decoding *or* source decoding (but usually not both, see also [22]). At the end of the previous section it was discussed that it is unclear how to divide residual correlation between source and channel decoding. Hence, it is just consequent to completely join source and channel decoding. If only one parameter is considered that is individually channel-encoded, and we are again interested in a good estimation in the minimum mean-squared error sense, we obtain [26,27]

$$\tilde{x}_k = \sum_{\forall i} \underbrace{\Pr(I_k = i | \tilde{z}_k, \tilde{z}_{k-1}, \ldots)}_{\text{index APP}} \cdot x_q(i) \tag{14}$$

with

$$\underbrace{\Pr(I_k = i | \tilde{z}_k, \tilde{z}_{k-1}, \ldots)}_{\text{index APP}} = A_k \cdot \underbrace{p(\tilde{z}_k | Z_k = CC(i))}_{\text{"channel" term}} \cdot$$

$$\sum_{\forall i'} \underbrace{\Pr(I_k = i | I_{k-1} = i')}_{\text{first-order Markov model}} \cdot \underbrace{\Pr(I_{k-1} = i' | \tilde{z}_{k-1}, \tilde{z}_{k-2}, \ldots)}_{\text{"old" index APP}} \tag{15}$$

where the notation $Z_k = CC(i)$ corresponds to the channel codeword Z_k that results from the quantizer index i. Note that the index APPs and the channel term, in contrast to (13), now contain the received channel words $\tilde{z}_k$ (instead of $\tilde{i}_k$) and the transmitted channel codeword Z_k (instead of $I_k = i$) that results from the channel code by some hypothesis i for the quantizer index. As long as we channel-encode the indices individually, there is no major difference in complexity compared with the method given by (13).[5] But in practice, we will have a set of quantizer indices (for the LPC coefficients, the lag information and others) that are jointly channel encoded. Therefore, the

---

[5]Some results for joint source-channel decoding for speech transmission by a CELP speech codec in the GSM half-rate channel are given in [25].

received channel codeword depends on the whole set of source-codec indices and, hence, we would have to compute the channel term in (15) for all possible bit combinations of all quantizer indices: an infeasible task if we have, e.g. 10 indices with a total of 200 data bits. Although these issues limit the applicability of (15), the formula provides an interesting and unconventional view of channel coding: its effect on the source decoder output $\tilde{x}_k$ is in no way different from other kinds of redundancies like the residual correlations of the quantizer indices. Both types of redundancy form weighting factors for the APPs of the data indices in (15). This insight into source-channel coding paved the way for iterative source-channel decoding.

Iterative source-channel decoding: In [27,29,65] iterative source-channel decoding is introduced. Although the sketch of the system in Figure 12 is similar to the systems in Figure 11, the theoretical concept is different: the transmitter is interpreted as a serially concatenated coding scheme; the constituent "codes" are formed by the implicit residual redundancies within the source encoder indices and by the explicit redundancy of the channel code.

The interpretation as a concatenated coding system calls for the use of the turbo principle [5,34] for decoding: as in all iterative decoding schemes so-called APP decoders (a-posteriori probability decoders) for both constituent "codes" must be available that are able to exchange extrinsic information on the data bits within the decoding iterations. The extrinsic information contains only the *new* part of information that is generated from *one* type of redundancy. While APP decoders for the frequently used binary convolutional channel codes are well known from literature [3,34], they can also be formulated on the basis of estimation-based decoding for the source-coding part of the system [27,28]. Among the realizable decoding schemes, iterative source-channel decoding works best for a given transmitter but at the same time it is the most complex approach, as several iterations of the component decoders are needed.

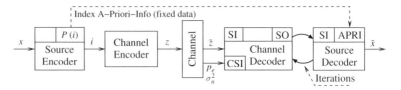

FIGURE 12 Iterative source-channel decoding with a soft-in (SI) soft-out (SO) channel decoder that requires channel state information (CSI) and a soft-in source decoder using a-priori information (APRI).

An intermediate discussion: In practice, it strongly depends on the encoding scheme if the application of advanced receiver structures such as iterative source-channel decoding is recommendable: if the source encoder works very well, it will leave only a small amount of redundancy in its output bits. Hence, it may be questionable if the effort for advanced joint decoding schemes really pays off as the potential quality gain might be small. Moreover, any decoding scheme can only be optimal for one given transmitter but *if* we are free to choose the encoder too, it is not really clear which scheme is optimal, especially in terms of auditory quality.

We recall from the earlier discussion of the separation principle that, ideally, the channel code makes available the amount of information that can be reliably transmitted over the channel. As the channel capacities in mobile radio are strongly time-variant, we actually have to vary the channel code rate, too. But if we are forced (e.g. by a standard) to use a fixed channel code rate we can only choose it for the "average case". This means, however, that the scheme is not optimal in many cases and that we also should consider more flexible systems including encoder modifications. In the discussion above we actually did that already, when we added a channel code to at least detect the errors that cannot be corrected and we used unequal error protection. Below we will discuss more elaborate schemes that can also be adapted to varying channels statistics.

Channel-optimized vector quantization: The idea of channel-optimized vector quantization (COVQ) [21,51,78] is to use the statistics of the source and the channel in the design of the source encoder (quantizer codebook) and in the encoding algorithm. COVQ is one of the few concepts which do not separate at all into source and channel coding. The scheme assumes a limited block length which is fixed for each source vector (i.e. fixed bit rate) but there are no further restrictions. Given the system constraints, COVQ does the best job possible with strong gains over source-optimized VQ.

Due to the complexity of encoding and the memory requirements for the codebook, COVQ is only applicable, if the number K of bits representing the quantizer indices is smaller than, e.g. 10. Therefore, COVQ cannot be used directly for speech, audio or image coding. In our setup, COVQ could be used to quantize the parameters of a source codec. Note that as long as the encoder and the decoder use the same codebooks, COVQ could be also used within an analysis-by-synthesis loop of an advanced CELP speech codec. If the channel conditions change, it is possible to adapt COVQ by use of an appropriate codebook that is designed for the current bit error rate. This requires, however, that the bit error rate is known at the transmitter, which is only realizable if the channel statistics change slowly (that is not the case for fast fading on a mobile radio link). Another drawback is that one would, in principle, need a new COVQ codebook for each channel condition, i.e. a huge amount of storage would be necessary. It was observed in [48] that the performance degradation due to moderately varied channel properties is small, so that a small set of codebooks is enough to cover the range of potential channel variations. Moreover, in [30] a simplification is introduced, which often allows to almost achieve the performance of COVQ while only one codebook and a channel-adaptive scaling factor for the fixed codevectors are required.

A major problem with parameter-individual COVQs is that at a bit rate of, say, one bit per sample, we could apply efficient error control coding for a long block of, e.g. 200 data bits. That would be much more efficient than the implicit error control implemented by each "small" parameter-individual COVQ. Therefore, COVQ will probably not find its way into speech coding with moderate block lengths, but there might be low-delay speech coding applications in which the block length is too small for efficient channel coding. An example is line-multiplexers for transatlantic telephony which use the low-delay CELP codec [46].

Adaptive rate-allocation techniques: From the discussion above it is clear that for varying channel conditions we should vary the bit rates for source and channel coding. As this rate adaptation requires signalling between the transmitter and the receiver, it cannot be carried out extremely fast, i.e. it is mainly useful to combat relatively slow fading in mobile radio.[6]

The basic idea is also clear from the discussion above: if the channel causes strong impairments, the source-coding bit rate is lowered and the number of redundancy bits used for channel coding is enlarged. This improves the error correction capabilities at the price of lower quality of source coding. If the channel noise is low, we choose a small number of redundancy bits and we increase the bit rate for source coding. As standardized for the adaptive multirate (AMR) system in GSM [15,16,20] and also for UMTS, the idea can be implemented by a set of coding schemes (for an introduction, see [40]).

---

[6]A way to combat *fast* fading is to use interleaving (which means increased delay) to spread the bits of a channel codeword over longer time periods, so that, in spite of the quickly time-varying channel, we obtain almost constant channel statistics after de-interleaving. The method is widely used in mobile radio.

Rate-allocation techniques can be combined with the other source-channel (de)coding methods described above, i.e. a good system will contain a mixture of rate-allocation techniques, bad frame handling, unequal error protection and possibly estimation-based/iterative source-channel decoders.

4 • Further Work, Results and Literature

4.1 Speech Coding

Good telephone quality will soon be achieved at $f_B = 4$ kbit/s, while good wide-band quality is available at $f_B = \ldots 12 \ldots$ kbit/s. Good telephone speech with $f_B = \ldots 2.4 \ldots$ kbit/s is currently studied. For instance, SM/HC ideas reappear in the MPEG-4 subsystem called "harmonic vector-excited coder" (HVXC); pitch-dependent waveform-interpolation concepts as proposed in [50] are investigated; models of the hearing process are used in the LP excitation [2]; also phoneme classification is applied [10].

4.2 Joint Source-Channel Coding

During the last decade, various methods have been derived that strongly improve the quality of coded speech transmission over noisy channels. The more elaborate methods such as estimation-based decoding will have to be implemented efficiently to make their potential available in practice. Techniques for rate adaptation, depending on the quality demands and on the current channel conditions, are important fields of future work. Another crucial aspect is the evolution from traditional circuit-switched to packet-switched transmission techniques, which are used, e.g. in the Internet.

4.3 Further Reading

A huge variety of literature, including books, are available. The choice given here is, of course, limited – and subjective. This holds also for the following recommendations:

A good entrance into the details of waveform-preserving coder approaches applied to speech (and video) is found in the "standard" [49], a broader overview on speech processing in [62], a less broad but more detailed and recent one in [77]. Two other good references for speech coding are [41,52]. The books also contain good introductory as well as more advanced aspects of error concealment and joint source-channel coding. A comprehensive treatment of channel coding can be found in [56]. All books and overview papers are, of course, rich sources for further references.

References

1. Alajaji, F., Phamdo, N., Fuja, T. (1996) Channel Codes that Exploit the Residual Redundancy in CELP-Encoded Speech. IEEE Trans. Speech Audio Process., **4**, 325–336

2. Ambikairajah, E., et al. (2001) Wideband-Speech and Audio Coding Using Gammatone Filterbanks. Proc. IEEE Int. Conf. Acoustics Speech Signal Process., Salt Lake City, **II**, 773–776
3. Bahl, L.R., Cocke, J., Jelinek, F., Raviv, J. (1974) Optimal Decoding of Linear Codes for Minimizing Symbol Error Rate. IEEE Trans. Inform. Theory, **20**, 284–287
4. Bell, C.G., Fujisaki, H., Heinz, J.M., Stevens, N.K., House, A.S. (1961) Reduction of Speech Spectra by Analysis-by-Synthesis Techniques. J. Acoust. Soc. Am., **33**, 1725–1736
5. Berrou, C., Glavieux, A. (1996) Near Optimum Error Correcting Coding and Decoding: Turbo-Codes. IEEE Trans. Commun., **44**, 1261–1271
6. Bessette, B., Jaervinen, K., et al. (2001) Techniques for High-Quality ACELP Coding of Wide-Band Speech. Proc. EUROSPEECH, Aalborg, **3**, 1993–1996
7. Brehm, H., Stammler, W. (1987) Description and Generation of Spherically-Invariant Speech-Model Signals. EURASIP Signal Process., **12**, 119–141
8. Carl, H. (1994) Examination of Different Speech-Coding Methods and an Application to Band-Width Enhancement for Narrow-Band Speech Signals (in German). Diss. Ruhr-Univ., Bochum.
9. Carl, H., Heute, U. (1994) Bandwidth Enhancement of Narrow-Band Speech Signals. Proc. EUSIPCO, Edinburgh, 1716–1719
10. Ehnert, W., Heute, U. (1997) Variable-Rate Speech Coding: Replacing Unvoiced Excitations by Linear-Prediction Residues of Different Phonemes. Proc. GRETSI, Grenoble, 993–996
11. Erdmann, Ch., Vary, P., et al. (2001) A Candidate Proposal for a 3GPP Adaptive Multi-rate Wide-Band Speech Codec. Proc. IEEE Int. Conf. Acoustics Speech Signal Process., Salt Lake City, **II**, 757–760
12. ETSI (1999) Rec. GSM 06.10: Digital Cellular Telecommunication System (phase 2+) (GSM); Full-Rate Speech Transcoding
13. ETSI (1999) Rec. GSM 06.20: Digital Cellular Telecommunications System (phase 2+) (GSM); Half-Rate Speech Transcoding
14. ETSI (1999) Rec. GSM 06.60: Digital Cellular Telecommunications System (phase 2+) (GSM); Enhanced Full-Rate (EFR) Speech Transcoding
15. ETSI (1999) Rec. AMR 06.90: Digital Cellular Telecommunications System: Adaptive Multi-rate (AMR) Speech Transcoding
16. ETSI (1999) Rec. GSM 05.03: Digital Cellular Telecommunications System (phase 2+) (GSM): Channel Coding
17. ETSI (1999) Rec. GSM 06.11: Digital Cellular Telecommunications System (phase 2+) (GSM): Substitution and Muting of Lost Frames for Full-Rate Speech Channels
18. ETSI (1999) Rec. 06.21: Digital Cellular Telecommunications System (phase 2+) (GSM): Substitution and Muting of Lost Frames for Half-Rate Speech Traffic Channels
19. ETSI (1999) Rec. 06.61: Digital Cellular Telecommunications System (phase 2+) (GSM): Substitution and Muting of Lost Frames for Enhanced Full-Rate (EFR) Speech Traffic Channels

20. ETSI (1998) Rec. 06.91: Digital Cellular Telecommunications System (phase 2+) (GSM): Substitution and Muting of Lost Frames for Adaptive Multi-rate (AMR) Speech Traffic Channels
21. Farvardin, N., Vaishampayan, V. (1991) On the Performance and Complexity of Channel-Optimized Vector Quantizers. IEEE Trans. Inform. Theory, **37**, 155–160
22. Fazel, T., Fuja, T. (2003) Robust Transmission of MELP-Compressed Speech: An Illustrative Example of Joint Source-Channel Decoding. IEEE Trans. Commun., **51**, 973–982
23. Fingscheidt, T., Vary, P. (1997) Robust Speech Decoding: A Universal Approach to Bit Error Concealment. Proc. IEEE Int. Conf. Acoust. Speech Signal Process., **3**, 1667–1670
24. Gluth, R., Guendel, L., Heute, U. (1986) ATC – A Candidate for Digital Mobile-Radio Telephony. Proc. Nordic Semin. Dig. Land-Mob. Rad. Commun., Stockholm, 230–235
25. Görtz, N. (1999) Joint Source-Channel Decoding by Channel-Coded Optimal Estimation (CCOE) for a CELP Speech Codec. Proc. EUROSPEECH, **3**, 1447–1450
26. Görtz, N. (2000) Joint Source-Channel Decoding by Channel Coded Optimal Estimation. Proc. 3rd ITG Conf. Source Channel Cod., VDE-Verlag, 267–272
27. Görtz, N. (2001) On the Iterative Approximation of Optimal Joint Source-Channel Decoding. IEEE J. Select. Areas Commun., **19**, 1662–1670
28. Görtz, N. (2003) Optimization of Bit-Mappings for Iterative Source-Channel Decoding. Proc. Int. Symp. Turbo Codes Relat. Top., 255–258
29. Görtz, N., Heute, U. (2000) Joint Source-Channel Decoding with Iterative Algorithms. Proc. EUSIPCO, **1**, 425–428
30. Görtz, N., Kliewer, J. (2003) Memory-Efficient Adaptation of Vector Quantizers to Time-Varying Channels. EURASIP Signal Process., **83**, 1519–1528
31. Gray, R.M. (1984) Vector Quantization. IEEE ASSP Mag., **1**, no. 2, 4–29
32. Grill, B., Edler, B., et al. (1998) Information Technology – Very Low Bitrate Audio-Visual Coding, Part 3: Audio. ISI/IEC FCD 14496–3, Subpart 1
33. Hagenauer, J. (1995) Source-Controlled Channel Decoding. IEEE Trans. Commun., **43**, 2449–2457
34. Hagenauer, J., Offer, E., Papke, L. (1996) Iterative Decoding of Binary Block and Convolutional Codes. IEEE Trans. Inform. Theory, **42**, 429–445
35. Halle, M., Stevens, N.K. (1959) Analysis by Synthesis. Proc. Semin. Inform. Theory Speech Compression **II**, paper no. D7
36. Heinen, S., Geiler, A., Vary, P. (1998) MAP Channel Decoding by Exploiting Multilevel Source A-Priori Knowledge. Proc. ITG-Fachtgg. "Cod. für Quelle, Kanal und Übertrgg.", 89–94
37. Hess, W. (1983) Pitch Determination of Speech Signals. Springer, Berlin

38. Heute, U. (1988) Medium-Rate Speech Coding – Trial of a Review. Speech Commun., **7**, 125–149
39. Heute, U. (1994) Speech Coding: Approaches, Trends, Standards (in German). Proc. ITG Conf. Source Channel Cod., Munich, 437–448
40. Hindelang, T., Kaindl, M., Hagenauer, J., Schmauz, M., Xu, W. (2000) Improved Channel Coding and Estimation for Adaptive Multirate (AMR) Speech Transmission. Proc. IEEE 51st Vehic. Technol. Conf., 1210–1214
41. Kleijn, W.B., Paliwal, K.K. (Editors) (1995) Speech Coding and Synthesis. Elsevier
42. Itakura, F. (1975) Line-Spectral Representation of Linear-Prediction Coefficients of Speech Signals. JASA, **57**, Suppl. S35A
43. ITU/CCITT (1988) Recommendation G.711: "Coding of Analogue Signals by Pulse-Code Modulation (PCM) of Voice-Frequencies"
44. ITU-T (1988) Recommendation G.722: "7 kHz Audio Coding within 64 kbit/s. Fascicle III.4, Blue Book, 269–341
45. ITU-T (1992) Recommendation G.726: "40,32,24,16 kbit/s Adaptive Differential Pulse-Code Modulation (ADPCM)"
46. ITU-T (1992) Recommendation G.728: "Coding of Speech at 16 kbit/s Using Low-Delay Code-Excited Linear Prediction"
47. ITU-T (1995) Recommendation G.729: "Coding of Speech at 8 kbit/s Using Conjugate-Structure Algebraic Code-Excited Linear Prediction (CS-ACELP)"
48. Jafarkhani, H., Farvardin, N. (2000) Design of Channel-Optimized Vector Quantizers in the Presence of Channel Mismatch. IEEE Trans. Commun., **4**, 118–124
49. Jayant, N.S., Noll, P. (1984) Digital Coding of Waveforms. Prentice Hall, Englewood Cliffs, NJ
50. Kleijn, B., Haagen, J. (1995) A Speech Coder Based on Decomposition of Characteristic Waveforms. Proc. IEEE Int. Conf. Acoustics, Speech, Signal Process., Detroit, 508–511
51. Kumazawa, H., Kasahara, M., Namekawa, T. (1984) A Construction of Vector Quantizers for Noisy Channels. Electron. Eng. Jpn, **67-B**, 39–47
52. Kondoz, A. (1994) Digital Speech for Low Bit Rate Communication Systems. Wiley
53. Lacroix, A. (2004) Speech-Production – Acoustics, Models, and Applications. In: Blauert, J. (Ed.): Communication Acoustics, Chapter 13. Springer, Berlin Heidelberg.
54. Liu, F.H., Ho, P., Cuperman, V. (1993) Joint Source and Channel Coding Using a Non-linear Receiver. Proc. IEEE Int. Conf. Commun., **3**, 1502–1507
55. Lazzari, V., Montagna., R., Sereno, D. (1988) Comparison of Two Speech Codecs for DMR Systems. Speech Commun., **7**, 193–207
56. Lin, S., Costello, D.J. (2nd edition 2004) Error Control Coding. Pearson Prentice-Hall
57. Linde, Y., Buzo, A., Gray, R.M. (1980) An Algorithm for Vector-Quantizer Design. IEEE Trans. Commun. **28**, 84–95

58. Max, J. (1960) Quantizing for Minimum Distortion. IRE Trans. Inform. Theory, **6**, 7–12
59. Miller, D.J., Park, M. (1998) A Sequence-Based Approximate MMSE Decoder for Source Coding over Noisy Channels Using Discrete Hidden Markov Models. IEEE Trans. Commun., **46**, 222–231
60. Noll, P., Zelinski, R. (1977) Adaptive Transform Coding of Speech Signals. IEEE Trans. Acoustics Speech Signal Process., **25**, 299–309
61. Noll, P. (1993) Wideband-Speech and Audio Coding. IEEE Commun. Mag., **Nov.**, 34–44
62. O'Shaughnessy, D. (1987) Speech Communication – Human and Machine. Addison-Wesley, New York
63. Paliwal, K.K., Atal, B.S. (1993) Efficient Vector Quantization of LPC Parameters at 24 bits/frame. IEEE Trans. Speech Audio Process., **1**, 3–14.
64. Paulus, J. (1997) Coding of Wide-Band Speech Signals at Low Data Rates (in German). Diss. RWTH, Aachen
65. Perkert, R., Kaindl, M., Hindelang, T. (2001) Iterative Source and Channel Decoding for GSM. Proc. IEEE Int. Conf. Acoustics Speech Signal Process., **4**, 2649–2652
66. Phamdo, N., Farvardin, N. (1994) Optimal Detection of Discrete Markov Sources over Discrete Memoryless Channels–Applications to Combined Source-Channel Coding. IEEE Trans. Inform. Theory, **40**, 186–193
67. Rabiner, L.R., Schafer, R.W. (1978) Digital Processing of Speech Signals. Prentice-Hall, Englewood Cliffs, NJ
68. Sayood, K., Borkenhagen, J.C. (1991) Use of Residual Redundancy in the Design of Joint Source/Channel Coders. IEEE Trans. Commun., **39**, 838–846
69. Schroeder, L.R., Atal, B.S. (1985) Code-Excited Linear Prediction (CELP): High-Quality Speech at Very Low Bit Rates. Proc. IEEE Int. Conf. Acoustics Speech Signal Process., Paris, **I**, 25.1.1–25.1.4.
70. Skoglund, M. (1999) Soft Decoding for Vector Quantization over Noisy Channels with Memory. IEEE Trans. Inform. Theory, **45**, 1293–1307
71. Spanias, A.S. (1994) Speech Coding: A Tutorial Review. Proc. IEEE **82**, 1541–1582
72. Thyssen, J., et al. (2001) EX-CELP: A Speech-Coding Paradigm. Proc. IEEE Int. Conf. Acoustics Speech Signal Process., Salt Lake City, **II**, 689–692
73. Thyssen, J., et al. (2001) A Candidate for the ITU-T 4 kbit/s Speech-Coding Standard. Proc. IEEE Int. Conf. Acoustics Speech Signal Process., Salt Lake City, **II**, 681–684
74. Trancoso, I.M., et al. (1988) Harmonic Coding – State of the Art and Future Trends. Speech Commun., **7**, 239–245
75. Tremain T.E. (1982) The Government Standard Linear-Predictive Coding: LPC-10. Speech Technol., **1**, 40–49
76. Varga, I. (2001) Standardization of the Adaptive Multi-rate Wideband Codec. Proc. ITG Conf. Source Channel Cod., Berlin, 341–346

77. Vary, P., Heute, U., Hess, W. (1998) Digitale Sprachsignalverarbeitung. Teubner, Stuttgart
78. Zeger, K., Gersho, A. (1998) Vector Quantizer Design for Memoryless Noisy Channels. Proc. IEEE Int. Conf. Commun., **3**, 1593–1597
79. Zwicker, Fastl (1990). Psychoacoustics – Facts and Models. Springer, Berlin Heidelberg

46

Echo Cancellation

Walter Kellermann

Multimedia Communications and Signal Processing, University Erlangen-Nuremberg, Erlangen, Germany, e-mail: wk@lnt.de

1 • Echo Cancellation and Suppression

Echoes as perceived with speech communication terminals in telecommunication networks typically result from two sources (Figure 1): Firstly, the so-called *electrical echoes* or *line echoes* are usually caused by hybrid circuits, as employed for connecting four-wire links and two-wire links [1]; secondly, *acoustic echoes* result from acoustic feedback from the loudspeakers into the microphones of the terminal equipment with widely varying feedback gain depending on the terminal and the acoustic scenario. In both cases, such echoes should be removed from the signal $x(n)$ that is transmitted to the far-end terminal. The signal $x(n)$ can generally be written as a sum of three components

$$x(n) = x_s(n) + x_v(n) + x_r(n), \qquad (1)$$

where in the acoustic echo scenario, $x_s(n)$ represents the component due to the local talker's original speech $s(n)$, $x_v(n)$ is the loudspeaker signal echo and $x_r(n)$ results from noise sources $r(n)$ in the acoustic environment. In the electrical echo scenario, $x_r(n)$ is often neglected and $x_s(n)$ corresponds to the speech from the local terminal. While *echo suppression* aims at attenuating the residual echo $x_v(n)$ within $z(n)$ by a time-varying gain or filtering of $y(n)$, *echo cancellation* aims at creating an echo estimate $\hat{x}_v(n)$

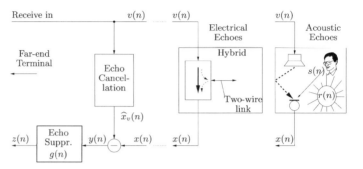

FIGURE 1 Cancellation and suppression of line echoes and acoustic echoes.

which minimizes $e(n) = x_v(n) - \hat{x}_v(n)$ subject to a given error criterion. Obviously, $e(n)$ is identical to y and as such can be directly observed if $x_s(n) = x_r(n) = 0$ can be assumed.

1.1 Echo Suppression

In its simplest version, an echo suppressor operates as a voice-activated switch: Whenever the far-end speaker is active, the echo suppressor applies zero gain $g(n) = 0$, so that feedback of $x_v(n)$ in $z(n)$ is completely suppressed. Obviously, this reduces full-duplex communication to a half-duplex mode, as during *double-talk*, i.e. when both $v(n)$ and $s(n)$ are active, the gain must be zero and the desired signal $s(n)$ will not be transmitted. To mitigate this undesirable effect, the gain control will aim at a compromise between attenuation of the echo $x_v(n)$ and gain for the desired signal component $x_s(n)$. Typically, short-time estimates for the levels of $x_v(n)$, $x_s(n)$ and $x_r(n)$ are extracted from $x(n)$ to support a proper choice of $g(n)$. For this, short-term variation of the speech signal's envelope, stationarity assumptions for the background noise $r(n)$ and correlation between $v(n)$ and $x_v(n)$ can be exploited. Fast and reliable detection of double-talk in $x(n)$ is crucial to subjective quality to avoid suppression of initial low-energy fricatives in $x_s(n)$, which is especially difficult for noisy signals, i.e. with high levels of $x_r(n)$. Besides linear gain control, nonlinear processing for echo suppression has also been introduced in various forms, most notably, centre-clipping [2]. Over the years, the control of echo suppression has reached a high level of sophistication to overcome its structural deficiencies [3]. As a common refinement in current systems, the gain control $g(n)$ is applied individually in frequency subbands or DFT (*Discrete Fourier Transform*) bins, especially for *residual echo suppression*, i.e. as a complement to acoustic echo cancellation (see below). For that, not only the above traditional echo suppression techniques but also the advanced noise reduction schemes (see below) are employed, handling residual echoes with similar techniques as background noise [4,3].

1.2 Echo Cancellation

Echo cancellation is commonly viewed as a system identification problem where the echo path is mostly modelled by a linear finite impulse response (FIR) filter.

For electrical echoes, the duration of the corresponding impulse reponse varies strongly, mostly depending on the delays for the hybrids where the echoes are caused. With an increasing amount of digital signal processing introducing extra delay in modern telecommunication networks, link delays are increasing and, in addition to the long-standing satellite links with one-way delays in the order of 250 ms, more and more links involve large delays, especially if packet-switched networks and mobile phones are involved. As the echo path of the actual hybrid is well modelled by a relatively short impulse response (covering not more than 50 ms), the impulse response modelling an entire link is usually sparse, i.e. many samples exhibit only small values.

The acoustic echo path is essentially determined by the reverberation in the local acoustic environment, which is commonly captured by the reverberation time T_{60}, which is the time interval for the sound level to decay by 60 dB in a room after the source is muted [5]. This means for attenuating the acoustic echo by x dB, roughly $L = x \cdot f_s \cdot T_{60}/60$ FIR filter coefficients are needed, where f_s is the sampling frequency [3]. Thus, in a well-damped office room, e.g. $L = 25 \cdot 8\,\text{kHz} \cdot 300\,\text{ms}/60 = 1,000$ coefficients are needed for basic telephone bandwidth if 25 dB echo attenuation should be achievable (Figure 2).

1.2.1 The System Identification Problem

The identification of a linear FIR filter with impulse response $\hat{h}_k(n)$ of length L, with $k = 0,\ldots,L-1$ indexing the impulse response samples, is commonly solved via minimizing a mean square error criterion[1]

$$\varepsilon\{e^2(n)\} \to \min, \qquad (2)$$

where ε is the expectation operator and

$$e(n) = \hat{h}_k(n) \times v(n) - x_v(n). \qquad (3)$$

---

[1]For notational convenience, only real-valued signals are considered here, the straightforward generalization to complex signals is given, e.g. in [6].

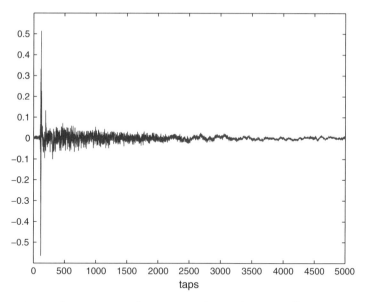

FIGURE 2 Impulse response of acoustic echo path in an office (5.5 m × 3 m × 2.8 m, $T_{60} \approx 300$ ms, sampling frequency $f_s = 12$ kHz).

Assuming (at least wide-sense) stationary processes $v(n), x(n)$ and a time-invariant impulse response $\hat{h}_k(n) = \hat{h}_k$, the solution is given by the Wiener filter [6]

$$\hat{\mathbf{h}} = \mathbf{R}_{vv}^{-1} \mathbf{r}_{xv} \qquad (4)$$

where $\hat{\mathbf{h}} = \left(\hat{h}_0, \ldots, \hat{h}_{L-1}\right)^T$ is the vector of the FIR filter coefficients, $\mathbf{R}_{vv}$ is the $L \times L$ autocorrelation matrix and $\mathbf{r}_{xv}$ is the $L \times 1$ crosscorrelation vector, with elements $\mathbf{R}_{vv,ij} = \varepsilon\{v(n-i)v(n-j)\}$ and $\mathbf{r}_{xv} = \varepsilon\{x(n)v(n-j)\}(i,j = 0, 1, \ldots, L-1)$, respectively $((\cdot)^T$ denotes transposition).

In the given context, no signal ensembles for computing expectations are available, but only single finite observations, which requires deterministic algorithms. Corresponding to Eq. 2, we formulate a least squares criterion as the weighted squared error signal $e^2(n)$ for a set of observable samples:

$$J(n) = \sum_{k=0}^{n} w(k,n) e^2(n-k) \rightarrow \min. \qquad (5)$$

Here, $w(k,n)$ denotes a windowing function with $0 \leq w(k,n) \leq 1$. This leads to a least squares solution of the form [6,7]

$$\hat{\mathbf{h}}(n) = \hat{\mathbf{R}}_{vv}^{-1}(n) \hat{\mathbf{r}}_{xv}(n), \qquad (6)$$

where the required estimates $\hat{\mathbf{R}}_{vv}(n)$, $\hat{\mathbf{r}}_{xv}(n)$ obviously depend on the window $w(k,n)$ in Eq. 5.

In the echo cancellation context, one has to account for the time-variance of the echo path, which especially for acoustic echoes may be drastic and unpredictable. Thus, *adaptive filters* are employed which try to identify the above optimum filters, by using update equations of the form:

$$\hat{\mathbf{h}}(n+1) = \hat{\mathbf{h}}(n) + \Delta\hat{\mathbf{h}}(n). \qquad (7)$$

For defining the update term, two philosophies are in common use: For one, gradient techniques iteratively approximate the minimum of the cost function, while others directly aim at the optimum solution at time n as given by Eq. 6.

For the well-known and widely studied *Least Mean Square (LMS) algorithm* [8,6], the update term $\Delta\hat{\mathbf{h}}(n)$ is obtained as the negated gradient of Eq. 5 for $w(k,n) = \delta(k)$ (with $\delta(k)$ as the unit impulse [9]), multiplied by a stepsize $\mu/2$, and reads with $\mathbf{v}(n) = (v(n),\ldots,v(n-L+1))^T$

$$\hat{\mathbf{h}}(n+1) = \hat{\mathbf{h}}(n) + \mu\mathbf{v}(n)e(n). \qquad (8)$$

This algorithm was suggested already in the first publication on acoustic echo cancellation [10] and its normalized version, the *Normalized Least Mean Square (NLMS) algorithm* ($\mu = \alpha/(\mathbf{v}^T(n)\mathbf{v}(n))$, $0 \leq \alpha \leq 2$) is still the most commonly used algorithm in real-world applications:

$$\hat{\mathbf{h}}(n+1) = \hat{\mathbf{h}}(n) + \alpha\frac{\mathbf{v}(n)e(n)}{\mathbf{v}^T(n)\mathbf{v}(n)}. \qquad (9)$$

The popularity of the NLMS algorithm is based not only on its simplicity and low complexity (roughly $2L$ multiplications/additions per sampling interval) but also on its robustness against changing signal statistics. The price for the simplicity and robustness of the LMS/NLMS-type algorithms is a relatively slow convergence for correlated signals $v(n)$, such as speech in the given context. Among the many variations of the LMS algorithms, the *Proportionate NLMS* algorithm [11,7] is of special interest for sparse impulse responses (as common for electrical echoes) because it allows for fast adaptation of the largest coefficients of the impulse response.

On the other hand, the *Recursive Least Squares (RLS) algorithm*[12] guarantees the optimum solution of Eq. 6 at each time instant n. Here, the update term recursively incorporates new observations $v(n)$ and $e(n)$ and adapts the solution to the changing cost function value $J(n)$. Choosing an exponential weighting, $w(k,n) = \lambda^{n-k}, \lambda < 1$, the update equation can be written as [6]:

$$\hat{\mathbf{h}}(n+1) = \hat{\mathbf{h}}(n) + \hat{\mathbf{R}}_{vv}^{-1}(n)\mathbf{v}(n)e(n), \qquad (10)$$

with

$$\hat{\mathbf{R}}_{vv}(n) = \lambda \hat{\mathbf{R}}_{vv}^{-1}(n-1) + v(n)\mathbf{v}^T(n). \qquad (11)$$

Although the inversion of the autocorrelation matrix at each sampling instant n is usually circumvented by the matrix inversion lemma [6], the computational complexity of the RLS is still of $\mathcal{O}(L^2)$ which is prohibitive for long FIR filters. The so-called *Fast RLS* algorithms or *Fast Transversal Filters* (*FTF*) exploit the transversal filter structure versions to further reduce computational load [12]. However, robustness is a crucial issue for RLS algorithms: Regularization for $\hat{\mathbf{R}}_{vv}(n)$ or its inverse has to account for strongly correlated signals (leading to a large eigenvalue spread) as well as for speech pauses, and, moreover, the choice of λ needs to balance tracking capability versus stability of convergence [13].

The *Affine Projection (AP) algorithm* [14] can be viewed as a compromise between LMS-type algorithms and RLS algorithms. The detrimental effect of signal correlation on LMS-type algorithms, as, e.g. given for autoregressive (AR) processes $v(n)$ such as speech, is removed by the AP algorithm up to a given AR process order K at the cost of an increased computational complexity of $\mathcal{O}(KL)$. As with the RLS algorithm, fast versions have been proposed [15,16], which require additional stabilization measures.

To reconcile the requirements of computational efficiency and fast and robust convergence for long adaptive FIR filters, a transformation of the filter adaptation in Eq. 10 seems attractive so that ideally the inversion of the autocorrelation matrix reduces to the inversion of a diagonal matrix. Obviously, the signal-dependent Karhunen–Loève transform [6] would be an ideal choice. In practice, signal-independent transforms into a frequency subband domain or into the DFT domain proved to be reasonably close and efficient approximations. Two classes of practically relevant concepts resulted from this idea:

Frequency subband filtering. The oversampled frequency subband approach [17,18,19] decomposes $v(n)$ and $x(n)$ into N frequency subbands (Figure 3) such that aliasing after downsampling by a factor $R < N$ in each subband is sufficiently small while overall analysis/synthesis filterbanks are still sufficiently close to

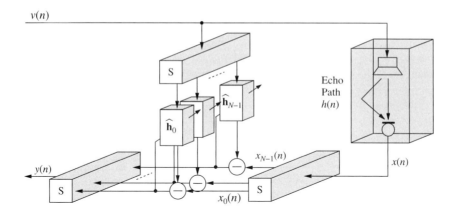

FIGURE 3 Frequency subband acoustic echo cancellation.

perfect reconstruction [20]. Then, adaptive subband filters can operate independently in each frequency subband. These filters are shorter than the equivalent fullband filter due to the reduced sampling rate and also assure a faster and more robust overall convergence for low-complexity adaptation algorithms such as the NLMS [17]. A significant computational gain is achieved over the corresponding fullband system due to the reduced sampling rate by R in the subbands, which implies both reduced clock rate for filtering and reduced filter length. In principle, all adaptive filtering algorithms for fullband signal processing can be implemented in each subband, with the additional advantage that data can be shared among different subbands, if desired, e.g. for control strategies. The filtering cost for analysis and synthesis is minimized if efficient uniform filterbanks based on polyphase structures [20] are used so that only the additional signal delay due to the filterbanks remains as a potential disadvantage in practical applications.

Frequency-Domain Adaptive Filtering (FDAF). This second class of transform-domain adaptive filtering is directly based on the DFT [21,22] and was initially motivated by the advantages of performing linear filtering by multiplication of DFT bins instead of convolving time-domain sequences (*Fast Convolution*). This computational advantage carries over to adaptive filtering and is accompanied by an almost-diagonalization of the autocorrelation matrix $\hat{\mathbf{R}}_{vv}$ in the update equation, Eq. 10, if an RLS-based formulation is used [23,7]. The block delay associated to processing successive DFT data frames can be reduced at the cost of increased computational load by using a large overlap of the DFT frames or by using *multidelay filters (MDF)* [24,25] (also,

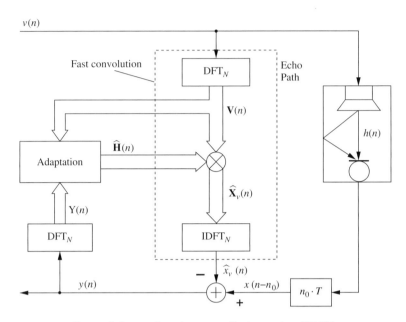

FIGURE 4 Acoustic echo cancellation using FDAF.

partitioned block frequency-domain adaptive filters (PBFDAF)), where the impulse response of the echo path model is partitioned into several segments, so that the DFT length is reduced. As a generalization to the MDF, the *extended multidelay filter (EMDF)* accounts for the signal correlation between partitions and thereby approximates both the low complexity and the fast convergence of the FDAF without the large delay, which is especially attractive for very high-order echo path models [26] (Figure 4).

1.2.2 Controlling the Adaptation

Using the above adaptive filtering algorithms, the echo path can only be identified as long as the signal components $x_s(n)$ and $x_r(n)$ do not interfere significantly with the adaptation, so that $y(n) = x(n) - \hat{x}_v(n)$ is essentially equal to the system identification error $e(n)$. However, typically permanent background noise $x_r(n)$ and intermittent and unpredictable local talker activity $x_s(n)$ call for additional control measures. Aside from adaptation under permanent noise, the distinction between a change of the echo path and the onset of local talker activity, which both lead to large signal values for $x(n)$, presents the main problem: In the first case, adaptation should be aggressive, whereas in the second case, it should be stalled. This so-called *double-talk detection* problem [23,7] can be addressed by deriving some estimate for the short-time crosscorrelation between $v(n)$ and

$x(n)$, and by a two-path model which compares the effect of the updated filter to the previous version [27]. For limiting divergence in case of detection errors, robust statistics has been introduced as an elegant and effective extension to various of the above adaptive filtering algorithms [23,7,28]. On the other hand, especially for (N)LMS-based algorithms empirical stepsize control methods for maximizing convergence speed while minimizing the risk of divergence have been developed to a high level of sophistication [3].

1.2.3 Complementing the Linear FIR Filter Model

The adaptive FIR filter will in general not be able to model the echo path perfectly at all times, not only because of limited tracking capability of the adaptation algorithms but also because the implemented filters may be too short, or the echo path is not strictly linear. Often, an additional echo suppression mechanism is implemented for the residual echo (see above). In case of severe nonlinearities in the echo path as they occur, e.g. in mobile telephone terminals due to saturated amplifiers and overloaded loudspeakers, efficient nonlinear echo path models are desirable [29,30,31].

1.3 Multichannel Acoustic Echo Cancellation

Since stereo reproduction has been invented, users appreciate multichannel sound reproduction. For hands-free telecommunications and human/machine dialogues, this introduces additional fundamental problems for acoustic echo cancellation, as is illustrated for the stereo case in Figure 5. Obviously, each of the microphones at the local terminal simultaneously records echoes of both loudspeaker signals $v_1(n), v_2(n)$. To identify the two echo path models $\hat{\mathbf{h}}_1(n), \hat{\mathbf{h}}_2(n)$ with only one error signal derived from $x_1(n)$ requires that $v_1(n)$ and $v_2(n)$ contain uncorrelated components [32,33]. In fact, the optimum stereo acoustic echo cancellation system can be written in the same way as Eq. 6 with the input data and filter coefficient vectors of the single-channel case replaced by a concatenation of the two-channel input data and filter coefficient vectors for the two channels, e.g. $\mathbf{v}(n) = (\mathbf{v}_1^T(n), \mathbf{v}_2^T(n))^T$. Accordingly, all the above adaptive filtering algorithms can be formulated for the P-channel case [7]. The usually strong crosscorrelation between the reproduction channel leads however to a much greater matrix condition number for the resulting $PL \times PL$ autocorrelation matrix $\mathbf{R_{vv}}$ which now also contains the crosscorrelations between the channels in the off-diagonal $L \times L$ block matrices. In the typical stereo communication scenario of Figure 5, the two reproduction channels are recorded from a single source, so that the microphone signals $v'_1(n), v'_2(n)$ are essentially convolutional products

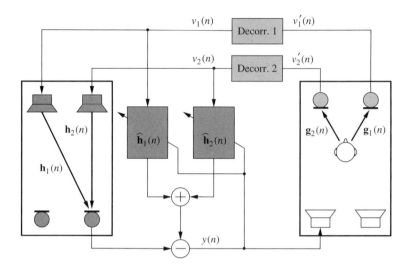

FIGURE 5 Scenario for stereo acoustic echo cancellation.

derived from a single signal source and therefore strongly correlated. To reduce this correlation, time-varying filtering, nonlinear distortion and/or injection of noise can be applied to $v'_1(n), v'_2(n)$, which must be different for each channel. Obviously, the cost is some distortion in the signals $v_1(n), v_2(n)$ which should be minimized. At the same time, this distortion should provide significant energy for uncorrelated components to sufficiently support the system identification. For the given options, time-varying allpass filtering [34] must be used diligently if spatial realism should not be impaired, noise injection should be limited to the masking thresholds of the auditory system [35] and nonlinear distortion should also remain inaudible [36].

For computationally inexpensive time-domain algorithms like the NLMS, the increased condition number for $\mathbf{R}_{vv}$ directly reduces the convergence speed. Thus, the demand for diagonalization of the correlation matrix becomes even stronger for multichannel systems, while at the same time the number of filter coefficients is multiplied by the channel number, so that an RLS implementation in the time-domain becomes increasingly unrealistic. As a consequence, both the frequency subband concept and the DFT domain approach become even more attractive and have been studied extensively [7, chapters 5–9]. While the subband approach has been refined to product maturity for the stereo teleconferencing application [7], the multichannel FDAF algorithm has been practically verified even for surround systems with five reproduction channels [37].

References

1. M.M. Sondhi and D.A. Berkley. Silencing echoes on the telephone network. *Proceedings of the IEEE*, 68(8):948–963, August 1980.
2. O.M.M. Mitchell and D.A. Berkley. A full-duplex echo suppressor using center-clipping. *Bell System Technical Journal*, 50(3): 1619–1630, May/June 1971.
3. E. Hänsler and G. Schmidt. *Acoustic Echo and Noise Control: A Practical Approach*. Wiley, Hoboken, NJ, 2004.
4. S. Gustafsson, R. Martin, and P. Vary. Combined acoustic echo cancellation and noise reduction for hands-free telephony. *Signal Processing*, 64(1):21–32, 1998.
5. H. Kuttruff. *Room Acoustics*. Spon Press, London, 4th edition, 2000.
6. S. Haykin. *Adaptive Filter Theory*. Prentice Hall, Upper Saddle River, NJ, 4th edition, 2002.
7. J. Benesty, T. Gänsler, D.R. Morgan, M.M. Sondhi, and S.L. Gay. *Advances in Network and Acoustic Echo Cancellation*. Springer, Berlin, 2001.
8. B. Widrow and M.E. Hoff. Jr. Adaptive switching circuits. In *IRE Wescon Conf. Rec., Part 4*, pp. 96–104, 1960.
9. A.V. Oppenheim and R.W. Schafer. *Digital Signal Processing*. Prentice Hall, Englewood Cliffs, NJ, 1975.
10. M.M. Sondhi. An adaptive echo canceler. *Bell System Technical Journal*, 46:497–511, March 1967.
11. D.L. Duttweiler. Proportionate normalized least-mean-squares adaptation in echo cancelers. *IEEE Transactions on Speech and Audio Processing*, 8(5):508–518, September 2000.
12. J.M. Cioffi and T. Kailath. Fast, recursive-least-squares transversal filters for adaptive filtering. *IEEE Transactions on Acoustics, Speech, and Signal Processing*, 32(2):304–337, April 1984.
13. D.T.M. Slock and T. Kailath. Numerically stable fast transversal filters for recursive least squares adaptive filtering. *IEEE Transactions on Signal Processing*, 39(1):92–114, January 1991.
14. K. Ozeki and T. Umeda. An adaptive filtering algorithm using an orthogonal projection to an affine subspace and its properties. *Electronic Communications Japan*, 67-A(5):19–27, January 1984.
15. M. Tanaka, Y. Kaneda, S. Makino, and J. Kojima. Fast projection algorithm and its step size control. In *Proceedings of Internation Conference on Acoustics, Speech, and Signal Processing*, pp. 945–948, Detroit, MI, May 1995. IEEE.
16. S.L. Gay and S. Tavathia. The fast affine projection algorithm. In *Proceedings of International Conference on Acoustics, Speech, and Signal Processing*, pp. 3023–3026, Detroit, MI, May 1995. IEEE.
17. W. Kellermann. Analysis and design of multirate systems for cancellation of acoustical echoes. In *Proceedings of International Conference on Acoustics, Speech, and Signal Processing*, pp. 2570–2573, New York, April 1988. IEEE.
18. M.M. Sondhi and W. Kellermann. Echo cancellation for speech signals. In S. Furui and M.M. Sondhi, editors, *Advances in*

Speech Signal Processing, chapter 11, pp. 327–356. Marcel Dekker, New York, 1991.
19. S. Gay and J. Benesty, editors. *Acoustic Signal Processing for Telecommunications*. Kluwer, Norwell, MA, 2000.
20. P.P. Vaidyanathan. *Multirate System and Filter Banks*. Prentice Hall, Englewood Cliffs, NJ, 1993.
21. D. Mansour and A.H. Gray Jr. Unconstrained frequency domain adaptive filter. *IEEE Transactions on Acoustics, Speech, and Signal Processing*, 30(5):726–734, October 1982.
22. J.J. Shynk. Frequency-domain and multirate adaptive filtering. *IEEE Signal Processing Magazine*, 9(1):14–37, January 1992.
23. S.L. Gay and J. Benesty, editors. *Acoustic Signal Processing for Telecommunication*. Kluwer, Boston, MA, 2000.
24. J.-S. Soo and K.K. Pang. Multidelay block frequency domain adaptive filter. *IEEE Transactions on Acoustics, Speech, and Signal Processing*, 38(2):373–376, February 1990.
25. E. Moulines, O.A. Amrane, and Y. Grenier. The generalized multi-delay adaptive filter: Structure and convergence analysis. *IEEE Transactions on Signal Processing*, 43(1):14–28, January 1995.
26. H. Buchner, J. Benesty, and W. Kellermann. An extended multi-delay filter: Fast low-delay algorithms for very high-order adaptive systems. In *Proceedings of International Conference on Acoustics, Speech, and Signal Processing*, pp. 385–388, Hongkong, China, April 2003 IEEE.
27. K. Ochiai, T. Araseki, and T. Ogihara. Echo canceler with two echo path models. *IEEE Transactions on Communications*, 25(6):589–595, June 1977.
28. H. Buchner, J. Benesty, and W. Kellermann. An outlier-robust extended multidelay filter with application to acoustic echo cancellation. In *Proceedings of International Workshop on Acoustic Echo and Noise Control*, pp. 19–22, Kyoto, Japan, September 2003.
29. A. Stenger and W. Kellermann. Adaptation of a memoryless preprocessor for nonlinear acoustic echo cancelling. *Signal Processing*, 80:1747–1760, Sep 2000.
30. F. Küch and W. Kellermann. Nonlinear line echo cancellation using a simplified second-order Volterra filter. In *Proceedings of International Conference on Acoustics, Speech, and Signal Processing*, Orlando, FL, May 2002. IEEE.
31. F. Kuech and W. Kellermann. Partitioned block frequency-domain adaptive second-order volterra filter. *IEEE Transactions on Signal Processing*, 53(2):564–575, February 2005.
32. M.M. Sondhi, D.R. Morgan, and J.L. Hall. Stereophonic echo cancellation: An overview of the fundamental problem. *IEEE Signal Processing Letters*, 2(8):148–151, August 1995.
33. J. Benesty, D.R. Morgan, and M.M. Sondhi. A better understanding and an improved solution to the specific problems of stereophonic acoustic echo cancellation. *IEEE Transactions on Speech and Audio Processing*, 6(2):156–165, March 1998.

34. Y. Joncours and K. Sugiyama. A stereo echo canceller with preprocessing for correct echo-path identification. In *Proceedings of International Conference on Acoustics, Speech, and Signal Processing*, pp. 3677–3680, Seattle, WA, 1998. IEEE.
35. T. Gänsler and P. Eneroth. Influence of audio coding on stereophonic acoustic echo cancellation. In *Proceedings of International Conference on Acoustics, Speech, and Signal Processing*, pp. 3649–3652, Seattle, WA, 1998. IEEE.
36. D.R. Morgan, J.L. Hall, and J. Benesty. Investigation of several types of nonlinearities for use in stereo acoustic echo cancellation. *IEEE Transactions on Speech and Audio Processing*, 9(6), September 2001.
37. H. Buchner, J. Benesty, and W. Kellermann. Multichannel frequency-domain adaptive filtering with application to acoustic echo cancellation. In J. Benesty and Y. Huang, editors, *Adaptive Signal Processing: Application to Real-World Problems*, chapter 3. Springer, Berlin, January 2003.

47

Noise Reduction and Interference Cancellation

Walter Kellermann

Multimedia Communications and Signal Processing, University Erlangen-Nuremberg, Erlangen, Germany, e-mail: wk@LNT.de

Noise reduction and interference cancellation differ structurally from echo cancellation insofar as no undisturbed reference signal (such as the loudspeaker signal for AEC) is available, neither for the desired signal nor for the unwanted noise or interference component. It can be viewed as a signal separation problem where estimates of desired and undesired signal components have to be found and separated from each other.

Based on the number of used sensor (i.e. microphone) signals $x_\nu(n)$, we distinguish between single-channel and multi-channel techniques. With single-channel techniques typically no reliable phase information for the various signal components can be extracted, whereas multichannel techniques implicitly or explicitly exploit phase differences between the various sensor signal components in order to perform spatial filtering or to separate signal components by their different spatial coherence.

1 • Single-Channel Techniques

Assuming a microphone signal $x(n)$ which contains the desired (speech) signal component $s(n)$ plus some unwanted noise or interference $r(n)$ ($x_s(n) = s(n), x_r(n) = r(n)$ in Figure 1 of Chapter 46), we seek an optimum filter for suppressing the noise $r(n)$ without distorting $s(n)$ unduly. If the optimization criterion is the minimum mean square error (MSE) between the filter output and the desired signal $s(n)$ (equivalent to maximum signal-to-noise power ratio (SNR)) for at least weakly stationary und mutually orthogonal signals $s(n)$, $r(n)$, Wiener filter theory [1] yields the corresponding optimum filter, which in its noncausal version is given in the frequency domain for discrete-time signals by

$$H_W(e^{j\omega}) = \frac{S_{ss}(e^{j\omega})}{S_{rr}(e^{j\omega}) + S_{ss}(e^{j\omega})} = 1 - \frac{S_{rr}(e^{j\omega})}{S_{rr}(e^{j\omega}) + S_{ss}(e^{j\omega})}$$

$$= 1 - \frac{S_{rr}(e^{j\omega})}{S_{xx}(e^{j\omega})}. \qquad (1)$$

Here, $S_{xx}(e^{j\omega})$, and $S_{ss}(e^{j\omega})$, and $S_{rr}(e^{j\omega})$ are the power spectral densities (PSDs) of $x(n)$, $s(n)$, and $r(n)$, respectively, and ω is the normalized frequency.

Alternatively, under the same assumptions for $s(n)$, $r(n)$, a spectral subtraction seems reasonable, $S_{ss}(e^{j\omega}) = S_{xx}(e^{j\omega}) - S_{rr}(e^{j\omega})$, to obtain the PSD of the desired speech [2]. Formulating this as a filtering operation yields

$$S_{ss}(e^{j\omega}) = \left(1 - \frac{S_{rr}(e^{j\omega})}{S_{xx}(e^{j\omega})}\right) \cdot S_{xx}(e^{j\omega}) =: |H_{SS}(e^{j\omega})|^2 \cdot S_{xx}(e^{j\omega}), \qquad (2)$$

so that obviously $|H_{SS}(e^{j\omega})| = \sqrt{H_W(e^{j\omega})}$. Integrating both versions, a *generalized Wiener filter* can be defined as follows:

$$H_{WG}(e^{j\omega}) := \left[1 - \left(\frac{S_{rr}(e^{j\omega})}{S_{xx}(e^{j\omega})}\right)^{\beta}\right]^{\alpha}. \qquad (3)$$

Aside from Wiener filtering ($\alpha = 1, \beta = 1$) and PSD subtraction ($\alpha = 0.5, \beta = 1$), this includes also the so-called spectral magnitude subtraction ($\alpha = 1, \beta = 0.5$) [3,4] and forms the basis for several other variations [5].

In noise reduction for speech enhancement, the PSDs $S_{rr}(e^{j\omega})$ and $S_{xx}(e^{j\omega})$ must be estimated on a short-term basis due to the nonstationarity of the speech signal $s(n)$, so that for a length-N data window taken at time n, only the discrete frequencies

$\omega_\nu = 2\pi\nu/N, \nu = 0,\ldots,N-1$, must be considered. For estimating the noise PSD, two methods are well established: Either it is estimated only during speech pauses, which requires voice activity detection, or it is continually estimated by tracking spectral minima [6]. In both cases, the desired noise estimates must be computed by averaging measured short-time PSDs over time, whereas the PSD estimate of $x(n)$ cannot be averaged in the same way, as the nonstationarity of $s(n)$ must be followed. This mismatch of time constants for the PSD estimation leads to the so-called *musical tones*, which are decisive for subjective quality. Usually some amount of noise power suppression is sacrificed to better avoid musical tones. Besides reducing α, β, the introduction of a minimum spectral gain $\hat{H}_{\min}(n, \omega_\nu)$ (*spectral floor*) is common, so that the time-varying estimated generalized Wiener filter can be written in the DFT domain as

$$\hat{H}_{WG}(n, \omega_\nu) = \max \left\{ \left[1 - \left(\frac{\hat{S}_{rr}(n, \omega_\nu)}{\hat{S}_{xx}(n, \omega_\nu)} \right)^\beta \right]^\alpha , \hat{H}_{\min}(n, \omega_\nu) \right\}, \quad (4)$$

where $\hat{S}(n, \omega_\nu)$ represents the νth DFT bin of the corresponding PSD estimate at time n. Obviously, the larger the $\hat{H}_{\min}(n, \omega_\nu)$, the less the noise attenuation.

While the above filters are all based on broadband optimization criteria, another widely used scheme directly minimizes the mean square spectral magnitude error for each DFT bin or its logarithm [7,8] assuming normally distributed DFT bin values for both speech and noise. This approach involves careful smoothing of SNR estimates and, with proper parametrization, has been found to be relatively robust against musical noise.

The optimum choice of the various parameters strongly depends on underlying assumptions for the noise component and for the acceptable level of musical tones. As a general tendency, the less stationary the noise is, the more fragile the noise reduction methods are. This, especially for human listeners, recommends larger values for $\hat{H}_{\min}$ to preserve the desired speech integrity and maintain naturalness of the background noise while accepting higher noise power levels.

Note also that the above filters (Eqs. (3) and (4)) are all linear-phase filters due to the even symmetry of the underlying PSD estimates. This implies that the phase of the estimated desired signal is still the phase of the measured noisy signal $x(n)$, except for some additive linear-phase term.

In addition to removing the noise PSD, significant research on speech enhancement was dedicated to exploiting structural knowledge about the speech signal to improve perceptual speech quality. One of the key ideas is here to exploit the harmonic structure of speech and emphasize spectral components corresponding to the pitch frequency and its higher harmonics [9] or the poles of an estimated all-pole model for speech production [10]. The latter can then also be connected to pattern matching techniques as common in speech recognition (see [11] and references therein).

2 • Multichannel Techniques

The advantages of multichannel techniques over single-channel techniques all originate from the exploitation of spatial diversity as given by the simultaneous sampling of sound fields at different locations. Essentially, the available spatial information provides another domain for separating desired from undesired signals aside from time and frequency.

According to the setup of Figure 1, the M microphone signals $x_\mu(n)$ contain differently filtered (or delayed) versions $x_{\mu,s}(n)$ of a desired source signal $s(n)$ and undesired noise components $x_{\mu,r}(n)$. The filtering by $\mathbf{w}_\mu(n)$ usually aims at extracting the desired signal from the set of microphone signals and at eliminating all undesired components $x_{\mu,r}(n)$ while allowing for some propagation and processing delay $n_0 \geq 0$, so that $y(n) \approx s(n - n_0)$. For this, ideally, all the desired source components should be time-aligned and added constructively, whereas all undesired components should be cancelled out.

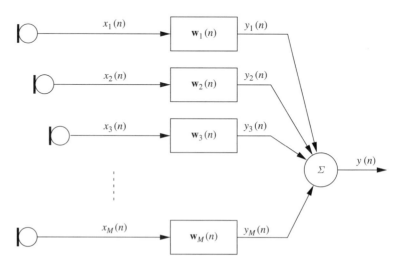

FIGURE 1 Multichannel recording by a microphone array for noise reduction and interference cancellation.

2.1 Data-Independent Beamforming

In a basic version, the differences between propagation delays for the direct paths from source to microphones are estimated – or assumed based on a priori knowledge – and then compensated by $\mathbf{w}_\mu$ prior to summation. It can be shown that with such a *delay and sum beamformer* the power of noise and interference which appears uncorrelated and with identical level at the M sensors can be suppressed by a factor M, commonly expressed as *white noise gain* of $10 \log_{10} M$ dB. For a given *look direction*, the filters $\mathbf{w}_\mu$ are here time invariant and only realize a (possibly fractional) delay, whereas in the more general case of a *filter and sum beamformer* they will also exhibit frequency-dependent and possibly time-variant magnitude characteristics.

A characteristic example for the spatial selectivity for a signal-independent *filter and sum beamformer* is shown in Figure 2. It shows the amplitude gain factor of a linear array using $M = 8$ sensors with a uniform spacing of $d = 4$ cm as a function of frequency and the angle of incidence of a plane wave (*far-field assumption*) in a horizontal half-plane. The look direction of the beamforming is toward *broadside*, that is, the array provides a distortionless response for plane waves impinging perpendicularly to the array axis (at 0 degrees), whereas plane waves from other directions are suppressed depending on their angle of incidence. In this example, some fundamental properties of beamforming for wideband audio signals using microphone arrays are illustrated (see also [12]): As beamforming corresponds to a spatial sampling, the spatial resolution is determined by the number of microphones and the spacing of the microphones relative to the signal wavelengths. Therefore, for a given array

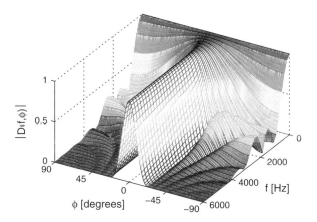

FIGURE 2 Spatial directivity of a linear array with filter and sum beamformer ($M = 8$ microphones with spacing 4 cm).

configuration, the spatial selectivity at low frequencies is less pronounced than for higher frequencies, as is clearly reflected in Figure 2. In order to avoid a second beam of increased sensitivity (*spatial aliasing*) at high frequencies in this half-plane, the microphone spacing may not exceed half a wavelength.

Due to the wideband nature of speech and audio signals which often span two to three orders of magnitude of wavelengths, traditional beamforming methods focussing on narrowband signals [13] have been extended to aim at frequency-independent spatial selectivity [14], which is reasonably well approximated in Figure 2 for the mainlobe for frequencies above 2,000 Hz.

In order to overcome the low spatial resolution at low frequencies, nested arrays can be used which realize different apertures with different spatial sampling for each octave of bandwidth [12]. Often, however, enlarging the aperture is not acceptable for the given application, so that differential (*superdirective*) beamforming becomes attractive which essentially relies on finite difference approximations of spatial derivatives of the sound field [15]. The approximation is valid as long as the acoustic wavelength is large relative to the sensor spacing and therefore can be used for conventional array geometries at the low frequencies. The price for the relatively small apertures is the sensitivity to spatially uncorrelated noise which so far limits the applicability for practical systems to third-order derivatives [15]. Another advanced concept for differential beamforming with small apertures (notably circular or spherical arrays) foresees a decomposition of the acoustic wavefield into eigenfunctions. Instead of the plain microphone signals, the contributions to the various modes are combined to achieve spatial selectivity, i.e. to realize a modal beamformer [16]. As an important advantage over classic conventional beamforming techniques, here, the spectral selectivity and the spatial selectivity can be designed independently.

2.2 Data-Dependent Beamforming

While the above beamformers are data-independent in that they do not exploit any knowledge about the actual signal properties and the positions of interferers or noise sources, data-dependent beamformers exploit such knowledge to optimally suppress noise and interference. Two optimization philosophies are generally followed: One is based on a mean square error (MSE) criterion aiming at maximizing SNR as a multichannel version of the single-channel Wiener filter. As opposed to the single-channel Wiener filter, the spatial domain is exploited to better estimate noise and interference or desired signal characteristics by measuring the spatial co-herence of the signals

[17,18]. In this multichannel Wiener filter approach, suppression of spatially incoherent noise can also efficiently be realized as a single-channel postfilter [19], with some unavoidable distortion of the desired signal as a consequence. The multichannel Wiener filter can also be the basis of subspace methods which essentially try to decompose the multichannel covariance matrices into two orthogonal subspaces for the desired signal and noise, respectively, by, e.g. using general singular value decomposition [18].

An alternative concept introduces a constraint in order to preserve the integrity of the desired signal, and, as in the popular case of the *linearly constrained minimum variance* (LCMV) beamformer [13], minimizes the output signal power subject to this constraint. As a special case, the *minimum variance distortionless response* (MVDR) beamformer demands a distortionless response for the assumed direction of the desired signal. To avoid the constrained optimization in practical applications, the *generalized sidelobe canceller* (GSC) [20] separates the problem into a signal-independent beamformer and a so-called blocking matrix for assuring the constraint, and an unconstrained optimization of the interference canceller (see Figure 3). The blocking matrix has to "block" all desired signal components, so that at its output only noise components appear, which are then used by the interference canceller to estimate the remaining noise component in the fixed beamformer's output. If desired signal components leak through the blocking matrix they will be interpreted as noise and subtracted from the fixed beamformer's output leading to highly undesirable cancellation of the desired signal. In practice, for robustness against source movements and acoustic echoes of the desired source signal, the blocking matrix is preferably adaptive, requiring a carefully designed adaptation control [21,22].

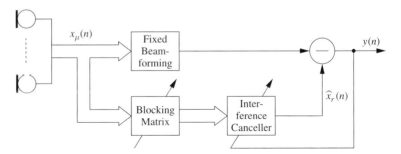

FIGURE 3 GSC with adaptive blocking matrix and adaptive interference canceller.

2.3 Adaptive Beamforming

In practical systems, signal-dependent optimum beamformers have to be adapted to the nonstationary charateristics of the involved speech and noise signals and the time-variant environment. The adaptive filtering algorithms as outlined for acoustic echo cancellation (from LMS to RLS, in time, frequency subband, and DFT domain) can directly be applied to beamforming once the the reference signals (filter inputs) and the desired signal have been defined. This is obvious for the multichannel interference canceller in the GSC, where the blocking matrix outputs play the same role as the loudspeaker inputs in the multichannel echo cancellation system, in the sense that they constitute reference input signals for identifying an optimum multichannel filter. Other adaptive beamformer algorithms face essentially the same estimation challenge as single-channel noise reduction: Reference signals that contain only noise (or desired signal) are difficult to obtain. However, if reliable spatial information on source directions is exploited properly, these references will always lead to better system performance than in the single-channel case [17,18,22].

2.4 Source Localization

If the above spatial filtering techniques require spatial knowledge on source positions which is not given a priori, localization algorithms are needed. A widely accepted classification [23] divides the current methods into "steered beamformers", "high-resolution spectral estimation", and "TDOA (*time difference of arrival*)-based methods". While "steered beamformers" essentially scan the spatial region of interest to measure the response power to localize a source, spectral estimation techniques may include statistical signal models and use eigenanalysis. TDOA-based methods are still dominant in the given applications: They rely on relative delays between two sensor signals which are typically estimated by the *generalized crosscorrelation* [24]. As a more advanced concept, the *adaptive eigenvalue decomposition* [25] extracts the TDOA estimates from estimated impulse responses for the propagation paths from the sources to the microphones and can be expected to perform better in moderately reverberant environments. In general, reflections and reverberation affect all of the above localization algorithms severely, as these signal components are acting as interference correlated to the desired component.

3 • Blind Signal Separation of Convolutive Mixtures

While the above multichannel techniques assume some knowledge about the position of one desired source and try to suppress signals from all other directions, blind source separation (BSS) effectively performs a blind beamforming by identfying linear filters that minimize statistical dependency between several desired signals [26–28]. Thereby, several desired source signals can be extracted simultaneously from the given convolutive mixtures at the microphones, which record a sum of all source signals each convolved with the corresponding acoustic impulse response from the source to the microphone.

Comparing the signal paths in the 2-by-2 case in Figure 4 to the GSC in Figure 3 the structural similarity can be illustrated: BSS can be viewed as performing interference cancellation for each output signal, with the respective other output being the interferer. A fundamental difference to known interference cancellers is given by the optimization criterion that is used for identifying the filters $\mathbf{w}_{ij}(n)$. Often, only second-order statistics are used, i.e. entries of crosscorrelation matrices between output values $y_i(n)$ are minimized, and useful solutions are found because the speech signals to be separated are nonwhite and nonstationary [27,29]. In the most general case, mutual information between the output channels is minimized, which then also requires higher order statistics to additionally exploit the nongaussianity of the signals [29]. Two families of algorithms developed for separation of convolutive speech mixtures can be distinguished: DFT-domain algorithms reduce the convolutive mixtures to scalar mixtures in each DFT bin and then solve the separation problem separately in each bin. While conceptually simple, the alignment of the separation results of several bins to the respective sources requires additional repair mechanisms, typically introducing an additional constraint [27,28]. On the other hand, time-domain algorithms fully acknowledge the convolutional character of the mixture, usually at the cost of greatly increased computational load. However, similar to fast convolution, some time-domain algorithms may be implemented in the DFT-domain with computational advantage and thereby become practicable [29].

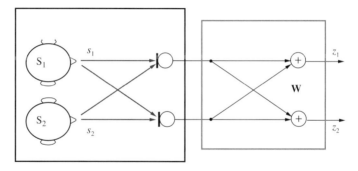

FIGURE 4 Blind source separation for the 2-by-2 case.

References

1. S. Haykin. *Adaptive Filter Theory*. Prentice Hall, Upper Saddle River, NJ, 4th ed., 2002.
2. S. Boll. Suppression of acoustic noise in speech using spectral subtraction. *IEEE Transactions on Acoustics, Speech, and Signal Processing*, 27:113–120, 1979.
3. S.L. Gay and J. Benesty (eds). *Acoustic Signal Processing for Telecommunication*. Kluwer, Boston, MA, 2000.
4. P. Vary. Noise suppression by spectral magnitude estimation – mechanism and theoretical limits. *Signal Processing*, 8(4):387–400, July 1985.
5. J.S. Lim (ed.). *Speech Enhancement*. Prentice Hall, Englewood Cliffs, NJ, 1983.
6. R. Martin. Spectral subtraction based on minimum statistics. *Signal Processing VII: Theories and Applications*, pp. 1182–1185, Edinburgh, Scotland, September 1994. EURASIP.
7. Y. Ephraim and D. Malah. Speech enhancement using a minimum meansquare error short-time spectral amplitude estimator. *IEEE Transactions on Acoustics, Speech, and Signal Processing*, 32: 1109–1121, 1984.
8. Y. Ephraim and D. Malah. Speech enhancement using a minimum mean-square error log-spectral amplitude estimator. *IEEE Transactions on Acoustics, Speech, and Signal Processing*, 33:443–445, 1985.
9. J.H. Chen and A. Gersho. Adaptive postfiltering for quality enhancement of coded speech. *IEEE Transactions on Speech and Audio Processing*, 3(1):59–71, January 1995.
10. J.S. Lim and A.V. Oppenheim. Enhancement and bandwidth compression of noisy speech. *Proceedings of the IEEE*, 67(12): 1286–1604, December 1979.
11. S.F. Boll. Speech enhancement in the 1980s: noise suppression with pattern matching. In M.M. Sondhi and S. Furui (eds). *Advances in Speech Signal Processing*, pp. 309–323. Marcel Dekker, New York, 1991.
12. J.L. Flanagan, D.A. Berkley, G.W. Elko, J.E. West, and M.M. Sondhi. Autodirective microphone systems. *Acustica*, 73:58–71, 1991.
13. B.D. Van Veen and K.M. Buckley. Beamforming: a versatile approach to spatial filtering. *IEEE ASSP Magazine*, 5(2):4–24, April 1988.
14. D.B. Ward, R.A. Kennedy, and R.C. Williamson. Constant directivity beamforming. In M.S. Brandstein and D. Ward (eds). *Microphone Arrays: Signal Processing Techniques and Applications*, chapter 13, pp. 3–18. Springer, Berlin, May 2001.
15. G. Elko. Differential microphone arrays. In Y. Huang and J. Benesty (eds). *Audio Signal Processing for Next-Generation Multimedia Communication Systems*, pp. 2–65. Kluwer, 2004.
16. J. Meyer and G. Elko. Spherical microphone arrays for 3d sound recording. In Y. Huang and J. Benesty (eds). *Audio Signal Processing*

for *Next-Generation Multimedia Communication Systems*, pp. 67–89. Kluwer, 2004.

17. J. Bitzer and K.U. Simmer. Superdirective microphone arrays. In M.S. Brandstein and D. Ward (eds). *Microphone Arrays: Signal Processing Techniques and Applications*, chapter 2, pp. 19–38. Springer, Berlin, May 2001.

18. S. Doclo and M. Moonen. Gsvd-based optimal filtering for multi-microphone speech enhancement. In M.S. Brandstein and D. Ward (eds). *Microphone Arrays: Signal Processing Techniques and Applications*, chapter 6, pp. 111–132. Springer, Berlin, May 2001.

19. K.U. Simmer, J. Bitzer, and C. Marro. Post-filtering techniques. In M.S. Brandstein and D. Ward (eds). *Microphone Arrays: Signal Processing Techniques and Applications*, chapter 3, pp. 39–60. Springer, Berlin, May 2001.

20. L.J. Griffiths and C.W. Jim. An alternative approach to linear constrained adaptive beamforming. *IEEE Transactions on Antennas and Propagation*, 30(1):27–34, January 1982.

21. O. Hoshuyama and A. Sugiyama. A robust adaptive beamformer with a blocking matrix using constrained adaptive filters. *Proceedings of the International Conference on Acoustics, Speech, and Signal Processing*, pp. 925–928. IEEE, 1996.

22. W. Herbordt and W. Kellermann. Adaptive beamforming for audio signal acquisition. In J. Benesty (ed.). *Adaptive Signal Processing: Application to Real-World Problems*. Springer, Berlin, January 2003.

23. J.H. DiBiase, H.F. Silverman, and M.S. Brandstein. Robust localization in reverberant rooms. In M.S. Brandstein and D. Ward (eds). *Microphone Arrays: Signal Processing Techniques and Applications*, chapter 8, pp. 156–180. Springer, Berlin, May 2001.

24. C.H. Knapp and G.C. Carter. The generalized correlation method for estimation of time delay. *IEEE Transactions on Acoustics, Speech, and Signal Processing*, 24(8):320–327, August 1976.

25. Y. Huang, J. Benesty, and G.W. Elko. Adaptive eigenvalue decomposition algorithm for realtime acoustic source localization system. *Proceedings of the International Conference on Acoustics, Speech, and Signal Processing*, Phoenix, AZ, USA. IEEE, March 1999.

26. J.-F. Cardoso. Blind signal separation: statistical principles. *Proceedings of the IEEE*, 86(10):2009–2025, October 1998.

27. L. Parra and C. Spence. Convolutive blind separation of non-stationary sources. *IEEE Transactions on Speech and Audio Processing*, 8(4):320–327, May 2000.

28. S. Makino. Blind source separation of convolutive mixtures of speech. In J. Benesty (ed.). *Adaptive Signal Processing: Application to Real-World Problems*. Springer, Berlin, January 2003.

29. H. Buchner, R. Aichner, and W. Kellermann. Blind source separation for convolutive mixtures: a unified treatment. In Y. Huang and J. Benesty (eds). *Audio Signal Processing for Next-Generation Multimedia Communication Systems*, chapter 10, pp. 255–296. Kluwer, Boston, MA, 2004.

48
Terminals and Their Influence on Communication Quality

Walter Kellermann

Multimedia Communications and Signal Processing, University Erlangen-Nuremberg, Erlangen, Germany, e-mail: wk@lnt.de

With the advent of affordable signal processing hardware, acoustic interfaces of many telecommunication terminals developed from loudspeakers and microphones with some gain control into complex systems integrating powerful signal processing for enhancement of communication quality. The requirements for signal processing to satisfy the human user vary strongly depending on the acoustic interface and the communication scenario. Several perceptually motivated criteria (which are also partially incorporated in recommendations issued by standardizing organizations, such as ITU, ETSI, ISO-MPEG) must be accounted for: Echoes in telecommunication networks should generally be suppressed as much as possible, and the later the echoes arrive, the more crucial is their cancellation/suppression [1]. The echo cancellation requirement can be somewhat relieved for double-talk situations, assuming that voice activity of a communication partner masks the echo of one's own voice.

For conversations among humans, signal delays due to processing and transmission should be minimized in order not to impair perceived communication quality. However, while until the 1980s only a few milliseconds were tolerated by the established standards for the mostly analog wired networks, the driving forces for digital mobile phones, low-bitrate video conferencing, and voice over IP pushed the limits to larger delays, and nowadays one-way delays on the order of 100 ms are widely accepted.

When eliminating noise from microphone signals, it is not just the amount of noise power which should be minimized, but ideally the signal characteristics of the original noise should be maintained. In any case, artifacts of the noise reduction method must be avoided, as they will call for the listener's special attention and are all the more noticeable, even if they are of little average power. In these cases, even adding some inconspicuous *comfort noise* [2] for masking such artifacts will improve perceptual quality. This technique can also be used for masking residual echoes. Moreover, perception and acceptance of undesired signal components from a remote communication partner depends also on the listening environment. On the other hand, with increasing noise level, a speaker will automatically raise his speech level (*Lombard effect*) and thereby enhance the SNR, which benefits the communication efficiency. If communication is supported by video or still images, the spatial alignment between visual and acoustic perception becomes necessary for user acceptance and, therefore, spatial audio is highly desirable for speech reproduction.

1 • Acoustic Interfaces

Obviously, the closer the microphones and loudspeakers are to the human's mouth and ears, respectively, the less problematic are the corresponding transmission channels and the less is the signal processing needed to assure acceptable communication quality. For instance, for headsets and telephone handsets the quality is essentially determined by the quality of the transducers and their integration in the housing.

For some handsets, especially for handhelds in mobile telephony, the acoustic coupling between loudspeaker and microphone, however, calls already for acoustic echo cancellation, and in case of low-voltage amplifiers and low-cost loudspeakers, the according echo path often exhibits significant nonlinearities [3,4].

In seamless hands-free communication terminals, as depicted for the general multichannel case in Figure 1, every microphone signal generally contains not only the desired signal but also its reverberation, plus additional noise and undesired interfering sources of the local environment, plus acoustic echoes from the loudspeaker(s) broadcasting the remote

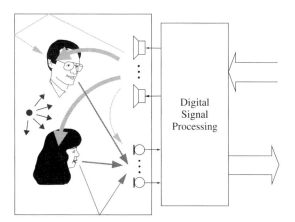

FIGURE 1 General scenario for seamless hands-free speech communication.

communication partners' signals. Along with increasing distance of the desired source(s), the gain for the desired signal(s) must be increased which automatically increases the recorded level of all undesired signal components as well. Similarly, for sound reproduction, the loudspeaker volume will be increased along with the distance between loudspeaker(s) and listener(s) which additionally increases the acoustic feedback into the microphones.

Signal processing systems should fulfill several tasks in the hands-free communication scenario: On the recording side, all desired sources should be captured separately without reverberation, the desired signals should be free from interference and noise, and, often, the source positions should be identified. As discussed above, the techniques of echo cancellation, noise reduction, beamforming, and blind source separation are capable of meeting these goals to a certain extent. So far, dereverberation remains the most difficult task: While beamforming will suppress some reflections from others than the desired direction and the filtering of blind source separation may or may not result in some dereverberation, the actually desired blind equalization of the acoustic transfer functions is still out of reach for realistic time-varying acoustic scenarios.

On the sound reproduction side, the listener, ideally, should hear a clean undistorted signal from the remote partner, while the ambient noise in the listening environment should be suppressed. With multichannel reproduction schemes, typically, spatial information for the remote sources should be represented as well. Current techniques mostly focus on the latter aspect, either by creating a *sweet spot* in the listening environment, where spatial information can be perceived correctly (stereo systems or extensions to five or more channels), or by binaural processing which can be used for sound reproduction via headsets and crosstalk-canceling loudspeaker systems [5]. Multichannel sound reproduction has also been extended to using dozens to hundreds of channels for synthesizing predescribed wavefields (*ambisonics* and *wavefield synthesis*) [6]. With these concepts, equalization of the listening room acoustics and active cancellation of noise sources and interferers in the listening environment may come into reach.

Considering typical terminals for hands-free telecommunication, the prevalent representatives are desktop telephones incorporating hands-free mode, hands-free telephones in cars, self-contained mobile teleconferencing terminals as used in normal offices especially equipped with teleconferencing rooms and, to a much lesser extent, teleconferencing equipment as used for auditoria [7].

While for the hands-free mode in standard telephone terminals mere echo suppression is still common, echo cancellation and noise reduction are almost always implemented in hands-free equipment for cars where the distance from the mouth to the microphone is often more than half a meter and the ambient noise level is high. Here, microphone arrays with up to four sensors are already in use allowing for multichannel noise reduction and beamforming. This is not only desirable for telecommunications, but also for speech dialogue systems which otherwise suffer severely from the high ambient noise levels in cars. So far, echo cancellation in cars is mostly implemented for monaural sound reproduction, but as most cars are equipped with high-quality stereo or multichannel reproduction systems, their use for telecommunications becomes highly desirable, although this calls for multichannel acoustic echo cancellation (see Sect. 46.1.3).

For mobile teleconferencing equipment, low-cost directional microphones with small apertures are desirable, and therefore differential microphone arrays [8] or just satellite microphones [9] are used which are placed close to the talkers. Satellite microphones for each participant are also often used for dedicated teleconferencing rooms, but here, microphone arrays with an aperture size on the order of the video screen size are already a realistic option to capture several talkers without requiring individual microphones. The advantage of microphone arrays becomes even more evident, when a large number of participants should be covered as, e.g., for people asking questions in a classroom or an auditorium.

If instead of single microphones a – possibly adaptive – microphone array is used to capture the various sources, a combination of AEC and beamforming becomes necessary, whose interactions have been studied in [10].

Especially in teleconferencing with video modality and multiple talkers on each side, multichannel sound reproduction is necessary to align the acoustic to the visual perception, which requires multichannel AEC either in its general or in some special form [9,11 (Chaps. 5,7)]. As the background noise level is generally very low in such environments and therefore residual echoes are well audible even at very low levels, residual echo suppression is often applied to assure echo attenuation beyond 30 dB which cannot be provided by echo cancellers alone.

References

1. S. Gay and J. Benesty, editors. *Acoustic Signal Processing for Telecommunications*. Kluwer, Norwell, MA, 2000.
2. E. Hänsler and G. Schmidt. *Acoustic Echo and Noise Control: A Practical Approach*. Wiley, Hoboken, NJ, 2004.
3. A. Stenger and W. Kellermann. Adaptation of a memoryless preprocessor for nonlinear acoustic echo cancelling. *Signal Processing*, 80:1747–1760, Sep 2000.
4. F. Küch and W. Kellermann. Nonlinear line echo cancellation using a simplified second-order Volterra filter. *Proceedings of International Conference on Acoustics, Speech, and Signal Processing*, Orlando, FL, USA, May 2002. IEEE.
5. C. Avendano. Virtual spatial sound. In Y. Huang and J. Benesty, editors, *Audio Signal Processing for Next-Generation Multimedia Communication Systems*, Chap. 13, pages 345–371. Kluwer, Boston, MA, 2004.
6. S. Spors, H. Teutsch, A. Kuntz, and R. Rabenstein. Sound field synthesis. In Y. Huang and J. Benesty, editors, *Audio Signal Processing for Next-Generation Multimedia Communication Systems*, Chap. 11, pages 323–344. Kluwer, Boston, MA, 2004.
7. J.L. Flanagan, D.A. Berkley, G.W. Elko, J.E. West, and M.M. Sondhi. Autodirective microphone systems. *Acustica*, 73:58–71, 1991.
8. P. Chu. Superdirective microphone array for a set-top videoconferencing system. *Proceedings of International Conference on Acoustics, Speech, and Signal Processing*, pages 235–238, Munich, Germany, April 1997. IEEE.
9. A. Nakagawa, S. Shimauchi, Y. Haneda, S. Aoki, and S. Makino. Channel-number-compressed multichannel acoustic echo canceller for high-presence teleconferencing system with large display. *Proceedings of International Conference on Acoustics, Speech, and Signal Processing*, pages 813–816, Istanbul, Turkey, June 2000. IEEE.
10. W. Kellermann. Acoustic echo cancellation for beamforming microphone arrays. In M.S. Brandstein and D. Ward, editors, *Microphone

Arrays: Signal Processing Techniques and Applications, Chap. 13, pages 281–306. Springer, Berlin, May 2001.
11. J. Benesty, T. Gänsler, D.R. Morgan, M.M. Sondhi, and S.L. Gay. *Advances in Network and Acoustic Echo Cancellation*. Springer, Berlin, 2001.

49
Networks and Their Influence on Communication Quality

H.W. Gierlich

HEAD acoustics GmbH, Herzogenrath, Germany

1 • PSTN (Public Switched Telephone Networks)

The dominating influences of traditional TDM-type networks on the communicational quality are

- Frequency-dependent attenuation in the analog parts of the network
- Delay
- Echo performance problems due to insufficient trans-hybrid loss of analog systems

In analog networks no synchronization problems occur; TDM-type networks work isochronously; in all situations the speech channel is associated to a physical channel in the network. The delays in the networks are typically well controlled due to proper transmission planning. Within the network of one operator, conversational speech quality problems due to high delay are unlikely as long as terrestrial transmission systems are used within the network. As soon as satellite transmission is used, e.g., for transcontinental communication, the communication dynamics may be impaired due to long delay which is typically around 240 ms one-way transmission time.

Factors impairing the listening speech quality are found in analog networks: the line characteristics (length and type of lines) have to be taken into account, since they may have impact on loudness ratings, frequency responses, noise, sidetone performance and trans-hybrid losses (see [5,11]).

When a delay greater than 15 ms is introduced in a connection, echo cancellation has to be provided. In the PSTN, echo cancellation is typically concentrated in international exchanges since within national or continental connections the network delay is typically below 15 ms. The echo cancellation is intended to cancel echoes produced by the hybrid of analog terminals. The maximum echo loss to be provided is generally determined by the maximum delay which may be inserted in a connection. However, most modern network echo cancellations conform to ITU-T Recommendation G. 168 [7] which assumes 250 ms one-way transmission time and provides adequate performance limits. If echo suppression is used (sometimes still found in older networks), instead of proper echo cancellation, the speech communication quality may be impaired significantly. Due to switching, the speech signal may be clipped partially, no double talk is possible and the background noise performance is significantly impaired by the network (see description in Section 44.3).

Different speech levels due, for example, to different national level settings or different attenuation in analog systems, can be expected and may degrade the performance of echo cancellers and other signal processing in the network and the terminals.

Speech coding is not typically found in PSTN networks, with the exception of DCME systems (digital circuit multiplication equipment, see [9]). If DCME is inserted in a PSTN network, all considerations concerning speech quality found in Section 44 apply.

2 • Packet-Based Transmission

Packet-based networks differ considerably from circuit-switched networks. In packet-based networks the communication quality may be influenced by

- Speech coding and packetization
- Delay which may be high and unpredictable
- Delay variation during a call
- Packet loss and packet loss concealment
- Echo performance problems due to insufficient echo cancellation
- VAD (voice activity detection) and comfort noise insertion

The (coded) input signal (speech) is segmented into packets typically of fixed length. Typical packet lengths are in the range 5–30 ms. Packetization itself introduces a minimum delay of the packet length introduced into the transmission. Packets from different sources (e.g., terminals or subscribers) are sequenced for transmission. This process, as well as any buffering or routing in a connection, may add additional delay. While in isochronous networks the delay of a connection is constant and predictable, the delay in IP networks may vary between different connections depending on network load, routing, bandwidth, switching and other effects. In addition, jitter may be introduced by the IP connection. Jitter occurs when the delays for different packets are different. Delay and jitter are not predictable in IP networks; therefore, they are described by statistical parameters, such as average delay (including delay distribution) and packet loss distribution.

At the IP network termination points (in gateways or the IP terminals) the receiver has to take care of the collection of the packets and their correct ordering. Jitter buffers are used in order to collect and sequence the packets received in such a way that an almost error-free transmission is achieved.

Due to the time-variant nature of IP networks (see Chapter 43.2.2) time-variant delay and jitter as well as packet loss have to be taken into account, and speech quality measurements have to be conducted for the various network conditions to be expected. It is advisable to make appropriate simulations in a lab environment in order to get an estimate of the speech quality range to be expected under real use conditions (see ITU-T Y. 1541 [10]).

In most packet-based systems, echo cancellation is provided, either in the terminal or in the network. In IP connections gateways have to be equipped with echo cancellers providing sufficient echo loss for all expected delays in a connection. The maximum echo loss to be provided is determined by the maximum delay which may be inserted in a connection. Since under worst-case conditions the maximum one-way transmission delay may be higher than 300 ms, the minimum echo loss which has to be provided is 55 dB [6]. Consequently, any IP terminal connected to an IP network has to provide this amount of echo loss. Different speech levels, due to computer terminals with insufficiently controlled loudness ratings or equipment with improperly implemented codecs for example, may degrade the performance of echo cancellers and other signal processing in an IP network. Speech coding especially in combination with packet loss and jitter may add further impairments to the transmitted speech signal. Speech coders with integrated packet loss concealment algorithms (PLC), e.g., [8], may improve the listening speech quality.

Frequently, silent periods, when the signal level is below a defined threshold, are recognized using voice activity detection (VAD) and are not transmitted in order to save bandwidth or power. At the far end, silent periods are typically replaced by low-level noise (comfort noise). In background noise situations, this procedure may lead to background noise modulations which may be annoying for the user. Furthermore, voice activity detection may cause clipping of the actual speech signal.

Further information about VoIP networks and configurations and QoS requirements can be found in [4] and [11].

3 • Mobile Networks

In mobile networks, the main parameters contributing to the communication speech quality are

- The radio channel and its transmission impairments
- Speech coding
- Delay introduced by the speech coding, the channel coding and the transmission in the mobile network
- Bit errors, frame loss and frame loss concealment
- Handover between different cells
- Echo performance of the echo cancellers used for interconnection to the PSTN and echo cancellation of the mobile terminal's echo
- Performance of additional signal processing introduced in the network and intended to automatically adjust signal levels or to reduce background noise

The radio network and its cellular structure is very sensitive to transmission errors. Disturbances introduced by the radio channel may result in bit errors or frame loss. As shown in Chapter 43.2.3, the different elements of the mobile network show a different sensitivity to transmission errors.

3.1 Listening Speech Quality

In the mobile network, the listening speech quality is limited by the codecs used [1,2]. However, even if no degradation by the mobile network is observed, the speech quality provided by the speech coder may be decreased due to environmental influences, especially background noise. Noise cancellation in mobile terminals may help to reduce the background noise, but will degrade the speech sound quality. Furthermore, the listening speech quality may be degraded by a number of different impairments introduced by the network. In the radio network, co-channel interference (C/I), adjacent channel interference (C/A), noise limitations (C/N) and other effects may lead to increasing bit error rates and to frame errors. Increasing BER and FER lead to less information about the coded speech and thus the listening speech quality will degrade. The various techniques used in the radio network to increase the quality of the radio environment also have impact on the listening speech quality. The advantage of DTX (discontinuous transmission) is the reduction of the interference in the radio channel since the signal is only transmitted during speech activity. This may lead to less BER and FER and thus to an increase of the listening quality of the speech transmitted. On the other hand, however, the speech quality may be impaired by speech clipping (due to non-ideal speech activity detection) and comfort noise insertion used at the far end in order to substitute the background noise (which is not transmitted). Frequency hopping may increase the listening speech quality since the effect of multipath fading and interference can be reduced. Dynamic power control may result in a further increase in the protection against fading dips and interference and therefore may contribute to less BER and, as a result of lower BER, to higher listening speech quality. The quality is always impaired during handover. When a user "leaves" one cell and enters another one (handover), the speech is interrupted briefly, and speech frames are used to transmit signaling information.

3.2 Conversational Speech Quality

Due to the delay introduced by mobile networks for source coding, channel coding and radio transmission, the conversation dynamics may be impaired, especially in connections where the mobile network is interconnected to other networks introducing additional transmission delay. Furthermore, due to the delay introduced by the mobile network, echo cancellation must be provided in the mobile switching centers to cancel echoes produced, for example, by analog PSTN connections. This echo cancellation is required to protect the mobile terminal user from echoes generated by the PSTN. To avoid audible acoustical echoes the echo loss provided by a mobile terminal has to be at least 46 dB [3]. Some mobile terminals, however, do not provide the required echo attenuation in all operational modes; therefore, in addition to the echo cancellers facing the PSTN, echo cancellers facing the mobile users' terminals are integrated into some mobile networks. These are intended to cancel acoustical echoes remaining from mobile terminals. It should be recognized that these echo cancellers have to operate on a non-linear and time-variant echo path. The complete radio channel including the speech codec is part of the echo path. Due to the mostly uncorrelated input and output signal (see Figure 1 in Chapter 50), the adaptive FIR filter is not suitable to model the echo path. As a consequence, these speech echo cancellers typically rely on the non-linear processor and operate as a type of echo suppressor. This operating mode leads to speech clipping during double talk and thus to an insufficient performance. Furthermore, strongly varying speech levels in the network may impair the efficiency of the echo cancellers.

In addition, a variety of non-linear and/or time-variant signal processing, such as AGC (automatic gain control) or compression techniques, may be used in mobile terminals in order to enhance the speech quality in the different acoustical conditions, but this processing may interact with the network signal processing (see Chapter 50).

References

1. 3GPP: TS 46.010: Full Rate Speech Encoding, Third Generation Partnership Project (3GPP), 2002
2. 3GPP: TS 46.051: GSM Enhanced Full Rate Speech Processing Functions: General Description, Third Generation Partnership Project (3GPP), 2002
3. ETSI EG 201 377-2: Specification and Measurement of Speech Transmission Quality; Part 2: Mouth-to-Ear Speech Transmission Quality including Terminals, 2004
4. ETSI TS 101 329-5: End-to-End Quality of Service in TIPHON Systems, Part 5: Quality of Service (QoS) Measurement Methodologies, 2002
5. ITU-T Recommendation G.107: The E-model, A Computational Model for Use in Transmission Planning, International Telecommunication Union, Geneva, 2003
6. ITU-T Recommendation G. 131: Control of Talker Echo, International Telecommunication Union, Geneva, 2000
7. ITU-T Recommendation G. 168: Digital Network Echo Cancellers, International Telecommunication Union, Geneva, 2004
8. ITU-T Recommendation G.711: Appendix I: A High Quality Low-Complexity Algorithm for Packet Loss Concealment with G.711, International Telecommunication Union, Geneva, 1999
9. ITU-T Recommendation G. 763: Digital Circuit Multiplication Equipment Using G.726 ADPCM and Digital Speech Interpolation, International Telecommunication Union, Geneva, 1998
10. ITU-T Recommnedation Y.1541: Network Performance Objectives for IP-Based Services, International Telecommunication Union, Geneva, 2001
11. Richards, D.L.: Telecommunication by Speech, Wiley, New York, 1973, ISBN 0 470 71949 4

50

Interaction of Terminals, Networks and Network Configurations

H.W. Gierlich

HEAD acoustics GmbH, Herzogenrath, Germany

The interaction of terminals and different network configurations may be many-fold and is discussed in this chapter only with reference to some examples. Typically, it is the task of standardization to avoid, or at least to minimize, interaction problems which might occur in different network configurations. Due to the liberalized telecommunications market, however, the variety of configurations found in today's telecommunication is increasing rapidly. Non-linear and time-variant signal processing is used in terminals as well as in networks and network components. Also different types of networks (analog, TDM, IP, mobile networks) are interconnected globally and different network providers may be involved in one connection.

Figure 1 gives an overview of a possible configuration which is used to explain different types of interaction problems which may occur.

1 • Tandeming of Speech Coders

Different types of speech coders may be used in different parts of the connection. Depending on the speech coding, the degradation of listening speech quality may vary significantly. In

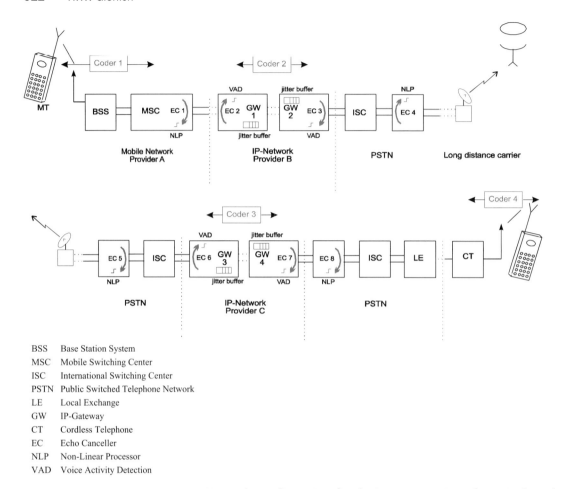

FIGURE 1 Example configuration for the interconnection of terminals and networks.

general, tandeming of all types of CELP coders using a parametric description of the speech signal (Chapter 45.2) will lead to stronger degradation as compared to the tandeming of waveform coders. Reference [1] gives an overview of the expected quality degradation for different types of tandemed coders. The influence of tandeming on the listening speech quality has to be derived subjectively. For transmission planning, the impairments are provided in terms of equipment impairment factors (Ie) based on the E-model [2]. The E-model assumes that impairments are additive on a psychological scale, but this is not really valid when tandeming different types of speech coders. Since this interaction problem is known, more effort is put into the standardization of tandem-free operation. However, due to the different topologies of networks interconnected and due to missing or improperly implemented signaling information about the type of coders used in a connection, such tandem-free operation can be realized typically only in one network. The realization of universally accepted signaling information about the type of coders seems to be not achievable.

2 • Delay

Figure 1 illustrates that when interconnecting multiple networks and different types of terminals, the transmission delay is increased with the insertion of each additional network and the type of signal processing. Although the individual components of a connection might well be within a specified range, the overall transmission delay might be unacceptably high. This is to be expected especially in cases where mobile and IP networks are interconnected. Therefore, in terminals as well as in networks, the general rule to keep the delay as low as technically possible should be followed during the design process. Otherwise, the communication quality might be impaired significantly so that successful interaction between the conversation partners is no longer possible. Furthermore, the increase of the delay increases the requirements on echo loss significantly [3].

3 • Tandeming of Speech Echo Cancellers

Echo cancellation may be found in many places in the network but the general rule should be that echo cancellation should be performed as closely as possible to the echo source itself. A four-wire network itself typically does not produce any echo, so the echo cancellation should be realized in the terminals which produce acoustical echo. During recent years, this has been more clearly recognized, and the requirements have been specified by standards [4]; however, it is still not guaranteed that a sufficient amount of echo loss is provided in all conversational situations. Today's telecommunications requirements often do not address the problem of time-variant behavior of echo cancellers in different conversational situations in the appropriate standards. This situation leads to the insertion of additional echo cancellers in networks, even where the echo cancellation should have been completely provided by the terminal itself (Figure 1, ECs 2, 4, 6).

One more reason for deployment of speech echo cancellers in a network is the analog network which, for the end point connections to the customer, uses two-wire connections. These end points are terminated by a hybrid [5] and the trans-hybrid loss is always limited, causing echo. Echo cancellation therefore has to be provided for these connections where the transmission delay exceeds about 15 ms. In the past, this was typically required only for lengthy international connections, and the echo cancellers were deployed only in the international exchanges [6,7]. With proper network planning, it was possible to avoid the use of echo cancellers in the shorter national networks, at least for smaller countries. In today's network configurations, these echo cancellers are still used, but the PSTN may be interconnected to other types of networks rather than directly to the conversation partner (Figure 1). When assuming that an IP network exists in the connection between the conversation partners, both gateways in the IP network are equipped with echo cancellers to reduce possible echoes from their end points. As seen in Figure 1, many echo cancellers (ECs 1, 3, 5, 7, 8 and ECs 2, 4, 6) are facing the same direction. Under this condition, only a proper design of the network speech echo canceller will avoid echo problems: The echo canceller closest to the echo source needs to sufficiently reduce the echo produced by the terminal under all conditions. The echo cancellers in the network facing the same direction should be mostly transparent. The ideal solution would be proper signaling between the networks for disabling the echo cancellers which, however, cannot be guaranteed. Therefore, a proper control of the "internal" echo cancellers is the only possibility to avoid effects like echo generation by the digital filter in the echo canceller (due to a "non-existent echo path"), temporal clipping (due to improperly activated non-linear processing) and others.

4 • Tandeming VAD and NLP

Yet another tandeming problem can be seen in the example shown in Figure 1. Voice activity detection in the gateways in combination with non-linear processing in the echo cancellers may lead to a variety of speech clipping effects. Different thresholds may lead to interrupted speech when low-level speech signals are transmitted. Different types of comfort noise insertion used in combination with different VAD- and NLP-implementations may lead to interrupted speech signals modulated with various types of comfort noise and – in the presence of background noise – also to background noise modulations. Such effects may be very annoying and should be avoided.

5 • Other Signal Processing

Recent standardizations [8] also describe the so-called voice enhancement devices in the network. The signal processing is intended to equalize speech levels found in a network, to reduce background noise which might have been picked up by the terminal and transmitted to the network, or even to enhance the "clarity" of the voice. The use of such equipment in a network may cause problems in cases where similar signal processing is already implemented in the terminal. For example, tandemed noise reduction will lead to speech sounding robotic and often to background noise modulation.

All the effects described are not properly modeled by the network planning tools used today [2]. Therefore, only careful transmission planning by experts which are aware of the problems described, in combination with careful system design, can avoid or minimize interaction problems.

References

1. ITU-T Recommendation G.113: Transmission Impairments due to Speech Processing, International Telecommunication Union, Geneva, 2001
2. ITU-T Recommendation G.107: The E-model, A Computational Model for Use in Transmission Planning, International Telecommunication Union, Geneva, 2003
3. ITU-T Recommendation G. 131: Control of Talker Echo, International Telecommunication Union, Geneva, 2000
4. ANSI-TIA/EIA 810-A: Transmission Requirements for Narrowband Voice over IP and Voice over PCM Digital Wireline Telephones, Dec. 19, 2000 ETSI EG 201 377-1: Specification and Measurement of Speech Transmission Quality; Part 1: Introduction to Objective Comparison Measurement Methods for One-Way Speech Quality Across Networks, 2003
5. Richards, D.L.: Telecommunication by Speech, Wiley, New York, 1973, ISBN 0 470 71949 4
6. Schütte, U.: Übertragungstechnische Planung von Netzen für Sprachdienste, Schiele & Schön, 1992, ISBN 379 4905 326
7. ITU-T Recommendation G.101: The Transmission Plan, International Telecommunication Union, Geneva, 2003
8. ITU-T Recommendation G. 169: Automatic Level Control Devices, International Telecommunication Union, Geneva, 1999

List of Important Abbreviations

A- Interface	Interface between the BSS and the MSC, manages the allocation of suitable radio resources to the MSs and mobility management
A-bis	BSS internal interface linking the BSC and a BTS
ACELP	Algebraic CELP (with special codebook, see CELP)
ACR	Absolute Category Rating
ADPCM	Adaptive Differential PCM (with adaptive prediction and quantization)
AHr,dt	Attenuation Range Receiving, Double Talk
AHs,dt	Attenuation Range Sending, Double Talk
AMR	Adaptive Multi-Rate coding
APC	Adaptive Predictive Coding (also "block-ADPCM")
APP	A-Posteriori Probability
ATC	Adaptive Transform Coding
A-ter:	BSS internal interface linking the BSC and a TRC
ATM	Asynchronous Transfer Mode
BB-RELP	Base-Band (i.e., low-frequency band) RELP (see RELP)
BER	Bit Error Rate
BFH	Bad Frame Handling
BFI	Bad Frame Indication
BSC	Base Station Controller
BTS	Base Transceiver Station
C/I	Co-channel interference
CCR	Comparison Category Rating
CELP	Code-Excited Linear Predictor/Prediction
COVQ	Channel-Optimized Vector Quantization
CRC	Cyclic Redundancy Check
CSI	Channel-State Information
DCR	Degradation Category Rating

DCT	Discrete Cosine Transformation
DFT	Discrete Fourier Transformation
DSP	Digital Signal Processor
DTX	Discontinuous Transmission
D-value	Relationship between direct sound field sensitivity and diffuse sound field sensitivity
EDGE	Enhanced Data-rate for GSM Evolution (an add-on to GSM)
EFR	Enhanced Full-Rate system (with the original GSM data rate and better quality)
EMBE	Enhanced MBE (see MBE)
ETSI	European Telecommunication Standardization Institution
eXCELP	Extended CELP (with classified codebooks, see CELP)
FER	Frame Error Rate
FIR	Finite (-length) Impulse Response (also "non-recursive") filter
FR	Full-Rate system (= the first standard GSM speech coding at 13 kb/s)
GSM	Global System for Mobile communication
HC	Harmonic Coding (= SM with a frequency grid, see SM)
HR	Half-Rate system (with half the original GSM data rate and less quality)
HVXC	Harmonic Vector-eXcited Coding (a HC / VQ combinaztion, see HC, CELP, VQ)
IMBE	Improved MBE (see MBE)
IP	Internet Protocol
ISCD	Iterative Source-Channel Decoding
ISO	International Standardization Organization
ITU	International Telecommunication Union (formerly: CCITT, Conseil Consultative de la Téléphonie et Télégraphie)
JLR	Junction Loudness Rating
LMS	Least Mean Squares (algorithm)
LP	Linear Prediction (= linear estimation of a new sample from n past samples)
LPC	Linear Predictive Coding (used generally for the filter inside all predictive coders or specifically for the LP-based VOCODER)
LSF	Line Spectral Frequencies (= transformed LPC coefficients)
LTP	Long-Term Prediction (with a large lag in the order of a "pitch" period)
MBE	Multi-Band Excitation (of an LP filter)
Mops	Mega (i.e., million) Operations per Second
MOS	Mean Opinion Score
MOS-LQO	Mean Opinion Score, Listening Quality Objective
MPE	Multi-Pulse Excitation (with a reduced number of entries on a general time grid)
MPEG	Moving-Picture Expert Group (also standardizing speech and audio systems)
MSC	Mobile Switching Center

MSE	Mean-Square Error
NLMS	Normalized Least Mean Squares (algorithm)
NLP	Non Linear Processor
OLR	Overall Loudness Rating
PCM	Pulse-Code Modulation (= analog-to-digital conversion, with linear or possibly non-linear quantization)
PLMN	Public Land Mobile Radio
PSD	Power spectral density
PSTN	Public Switched Telephone Network
R	Rating Factor, used in the E-model for transmission planning
RELP	Residual (signal)-Excited Linear Predictor
RLR	Receiving Loudness Rating
RLS	Recursive Least Squares (algorithm)
RPE	Regular-Pulse Excitation (with a reduced number of entries on a regular time grid)
SBC	Sub-Band Coding
SIRP	Spherically Invariant Random Process
SISO	Soft-In/Soft-Out
SLR	Sending Loudness Rating
SM	Sinusoidal Modeling (of signals by sums of general sinusoids)
SNR	Signal-to-Noise (power) Ratio (usually on a log. scale, in dB)
SOVA	Soft-Output Viterbi Algorithm
TC	Transform Coding
TDM	Time Division Multiplex
TELR	Talker Echo Loudness Rating
TELRdt	Talker Echo Loudness Rating, Double Talk
TRC	Transcoder Controller
UEP	Unequal Error Protection
UMTS	Universal Mobile Telecommunication System
VAD	Voice Activity Detection
VOCODER	Voice CODER with parametric speech description and synthesis
VoIP	Voice over Internet Protocol
VQ	Vector Quantization (also "block quantization")
WB	Wide-Band (speech, with ~7 kHz bandwidth rather than 3.5 kHz)

Printed in the United States of America